MAXWELL ELECTRODYNAMICS AND BOSON FIELDS IN SPACES OF CONSTANT CURVATURE

CONTEMPORARY FUNDAMENTAL PHYSICS

VALERIY V. DVOEGLAZOV - SERIES EDITOR

UNIVERSITY DE ZACATECAS

ZACATECAS, MEXICO

CONTEMPORARY FUNDAMENTAL PHYSICS

MAXWELL ELECTRODYNAMICS AND BOSON FIELDS IN SPACES OF CONSTANT CURVATURE

E. M. OVSIYUK

V. V. KISEL

AND

V. M. RED'KOV

nova
publishers

New York

Library of Congress Cataloging-in-Publication Data

Maxwell electrodynamics and Boson fields in spaces of constant curvature / [edited by] E.M. Ovsiyuk, V.V. Kisel and V. M. Red'kov (Institute of Physics, Minsk, Belarus and others).
    pages cm
  Includes bibliographical references and index.
  ISBN 978-1-62618-891-4 (hardcover)
 1. Boundary value problems. 2. Maxwell equation. 3. Electrodynamics. 4. Quantum theory. I. Ovsiyuk, E. M., editor of compilation. II. Kisel, V. V., editor of compilation. III. Redkov, V. M., editor of compilation.
  QC20.7.B6M39 2013
  537'.1--dc23
                            2013017669

Published by Nova Science Publishers, Inc. † New York

Contents

Introduction

Maxwell electrodynamics plays a special role in physics, because it helped establish the idea for presenting the concept of fields mediating interaction that ultimately became physical reality. The search for symmetry properties in the Maxwell equations with respect to the moving the reference frames led Lorentz, Poincaré and Einstein [1–3] to the creation of Special Relativity.

Maxwell theory turned out to be the first relativistically invariant field model, and it became a starting point to extend such an approach to all other physical fields and forces [4–17].

In particular, after recognizing the role of relativistic symmetry, a problem of relativistic theory for gravitational interaction arose. Evidently, the Newton approach to gravitation though proved its physical status through great successes, was nonrelativistic – according to it the gravitational forces were transmitted instantly. The first known attempt by Poincaré to make this connection was not successful. However, Poinaré had formulated an idea that instead of using the language of forces it would be better to try the language of geometry. This initial almost philosophic idea was realized in full in General relativity [18–20]. In general relativity, the old idea of universal gravitational interaction (Newtonian attraction) was transformed to a new concept: all physical objects, particles and fields, are subject to geometrical properties of the background space-time that describe (explain or simulate) the old gravitational forces.

In this broad and much more difficult analysis subject – particles and fields in presence of non-trivial geometrical (gravitational) background – the role of classical electrodynamics still remains enormous. The main idea of this book is to demonstrate and explore this broad and interesting topic.

It should be mentioned here that there are excellent textbooks for those who have further interest on the study of Special and General Relativity and field theory in curved space-time models. For instance, some specfic resources regarding this topic are: Pauli [21], Eddington [22], Weyl [23], Tolman [24] Fock [25] Schrödinger [26, 27], Schmutzer [28] Mitckevich [29] Weinberg [30] Misner–Thorne–Wheeler [31] Birrell-Davies [32] Penrose–Rindler [33], Chandraseckhar [34], Galtcov [36]. However, only a small number of these resources deal with the role of curved geometry in the quantum mechanics itself (for example see the book by Gorbatcevich [35]).

This book provides a detailed analytical treatment of exact solutions that regard various problems for classical electrodynamics, Boson field theory, and quantum mechanics in simplest non-Euclidean space-time models, open Bolyai and Lobachevsky space H_3 and closed Riemann space S_3, and (anti) de Sitter space-times. The main objective is to focus on the related new themes created by non-vanishing curvature in the following topics: *electrodynamics in curved space-time and modeling of the media, Majorana–Oppenheimer approach in curved space time, spin 1 field theory, tetrad based Duffin–Kemmer-Petiau formalism, Schrödinger–Pauli limit, Dirac–Kähler particle, spin 2 field, anomalous magnetic moment, plane wave, cylindrical, and spherical solutions, spin 1 particle in a magnetic field, spin 1 field and cosmological radiation in de Sitter space-time, electromagnetic field and Schwarzschild black hole.*

The topics are of fundamental significance in this field of study, and are even beyond their possible experimental testing.

We have tried to adhere to the general notation used in the well-known course of theoretical physics by Landau et al., [37–39]; for other special techniques related to field theory in curved space-time we follow the treatment given in [40–42].

We are grateful to our colleagues in B.I. Stepanov Institute of Physics, the National Academy of Sciences in Belarus, Belorussian State University, and Mozyr State Pedagogical University: L.M. Tomil'chik, Yu.A. Kurochkin, E.A. Tolkachev, I.S. Satsunkevich, Yu.P. Vably, V.V. Kudryashov, S.Yu. Sakovich, G.G. Krylov, G.V. Grushevskaya, I.N. Kralevich, I.N. Koval'chuk, V.S. Savenko, and V.V. Shepelevich, for advice and help when working on this manuscript.

Authors: Elena M. Ovsiyuk, Vasiliy V. Kisel, and Victor M. Red'kov

Chapter 1

Maxwell Equations, Geometry Effects on Constitutive Relations

1.1 Introduction

There is a scope of details in the analysis to the problem of constitutive relations in Maxwell electrodynamics. These details involve their possible form and its role in physical manifestation of electromagnetic fields. Also, in its behavior under the motion of the reference frame and its connection with Special Relativity theory, interplay between electrodynamic constitutive relations and gravity theory, and so on. There is a long history on how this is tightly linked with the problem of electromagnetic energy-momentum tensor in matter.

Note, it seemed that Gordon [44] was first largely interested in trying to describe dielectric media by an "effective metrics". That is, Gordon tried to use a gravitational field to mimic a dielectric medium. The idea was taken up and developed by Tamm and Mandel'stam [45], [46]. A noticeable contribution was one of the "problems" in Landau and Lifshitz book [37] (1935).

We have located extensive interest to this very problem. We can cite a huge list of literature that has been published on this issue: Minkowski [6], Kottler [43], Dantzig [47], Gordon [44], Tamm and Mandel'stam [45, 46], Landau and Lifshitz [37], Schrödinger [48, 52], Tonnelat [49], Skrotskii [54], Pham Mau Quan [56, 58], Balazs [55, 59], Tomil'chik and Fedorov [60], Dzyaloshinskii [61], Plebanski [62], Winterberg [64], de Felice [66], Post [63], O'Dell [65], Bolotowskij and Stoliarov [67], Veselago [68–71], Fedorov [72], Barykin et al. [73], Berezin et al. [76, 77], Turov [74], Schleich [75], Antoci and Minich [78–81], Weiglhofer [82,83], Hillion [84], Lakhtakia et al. [85–90], Leonhardt and Piwnicki [92–98], Novello et al. [99–102], Lange and Raab [103, 104], Piwnicki [105], Brevik and Halnes [106], Vinogradov [107], Barkovsky et al. [108, 109], Obukhov et al. [110–112], Nandi [113], Vlokh et al. [114–118], Matagne [119], Barcelro [121], Boonserm et al. [120], Delphenich [122–124], Dereli et al. [125, 126].

1.2   Riemannian Geometry and Maxwell Theory

Let us start with the Maxwell equations in Minkowski space – in vector notation they are [50]

$$\text{div } \mathbf{B} = 0 \,, \qquad \text{rot } \mathbf{E} = -\frac{\partial \mathbf{B}}{\partial t} \,,$$

$$\epsilon\epsilon_0 \text{ div } \mathbf{E} = \rho \,, \quad \frac{1}{\mu\mu_0} \text{ rot } \mathbf{B} = \mathbf{J} + \epsilon\epsilon_0 \frac{\partial \mathbf{E}}{\partial t} \,, \tag{1.2.1}$$

wherein SI units are implemented. Through the use of material (constitutive) relations

$$\mathbf{H} = \frac{\mathbf{B}}{\mu\mu_0} \,, \qquad \mathbf{D} = \epsilon\epsilon_0 \, \mathbf{E} \tag{1.2.2}$$

Eqs. (1.2.1) can be written in terms of 4-vectors

$$\text{div } c\mathbf{B} = 0 \,, \quad \text{rot } \mathbf{E} = -\frac{\partial c\mathbf{B}}{\partial x^0} \,,$$

$$\text{div } \mathbf{D} = j^0, \quad \text{rot } \frac{\mathbf{H}}{c} = \frac{\mathbf{J}}{c} + \frac{\partial \mathbf{D}}{\partial x^0} \,, \tag{1.2.3}$$

where $x^0 = ct$. In terms of two electromagnetic tensors we represent the electric displacement $\mathbf{D}$ and the magnetic field $\mathbf{H}$ by the antisymmetric tensor H^{ik}, while the electric $\mathbf{E}$ and the magnetic induction $\mathbf{B}$ are accounted for by the tensor F^{ik}

$$(F^{\alpha\beta}) = \begin{vmatrix} 0 & -E^1 & -E^2 & -E^3 \\ E^1 & 0 & -cB^3 & cB^2 \\ E^2 & cB^3 & 0 & -cB^1 \\ E^3 & -cB^2 & cB^1 & 0 \end{vmatrix} \,,$$

$$(H^{\alpha\beta}) = \begin{vmatrix} 0 & -D^1 & -D^2 & -D^3 \\ D^1 & 0 & -H^3/c & H^2/c \\ D^2 & H^3/c & 0 & -H^1/c \\ D^3 & -H^2/c & H^1/c & 0 \end{vmatrix} \,,$$

$$E^i = -E_i \,, \; D^i = -D_i \,, \qquad B^i = +B_i \,, \; H^i = +H_i \,, \qquad j^a = (\rho, \mathbf{J}/c) \,;$$

then Eqs. (1.2.3) take the form

$$\partial_a F_{bc} + \partial_b F_{ca} + \partial_c F_{ab} = 0 \,, \qquad \partial_b H^{ba} = j^a. \tag{1.2.4}$$

For a vacuum case, the constitutive relations are

$$\mathbf{D} = \epsilon_0 \mathbf{E} = (D^i) \,, \qquad \mathbf{H} = \frac{1}{\mu_0} \mathbf{B} = (H^i) \,,$$

Now, to consider the tensor form as a trivial linear identity

$$H^{ab}(x) = \epsilon_0 \, F^{ab}(x) \,, \tag{1.2.5}$$

and Eqs. (1.2.4) then becomes

$$\partial_a F_{bc} + \partial_b F_{ca} + \partial_c F_{ab} = 0 \,, \qquad \partial_b F^{ba} = \frac{1}{\epsilon_0}\, j^a \,. \tag{1.2.6}$$

The situation is quite different with a non-vacuum case. Even in the simplest case of a uniform media, the procedure relativizing the constitutive equations

$$\mathbf{D} = \epsilon_0 \epsilon \mathbf{E} = (D^i) \,, \qquad \mathbf{H} = \frac{1}{\mu_0 \mu}\, \mathbf{B} = (H^i)$$

requires a subsidiary (4×4)-matrix with presumed properties of a 2-rank tensor:

$$\eta^{am} = \sqrt{\epsilon}\ \begin{vmatrix} 1/k & 0 & 0 & 0 \\ 0 & -k & 0 & 0 \\ 0 & 0 & -k & 0 \\ 0 & 0 & 0 & -k \end{vmatrix} \,,$$

$$k = \frac{1}{\sqrt{\epsilon\mu}} \,, \qquad H^{ab} = \epsilon_0\, \eta^{am} \eta^{bn}\, F_{mn} \,. \tag{1.2.7}$$

We may postulate a class of linear inhomogeneous electromagnetic media characterized by a 4-index constitutive object [45, 46]:

$$H^{ab}(x) = \epsilon_0\, \Delta^{abmn}(x)\, F_{mn}(x) \,, \tag{1.2.8}$$

relativistic symmetry presume that the quantity Δ^{abmn} to be a 4-rank tensor under the Lorentz group, with evident symmetry constrains:

$$\Delta^{abmn}(x) = -\Delta^{bamn}(x) = -\Delta^{abnm}(x) \,, \tag{1.2.9}$$

so $\Delta^{abmn}(x)$ depends upon 36 parameters. The constitutive object $\Delta^{abmn}(x)$ bears a resemblance to the curvature Riemann tensor; so then, this can provide us with a classifying problem. The constitutive object $\Delta^{abmn}(x)$ can include differential operators as well – see [72, 74, 107].

When extending the Maxwell theory to a space-time with non-Euclidean geometry, which can describe gravity according to General Relativity; one must change previous equations to a more general form – for simplicity, let us start with the most simple case of vacuum Maxwell equations

$$\nabla_\alpha f_{\beta\gamma} + \nabla_\beta f_{\gamma\alpha} + \nabla_\gamma f_{\alpha\beta} = 0 \,,$$

$$\nabla_\beta h^{\beta\alpha} = j^\alpha \,, \qquad h_{\alpha\beta} = \epsilon_0\, f_{\alpha\beta} \,; \tag{1.2.10}$$

to distinguish formulas referring to a flat and curved models let us use small letters to designate electromagnetic tensors in curved model, f_{ab} and h^{ab}.

1.3   Maxwell Equations in Riemannian Space-Time and Media

Let us discuss in detail by considering the (vacuum) Maxwell equations in a curved space-time as Maxwell equations in flat space-time but then specified for an effective medium, the properties of which are determined by metrical structure of the initial curved model $g_{\alpha\beta}(x)$. Let us restrict ourselves to the case of curved space-time models which are parameterized by some quasi-Cartesian coordinate system x^a.

Vacuum Maxwell equations in a Riemannian space-time, parameterized by some quasi-Cartesian coordinates can be brought to the form [37]

$$\partial_a f_{bc} + \partial_b f_{ca} + \partial_c f_{ab} = 0 \ ,$$

$$\frac{1}{\sqrt{-g}} \ \partial_b \ \sqrt{-g} \ f^{ba} = \frac{1}{\epsilon_0} j^a \ . \tag{1.3.1}$$

Just by the form in Eqs.(1.3.1) gives rise to the line of thinking that Maxwell theory can have a pre-metric status. Indeed, one can immediately see that when introducing new (formal) variables

$$\sqrt{-g} \ j^a \longrightarrow j^a \ , \qquad f_{ab} \longrightarrow F_{ab} \ ,$$

$$\epsilon_0 \ \sqrt{-g} \ g^{am}(x)g^{bn}(x) \ f_{mn}(x) \longrightarrow H^{ba} \ , \tag{1.3.2}$$

Equations (1.3.1) in the curved space can be re-written as Maxwell equations in flat space but in a medium:

$$\partial_a F_{bc} + \partial_b F_{ca} + \partial_c F_{ab} = 0 \ , \qquad \partial_b \ H^{ba} = \frac{1}{\epsilon_0} j^a \ . \tag{1.3.3}$$

Relations playing the role of constitutive equations are determined by the metrical structure

$$H^{\beta\alpha}(x) = \epsilon_0 \left[\sqrt{-g(x)} \ g^{\alpha\rho}(x)g^{\beta\sigma}(x) \right] F_{\rho\sigma}(x) \ . \tag{1.3.4}$$

There exists one special case: namely, if $g(x)$ does not depend on coordinates. In fact, then the factor $\sqrt{-g(x)} = \sqrt{-g}$ can be omitted from the previous formulas.

1.4   Metrical Tensor $g_{\alpha\beta}(x)$ and Constitutive Relations

In this section, let us consider the material equations for electromagnetic fields which are generated by metrical structure of the curved space-time model. Consider the case of arbitrary metrical tensor

$$g_{\alpha\beta}(x) = \begin{vmatrix} g_{00} & g_{01} & g_{02} & g_{03} \\ g_{01} & g_{11} & g_{12} & g_{13} \\ g_{02} & g_{12} & g_{22} & g_{23} \\ g_{03} & g_{13} & g_{23} & g_{33} \end{vmatrix} . \tag{1.4.1}$$

We can obtain a 3-dimensional form of relation in Eqs. (1.3.4). Their general structure should be as follows

$$D^i = \epsilon_0 \ \epsilon^{ik}(x) \ E_k + \epsilon_0 c \ \alpha^{ik}(x) \ B_k \ ,$$

$$H^i = \epsilon_0 c \ \beta^{ik}(x) \ E_k + \mu_0^{-1} \ (\mu^{-1})^{ik}(x) \ B_k \ . \tag{1.4.2}$$

By interpreting Eqs. (1.4.2), one should give attention to the different positions of the vector index in D^i, H^i and E_k and remember the accepted agreement

$$\mathbf{E} = (E^i) , \quad \mathbf{D} = (D^i) , \quad \mathbf{B} = (B^i) = (B_i) , \quad \mathbf{H} = (H^i) = (H_i) .$$

Four (constitutive) dimensionless (3×3)-matrices (remember on 36 parameters)

$$\epsilon^{ik}(x), \; \alpha^{ik}(x), \; \beta^{ik}(x), \; (\mu^{-1})^{ik}(x)$$

are not independent; because generally speaking, they are bilinear functions of only 10 components of the symmetrical tensor $g_{\alpha\beta}(x)$. After quite simple calculation, one produces expressions for four matrices:

$$\epsilon^{ik}(x) = \sqrt{-g} \left(g^{00}(x) \, g^{ik}(x) - g^{0i}(x) \, g^{0k}(x) \right) ,$$

$$(\mu^{-1})^{ik}(x) = \frac{1}{2} \sqrt{-g} \; \epsilon_{imn} \; g^{ml}(x) g^{nj}(x) \; \epsilon_{ljk} ,$$

$$\alpha^{ik}(x) = +\sqrt{-g} \; g^{ij}(x) \, g^{0l}(x) \; \epsilon_{ljk} ,$$

$$\beta^{ik}(x) = -\sqrt{-g} \; g^{0j}(x) \; \epsilon_{jil} \; g^{lk}(x) . \tag{1.4.3}$$

The tensor $\epsilon^{ik}(x)$ is evidently symmetrical; it is easy to demonstrate the same property for $(\mu^{-1})^{ik}(x)$. Indeed,

$$(\mu^{-1})^{ki}(x) = \frac{1}{2} \; \epsilon_{kmn} g^{ml}(x) g^{nj}(x) \epsilon_{lji} ,$$

allowing for changes in mute indices, $m \leftrightarrow j, n \leftrightarrow l$, we get

$$(\mu^{-1})^{ki}(x) = \frac{1}{2} \; \epsilon_{kjl} g^{jn}(x) g^{lm}(x) \epsilon_{nmi}$$

$$= \epsilon_{imn} g^{lm}(x) g^{jn}(x) \epsilon_{ljk} = (\mu^{-1})^{ik}(x) .$$

In the same manner, one can prove the identity $\beta^{ki}(x) = +\alpha^{ik}$. Indeed,

$$\beta^{ki} = -g^{0j}(x) \epsilon_{jkl} g^{li}(x) = g^{il}(x) g^{0j}(x) \epsilon_{jlk} = +\alpha^{ik} .$$

These tensors obey special symmetry conditions:

$$\epsilon^{ik}(x) = +\epsilon^{ki}(x) ,$$

$$(\mu^{-1})^{ik}(x) = +(\mu^{-1})^{ki}(x) , \qquad \beta^{ki}(x) = \alpha^{ik}(x) ; \tag{1.4.4}$$

which mean that the (6×6)-matrix defining material (constitutive) equations

$$\begin{vmatrix} D^i(x) \\ H^i(x) \end{vmatrix} = \begin{vmatrix} \epsilon_0 \epsilon^{ik}(x) & \epsilon_0 c \alpha^{ik}(x) \\ \epsilon_0 c \beta^{ik}(x) & \mu_0^{-1} (\mu^{-1})^{ik}(x) \end{vmatrix} \begin{vmatrix} E_k(x) \\ B_k(x) \end{vmatrix} \tag{1.4.5}$$

is a symmetrical matrix.

It is convenient to introduce $(3+1)$-splitting for the metrical tensor $g^{\alpha\beta}(x)$

$$g^{\alpha\beta}(x) = \left| \begin{array}{cc} g^{00} & (g^{0i}) = \bar{g} \\ (g^{i0}) = \bar{g} & (g^{ik}) = g \end{array} \right| ,$$

$$(\bar{g}^{\times})_{jk} \equiv g^{0l}(x)\epsilon_{ljk} = g^l(x)\epsilon_{ljk} \tag{1.4.6}$$

tensors $(\epsilon^{ik}), (\alpha^{ik}), (\beta^{ik})$ can be written in the form

$$\epsilon(x) = \sqrt{-g}\,[\,g^{00}(x)g(x) - \bar{g}(x)\bullet\bar{g}(x)\,]\,,$$

$$\alpha(x) = \sqrt{-g}\,g(x)\,\bar{g}^{\times}(x)\,, \qquad \beta(x) = -\sqrt{-g}\,\bar{g}^{\times}(x)\,g(x)\,. \tag{1.4.7}$$

Also, one can produce a more convenient representation for $(\mu^{-1})^{ik}(x)$. Indeed, with notation $(\tau_i)_{mn} = \epsilon_{imn}$, we get

$$(\mu^{-1})^{ik}(x) = -\frac{1}{2}\,\sqrt{-g}\,\mathrm{Sp}\,[\,\tau_i\,g(x)\,\tau_k\,g(x)\,]\,. \tag{1.4.8}$$

Metrical tensors are very physically interesting within the context of General relativity [37], in that they have a quasi-diagonal structure and correspondingly the effective constitutive relations are much more simplified:

$$g^{\alpha\beta}(x) = \left| \begin{array}{cccc} g^{00} & 0 & 0 & 0 \\ 0 & g^{11} & g^{12} & g^{13} \\ 0 & g^{21} & g^{22} & g^{23} \\ 0 & g^{31} & g^{32} & g^{33} \end{array} \right| , \qquad \alpha(x) = 0\,, \qquad \beta(x) = 0\,,$$

$$\epsilon(x) = \sqrt{-g}\,g^{00}(x)g(x)\,, \qquad (\mu^{ik})(x) = -\frac{1}{2}\,\sqrt{-g}\,\mathrm{Sp}\,[\,\tau_i\,g(x)\,\tau_k\,g(x)\,]\,. \tag{1.4.9}$$

Explicit expression for tensors $\epsilon^{ik}(x)$ and $(\mu^{-1})^{ik}(x)$ given by Eqs.(1.4.9) are:

$$(\epsilon^{ik}) = \sqrt{-g}\,g^{00}\left| \begin{array}{ccc} g^{11} & g^{12} & g^{13} \\ g^{21} & g^{22} & g^{23} \\ g^{31} & g^{32} & g^{33} \end{array} \right| , \qquad ((\mu^{-1})^{ik}) = \sqrt{-g}\left| \begin{array}{ccc} G^{11} & G^{12} & G^{13} \\ G^{21} & G^{22} & G^{23} \\ G^{31} & G^{32} & G^{33} \end{array} \right| ,$$

$$\tag{1.4.10}$$

where $G^{ik}(x)$ stands for (algebraic) co-factor to the element $g^{ik}(x)$

$$G^{ik}(x) = (-1)^{j+k}M^{ik}(x)$$

$$= \left| \begin{array}{ccc} (g^{22}g^{33} - g^{23}g^{32}) & (g^{31}g^{23} - g^{21}g^{33}) & (g^{21}g^{32} - g^{22}g^{31}) \\ (g^{32}g^{13} - g^{33}g^{12}) & (g^{33}g^{11} - g^{31}g^{13}) & (g^{31}g^{12} - g^{32}g^{11}) \\ (g^{12}g^{23} - g^{13}g^{22}) & (g^{13}g^{21} - g^{11}g^{23}) & (g^{11}g^{22} - g^{12}g^{21}) \end{array} \right| . \tag{1.4.11}$$

Therefore, two constitutive matrices $\epsilon(x)$ and $\mu^{-1}(x)$ are not independent and obey identity

$$\epsilon(x)\mu^{-1}(x) = \frac{-g^{00}g_{00}}{\det g_{ik}}\left| \begin{array}{ccc} g^{11} & g^{12} & g^{13} \\ g^{21} & g^{22} & g^{23} \\ g^{31} & g^{32} & g^{33} \end{array} \right|$$

$$\times \begin{vmatrix} (g^{22}g^{33} - g^{23}g^{32}) & (g^{31}g^{23} - g^{21}g^{33}) & (g^{21}g^{32} - g^{22}g^{31}) \\ (g^{32}g^{13} - g^{33}g^{12}) & (g^{33}g^{11} - g^{31}g^{13}) & (g^{31}g^{12} - g^{32}g^{11}) \\ (g^{12}g^{23} - g^{13}g^{22}) & (g^{13}g^{21} - g^{11}g^{23}) & (g^{11}g^{22} - g^{12}g^{21}) \end{vmatrix} = -I . \qquad (1.4.12)$$

Thus, the most physical in the context of General relativity metrical tensors with a quasi-diagonal structure effectively describe the media with the following constitutive relations (the minus sign given in this example may be taken away by changing the notation)

$$\mathbf{D} = -\epsilon_0 \, \epsilon(x) \, \mathbf{E} , \qquad \mathbf{B} = \mu_0 \mu(x) \, \mathbf{H} , \qquad \mu(x) = -\epsilon(x) ,$$

$$(\epsilon^{ik})(x) = \sqrt{-g(x)} \, g^{00}(x) \begin{vmatrix} g^{11}(x) & g^{12}(x) & g^{13}(x) \\ g^{21}(x) & g^{22}(x) & g^{23}(x) \\ g^{31}(x) & g^{32}(x) & g^{33}(x) \end{vmatrix} . \qquad (1.4.13)$$

Turning again to the general case of arbitrary metrical tensor, four constitutive tensors may be given explicitly in terms of metrical tensor $g^{\alpha\beta}(x)$ (we use notation $g^i(x) = g^{i0}(x)$)

$$[\, \epsilon^{ik}(x) \,] = \sqrt{-g} \, g^{00} \begin{vmatrix} g^{11} & g^{12} & g^{13} \\ g^{21} & g^{22} & g^{23} \\ g^{31} & g^{32} & g^{33} \end{vmatrix} - \sqrt{-g} \begin{vmatrix} g^1 g^1 & g^1 g^2 & g^1 g^3 \\ g^2 g^1 & g^2 g^2 & g^2 g^3 \\ g^3 g^1 & g^3 g^2 & g^3 g^3 \end{vmatrix} ,$$

$$(\mu^{-1})^{ik}(x) = (\sqrt{-g} \begin{vmatrix} (g^{22}g^{33} - g^{23}g^{32}) & (g^{31}g^{23} - g^{21}g^{33}) & (g^{21}g^{32} - g^{22}g^{31}) \\ (g^{32}g^{13} - g^{33}g^{12}) & (g^{33}g^{11} - g^{31}g^{13}) & (g^{31}g^{12} - g^{32}g^{11}) \\ (g^{12}g^{23} - g^{13}g^{22}) & (g^{13}g^{21} - g^{11}g^{23}) & (g^{11}g^{22} - g^{12}g^{21}) \end{vmatrix} ,$$

$$\alpha^{ik}(x) = \sqrt{-g} \begin{vmatrix} (-g^{12}g^3 + g^{13}g^2) & (g^{11}g^3 - g^{13}g^1) & (-g^{11}g^2 + g^{12}g^1) \\ (-g^{22}g^3 + g^{23}g^2) & (g^{21}g^3 - g^{23}g^1) & (-g^{21}g^2 + g^{22}g^1) \\ (-g^{32}g^3 + g^{33}g^2) & (g^{31}g^3 - g^{33}g^1) & (-g^{31}g^2 + g^{32}g^1) \end{vmatrix} = \beta^{ki}(x) .$$

$$(1.4.14)$$

1.5   Inverse Material Equations

Estabishing the constitutive equations as follows

$$H^{\rho\sigma}(x) = \epsilon_0 \, \sqrt{-g(x)} \, g^{\rho\alpha}(x) g^{\sigma\beta}(x) \, F_{\alpha\beta}(x) ,$$

in 3-vector form, they are

$$D^i = \epsilon_0 \epsilon^{ik} \, E_k + \epsilon_0 c \, \alpha^{ik} \, B_k ,$$

$$H^i = \epsilon_0 c \, \beta^{ik} \, E_k + \frac{1}{\mu_0} (\mu^{-1})^{ik} \, B_k . \qquad (1.5.1)$$

Let us now derive the inverse of this

$$F_{\rho\sigma} = \frac{1}{\epsilon_0} \frac{1}{\sqrt{-g(x)}} g_{\rho\alpha}(x) g_{\sigma\beta}(x) \, H^{\alpha\beta} ,$$

$$E_i = \frac{1}{\epsilon_0} (\epsilon^{-1})_{ik} \, D^k + \frac{1}{\epsilon_0 c} \, (\alpha^{-1})_{ik} \, H^k ,$$

$$B_i = \frac{1}{\epsilon_0 c} \left(\beta^{-1} \right)_{ik} D^k + \mu_0 \mu_{ik} H^k . \tag{1.5.2}$$

By symmetry reason, one does not need to make any substantially new calculation. In addition, as given in Section **1.4**, the expressions for inverse constitutive tensors are:

$$\left(\epsilon^{-1} \right)_{ik} = \frac{1}{\sqrt{-g(x)}} \left[g_{00}(x) g_{ik}(x) - g_i(x) g_k(x) \right] ,$$

$$\mu_{ik} = \frac{1}{2} \frac{1}{\sqrt{-g(x)}} \epsilon_{imn} g_{ml}(x) g_{nj}(x) \epsilon_{ljk} ,$$

$$\left(\alpha^{-1} \right)_{ik} = + \frac{1}{\sqrt{-g(x)}} g_{ij}(x) g_l(x) \epsilon_{ljk} ,$$

$$\left(\beta^{-1} \right)_{ik} = - \frac{1}{\sqrt{-g(x)}} g_j(x) \epsilon_{jil} g_{lk}(x) , \tag{1.5.3}$$

with symmetries

$$\left(\epsilon^{-1} \right)_{ik}(x) = + \left(\epsilon^{-1} \right)_{ki}(x) ,$$

$$\mu_{ik}(x) = + \mu_{ki}(x) ,$$

$$\left(\beta^{-1} \right)_{ki}(x) = + \left(\alpha^{-1} \right)_{ik} . \tag{1.5.4}$$

Therefore, the (6×6)-matrix determining inverse material equations is symmetrical

$$\left| \begin{array}{c} E_k \\ B_k \end{array} \right| = \left| \begin{array}{cc} \epsilon_0^{-1} (\epsilon^{-1})_{kl} & \epsilon_0^{-1} c^{-1} (\alpha^{-1})_{kl} \\ \epsilon_0^{-1} c^{-1} (\beta^{-1})_{kl} & \mu_0 \mu_{kl} \end{array} \right| \left| \begin{array}{c} D^l \\ H^l \end{array} \right| . \tag{1.5.5}$$

1.6   Geometric Simulating of Inhomogeneous Media

Let us consider a special form of the diagonal metrical tensor

$$g_{\alpha\beta} = \left| \begin{array}{cccc} a^2(x) & 0 & 0 & 0 \\ 0 & -b^2(x) & 0 & 0 \\ 0 & 0 & -b^2(x) & 0 \\ 0 & 0 & 0 & -b^2(x) \end{array} \right| . \tag{1.6.1}$$

Relations in Eqs.(1.4.13) take the form

$$\mathbf{D} = -\epsilon_0 \, \epsilon(x) \, \mathbf{E} , \quad \mathbf{B} = \mu_0 \mu(x) \mathbf{H} ,$$

$$\mu(x) = -\epsilon(x) = \frac{b(x)}{a(x)} \, I . \tag{1.6.2}$$

For a general diagonal anisotropic metrical tensor

$$g_{\alpha\beta} = \left| \begin{array}{cccc} a^2(x) & 0 & 0 & 0 \\ 0 & -b_1^2(x) & 0 & 0 \\ 0 & 0 & -b_2^2(x) & 0 \\ 0 & 0 & 0 & -b_3^2(x) \end{array} \right| , \tag{1.6.3}$$

relations in Eqs.(1.4.13) take the form

$$\mathbf{D} = -\epsilon_0\, \epsilon(x)\, \mathbf{E}\ , \qquad \mathbf{B} = \mu_0 \mu(x) \mathbf{H}\ ,$$

$$\mu(x) = -\epsilon(x) = \begin{vmatrix} b_2 b_3/ab_1 & 0 & 0 \\ 0 & b_3 b_1/ab_2 & 0 \\ 0 & 0 & b_1 b_2/ab_3 \end{vmatrix}. \tag{1.6.4}$$

1.7   Geometrical Simulating of the Uniform Media

Let us consider one special form for the metrical tensor:

$$g_{\alpha\beta} = \begin{vmatrix} a^2 & 0 & 0 & 0 \\ 0 & -b^2 & 0 & 0 \\ 0 & 0 & -b^2 & 0 \\ 0 & 0 & 0 & -b^2 \end{vmatrix}, \tag{1.7.1}$$

where a^2 and b^2 are taken as arbitrary (positive) numerical parameters. This is a special case mentioned in connection with Eq.(1.3.2): if $g(x)$ does not depend on coordinates; then, in fact, the factor $\sqrt{-g}$ can be omitted from the formulas. In this conclusion, we get the material equations generated by that geometry

$$(\epsilon^{ik}) = \frac{1}{a^2 b^2}\begin{vmatrix} -1 & 0 & 0 \\ 0 & -1 & 0 \\ 0 & 0 & -1 \end{vmatrix}, \qquad ((\mu^{-1})^{ik}) = \frac{1}{b^4}\begin{vmatrix} 1 & 0 & 0 \\ 0 & 1 & 0 \\ 0 & 0 & 1 \end{vmatrix}, \tag{1.7.2}$$

or differently

$$D^i = -\frac{\epsilon_0}{a^2 b^2}\, E_i\ , \qquad H^i = \frac{1}{\mu_0 b^4}\, B_i\ , \tag{1.7.3}$$

from which it follows

$$b^2 = \sqrt{\mu}\ , \qquad a^2 = \frac{1}{\epsilon}\frac{1}{\sqrt{\mu}}\ . \tag{1.7.4}$$

Corresponding metrical tensor of Eq.(1.7.1) is

$$g_{\alpha\beta} = \frac{1}{\sqrt{\epsilon}}\begin{vmatrix} 1/\sqrt{\epsilon\mu} & 0 & 0 & 0 \\ 0 & -\sqrt{\epsilon\mu} & 0 & 0 \\ 0 & 0 & -\sqrt{\epsilon\mu} & 0 \\ 0 & 0 & 0 & -\sqrt{\epsilon\mu} \end{vmatrix}. \tag{1.7.5}$$

It should be mentioned that in two very (geometrically) peculiar cases we get modeling media with negative values of ϵ and μ:

$$
\begin{array}{ccc}
 & a^2 > 0 & a^2 < 0 \\
b^2 > 0 & \text{ordinary} & \text{special} \\
b^2 < 1 & \text{special} & \text{ordinary}
\end{array}
\qquad . \tag{1.7.6}
$$

Such peculiar possibilities will exist in all other cases as considered next.

1.8   Geometrical Modeling of Anisotropic Uniform Media

Let us extend the previous analysis and consider another metrical tensor

$$
g_{\alpha\beta} =
\begin{vmatrix}
a^2 & 0 & 0 & 0 \\
0 & -b_1^2 & 0 & 0 \\
0 & 0 & -b_2^2 & 0 \\
0 & 0 & 0 & -b_3^2
\end{vmatrix},
\tag{1.8.1}
$$

where a^2, b_i^2 are arbitrary numerical parameters. The constitutive equations generated by that geometry are

$$
D^i = \epsilon_0 \epsilon^{ik} E_k , \qquad H^i = \mu_0^{-1} \mu^{ik} B_k ,
$$

$$
(\epsilon^{ik}) = a^{-2}
\begin{vmatrix}
-b_1^{-2} & 0 & 0 \\
0 & -b_2^{-2} & 0 \\
0 & 0 & -b_3^{-2}
\end{vmatrix},
$$

$$
(\mu^{ik}) =
\begin{vmatrix}
b_2^{-2}b_3^{-2} & 0 & 0 \\
0 & b_3^{-2}b_1^{-2} & 0 \\
0 & 0 & b_1^{-2}b_2^{-2}
\end{vmatrix},
\tag{1.8.2}
$$

or differently

$$
D^1 = -\frac{\epsilon_0 E_1}{a^2 b_1^2} , \qquad D^2 = -\frac{\epsilon_0 E_2}{a^2 b_2^2} , \qquad D^3 = -\frac{\epsilon_0 E_3}{a^2 b_3^2} ,
$$

$$
H^1 = \frac{B_1}{\mu_0 b_2^2 b_3^2} , \qquad H^2 = \frac{B_2}{\mu_0 b_3^2 b_1^2} , \qquad H^3 = \frac{B_3}{\mu_0 b_1^2 b_2^2} .
$$

These equations should be compared with

$$
D^1 = -\epsilon_0 \epsilon_1 E_1 , \qquad D^2 = -\epsilon_0 \epsilon_2 E_2 , \qquad D^3 = -\epsilon_0 \epsilon_3 E_3 ,
$$

$$
H^1 = \frac{B_1}{\mu_0 \mu_1} , \qquad H^2 = \frac{B_2}{\mu_0 \mu_2} , \qquad H^3 = \frac{B_3}{\mu_0 \mu_3} ,
$$

from which it follows

$$
\epsilon_1 = \frac{1}{a^2 b_1^2} , \qquad \epsilon_2 = \frac{1}{a^2 b_2^2} , \qquad \epsilon_3 = \frac{1}{a^2 b_3^2} ,
$$

$$
\mu_1 = b_2^2 \, b_3^2 , \qquad \mu_2 = b_3^2 \, b_1^2 , \qquad \mu_3 = b_1^2 \, b_2^2 .
\tag{1.8.3}
$$

One can readily obtain

$$
\frac{\mu_1}{\epsilon_1} = \frac{\mu_2}{\epsilon_2} = \frac{\mu_3}{\epsilon_3} = (a^2 \, b_1^2 \, b_2^2 \, b_3^2) = -g ,
$$

$$
-g = \sqrt{\frac{\mu_1^2 + \mu_2^2 + \mu_3^2}{\epsilon_1^2 + \epsilon_2^2 + \epsilon_2^2}} ,
$$

$$
\frac{\mu_i}{\sqrt{\mu_1^2 + \mu_2^2 + \mu_3^2}} = \frac{\epsilon_i}{\sqrt{\epsilon_1^2 + \epsilon_2^2 + \epsilon_2^2}} .
\tag{1.8.4}
$$

The latter means that one may use four independent parameters, ϵ, μ, n_i:

$$\epsilon_i = \epsilon \, n_i \,, \qquad \mu_i = \mu \, n_i, \qquad \mathbf{n}^2 = 1 \,. \qquad (1.8.5)$$

One can readily express b_i^2 in terms of μ_i:

$$b_1^2 = \sqrt{\frac{\mu_2 \mu_3}{\mu_1}} = \sqrt{\mu} \, \sqrt{\frac{n_2 n_3}{n_1}} \,,$$

$$b_2^2 = \sqrt{\frac{\mu_3 \mu_1}{\mu_2}} = \sqrt{\mu} \, \sqrt{\frac{n_3 n_1}{n_2}} \,,$$

$$b_3^2 = \sqrt{\frac{\mu_1 \mu_2}{\mu_3}} = \sqrt{\mu} \, \sqrt{\frac{n_1 n_2}{n_3}} \,. \qquad (1.8.6)$$

In turn, from $a^2 b_1^2 \, b_2^2 \, b_3^2 = \mu/\epsilon$ it follows

$$a^2 = \frac{\mu}{\epsilon} \, \frac{1}{b_1^2 b_2^2 b_3^2} = \frac{1}{\epsilon\sqrt{\mu}} \, \frac{1}{\sqrt{n_1 n_2 n_3}} \,. \qquad (1.8.7)$$

The formulas in Eqs. (1.8.6) and (1.8.7) provide us with (anisotropic) extension of the previous (isotropic) metrical tensor

$$g_{ab}(x) = \frac{1}{\sqrt{\epsilon}} \begin{vmatrix} \frac{1}{\sqrt{\epsilon\mu}} \frac{1}{\sqrt{n_1 n_2 n_3}} & 0 & 0 & 0 \\ 0 & -\sqrt{\epsilon\mu} \sqrt{\frac{n_2 n_3}{n_1}} & 0 & 0 \\ 0 & 0 & -\sqrt{\epsilon\mu} \sqrt{\frac{n_3 n_1}{n_2}} & 0 \\ 0 & 0 & 0 & -\sqrt{\epsilon\mu} \sqrt{\frac{n_1 n_2}{n_3}} \end{vmatrix} \,.$$

$$(1.8.8)$$

Here, again, we can produce geometrically very peculiar media with negative values for ϵ^{ik}, μ^{ik} (see [68–71]) when using different combinations of signs for the parameters a^2, b_1^2, b_2^2, b_3^2.

1.9 The Moving Medium and Anisotropy

One other example, is somewhat more involved, an effective anisotropic medium is provided by the material equations with a uniform media for the case with a moving observer (following the initial Minkowski investigation [6] this problem was considered and discussed in literature by many authors). The intitial starting point in Minkowski's approach to electrodynamics constitutive relations explicitly depends on the 4-velocity of motion in the reference frame under a media. Gordon [44], Tamm and Mandel'stam [45, 46], noticed that this observer dependent constitutive relations can be expressed with the help of an effective metric:

$$H^{ab}(x) = \epsilon_0 \, \Delta^{abmn} \, F_{mn} \,,$$

$$\Delta^{abmn} = \epsilon_0 \frac{1}{\sqrt{\mu}} [\, g^{am} + (\epsilon\mu - 1)u^a u^m \,] \frac{1}{\sqrt{\mu}} [\, g^{bn} + (\epsilon\mu - 1)u^b u^n \,] \,, \qquad (1.9.1)$$

here $g^{ab} = \mathrm{diag}(+1, -1, -1, -1)$. Corresponding constitutive 3-dimensional tensors are (let it be $\epsilon\mu - 1 = \gamma$)

$$
\epsilon^{ik} = \frac{1}{\mu} \begin{vmatrix} (-1 + \gamma u^1 u^1 - \gamma u^0 u^0) & \gamma u^1 u^2 & \gamma u^1 u^3 \\ \gamma u^1 u^2 & (-1 + \gamma u^2 u^2 - \gamma u^0 u^0) & \gamma u^2 u^3 \\ \gamma u^3 u^1 & \gamma u^3 u^2 & (-1 + \gamma u^3 u^3 - \gamma u^0 u^0) \end{vmatrix} ,
$$

$$
(\mu^{-1})^{ik} = \frac{1}{\mu} \begin{vmatrix} (1 - \gamma u^2 u^2 - \gamma u^3 u^3) & \gamma u^1 u^2 & \gamma u^1 u^3 \\ \gamma u^1 u^2 & (1 - \gamma u^3 u^3 - \gamma u^1 u^1) & \gamma u^2 u^3 \\ \gamma u^3 u^1 & \gamma u^3 u^2 & (1 - \gamma u^1 u^1 - \gamma u^2 u^2) \end{vmatrix} ,
$$

$$
\alpha^{ik} = \frac{1}{\mu} \begin{vmatrix} 0 & -\gamma u^0 u^3 & +\gamma u^0 u^2 \\ +\gamma u^0 u^3 & 0 & -\gamma u^0 u^1 \\ -\gamma u^0 u^2 & +\gamma u^0 u^1 & 0 \end{vmatrix} ,
$$

$$
\beta^{ik} = \frac{1}{\mu} \begin{vmatrix} 0 & +\gamma u^0 u^3 & -\gamma u^0 u^2 \\ -\gamma u^0 u^3 & 0 & +\gamma u^0 u^1 \\ +\gamma u^0 u^2 & -\gamma u^0 u^1 & 0 \end{vmatrix} . \tag{1.9.2}
$$

Let us now deduce the 3-dimensional vector form of these relations. For the vector D^i, we have

$$
D^1 = \frac{\epsilon_0}{\mu} \left[(-1 + \gamma u^1 u^1 - \gamma u^0 u^0) E_1 + \gamma u^1 u^2 E_2 + \gamma u^1 u^3 E_3 \right]
$$

$$
+ \frac{\epsilon_0 c}{\mu} (-\gamma u^0 u^3 B_2 + \gamma u^0 u^2 B_3) ,
$$

$$
D^2 = \frac{\epsilon_0}{\mu} \left[+\gamma u^1 u^2 E_1 + (-1 + \gamma u^2 u^2 - \gamma u^0 u^0) E_2 + \gamma u^2 u^3 E_3 \right]
$$

$$
+ \frac{\epsilon_0 c}{\mu} (\gamma u^0 u^3 B_1 - \gamma u^0 u^1 B_3) ,
$$

$$
D^1 = \frac{\epsilon_0}{\mu} \left[+\gamma u^1 u^3 E_1 + \gamma u^2 u^3 E_2 + (-1 + \gamma u^3 u^3 - \gamma u^0 u^0) E_3 \right]
$$

$$
+ \frac{\epsilon_0 c}{\mu} (-\gamma u^0 u^2 B_1 + \gamma u^0 u^1 B_2) ,
$$

and further

$$
D^1 = -\frac{\epsilon_0}{\mu} E_1 + \frac{\epsilon_0 \gamma}{\mu} [-u^0 u^0 E_1 + (u^1 E_1 + u^2 E_2 + u^3 E_3) u^1] + \frac{\epsilon_0 c \gamma}{\mu} u^0 (u^2 B_3 - u^3 B_2),
$$

$$
D^2 = -\frac{\epsilon_0}{\mu} E_2 + \frac{\epsilon_0 \gamma}{\mu} [-u^0 u^0 E_2 + (u^1 E_1 + u^2 E_2 + u^3 E_3) u^2] + \frac{\epsilon_0 c \gamma}{\mu} u^0 (u^3 B_1 - u^1 B_3),
$$

$$
D^3 = -\frac{\epsilon_0}{\mu} E_3 + \frac{\epsilon_0 \gamma}{\mu} [-u^0 u^0 E_3 + (u^1 E_1 + u^2 E_2 + u^3 E_3) u^3] + \frac{\epsilon_0 c \gamma}{\mu} u^0 (u^1 B_2 - u^2 B_1).
$$

By using this notation (we use dimensionless velocity variable $V = v/c$)

$$
u^0 = \frac{1}{\sqrt{1 - V^2}} , \qquad u^i = \frac{V^i}{\sqrt{1 - V^2}}
$$

previous relations look as follows:

$$D^1 = -\frac{\epsilon_0}{\mu}E_1 + \frac{\epsilon_0\gamma}{\mu}\frac{[-E_1 + (V^1E_1 + V^2E_2 + V^3E_3)V^1]}{1-V^2} + \frac{\epsilon_0 c\gamma}{\mu}\frac{(V^2B_3 - V^3B_2)}{1-V^2}\ ,$$

$$D^2 = -\frac{\epsilon_0}{\mu}E_2 + \frac{\epsilon_0\gamma}{\mu}\frac{[-E_1 + (V^1E_1 + V^2E_2 + V^3E_3)\,V^2]}{1-V^2} + \frac{\epsilon_0 c\gamma}{\mu}\frac{(V^3B_1 - V^1B_3)}{1-V^2}\ ,$$

$$D^3 = -\frac{\epsilon_0}{\mu}E_3 + \frac{\epsilon_0\gamma}{\mu}\frac{[-E_3 + (V^1E_1 + V^2E_2 + V^3E_3)\,V^3]}{1-V^2} + \frac{\epsilon_0 c\gamma}{\mu}\frac{(V^1B_2 - V^2B_1)}{1-V^2}\ ,$$

or in the vector form, are shown to be:

$$\mathbf{D} = \frac{\epsilon_0}{\mu}\mathbf{E} + \frac{\epsilon_0\gamma}{\mu}\frac{\mathbf{E} - (\mathbf{VE})\mathbf{V}}{1-V^2} + \frac{\epsilon_0 c\gamma}{\mu}\frac{\mathbf{V}\times\mathbf{B}}{1-V^2}\ . \tag{1.9.3}$$

Analogously, we should consider three relations for H^i:

$$H_1 = \frac{1}{\mu_0\mu}[(1 - \gamma u^2 u^2 - \gamma u^3 u^3)B_1 + \gamma u^1 u^2 B_2 + \gamma u^1 u^3 B_3] + \frac{\epsilon_0 c}{\mu}(\gamma u^0 u^3 E_2 - \gamma u^0 u^2 E_3)\ ,$$

$$H_2 = \frac{1}{\mu_0\mu}[\gamma u^1 u^2 B_1 + (1 - \gamma u^3 u^3 - \gamma u^1 u^1)B_2 + \gamma u^2 u^3 B_3] + \frac{\epsilon_0 c}{\mu}(\gamma u^0 u^3 E_2 - \gamma u^0 u^2 E_3)\ ,$$

$$H_3 = \frac{1}{\mu_0\mu}[\gamma u^3 u^1 B_1 + \gamma u^3 u^2 B_3 + (1 - \gamma u^1 u^1 - \gamma u^2 u^2)B_3] + \frac{\epsilon_0 c}{\mu}(\gamma u^0 u^2 E_1 - \gamma u^0 u^1 E_2)\ ,$$

or further

$$H_1 = \frac{1}{\mu_0\mu}B_1 + \frac{\gamma}{\mu_0\mu}(-u^2 u^2 B_1 - u^3 u^3 B_1 + u^1 u^2 B_2 + u^1 u^3 B_3) + \frac{\epsilon_0 c\gamma}{\mu}u^0(u^3 E_2 - u^2 E_3)\ ,$$

$$H_2 = \frac{1}{\mu_0\mu}B_2 + \frac{\gamma}{\mu_0\mu}(+u^1 u^2 B_1 - u^3 u^3 B_2 - u^1 u^1 B_2 + u^2 u^3 B_3) + \frac{\epsilon_0 c\gamma}{\mu}u^0(u^3 E_2 - u^2 E_3)\ ,$$

$$H_3 = \frac{1}{\mu_0\mu}B_3 + \frac{\gamma}{\mu_0\mu}(+u^3 u^1 B_1 + u^3 u^2 B_2 - u^1 u^1 B_3 - u^2 u^2 B_3) + \frac{\epsilon_0 c\gamma}{\mu}u^0(u^2 E_1 - u^1 E_2)\ ;$$

these relations can be rewritten in a vector form as follows:

$$\mathbf{H} = \frac{1}{\mu_0\mu}\mathbf{B} + \frac{\gamma}{\mu_0\mu}\frac{\mathbf{V}\times(\mathbf{V}\times\mathbf{B}))}{1-V^2} + \frac{\epsilon_0 c\,\gamma}{\mu}\frac{\mathbf{V}\times\mathbf{E}}{1-V^2}\ . \tag{1.9.4}$$

Relations in Eqs. (1.9.3) and (1.9.4) provide us with 3-dimensional vector form of constitutive relations in a media moving with velocity $\mathbf{V}$. The motion is effectively equivalent to an anisotropic media.

1.10   Effective Constitutive Equations Generated by Riemann Spherical Geometry

A 3-dimensional space of constant positive curvature, S_3, has many applications in physical problems. The most simple realization of this model is given by three-sphere in 4-space:

$$V_4^2 + V_1^2 + V_2^2 + V_3^2 = 1 \; ,$$

R is a curvature radius and used as a unit of length. These four coordinates are connected with those being the most used quasi-spherical relations:

$$x^0 = ct/R \; , \quad V_4 = \cos\chi \; , \quad V_i = \sin\chi \, n_i \; ,$$

$$n_i = (\sin\theta\cos\phi, \sin\theta\sin\phi, \sin\theta) \; ,$$

$$dS^2 = dx_0^2 - d\chi^2 - \sin^2\chi \, (d\theta^2 + \sin^2\theta d\phi^2) \; . \tag{1.10.1}$$

Also, interesting ones, are conformally-flat coordinates

$$y^0 = ct/R \; , \quad y^i = \frac{2V_i}{1+V_4} = 2\tan\chi/2 \, n_i \; ,$$

$$dS^2 = dt^2 - (1 + \frac{y^2}{4})^{-2}(y_1^2 + dy_2^2 + dy_3^2) \; ; \tag{1.10.2}$$

and quasi-Cartesian coordinates (they parameterize only elliptical model, the space of orthogonal group $SO(3)$)

$$x^i = \frac{y^i}{1 - y^2/4} = \tan\chi \, n_i = \frac{V_i}{V_4} \; ,$$

$$dS^2 = dx_0^2 - g_{jk}dx^j dx^k \; , \quad \chi \in [0, \pi/2] \; ;$$

$$g_{jk} = -\left(\frac{\delta_{jk}}{1+x^2} - \frac{x^j x^k}{(1+x^2)^2}\right) \; , \qquad g^{kl} = -(1+x^2)(\delta_{kl} + x^k x^l) \; , \tag{1.10.3}$$

calculating the determinant

$$\sqrt{-\det(g_{\alpha\beta})} = \frac{1}{(1+x^2)^{3/2}} \; . \tag{1.10.4}$$

For effective dielectric tensor $\epsilon^{ik}(x)$, we have

$$\epsilon^{ik}(x) = \begin{vmatrix} 1+x^1x^1 & x^1x^2 & x^1x^3 \\ x^1x^2 & 1+x^2x^2 & x^2x^3 \\ x^3x^1 & x^3x^2 & 1+x^3x^3 \end{vmatrix} . \tag{1.10.5}$$

For effective magnetic tensor $(\mu^{-1})^{ik}(x)$, we have

$$(\mu^{-1})^{ik}(x) = \sqrt{1+x^2} \begin{vmatrix} (1+x^2x^2+x^3x^3) & -x^1x^2 & -x^1x^3 \\ -x^2x^1 & (1+x^3x^3+x^1x^1) & -x^2x^3 \\ -x^3x^1 & -x^3x^2 & (1+x^1x^1+x^2x^2) \end{vmatrix} . \tag{1.10.6}$$

It is easily verified by direct calculation that dielectric tensor $(-\epsilon(x))$ and tensor $\mu^{-1}(x)$ are the inverse to each other

$$-\epsilon(x)\,\mu^{-1}(x) = I, \quad \text{or} \quad \mu(x) = -\epsilon(x) \; . \tag{1.10.7}$$

1.11   Effective Material Equations Generated by Lobachevsky Geometry

3-Dimensional space of constant negative curvature, H_3, has many applications in physical problems. The most simple realization of this model is given by pseudo-sphere in 4-space:

$$u_4^2 - u_1^2 - u_2^2 - u_3^2 = 1 .$$

These four coordinate are connected with quasi-spherical ones be relations:

$$\chi \in [0, +\infty) , \quad u_4 = \cosh \chi, \qquad w_i = \sinh \chi \, n_i ,$$

$$x^0 = ct/R, \quad n_i = (\sin \theta \cos \phi, \sin \theta \sin \phi, \sin \theta) ,$$

$$dS^2 = dx_0^2 - d\chi^2 - \sinh^2\chi \, (d\theta^2 + \sin^2 \theta d\phi^2) . \tag{1.11.1}$$

Frequently used coordinates are conformally-flat

$$y^0 = ct/R , \quad y^i = \frac{2u_i}{1 + u_4} = 2 \tanh \chi/2 \, n_i ,$$

$$dS^2 = dy_0^2 - (1 - \frac{y^2}{4})^{-2} \, (dy_1^2 + dy_2^2 + dy_3^2) , \tag{1.11.2}$$

and so are the quasi-Cartesian type, as shown:

$$x^i = \frac{y^i}{1 + y^2/4} = \tanh \chi \, n_i = \frac{w_i}{w_4} ,$$

$$dS^2 = c^2 dt^2 - (\frac{\delta_{jk}}{1 - x^2} + \frac{x^j x^k}{(1 - x^2)^2})dx^j dx^j ,$$

$$g_{\alpha\beta} = \begin{vmatrix} 1 & 0 \\ 0 & g_{jk} \end{vmatrix} , \quad g_{jk} = -(\frac{\delta_{jk}}{1 - x^2} + \frac{x^j x^k}{(1 - x^2)^2}) ,$$

$$g^{\alpha\beta} = \begin{vmatrix} 1 & 0 \\ 0 & g^{kl} \end{vmatrix} , \quad g^{kl} = -(1 - x^2)(\delta_{kl} - x^k x^l) , \tag{1.11.3}$$

with the determinant

$$\sqrt{-\det (g_{\alpha\beta})} = \frac{1}{(1 - x^2)^{3/2}} . \tag{1.11.4}$$

For effective dielectric tensor $\epsilon^{ik}(x)$, we have

$$\epsilon(x) = -\frac{1}{\sqrt{1 - x^2}} \begin{vmatrix} 1 - x^1 x^1 & -x^1 x^2 & -x^1 x^3 \\ -x^1 x^2 & 1 - x^2 x^2 & -x^2 x^3 \\ -x^3 x^1 & -x^3 x^2 & 1 - x^3 x^3 \end{vmatrix} . \tag{1.11.5}$$

For effective magnetic tensor $\mu^{ik}(x)$, we have

$$(\mu^{-1}(x) = \sqrt{1 - x^2} \begin{vmatrix} (1 - x^2 x^2 - x^3 x^3) & x^1 x^2 & x^1 x^3 \\ x^2 x^1 & (1 - x^3 x^3 - x^1 x^1) & x^2 x^3 \\ x^3 x^1 & x^3 x^2 & (1 - x^1 x^1 - x^2 x^2) \end{vmatrix} .$$

$$\tag{1.11.6}$$

It is easily verified by direct calculation that dielectric tensor $(-\epsilon(x))$ and tensor $\mu^{-1}(x)$ are inverse to each other $\epsilon(x) \, \mu^{-1}(x) = -I$.

1.12 Geometry Effect on Material Equations in Arbitrary Linear Media

Previously, we started with Maxwell equations in vacuum and changed them to generally covariant in Riemannian space-time. At this, vacuum material equations $H_{\alpha\beta} = \epsilon_0\, F_{\alpha\beta}$, due to presence of metrical tensor $g^{\rho\alpha}(x)$ gave us the modified material equations

$$H^{\rho\sigma}(x) = \sqrt{-g}\; g^{\rho\alpha}(x)g^{\sigma\beta}(x)\; \epsilon_0\, F_{\alpha\beta}(x)\;. \qquad (1.12.1)$$

For the first generalization, let us start with Maxwell equations in a uniform media; beginning with material equations for the uniform media

$$H_{\alpha\beta}(x) = \epsilon_0\eta_\alpha{}^a\eta_\beta{}^b\, F_{ab}(x) \qquad (1.12.2)$$

we will arrive at

$$H^{\rho\sigma}(x) = \sqrt{-g}\; g^{\rho\alpha}(x)g^{\sigma\beta}(x)\; \epsilon_0\eta_\alpha{}^a\eta_\beta{}^b F_{ab}(x)\;. \qquad (1.12.3)$$

Using this notation $\hat{F}_{\alpha\beta}(x) = \eta_\alpha{}^a\,\eta_\beta{}^b\, F_{ab}(x)$, they are written as follows:

$$H^{\rho\sigma}(x)(x) = \sqrt{-g}\; g^{\rho\alpha}(x)g^{\sigma\beta}(x)\; \epsilon_0\, \hat{F}_{\alpha\beta}(x)\;,$$

$$\hat{F}_{\alpha\beta}(x) = \begin{vmatrix} 0 & \epsilon F_{0i} \\ \epsilon F_{i0} & \mu^{-1}F_{ik} \end{vmatrix}. \qquad (1.12.4)$$

One should not make any additional calculation, instead it suffices to make one formal change $F_{\alpha\beta}(x) \Longrightarrow \hat{F}_{\alpha\beta}(x)$, and now material equations are

$$D^i = \epsilon_0\epsilon\, \epsilon^{ik}(x)\, E_k + \epsilon_0\epsilon c\, \alpha^{ik}(x)\, B_k\;,$$

$$H^i = \epsilon_0\epsilon c\, \beta^{ik}(x)\, E_k + \frac{1}{\mu_0\mu}\, (\mu^{-1})^{ik}(x)\, B_k\;. \qquad (1.12.5)$$

These relations provide us with material equations for uniform media modified by Riemannian geometry of background space-time. It is easily to make one other extension: let us start with an anisotropic media in Minkowski space

$$D_i = \epsilon_0\, \epsilon_{(0)kl}E_l\;, \qquad H_i = \mu_0^{-1}\, \mu_{(0)kl}^{-1}B_k\;; \qquad (1.12.6)$$

they will be modified into

$$D^i = \epsilon_0\, \epsilon^{ik}(x)\epsilon_{(0)kl}\, E_l + \epsilon_0 c\, \alpha^{ik}(x)\mu_{(0)kl}^{-1}\, B_l\;,$$

$$H^i = \epsilon_0 c\, \beta^{ik}(x)\epsilon_{(0)kl}\, E_l + \mu_0^{-1}\, (\mu^{-1})^{ik}(x)\mu_{(0)kl}\, B_l\;. \qquad (1.12.7)$$

And now, for the final extension: let us start with arbitrary (linear) media when material equations are determined by 4-rank tensor $H_{\alpha\beta}(x) = \epsilon_0\, \Delta_{\alpha\beta}{}^{ab}\, F_{ab}(x)$, from which Riemannian geometry will generate the following:

$$H^{\rho\sigma}(x)(x) = \sqrt{-g}\; g^{\rho\alpha}(x)g^{\sigma\beta}(x)\epsilon_0\Delta_{\alpha\beta}{}^{ab}F_{ab}(x)\;, \qquad (1.12.8)$$

or in 3-dimensional form

$$D^i = \epsilon_0 \left[\epsilon^{ik}(x)\, \epsilon_{(0)kl} + \alpha^{ik}(x)\beta_{(0)kl} \right] E_l + \epsilon_0 c \left[\epsilon^{ik}(x)\alpha_{(0)kl} + \alpha^{ik}(x)\mu^{-1}_{(0)kl} \right] B_l \,,$$

$$H^i = \epsilon_0 c \left[\beta^{ik}(x)\epsilon_{(0)kl} + (\mu^{-1})^{ik}(x)\beta_{(0)kl} \right] E_l + \frac{1}{\mu_0} \left[\beta^{ik}(x)\alpha_{(0)kl} + (\mu^{-1})^{ik}(x)\mu^{-1}_{(0)kl} \right] B_l,$$

or in a matrix form (with no indices)

$$\mathbf{D} = \epsilon_0 \left[\epsilon(x)\, \epsilon_{(0)} + \alpha(x)\beta_{(0)} \right] \mathbf{E} + \epsilon_0 c \left[\epsilon(x)\alpha_{(0)} + \alpha(x)\mu^{-1}_{(0)} \right] \mathbf{B} \,,$$

$$\mathbf{H} = \epsilon_0 c \left[\beta(x)\epsilon_{(0)} + \mu^{-1}(x)\beta_{(0)} \right] \mathbf{E} + \frac{1}{\mu_0} \left[\beta(x)\alpha_{(0)} + \mu^{-1}(x)\mu^{-1}_{(0)} \right] \mathbf{B} \,.$$

$$(1.12.9)$$

These formulas can be read symbolically:

$$\hat{\epsilon} = \epsilon(x)\, \epsilon_{(0)} + \alpha(x)\beta_{(0)} \,, \qquad \hat{\alpha} = \epsilon(x)\alpha_{(0)} + \alpha(x)\mu^{-1}_{(0)} \,,$$

$$\hat{\beta} = \beta(x)\epsilon_{(0)} + \mu^{-1}(x)\beta_{(0)} \,, \qquad \hat{\mu}^{-1} = \beta(x)\alpha_{(0)} + \mu^{-1}(x)\mu^{-1}_{(0)} \,. \qquad (1.12.10)$$

For instance, if starting material equations have only diagonal blocks

$$\epsilon_{(0)} \,, \qquad \alpha_{(0)} = 0 \,, \qquad \beta_{(0)} = 0 \,, \qquad \mu^{-1}_{(0)} \,,$$

the relations in Eqs. (1.12.10) become simpler:

$$\hat{\epsilon} = \epsilon(x)\, \epsilon_{(0)}, \qquad \hat{\alpha} = +\alpha(x)\, \mu_{(0)} \,,$$

$$\hat{\beta} = \beta(x)\epsilon_{(0)} \,, \qquad \hat{\mu}^{-1} = \mu(x)\, \mu^{-1}_{(0)} \,. \qquad (1.12.11)$$

1.13   Example: A Plane Wave in the Lobachevsky Space, Simulating a Medium

Let us construct some simple solutions for the plane wave type with the Maxwell equations in 3-dimensional Lobachevsky space model when using horospherical coordinates. We start with

$$(I) \qquad \partial_\alpha F_{\beta\gamma} + \partial_\beta F_{\gamma\alpha} + \partial_\gamma F_{\alpha\beta} = 0 \,,$$

$$(II) \qquad \frac{1}{\sqrt{-g}}\, \partial_\beta \sqrt{-g}\, F^{\beta\alpha} = \epsilon_0^{-1} j^\alpha \,, \qquad (1.13.1)$$

where $g(x) = \det [g_{\alpha\beta}(x)] < 0$. Eqs.(I) are equivalent to

$$(123) \qquad \partial_1 F_{23} + \partial_2 F_{31} + \partial_3 F_{12} = 0 \,,$$

$$(012) \qquad \partial_0 F_{12} + \partial_1 F_{20} + \partial_2 F_{01} = 0 \,,$$

$$(023) \qquad \partial_0 F_{23} + \partial_2 F_{30} + \partial_3 F_{02} = 0 \,,$$

$$(031) \qquad \partial_0 F_{31} + \partial_3 F_{10} + \partial_1 F_{03} = 0 \; . \qquad (1.13.2)$$

In turn, Eqs.(II) in the Eqs. (1.13.1) are equivalent to

$$\frac{1}{\sqrt{-g}} \partial_1 \sqrt{-g} \, F^{10} + \frac{1}{\sqrt{-g}} \partial_2 \sqrt{-g} \, F^{20} + \frac{1}{\sqrt{-g}} \partial_3 \sqrt{-g} \, F^{30} = \epsilon_0^{-1} j^0 \; ,$$

$$\frac{1}{\sqrt{-g}} \frac{\partial}{\partial x^0} \sqrt{-g} \, F^{01} - \frac{1}{\sqrt{-g}} \frac{\partial}{\partial x^2} \sqrt{-g} \, F^{12} + \frac{1}{\sqrt{-g}} \frac{\partial}{\partial x^3} \sqrt{-g} \, F^{31} = \epsilon_0^{-1} j^1 \; ,$$

$$\frac{1}{\sqrt{-g}} \frac{\partial}{\partial x^0} \sqrt{-g} \, F^{02} + \frac{1}{\sqrt{-g}} \frac{\partial}{\partial x^1} \sqrt{-g} \, F^{12} - \frac{1}{\sqrt{-g}} \frac{\partial}{\partial x^3} \sqrt{-g} \, F^{23} = \epsilon_0^{-1} j^2 \; ,$$

$$\frac{1}{\sqrt{-g}} \frac{\partial}{\partial x^0} \sqrt{-g} \, F^{03} - \frac{1}{\sqrt{-g}} \frac{\partial}{\partial x^1} \sqrt{-g} \, F^{31} + \frac{1}{\sqrt{-g}} \frac{\partial}{\partial x^2} \sqrt{-g} \, F^{23} = \epsilon_0^{-1} j^3 \; .$$

$$(1.13.3)$$

We will specify these equations in the absence of any sources and for horospherical coordinates in the hyperbolic model H_3 (see in [42]):

$$u_1 = R e^{-Z} \cos \phi \; , \qquad u_2 = R e^{-Z} \sin \phi \; ,$$

$$u_3 = \sinh Z + \frac{1}{2} R^2 e^{-Z} = \frac{1}{2} \left[e^{+Z} + (R^2 - 1) e^{-Z} \right] \; ,$$

$$u_0 = \cosh Z + \frac{1}{2} R^2 e^{-Z} = \frac{1}{2} \left[e^{+Z} + (R^2 + 1) e^{-Z} \right] \; ,$$

$$dS^2 = c^2 dt^2 - e^{-2Z} dR^2 - e^{-2Z} R^2 d\phi^2 - dZ^2 \; . \qquad (1.13.4)$$

In the limit of a vanishing curvature, the given coordinates reduce to cylindric ones. In using of notation

$$\frac{ct}{\rho} = x^0 \; , \qquad r = \frac{R}{\rho} = x^1 \; , \qquad \phi = x^2 \; , \qquad z = \frac{Z}{\rho} = x^3 \; ,$$

where ρ is the curvature radius, Eqs. (1.13.2) and (1.13.3) give

$$\partial_1 F_{23} + \partial_2 F_{31} + \partial_3 F_{12} = 0 \; ,$$

$$\partial_0 F_{12} + \partial_1 F_{20} + \partial_2 F_{01} = 0 \; ,$$

$$\partial_0 F_{23} + \partial_2 F_{30} + \partial_3 F_{02} = 0 \; ,$$

$$\partial_0 F_{31} + \partial_3 F_{10} + \partial_1 F_{03} = 0 \; , \qquad (1.13.5)$$

$$\frac{1}{r} \partial_r r F^{10} + \partial_\phi F^{20} + e^{2z} \partial_z e^{-2z} F^{30} = 0 \; ,$$

$$\partial_0 F^{01} - \partial_\phi F^{12} + e^{2z} \partial_z e^{-2z} F^{31} = 0 \; ,$$

$$\partial_0 F^{02} + \frac{1}{r} \partial_r r F^{12} - e^{2z} \partial_z e^{-2z} F^{23} = 0 \; ,$$

$$\partial_0 F^{03} - \frac{1}{r} \partial_r r F^{31} + \partial_\phi F^{23} = 0 \; . \qquad (1.13.6)$$

Here, let us impose simplicity restrictions

$$F_{01} = 0 \, , \qquad F_{02} = F_{0\phi}(r, z) \neq 0 \, , \qquad F_{03} = 0 \, , \tag{1.13.7}$$

Eqs. (1.13.5) and (1.13.6) become simpler

$$\partial_r \, F_{23} + \partial_z \, F_{12} = 0 \, , \ \partial_0 \, F_{12} + \partial_r \, F_{20} = 0 \, , \ \partial_0 \, F_{23} + \partial_z \, F_{02} = 0 \, , \tag{1.13.8}$$

$$F_{31} = 0 \, , \qquad \partial_0 F^{02} + \frac{1}{r}\partial_r r F^{12} - e^{2z}\partial_z e^{-2z} F^{23} = 0 \, . \tag{1.13.9}$$

Note that for Eqs. (1.13.8), the first equation follows from both the 2-nd and 3-rd equations, thus allowing only three equations to be independent:

$$\partial_0 F_{12} = \partial_r \, F_{02} \, , \qquad \partial_0 F_{23} = -\partial_z \, F_{02} \, , \tag{1.13.10}$$

$$\partial_0^2 g^{22} \, F_{02} + \frac{1}{r}\partial_r r \, g^{11}g^{22} \, \partial_0 F_{12} - e^{2z}\partial_z e^{-2z} g^{22} g^{33}\partial_0 F_{23} = 0 \, . \tag{1.13.11}$$

Differentiating the third equation on x_0 and by taking into account Eqs.(1.13.10), we get

$$-\frac{e^{2z}}{r^2}\partial_0^2 F_{02} + \frac{1}{r}\partial_r \frac{e^{4z}}{r} \, \partial_r F_{02} + \frac{e^{2z}}{r^2}\partial_z^2 F_{02} = 0 \, . \tag{1.13.12}$$

Let us search for solutions in the form of a plane wave

$$F_{02} = E(r) \, \cos(k_0 x^0 - kz + \beta) \, ; \tag{1.13.13}$$

where dimensionless quantities are used

$$k_0 = \frac{\omega\rho}{c} \, , \qquad k = \pm \, \frac{\omega\rho}{c} = \pm k_0 \, , \tag{1.13.14}$$

Eqs. (1.13.12) led to

$$\frac{k_0^2 e^{2z}}{r^2} \, E + \frac{e^{4z}}{r}\frac{d}{dr}\frac{1}{r}\frac{d}{dr} \, E - \frac{k_0^2 e^{2z}}{r^2} \, E = 0 \, , \tag{1.13.15}$$

or

$$\frac{d}{dr}\frac{1}{r}\frac{d}{dr} \, E = 0 \tag{1.13.16}$$

with solutions

$$E = \text{const} \, , \qquad E = \text{const} \, r^2; \tag{1.13.17}$$

Equation (1.13.16) coincides with a similar equation that arises in flat space; correspondingly its solutions coincides with the established known ones found in the flat Minkowski space. What is most interesting, here, is the solution for $E = \text{const}$, so that

$$F_{02} = E \, \cos(k_0 x^0 - kz + \beta) \, . \tag{1.13.18}$$

From Eq. (1.13.10), we readily find F_{12} and F_{23}:

$$\partial_0 F_{12} = \partial_r \, F_{02} \, , \qquad \partial_0 F_{23} = -\partial_z \, F_{02} \Longrightarrow$$

$$F_{12} = 0 , \qquad F_{23} = E \frac{k}{k_0} \cos(k_0 x^0 - kz + \beta) . \qquad (1.13.19)$$

In the electromagnetic field of Eqs. (1.13.18) and (1.13.19), it corresponds then for the 4-potential

$$A_\phi = A_2 = \frac{E}{k_0} \sin(k_0 x^0 - kz + \beta) . \qquad (1.13.20)$$

It is exactly coincides with an expression for the 4-potential for the simplest cylindric wave in Minkowski space. In taking $\beta = 0, \pi/2$, we get two linearly independent solutions of the Maxwell equations.

This solution obtained in horospheric coordinate can also be translated to quasi-Cartesian coordinates in H_3:

$$(x^1, x^2, x^3) = (r, \phi, z) \qquad \Longrightarrow \qquad y^i = \frac{u_i}{u_0} = (q_1, q_2, q_3) ,$$

$$y^1 = \frac{2r \cos \phi}{e^{2z} + r^2 + 1} , \qquad y^2 = \frac{2r \sin \phi}{e^{2z} + r^2 + 1} , \qquad y^3 = \frac{e^{2z} + r^2 - 1}{e^{2z} + r^2 + 1} . \qquad (1.13.21)$$

Allowing for the relations

$$A_i(y) = \frac{\partial x^j}{\partial y^i} A_j = \frac{\partial \phi}{\partial y^i} A_\phi(x) ,$$

$$\frac{y^2}{y^1} = \tan \phi , \qquad \frac{1}{\cos^2 \phi} = \frac{(y^1)^2 + (y^2)^2}{(y^1)^2} ,$$

$$\frac{(y^1)^2 + (y^2)^2}{(y^1)^2} \frac{\partial \phi}{\partial y^1} = -\frac{y^2}{(y^1)^2} \qquad \Longrightarrow \qquad \frac{\partial \phi}{\partial y^1} = -\frac{y^2}{(y^1)^2 + (y^2)^2} ,$$

$$\frac{(y^1)^2 + (y^2)^2}{(y^1)^2} \frac{\partial \phi}{\partial y^2} = \frac{1}{y^1} \qquad \Longrightarrow \qquad \frac{\partial \phi}{\partial y^2} = \frac{y^1}{(y^1)^2 + (y^2)^2} , \qquad (1.13.22)$$

we arrive at (for simplicity, we use notation $y^j = q_j$)

$$A_1(q) = -\frac{q_2}{q_1^2 + q_2^2} A \sin(\omega t - kz + \beta) ,$$

$$A_2(q) = \frac{q_1}{q_1^2 + q_2^2} A \sin(\omega t - kz + \beta) , \qquad A_3(q) = 0 . \qquad (1.13.23)$$

Allowing for formulas

$$e^z = \frac{\sqrt{1 - q_1^2 - q_2^2 - q_3^2}}{1 - q_3} \qquad \Longrightarrow \qquad z = \ln \frac{\sqrt{1 - q_1^2 - q_2^2 - q_3^2}}{1 - q_3} , \qquad (1.13.24)$$

we obtain the following explicit form for these solutions: $A_3(q) = 0$ and

$$A_1(q) = -\frac{q_2}{q_1^2 + q_2^2} A \sin \left(\omega t - k \ln \frac{\sqrt{1 - q_1^2 - q_2^2 - q_3^2}}{1 - q_3} + \beta \right) ,$$

$$A_2(q) = \frac{q_1}{q_1^2 + q_2^2} \, A \sin\left(\omega t - k \ln \frac{\sqrt{1 - q_1^2 - q_2^2 - q_3^2}}{1 - q_3} + \beta\right).$$

$$(1.13.25)$$

It should be noted that the very similar and trivial form of solutions in horospherical coordinates can also turn to be much more complex for those in quasi-Cartesian. Let us specify electromagnetic tensor for that solution (compare this with Eqs. (1.13.18) and (1.13.19))

$$\varphi = \left(\omega t - k \ln \frac{\sqrt{1 - q^2}}{1 - q_3} + \beta\right), \qquad q^2 = q_1^2 + q_2^2 + q_3^2 \, ,$$

$$E_1 = F_{01} = -\frac{q_2 \, \omega}{q_1^2 + q_2^2} \, A \cos\varphi \, , \qquad E_2 = F_{02} = \frac{q_1 \, \omega}{q_1^2 + q_2^2} \, A \cos\varphi \, ,$$

$$E_3 = F_{03} = 0 \, , \qquad -B_3 = F_{12} = Ak \frac{1}{1 - q^2} \, \cos\varphi \, ,$$

$$-B_1 = F_{23} = Ak \frac{q_1}{(q_1^2 + q_2^2)} \frac{(1 - q_3 - q_1^2 - q_2^2)}{(1 - q_3)(1 - q^2)} \, \cos\varphi = Ak \frac{q_1}{(q_1^2 + q_2^2)} \, F \, \cos\varphi \, ,$$

$$-B_2 = F_{31} = Ak \frac{q_2}{(q_1^2 + q_2^2)} \frac{(1 - q_3 - q_1^2 - q_2^2)}{(1 - q_3)(1 - q^2)} \, \cos\varphi = Ak \frac{q_2}{(q_1^2 + q_2^2)} \, F \, \cos\varphi \, ;$$

$$(1.13.26)$$

where this notation is used

$$F = \frac{(1 - q_3 - q_1^2 - q_2^2)}{(1 - q_3)(1 - q^2)} \, .$$

In this wave, electric vector is oriented along the circle in the plane $1 - 2$; whereas magnetic vector has projection on axis 3 and radial direction. The density of energy flow $\mathbf{S} = \mathbf{E} \times \mathbf{B}$ is given by

$$S_1 = E_2 B_3 = -A^2 k\omega \, \frac{q_1}{q_1^2 + q_2^2} \frac{1}{1 - q^2} \frac{1 + \cos 2\varphi}{2} \, ,$$

$$S_2 = -E_1 B_3 = -A^2 k\omega \, \frac{q_2}{q_1^2 + q_2^2} \frac{1}{1 - q^2} \frac{1 + \cos 2\varphi}{2} \, ,$$

$$S_3 = E_1 B_2 - E_2 B_1 = \frac{A^2 \omega k}{(q_1^2 + q_2^2)} \, F \, \frac{1 + \cos 2\varphi}{2} \, ; \qquad (1.13.27)$$

that is the vector $\mathbf{S}$ has the structure $\mathbf{S} = \mathbf{S}_{\parallel} + \mathbf{S}_{\perp}$. Let us detail the surface of fixed phase

$$\omega t - k \ln \frac{\sqrt{1 - q_1^2 - q_2^2 - q_3^2}}{1 - q_3} = \lambda = \omega t_0 \, ,$$

whence it follows

$$\frac{1 - q_1^2 - q_2^2 - q_3^2}{(1 - q_3)^2} = a^2(t) \, , \quad a^2(t) = e^{2\omega(t - t_0)} \, . \qquad (1.13.28)$$

In canonical form it is recognized as ellipsoid equation

$$\left(q_3 - \frac{a^2}{1+a^2}\right)^2 + \frac{q_1^2}{1+a^2} + \frac{q_2^2}{1+a^2} = \frac{1}{(1+a^2)^2} \ . \tag{1.13.29}$$

In the moment $t = t_0$, the parameter $a = 1$

$$\left(q_3 - \frac{1}{2}\right)^2 + \frac{q_1^2}{2} + \frac{q_2^2}{2} = \frac{1}{4} \ .$$

When $a \to 0$ ($t \to -\infty$) we have

$$q_3^2 + q_1^2 + q_2^2 = 1 \ .$$

When $a \to \infty$ ($t \to +\infty$) we will have

$$(q_3 - 1)^2 + \frac{q_1^2}{\infty^2} + \frac{q_2^2}{\infty^2} = \frac{1}{\infty^4} \ .$$

This simple solution can be considered as an exact solution for the Maxwell equations in Minkowski space-time, but in the special effective medium (see Section **1.11**).

1.14   The Case of Arbitrary Metric, Additional Consideration

For the general case, four 3×3 constitutive tensors contain $6 + 6 + 9 = 21$ independent functions, but defining metrical tensor $g^{\alpha\beta}(x)$ is fixed by 10 functions. Here, we must assume additional constrains on these constitutive tensors.

In the first place, note the evident identity

$$\mathrm{Sp}\ \alpha = \mathrm{Sp}\ \beta = 0 \ . \tag{1.14.1}$$

Besides, we can now see two identities

$$\mathbf{g}\,\beta = 0 \quad \Longrightarrow \quad (g^{01}, g^{02}, g^{03}) \begin{vmatrix} \beta^{11} & \beta^{12} & \beta^{13} \\ \beta^{21} & \beta^{22} & \beta^{23} \\ \beta^{31} & \beta^{32} & \beta^{33} \end{vmatrix} = 0 \ ; \tag{1.14.2a}$$

$$\alpha\,\mathbf{g}^+ = 0 \quad \Longrightarrow \quad \begin{vmatrix} \alpha^{11} & \alpha^{12} & \alpha^{13} \\ \alpha^{21} & \alpha^{22} & \alpha^{23} \\ \alpha^{31} & \alpha^{32} & \alpha^{33} \end{vmatrix} \begin{vmatrix} g^{01} \\ g^{02} \\ g^{03} \end{vmatrix} = 0 \ , \tag{1.14.2b}$$

these identities are equivalent to each other.

To proceed, let us rewrite expressions for $\epsilon^{ik}(x)$ and $(\mu^{-1})^{ik}(x)$ as follows

$$\frac{1}{g^{00}(x)} \left(\frac{\epsilon^{ik}(x)}{\sqrt{-G(x)}} + (\mathbf{g}(x) \bullet \mathbf{g}(x))^{ik} \right) = g^{ik}(x) \ ,$$

$$\frac{(\mu^{-1})^{ik}(x)}{\sqrt{-G(x)}} = \det\left(g^{ik}(x)\right)\left(g^{ik}(x)\right)^{-1} = \frac{1}{\det\left(g_{ik}(x)\right)}\ g_{ik}(x)$$

$$\implies \quad \mu^{ik}(x)\,\sqrt{-G(x)} = \det\left(g_{ik}(x)\right)\,g^{ik}(x)\ ; \qquad (1.14.3a)$$

from this, we derive

$$\frac{1}{g^{00}(x)}\left(\frac{\epsilon^{ik}(x)}{\sqrt{-G(x)}} + (\mathbf{g}(x)\bullet\mathbf{g}(x))^{ik}\right) = \frac{\mu^{ik}(x)\sqrt{-G(x)}}{\det\left(g_{ik}(x)\right)}\ , \qquad (1.14.3b)$$

or differently

$$\epsilon^{ik}(x) + \sqrt{-G(x)}\,[\mathbf{g}(x)\bullet\mathbf{g}(x)]^{ik} = -g^{00}(x)\,G(x)\,\frac{\mu^{ik}(x)}{\det\left(g_{ik}(x)\right)}\ . \qquad (1.14.3c)$$

This relationship provides us with an important additional constraint on $\epsilon^{ik}(x)$ and $\mu^{ik}(x)$. For the special form in the case of the metrical tensor

$$\mathbf{g}(x) = 0\ , \qquad G(x) = g_{00}\,\det\left[g_{ij}(x)\right]\ ; \qquad (1.14.4a)$$

we obtain an already known result

$$\epsilon^{ik}(x) = -\frac{1}{g_{00}(c)}\,G(x)\,\frac{1}{\det\left[g_{ik})x)\right]}\,\mu^{ik}(x) = -\mu^{ik}(x)\ . \qquad (1.14.4b)$$

Chapter 2

Electrodynamics, Complex Rotation Group, Media, Gravity

2.1 Introduction

Special Relativity arose from the study of the symmetry properties for the Maxwell equations with respect to motion of references frames: Lorentz [1], Poincaré [2], Einstein [3]. Naturally, an analysis of the Maxwell equations with respect to Lorentz transformations was the first objects of relativity theory. After discovering the relativistic equation for a particle with spin 1/2 – Dirac [127] – much work was done to study spinor and vectors within the Lorentz group theory: Möglich [129], Ivanenko – Landau [130], Neumann [131], van der Waerden [132], Juvet [323]. This established that any quantity which transforms linearly under Lorentz transformations is a spinor. For that reason spinor quantities are considered as fundamental in quantum field theory and basic equations for such quantities should be written in a spinor form. A spinor formulation of Maxwell equations was studied by Laporte and Uhlenbeck [138], also see Rumer [149]. In 1931, Majorana [140] and Oppenheimer [139] proposed to consider the Maxwell theory of electromagnetism as wave mechanics of the photon. They introduced a complex 3-vector wave function which satisfied the massless Dirac-like equations. Before Majorana and Oppenheimer, the most crucial steps were made by Silberstein [5], who showed the possibility to have formulated Maxwell equation in terms of complex 3-vector entities. Silberstein in his second paper [5] writes that the complex form of Maxwell equations has been known before; where he refers there to the second volume of the lecture notes on the differential equations of mathematical physics by Riemann that were edited and published by H. Weber in 1901 [128]. This is not a widely used fact as noted by Bialynicki-Birula [220].

Maxwell equations in the matrix Dirac-like form has been considered for a long time by many authors. Yet, the interest in the Majorana–Oppenheimer formulation of electrodynamics has grown in recent years:

Luis de Broglie [141–144], Mercier [145], Petiau [146], Proca [147, 148], Duffin [150], Kemmer [151–153], Bhabha [154], Belinfante [155, 156], Taub [157], Sakata and Taketani [160], Erikson [161], Schrödinger [48, 158, 159], Tonnelat [49], Heitler [162], Harish-Chandra [164, 165], Hoffmann [166], Utiyama [167], Schouten [168], Mercier [169], de Broglie and Tonnelat [170], Gupta [171], Bleuler [172], Brulin and Hjalmars [173], Rosen [174], Fujiwara [175], Gürsey [176], Gupta [177], Lichnerowicz [178], Ohmura [179], Borgardt [180, 181], Fedorov [182], Kuohsien [183], Bludman [184], Good [185], Moses [186–188], Silveira [189, 190], Lomont [191], Kibble [192], Post [63], Bogush and Fedorov [193], Sachs and Schwebel [194], Ellis [195], Beckers and Pirotte [196], Casanova [197], Carmeli [198], Weingarten [199], Mignani, Recami, and Baldo [200], Frankel [202], Jackson [203], Fushchich [204], Edmonds [205], Strazhev and Tomil'chik [247], Jena et al. [206], Venuri [207], Chow [208], Fushchich and Nikitin [209], Cook [210, 211], Giannetto [213], Nunez et al. [214], Kidd et al. [215], Recami [216], Krivsky and Simulik [217], Hillion [218], Inagaki [219], Bialynicki-Birula [220–222], Sipe [223], Ghose [224], Gersten [225], Esposito [226], Torres del Castill and Mercado-Perez [227], Dvoeglazov [228–230], Gsponer [231], Ivezic [232–236], Donev and Tashkova [237, 238], Kravchenko [239], Varlamov [240], Khan [241], Armour [242].

Our treatment shall have quite a definite accent in this study: the main attention is focused on the technical aspect of classical electrodynamics based on the theory of rotation complex group SO(3.C) (isomorphic to the Lorentz group – see Kurşunoǧlu [243], Macfarlane [244, 245], Fedorov [246]).

2.2 Complex Matrix Form for the Maxwell Theory in Vacuum

Let us start with Maxwell equations in a uniform (ϵ, μ)-media in presence of external sources:

$$(F^{ab}) \qquad \text{div } c\mathbf{B} = 0 \,, \qquad \text{rot } \mathbf{E} = -\frac{\partial c\mathbf{B}}{\partial ct} \,,$$

$$(H^{ab}) \qquad \text{div } \mathbf{E} = \frac{\rho}{\epsilon\epsilon_0}, \qquad \text{rot } c\mathbf{B} = \mu\mu_0 c\mathbf{J} + \epsilon\mu\frac{\partial \mathbf{E}}{\partial ct} \,. \qquad (2.2.1)$$

Through the use of usual notation with a current 4-vector

$$j^a = (\rho, \mathbf{J}/c) \,, \qquad c^2 = \frac{1}{\epsilon_0\mu_0} \,, \qquad (2.2.2)$$

Eqs. (2.2.1) reads

$$\text{div } c\mathbf{B} = 0 \,, \qquad \text{rot } \mathbf{E} = -\frac{\partial c\mathbf{B}}{\partial ct} \,,$$

$$\text{div } \mathbf{E} = \frac{\rho}{\epsilon_0}, \qquad \text{rot } c\mathbf{B} = \frac{\mathbf{j}}{\epsilon_0} + \frac{\partial \mathbf{E}}{\partial ct} \,, \qquad (2.2.3)$$

or in an explicit component form (let $x_0 = ct, \ \partial_0 = c\,\partial_t$)

$$\partial_1 cB^1 + \partial_2 cB^2 + \partial_3 cB^3 = 0 \,,$$

$$\partial_2 E^3 - \partial_3 E^2 + \partial_0 cB^1 = 0 \,,$$

$$\partial_3 E^1 - \partial_1 E^3 + \partial_0 cB^2 = 0 \,,$$

$$\partial_1 E^2 - \partial_2 E^1 + \partial_0 cB^3 = 0 \,,$$

$$\partial_1 E^1 + \partial_2 E^2 + \partial_3 E^3 = j^0/\epsilon_0 \,,$$

$$\partial_2 cB^3 - \partial_3 cB^2 - \partial_0 E^1 = j^1/\epsilon_0 \, ,$$

$$\partial_3 cB^1 - \partial_1 cB^3 - \partial_0 E^2 = j^2/\epsilon_0 \, ,$$

$$\partial_1 cB^2 - \partial_2 cB^1 - \partial_0 E^3 = j^3/\epsilon_0 \, . \tag{2.2.4}$$

Let us introduce a 3-dimensional complex vector

$$\psi^k = E^k + icB^k \, , \tag{2.2.5}$$

with the help of it previous equations (2.2.4) can be combined into

$$\partial_1 \Psi^1 + \partial_2 \Psi^0 + \partial_3 \Psi^3 = j^0/\epsilon_0 \, ,$$

$$-i\partial_0 \psi^1 + (\partial_2 \psi^3 - \partial_3 \psi^2) = i\, j^1/\epsilon_0 \, ,$$

$$-i\partial_0 \psi^2 + (\partial_3 \psi^1 - \partial_1 \psi^3) = i\, j^2/\epsilon_0 \, ,$$

$$-i\partial_0 \psi^3 + (\partial_1 \psi^2 - \partial_2 \psi^1) = i\, j^3/\epsilon_0 \, . \tag{2.2.6}$$

These four relations can be rewritten in a matrix form using a 4-dimensional column ψ with one additional zero-element:

$$\left(-i\partial_0 \begin{vmatrix} a_0 & 0 & 0 & 0 \\ a_1 & 1 & 0 & 0 \\ a_2 & 0 & 1 & 0 \\ a_3 & 0 & 0 & 1 \end{vmatrix} + \partial_1 \begin{vmatrix} b_0 & 1 & 0 & 0 \\ b_1 & 0 & 0 & 0 \\ b_2 & 0 & 0 & -1 \\ b_3 & 0 & 1 & 0 \end{vmatrix} + \partial_2 \begin{vmatrix} c_0 & 0 & 1 & 0 \\ c_1 & 0 & 0 & 1 \\ c_2 & 0 & 0 & 0 \\ c_3 & -1 & 0 & 0 \end{vmatrix} \right.$$

$$\left. + \partial_3 \begin{vmatrix} d_0 & 0 & 0 & 1 \\ d_1 & 0 & -1 & 0 \\ d_2 & 1 & 0 & 0 \\ d_3 & 0 & 0 & 0 \end{vmatrix} \right) \begin{vmatrix} 0 \\ \psi^1 \\ \psi^2 \\ \psi^3 \end{vmatrix} = \frac{1}{\epsilon_0} \begin{vmatrix} j^0 \\ i\, j^1 \\ i\, j^2 \\ i\, j^3 \end{vmatrix} .$$

Here, there arise four ambiguously determined matrices (numerical parameters a_k, b_k, c_k, d_k are arbitrary):

$$(-i\alpha^0 \partial_0 + \alpha^j \partial_j)\Psi = J \, , \qquad \Psi = \begin{vmatrix} 0 \\ \psi^1 \\ \psi^2 \\ \psi^3 \end{vmatrix} , \qquad \alpha^0 = \begin{vmatrix} a_0 & 0 & 0 & 0 \\ a_1 & 1 & 0 & 0 \\ a_2 & 0 & 1 & 0 \\ a_3 & 0 & 0 & 1 \end{vmatrix} ,$$

$$\alpha^1 = \begin{vmatrix} b_0 & 1 & 0 & 0 \\ b_1 & 0 & 0 & 0 \\ b_2 & 0 & 0 & -1 \\ b_3 & 0 & 1 & 0 \end{vmatrix} , \alpha^2 = \begin{vmatrix} c_0 & 0 & 1 & 0 \\ c_1 & 0 & 0 & 1 \\ c_2 & 0 & 0 & 0 \\ c_3 & -1 & 0 & 0 \end{vmatrix} , \alpha^3 = \begin{vmatrix} d_0 & 0 & 0 & 1 \\ d_1 & 0 & -1 & 0 \\ d_2 & 1 & 0 & 0 \\ d_3 & 0 & 0 & 0 \end{vmatrix} . \tag{2.2.7}$$

Consider the products for these matrices. By taking into account

$$(\alpha^0)^2 = \begin{vmatrix} a_0 & 0 & 0 & 0 \\ a_1 & 1 & 0 & 0 \\ a_2 & 0 & 1 & 0 \\ a_3 & 0 & 0 & 1 \end{vmatrix} \begin{vmatrix} a_0 & 0 & 0 & 0 \\ a_1 & 1 & 0 & 0 \\ a_2 & 0 & 1 & 0 \\ a_3 & 0 & 0 & 1 \end{vmatrix} = \begin{vmatrix} a_0 a_0 & 0 & 0 & 0 \\ a_1 a_0 + a_1 & 1 & 0 & 0 \\ a_2 a_0 + a_2 & 0 & 1 & 0 \\ a_3 a_0 + a_3 & 0 & 0 & 1 \end{vmatrix} ,$$

let us require the identity for $(\alpha^0)^2 = +I$, so that,

$$a_0 a_0 = 1 \,, \quad a_1 a_0 + a_1 = 0 \,, \quad a_2 a_0 + a_2 = 0 \,, \quad a_3 a_0 + a_3 = 0 \,;$$

the most simple solution is

$$a_0 = \pm 1 \,, \qquad a_j = 0 \,, \qquad \alpha^0 = \begin{vmatrix} \pm 1 & 0 & 0 & 0 \\ 0 & 1 & 0 & 0 \\ 0 & 0 & 1 & 0 \\ 0 & 0 & 0 & 1 \end{vmatrix} . \qquad (2.2.8)$$

In the same manner, starting from

$$(\alpha^1)^2 = \begin{vmatrix} b_0^2 + b_1 & b_0 & 0 & 0 \\ b_1 b_0 & b_1 & 0 & 0 \\ b_2 b_0 - b_3 & b_2 & -1 & 0 \\ b_3 b_0 - b_2 & b_3 & 0 & -1 \end{vmatrix} ,$$

let us require the identity of $(\alpha^1)^2 = -I$, the most simple solution looks

$$b_0 = 0 \,, \qquad b_1 = -1 \,, \qquad b_2 = 0 \,, \qquad b_3 = 0 \,,$$

$$\alpha^1 = \begin{vmatrix} 0 & 1 & 0 & 0 \\ -1 & 0 & 0 & 0 \\ 0 & 0 & 0 & -1 \\ 0 & 0 & 1 & 0 \end{vmatrix} . \qquad (2.2.9)$$

In turn, from relation

$$(\alpha^2)^2 = \begin{vmatrix} c_0 c_0 + c_2 & 0 & c_0 & 0 \\ c_1 c_0 + c_3 & -1 & c_1 & 0 \\ c_2 c_0 & 0 & c_2 & 0 \\ c_3 c_0 - c_1 & 0 & c_3 & -1 \end{vmatrix} = -I$$

it follows

$$c_0 = 0 \,, \qquad c_1 = 0 \,, \qquad c_2 = -1 \,, \qquad c_3 = 0 \,,$$

$$\alpha^2 = \begin{vmatrix} 0 & 0 & 1 & 0 \\ 0 & 0 & 0 & 1 \\ -1 & 0 & 0 & 0 \\ 0 & -1 & 0 & 0 \end{vmatrix} , \qquad (\alpha^2)^2 = -I \,. \qquad (2.2.10)$$

And from

$$(\alpha^3)^2 = \begin{vmatrix} d_0 d_0 + d_3 & 0 & 0 & d_0 \\ d_1 d_0 - d_2 & -1 & 0 & 0 \\ d_2 d_0 + d_1 & 0 & -1 & d_2 \\ d_3 d_0 & 0 & 0 & d_3 \end{vmatrix} = -I$$

we get

$$d_0 = 0 \,, \qquad d_1 = 0 \,, \qquad d_2 = 0 \,, \qquad d_3 = -1 \,,$$

$$\alpha^3 = \begin{vmatrix} 0 & 0 & 0 & 1 \\ 0 & 0 & -1 & 0 \\ 0 & 1 & 0 & 0 \\ -1 & 0 & 0 & 0 \end{vmatrix} , \qquad (\alpha^3)^2 = -I . \qquad (2.2.11)$$

For other products, we have

$$\alpha^1 \alpha^2 = \begin{vmatrix} 0 & 1 & 0 & 0 \\ -1 & 0 & 0 & 0 \\ 0 & 0 & 0 & -1 \\ 0 & 0 & 1 & 0 \end{vmatrix} \begin{vmatrix} 0 & 0 & 1 & 0 \\ 0 & 0 & 0 & 1 \\ -1 & 0 & 0 & 0 \\ 0 & -1 & 0 & 0 \end{vmatrix} = \begin{vmatrix} 0 & 0 & 0 & 1 \\ 0 & 0 & -1 & 0 \\ 0 & 1 & 0 & 0 \\ -1 & 0 & 0 & 0 \end{vmatrix} = +\alpha^3 ,$$

$$\alpha^2 \alpha^1 = \begin{vmatrix} 0 & 0 & 1 & 0 \\ 0 & 0 & 0 & 1 \\ -1 & 0 & 0 & 0 \\ 0 & -1 & 0 & 0 \end{vmatrix} \begin{vmatrix} 0 & 1 & 0 & 0 \\ -1 & 0 & 0 & 0 \\ 0 & 0 & 0 & -1 \\ 0 & 0 & 1 & 0 \end{vmatrix} = \begin{vmatrix} 0 & 0 & 0 & -1 \\ 0 & 0 & 1 & 0 \\ 0 & -1 & 0 & 0 \\ 1 & 0 & 0 & 0 \end{vmatrix} = -\alpha^3 ,$$

$$\alpha^1 \alpha^2 = -\alpha^2 \alpha^1 = \alpha^3 , \qquad (2.2.12)$$

$$\alpha^2 \alpha^3 = \begin{vmatrix} 0 & 0 & 1 & 0 \\ 0 & 0 & 0 & 1 \\ -1 & 0 & 0 & 0 \\ 0 & -1 & 0 & 0 \end{vmatrix} \begin{vmatrix} 0 & 0 & 0 & 1 \\ 0 & 0 & -1 & 0 \\ 0 & 1 & 0 & 0 \\ -1 & 0 & 0 & 0 \end{vmatrix} = \begin{vmatrix} 0 & 1 & 0 & 0 \\ -1 & 0 & 0 & 0 \\ 0 & 0 & 0 & -1 \\ 0 & 0 & 1 & 0 \end{vmatrix} = \alpha^1 ,$$

$$\alpha^3 \alpha^2 = \begin{vmatrix} 0 & 0 & 0 & 1 \\ 0 & 0 & -1 & 0 \\ 0 & 1 & 0 & 0 \\ -1 & 0 & 0 & 0 \end{vmatrix} \begin{vmatrix} 0 & 0 & 1 & 0 \\ 0 & 0 & 0 & 1 \\ -1 & 0 & 0 & 0 \\ 0 & -1 & 0 & 0 \end{vmatrix} = \begin{vmatrix} 0 & -1 & 0 & 0 \\ 1 & 0 & 0 & 0 \\ 0 & 0 & 0 & 1 \\ 0 & 0 & -1 & 0 \end{vmatrix} = -\alpha^1 ,$$

$$\alpha^2 \alpha^3 = -\alpha^3 \alpha^2 = \alpha^1 , \qquad (2.2.13)$$

and also

$$\alpha^3 \alpha^1 = -\alpha^1 \alpha^3 = \alpha^2 . \qquad (2.2.14)$$

Let us turn to $\alpha^0 \alpha^i$:

$$\delta = \pm 1, \quad \alpha^0 \alpha^1 = \begin{vmatrix} \delta & 0 & 0 & 0 \\ 0 & 1 & 0 & 0 \\ 0 & 0 & 1 & 0 \\ 0 & 0 & 0 & 1 \end{vmatrix} \begin{vmatrix} 0 & 1 & 0 & 0 \\ -1 & 0 & 0 & 0 \\ 0 & 0 & 0 & -1 \\ 0 & 0 & 1 & 0 \end{vmatrix} = \begin{vmatrix} 0 & \delta & 0 & 0 \\ -1 & 0 & 0 & 0 \\ 0 & 0 & 0 & -1 \\ 0 & 0 & 1 & 0 \end{vmatrix} ,$$

$$\delta = \pm 1, \quad \alpha^1 \alpha^0 = \begin{vmatrix} 0 & 1 & 0 & 0 \\ -1 & 0 & 0 & 0 \\ 0 & 0 & 0 & -1 \\ 0 & 0 & 1 & 0 \end{vmatrix} \begin{vmatrix} \delta & 0 & 0 & 0 \\ 0 & 1 & 0 & 0 \\ 0 & 0 & 1 & 0 \\ 0 & 0 & 0 & 1 \end{vmatrix} = \begin{vmatrix} 0 & 1 & 0 & 0 \\ -\delta & 0 & 0 & 0 \\ 0 & 0 & 0 & -1 \\ 0 & 0 & 1 & 0 \end{vmatrix} .$$

It should be noted that only at $\delta = +1$, we have a simple commutation rule:

$$\alpha^0 = I , \qquad \alpha^i \alpha^0 = \alpha^0 \alpha^i = \alpha^i . \qquad (2.2.15)$$

Thus, eight Maxwell equations are presented in matrix form:

$$(-i\partial_0 + \alpha^j \partial_j)\Psi = J, \qquad \Psi = \begin{vmatrix} 0 \\ \psi^1 \\ \psi^2 \\ \psi^3 \end{vmatrix}, \qquad J = \frac{1}{\epsilon_0} \begin{vmatrix} j^0 \\ i\,j^1 \\ i\,j^2 \\ i\,j^3 \end{vmatrix},$$

$$\alpha^1 = \begin{vmatrix} 0 & 1 & 0 & 0 \\ -1 & 0 & 0 & 0 \\ 0 & 0 & 0 & -1 \\ 0 & 0 & 1 & 0 \end{vmatrix}, \quad \alpha^2 = \begin{vmatrix} 0 & 0 & 1 & 0 \\ 0 & 0 & 0 & 1 \\ -1 & 0 & 0 & 0 \\ 0 & -1 & 0 & 0 \end{vmatrix}, \quad \alpha^3 = \begin{vmatrix} 0 & 0 & 0 & 1 \\ 0 & 0 & -1 & 0 \\ 0 & 1 & 0 & 0 \\ -1 & 0 & 0 & 0 \end{vmatrix},$$

$$(\alpha^1)^2 = -I, \qquad (\alpha^2)^2 = -I, \qquad (\alpha^2)^2 = -I,$$

$$\alpha^1\alpha^2 = -\alpha^2\alpha^1 = \alpha^3, \qquad \alpha^2\alpha^3 = -\alpha^3\alpha^2 = \alpha^1, \qquad \alpha^3\alpha^1 = -\alpha^1\alpha^3 = \alpha^2.$$

$$(2.2.16)$$

Such a complex 4-dimensional matrix form of Maxwell theory can readily provide us with a real 8-dimensional matrix form (see in [209]). Indeed, we have two conjugated equations:

$$(-i\partial_0 + \alpha^j\partial_j)\Psi = J, \qquad (+i\partial_0 + \alpha^j\partial_j)\Psi^* = J^*, \qquad (2.2.17)$$

$$E = \frac{\psi + \psi^*}{2}, \qquad B = \frac{\psi - \psi^*}{2i}. \qquad (2.2.18)$$

By summing these equations and then subtracting them, we get:

$$-i\partial_0\Psi + \alpha^j\partial_j\Psi + i\partial_0\Psi^* + \alpha^j\partial_j\Psi^* = J + J^*,$$

$$-i\partial_0\Psi + \alpha^j\partial_j\Psi - i\partial_0\Psi^* - \alpha^j\partial_j\Psi^* = J - J^*,$$

or further

$$\partial_0\frac{\Psi - \Psi^*}{2i} + \alpha^j\partial_j\frac{\Psi + \Psi^*}{2} = \frac{J + J^*}{2},$$

$$-\partial_0\frac{\Psi + \Psi^*}{2} + \alpha^j\partial_j\frac{\Psi - \Psi^*}{2i} = \frac{J - J^*}{2i};$$

that is

$$\partial_0 B + \alpha^j\partial_j E = \mathrm{Re}\,(J),$$

$$-\partial_0 E + \alpha^j\partial_j B = \mathrm{Im}\,(J).$$

These equations can be presented in a matrix 8-dimensional form

$$\begin{vmatrix} \alpha^j\partial_j & \partial_0 \\ -\partial_0 & \alpha^j\partial_j \end{vmatrix} \begin{vmatrix} E \\ B \end{vmatrix} = \begin{vmatrix} \mathrm{Re}\,(J) \\ \mathrm{Im}\,(J) \end{vmatrix},$$

$$(\Gamma^0\partial_0 + \Gamma^i\partial_i)\Psi = J, \qquad (2.2.19)$$

where

$$\Gamma^0 = \begin{vmatrix} 0 & I \\ -I & 0 \end{vmatrix}, \qquad \Gamma^i = \begin{vmatrix} \alpha^i & 0 \\ 0 & \alpha^i \end{vmatrix}, \qquad (2.2.20)$$

with the properties

$$(\Gamma^0)^2 = -I , \qquad (\Gamma^1)^2 = -I , \qquad (\Gamma^1)^2 = -I , \qquad (\Gamma^1)^2 = -I ,$$

$$\Gamma^1\Gamma^2 = -\Gamma^2\Gamma^1 = \Gamma^3 , \qquad \Gamma^2\Gamma^3 = -\Gamma^3\Gamma^2 = \Gamma^1 , \qquad \Gamma^3\Gamma^1 = -\Gamma^1\Gamma^3 = \Gamma^2 ,$$

$$\Gamma^0\Gamma^i = \Gamma^i\Gamma^0 = \begin{vmatrix} 0 & \alpha^i \\ -\alpha^i & 0 \end{vmatrix} \neq \Gamma^i . \tag{2.2.21}$$

Now, let us turn again to 4-dimensional complex form as it is more compact. Let us consider the problem of relativistic invariance for this equation. The lack of manifest invariance of a 3-vector complex form using the Maxwell theory has been intensively discussed in various aspects by Ivezic [232–236].

Let us start with relations

$$(-i\partial_0 + \alpha^j\partial_j)\Psi = J , \qquad \Psi = \begin{vmatrix} 0 \\ \psi^1 \\ \psi^2 \\ \psi^3 \end{vmatrix} , \qquad J = \frac{1}{\epsilon_0}\begin{vmatrix} j^0 \\ i\,j^1 \\ i\,j^2 \\ i\,j^3 \end{vmatrix} .$$

Arbitrary Lorentz transformation over the function Ψ is given by (take notice that one may introduce four undefined parameters $s_0, ..., s_3$, but we will take $s_0 = 1, s_j = 0$)

$$S = \begin{vmatrix} s_0 & 0 & 0 & 0 \\ s_1 & \cdot & \cdot & \cdot \\ s_2 & \cdot & O(k) & \cdot \\ s_3 & \cdot & \cdot & \cdot \end{vmatrix} , \qquad \Psi' = S\Psi , \qquad \Psi = S^{-1}\Psi' , \tag{2.2.22}$$

where $O(k)$ stands for a (3×3)-rotation complex matrix from $SO(3.C)$, isomorphic to the Lorentz group – for more details see in the book by Fedorov [246] (and later explained in this present text). Equation for a primed function Ψ' is

$$\left(-i\partial_0 + S\alpha^j S^{-1}\partial_j\right)\Psi' = S\,J . \tag{2.2.23}$$

When working with matrices α^j, we will use vectors $\mathbf{e}_i$ and (3×3)-matrices τ_i, then the structure $S\alpha^j S^{-1}$ is

$$S\alpha^j S^{-1} = \begin{vmatrix} 1 & 0 \\ 0 & O(k) \end{vmatrix} \begin{vmatrix} 0 & \mathbf{e}_j \\ -\mathbf{e}_j^t & \tau_j \end{vmatrix} \begin{vmatrix} 1 & 0 \\ 0 & O^{-1}(k) \end{vmatrix}$$

$$= \begin{vmatrix} 0 & \mathbf{e}_j O^{-1}(k) \\ -O(k)\mathbf{e}_j^t & O(k)\tau_j O^{-1}(k) \end{vmatrix} .$$

This relationship can be rewritten with the help of the indices and properties of orthogonal complex rotation group; given as follows:

$$S\alpha^j S^{-1} = \begin{vmatrix} 0 & \delta_{ji}O^{-1}(k)_{in} \\ -O(k)_{ni}\delta_{ij} & \tau_n O(k)_{nj} \end{vmatrix} = \begin{vmatrix} 0 & O(k)_{nj} \\ -O(k)_{nj} & \tau_n O(k)_{nj} \end{vmatrix}$$

$$= \begin{vmatrix} 0 & (e_m)_n O(k)_{mj} \\ -(e_m)_n O(k)_{mj} & \tau_m O(k)_{mj} \end{vmatrix} = \alpha^m O_{mj}(k) . \tag{2.2.24}$$

Therefore, after this transformation the matrix Maxwell equation is brought to

$$(-i\partial_0 + S\alpha^j S^{-1}\partial_j)\Psi' = SJ ,$$

$$(-i\partial_0 + \alpha^m O_{mj}\partial_j)\Psi' = SJ , \qquad O_{mj}\partial_j = \partial'_m ,$$

or

$$(-i\partial_0 + \alpha^m \partial'_m)\Psi' = SJ . \tag{2.2.25}$$

Let us specify, and we can verify this at the same time in the results for this particularily simple case: when S looks as follows (to a real value) a corresponds Euclidean rotation, to an imaginary value a corresponds to a Lorentzian rotation:

Euclidean rotation $(1-2)$,

$$S\alpha^1 S^{-1} = \begin{vmatrix} 1 & 0 & 0 & 0 \\ 0 & \cos a & -\sin a & 0 \\ 0 & \sin a & \cos a & 0 \\ 0 & 0 & 0 & 1 \end{vmatrix} \begin{vmatrix} 0 & 1 & 0 & 0 \\ -1 & 0 & 0 & 0 \\ 0 & 0 & 0 & -1 \\ 0 & 0 & 1 & 0 \end{vmatrix} \begin{vmatrix} 1 & 0 & 0 & 0 \\ 0 & \cos a & +\sin a & 0 \\ 0 & -\sin a & \cos a & 0 \\ 0 & 0 & 0 & 1 \end{vmatrix}$$

$$= \begin{vmatrix} 0 & \cos a & \sin a & 0 \\ -\cos a & 0 & 0 & \sin a \\ -\sin a & 0 & 0 & -\cos a \\ 0 & -\sin a & \cos a & 0 \end{vmatrix} = \cos a \, \alpha^1 + \sin a \, \alpha^2 = \alpha^j O_{j1} . \tag{2.2.26}$$

$$S\alpha^2 S^{-1} = \begin{vmatrix} 1 & 0 & 0 & 0 \\ 0 & \cos a & -\sin a & 0 \\ 0 & \sin a & \cos a & 0 \\ 0 & 0 & 0 & 1 \end{vmatrix} \begin{vmatrix} 0 & 0 & 1 & 0 \\ 0 & 0 & 0 & 1 \\ -1 & 0 & 0 & 0 \\ 0 & -1 & 0 & 0 \end{vmatrix} \begin{vmatrix} 1 & 0 & 0 & 0 \\ 0 & \cos a & \sin a & 0 \\ 0 & -\sin a & \cos a & 0 \\ 0 & 0 & 0 & 1 \end{vmatrix}$$

$$= \begin{vmatrix} 0 & -\sin a & \cos a & 0 \\ \sin a & 0 & 0 & \cos a \\ -\cos a & 0 & 0 & \sin a \\ 0 & -\cos a & -\sin a & 0 \end{vmatrix} = -\sin a \, \alpha^1 + \cos a \, \alpha^2 = \alpha^j O_{j2} .$$

$$\tag{2.2.27}$$

$$S\alpha^3 S^{-1} = \begin{vmatrix} 1 & 0 & 0 & 0 \\ 0 & \cos a & -\sin a & 0 \\ 0 & \sin a & \cos a & 0 \\ 0 & 0 & 0 & 1 \end{vmatrix} \begin{vmatrix} 0 & 0 & 0 & 1 \\ 0 & 0 & -1 & 0 \\ 0 & 1 & 0 & 0 \\ -1 & 0 & 0 & 0 \end{vmatrix} \begin{vmatrix} 1 & 0 & 0 & 0 \\ 0 & \cos a & \sin a & 0 \\ 0 & -\sin a & \cos a & 0 \\ 0 & 0 & 0 & 1 \end{vmatrix}$$

$$= \begin{vmatrix} 0 & 0 & 0 & 1 \\ 0 & 0 & -1 & 0 \\ 0 & 1 & 0 & 0 \\ -1 & 0 & 0 & 0 \end{vmatrix} = \alpha^3 = \alpha^j O_{j3} . \tag{2.2.28}$$

Now, one should give special attention to the fact that the symmetry properties given in Eqs.(2.2.25) look satisfactory only for real values of parameter a – in this case it describes symmetry of the Maxwell equations under Euclidean rotations. However, if the values of a are imaginary, then this transformation S gives a Lorentzian boost:

$$ a = ib \, , \qquad b^* = b \, , $$

$$ \sin a = \frac{e^{ia} - e^{-ia}}{2i} = \frac{e^{-b} - e^{b}}{2i} = i \sinh b \, , $$

$$ \cos a = \frac{e^{ia} + e^{-ia}}{2} = \frac{e^{-b} + e^{b}}{2} = \cosh b \, , $$

$$ S(a = ib) = \begin{vmatrix} 1 & 0 & 0 & 0 \\ 0 & \cosh b & -i \sinh b & 0 \\ 0 & i \sinh b & \cosh b & 0 \\ 0 & 0 & 0 & 1 \end{vmatrix} \, , \tag{2.2.29} $$

and the formulas in Eqs.(2.2.26) will take the form

$$ S\alpha^1 S^{-1} = \cosh b \, \alpha^1 + i \sinh b \, \alpha^2 \, , $$

$$ S\alpha^2 S^{-1} = -i \sinh b \, \alpha^1 + \cosh b \, \alpha^2 \, , \qquad S\alpha^3 S^{-1} = \alpha^3 \, . \tag{2.2.30} $$

Correspondingly, the Maxwell matrix equation after the transformation of Eqs. (2.2.29) and (2.2.30) will look asymmetric

$$ \left[\, (-i\partial_0 + \alpha^3 \partial_3) + (\cosh b \, \alpha^1 + i \sinh b \, \alpha^2) \, \partial_2 \right. $$

$$ \left. + (-i \sinh b \, \alpha^1 + \cosh b \, \alpha^2) \, \partial_3 \, \right] \Psi' = SJ \, . \tag{2.2.31} $$

One can note the identity for

$$ (\cosh b - i \sinh b \, \alpha^3)(-i\partial_0 + \alpha^3 \partial_3) $$

$$ = -i(\cosh b \, \partial_0 - \sinh b \, \partial_3) + \alpha^3(-\sinh b \, \partial_0 + \cosh b \, \partial_3) = -i\partial_0' + \alpha^3 \partial_3' \, , \tag{2.2.32} $$

where derivatives are changed in accordance with the Lorentzian boost rule:

$$ \cosh b \, \partial_0 - \sinh b \, \partial_3 = \partial_0' \, , \qquad -\sinh b \, \partial_0 + \cosh b \, \partial_3 = \partial_3' \, . $$

It remains to determine the action of the operator

$$ \Delta = \cosh b - i \, \text{insh} \, b \, \alpha^3 \tag{2.2.33} $$

on two other terms in Eqs. (2.2.31). One might expect two relations:

$$ (\cosh b - i \sinh b \, \alpha^3)(\cosh b \, \alpha^1 + i \sinh b \, \alpha^2) = \alpha^2 \, , $$

$$ (\cosh b - i \sinh b \, \alpha^3)(-i \sinh b \, \alpha^1 + \cosh b \, \alpha^2) = \alpha^3 \, . \tag{2.2.34} $$

As easily verified, these indeed hold. We should determine the term $\Delta S\, J$. First, we have

$$
S\,J = \epsilon_0^{-1}
\begin{vmatrix}
1 & 0 & 0 & 0 \\
0 & \cosh b & -i\sinh b & 0 \\
0 & i\sinh b & \cosh b & 0 \\
0 & 0 & 0 & 1
\end{vmatrix}
\begin{vmatrix}
j^0 \\
ij^1 \\
ij^2 \\
ij^3
\end{vmatrix}
=
\begin{vmatrix}
j^0 \\
i\cosh b\, j^1 + \sinh b\, j^2 \\
-\sinh b\, j^1 + i\cosh b\, j^2 \\
ij^3
\end{vmatrix},
$$

and then

$$
\Delta S\,J = \epsilon_0^{-1}
\begin{vmatrix}
\cosh b & 0 & 0 & -i\sinh b \\
0 & \cosh b & i\sinh b & 0 \\
0 & -i\sinh b & \cosh b & 0 \\
i\sinh b & 0 & 0 & \cosh b
\end{vmatrix}
\begin{vmatrix}
j^0 \\
i\cosh b\, j^1 + \sinh b\, j^2 \\
-\sinh b\, j^1 + i\cosh b\, j^2 \\
ij^3
\end{vmatrix}
$$

$$
=
\begin{vmatrix}
\cosh b\, j^0 + \sinh b\, j^3 \\
i\, j^1 \\
i\, j^2 \\
i(\sinh b\, j^0 + \cosh b\, j^3)
\end{vmatrix}; \tag{2.2.35}
$$

the right-hand side of Eqs. (2.2.35) is what we need. Thus, the symmetry of the matrix Maxwell equation under the Lorentzian boost in the plane $(0-3)$ is described by relations:

Pseudo-Euclidean rotation $(0-3)$

$$
S(b) =
\begin{vmatrix}
1 & 0 & 0 & 0 \\
0 & \cosh b & -i\sinh b & 0 \\
0 & i\sinh b & \cosh b & 0 \\
0 & 0 & 0 & 1
\end{vmatrix},
\qquad
\Delta(b) = \cosh b - i\sinh b\, \alpha^3 ,
$$

$$
\Delta(b)\,(-i\partial_0 + S\alpha^j S^{-1}\partial_j)\,\Psi' = \Delta S\,J \equiv J' \quad \Longrightarrow \quad (-i\partial_0' + \alpha^j \partial_j')\Psi' = J' ,
$$

$$
\cosh b\, \partial_0 - \sinh b\, \partial_3 = \partial_0' , \qquad -\sinh b\, \partial_0 + \cosh b\, \partial_3 = \partial_3' . \tag{2.2.36}
$$

The symmetry properties of this equation under two other boosts are similar.

Pseudo-Euclidean rotation $(0-1)$

$$
S(a=b) =
\begin{vmatrix}
1 & 0 & 0 & 0 \\
0 & 0 & 0 & 0 \\
0 & 0 & \cosh b & i\sinh b \\
0 & 0 & -i\sinh b & \cosh b
\end{vmatrix},
\qquad
\Delta(b) = \cosh b + i\sinh b\, \alpha^1 ,
$$

$$
\Delta(b)\,(-i\partial_0 + S\alpha^j S^{-1}\partial_j)\Psi' = \Delta S\,J \equiv J' \quad \Longrightarrow \quad (-i\partial_0' + \alpha^j \partial_j')\Psi' = J' ,
$$

$$
\cosh b\, \partial_0 + \sinh b\, \partial_1 = \partial_0' , \qquad \sinh b\, \partial_0 + \cosh b\, \partial_1 = \partial_3' . \tag{2.2.37}
$$

Pseudo-Euclidean rotation $(0-2)$

$$
S(ib) =
\begin{vmatrix}
1 & 0 & 0 & 0 \\
0 & \cosh b & 0 & i\sinh b \\
0 & 0 & 1 & 0 \\
0 & -i\sinh b & 0 & \cosh b
\end{vmatrix},
\qquad
\Delta(b) = \cosh b - i\sinh b\, \alpha^2 ,
$$

$$\Delta(b)\,(-i\partial_0 + S\alpha^j S^{-1}\partial_j)\Psi' = \Delta S\,J \equiv J' \qquad \Longrightarrow \qquad (-i\partial'_0 + \alpha^j \partial'_j)\Psi' = J'\,,$$

$$\cosh b\,\partial_0 - \sinh b\,\partial_2 = \partial'_0\,, \qquad -\sinh b\,\partial_0 + \cosh b\,\partial_2 = \partial'_3\,. \tag{2.2.38}$$

In the general case, one can think that for an arbitrary oriented boost the operator Δ should be in the form:

$$\Delta = \Delta_\alpha = \cosh b - i\,\sinh b\,n_j\,\alpha^j\,.$$

Now, let us verify this. To obtain mathematical description of that boost we will start with the known parametrization of the real 3-dimension group [246])

$$O(c) = I + 2\,[\,c_0\,\mathbf{c}^\times + (\mathbf{c}^\times)^2\,]\,, \qquad (\mathbf{c}^\times)_{kl} = -\epsilon_{klj}\,c_j\,,$$

$$O(c) = \begin{vmatrix} 1 - 2(c_2^2 + c_3^2) & -2c_0c_3 + 2c_1c_2 & +2c_0c_2 + 2c_1c_3 \\ +2c_0c_3 + 2c_1c_2 & 1 - 2(c_3^2 + c_1^2) & -2c_0c_1 + 2c_2c_3 \\ -2c_0c_2 + 2c_1c_3 & +2c_0c_1 + 2c_2c_3 & 1 - 2(c_1^2 + c_2^2) \end{vmatrix}\,. \tag{2.2.39}$$

For instance, rotation in the plane $(1-2)$ is given by

$$O(1-2) = O(c_0, 0, 0, c_3) = \begin{vmatrix} 1 - 2c_3^2 & -2c_0c_3 & 0 \\ +2c_0c_3 & 1 - 2c_3^2 & 0 \\ 0 & 0 & 1 \end{vmatrix} = \begin{vmatrix} \cos a & -\sin a & 0 \\ \sin a & \cos a & 0 \\ 0 & 0 & 1 \end{vmatrix}\,,$$

that is

$$c_0 = \cos\frac{a}{2}\,, \qquad c_3 = \sin\frac{a}{2}\,. \tag{2.2.40}$$

Transition to a Lorentzian boost is achieved by the formal change:

$$c_0 \implies \cosh\frac{b}{2}\,, \qquad c_3 \implies i\,\sinh\frac{b}{2}\,,$$

$$O(1-2) \implies \begin{vmatrix} 1 - 2c_3^2 & -2c_0c_3 & 0 \\ +2c_0c_3 & 1 - 2c_3^2 & 0 \\ 0 & 0 & 1 \end{vmatrix} = \begin{vmatrix} \cosh b & -i\,\sinh b & 0 \\ i\,\sinh b & \cosh b & 0 \\ 0 & 0 & 1 \end{vmatrix}\,. \tag{2.2.41}$$

In general, for this case such a transition is achieved by the change

$$c_0 \implies \cosh\frac{b}{2}\,, \ c_j \implies i\,\sinh\frac{b}{2}\,n_j\,, \ n_jn_j = 1\,, \qquad O(b, \mathbf{n})$$

$$= \begin{vmatrix} 1 + 2\sinh^2\frac{b}{2}(n_2^2 + n_3^2) & -i\,\sinh b n_3 - 2\sinh^2\frac{b}{2}n_1n_2 & i\,\sinh b n_2 - 2\sinh^2\frac{b}{2}n_1n_3 \\ i\,\sinh b n_3 - 2\sinh^2\frac{b}{2}n_1n_2 & 1 + 2\sinh^2\frac{b}{2}(n_3^2 + n_1^2) & -i\,\sinh b n_1 - 2\sinh^2\frac{b}{2}n_2n_3 \\ -i\,\sinh b n_2 - 2\sinh^2\frac{b}{2}n_1n_3 & i\,\sinh b n_1 - 2\sinh^2\frac{b}{2}n_2n_3 & 1 + 2\sinh^2\frac{b}{2}(n_1^2 + n_2^2) \end{vmatrix}$$

$$\tag{2.2.42}$$

from this point, when keeping in mind the formula $1 - \cosh b = -2\sinh^2(b/2)$, we arrive at

$$f \equiv (1 - \cosh b)\,, \qquad O(b, \mathbf{n})$$

$$= \begin{vmatrix} 1 - f(n_2^2 + n_3^2) & -i \sinh b \; n_3 + f n_1 n_2 & i \sinh b \; n_2 + f n_1 n_3 \\ i \sinh b \; n_3 + f n_1 n_2 & 1 - f(n_3^2 + n_1^2) & -i \sinh b \; n_1 + f n_2 n_3 \\ -i \sinh b \; n_2 + f n_1 n_3 & i \sinh b \; n_1 + f n_2 n_3 & 1 - f(n_1^2 + n_2^2) \end{vmatrix} .$$

$$(2.2.43)$$

One may specify particular cases:

$$O(b, (1,0,0)) = \begin{vmatrix} 1 & 0 & 0 \\ 0 & \cosh b & -i \sinh b \\ 0 & i \sinh b & \cosh b \end{vmatrix} ,$$

$$O(b, (0,1,0)) = \begin{vmatrix} \cosh b & 0 & i \sinh b \\ 0 & 1 & 0 \\ -i \sinh b & 0 & \cosh b \end{vmatrix} ,$$

$$O(b, (0,0,1)) = \begin{vmatrix} \cosh b & -i \sinh b & 0 \\ i \sinh b & \cosh b & 0 \\ 0 & 0 & 1 \end{vmatrix} . \qquad (2.2.44)$$

We need to examine relation

$$\Delta(b, \mathbf{n}) \; (-i \partial_0 + \alpha^i O_{ij}(b, \mathbf{n}) \partial_j) \; \Psi' = \Delta(b, \mathbf{n}) S J , \qquad (2.2.45)$$

where

$$\Delta = \cosh b - i \sinh b \; n_1 \alpha^1 - i \sinh b \; n_2 \alpha^2 - i \sinh b \; n_3 \alpha^3 .$$

Allowing for these identities:

$$(\cosh b - i \sinh b \; n_1 \alpha^1 - i \sinh b \; n_2 \alpha^2 - i \sinh b \; n_3 \alpha^3) \; \alpha^1$$

$$= \cosh b \; \alpha^1 + i \sinh b \; n_1 + i \sinh b \; n_2 \; \alpha^3 - i \sinh b \; n_3 \; \alpha^2 ,$$

$$(\cosh b - i \sinh b \; n_1 \alpha^1 - i \sinh b \; n_2 \alpha^2 - i \sinh b \; n_3 \alpha^3) \; \alpha^2$$

$$= \cosh b \; \alpha^2 - i \sinh b \; n_1 \; \alpha^3 + i \sinh b \; n_2 + i \sinh b \; n_3 \alpha^1 ,$$

$$(\cosh b - i \sinh b \; n_1 \alpha^1 - i \sinh b \; n_2 \alpha^2 - i \sinh b \; n_3 \alpha^3) \; \alpha^3$$

$$= \cosh b \; \alpha^3 + i \sinh b \; n_1 \; \alpha^2 - i \sinh b \; n_2 \alpha^1 + i \sinh b \; n_3 , \qquad (2.2.46)$$

Equation (2.2.45) takes the form

$$[\; (-i \; (\cosh b - i \sinh b \; n_1 \alpha^1 - i \sinh b \; n_2 \alpha^2 - i \sinh b \; n_3 \alpha^3) \; \partial_0$$

$$+ (\cosh b \; \alpha^1 + i \sinh b \; n_1 + i \sinh b \; n_2 \; \alpha^3 - i \sinh b \; n_3 \; \alpha^2) O_{1i} \partial_i$$

$$+ (\cosh b \; \alpha^2 - i \sinh b \; n_1 \; \alpha^3 + i \sinh b \; n_2 + i \sinh b \; n_3 \alpha^1) O_{2i} \partial_i$$

$$+ (\cosh b \; \alpha^3 + i \sinh b \; n_1 \; \alpha^2 - i \sinh b \; n_2 \alpha^1 + i \sinh b \; n_3) O_{3i} \partial_i \;] \; \Psi' = \Delta S J .$$

After some regrouping, we get the terms

$$(-i \; A_0 + \alpha^1 A_1 + \alpha^2 A_2 + \alpha^3 A_3) \; \Psi$$

$$= [-i (\cosh b \, \partial_0 - \sinh b \, n_1 \, O_{1i} \, \partial_i - \sinh b \, n_2 \, O_{2i}\partial_i - \sinh b \, n_3 \, O_{3i} \, \partial_i)$$

$$+ \alpha^1 (-\sinh b \, n_1 \, \partial_0 + \cosh b \, O_{1i} \, \partial_i - i \sinh b \, n_2 \, O_{3i}\partial_i + i \sinh b \, n_3 \, O_{2i} \, \partial_i)$$

$$+ \alpha^2 (-\sinh b \, n_2 \, \partial_0 + \cosh b \, O_{2i} \, \partial_i - i \sinh b \, n_3 \, O_{1i} \, \partial_i + i \sinh b \, n_1 \, O_{3i}\partial_i)$$

$$+ \alpha^3 (-\sinh b \, n_3 \, \partial_0 + \cosh b \, O_{3i} \, \partial_i - i \sinh b \, n_1 \, O_{2i} \, \partial_i + i \sinh b \, n_2 \, O_{1i}\partial_i)] \, \Psi'$$

$$= \Delta SJ . \tag{2.2.47}$$

Let us calculate the coefficient at $\alpha^0 = -i \, I$:

$$A_0 = \cosh b \, \partial_0 - \sinh b \, n_1 \, O_{1i} \, \partial_i - \sinh b \, n_2 \, O_{2i}\partial_i - \sinh b \, n_3 \, O_{3i} \, \partial_i$$

$$= \cosh b \, \partial_0$$

$$-\sinh b \, (n_1 O_{11} + n_2 O_{21} + n_3 O_{31}) \, \partial_1$$

$$-\sinh b \, (n_1 O_{12} + n_2 O_{22} + n_3 O_{32}) \, \partial_2$$

$$-\sinh b \, (n_1 O_{13} + n_2 O_{23} + n_3 O_{33}) \, \partial_3 . \tag{2.2.48}$$

Allowing for these identities

$$(n_1 O_{11} + n_2 O_{21} + n_3 O_{31}) = n_1[1 - (1 - \cosh b)(n_2^2 + n_3^2)]$$

$$+n_2[i \sinh b \, n_3 + (1 - \cosh b)n_1 n_2] + n_3[-i \sinh b \, n_2 + (1 - \cosh b)n_1 n_3]$$

$$= n_1[1 - (1 - \cosh b)(n_2^2 + n_3^2)] + (1 - \cosh b)n_1(n_2^2 + n_3^2) = n_1 ,$$

$$(n_1 O_{12} + n_2 O_{22} + n_3 O_{32}) = n_1[-i \sinh b \, n_3 + (1 - \cosh b)n_1 n_2]$$

$$+n_2[1 - (1 - \cosh b)(n_3^2 + n_1^2)] + n_3[i \sinh b \, n_1 + (1 - \cosh b)n_2 n_3]$$

$$= (1 - \cosh b)(n_1^2 + n_3^2)n_2 + n_2[1 - (1 - \cosh b)(n_3^2 + n_1^2)] = n_2 ,$$

$$(n_1 O_{13} + n_2 O_{23} + n_3 O_{33}) = n_1[+i \sinh b \, n_2 + (1 - \cosh b)n_1 n_3]$$

$$+n_2[-i \sinh b \, n_1 + (1 - \cosh b)n_2 n_3] + n_3[1 - (1 - \cosh b)(n_1^2 + n_2^2)]$$

$$= (1 - \cosh b)(n_1^2 + n_2^2)n_3 + n_3[1 - (1 - \cosh b)(n_1^2 + n_2^2)] = n_3 ,$$

we get

$$A_0 = \cosh b \, \partial_0 - \sinh b \, (n_1 \partial_1 + n_2 \partial_2 + n_3 \partial_3) . \tag{2.2.49}$$

Now calculate A_1:

$$A_1 = -\sinh b \, n_1 \, \partial_0 + \cosh b \, O_{1i} \, \partial_i - i \sinh b \, n_2 \, O_{3i}\partial_i + i \sinh b \, n_3 \, O_{2i} \, \partial_i$$

$$= -\sinh b \, n_1 \, \partial_0$$

$$+(\cosh b \, O_{11} - i \sinh b \, n_2 \, O_{31} + i \sinh b \, n_3 \, O_{21}) \, \partial_1$$

$$+(\cosh b \, O_{12} - i \sinh b \, n_2 \, O_{32} + i \sinh b \, n_3 \, O_{22}) \, \partial_2$$

$$+(\cosh b \, O_{13} - i \sinh b \, n_2 \, O_{33} + i \sinh b \, n_3 \, O_{23}) \, \partial_3 . \tag{2.2.50}$$

Allowing for these identities

$$\cosh b\, O_{11} - i \sinh b\, n_2\, O_{31} + i \sinh b\, n_3\, O_{21}$$

$$= \cosh b[1 - (1-\cosh b)(n_2^2 + n_3^2)]$$
$$-i \sinh b\, n_2[-i \sinh b\, n_2 + (1-\cosh b)n_1 n_3]$$
$$+i \sinh b\, n_3[i \sinh b\, n_3 + (1-\cosh b)n_1 n_2]$$
$$= \cosh b\, [1 - (1-\cosh b)(n_2^2 + n_3^2)] - \sinh^2 b\, (n_2^2 + n_3^2)$$
$$= \cosh b - \cosh b\, (1-\cosh b)(1 - n_1^2) - \sinh^2 b(1 - n_1^2)$$
$$= 1 + (\cosh b - 1)n_1 n_1\ ,$$

$$\cosh b\, O_{12} - i \sinh b\, n_2\, O_{32} + i \sinh b\, n_3\, O_{22}$$

$$= \cosh b\, [-i \sinh b\, n_3 + (1-\cosh b)n_1 n_2]$$
$$-i \sinh b\, n_2\, [i \sinh b\, n_1 + (1-\cosh b)n_2 n_3]$$
$$+i \sinh b\, n_3\, [1 - (1-\cosh b)(n_3^2 + n_1^2)]$$
$$= -i\ \sinh b\, [\, \cosh b\, n_3 + (1-\cosh b)\, n_2^2 n_3 - n_3 + n_3(1-\cosh b)(n_1^2 + n_3^2)\,]$$
$$+(\cosh b - 1)n_1 n_2 = (\cosh b - 1)n_1 n_2\ ,$$

$$\cosh b\, O_{13} - i \sinh b\, n_2\, O_{33} + i \sinh b\, n_3\, O_{23}$$

$$= \cosh b\, [i \sinh b\, n_2 + (1-\cosh b)n_1 n_3] - i \sinh b\, n_2\, [1 - (1-\cosh b)(n_1^2 + n_2^2)]$$
$$+i \sinh b\, n_3\, [i \sinh b\, n_1 + (1-\cosh b)n_2 n_3]$$
$$= -i \sinh b\, [-\cosh b\, n_2 + n_2 - (1-\cosh b)n_2\, (n_1^2 + n_2^2) - (1-\cosh b)\, n_2 n_3^2\,]$$
$$-(\cosh b - 1)n_1 n_3 = (\cosh b - 1)\, n_1 n_3\ ,$$

for A_1 we get
$$A_1 = -\sinh b\, n_1\, \partial_0$$
$$+[\, 1 + (\cosh b - 1)n_1 n_1\,]\, \partial_1 + (\cosh b - 1)n_1 n_2\, \partial_2 + (\cosh b - 1)n_1 n_3\, \partial_3\ ,$$

that is

$$A_1 = -\sinh b\, n_1\, \partial_0 + [\, \partial_1 + (\cosh b - 1)n_1(n_1\partial_1 + n_2\partial_2 + n_3\partial_3)\,]\ . \qquad (2.2.51)$$

Now calculate A_2:
$$A_2 = -\sinh b\, n_2\, \partial_0$$
$$+\cosh b\, O_{2i}\, \partial_i - i \sinh b\, n_3\, O_{1i}\, \partial_i + i \sinh b\, n_1\, O_{3i}\partial_i$$
$$= \sinh b\, n_2\, \partial_0$$
$$+(\cosh b\, O_{21} - i \sinh b\, n_3\, O_{11} + i \sinh b\, n_1\, O_{31})\, \partial_1$$
$$+(\cosh b\, O_{22} - i \sinh b\, n_3\, O_{12} + i \sinh b\, n_1\, O_{32})\, \partial_2$$

$$+(\cosh b \; O_{23} - i \sinh b \; n_3 \; O_{13} + i \sinh b \; n_1 \; O_{33}) \; \partial_3 \; . \qquad (2.2.52)$$

Allowing for these identities

$$\cosh b \; O_{21} \; - i \sinh b \; n_3 \; O_{11} + i \sinh b \; n_1 \; O_{31}$$

$$= \cosh b \; [i \sinh b \; n_3 + (1 - \cosh b)n_1 n_2]$$

$$- i \sinh b \; n_3 \; [1 - (1 - \cosh b)(n_2^2 + n_3^2)]$$

$$+ i \sinh b \; n_1 \; [-i \sinh b \; n_2 + (1 - \cosh b)n_1 n_3]$$

$$= -i \sinh b \; [\; -\cosh b \; n_3 + n_3 - (1 - \cosh b)n_3(n_2^2 + n_3^2) - (1 - \cosh b)n_1^2 n_3 \;]$$

$$+ (\cosh b - 1)n_2 n_1 = (\cosh b - 1)n_2 n_1 \; ,$$

$$\cosh b \; O_{22} - i \sinh b \; n_3 \; O_{12} + i \sinh b \; n_1 \; O_{32}$$

$$= \cosh b \; [1 - (1 - \cosh b)(n_3^2 + n_1^2)]$$

$$- i \sinh b \; n_3 \; [-i \sinh b \; n_3 + (1 - \cosh b)n_1 n_2]$$

$$+ i \sinh b \; n_1 \; [i \sinh b \; n_1 + (1 - \cosh b)n_2 n_3]$$

$$= \cosh b \; [1 - (1 - \cosh b)(n_3^2 + n_1^2)] - \sinh b \; n_3 \sinh b \; n_3 - \sinh b \; n_1 \sinh b \; n_1$$

$$= \cosh b \; [1 - (1 - \cosh b)(1 - n_2^2)] - \sinh^2 b \; (1 - n_2^2) = 1 - (\cosh b - 1)n_2 n_2 \; ,$$

$$\cosh b \; O_{23} - i \sinh b \; n_3 \; O_{13} + i \sinh b \; n_1 \; O_{33}$$

$$= \cosh b \; [-i \sinh b \; n_1 + (1 - \cosh b)n_2 n_3]$$

$$- i \sinh b \; n_3 \; [\sinh b \; n_2 + (1 - \cosh b)n_1 n_3]$$

$$+ i \sinh b \; n_1 \; [1 - (1 - \cosh b)(n_1^2 + n_2^2)]$$

$$= -i \sinh b [\cosh b \; n_1 + (1 - \cosh b) \; n_1 n_3^2 - n_1 + (1 - \cosh b)n_1(n_1^2 + n_2^2)]$$

$$+ (\cosh b - 1)n_3 n_2 = (\cosh b - 1)n_3 n_2 \; ,$$

for A_2, we get

$$A_2 = -\sinh b \; n_2 \; \partial_0 + (\cosh b - 1)n_2 n_1 \; \partial_1$$

$$+ [\; 1 - (\cosh b - 1)n_2 n_2 \;] \; \partial_2 + (\cosh b - 1)n_2 n_3 \; \partial_3 \; ,$$

that is

$$A_2 = -\sinh b \; n_2 \; \partial_0 + [\; \partial_2 + (\cosh b - 1)n_2(n_1 \partial_1 + n_2 \partial_2 + n_3 \partial_3) \;] \; . \qquad (2.2.53)$$

Calculating A_3

$$A_3 = -\sinh b \; n_3 \; \partial_0 + \cosh b \; O_{3i} \; \partial_i - i \sinh b \; n_1 \; O_{2i} \; \partial_i + i \sinh b \; n_2 \; O_{1i} \partial_i \qquad (2.2.54)$$

we get

$$A_3 = -\sinh b \; n_3 \; \partial_0 + [\partial_3 + (\cosh b - 1)n_2(n_1 \partial_1 + n_2 \partial_2 + n_3 \partial_3)] \; . \qquad (2.2.55)$$

Therefore, under general Lorentzian boost (take note that $x^a = (x^0, \mathbf{x})$ is a contra-variant 4-vector, whereas ∂_a is a covariant one)

$$t' = \cosh \beta \, t + \sinh \beta \, \mathbf{n} \, \mathbf{x} \, ,$$

$$\mathbf{x}' = +\mathbf{n} \sinh \beta \, t + \mathbf{x} + (\cosh \beta - 1) \, \mathbf{n} \, (\mathbf{n}\mathbf{x}) \, ; \qquad (2.2.56)$$

the matrix Maxwell equation transforms according to

$$(A_0 + \alpha^1 A_1 + \alpha^2 A_2 + \alpha^3 A_3) \, \Psi' = \Delta S J \, ,$$

$$\partial_0' \equiv A_0 = \cosh b \, \partial_0 - \sinh b \, (n_1 \partial_1 + n_2 \partial_2 + n_3 \partial_3) \, ,$$

$$\partial_1' \equiv A_1 = -\sinh b \, n_1 \, \partial_0 + [\partial_1 + (\cosh b - 1)n_1(n_1 \partial_1 + n_2 \partial_2 + n_3 \partial_3)] \, ,$$

$$\partial_2' \equiv A_2 = -\sinh b \, n_2 \, \partial_0 + [\partial_2 + (\cosh b - 1)n_2(n_1 \partial_1 + n_2 \partial_2 + n_3 \partial_3)] \, ,$$

$$\partial_3' \equiv A_3 = -\sinh b \, n_3 \, \partial_0 + [\partial_3 + (\cosh b - 1)n_2(n_1 \partial_1 + n_2 \partial_2 + n_3 \partial_3)] \, ;$$

$$(2.2.57)$$

so that this transformed equation reads

$$(\partial_0' + \alpha^1 \partial_1' + \alpha^2 \partial_2' + \alpha^3 \partial_3') \, \Psi' = \Delta S J \, . \qquad (2.2.58)$$

What remains to be examined is the term $\Delta S J$ in the right-hand side. Using

$$\Delta = \cosh b - i \sinh b \, n_1 \alpha^1 - i \sinh b \, n_2 \alpha^2 - i \sinh b \, n_3 \alpha^3$$

$$= \begin{vmatrix} \cosh b & -i\sinh b n_1 & -i\sinh b n_2 & -i\sinh b n_3 \\ i\sinh b n_1 & \cosh b & i\sinh b n_3 & -i\sinh b n_2 \\ i\sinh b n_2 & -i\sinh b n_3 & \cosh b & i\sinh b n_1 \\ i\sinh b n_3 & i\sinh b n_2 & -i\sinh b n_1 & \cosh b \end{vmatrix} , \qquad (2.2.59)$$

and (for a time, the factor ϵ_0^{-1} is omitted; remember the notation $f \equiv 1 - \cosh b$):

$$S J$$

$$= \begin{vmatrix} 1 & 0 & 0 & 0 \\ 0 & 1 - f(n_2^2 + n_3^2) & -i\sinh b n_3 + f n_1 n_2 & i\sinh b \, n_2 + f n_1 n_3 \\ 0 & i\sinh b n_3 + f n_1 n_2 & 1 - f(n_3^2 + n_1^2) & -i\sinh b n_1 + f n_2 n_3 \\ 0 & -i\sinh b n_2 + f n_1 n_3 & i\sinh b n_1 + f n_2 n_3 & 1 - f(n_1^2 + n_2^2) \end{vmatrix}$$

$$\times \begin{vmatrix} j^0 \\ i j^1 \\ i j^2 \\ i j^2 \end{vmatrix}$$

$$= \begin{vmatrix} j^0 \\ i j^1[1 - f(n_2^2 + n_3^2)] + i j^2[-i\sinh b n_3 + f n_1 n_2] + i j^3[i\sinh b n_2 + f n_1 n_3] \\ i j^1[i\sinh b n_3 + f n_1 n_2] + i j^2[1 - f(n_3^2 + n_1^2)] + i j^3[-i\sinh b n_1 + f n_2 n_3] \\ i j^1[-i\sinh b n_2 + f n_1 n_3] + i j^2[i\sinh b n_1 + f n_2 n_3] + i j^3[1 - f(n_1^2 + n_2^2)] \end{vmatrix}$$

$$(2.2.60)$$

for ΔSJ, we can then produce

$$\Delta SJ = \begin{vmatrix} j'^0 \\ ij'^1 \\ ij'^2 \\ ij'^3 \end{vmatrix} .$$

$$(2.2.61)$$

First, let us consider

$$j'^0 = \cosh b \, j^0 + \sinh b \, \{ j^1 n_1 [1 - (1 - \cosh b)(n_2^2 + n_3^2)]$$

$$+ j^2 \, n_1 \, [-i \sinh b \, n_3 + (1 - \cosh b)n_1 n_2] + j^3 n_1 \, [i \sinh b \, n_2 + (1 - \cosh b)n_1 n_3]$$

$$+ j^1 n_2 \, [i \sinh b \, n_3 + (1 - \cosh b)n_1 n_2] + j^2 n_2 \, [1 - (1 - \cosh b)(n_3^2 + n_1^2)]$$

$$+ j^3 n_2 \, [-i \sinh b \, n_1 + (1 - \cosh b)n_2 n_3] + j^1 n_3 \, [-i \sinh b \, n_2 + (1 - \cosh b)n_1 n_3]$$

$$+ j^2 n_3 \, [i \sinh b \, n_1 + (1 - \cosh b)n_2 n_3] + j^3 n_3 \, [1 - (1 - \cosh b)(n_1^2 + n_2^2)] \} \, ,$$

that is

$$j'^0 = \cosh b \, j^0 + \sinh b \, (\, n_1 \, j^1 + n_2 \, j^2 + n_3 \, j^3 \,)$$

$$= \cosh b \, j^0 + \sinh b \, (\mathbf{nj}) \, .$$

$$(2.2.62)$$

Now, let us calculate the term

$$ij'^1 = +i \sinh b \, n_1 \, j^0 + \cosh b \{ ij^1 \, [1 - (1 - \cosh b)(n_2^2 + n_3^2)]$$

$$+ ij^2 \, [-i \sinh b \, n_3 + (1 - \cosh b)n_1 n_2] + ij^3 \, [i \sinh b \, n_2 + (1 - \cosh b)n_1 n_3] \}$$

$$+ i \sinh b n_3 \{ ij^1 \, [+i \sinh b \, n_3 + (1 - \cosh b)n_1 n_2] + ij^2 \, [1 - (1 - \cosh b)(n_3^2 + n_1^2)]$$

$$+ ij^3 \, [-i \sinh b \, n_1 + (1 - \cosh b)n_2 n_3] \} - i \sinh b n_2 \{ ij^1 \, [-i \sinh b \, n_2 + (1 - \cosh b)n_1 n_3]$$

$$+ ij^2 \, [i \sinh b \, n_1 + (1 - \cosh b)n_2 n_3] + ij^3 [1 - (1 - \cosh b)(n_1^2 + n_2^2)] \} \, ,$$

that is

$$j'^1 = +\sinh b \, n_1 \, j^0 + j^1 + (\cosh b - 1)n_1(n_1 j^1 + n_2 j^2 + n_3 j^3) \, .$$

$$(2.2.63)$$

It is evidently the first component of the vector formula (see Eqs.(2.2.56))

$$\mathbf{j}' = +\sinh b \, \mathbf{n} \, j^0 + \mathbf{j} + (\cosh b - 1) \, \mathbf{n} \, (\mathbf{nj}) \, .$$

$$(2.2.64)$$

Let us calculate the term j'^2:

$$j'^2 = +i \sinh b \, n_2 j^0 - i \sinh b \, n_3 \{ ij^1 [1 - (1 - \cosh b)(n_2^2 + n_3^2)]$$

$$+ ij^2 [-i \sinh b \, n_3 + (1 - \cosh b)n_1 n_2] + ij^3 [i \sinh b \, n_2 + (1 - \cosh b)n_1 n_3] \}$$

$$+ \cosh b \{ ij^1 [i \sinh b \, n_3 + (1 - \cosh b)n_1 n_2] + ij^2 [1 - (1 - \cosh b)(n_3^2 + n_1^2)]$$

$$+ ij^3 [-i \sinh b \, n_1 + (1 - \cosh b)n_2 n_3] \}$$

$$+i \sinh b n_1 \{ij^1[-i \sinh b\, n_2 + (1-\cosh b)n_1 n_3]$$

$$+ij^2[i \sinh b\, n_1 + (1-\cosh b)n_2 n_3] + ij^3[1-(1-\cosh b)(n_1^2 + n_2^2)]\}\,,$$

that is

$$j'^2 = +\sinh b\, n_2\, j^0 + j^2 + (\cosh b - 1)n_2(n_1 j^1 + n_2 j^2 + n_3 j^3)\,.$$

$$(2.2.65)$$

In this same way, we prove the formula

$$j'^3 = +\sinh b\, n_3\, j^0 + j^3 + (\cosh b - 1)n_3(n_1 j^1 + n_2 j^2 + n_3 j^3)\,. \qquad (2.2.66)$$

Thus, the matrix Maxwell equation

$$(-i\,\partial_0 + \alpha^i \partial_i)\,\Psi = J$$

is invariant under an arbitrary Lorentzian boost:

$$S(ib, \mathbf{n}) = \begin{vmatrix} 1 & 0 \\ 0 & O(ib, \mathbf{n}) \end{vmatrix}\,,$$

$$t' = \cosh \beta\, t + \sinh \beta\, \mathbf{n}\, \mathbf{x}\,, \qquad \mathbf{x}' = +\mathbf{n} \sinh \beta\, t + \mathbf{x} + (\cosh \beta - 1)\,\mathbf{n}\,(\mathbf{n}\mathbf{x})\,,$$

$$\Delta(-i\,\partial_0 + S\alpha^i S^{-1}\partial_i)\,S\Psi = \Delta SJ \qquad \Longrightarrow \qquad (\partial_0' + \alpha^i \partial_i')\,\Psi' = J'\,;$$

$$(2.2.67)$$

the derivatives and current will transform by the rules:

$$\partial_0' = \cosh b\, \partial_0 - \sinh b\, (\mathbf{n}\nabla)\,,$$

$$\nabla' = -\sinh b\, \mathbf{n}\, \partial_0 + [\nabla + (\cosh b - 1)\mathbf{n}(\mathbf{n}\nabla)\,,$$

$$j'^0 = \cosh b\, j^0 + \sinh b\, (\mathbf{n}\mathbf{j})\,,$$

$$\mathbf{j}' = +\sinh b\, \mathbf{n}\, j^0 + \mathbf{j} + (\cosh b - 1)\,\mathbf{n}\,(\mathbf{n}\mathbf{j})\,. \qquad (2.2.68)$$

Invariance of the matrix equation under Euclidean rotations is achieved in a simpler way:

$$(-i\,\partial_0 + \alpha^i \partial_i)\,\Psi = J\,,$$

$$S(a, \mathbf{n}) = \begin{vmatrix} 1 & 0 \\ 0 & O(a, \mathbf{n}) \end{vmatrix}\,, \qquad t' = t\,, \qquad \mathbf{x}' = R(a, \mathbf{n})\mathbf{x}\,,$$

$$(-i\,\partial_0 + S\alpha^i S^{-1}\partial_i)\,S\Psi = SJ \qquad \Longrightarrow \qquad (-i\,\partial_0' + \alpha^i \partial_i')\,\Psi' = J'\,, \qquad (2.2.69)$$

$$\partial_0' = \partial_0\,, \qquad \nabla' = R(a, -\mathbf{n})\nabla\,,$$

$$j'^0 = j^0\,, \qquad \mathbf{j}' = R(a, \mathbf{n})\mathbf{j}\,. \qquad (2.2.70)$$

2.3   Maxwell Equations in a Uniform Medium, Modified Lorentz Symmetry

Let us start with Maxwell equations in a uniform medium:

$$(F^{ab}) \qquad \text{div } c\mathbf{B} = 0 \;, \qquad \text{rot } \mathbf{E} = -\frac{\partial c\mathbf{B}}{\partial ct} \;,$$

$$(H^{ab}) \qquad \text{div } \mathbf{E} = \frac{\rho}{\epsilon\epsilon_0}, \qquad \text{rot } c\mathbf{B} = \mu\mu_0 c\mathbf{J} + \epsilon\mu\frac{\partial \mathbf{E}}{\partial ct} \;. \qquad (2.3.1)$$

The coefficient $\epsilon\mu$ can be factorized by

$$\epsilon\mu = \sqrt{\epsilon\mu}\,\sqrt{\epsilon\mu} = \frac{1}{k^2} \;, \qquad c' = \frac{1}{\sqrt{\epsilon_0\epsilon\mu_0\mu}} = k\,c \;; \qquad (2.3.2)$$

and this previous system may be presented differently

$$\text{div } kc\mathbf{B} = 0 \;, \qquad \text{rot } \mathbf{E} = -\frac{\partial kc\mathbf{B}}{\partial kct} \;,$$

$$\text{div } \mathbf{E} = \frac{\rho}{\epsilon\epsilon_0} \;, \qquad \text{rot } kc\mathbf{B} = \frac{1}{\epsilon\epsilon_0}\frac{\mathbf{J}}{kc} + \frac{\partial \mathbf{E}}{\partial kct} \;. \qquad (2.3.3)$$

By introducing these variables

$$x^a = (x^0 = kct \;, \quad x^i) \;, \qquad j^a = (j^0 = \rho, \mathbf{j} = \frac{\mathbf{J}}{kc}), \qquad (2.3.4)$$

the previous equation can be rewritten as

$$\text{div } kc\mathbf{B} = 0 \;, \qquad \text{rot } \mathbf{E} = -\frac{\partial kc\mathbf{B}}{\partial x^0} \;;$$

$$\text{div } \mathbf{E} = \frac{1}{\epsilon\epsilon_0}\,j^0, \qquad \text{rot } kc\mathbf{B} = \frac{1}{\epsilon\epsilon_0}\mathbf{j} + \frac{\partial \mathbf{E}}{\partial x^0} \;, \qquad (2.3.5)$$

or in explicit component form

$$\partial_1 c'B^1 + \partial_2 c'B^2 + \partial_3 c'B^3 = 0 \;,$$

$$\partial_2 E^3 - \partial_3 E^2 + \partial_0 c'B^1 = 0 \;,$$

$$\partial_3 E^1 - \partial_1 E^3 + \partial_0 c'B^2 = 0 \;,$$

$$\partial_1 E^2 - \partial_2 E^1 + \partial_0 c'B^3 = 0 \;,$$

$$\partial_1 E^1 + \partial_2 E^2 + \partial_3 E^3 = j^0/\epsilon\epsilon_0 \;,$$

$$\partial_2 cB^3 - \partial_3 cB^2 - \partial_0 E^1 = j^1/\epsilon\epsilon_0 \;,$$

$$\partial_3 cB^1 - \partial_1 cB^3 - \partial_0 E^2 = j^2/\epsilon\epsilon_0 \;,$$

$$\partial_1 cB^2 - \partial_2 cB^1 - \partial_0 E^3 = j^3/\epsilon\epsilon_0 \;. \qquad (2.3.6)$$

Equations (2.3.5) and (2.3.6) formally differ from Eqs. (2.2.4) only in one parametric change $c \Longrightarrow c' = kc$ (and also $\epsilon_0 \Longrightarrow \epsilon_0\epsilon$); therefore, all analysis performed in Section **2.1** is applicable here:

$$\psi^k = E^k + ic'B^k \; ; \tag{2.3.7}$$

Eqs. (2.3.6) are combined into

$$\partial_1\Psi^1 + \partial_2\Psi^0 + \partial_3\Psi^3 = j^0/\epsilon\epsilon_0 \; ,$$
$$-i\partial_0\psi^1 + (\partial_2\psi^3 - \partial_3\psi^2) = i\,j^1/\epsilon\epsilon_0 \; ,$$
$$-i\partial_0\psi^2 + (\partial_3\psi^1 - \partial_1\psi^3) = i\,j^2/\epsilon\epsilon_0 \; ,$$
$$-i\partial_0\psi^3 + (\partial_1\psi^2 - \partial_2\psi^1) = i\,j^3/\epsilon\epsilon_0 \; . \tag{2.3.8}$$

These are rewritten in the matrix form

$$\left(-i\partial_0 + \alpha^i\partial_i\right)\Psi = J \; ; \tag{2.3.9}$$

it involves the same old matrices.

The given matrix form of the Maxwell theory in a uniform media proves the existence of symmetry in the theory under a modified Lorentz group (see Rosen [174]) where instead of using the vacuum speed of light we are using the modified speed:

$$c' = kc \; , \qquad c = \frac{1}{\sqrt{\epsilon_0\mu_0}} \; , \qquad k = \frac{1}{\sqrt{\epsilon\mu}} \; .$$

2.4   Dual Symmetry of the Maxwell Equations

Let us consider the known dual symmetry of Maxwell theory in matrix formalism, starting from Maxwell equations without sources:

$$\left(-i\partial_0 + \alpha^j\partial_j\right) \left| \begin{matrix} 0 \\ \mathbf{E} + ic\mathbf{B} \end{matrix} \right| = \left| \begin{matrix} 0 \\ \mathbf{0} \end{matrix} \right| . \tag{2.4.1}$$

It is evident that there exists a simple transform, multiplication by imaginary i, with the following properties:

$$\Psi^D = i\,\Psi \; , \; \Psi = -i\,\Psi^D \; , \qquad \left(-i\partial_0 + \alpha^j\partial_j\right)\Psi^D = 0 \; ,$$

$$\Psi^D = \left| \begin{matrix} 0 \\ i\mathbf{E} - c\mathbf{B} \end{matrix} \right| = \left| \begin{matrix} 0 \\ \mathbf{E}^D + ic\mathbf{B}^D \end{matrix} \right| ,$$

$$\mathbf{E}^D = -c\mathbf{B} \; , \qquad c\mathbf{B}^D = +\mathbf{E} \; . \tag{2.4.2}$$

It is the dual transformation of the electromagnetic field. Two points should be clarified.

I. This transformation is not a symmetry operation in presence of external sources (for details see [247]). Indeed, in this case we have

$$\left(-i\partial_0 + \alpha^j\partial_j\right) \left| \begin{matrix} 0 \\ \mathbf{E} + ic\mathbf{B} \end{matrix} \right| = \frac{1}{\epsilon_0} \left| \begin{matrix} \rho \\ i\mathbf{j} \end{matrix} \right| , \tag{2.4.3}$$

and further

$$\Psi^D = i\,\Psi\,, \qquad (-i\partial_0 + \alpha^j\partial_j)\,\Psi^D = \frac{1}{\epsilon_0}\begin{vmatrix} i\rho \\ -\mathbf{j} \end{vmatrix},$$

$$\Psi^D = \begin{vmatrix} 0 \\ i\mathbf{E} - c\mathbf{B} \end{vmatrix} = \begin{vmatrix} 0 \\ \mathbf{E}^D + ic\mathbf{B}^D \end{vmatrix},$$

$$\mathbf{E}^D = -c\mathbf{B}\,, \qquad c\mathbf{B}^D = +\mathbf{E}\,. \tag{2.4.4}$$

II. To save the situation one can extend the Maxwell equations by introducing magnetic sources [247]:

$$(-i\partial_0 + \alpha^j\partial_j)\begin{vmatrix} 0 \\ \mathbf{E} + ic\mathbf{B} \end{vmatrix} = \frac{1}{\epsilon_0}\begin{vmatrix} \rho_e + i\rho_m \\ i\mathbf{j}_e + \mathbf{j}_m \end{vmatrix}, \tag{2.4.5}$$

which permits us to consider the dual transformation as a symmetry

$$\Psi^D = i\,\Psi\,, \qquad \Psi = -i\,\Psi^D\,,$$

$$(-i\partial_0 + \alpha^j\partial_j)\,\Psi^D = \frac{1}{\epsilon_0}\begin{vmatrix} -\rho_m + i\rho_e \\ i\mathbf{j}_m - \mathbf{j}_e \end{vmatrix},$$

$$\Psi^D = \begin{vmatrix} 0 \\ i\mathbf{E} - c\mathbf{B} \end{vmatrix} = \begin{vmatrix} 0 \\ \mathbf{E}^D + ic\mathbf{B}^D \end{vmatrix},$$

$$\mathbf{E}^D = -c\mathbf{B}\,, \qquad c\mathbf{B}^D = +\mathbf{E}\,,$$

$$\rho_e^D = -\rho_m, \qquad \mathbf{j}_e^D = +\mathbf{j}_m\,, \tag{2.4.6}$$

$$\rho_m^D = +\rho_e\,, \qquad \mathbf{j}_m^D = -\mathbf{j}_e\,.$$

In a real representation the Eqs. (2.4.6) will look

$$\partial_0\frac{\Psi - \Psi^*}{2i} + \alpha^j\partial_j\frac{\Psi + \Psi^*}{2} = \frac{J + J^*}{2}\,,$$

$$\partial_0\frac{\Psi + \Psi^*}{2} + \alpha^j\partial_j\frac{\Psi - \Psi^*}{2i} = \frac{J - J^*}{2i}\,,$$

$$\mathrm{Re}\,(J) = \frac{1}{\epsilon_0}\begin{vmatrix} \rho_e \\ \mathbf{j}_m \end{vmatrix}\,, \qquad \mathrm{Im}\,(J) = \frac{1}{\epsilon_0}\begin{vmatrix} \rho_m \\ \mathbf{j}_e \end{vmatrix}\,,$$

that is

$$\partial_0 B + \alpha^j\partial_j E = \mathrm{Re}\,(J)\,,$$

$$-\partial_0 E + \alpha^j\partial_j B = \mathrm{Im}\,(J)\,. \tag{2.4.7}$$

Eqs. (2.4.7) in explicit form are

$$\begin{vmatrix} 0 \\ \partial_0 cB^1 \\ \partial_0 cB^2 \\ \partial_0 cB^3 \end{vmatrix} + \begin{vmatrix} 0 & \partial_1 & \partial_2 & \partial_3 \\ -\partial_1 & 0 & -\partial_3 & \partial_2 \\ -\partial_2 & \partial_3 & 0 & -\partial_1 \\ -\partial_3 & -\partial_2 & \partial_1 & 0 \end{vmatrix}\begin{vmatrix} 0 \\ E^1 \\ E^2 \\ E^3 \end{vmatrix} = \frac{1}{\epsilon_0}\begin{vmatrix} \rho_e \\ \mathbf{j}_m \end{vmatrix}\,,$$

$$\begin{vmatrix} 0 \\ -\partial_0 E^1 \\ -\partial_0 E^2 \\ -\partial_0 E^3 \end{vmatrix} + \begin{vmatrix} 0 & \partial_1 & \partial_2 & \partial_3 \\ -\partial_1 & 0 & -\partial_3 & \partial_2 \\ -\partial_2 & \partial_3 & 0 & -\partial_1 \\ -\partial_3 & -\partial_2 & \partial_1 & 0 \end{vmatrix} \begin{vmatrix} 0 \\ cB^1 \\ cB^2 \\ cB^3 \end{vmatrix} = \frac{1}{\epsilon_0} \begin{vmatrix} \rho_m \\ \mathbf{j}_e \end{vmatrix} ,$$

or in a vector notation

$$\operatorname{div} \mathbf{E} = \frac{\rho_e}{\epsilon_0} , \qquad \operatorname{rot} \mathbf{E} = +\frac{\mathbf{j}_m}{\epsilon_0} - \frac{\partial c\mathbf{B}}{\partial ct} ,$$

$$\operatorname{div} c\mathbf{B} = \frac{\rho_m}{\epsilon_0} , \qquad \operatorname{rot} c\mathbf{B} = \frac{\mathbf{j}_e}{\epsilon_0} + \frac{\partial \mathbf{E}}{\partial ct} . \tag{2.4.8}$$

One addition should be made: the well-known continuous dual symmetry looks as a phase transformation over complex variables:

$$e^{i\chi} \left(\mathbf{E} + ic\,\mathbf{B} \right) , \qquad e^{-i\chi} \left(i\mathbf{j}_e + \mathbf{j}_m \right) , \qquad e^{-i\chi} \left(\rho_e + i\rho_m \right) . \tag{2.4.9}$$

2.5 On a Matrix Form of Minkowski Electrodynamics in Moving Bodies

In agreement with the Minkowski approach [6], in the presence of a uniform media we should introduce two electromagnetic tensors F^{ab} and H^{ab} that transform independently under the Lorentz group. At this, the known constitutive (or material) relations change their form in the moving reference frame.

Initial Maxwell equations (in the "rest reference frame") are

$$F^{ab} = (\mathbf{E}, \ c\mathbf{B})$$

$$\operatorname{div} \mathbf{B} = 0 , \qquad \operatorname{rot} \mathbf{E} = -\frac{\partial c\mathbf{B}}{\partial ct} , \tag{2.5.1}$$

$$H^{ab} = (\mathbf{D}, \ \mathbf{H}/c)$$

$$\operatorname{div} \mathbf{D} = \rho , \qquad \operatorname{rot} \frac{\mathbf{H}}{c} = \frac{\mathbf{J}}{c} + \frac{\partial \mathbf{D}}{\partial ct} , \tag{2.5.2}$$

with the constitutive relations

$$\mathbf{D} = \epsilon_0 \epsilon \, \mathbf{E} , \qquad \mathbf{H} = \frac{1}{\mu \mu_0} \mathbf{B} . \tag{2.5.3}$$

In the Minkowski approach, quantities with simple transformation laws under the Lorentz group are

$$\mathbf{f} = \mathbf{E} + ic\mathbf{B} , \qquad \mathbf{h} = \frac{1}{\epsilon_0} \left(\mathbf{D} + i\mathbf{H}/c \right) ,$$

$$j^a = (j^0 = \rho, \ \mathbf{j} = \mathbf{J}/c) ; \tag{2.5.4}$$

where $\mathbf{f}, \mathbf{h}$ are the complex 3-vector under the complex orthogonal group $SO(3.C)$, the latter is isomorphic to the Lorentz group. One can combine Eqs. (2.5.1) and (2.5.2) into following ones

$$\operatorname{div} \left(\frac{\mathbf{D}}{\epsilon_0} + i \, c\mathbf{B} \right) = \frac{1}{\epsilon_0} \rho ,$$

$$-i\partial_0(\frac{\mathbf{D}}{\epsilon_0} + ic\mathbf{B}) + \text{rot } (\mathbf{E} + i\frac{\mathbf{H}/c}{\epsilon_0}) = \frac{i}{\epsilon_0}\,\mathbf{j}\,. \qquad (2.5.5)$$

By taking into account the relationships

$$\mathbf{E} = \frac{\mathbf{f} + \mathbf{f}^*}{2}\,, \qquad i\,\frac{\mathbf{H}/c}{\epsilon_0} = \frac{\mathbf{h} - \mathbf{h}^*}{2}\,,$$

$$\frac{\mathbf{D}}{\epsilon_0} = \frac{\mathbf{h} + \mathbf{h}^*}{2}\,, \qquad ic\mathbf{B} = \frac{\mathbf{f} - \mathbf{f}^*}{2}\,, \qquad (2.5.6)$$

Eqs. (2.5.5) can be rewritten in the form

$$\text{div } (\frac{\mathbf{h} + \mathbf{h}^*}{2} + \frac{\mathbf{f} - \mathbf{f}^*}{2}) = \frac{1}{\epsilon_0}\,\rho\,,$$

$$-i\partial_0(\frac{\mathbf{h} + \mathbf{h}^*}{2} + \frac{\mathbf{f} - \mathbf{f}^*}{2}) + \text{rot } (\frac{\mathbf{f} + \mathbf{f}^*}{2} + \frac{\mathbf{h} - \mathbf{h}^*}{2}) = \frac{i}{\epsilon_0}\,\mathbf{j}\,. \qquad (2.5.7)$$

Let us introduce two quantities

$$\mathbf{M} = \frac{\mathbf{h} + \mathbf{f}}{2}\,, \qquad \mathbf{N} = \frac{\mathbf{h}^* - \mathbf{f}^*}{2}\,, \qquad (2.5.8)$$

which are different 3-vectors under the group $SO(3.C)$:

$$\mathbf{M}' = O\,\mathbf{M}\,, \qquad \mathbf{N}' = O^*\,\mathbf{N}\,. \qquad (2.5.9)$$

Evidently, with respect to Euclidean rotations, the identity $O^* = O$ holds; instead for Lorentzian boosts we have $O^* = O^{-1}$.

In terms of $\mathbf{M}, \mathbf{N}$, Eqs. (2.5.7) will be

$$\text{div } \mathbf{M} + \text{div } \mathbf{N} = \frac{1}{\epsilon_0}\,\rho\,,$$

$$-i\partial_0\mathbf{M} + \text{rot } \mathbf{M} - i\partial_0\mathbf{N} - \text{rot } \mathbf{N} = \frac{i}{\epsilon_0}\,\mathbf{j}\,, \qquad (2.5.10)$$

or in a matrix form

$$(-i\partial_0 + \alpha^i\partial_i)\,M + (-i\partial_0 + \beta^i\partial_i)\,N = J\,, \qquad (2.5.11)$$

where

$$M = \begin{vmatrix} 0 \\ \mathbf{M} \end{vmatrix}\,, \qquad N = \begin{vmatrix} 0 \\ \mathbf{N} \end{vmatrix}\,, \qquad J = \frac{1}{\epsilon_0}\begin{vmatrix} \rho \\ i\,\mathbf{j} \end{vmatrix}\,.$$

The matrices α^i and β^i are taken in the form

$$\alpha^1 = \begin{vmatrix} 0 & 1 & 0 & 0 \\ -1 & 0 & 0 & 0 \\ 0 & 0 & 0 & -1 \\ 0 & 0 & 1 & 0 \end{vmatrix}\,, \quad \alpha^2 = \begin{vmatrix} 0 & 0 & 1 & 0 \\ 0 & 0 & 0 & 1 \\ -1 & 0 & 0 & 0 \\ 0 & -1 & 0 & 0 \end{vmatrix}\,, \quad \alpha^3 = \begin{vmatrix} 0 & 0 & 0 & 1 \\ 0 & 0 & -1 & 0 \\ 0 & 1 & 0 & 0 \\ -1 & 0 & 0 & 0 \end{vmatrix}\,,$$

$$\beta^1 = \begin{vmatrix} 0 & 1 & 0 & 0 \\ -1 & 0 & 0 & 0 \\ 0 & 0 & 0 & 1 \\ 0 & 0 & -1 & 0 \end{vmatrix}, \quad \beta^2 = \begin{vmatrix} 0 & 0 & 1 & 0 \\ 0 & 0 & 0 & -1 \\ -1 & 0 & 0 & 0 \\ 0 & 1 & 0 & 0 \end{vmatrix}, \quad \beta^3 = \begin{vmatrix} 0 & 0 & 0 & 1 \\ 0 & 0 & 1 & 0 \\ 0 & -1 & 0 & 0 \\ -1 & 0 & 0 & 0 \end{vmatrix}.$$

$$(2.5.12)$$

It should be noted that the main idea given by Minkowski is to divide the Maxwell equations into equations for tensors F^{ab} and H^{ab} by transforming them independently under the Lorentz group – this should correspond to immiscible equations for $\mathbf{f}, \mathbf{f}^*$ and $\mathbf{h}, \mathbf{h}^*$, respectively. Such a form can be found easily. Indeed, let us start with Eqs. (2.5.7) and use its conjugated form:

$$\mathrm{div}\,(\frac{\mathbf{h} + \mathbf{h}^*}{2} + \frac{\mathbf{f} - \mathbf{f}^*}{2}) = \frac{1}{\epsilon_0}\,\rho\,, \qquad \mathrm{div}\,(\frac{\mathbf{h} - \mathbf{h}^*}{2} + \frac{\mathbf{f} - \mathbf{f}^*}{2}) = \frac{1}{\epsilon_0}\,\rho\,;$$

$$-i\partial_0(\frac{\mathbf{h} + \mathbf{h}^*}{2} + \frac{\mathbf{f} - \mathbf{f}^*}{2}) + \mathrm{rot}\,(\frac{\mathbf{f} + \mathbf{f}^*}{2} + \frac{\mathbf{h} - \mathbf{h}^*}{2}) = \frac{i}{\epsilon_0}\,\mathbf{j}\,,$$

$$+i\partial_0(\frac{\mathbf{h} + \mathbf{h}^*}{2} - \frac{\mathbf{f} - \mathbf{f}^*}{2}) + \mathrm{rot}\,(\frac{\mathbf{f} + \mathbf{f}^*}{2} - \frac{\mathbf{h} - \mathbf{h}^*}{2}) = -\frac{i}{\epsilon_0}\,\mathbf{j}\,.$$

From these deductions, it follows

$$\mathrm{div}\,\frac{\mathbf{h} + \mathbf{h}^*}{2} = \frac{1}{\epsilon_0}\,\rho\,, \qquad \mathrm{div}\,\frac{\mathbf{f} - \mathbf{f}^*}{2} = 0\,;$$

$$-i\partial_0\frac{\mathbf{f} - \mathbf{f}^*}{2} + \mathrm{rot}\,\frac{\mathbf{f} + \mathbf{f}^*}{2} = 0\,, \qquad -i\partial_0\frac{\mathbf{h} + \mathbf{h}^*}{2} + \mathrm{rot}\,\frac{\mathbf{h} - \mathbf{h}^*}{2} = \frac{i}{\epsilon_0}\,\mathbf{j}\,.$$

Thus, in agreement with the Minkowski principle, a required form is

$$\mathrm{div}\,\frac{\mathbf{f} - \mathbf{f}^*}{2} = 0\,, \qquad -i\partial_0\frac{\mathbf{f} - \mathbf{f}^*}{2} + \mathrm{rot}\,\frac{\mathbf{f} + \mathbf{f}^*}{2} = 0\,, \qquad (2.5.13)$$

$$\mathrm{div}\,\frac{\mathbf{h} + \mathbf{h}^*}{2} = \rho\,, \qquad -i\partial_0\frac{\mathbf{h} + \mathbf{h}^*}{2} + \mathrm{rot}\,\frac{\mathbf{h} - \mathbf{h}^*}{2} = \frac{i}{\epsilon_0}\,\mathbf{j}\,. \qquad (2.5.14)$$

The quantities entering these equations behave themselves under the Lorentz group in accordance with the rules:

$$\mathbf{f}' = O\,\mathbf{f}\,, \qquad \mathbf{f}^{*'} = O^*\,\mathbf{f}^*\,,$$

$$\mathbf{h}' = O\,\mathbf{h}\,, \qquad \mathbf{h}^{*'} = O^*\,\mathbf{h}^*\,.$$

When comparing Eqs. (2.5.13) and (2.5.14) with Eqs. (2.5.10) and (2.5.11):

$$\mathrm{div}\,\mathbf{M} + \mathrm{div}\,\mathbf{N} = \frac{1}{\epsilon_0}\,\rho\,,$$

$$-i\partial_0\mathbf{M} + \mathrm{rot}\,\mathbf{M} - i\partial_0\mathbf{N} - \mathrm{rot}\,\mathbf{N} = \frac{i}{\epsilon_0}\,\mathbf{j}\,,$$

$$(-i\partial_0 + \alpha^i\partial_i)M + (-i\partial_0 + \beta^i\partial_i)N = J\,,$$

one can immediately conclude from this comparison that due do

$$\operatorname{div} \mathbf{h} + \operatorname{div} \mathbf{h}^* = \frac{1}{\epsilon_0} \, \rho \, ,$$

$$-i\partial_0 \mathbf{h} + \operatorname{rot} \mathbf{h} - i\partial_0 \mathbf{h}^* - \operatorname{rot} \mathbf{h}^* = \frac{i}{\epsilon_0} \, \mathbf{j} \, , \qquad (2.5.15)$$

$$\operatorname{div} \mathbf{f} - \operatorname{div} \mathbf{f}^* = 0 \, ,$$

$$-i\partial_0 \mathbf{f} + \operatorname{rot} \mathbf{f} + i\partial_0 \mathbf{f}^* + \operatorname{rot} \mathbf{f}^* = 0 \qquad (2.5.16)$$

the matrix forms we need are

$$(-i\partial_0 + \alpha^i \partial_i) h + (-i\partial_0 + \beta^i \partial_i) h^* = J \, , \qquad (2.5.17)$$

$$(-i\partial_0 + \alpha^i \partial_i) f - (-i\partial_0 + \beta^i \partial_i) f^* = 0 \, . \qquad (2.5.18)$$

Below, we will work with a more simple form according to Eq. (2.5.11).

2.6   Minkowski Constitutive Relations in a Complex 3-Vector Form

Symmetry of the matrix equation under the Lorentz transformations must exist because symmetry exists for these equations written in the ordinary tensor form. Now, let us examine how these constitutive relations behave under the Lorentz transformations.

Let us start with these relations in the rest reference frame

$$\mathbf{D} = \epsilon_0 \epsilon \, \mathbf{E} \, , \qquad \frac{\mathbf{H}}{c} = \frac{1}{\mu_0 \mu} \frac{1}{c^2} \, c\mathbf{B} = \frac{\epsilon_0}{\mu} \, c\mathbf{B} \, . \qquad (2.6.1)$$

Allowing for Eq. (2.5.6), then Eqs. (2.6.1) can be rewritten as

$$\frac{\mathbf{h} + \mathbf{h}^*}{2} = \epsilon \, \frac{\mathbf{f} + \mathbf{f}^*}{2} \, , \qquad \frac{\mathbf{h} - \mathbf{h}^*}{2} = \frac{1}{\mu} \frac{\mathbf{f} - \mathbf{f}^*}{2} \, ; \qquad (2.6.2)$$

from whence it follows

$$2\mathbf{h} = (\epsilon + \frac{1}{\mu}) \, \mathbf{f} + (\epsilon - \frac{1}{\mu}) \, \mathbf{f}^* \, ,$$

$$2\mathbf{h}^* = (\epsilon + \frac{1}{\mu}) \, \mathbf{f}^* + (\epsilon - \frac{1}{\mu}) \, \mathbf{f} \, . \qquad (2.6.3)$$

This is a complex form of the constitutive relations as in Eq. (2.6.1). It should be noted that Eqs. (2.6.2) can be resolved under $\mathbf{f}, \, \mathbf{f}^*$ as well:

$$2\mathbf{f} = (\frac{1}{\epsilon} + \mu) \, \mathbf{h} + (\frac{1}{\epsilon} - \mu) \, \mathbf{h}^* \, ,$$

$$2\mathbf{f}^* = (\frac{1}{\epsilon} + \mu) \, \mathbf{h}^* + (\frac{1}{\epsilon} - \mu) \, \mathbf{h} \, ; \qquad (2.6.4)$$

these are the same constitutive equations as in Eqs. (2.6.3) but in other form.

Now let us allow for the Lorentz transformations:

$$\mathbf{f}' = O\,\mathbf{f}\ , \qquad \mathbf{f}'^* = O^*\,\mathbf{f}^*\ , \qquad \mathbf{h}' = O\,\mathbf{h}\ , \qquad \mathbf{h}'^* = O^*\,\mathbf{h}^*\ ;$$

then Eqs. (2.6.2) will become

$$\frac{O^{-1}\mathbf{h}' + (O^{-1})^*\mathbf{h}^{*'}}{2} = \epsilon\,\frac{O^{-1}\mathbf{f}' + (O^{-1})^*\mathbf{f}^{*'}}{2}\ ,$$

$$\frac{O^{-1}\mathbf{h}' - (O^{-1})^*\mathbf{h}^{*'}}{2} = \frac{1}{\mu}\,\frac{O^{-1}\mathbf{f}' - (O^{-1})^*\mathbf{f}^{*'}}{2}\ .$$

Multiplying both equations by O and summing (or subtracting) the results, we get

$$\mathbf{h}' = \epsilon\,\frac{\mathbf{f}' + O(O^{-1})^*\mathbf{f}^{*'}}{2} + \frac{1}{\mu}\,\frac{\mathbf{f}' - O(O^{-1})^*\mathbf{f}^{*'}}{2}\ , \tag{2.6.5}$$

$$\mathbf{h}^{*'} = \epsilon\,\frac{O^*O^{-1}\mathbf{f}' + \mathbf{f}^{*'}}{2} - \frac{1}{\mu}\,\frac{O^*O^{-1}\mathbf{f}' - \mathbf{f}^{*'}}{2}\ . \tag{2.6.6}$$

Eqs. (2.6.5) and (2.6.6) can be presented as

$$2\mathbf{h}' = (\epsilon + \frac{1}{\mu})\,\mathbf{f}' + (\epsilon - \frac{1}{\mu})\,O(O^{-1})^*\,\mathbf{f}'^*\ ,$$

$$2\mathbf{h}'^* = (\epsilon + \frac{1}{\mu})\,\mathbf{f}'^* + (\epsilon - \frac{1}{\mu})\,O^*O^{-1}\,\mathbf{f}'\ . \tag{2.6.7}$$

Analogously, starting from Eqs. (2.6.4) we can produce

$$2\mathbf{f}' = (\frac{1}{\epsilon} + \mu)\,\mathbf{h}' + (\frac{1}{\epsilon} - \mu)\,O(O^{-1})^*\,\mathbf{h}'^*\ ,$$

$$2\mathbf{f}'^* = (\frac{1}{\epsilon} + \mu)\,\mathbf{h}'^* + (\frac{1}{\epsilon} - \mu)\,O^*O^{-1}\,\mathbf{h}'\ . \tag{2.6.8}$$

Equations (2.6.7) and (2.6.8) represent the constitutive relations after changing the reference frame. At this point, one should be able to distinguish between two cases: Euclidean rotation and Lorentzian boosts.

Indeed, for any Euclidean rotations

$$O^* = O, \qquad \Longrightarrow \qquad O(O^{-1})^* = I, \qquad O^*O^{-1} = I\ ;$$

which, therefore Eqs. (2.6.7) and (2.6.8) take the form of Eqs. (2.6.3) and (2.6.4); in other words, at Euclidean rotations the constitutive relations do not change their form.

However, for any pseudo-Euclidean rotations (Lorentzian boosts)

$$O^* = O^{-1} \qquad \Longrightarrow \qquad O(O^{-1})^* = 0^2, \qquad O^*O^{-1} = O^{*2}\ ;$$

and Eqs. (2.6.7) and (2.6.8) look

$$2\mathbf{h}' = (\epsilon + \frac{1}{\mu})\,\mathbf{f}' + (\epsilon - \frac{1}{\mu})\,O^2\,\mathbf{f}'^*\ ,$$

$$2\mathbf{h}'^* = (\epsilon + \frac{1}{\mu}) \, \mathbf{f}'^* + (\epsilon - \frac{1}{\mu}) \, O^2 \, \mathbf{f}' \; ; \tag{2.6.9}$$

$$2\mathbf{f}' = (\frac{1}{\epsilon} + \mu) \, \mathbf{h}' + (\frac{1}{\epsilon} - \mu) \, O^{*2} \, \mathbf{h}'^* \, ,$$

$$2\mathbf{f}'^* = (\frac{1}{\epsilon} + \mu) \, \mathbf{h}'^* + (\frac{1}{\epsilon} - \mu) \, O^{*2} \, \mathbf{h}' \, . \tag{2.6.10}$$

In complex 3-vector form these relations seem to be shorter than in real 3-vector form:

$$\mathbf{D}' + i\frac{\mathbf{H}'}{c} = \epsilon_0\epsilon \, \frac{(\mathbf{E}' + ic\mathbf{B}') + O(O^{-1})^*(\mathbf{E}' - ic\mathbf{B}')}{2}$$

$$+\frac{\epsilon_0}{\mu} \, \frac{(\mathbf{E}' + ic\mathbf{B}') - O(O^{-1})^*(\mathbf{E}' - ic\mathbf{B}')}{2} \, ,$$

$$\mathbf{D}' - i\frac{\mathbf{H}'}{c} = \epsilon_0\epsilon \, \frac{(\mathbf{E}' - ic\mathbf{B}') + O^*O^{-1}(\mathbf{E}' + ic\mathbf{B}')}{2}$$

$$+\frac{\epsilon_0}{\mu} \, \frac{(\mathbf{E}' - ic\mathbf{B}') - O^*O^{-1}(\mathbf{E}' + ic\mathbf{B}')}{2} \, ,$$

or

$$2\mathbf{D}' = \epsilon_0\epsilon \, [\, \mathbf{E}' + \frac{O(O^{-1})^* + O^*O^{-1}}{2} \, \mathbf{E}' + \frac{O(O^{-1})^* - O^*O^{-1}}{2i} \, c\mathbf{B}' \,]$$

$$+\frac{\epsilon_0}{\mu} \, [\, \mathbf{E}' - \frac{O(O^{-1})^* + O^*O^{-1}}{2} \, \mathbf{E}' - \frac{O(O^{-1})^* - O^*O^{-1}}{2i} \, c\mathbf{B}' \,] \, ,$$

$$\tag{2.6.11}$$

$$2\frac{\mathbf{H}'}{c} = \epsilon_0\epsilon \, [\, c\mathbf{B}' - \frac{O(O^{-1})^* + O^*O^{-1}}{2} \, c\mathbf{B}' + \frac{O(O^{-1})^* - O^*O^{-1}}{2i} \, \mathbf{E}' \,]$$

$$+\frac{\epsilon_0}{\mu} \, [\, c\mathbf{B}' + \frac{O(O^{-1})^* + O^*O^{-1}}{2} \, c\mathbf{B}' - \frac{O(O^{-1})^* - O^*O^{-1}}{2i} \, \mathbf{E}' \,] \, .$$

$$\tag{2.6.12}$$

These are the constitutive equations for electromagnetic field in a uniform media according to Minkowski in a moving reference frame. For any Euclidean rotation, $O^* = O$, and Eqs. (2.6.11) and (2.6.12) coincide with the initial ones

$$\mathbf{D}' = \epsilon_0\epsilon \, \mathbf{E}' \, , \qquad \mathbf{H}'/c = \epsilon_0 \frac{1}{\mu} \, c\mathbf{B}' \; \Longrightarrow \; \mathbf{H}' = \frac{1}{\mu_0\mu} \, \mathbf{B}' \, .$$

For any Lorentzian boosts $O^* = O^{-1}$, Eqs. (2.6.11) and (2.6.12) will read as:

$$2\mathbf{D}' = \epsilon_0\epsilon \, [\, (I + \frac{OO + O^*O^*}{2}) \, \mathbf{E}' + \frac{OO - O^*O^*}{2i} \, c\mathbf{B}' \,]$$

$$+\frac{\epsilon_0}{\mu} \, [\, (I - \frac{OO + O^*O^*}{2}) \, \mathbf{E}' - \frac{OO - O^*O^*}{2i} \, c\mathbf{B}' \,] \, ,$$

$$2\mathbf{H}'/c = \epsilon_0 \epsilon \left[\left(I - \frac{OO + O^*O^*}{2} \right) c\mathbf{B}' + \frac{OO - O^*O^*}{2i} \mathbf{E}' \right]$$
$$+ \frac{\epsilon_0}{\mu} \left[\left(I + \frac{OO + O^*O^*}{2} \right) c\mathbf{B}' - \frac{OO - O^*O^*}{2i} \mathbf{E}' \right]. \qquad (2.6.13)$$

They can be written differently

$$\mathbf{D}' = \frac{\epsilon_0}{2} \left\{ \left[\left(\epsilon + \frac{1}{\mu} \right) + \left(\epsilon - \frac{1}{\mu} \right) \operatorname{Re} O^2 \right] \mathbf{E}' + \left(\epsilon - \frac{1}{\mu} \right) \operatorname{Im} O^2 \, c\mathbf{B}' \right\},$$

$$\frac{\mathbf{H}'}{c} = \frac{\epsilon_0}{2} \left\{ \left[\left(\epsilon + \frac{1}{\mu} \right) - \left(\epsilon - \frac{1}{\mu} \right) \operatorname{Re} O^2 \right] c\mathbf{B}' + \left(\epsilon - \frac{1}{\mu} \right) \operatorname{Im} O^2 \, \mathbf{E}' \right\}.$$

$$(2.6.14)$$

Lorentzian complex vector boosts are given by the matrix (remembering $f = 1 - \cosh b$)

$$O = O(b, \mathbf{n})$$

$$= \begin{vmatrix} 1 - f(n_2^2 + n_3^2) & -i \sinh b\, n_3 + f n_1 n_2 & i \sinh b\, n_2 + f n_1 n_3 \\ i \sinh b\, n_3 + f n_1 n_2 & 1 - f(n_3^2 + n_1^2) & -i \sinh b\, n_1 + f n_2 n_3 \\ -\sinh b\, n_2 + f n_1 n_3 & i \sinh b\, n_1 + f n_2 n_3 & 1 - f(n_1^2 + n_2^2) \end{vmatrix}. \quad (2.6.15)$$

The square O^2 is (let $F = 1 - \cosh 2b$)

$$O^2 = \begin{vmatrix} \cosh 2b + (1 - \cosh 2b)n_1^2 & F n_1 n_2 - i \sinh 2b\, n_3 & F n_3 n_1 + i \sinh 2b\, n_2 \\ F n_1 n_2 + i \sinh 2b\, n_3 & \cosh 2b + F n_2^2 & F n_2 n_3 - i \sinh 2b\, n_1 \\ F n_3 n_1 - i \sinh 2b\, n_2 & F n_2 n_3 + i \sinh 2b\, n_1 & \cosh 2b + F n_3^2 \end{vmatrix}.$$

$$(2.6.16)$$

We have obtained the result that must be expected: Eq. (2.6.16) differs from Eq. (2.6.15) only in one change $b \longrightarrow 2b$. These expressions for O^2 ought to be substituted into the formulas:

$$2\mathbf{h}' = \left(\epsilon + \frac{1}{\mu} \right) \mathbf{f}' + \left(\epsilon - \frac{1}{\mu} \right) O^2 \mathbf{f}'^*,$$

$$2\mathbf{h}'^* = \left(\epsilon + \frac{1}{\mu} \right) \mathbf{f}'^* + \left(\epsilon - \frac{1}{\mu} \right) O^2 \mathbf{f}',$$

$$2\mathbf{f}' = \left(\frac{1}{\epsilon} + \mu \right) \mathbf{h}' + \left(\frac{1}{\epsilon} - \mu \right) O^{*2} \mathbf{h}'^*,$$

$$2\mathbf{f}'^* = \left(\frac{1}{\epsilon} + \mu \right) \mathbf{h}'^* + \left(\frac{1}{\epsilon} - \mu \right) O^{*2} \mathbf{h}',$$

or

$$\mathbf{D}' = \frac{\epsilon_0}{2} \left\{ \left[\left(\epsilon + \frac{1}{\mu} \right) + \left(\epsilon - \frac{1}{\mu} \right) \operatorname{Re} O^2 \right] \mathbf{E}' + \left(\epsilon - \frac{1}{\mu} \right) \operatorname{Im} O^2 \, c\mathbf{B}' \right\},$$

$$\frac{\mathbf{H}'}{c} = \frac{\epsilon_0}{2} \left\{ \left[\left(\epsilon + \frac{1}{\mu} \right) - \left(\epsilon - \frac{1}{\mu} \right) \operatorname{Re} O^2 \right] c\mathbf{B}' + \left(\epsilon - \frac{1}{\mu} \right) \operatorname{Im} O^2 \, \mathbf{E}' \right\}.$$

All formulas become much more simple for rotations in planes $(0-1), (0-2), (0-3)$:

$$\mathbf{n} = (1, 0, 0), \qquad O^2 = \begin{vmatrix} 1 & 0 & 0 \\ 0 & \cosh 2b & +i \sinh 2b \\ 0 & -i \sinh 2b & \cosh 2b \end{vmatrix} ,$$

$$\mathbf{n} = (0, 1, 0), \qquad O^2 = \begin{vmatrix} \cosh 2b & 0 & -i \sinh 2b \\ 0 & 1 & 0 \\ +i \sinh 2b & 0 & \cosh 2b \end{vmatrix} ,$$

$$\mathbf{n} = (0, 0, 1), \qquad O^2 = \begin{vmatrix} \cosh 2b & +i \sinh 2b & 0 \\ -i \sinh 2b & \cosh 2b & 0 \\ 0 & 0 & 1 \end{vmatrix} .$$

$$(2.6.17)$$

The previous result can be easily extended to more general medias. Here, we will restrict ourselves to linear medias specifically. Indeed, arbitrary linear media is characterized by the following constitutive equations:

$$\mathbf{D} = \epsilon_0 \, \epsilon(x) \, \mathbf{E} + \epsilon_0 c \, \alpha(x) \, \mathbf{B} ,$$

$$\mathbf{H} = \epsilon_0 c \, \beta(x) \, \mathbf{E} + \frac{1}{\mu_0} \, \mu(x) \, \mathbf{B} , \qquad (2.6.18)$$

where $\epsilon(x), \mu(x), \alpha(x), \beta(x)$ are (3×3)-dimensionless matrices. Eqs. (2.6.18) should be rewritten in terms of complex vectors $\mathbf{f}, \mathbf{h}$:

$$\frac{\mathbf{h} + \mathbf{h}^*}{2} = \epsilon(x) \, \frac{\mathbf{f} + \mathbf{f}^*}{2} + \alpha(x) \, \frac{\mathbf{f} - \mathbf{f}^*}{2i} ,$$

$$\frac{\mathbf{h} - \mathbf{h}^*}{2i} = \beta(x) \, \frac{\mathbf{f} + \mathbf{f}^*}{2} + \mu(x) \, \frac{\mathbf{f} - \mathbf{f}^*}{2i} . \qquad (2.6.19)$$

From Eqs. (2.6.18) and (2.6.19), it follows

$$\mathbf{h} = [\, (\epsilon(x) + \mu(x)) + i(\beta(x) - \alpha(x)) \,] \, \mathbf{f}$$

$$+ [\, (\epsilon(x) - \mu(x)) + i(\beta(x) + \alpha(x)) \,] \, \mathbf{f}^* ,$$

$$\mathbf{h}^* = [\, (\epsilon(x) + \mu(x)) - i(\beta(x) - \alpha(x)) \,] \, \mathbf{f}^*$$

$$+ [\, (\epsilon(x) - \mu(x)) - i(\beta(x) + \alpha(x)) \,] \, \mathbf{f} . \qquad (2.6.20)$$

Under Lorentz transformations, the relations of Eq. (2.6.20) will take the form

$$O^{-1}\mathbf{h}' = [(\epsilon(x) + \mu(x)) + i(\beta(x) - \alpha(x))] \, O^{-1}\mathbf{f}'$$

$$+ [(\epsilon(x) - \mu(x)) + i(\beta(x) + \alpha(x))] \, (O^{-1})^* \mathbf{f}'^* ,$$

$$(O^{-1})^* \mathbf{h}'^* = [(\epsilon(x) + \mu(x)) - i(\beta(x) - \alpha(x))] \, (O^{-1})^* \mathbf{f}'^*$$

$$+ [(\epsilon(x) - \mu(x)) - i(\beta(x) + \alpha(x))] \, (O^{-1})\mathbf{f}' , \qquad (2.6.21)$$

or

$$\mathbf{h}' = \epsilon_0 \left[(\epsilon(x) + \mu(x)) + i(\beta(x) - \alpha(x)) \right] \mathbf{f}'$$
$$+ \left[(\epsilon(x) - \mu(x)) + i(\beta(x) + \alpha(x)) \right] [O(O^{-1})^*] \mathbf{f}'^*,$$

$$\mathbf{h}'^* = \epsilon_0 \left[(\epsilon(x) + \mu(x)) - i(\beta(x) - \alpha(x)) \right] \mathbf{f}'^*$$
$$+ [(\epsilon(x) - \mu(x)) - i(\beta(x) + \alpha(x))] [O^*(O^{-1})] \mathbf{f}'. \tag{2.6.22}$$

For Euclidean rotation, we have

$$[O(O^{-1})^*] = I, \qquad [O^*(O^{-1})] = I ,$$

and the constitutive relations preserve their form. For Lorentz boosts, we have

$$[O(O^{-1})^*] = O^2, \qquad [O^*(O^{-1})] = O^{*2} ,$$

and the constitutive equations in a moving reference frame are

$$\mathbf{h}' = \left[(\epsilon(x) + \mu(x)) + i\,(\beta(x) - \alpha(x)) \right] \mathbf{f}'$$

$$+ \left[(\epsilon(x) - \mu(x)) + i\,(\beta(x) + \alpha(x)) \right] O^2 \mathbf{f}'^* ,$$

$$\mathbf{h}'^* = \left[(\epsilon(x) + \mu(x)) - i\,(\beta(x) - \alpha(x)) \right] \mathbf{f}'^*$$

$$+ \left[(\epsilon(x) - \mu(x)) - i\,(\beta(x) + \alpha(x)) \right] O^{*2} \mathbf{f}'. \tag{2.6.23}$$

These are the constitutive relations for arbitrary linear media in the moving reference frame.

2.7 Symmetry Properties of the Matrix Equation in the Media

Maxwell equations in media are written in two groups

$$F^{ab}, \qquad \operatorname{div} \mathbf{f} - \operatorname{div} \mathbf{f}^* = 0 ,$$

$$-i\partial_0 \mathbf{f} + \operatorname{rot} \mathbf{f} + i\partial_0 \mathbf{f}^* + \operatorname{rot} \mathbf{f}^* = 0 ; \tag{2.7.1}$$

$$H^{ab}, \qquad \operatorname{div} \mathbf{h} + \operatorname{div} \mathbf{h}^* = \frac{1}{\epsilon_0}\rho ,$$

$$-i\partial_0 \mathbf{h} + \operatorname{rot} \mathbf{h} - i\partial_0 \mathbf{h}^* - \operatorname{rot} \mathbf{h}^* = i\,\frac{1}{\epsilon_0}\mathbf{j} . \tag{2.7.2}$$

Their matrix forms are respectively given as

$$(-i\partial_0 + \alpha^i \partial_i)f - (-i\partial_0 + \beta^i \partial_i)f^* = 0 , \tag{2.7.3}$$

and

$$(-i\partial_0 + \alpha^i \partial_i)h + (-i\partial_0 + \beta^i \partial_i)h^* = J . \tag{2.7.4}$$

Maxwell equations can be presented in a more short form (see Eqs. (2.5.10) and (2.5.11)):

$$\operatorname{div} \mathbf{M} + \operatorname{div} \mathbf{N} = \frac{1}{\epsilon_0}\rho \,,$$

$$-i\partial_0 \mathbf{M} + \operatorname{rot} \mathbf{M} - i\partial_0 \mathbf{N} - \operatorname{rot} \mathbf{N} = \frac{i}{\epsilon_0}\mathbf{j} \,; \tag{2.7.5}$$

with the corresponding matrix form

$$(-i\partial_0 + \alpha^i \partial_i)\, M + (-i\partial_0 + \beta^i \partial_i)\, N = J \,. \tag{2.7.6}$$

The matrices α^i and β^i are given by

$$\alpha^1 = \begin{vmatrix} 0 & 1 & 0 & 0 \\ -1 & 0 & 0 & 0 \\ 0 & 0 & 0 & -1 \\ 0 & 0 & 1 & 0 \end{vmatrix}, \quad \alpha^2 = \begin{vmatrix} 0 & 0 & 1 & 0 \\ 0 & 0 & 0 & 1 \\ -1 & 0 & 0 & 0 \\ 0 & -1 & 0 & 0 \end{vmatrix}, \quad \alpha^3 = \begin{vmatrix} 0 & 0 & 0 & 1 \\ 0 & 0 & -1 & 0 \\ 0 & 1 & 0 & 0 \\ -1 & 0 & 0 & 0 \end{vmatrix},$$

$$\beta^1 = \begin{vmatrix} 0 & 1 & 0 & 0 \\ -1 & 0 & 0 & 0 \\ 0 & 0 & 0 & 1 \\ 0 & 0 & -1 & 0 \end{vmatrix}, \quad \beta^2 = \begin{vmatrix} 0 & 0 & 1 & 0 \\ 0 & 0 & 0 & -1 \\ -1 & 0 & 0 & 0 \\ 0 & 1 & 0 & 0 \end{vmatrix}, \quad \beta^3 = \begin{vmatrix} 0 & 0 & 0 & 1 \\ 0 & 0 & 1 & 0 \\ 0 & -1 & 0 & 0 \\ -1 & 0 & 0 & 0 \end{vmatrix}. \tag{2.7.7}$$

We now will consider Eq. (2.7.6). The terms with α^j matrices were previously examined, where the terms with β^j matrices are new. We restrict ourselves to demonstrate the Lorentz symmetry of Eq. (2.7.6) under two of the simplest transformations.

First, let us consider the Euclidean rotation in the plane $(1-2)$. Here, we examine additionally only the term with β-matrices:

$$S\beta^1 S^{-1} = \begin{vmatrix} 1 & 0 & 0 & 0 \\ 0 & \cos a & \sin a & 0 \\ 0 & -\sin a & \cos a & 0 \\ 0 & 0 & 0 & 1 \end{vmatrix} \begin{vmatrix} 0 & 1 & 0 & 0 \\ -1 & 0 & 0 & 0 \\ 0 & 0 & 0 & 1 \\ 0 & 0 & -1 & 0 \end{vmatrix} \begin{vmatrix} 1 & 0 & 0 & 0 \\ 0 & \cos a & -\sin a & 0 \\ 0 & \sin a & \cos a & 0 \\ 0 & 0 & 0 & 1 \end{vmatrix}$$

$$= \begin{vmatrix} 0 & \cos a & -\sin a & 0 \\ -\cos a & 0 & 0 & \sin a \\ \sin a & 0 & 0 & \cos a \\ 0 & -\sin a & -\cos a & 0 \end{vmatrix} = \cos a\, \beta^1 - \sin a\, \beta^2 = \beta^j O_{j1}\,, \tag{2.7.8}$$

$$S\beta^2 S^{-1} = \begin{vmatrix} 1 & 0 & 0 & 0 \\ 0 & \cos a & \sin a & 0 \\ 0 & -\sin a & \cos a & 0 \\ 0 & 0 & 0 & 1 \end{vmatrix} \begin{vmatrix} 0 & 0 & 1 & 0 \\ 0 & 0 & 0 & -1 \\ -1 & 0 & 0 & 0 \\ 0 & 1 & 0 & 0 \end{vmatrix} \begin{vmatrix} 1 & 0 & 0 & 0 \\ 0 & \cos a & -\sin a & 0 \\ 0 & \sin a & \cos a & 0 \\ 0 & 0 & 0 & 1 \end{vmatrix}$$

$$= \begin{vmatrix} 0 & \sin a & \cos a & 0 \\ -\sin a & 0 & 0 & -\cos a \\ -\cos a & 0 & 0 & \sin a \\ 0 & \cos a & -\sin a & 0 \end{vmatrix} = \sin a\, \beta^1 + \cos a\, \beta^2 = \beta^j O_{j2}\,, \tag{2.7.9}$$

$$S\beta^3 S^{-1} = \begin{vmatrix} 1 & 0 & 0 & 0 \\ 0 & \cos a & \sin a & 0 \\ 0 & -\sin a & \cos a & 0 \\ 0 & 0 & 0 & 1 \end{vmatrix} \begin{vmatrix} 0 & 0 & 0 & 1 \\ 0 & 0 & 1 & 0 \\ 0 & -1 & 0 & 0 \\ -1 & 0 & 0 & 0 \end{vmatrix} \begin{vmatrix} 1 & 0 & 0 & 0 \\ 0 & \cos a & -\sin a & 0 \\ 0 & \sin a & \cos a & 0 \\ 0 & 0 & 0 & 1 \end{vmatrix}$$

$$= \begin{vmatrix} 0 & 0 & 0 & 1 \\ 0 & 0 & 1 & 0 \\ 0 & -1 & 0 & 0 \\ -1 & 0 & 0 & 0 \end{vmatrix} = \beta^3 = \beta^j O_{j3} . \qquad (2.7.10)$$

Therefore, we conclude that Eq. (2.7.6) is symmetrical under Euclidean rotations in accordance with the relations

$$(-i\partial_0 + S\alpha^i S^{-1}\partial_i)\, M' + (-i\partial_0 + S\beta^i S^{-1}\partial_i)\, N' = +SJ$$

$$\implies \quad (-i\partial_0 + \alpha^i \partial_i')\, M' + (-i\partial_0 + \beta^i \partial_i')\, N' = +J' . \qquad (2.7.11)$$

For the Lorentz boost in the plane $(0-3)$, we have

$$M' = SM , \qquad N' = S^* N = S^{-1} N , \qquad S^* = S^{-1} ;$$

and Eq. (2.7.6) takes the form (note that the additional transformation $\Delta = \Delta_{(\alpha)}$ is combined in terms of α^j)

$$\Delta_{(\alpha)} S \left[(-i\partial_0 + \alpha^i \partial_i)\, S^{-1} M' + (-i\partial_0 + \beta^i \partial_i)\, SN' \right] = \Delta SJ , \qquad (2.7.12)$$

or

$$\Delta_{(\alpha)} \left[(-i\partial_0 + S\alpha^i S^{-1}\partial_i)\, M' + S^2(-i\partial_0 + S^{-1}\beta^i S\partial_i)\, N' \right] = J' , \qquad (2.7.13)$$

and further

$$(-i\partial_0' + \alpha^i \partial_i')\, M' + \Delta_{(\alpha)} S^2(-i\partial_0 + S^{-1}\beta^i S\partial_i)\, N' = J' . \qquad (2.7.14)$$

It remains to prove the relationship

$$\Delta_{(\alpha)} S^2 \, (-i\partial_0 + S^{-1}\beta^i S\partial_i)\, N' = (-i\partial_0' + \beta^i \partial_i')\, N' . \qquad (2.7.15)$$

By simplicity, one can with reasonably expect two identities:

$$\Delta_{(\alpha)} S^2 = \Delta_{(\beta)} \qquad \Longleftrightarrow \qquad \Delta_{(\alpha)} S = \Delta_{(\beta)} S^{-1} , \qquad (2.7.16)$$

and

$$\Delta_{(\beta)}(-i\partial_0 + S^{-1}\beta^i S\partial_i)\, N' = (-i\partial_0' + \beta^i \partial_i')\, N' .$$

$$(2.7.17)$$

Let us prove them for a Lorentzian boost in the plane $(0-3)$:

$$S = \begin{vmatrix} 1 & 0 & 0 & 0 \\ 0 & \cosh b & -i\sinh b & 0 \\ 0 & i\sinh b & \cosh b & 0 \\ 0 & 0 & 0 & 1 \end{vmatrix}, \qquad S^{-1} = \begin{vmatrix} 1 & 0 & 0 & 0 \\ 0 & \cosh b & -i\sinh b & 0 \\ 0 & i\sinh b & \cosh b & 0 \\ 0 & 0 & 0 & 1 \end{vmatrix}.$$

By allowing for Eqs. (2.7.8) – (2.7.10) and with the use the method of simple change, we readily get

$$S^{-1}\beta^1 S = \cosh b\,\beta^1 - i\sinh b\,\beta^2 = \beta^j O_{j1}^{-1},$$

$$S^{-1}\beta^2 S = i\sinh b\,\beta^1 + \cosh b\,\beta^2 = \beta^j O_{j2}^{-1},$$

$$S^{-1}\beta^3 S = \beta^3 = \beta^j O_{j3}^{-1}. \qquad (2.7.18)$$

To verify identity

$$\Delta_{(\alpha)}S = \Delta_{(\beta)}S^{-1},$$

or

$$(\cosh b - i\sinh b\,\alpha^3)S = (\cosh b - i\sinh b\,\beta^3)S^{-1},$$

let us calculate separately the left-hand part

$$(\cosh b - i\sinh b\,\alpha^3)S$$

$$= \begin{vmatrix} \cosh b & 0 & 0 & -i\sinh b \\ 0 & \cosh b & i\sinh b & 0 \\ 0 & -i\sinh b & \cosh b & 0 \\ i\sinh b & 0 & 0 & \cosh b \end{vmatrix} \begin{vmatrix} 1 & 0 & 0 & 0 \\ 0 & \cosh b & -i\sinh b & 0 \\ 0 & i\sinh b & \cosh b & 0 \\ 0 & 0 & 0 & 1 \end{vmatrix}$$

$$= \begin{vmatrix} \cosh b & 0 & 0 & -i\sinh b \\ 0 & 1 & 0 & 0 \\ 0 & 0 & 1 & 0 \\ i\sinh b & 0 & 0 & \cosh b \end{vmatrix};$$

and the right-hand part

$$(\cosh b - i\sinh b\,\beta^3)S^{-1}$$

$$= \begin{vmatrix} \cosh b & 0 & 0 & -i\sinh b \\ 0 & \cosh b & -i\sinh b & 0 \\ 0 & i\sinh b & \cosh b & 0 \\ i\sinh b & 0 & 0 & \cosh b \end{vmatrix} \begin{vmatrix} 1 & 0 & 0 & 0 \\ 0 & \cosh b & i\sinh b & 0 \\ 0 & -i\sinh b & \cosh b & 0 \\ 0 & 0 & 0 & 1 \end{vmatrix}$$

$$= \begin{vmatrix} \cosh b & 0 & 0 & -i\sinh b \\ 0 & 1 & 0 & 0 \\ 0 & 0 & 1 & 0 \\ i\sinh b & 0 & 0 & \cosh b \end{vmatrix}.$$

They coincide with each other, so Eq. (2.7.16) holds. Now to prove the relation in Eq. (2.7.17). By allowing for the properties in the β–matrices

$$(\beta^0)^2 = -I, \qquad (\beta^1)^2 = -I, \qquad \beta^1\beta^2 = -\beta^3, \qquad \beta^2\beta^1 = +\beta^3,$$

we readily find

$$\Delta_{(\beta)} \left(-i\partial_0 + S^{-1}\beta^i S\partial_i\right) N' = (\cosh b - i \sinh b \, \beta^3) \left[\, -i\partial_0 + \beta^3\partial_3 \right.$$

$$+(\cosh b \, \beta^1 - i \sinh b \, \beta^2) \, \partial_1 + (i \sinh b\beta^1 + \cosh b \, \beta^2) \, \partial_2 \left]\, N' \right.$$

$$= \left[\, -i(\cosh b \, \partial_0 - \sinh b \, \partial_3) + \beta^3 \, (-\sinh b \, \partial_0 + \cosh b \, \partial_3) + \beta^1 \, \partial_1 + \beta^2 \, \partial_2 \,\right] N' \,,$$

$$(2.7.19)$$

that is

$$\Delta_{(\beta)}(-i\partial_0 + S^{-1}\beta^i S\partial_i) \, N' = (-i\partial_0' + \beta^1\partial_1 + \beta^2\partial_2 + \beta^3\partial_3') \, N' \,; \qquad (2.7.20)$$

the relations in Eqs. (2.7.16) and (2.7.17) hold. Thus, the symmetry of the matrix Maxwell equation in media under the Lorentz group is proven.

2.8   Maxwell Theory, Dirac Matrices and Electromagnetic 4-vectors

Let us shortly discuss two points relevant to the previous matrix formulation of the Maxwell theory. First, let us write down explicit form for Dirac matrices in spinor basis:

$$\gamma^0 = \begin{vmatrix} 0 & 0 & 1 & 0 \\ 0 & 0 & 0 & 1 \\ 1 & 0 & 0 & 0 \\ 0 & 1 & 0 & 0 \end{vmatrix}, \qquad \gamma^5 = -i\gamma^0\gamma^1\gamma^2\gamma^3 = \begin{vmatrix} -1 & 0 & 0 & 0 \\ 0 & -1 & 0 & 0 \\ 0 & 0 & 1 & 0 \\ 1 & 0 & 0 & 1 \end{vmatrix},$$

$$\gamma^1 = \begin{vmatrix} 0 & 0 & 0 & -1 \\ 0 & 0 & -1 & 0 \\ 0 & 1 & 0 & 0 \\ 1 & 0 & 0 & 0 \end{vmatrix}, \gamma^2 = \begin{vmatrix} 0 & 0 & 0 & i \\ 0 & 0 & -i & 0 \\ 0 & -i & 0 & 0 \\ i & 0 & 0 & 0 \end{vmatrix}, \gamma^3 = \begin{vmatrix} 0 & 0 & -1 & 0 \\ 0 & 0 & 0 & 1 \\ 1 & 0 & 0 & 0 \\ 0 & -1 & 0 & 0 \end{vmatrix}.$$

By keeping in mind the expressions for α^i, β^i, we immediately see the identities

$$\alpha^1 = i\gamma^0\gamma^2, \qquad \alpha^2 = \gamma^0\gamma^5, \qquad \alpha^3 = i\gamma^5\gamma^2 \,,$$

$$\beta^1 = -\gamma^3\gamma^1, \qquad \beta^2 = -\gamma^3, \qquad \beta^3 = -\gamma^1 \,, \qquad\qquad (2.8.1)$$

so the Maxwell matrix equation in the media takes the form

$$(-i\partial_0 + i\gamma^0\gamma^2\partial_1 + \gamma^0\gamma^5\partial_2 + i\gamma^5\gamma^2\partial_3) \, M$$

$$+ (-i\partial_0 - \gamma^3\gamma^1\partial_1 - \gamma^3\partial_2 - \gamma^1\partial_3) \, N = J \,. \qquad (2.8.2)$$

This Dirac matrix-based form does not seem to be very useful when applying it to the Maxwell theory, because there is not much similarity with ordinary Dirac equation (though that analogy was often discussed in the literature).

Below, for simplicity, let us consider the vacuum case. We can start from electromagnetic 2-tensor $F_{\alpha\beta}$ and the dual tensor to it $\tilde{F}_{\rho\sigma}$:

$$\tilde{F}_{\rho\sigma} = \frac{1}{2} \, \epsilon_{\rho\sigma\alpha\beta}F^{\alpha\beta} \,, \qquad F_{\alpha\beta} = -\frac{1}{2} \, \epsilon_{\alpha\beta\rho\sigma}\tilde{F}^{\rho\sigma} \,,$$

let us introduce two electromagnetic 4-vectors (below u^α is any 4-vector, which in general may not coincide with 4-velocity of a reference frame)

$$e^\alpha = u_\beta F^{\alpha\beta} , \qquad b^\alpha = u_\beta \tilde{F}^{\alpha\beta} , \qquad u^\alpha u_\alpha = 1 ; \qquad (2.8.3)$$

Where the inverse formulas are

$$F^{\alpha\beta} = (e^\alpha u^\beta - e^\beta u^\alpha) - \epsilon^{\alpha\beta\rho\sigma} b_\rho u_\sigma ,$$

$$\tilde{F}^{\alpha\beta} = (b^\alpha u^\beta - b^\beta u^\alpha) + \epsilon^{\alpha\beta\rho\sigma} e_\rho u_\sigma . \qquad (2.8.4)$$

Such electromagnetic 4-vectors are always presented in literature by the electrodynamics of moving bodies, and from the very beginning of the relativistic tensor form of electrodynamics – see Minkowski [6], Gordon [44], Tamm and Mandel'stam [45, 46]; for instance see Yépez, Brito, and Vargas [214]. The interest in these field variables was renewed after the publishing of the Esposito paper [226] in 1998.

In the 3-dimensional notation

$$E^1 = -E_1 = F^{10} , \qquad cB^1 = cB_1 = \tilde{F}^{10} = -F_{23}, \qquad \text{and so on}$$

the formulas in Eqs. (2.8.3) take the form

$$e^0 = \mathbf{u}\,\mathbf{E} , \qquad \mathbf{e} = u^0\,\mathbf{E} + c\,\mathbf{u} \times \mathbf{B} ,$$

$$b^0 = c\,\mathbf{u}\,\mathbf{B} , \qquad \mathbf{b} = c\,u^0\,\mathbf{B} - \mathbf{u} \times \mathbf{E} , \qquad (2.8.5)$$

or symbolically $(e, b) = U(u)\,(\mathbf{E}, \mathbf{B})$; and inverse the formulas of Eqs. (2.8.4) look

$$\mathbf{E} = \mathbf{e}\,u^0 - e^0\,\mathbf{u} + \mathbf{b} \times \mathbf{u} , \qquad c\,\mathbf{B} = \mathbf{b}\,u^0 - b^0\,\mathbf{u} - \mathbf{e} \times \mathbf{u} , \qquad (2.8.6)$$

or in symbolical form $(\mathbf{E}, \mathbf{B}) = U^{-1}(u)\,(e, b)$.

This possibility is often used to produce a special form of the Maxwell equations. Let us start with

$$\partial_\alpha F_{\beta\gamma} + \partial_\beta F_{\gamma\alpha} + \partial_\gamma F_{\alpha\beta} = 0 , \qquad \partial_\alpha F^{\alpha\beta} = \epsilon_0^{-1} j^\beta ,$$

or differently with the help of the dual tensor:

$$\partial_\beta \tilde{F}^{\beta\alpha} = 0 , \qquad \partial_\alpha F^{\alpha\beta} = \epsilon_0^{-1} j^\beta . \qquad (2.8.7)$$

These can be transformed to variables e^α, b^α:

$$\partial_\alpha(b^\alpha u^\beta - b^\beta u^\alpha + \epsilon^{\alpha\beta\rho\sigma} e_\rho u_\sigma) = 0 ,$$

$$\partial_\alpha(e^\alpha u^\beta - e^\beta u^\alpha - \epsilon^{\alpha\beta\rho\sigma} b_\rho u_\sigma) = \epsilon_0^{-1} j^\beta . \qquad (2.8.8)$$

Then can be combined into the equations for complex field function

$$\Phi^\alpha = e^\alpha + i b^\alpha , \qquad \partial_\alpha\,[\,\Phi^\alpha u^\beta - \Phi^\beta u^\alpha + i\epsilon^{\alpha\beta\rho\sigma}\Phi_\rho u_\sigma\,] = \epsilon_0^{-1} j^\beta ,$$

or differently

$$\partial_\alpha\,[\,\delta^\alpha_\gamma u^\beta - \delta^\beta_\gamma u^\alpha + i\epsilon^{\alpha\beta\rho\sigma} g_{\rho\gamma} u_\sigma\,]\,\Phi^\gamma = \epsilon_0^{-1} j^\beta . \qquad (2.8.9)$$

This is Esposito's representation [226] of the Maxwell equations. One can introduce four matrices, functions of 4-vector u^α:

$$(\Gamma^\alpha)^\beta{}_\gamma = \delta^\alpha_\gamma u^\beta - \delta^\beta_\gamma u^\alpha + i\epsilon^{\alpha\beta\rho\sigma} g_{\rho\gamma} u_\sigma \,, \tag{2.8.10}$$

then Eq. (2.8.9) becomes

$$\partial_\alpha (\Gamma^\alpha)^\beta{}_\gamma \, \Phi^\gamma = \epsilon_0^{-1} j^\beta \,, \qquad \text{or} \qquad \Gamma^\alpha \partial_\alpha \Phi = \epsilon_0^{-1} j \,. \tag{2.8.11}$$

In the 'rest reference frame' when $u^\alpha = (1,0,0,0)$, the matrices Γ^α become simpler and $\Phi = \Psi$:

$$\Gamma^0 = \begin{vmatrix} 0 & 0 & 0 & 0 \\ 0 & -1 & 0 & 0 \\ 0 & 0 & -1 & 0 \\ 0 & 0 & 0 & -1 \end{vmatrix}, \qquad \Gamma^1 = \begin{vmatrix} 0 & 1 & 0 & 0 \\ 0 & 0 & 0 & 0 \\ 0 & 0 & 0 & i \\ 0 & 0 & -i & 0 \end{vmatrix},$$

$$\Gamma^2 = \begin{vmatrix} 0 & 0 & 1 & 0 \\ 0 & 0 & 0 & -i \\ 0 & 0 & 0 & 0 \\ 0 & i & 0 & 0 \end{vmatrix}, \qquad \Gamma^3 = \begin{vmatrix} 0 & 0 & 0 & 1 \\ 0 & 0 & i & 0 \\ 0 & -i & 0 & 0 \\ 0 & 0 & 0 & 0 \end{vmatrix}.$$

and Eq. (2.8.11) can take the form

$$\begin{vmatrix} 0 & \partial_1 & \partial_2 & \partial_3 \\ 0 & -\partial_0 & i\partial_3 & -i\partial_2 \\ 0 & -i\partial_3 & -\partial_0 & i\partial_1 \\ 0 & i\partial_2 & -i\partial_1 & -\partial_0 \end{vmatrix} \begin{vmatrix} 0 \\ E^1 + icB^1 \\ E^2 + icB^2 \\ E^3 + icB^3 \end{vmatrix} = \epsilon_0^{-1} \begin{vmatrix} \rho \\ j^1 \\ j^2 \\ j^3 \end{vmatrix} = \epsilon_0^{-1} j \,, \tag{2.8.12}$$

or

$$\operatorname{div}(\mathbf{E} + ic\mathbf{B}) = \epsilon_0^{-1} \rho \,,$$

$$-\partial_0(\mathbf{E} + ic\mathbf{B}) - i \operatorname{rot}(\mathbf{E} + ic\mathbf{B}) = \epsilon_0^{-1} \mathbf{j} \,.$$

From here, we get equations

$$\operatorname{div} c\mathbf{B} = 0 \,, \qquad \operatorname{rot} \mathbf{E} = -\frac{\partial c\mathbf{B}}{\partial ct} \,,$$

$$\operatorname{div} \mathbf{E} = \frac{\rho}{\epsilon_0}, \qquad \operatorname{rot} c\mathbf{B} = \frac{\mathbf{j}}{\epsilon_0} + \frac{\partial \mathbf{E}}{\partial ct} \,,$$

which coincides with the Maxwell equations. Relations in Eq. (2.8.12) corresponds to a special choice of α-matrices:

$$\beta(-i\alpha^0) = \Gamma^0 \,, \quad \beta \, \alpha^j = \Gamma^j \,, \qquad \text{where} \qquad \beta = \begin{vmatrix} 1 & 0 & 0 & 0 \\ 0 & -i & 0 & 0 \\ 0 & 0 & -i & 0 \\ 0 & 0 & 0 & -i \end{vmatrix}. \tag{2.8.13}$$

Esposito's representation of the Maxwell equation at any 4-vector u^α can be easily related to the matrix equation of Riemann–Silberstein–Majorana–Oppenheimer:

$$(-i\alpha^0 \partial_0 + \alpha^j \partial_j)\Psi = J \,, \qquad (2.8.14)$$

indeed

$$(-i\alpha^0 \partial_0 + \alpha^j \partial_j)U^{-1}\,(U\Psi) = J \,,$$

$$-i\alpha^0 \, U^{-1} = \beta \, \Gamma^0 \,, \qquad \alpha^j \, U^{-1} = \beta \, \Gamma^j \,, \qquad U\Psi = \Phi \,,$$

$$\beta \, (\Gamma^0 \partial_0 + \Gamma^j \partial_j) \, \Phi = J \,, \qquad \beta^{-1} J = \epsilon_0^{-1}(j^a) \,,$$

$$(\Gamma^0 \partial_0 + \Gamma^j \partial_j)\Phi = \epsilon_0^{-1} \, j \,. \qquad (2.8.15)$$

Equation (2.8.15) is the matrix representation of the Maxwell equations in Esposito's form

$$\partial_\alpha \left[\, \delta^\alpha_\gamma u^\beta - \delta^\beta_\gamma u^\alpha + i\epsilon^{\alpha\beta\rho\sigma} g_{\rho\gamma} u_\sigma \, \right] \Phi^\gamma = \epsilon_0^{-1} j^\beta \,. \qquad (2.8.16)$$

Evidently, Eqs. (2.8.14) and (2.8.16) are equivalent to each other. There is no grounds here to consider the form of Eq. (2.8.16) as obtained through the trivial use of the identity $I = U^{-1}(u)U(u)$ as existing in a certain essentially profound sense. Our point of view contrasts with the claim by Ivezic [232–236] that Eq. (2.8.16) has a status of a true Maxwell equation in a moving reference frame (at this u^α is identified with 4-velocity).

2.9   Maxwell Matrix Equation in a Curved Space-Time, in Absence of Media

Now, the main question is to query the previous Maxwell matrix equation for the absence of media (consider the no-media case)

$$(\alpha^0 \partial_0 + \alpha^j \partial_j) \, \Psi = J \,, \qquad \alpha^0 = -iI \,,$$

$$\Psi = \left| \begin{array}{c} 0 \\ \mathbf{E} + ic\mathbf{B} \end{array} \right| \,, \qquad J = \frac{1}{\epsilon_0} \left| \begin{array}{c} \rho \\ i\mathbf{j} \end{array} \right| ; \qquad (2.9.1)$$

this can be generalized in the case of a curved space-time background. We should expect the existence of an extended equation in the general frame of Tetrode–Weyl–Fock–Ivanenko tetrad approach [40]. Such an equation might be in the following form

$$\alpha^\rho(x) \left[\, \partial_\rho + A_\rho(x) \, \right] \Psi(x) = J(x) \,,$$

$$\alpha^\rho(x) = \alpha^c \, e^\rho_{(c)}(x) \,, \qquad A_\rho(x) = \frac{1}{2} j^{ab} \, e^\beta_{(a)} \, \nabla_\rho e_{(n)\beta} ; \qquad (2.9.2)$$

j^{ab} stands for generators of 3-vector field under complex orthogonal group $SO(3.C)$, their explicit form shall be given later. Tetrad represents four covariant vectors related to metric tensor through a bilinear function

$$g_{\alpha\beta}(x) = \eta^{ab} e_{(a)\alpha} e_{(b)\beta} = e_{(0)\alpha} e_{(0)\beta} - e_{(1)\alpha} e_{(1)\beta} - e_{(2)\alpha} e_{(2)\beta} - e_{(3)\alpha} e_{(3)\beta} ; \qquad (2.9.3)$$

where all the tetrads referred to by local Lorentz transformations correspond to the same metric $g_{\alpha\beta}(x)$:

$$e'_{(a)\alpha}(x) = L_a{}^b(x) \, e_{(b)\alpha}(x) \, . \tag{2.9.4}$$

Equation (2.9.2) can be rewritten as

$$\alpha^c \left(e^\rho_{(c)} \partial_\rho + \frac{1}{2} j^{ab} \gamma_{abc} \right) \Psi = J(x) \, , \tag{2.9.5}$$

where Ricci rotation coefficients are used

$$\gamma_{bac} = -\gamma_{abc} = -e_{(b)\beta;\alpha} e^\beta_{(a)} e^\alpha_{(c)} \, . \tag{2.9.6}$$

Regarding Eqs. (2.9.1) and (2.9.2), one should expect symmetry properties for the equation under local gauge transformations:

$$\Psi'(x) = S(x)\Psi(x) \, , \qquad e'_{(a)\alpha}(x) = L_a{}^b(x) \, e_{(b)\alpha}(x) \, ,$$

$$\alpha^\rho(x) \left[\partial_\rho + A_\rho(x) \right] \Psi(x) = J(x) \quad \Longrightarrow$$

$$\alpha'^\rho(x) \left[\partial_\rho + A'_\rho(x) \right] \Psi'(x) = J'(x) \, . \tag{2.9.7}$$

Next, we will separately consider Euclidean and Lorentzian tetrad rotations.

In the case of Euclidean rotations, we may expect the following symmetry:

$$S = S[a(x), \mathbf{n}(x)] \, ,$$

$$\Psi' = S\Psi \, , \qquad \Psi = S^{-1}\Psi', \qquad SJ(x) = J' \, ,$$

$$S\alpha^\rho S^{-1} \left(\partial_\rho + SA_\rho S^{-1} + S\partial_\rho S^{-1} \right) \Psi'(x) = SJ(x) \, ,$$

$$S\alpha^\rho S^{-1} = \alpha'^\rho \, , \qquad SA_\rho S^{-1} + S\partial_\rho S^{-1} = A'_\rho \, . \tag{2.9.8}$$

In the case of Lorentzian rotations, we may expect other symmetry realized in accordance with relations

$$S = S[ib(x), \mathbf{n}(x)] \, , \quad \Delta = \Delta[ib(x), \mathbf{n}(x)] \, ,$$

$$\Psi' = S\Psi \, , \qquad \Psi = S^{-1}\Psi', \qquad \Delta \, SJ(x) = J' \, ,$$

$$\Delta S\alpha^\rho S^{-1} \left(\partial_\alpha + SA_\alpha S^{-1} + S\partial_\alpha S^{-1} \right) \Psi'(x) = \Delta SJ(x) \, ,$$

$$\Delta \, S\alpha^\rho S^{-1} = \alpha'^\rho \, , \qquad SA_\alpha S^{-1} + S\partial_\alpha S^{-1} = A'_\alpha \, . \tag{2.9.9}$$

Symmetry properties of the local matrices $\alpha^\rho(x)$ can be straightforwardly found by the base analysis performed for the flat Minkowski space. Indeed, for local Euclidean rotations, the rule for $S\alpha^\rho(x)S^{-1}$ is

$$S\alpha^\rho S^{-1} = S\alpha^0 e^\rho_{(0)} S^{-1} + S\alpha^l e^\rho_{(l)} S^{-1}$$

$$= \alpha^0 e^\rho_{(0)} + \alpha^k O_{kl} e^\rho_{(l)} = \alpha^0 e^\rho_{(0)} + \alpha^k e'^\rho_{(k)} = \alpha'^\rho \, . \tag{2.9.10}$$

For local Lorentzian rotations, we can easily prove a symmetry relation:

$$\Delta\, S\alpha^\rho(x)S^{-1} = \Delta\, S\alpha^a e^\rho_{(a)}S^{-1}$$

$$= [\Delta\, S\alpha^a S^{-1}]\, e^\rho_{(a)} = \alpha^b L_b{}^a\, e^\rho_{(a)} = \alpha^b e'^\rho_{(b)} = \alpha'^\rho(x)\;. \qquad (2.9.11)$$

The transformation law for the complex 3-vector connection $A_\rho(x)$

$$S A_\rho S^{-1} + S\partial_\rho S^{-1} = A'_\rho$$

will be proved in Section **2.10**.

2.10 On Tetrad Transformation Law for complex 3-Vector connection $A_\alpha(x)$

First, let us write down six elementary rotations from the local group $SO(3.C)$

$$S_{23} = \begin{vmatrix} 1 & 0 & 0 & 0 \\ 0 & 1 & 0 & 0 \\ 0 & 0 & \cos a & -\sin a \\ 0 & 0 & \sin a & \cos a \end{vmatrix}, \qquad S_{01} = \begin{vmatrix} 1 & 0 & 0 & 0 \\ 0 & 0 & 0 & 0 \\ 0 & 0 & \cosh b & -i\sinh b \\ 0 & 0 & +i\sinh b & \cosh b \end{vmatrix},$$

$$S^1 = j^{23} = \begin{vmatrix} 0 & 0 \\ 0 & \tau_1 \end{vmatrix}, \qquad N^2 = j^{01} = +i\begin{vmatrix} 0 & 0 \\ 0 & \tau_1 \end{vmatrix},$$

$$S_{31} = \begin{vmatrix} 1 & 0 & 0 & 0 \\ 0 & \cos a & 0 & \sin a \\ 0 & 0 & 1 & 0 \\ 0 & -\sin a & 0 & \cos a \end{vmatrix}, \qquad S_{02} = \begin{vmatrix} 1 & 0 & 0 & 0 \\ 0 & \cosh b & 0 & +i\sinh b \\ 0 & 0 & 0 & 0 \\ 0 & -i\sinh b & 0 & \cosh b \end{vmatrix},$$

$$S^2 = j^{31} = \begin{vmatrix} 0 & 0 \\ 0 & \tau_2 \end{vmatrix}, \qquad N^2 = j^{02} = +i\begin{vmatrix} 0 & 0 \\ 0 & \tau_2 \end{vmatrix},$$

$$S_{12} = \begin{vmatrix} 1 & 0 & 0 & 0 \\ 0 & \cos a & -\sin a & 0 \\ 0 & \sin a & \cos a & 0 \\ 0 & 0 & 0 & 1 \end{vmatrix}, \qquad S_{03} = \begin{vmatrix} 1 & 0 & 0 & 0 \\ 0 & \cosh b & -i\sinh b & 0 \\ 0 & +i\sinh b & \cosh b & 0 \\ 0 & 0 & 0 & 1 \end{vmatrix},$$

$$S^3 = j^{12} = \begin{vmatrix} 0 & 0 \\ 0 & \tau_3 \end{vmatrix}, \qquad N^3 = j^{03} = +i\begin{vmatrix} 0 & 0 \\ 0 & \tau_3 \end{vmatrix};$$

these elementary rotations obey the commutative relations:

$$S^1 S^2 - S^2 S^1 = S^3\;, \qquad N^1 N^2 - N^2 N^1 = -S^3\;, \qquad S^1 N^2 - N^2 S^1 = +N^3\;;$$

and remaining rotations are written by cyclic symmetry.

Now, let us turn to some properties in the connection $A_\alpha(x)$:

$$A_\alpha(x) = \frac{1}{2} j^{ab} \, e^\beta_{(a)} \, \nabla_\alpha e_{(b)\beta}$$

$$= S^1 e^\beta_{(2)} \, \nabla_\alpha e_{(3)\beta} + S^2 e^\beta_{(3)} \, \nabla_\alpha e_{(2)\beta} + S^3 e^\beta_{(1)} \, \nabla_\alpha e_{(2)\beta}$$

$$+ N^1 e^\beta_{(0)} \, \nabla_\alpha e_{(1)\beta} + N^2 e^\beta_{(0)} \, \nabla_\alpha e_{(2)\beta} + N^3 e^\beta_{(0)} \, \nabla_\alpha e_{(3)\beta} \,. \qquad (2.10.1)$$

By keeping in mind the identity $N_k = +iS_k$, and through introducing new complex variables

$$A_{(1)\alpha} = e^\beta_{(2)} \, \nabla_\alpha e_{(3)\beta} + i \, e^\beta_{(0)} \, \nabla_\alpha e_{(1)\beta} \,,$$

$$A_{(2)\alpha} = e^\beta_{(3)} \, \nabla_\alpha e_{(1)\beta} + i \, e^\beta_{(0)} \, \nabla_\alpha e_{(2)\beta} \,,$$

$$A_{(3)\alpha} = e^\beta_{(1)} \, \nabla_\alpha e_{(2)\beta} + i \, e^\beta_{(0)} \, \nabla_\alpha e_{(3)\beta} \,, \qquad (2.10.2)$$

one can read the previous connection as

$$A_\alpha(x) = S^k A_{(k)\alpha} \,. \qquad (2.10.3)$$

When we use the notation

$$A_\alpha(x) = \frac{1}{2} j^{ab} \, e^\beta_{(a)} \, \nabla_\alpha e_{(b)\beta} = \frac{1}{2} j^{ab} \, A_{(a)(b)\alpha} \,, \qquad A_{(a)(b)\alpha} = -A_{(b)(a)\alpha} \,, \qquad (2.10.4)$$

this definition for $A_{(k)\alpha}$ can be rewritten differently:

$$A_{(1)\alpha} = A_{(2)(3)\alpha} + iA_{(0)(1)\alpha} \,,$$

$$A_{(2)\alpha} = A_{(3)(1)\alpha} + iA_{(0)(2)\alpha} \,,$$

$$A_{(3)\alpha} = A_{(1)(2)\alpha} + iA_{(0)(3)\alpha} \,. \qquad (2.10.5)$$

In other words, the 3-quantity $A_{(k)\alpha}$ with respect to 3-index (k) is constructed in terms of "tensor" $A_{(a)(b)\alpha}$ by the same rule as used in constructing the 3-dimensional complex vector $-i(\mathbf{E} + ic\mathbf{B})$ in the terms of components for tensor F_{ab}.

Let us show that such 3-dimensional complex vectors can be built in terms of a skew-symmetric 2-rank real tensor through a simple algebraic construction:

$$\mathrm{Sp}\,(\sigma^k \, \bar\sigma^a \sigma^b A_{ab}) \,, \qquad \mathrm{Sp}\,(\sigma^k \, \sigma^a \bar\sigma^b \sigma^a A_{ab}) \,,$$

$$\sigma^a = (I, +\sigma^k) \,, \qquad \bar\sigma^a = (I, -\sigma^k) \,,$$

$$\sigma^1 = \begin{vmatrix} 0 & 1 \\ 1 & 0 \end{vmatrix}, \, \sigma^2 = \begin{vmatrix} 0 & -i \\ i & 0 \end{vmatrix}, \, \sigma^3 = \begin{vmatrix} 1 & 0 \\ 0 & -1 \end{vmatrix}; \qquad (2.10.6)$$

and further

$$\frac{1}{2} \, \bar\sigma^a \sigma^b A_{ab} = (\bar\sigma^2 \sigma^3 A_{23} + \bar\sigma^0 \sigma^1 A_{01})$$

$$+ (\bar\sigma^3 \sigma^1 A_{31} + \bar\sigma^0 \sigma^2 A_{02}) + (\bar\sigma^1 \sigma^2 A_{12} + \bar\sigma^0 \sigma^3 A_{03})$$

$$= -i[\,\sigma^1 (A_{23} + iA_{01}) + \sigma^2 (A_{31} + iA_{02}) + \sigma^3 (A_{12} + iA_{03})\,] \,; \qquad (2.10.7)$$

$$\frac{1}{2}\,\sigma^a\bar{\sigma}^b A_{ab} = (\sigma^2\bar{\sigma}^3 A_{23} + \sigma^0\bar{\sigma}^1 A_{01})$$

$$+(\sigma^3\bar{\sigma}^1 A_{31} + \sigma^0\bar{\sigma}^2 A_{02}) + (\sigma^1\bar{\sigma}^2 A_{12} + \sigma^0\bar{\sigma}^3 A_{03})$$

$$= -i[\,\sigma^1(A_{23} - iA_{01}) + \sigma^2(A_{31} - iA_{02}) + \sigma^3(A_{12} - iA_{03})\,]\,. \tag{2.10.8}$$

Therefore, we have two decompositions

$$\frac{i}{2}\,\bar{\sigma}^a\sigma^b A_{(a)(b)\alpha} = \sigma^k A_{(k)\alpha}\,, \qquad \frac{i}{2}\,\sigma^a\bar{\sigma}^b A_{(a)(b)\alpha} = \sigma^k A^*_{(k)\alpha}\,. \tag{2.10.9}$$

From this, it follows covariant formulas for $A_{(k)\alpha}$ and $A^*_{(k)\alpha}$:

$$A_{(k)\alpha} = \frac{i}{4}\,\mathrm{Sp}\,[\bar{\sigma}^a\sigma^b A_{(a)(b)\alpha}]\,, \qquad A^*_{(k)\alpha} = \frac{i}{4}\mathrm{Sp}\,[\sigma_k\sigma^a\bar{\sigma}^b A_{(a)(b)\alpha}]\,. \tag{2.10.10}$$

Now, when starting from relations between any two tetrads by a local Lorentz transformation:

$$e^{'\alpha}_{(a)} = L_a{}^b e^{\alpha}_{(b)}\,, \qquad e^{\alpha}_{(a)} = (L^{-1})_a{}^b e^{'\alpha}_{(b)}\,, \tag{2.10.11}$$

let us derive a rule to transform 3-vector connection when we change the tetrad:

$$A_{(a)(b)\alpha} = e^{\beta}_{(a)}\,\nabla_\alpha e_{(b)\beta} = (L^{-1})_a{}^m e^{'\beta}_{(m)}\nabla_\alpha (L^{-1})_b{}^n e'_{(n)\beta}$$

$$= (L^{-1})_a{}^m e^{'\beta}_{(m)}(L^{-1})_b{}^n\,\nabla_\alpha e'_{(n)\beta} + (L^{-1})_a{}^m e^{'\beta}_{(m)}\,\frac{\partial (L^{-1})_b{}^n}{\partial x^\alpha}\,e'_{(n)\beta}\,,$$

that is

$$A_{(a)(b)\alpha} = (L^{-1})_a{}^m (L^{-1})_b{}^n\,A'_{(m)(n)\alpha} + (L^{-1})_a{}^m\,g_{(m)(n)}\,\frac{\partial (L^{-1})_b{}^n}{\partial x^\alpha}\,. \tag{2.10.12}$$

Let us act on this relation from the left by an operator $\frac{i}{4}\,\mathrm{Sp}\,[\sigma_k\bar{\sigma}^a\sigma^b\,...]$; this results in

$$A_{(k)\alpha} = \frac{i}{4}\,\mathrm{Sp}\left[\sigma_k\bar{\sigma}^a\sigma^b A_{(a)(b)\alpha}\right]$$

$$= \frac{i}{4}\,\mathrm{Sp}\left[\sigma_k\bar{\sigma}^a\sigma^b(L^{-1})_a{}^m(L^{-1})_b{}^n\,A'_{(m)(n)\alpha}\right]$$

$$+\frac{i}{4}\,\mathrm{Sp}\left[\sigma_k\bar{\sigma}^a\sigma^b(L^{-1})_a{}^m\,g_{(m)(n)}\,\frac{\partial (L^{-1})_b{}^n}{\partial x^\alpha}\right]\,. \tag{2.10.13}$$

One may expect Eqs. (2.10.13) to be equivalent to

$$A_{(k)\alpha} = O^{-1}_{kn} A'_{(n)\alpha} + \frac{i}{4}\mathrm{Sp}\left[\sigma_k\bar{\sigma}^a\sigma^b(L^{-1})_a{}^m\,g_{(m)(n)}\,\frac{\partial}{\partial x^\alpha}(L^{-1})_b{}^n\right]\,. \tag{2.10.14}$$

This is so, if an identity holds

$$\frac{i}{4}\mathrm{Sp}\left[\sigma_k\bar{\sigma}^a\sigma^b(L^{-1})_a{}^m(L^{-1})_b{}^n\,A'_{(m)(n)\alpha}\right] = O^{-1}_{kl} A'_{(l)\alpha}\,. \tag{2.10.15}$$

By reason of symmetry, with respect to indices $0, 1, 2, 3$; it will be sufficient to check Eq. (2.10.15) with the use of several particular cases for Euclidean and Lorentzian rotations.

First, let us consider $(1 - 2)$-rotation

$$L^{-1} = \begin{vmatrix} 1 & 0 & 0 & 0 \\ 0 & \cos a & \sin a & 0 \\ 0 & -\sin a & \cos a & 0 \\ 0 & 0 & 0 & 1 \end{vmatrix}, \qquad (2.10.16)$$

and construct the quantity (index α is omitted and $\Lambda = L^{-1}$)

$$\frac{i}{4} \, \bar{\sigma}^a \sigma^b \, \Lambda_a{}^m \Lambda_b{}^n \, A'_{mn}$$

$$= \frac{1}{2} \{ \ \sigma^1 (\Lambda_2{}^m \Lambda_3{}^n \, A'_{mn} + i\Lambda_0{}^m \Lambda_1{}^n \, A'_{mn})$$

$$+ \sigma^2 (\Lambda_3{}^m \Lambda_1{}^n \, A'_{mn} + i\Lambda_0{}^m \Lambda_2{}^n \, A'_{mn})$$

$$+ \sigma^3 (\Lambda_1{}^m \Lambda_2{}^n \, A'_{mn} + i\Lambda_0{}^m \Lambda_3{}^n \, A'_{mn}) \ \}$$

$$= \frac{1}{2} \, \sigma^1 \, [\Lambda_2^2 \Lambda_3^3 A'_{23} - \Lambda_2^1 \Lambda_3^3 A'_{31} + i\Lambda_0^0 \Lambda_1^1 A'_{01} + i\Lambda_0^0 \Lambda_1^2 A'_{02}]$$

$$+ \frac{1}{2} \, \sigma^2 \, [-\Lambda_3^3 \Lambda_1^2 A'_{23} + \Lambda_3^3 \Lambda_1^1 A'_{31} + i\Lambda_0^0 \Lambda_2^1 \, A'_{01} + i\Lambda_0^0 \Lambda_2^2 A'_{02}]$$

$$+ \frac{1}{2} \, \sigma^3 \, [(\Lambda_1^1 \Lambda_2^2 - \Lambda_1^2 \Lambda_2^1) A'_{12} + i \, \Lambda_0^0 \Lambda_3^3 A'_{03}]$$

$$= \frac{1}{2} \, \sigma^1 \, (\cos a A'_{23} + \sin a A'_{31} + i \cos a A'_{01} + i \sin a A'_{02})$$

$$+ \frac{1}{2} \, \sigma^2 \, (-\sin a A'_{23} + \cos a A'_{31} - i \sin a \, A'_{01} + i \cos a \, A_{02})' + \frac{1}{2} \, \sigma^3 (A'_{12} + iA'_{03})$$

$$= \frac{1}{2} \, \sigma^1 [\cos a \, (A'_{23} + iA'_{01}) + \sin a \, (A'_{31} + iA'_{02})]$$

$$+ \frac{1}{2} \, \sigma^2 [\cos a \, (A'_{31} + iA'_{02}) - \sin a \, (A'_{23} + iA'_{01})] - \frac{i}{2} \, \sigma^3 A'_3$$

$$= \frac{1}{2} [\sigma^1 \, (\cos a \, A_1 + \sin a \, A_2) + \sigma^2 \, (\cos a A'_2 - \sin a A'_1) + \sigma^3 A'_3] .$$

Thus,

$$\frac{i}{4} \, \bar{\sigma}^a \sigma^b \, \Lambda_a{}^m \Lambda_b{}^n \, A'_{mn}$$

$$= \frac{1}{2} [\sigma^1 \, (\cos a \, A'_1 + \sin a \, A'_2) + \sigma^2 \, (\cos a A'_2 - \sin a A'_1) + \sigma^3 A'_3] ,$$

and for

$$\frac{i}{4} \mathrm{Sp} \, \sigma^k \, \bar{\sigma}^a \sigma^b \Lambda_a{}^m \Lambda_b{}^n \, A'_{mn}$$

we get

$$\frac{i}{4}\mathrm{Sp}\,\sigma^1\,\bar{\sigma}^a\sigma^b\Lambda_a{}^m\Lambda_b{}^n\,A'_{mn} = \cos a\,A'_1 + \sin a\,A'_2\,,$$

$$\frac{i}{4}\mathrm{Sp}\,\sigma^2\,\bar{\sigma}^a\sigma^b\Lambda_a{}^m\Lambda_b{}^n\,A'_{mn} = \cos a\,A'_2 - \sin a\,A'_1\,,$$

$$\frac{i}{4}\mathrm{Sp}\,\sigma^3\,\bar{\sigma}^a\sigma^b\Lambda_a{}^m\Lambda_b{}^n\,A'_{mn} = A'_3\,, \qquad (2.10.17)$$

which is a particular case for the general relation in Eqs. (2.10.15).

Analogous calculation can be done for Lorentzian rotation in the plane $(0-3)$; by starting with

$$L^{-1} = \begin{vmatrix} \cosh b & 0 & 0 & \sinh b \\ 0 & 1 & 0 & 0 \\ 0 & 0 & 1 & 0 \\ \sinh b & 0 & 0 & \cosh b \end{vmatrix} \qquad (2.10.18)$$

consider the quantity (again index α is omitted and $\Lambda = L^{-1}$)

$$\frac{i}{4}\,\bar{\sigma}^a\sigma^b\,\Lambda_a{}^m\Lambda_b{}^n\,A'_{mn}$$

$$= \frac{1}{2}\{\,\sigma^1(\Lambda_2{}^m\Lambda_3{}^n\,A'_{mn} + i\Lambda_0{}^m\Lambda_1{}^n\,A'_{mn})$$

$$+ \sigma^2(\Lambda_3{}^m\Lambda_1{}^n\,A'_{mn} + i\Lambda_0{}^m\Lambda_2{}^n\,A'_{mn})$$

$$+ \sigma^3(\Lambda_1{}^m\Lambda_2{}^n\,A'_{mn} + i\Lambda_0{}^m\Lambda_3{}^n\,A'_{mn})\,\}$$

$$= \frac{1}{2}\,\sigma^1\,[\,-\Lambda_2^2\Lambda_3^0 A'_{02} + \Lambda_2^2\Lambda_3^3 A'_{23} + i\,(\Lambda_0^0\Lambda_1^1 A'_{01} + \Lambda_0^3\Lambda_1^1 A'_{31})\,]$$

$$+ \frac{1}{2}\,\sigma^2\,[\,\Lambda_3^0\Lambda_1^1 A'_{01} + \Lambda_3^3\Lambda_1^1 A'_{31} i\,(\Lambda_0^0\Lambda_2^2\,A'_{02} - \Lambda_0^3\Lambda_2^2 A'_{23})\,]$$

$$+ \frac{1}{2}\,\sigma^3\,[\,\Lambda_1^1\Lambda_2^2 A'_{12} + i\,(\Lambda_0^0\Lambda_3^3 - \Lambda_0^3\Lambda_3^0)A'_{03}\,]$$

$$= \frac{1}{2}\,\sigma^1\,(-\sinh b\,A'_{02} + \cosh b\,A'_{23} + i\cosh b\,A'_{01} + i\sinh b\,A'_{31})$$

$$+ \frac{1}{2}\,\sigma^2\,(\sinh b\,A'_{01} + \cosh b\,A'_{31} + i\cosh b\,A'_{02} - i\sinh b\,A'_{23})$$

$$+ \frac{1}{2}\,\sigma^3\,[A'_{12} + i\,(\cosh b\cosh b - \sinh b\sinh b\,)A'_{03}]$$

$$= \frac{1}{2}\,\sigma^1\,[\cosh b\,(A'_{23} + i\,A'_{01}) + i\sinh b\,(A'_{31} + iA'_{02})]$$

$$+ \frac{1}{2}\,\sigma^2\,[-i\sinh b\,(A'_{23} + i\,A'_{01}) + \cosh b\,(A'_{31} + i\,A'_{02})]$$

$$+ \frac{1}{2}\,\sigma^3\,(A'_{12} + i\,A'_{03})\,,$$

that is

$$\frac{i}{4} \, \bar\sigma^a \sigma^b \, \Lambda_a{}^m \Lambda_b{}^n \, A'_{mn}$$

$$= \frac{1}{2} \, \sigma^1 \left[\cosh b \, (A'_{23} + i \, A'_{01}) + i \, \sinh b \, (A'_{31} + i A'_{02}) \right]$$

$$+ \frac{1}{2} \, \sigma^2 \left[-i \, \sinh b \, (A'_{23} + i \, A'_{01}) + \cosh b \, (A'_{31} + i \, A'_{02}) \right]$$

$$+ \frac{1}{2} \, \sigma^3 \, (A'_{12} + i \, A'_{03}) \, ,$$

from where it follows

$$\frac{i}{4} \mathrm{Sp} \, \sigma^1 \, \bar\sigma^a \sigma^b \Lambda_a{}^m \Lambda_b{}^n \, A'_{mn} = \cosh b \, A'_1 + i \, \sinh b \, A'_2 \, ,$$

$$\frac{i}{4} \mathrm{Sp} \, \sigma^2 \, \bar\sigma^a \bar\sigma^b \Lambda_a{}^m \Lambda_b{}^n \, A'_{mn} = -i \, \sinh b \, B'_1 + \cosh b \, B'_2 \, ,$$

$$\frac{i}{4} \mathrm{Sp} \, \sigma^3 \, \bar\sigma^a \sigma^b \Lambda_a{}^m \Lambda_b{}^n \, A'_{mn} = +B'_3 \, , \qquad (2.10.19)$$

which is a particular case for the general relation in Eqs.(2.10.15).

The following general relations specify isomorphism between Lorentz group and 3-dimensional orthogonal group $SO(3.C)$:

$$\psi_k = \frac{i}{4} \, \mathrm{Sp} \, (\sigma_k \bar\sigma^a \sigma^b \psi_{ab}) \, , \qquad \psi_k^* = \frac{i}{4} \mathrm{Sp} \, (\sigma_k \sigma^a \bar\sigma^b \Psi_{ab}) \, ,$$

$$\frac{i}{4} \, \mathrm{Sp} \left[\sigma_k \bar\sigma^a \sigma^b L_a{}^m L_b{}^n \, \psi_{mn} \right] = O_{kl} \psi_l \, ,$$

$$\frac{i}{4} \, \mathrm{Sp} \left[\sigma_k \sigma^a \bar\sigma^b L_a{}^m L_b{}^n \, \psi_{mn} \right] = O_{kl}^* \psi_l^* \, . \qquad (2.10.20)$$

Now, we are ready to prove the following relationships:

$$O A_\rho O^{-1} + O \partial_\rho O^{-1} = A'_\rho \, . \qquad (2.10.21)$$

In taking into account the linear decomposition

$$A_\alpha = A_{(k)\alpha} \, \tau_k \, ,$$

Eq. (2.10.21) can be rewritten as

$$O \tau^k A_{(k)\alpha} O^{-1} + O \partial_\alpha O^{-1} = \tau^k A'_{(k)\alpha} \, ,$$

or

$$\tau^l O_{lk} A_{(k)\alpha} + O \partial_\alpha O^{-1} = \tau^k A'_{(k)\alpha} \, . \qquad (2.10.22)$$

Substituting expression for $A_{(k)\alpha}$ through $A'_{(k)\alpha}$ (see Eq. (2.10.14))

$$A_{(k)\alpha} = O_{kn}^{-1} A'_{(n)\alpha} + \frac{i}{4} \, \mathrm{Sp} \left[\, (\sigma_k \bar\sigma^a \sigma^b (L^{-1})_a{}^m \, g_{(m)(n)} \, \frac{\partial}{\partial x^\alpha} (L^{-1})_b{}^n \, \right] ;$$

we get

$$\tau^l O_{lk} \left[O_{kn}^{-1} A'_{(n)\alpha} + \frac{i}{4} \mathrm{Sp} \left[\sigma_k \bar{\sigma}^a \sigma^b (L^{-1})_a{}^m \ g_{(m)(n)} \ \frac{\partial}{\partial x^\alpha} (L^{-1})_b{}^n \right] \right]$$

$$+ O \partial_\alpha O^{-1} = \tau^k A'_{(k)\alpha} \ .$$

From this, the relationship must hold:

$$\tau^l O_{lk} \frac{i}{2} \mathrm{Sp} \left[\sigma_k \bar{\sigma}^a \sigma^b (L^{-1})_a{}^m \ g_{(m)(n)} \ \frac{\partial}{\partial x^\alpha} (L^{-1})_b{}^n \right] + O \partial_\alpha O^{-1} = 0 \ , \qquad (2.10.23)$$

or differently

$$\tau^l O_{lk} \frac{i}{4} \mathrm{Sp} \left[\sigma_k \bar{\sigma}^a \sigma^b C_{ab,\alpha} \right] + O \partial_\alpha O^{-1} = 0 \ , \qquad (2.10.24)$$

where

$$C_{ab,\alpha} = (L^{-1})_a{}^m \ g_{(m)(n)} \ \frac{\partial}{\partial x^\alpha} (L^{-1})_b{}^n \ . \qquad (2.10.25)$$

It suffices to prove relationships of Eq. (2.10.23) for the simplest transformations. For Euclidean rotation $(1 - 2)$:

$$L^{-1} = \begin{vmatrix} 1 & 0 & 0 & 0 \\ 0 & \cos a & \sin a & 0 \\ 0 & -\sin a & \cos a & 0 \\ 0 & 0 & 0 & 1 \end{vmatrix},$$

$$\partial_\alpha L^{-1} = \begin{vmatrix} 0 & 0 & 0 & 0 \\ 0 & -\sin a & \cos a & 0 \\ 0 & -\cos a & -\sin a & 0 \\ 0 & 0 & 0 & 0 \end{vmatrix} \partial_\alpha a \ , \qquad (2.10.26)$$

$$C_{ab,\alpha} = (L^{-1})_a{}^m \ g_{(m)(n)} \ \frac{\partial}{\partial x^\alpha} (L^{-1})_b{}^n$$

$$= \begin{vmatrix} 1 & 0 & 0 & 0 \\ 0 & \cos a & \sin a & 0 \\ 0 & -\sin a & \cos a & 0 \\ 0 & 0 & 0 & 1 \end{vmatrix} \begin{vmatrix} 0 & 0 & 0 & 0 \\ 0 & \sin a & \cos a & 0 \\ 0 & -\cos a & \sin a & 0 \\ 0 & 0 & 0 & 0 \end{vmatrix} \partial_\alpha a = \begin{vmatrix} 0 & 0 & 0 & 0 \\ 0 & 0 & 1 & 0 \\ 0 & -1 & 0 & 0 \\ 0 & 0 & 0 & 0 \end{vmatrix} \partial_\alpha a \ ,$$

$$(2.10.27)$$

and further

$$\frac{i}{4} \bar{\sigma}^a \sigma^b C_{ab,\alpha} = \frac{1}{2} \sigma^3 \ \partial_\alpha a \ , \qquad \frac{i}{4} \mathrm{Sp} \left[\sigma^k \bar{\sigma}^a \sigma^b C_{ab,\alpha} \right] = \begin{vmatrix} 0 \\ 0 \\ 1 \end{vmatrix} \partial_\alpha a \ ;$$

that is

$$O_{lk} = \begin{vmatrix} \cos a & -\sin a & 0 \\ \sin a & \cos a & 0 \\ 0 & 0 & 1 \end{vmatrix}, \qquad \tau^l O_{lk} \frac{i}{4} \mathrm{Sp} \left[\sigma^k \bar{\sigma}^a \sigma^b C_{ab,\alpha} \right] = + \tau^3 \ \partial_\alpha a \ .$$

$$(2.10.28)$$

Comparing it with

$$O\partial_\alpha O^{-1} = \begin{vmatrix} \cos a & -\sin a & 0 \\ \sin a & \cos a & 0 \\ 0 & 0 & 1 \end{vmatrix} \begin{vmatrix} -\sin a & \cos a & 0 \\ -\cos a & -\sin a & 0 \\ 0 & 0 & 0 \end{vmatrix} \partial_\alpha a$$

$$= \begin{vmatrix} 0 & 1 & 0 \\ -1 & 0 & 0 \\ 0 & 0 & 0 \end{vmatrix} \partial_\alpha = -\tau^3 \, \partial_\alpha a \,, \qquad (2.10.29)$$

we can conclude that Eq. (2.10.24) holds.

Now, in the same manner, let us consider Lorentzian rotation in the plane $(0-3)$:

$$L^{-1} = \begin{vmatrix} \cosh b & 0 & 0 & \sinh b \\ 0 & 1 & 0 & 0 \\ 0 & 0 & 1 & 0 \\ \sinh b & 0 & 0 & \cosh b \end{vmatrix}, \qquad \partial_\alpha L^{-1} = \begin{vmatrix} \sinh b & 0 & 0 & \cosh b \\ 0 & 1 & 0 & 0 \\ 0 & 0 & 1 & 0 \\ \cosh b & 0 & 0 & \sinh b \end{vmatrix} \partial_\alpha b \,,$$

$$(2.10.30)$$

$$C_{ab,\alpha} = (L^{-1})_a{}^m \; g_{(m)(n)} \; \frac{\partial}{\partial x^\alpha} (L^{-1})_b{}^n$$

$$= \begin{vmatrix} \cosh b & 0 & 0 & \sinh b \\ 0 & 1 & 0 & 0 \\ 0 & 0 & 1 & 0 \\ \sinh b & 0 & 0 & \cosh b \end{vmatrix} \begin{vmatrix} \sinh b & 0 & 0 & \cosh b \\ 0 & 0 & 0 & 0 \\ 0 & 0 & 0 & 0 \\ -\cosh b & 0 & 0 & -\sinh b \end{vmatrix} \partial_\alpha b = \begin{vmatrix} 0 & 0 & 0 & 1 \\ 0 & 0 & 0 & 0 \\ 0 & 0 & 0 & 0 \\ -1 & 0 & 0 & 0 \end{vmatrix} \partial_\alpha b \,,$$

$$(2.10.31)$$

and further

$$\frac{i}{4} \bar\sigma^a \sigma^b C_{ab,\alpha} = +\frac{i}{2}\sigma^3 \, \partial_\alpha b \,, \qquad \frac{i}{4}\text{Sp} \, [\, \sigma^k \bar\sigma^a \sigma^b C_{ab,\alpha} \,] = +i \begin{vmatrix} 0 \\ 0 \\ 1 \end{vmatrix} \partial_\alpha b \,;$$

that is

$$O_{lk} = \begin{vmatrix} \cosh b & -i\sinh b & 0 \\ +i\sinh b & \cosh b & 0 \\ 0 & 0 & 1 \end{vmatrix}, \qquad \tau^l O_{lk} \frac{i}{4}\text{Sp} \, [\, \sigma^k \bar\sigma^a \sigma^b C_{ab,\alpha} \,] = +i\,\tau^3 \, \partial_\alpha b \,;$$

$$(2.10.32)$$

comparing it with

$$O\partial_\alpha O^{-1} = \begin{vmatrix} \cosh b & -i\sinh b & 0 \\ +i\sinh b & \cosh b & 0 \\ 0 & 0 & 1 \end{vmatrix} \begin{vmatrix} \sinh b & +i\cosh b & 0 \\ -i\cosh b & \sinh b & 0 \\ 0 & 0 & 0 \end{vmatrix} \partial_\alpha b = -i\tau^3 \partial_\alpha b \,,$$

$$(2.10.33)$$

we can conclude that Eq. (2.10.24) holds.

Thus, generally the covariant Maxwell matrix equation in a Riemannian space-time

$$\alpha^\rho(x) \left[\partial_\rho + A_\rho(x) \right] \Psi(x) = J(x) \,, \qquad (2.10.34)$$

possesses all the needed symmetry properties under local tetrad transformations and is therefore correct.

2.11   Matrix Equation in a Curved Space in the Presence of Media

Now, we can extend the Maxwell matrix equation in media to a curved space-time background; starting from the equation

$$\left(-i\partial_0 + \alpha^i \partial_i \right) M + \left(-i\partial_0 + \beta^i \partial_i \right) N = J \,,$$

$$M' = SM \,, \qquad N' = S^* N \,; \qquad (2.11.1)$$

we may propose the following

$$\alpha_\rho(x)(i\partial_\rho + A_\rho) M + \beta_\rho(x)(i\partial_\rho + B_\rho) N = J \,, \qquad (2.11.2)$$

where $A(x)$, $B_\rho = A^*(x)$ stand connections related to the fields $M(x)$ and $N(x)$, respectively. We will, again, consider separately Euclidean and Lorentzian tetrad rotations.

In the case of Euclidean rotations, we can expect the following symmetry:

$$S^* = S \,, \qquad S(x)J(x) = J'(x) \,,$$

$$M'(x) = S(x)M(x) \,, \qquad N'(x) = S(x)N(x) \,, \qquad (2.11.3)$$

$$S\alpha^\rho S^{-1} \left(\partial_\rho + SA_\rho S^{-1} + S\partial_\rho S^{-1} \right) M'(x)$$

$$+ S\beta^\rho S^{-1} \left(\partial_\rho + SB_\rho S^{-1} + S\partial_\rho S^{-1} \right) N'(x) = SJ(x) \,, \qquad (2.11.4)$$

$$S\alpha^\rho S^{-1} = \alpha'^\rho \,, \qquad S\beta^\rho S^{-1} = \beta'^\rho \,,$$

$$SA_\rho S^{-1} + S\partial_\rho S^{-1} = A'_\rho \,, \qquad SB_\rho S^{-1} + S\partial_\rho S^{-1} = B'_\rho \,. \qquad (2.11.5)$$

In the case of Lorentzian rotations, we can expect other symmetry realized in accordance with the following relations

$$S^* = S^{-1} \,, \qquad \Delta_\alpha(x) \,, \qquad \Delta_\alpha(x) \, S(x) \, J(x) = J' \,,$$

$$M(x)' = S(x)M(x) \,, \qquad N'(x) = S^*(x)N'(x) = S^{-1}(x)N'(x) \,, \qquad (2.11.6)$$

$$\Delta_\alpha S\alpha^\rho S^{-1} \left(\partial_\alpha + SA_\alpha S^{-1} + S\partial_\alpha S^{-1} \right) M'(x)$$

$$+ \Delta_\alpha S^2 \, S^{-1} \beta^\rho S \left(\partial_\alpha + S^{-1} B_\alpha S + S^{-1}\partial_\alpha S \right) N'(x) = \Delta SJ(x) \,, \qquad (2.11.7)$$

$$\Delta_\alpha \, S\alpha^\rho S^{-1} = \alpha'^\rho \,, \qquad SA_\alpha S^{-1} + S\partial_\alpha S^{-1} = A'_\alpha \,,$$

$$\Delta_\alpha S^2 \, S^{-1} \beta^\rho S = \beta'^\rho \,, \qquad S^{-1} B_\alpha S + S^{-1} \partial_\alpha S = B'_\alpha \,. \tag{2.11.8}$$

In addition to our calculation performed in Sections **2.9** and **2.10**, we need to consider only the relations that involve additional matrices β^ρ and connection B_ρ.

For Euclidean rotations:

$$S \beta^\rho S^{-1} = S \beta^0 e^\rho_{(0)} S^{-1} + S \beta^l e^\rho_{(l)} S^{-1}$$

$$= \beta^0 e^\rho_{(0)} + \beta^k O_{kl} e^\rho_{(l)} = \beta^0 e^\rho_{(0)} + \beta^k e'^\rho_{(k)} = \beta'^\rho \,. \tag{2.11.9}$$

For local Lorentzian rotations (see Eqs. (2.7.15)–(2.7.17))

$$\Delta S^2 \, S^{-1} \beta^\rho(x) S = \Delta S^2 \, S^{-1} \beta^a e^\rho_{(a)} S$$

$$= [\Delta S^2 \, (S^{-1} \alpha^a S)] \, e^\rho_{(a)} = \alpha^b L_b{}^a \, e^\rho_{(a)} = \beta^b e'^\rho_{(b)} = \beta'^\rho(x) \,. \tag{2.11.10}$$

Transformation laws for two connections

$$S A_\rho S^{-1} + S \partial_\rho S^{-1} = A'_\rho \,, \qquad S^{-1} B_\rho S + S^{-1} \partial_\rho S = B_\rho \tag{2.11.11}$$

are complex conjugated relations, due to the identities

$$S^{-1} = S^* \,, \qquad S = (S^*)^{-1}, \qquad (B_\alpha)^* = A_\alpha.$$

Although, we do not need any additional calculation to prove the second relation in Eq. (2.11.11).

2.12   Matrix Equation in Explicit Component Form

Next, we will derive tensor generally covariant Maxwell equations using the matrix form

$$\alpha^c \left(e^\rho_{(c)} \partial_\rho + \frac{1}{2} j^{ab} \gamma_{abc} \right) \Psi = J(x) \,,$$

$$\alpha^0 = -iI \,, \quad \Psi = \left| \begin{array}{c} 0 \\ \mathbf{E} + ic\mathbf{B} \end{array} \right| \,, \quad J = \frac{1}{\epsilon_0} \left| \begin{array}{c} \rho \\ i\mathbf{j} \end{array} \right| \,. \tag{2.12.1}$$

A more detailed form for this is

$$-i \left(e^\rho_{(0)} \partial_\rho + \frac{1}{2} j^{ab} \gamma_{ab0} \right) \Psi + \alpha^k \left(e^\rho_{(k)} \partial_\rho + \frac{1}{2} j^{ab} \gamma_{abk} \right) \Psi = J(x) \,. \tag{2.12.2}$$

When taking this in mind, we have

$$\frac{1}{2} j^{ab} \gamma_{ab0} = [s^1(\gamma_{230} + i\gamma_{010}) + s^2(\gamma_{310} + i\gamma_{020}) + s^3(\gamma_{120} + i\gamma_{030})] \,,$$

$$\frac{1}{2} j^{ab} \gamma_{abk} = [s^1(\gamma_{23k} + i\gamma_{01k}) + s^2(\gamma_{31k} + i\gamma_{02k}) + s^3(\gamma_{12k} + i\gamma_{03k})] \,, \tag{2.12.3}$$

and by introducing the notation

$$e^\rho_{(0)} \partial_\rho = \partial_{(0)}, \qquad e^\rho_{(k)} \partial_\rho = \partial_{(k)} \,,$$

$$(\gamma_{01a}, \gamma_{02a}, \gamma_{03a}) = \mathbf{v}_a \ , \ (\gamma_{23a}, \gamma_{31a}, \gamma_{12a}) = \mathbf{p}_a \ , \ a = 0, 1, 2, 3 \ , \tag{2.12.4}$$

Equation (2.12.2) can be transformed to the form

$$-i \left[\, \partial_{(0)} + \mathbf{s}(\mathbf{p}_0 + i\mathbf{v}_0 \,) \, \right] \Psi + \alpha^k \left[\, \partial_{(k)} + \mathbf{s}(\mathbf{p}_k + i\mathbf{v}_k \,) \, \right] \Psi = J(x) \ ,$$

or

$$(\, \alpha^k \, \partial_{(k)} + \mathbf{s}\mathbf{v}_0 + \alpha^k \, \mathbf{s}\mathbf{p}_k \,) \left| \begin{array}{c} 0 \\ \mathbf{E} + ic\mathbf{B} \end{array} \right|$$

$$-i \, (\, \partial_{(0)} + \mathbf{s}\mathbf{p}_0 - \alpha^k \mathbf{s}\mathbf{v}_k) \left| \begin{array}{c} 0 \\ \mathbf{E} + ic\mathbf{B} \end{array} \right| = \frac{1}{\epsilon_0} \left| \begin{array}{c} \rho \\ i\,\mathbf{j} \end{array} \right| \ . \tag{2.12.5}$$

Let us divide Eq. (2.12.5) into real and imaginary parts:

$$(\, \alpha^k \, \partial_{(k)} + \mathbf{s}\mathbf{v}_0 + \alpha^k \, \mathbf{s}\mathbf{p}_k \,) \left| \begin{array}{c} 0 \\ \mathbf{E} \end{array} \right| + (\, \partial_{(0)} + \mathbf{s}\mathbf{p}_0 - \alpha^k \mathbf{s}\mathbf{v}_k) \left| \begin{array}{c} 0 \\ c\mathbf{B} \end{array} \right| = \frac{1}{\epsilon_0} \left| \begin{array}{c} \rho \\ 0 \end{array} \right| \ ,$$

$$\tag{2.12.6}$$

$$(\, \alpha^k \, \partial_{(k)} + \mathbf{s}\mathbf{v}_0 + \alpha^k \, \mathbf{s}\mathbf{p}_k \,) \left| \begin{array}{c} 0 \\ c\mathbf{B} \end{array} \right| - (\, \partial_{(0)} + \mathbf{s}\mathbf{p}_0 - \alpha^k \mathbf{s}\mathbf{v}_k) \left| \begin{array}{c} 0 \\ \mathbf{E} \end{array} \right| = \frac{1}{\epsilon_0} \left| \begin{array}{c} 0 \\ \mathbf{j} \end{array} \right| \ .$$

$$\tag{2.12.7}$$

Now, allowing for identities

$$\alpha^k \, \partial_{(k)} \left| \begin{array}{c} 0 \\ \mathbf{E} \end{array} \right| = \left| \begin{array}{c} \partial_{(k)} E_k \\ \partial_{(2)} E_3 - \partial_{(3)} E_2 \\ \partial_{(3)} E_1 - \partial_{(1)} E_3 \\ \partial_{(1)} E_2 - \partial_{(2)} E_1 \end{array} \right| \ , \qquad \mathbf{s}\mathbf{v}_0 \left| \begin{array}{c} 0 \\ \mathbf{E} \end{array} \right| = \left| \begin{array}{c} 0 \\ v_{20} E_3 - v_{30} E_2 \\ v_{30} E_1 - v_{10} E_3 \\ v_{10} E_2 - v_{20} E_1 \end{array} \right| \ ,$$

and

$$\alpha^k \, \mathbf{s}\mathbf{p}_k = \alpha^1 (S^1 p_{11} + S^2 p_{21} + S^3 p_{31})$$

$$+\alpha^2 (S^1 p_{12} + S^2 p_{22} + S^3 p_{32}) + \alpha^3 (S^1 p_{13} + S^2 p_{23} + S^3 p_{33})$$

$$= \left| \begin{array}{cccc} 0 & p_{32} - p_{23} & -p_{31} + p_{13} & p_{21} - p_{12} \\ 0 & -p_{22} - p_{33} & p_{12} & p_{13} \\ 0 & p_{21} & -p_{11} - p_{33} & p_{23} \\ 0 & p_{31} & p_{32} & -p_{11} - p_{22} \end{array} \right| \ ,$$

$$\alpha^k \, \mathbf{s}\mathbf{p}_k \left| \begin{array}{c} 0 \\ \mathbf{E} \end{array} \right| = \left| \begin{array}{c} -(p_{23} - p_{32})E_1 - (p_{31} - p_{13})E_2 - (p_{12} - p_{21})E_3 \\ -(p_{22} + p_{33})E_1 + p_{12}E_2 + p_{13}E_3 \\ p_{21}E_1 - (p_{11} + p_{33})E_2 + p_{23}E_3 \\ p_{31}E_1 + p_{32}E_2 - (p_{11} + p_{22})E_3 \end{array} \right| \ ,$$

Eqs. (2.12.6) and (2.12.7) take the form

$$\left(\left| \begin{array}{c} \partial_{(k)} E_k \\ \partial_{(2)} E_3 - \partial_{(3)} E_2 \\ \partial_{(3)} E_1 - \partial_{(1)} E_3 \\ \partial_{(1)} E_2 - \partial_{(2)} E_1 \end{array} \right| + \left| \begin{array}{c} 0 \\ v_{20} E_3 - v_{30} E_2 \\ v_{30} E_1 - v_{10} E_3 \\ v_{10} E_2 - v_{20} E_1 \end{array} \right| \right.$$

$$+ \begin{vmatrix} -(p_{23}-p_{32})E_1 - (p_{31}-p_{13})E_2 - (p_{12}-p_{21})E_3 \\ -(p_{22}+p_{33})E_1 + p_{12}E_2 + p_{13}E_3 \\ p_{21}E_1 - (p_{11}+p_{33})E_2 + p_{23}E_3 \\ p_{31}E_1 + p_{32}E_2 - (p_{11}+p_{22})E_3 \end{vmatrix} \Bigg)$$

$$+ \left(\partial_{(0)} \begin{vmatrix} 0 \\ cB_{1)} \\ cB_2 \\ cB_3 \end{vmatrix} + \begin{vmatrix} 0 \\ p_{20}cB_3 - p_{30}cB_2 \\ p_{30}cB_1 - p_{10}cB_3 \\ p_{10}cB_2 - p_{20}cB_1 \end{vmatrix} \right.$$

$$- \begin{vmatrix} -(v_{23}-v_{32})cB_1 - (v_{31}-v_{13})cB_2 - (v_{12}-v_{21})cB_3 \\ -(v_{22}+v_{33})cB_1 + v_{12}cB_2 + v_{13}cB_3 \\ v_{21}cB_1 - (v_{11}+v_{33})cB_2 + v_{23}cB_3 \\ v_{31}cB_1 + v_{32}cB_2 - (v_{11}+v_{22})cB_3 \end{vmatrix} \Bigg) = \frac{1}{\epsilon_0} \begin{vmatrix} \rho(x) \\ 0 \end{vmatrix} ,$$

$$(2.12.8)$$

and

$$\left(\begin{vmatrix} \partial_{(k)}cB_k \\ \partial_{(2)}cB_3 - \partial_{(3)}cB_2 \\ \partial_{(3)}cB_1 - \partial_{(1)}cB_3 \\ \partial_{(1)}cB_2 - \partial_{(2)}cB_1 \end{vmatrix} + \begin{vmatrix} 0 \\ v_{20}cB_3 - v_{30}cB_2 \\ v_{30}cB_1 - v_{10}cB_3 \\ v_{10}cB_2 - v_{20}cB_1 \end{vmatrix} \right.$$

$$+ \begin{vmatrix} -(p_{23}-p_{32})cB_1 - (p_{31}-p_{13})cB_2 - (p_{12}-p_{21})cB_3 \\ -(p_{22}+p_{33})cB_1 + p_{12}cB_2 + p_{13}cB_3 \\ p_{21}cB_1 - (p_{11}+p_{33})cB_2 + p_{23}cB_3 \\ p_{31}cB_1 + p_{32}cB_2 - (p_{11}+p_{22})cB_3 \end{vmatrix} \Bigg)$$

$$- \left(\partial_{(0)} \begin{vmatrix} 0 \\ E_1 \\ E_2 \\ E_3 \end{vmatrix} + \begin{vmatrix} 0 \\ p_{20}E_3 - p_{30}E_2 \\ p_{30}E_1 - p_{10}E_3 \\ p_{10}E_2 - p_{20}E_1 \end{vmatrix} \right.$$

$$- \begin{vmatrix} -(v_{23}-v_{32})E_1 - (v_{31}-v_{13})E_2 - (v_{12}-v_{21})E_3 \\ -(v_{22}+v_{33})E_1 + v_{12}E_2 + v_{13}E_3 \\ v_{21}E_1 - (v_{11}+v_{33})E_2 + v_{23}E_3 \\ v_{31}E_1 + v_{32}E_2 - (v_{11}+v_{22})E_3 \end{vmatrix} \Bigg) = \frac{1}{\epsilon_0} \begin{vmatrix} 0 \\ \mathbf{j} \end{vmatrix} .$$

$$(2.12.9)$$

Here, we write down them in explicit component form (combining in pairs from Eqs. (2.12.8) and (2.12.9)):

$$\partial_{(k)}E_k - [(p_{23}-p_{32})E_1 + (p_{31}-p_{13})E_2 + (p_{12}-p_{21})E_3]$$

$$+[(v_{23}-v_{32})cB_1 + (v_{31}-v_{13})cB_2 + (v_{12}-v_{21})cB_3] = \frac{1}{\epsilon_0}\rho , \qquad (2.12.10)$$

$$\partial_{(k)}cB_k - [(p_{23}-p_{32})cB_1 + (p_{31}-p_{13})cB_2 + (p_{12}-p_{21})cB_3]$$

$$-[(v_{23} - v_{32})E_1 + (v_{31} - v_{13})E_2 + (v_{12} - v_{21})E_{(3)}] = 0 \,, \qquad (2.12.11)$$

$$(\partial_{(2)}E_3 - \partial_{(3)}E_2) + (v_{20}E_3 - v_{30}E_2) + [-(p_{22} + p_{33})E_1 + p_{12}E_2 + p_{13}E_3]$$
$$+\partial_{(0)}cB_1 + (p_{20}cB_3 - p_{30}cB_2) - [-(v_{22} + v_{33})cB_1 + v_{12}cB_2 + v_{13}cB_3] = 0 \,, \quad (2.12.12)$$

$$(\partial_{(2)}cB_3 - \partial_{(3)}cB_2) + (v_{20}cB_3 - v_{30}cB_2) + [-(p_{22} + p_{33})cB_1 + p_{12}cB_2 + p_{13}cB_3]$$
$$-\partial_{(0)}E_1 - (p_{20}E_3 - p_{30}E_2) + [-(v_{22} + v_{33})E_1 + v_{12}E_2 + v_{13}E_3] = \frac{1}{\epsilon_0}j^1 \,, \quad (2.12.13)$$

$$(\partial_{(3)}E_1 - \partial_{(1)}E_3) + (v_{30}E_1 - v_{10}E_3) + [p_{21}E_1 - (p_{11} + p_{33})E_2 + p_{23}E_3]$$
$$+\partial_{(0)}cB_2 + (p_{30}cB_1 - p_{10}cB_1) - [v_{21}cB_1 - (v_{11} + v_{33})cB_2 + v_{23}cB_3] = 0 \,, \quad (2.12.14)$$

$$(\partial_{(3)}cB_1 - \partial_{(0)}cB_3) + (v_{30}cB_1 - v_{10}cB_3) + [+p_{31}cB_1 - (p_{11} + p_{33})cB_2 + p_{23}cB_3]$$
$$-\partial_{(0)}E_2 - (p_{30}E_1 - p_{10}E_3) + [v_{21}E_1 - (v_{11} + v_{33})E_2 + v_{23}E_3] = \frac{1}{\epsilon_0}j^2 \,, \quad (2.12.15)$$

$$(\partial_{(1)}E_2 - \partial_{(2)}E_1) + (v_{10}E_2 - v_{20}E_1) + [p_{31}E_1 + p_{32}E_2 - (p_{11} + p_{22})E_3]$$
$$+\partial_{(0)}cB_3 + (p_{10}cB_2 - p_{20}cB_1) - [v_{31}cB_1 + v_{32}cB_2 - (v_{11} + v_{22})cB_3] = 0 \,, \quad (2.12.16)$$

$$(\partial_{(1)}cB_2 - \partial_{(2)}cB_1) + (v_{10}cB_2 - v_{20}cB_1) + [+p_{31}cB_1 + p_{32}cB_2 - (p_{11} + p_{22})cB_3]$$
$$-\partial_{(0)}E_3 - (p_{10}E_2 - p_{20}E_1) + [v_{31}E_1 + v_{32}E_2 - (v_{11} + v_{22})E_3] = \frac{1}{\epsilon_0}j^3 \,. \quad (2.12.17)$$

We have estabished a rather complicated system of eight equations. In the next Section, we will prove its equivalence to tensor generally covariant Maxwell equations.

2.13   Relations Between Matrix and Tensor Equations

For covariant tensor Maxwell equations

$$\nabla^\alpha F^{\beta\gamma} + \nabla^\beta F^{\gamma\alpha} + \nabla^\gamma F^{\alpha\beta} = 0 \,, \qquad \nabla_\beta F^{\beta\alpha} = \frac{1}{\epsilon_0}\,j^\alpha \qquad (2.13.1)$$

let us introduce tetrad field variables:

$$\nabla^\alpha \,(e^\beta_{(b)}e^\gamma_{(c)}F^{(b)(c)}) + \nabla^\beta \,(e^\gamma_{(c)}e^\alpha_{(a)}F^{(c)(a)}) + \nabla^\gamma \,(e^\alpha_{(a)}e^\beta_{(b)}F^{(a)(b)}) = 0 \,,$$

$$\nabla_\beta \, e^\beta_{(b)} e^\alpha_{(a)} F^{(b)(a)} = \frac{1}{\epsilon_0} \, e^\alpha_{(a)} j^{(a)} \, ,$$

or

$$g^{\alpha\rho}\nabla_\rho \, (e^\beta_{(b)} e^\gamma_{(c)} F^{(b)(c)}) + g^{\beta\rho}\nabla_\rho \, (e^\gamma_{(c)} e^\alpha_{(a)} F^{(c)(a)}) + g^{\gamma\rho}\nabla_\rho \, (e^\alpha_{(a)} e^\beta_{(b)} F^{(a)(b)}) = 0 \, ,$$

$$\nabla_\beta e^\beta_{(b)} e^\alpha_{(a)} F^{(b)(a)} = \frac{1}{\epsilon_0} \, e^\alpha_{(a)} j^{(a)} \, .$$

Given the known properties of the covariant derivative, we get

$$g^{\alpha\rho} \, e^\beta_{(b);\rho} \, e^\gamma_{(c)} \, F^{(b)(c)} + g^{\alpha\rho} \, e^\beta_{(b)} \, e^\gamma_{(c);\rho} \, F^{(b)(c)} + g^{\alpha\rho} \, e^\beta_{(b)} e^\gamma_{(c)} \, \partial_\rho F^{(b)(c)}$$

$$+g^{\beta\rho} \, e^\gamma_{(c);\rho} \, e^\alpha_{(a)} \, F^{(c)(a)} + g^{\beta\rho} \, e^\gamma_{(c)} \, e^\alpha_{(a);\rho} \, F^{(c)(a)} + g^{\beta\rho} e^\gamma_{(c)} e^\alpha_{(a)} \, \partial_\rho \, F^{(c)(a)}$$

$$+g^{\gamma\rho} \, e^\alpha_{(a);\rho} \, e^\beta_{(b)} \, F^{(a)(b)} + g^{\gamma\rho} \, e^\alpha_{(a)} \, \nabla_\rho e^\beta_{(b);\rho} \, F^{(a)(b)} + g^{\gamma\rho} \, e^\alpha_{(a)} \, e^\beta_{(b)} \, \partial_\rho F^{(a)(b)} = 0 \, ,$$

$$(2.13.2)$$

$$e^\beta_{(b);\beta} \, e^\alpha_{(a)} \, F^{(b)(a)} + e^\beta_{(b)} \, e^\alpha_{(a);\beta} \, F^{(b)(a)} + e^\beta_{(b)} \, e^\alpha_{(a)} \, \partial_\beta F^{(b)(a)} = \frac{1}{\epsilon_0} \, e^\alpha_{(a)} j^{(a)} \, .$$

$$(2.13.3)$$

Recall the definition for Ricci rotation coefficients:

$$\gamma_{abc} = -e_{(a)\alpha;\beta} \, e^\alpha_{(b)} e^\beta_{(c)} \, ,$$

let us multiply Eq. (2.13.2) by $e_{\alpha(n)} e_{\beta(m)} e_{\gamma(l)}$:

$$e_{\alpha(n)} e_{\beta(m)} e_{\gamma(l)} g^{\alpha\rho} e^\beta_{(b);\rho} e^\gamma_{(c)} F^{(b)(c)} + e_{\alpha(n)} e_{\beta(m)} e_{\gamma(l)} g^{\alpha\rho} e^\beta_{(b)} e^\gamma_{(c);\rho} F^{(b)(c)}$$

$$+e_{\alpha(n)} e_{\beta(m)} e_{\gamma(l)} g^{\alpha\rho} e^\beta_{(b)} e^\gamma_{(c)} \partial_\rho F^{(b)(c)}$$

$$+e_{\alpha(n)} e_{\beta(m)} e_{\gamma(l)} g^{\beta\rho} e^\gamma_{(c);\rho} e^\alpha_{(a)} F^{(c)(a)} + e_{\alpha(n)} e_{\beta(m)} e_{\gamma(l)} g^{\beta\rho} e^\gamma_{(c)} e^\alpha_{(a);\rho} F^{(c)(a)}$$

$$+e_{\alpha(n)} e_{\beta(m)} e_{\gamma(l)} g^{\beta\rho} e^\gamma_{(c)} e^\alpha_{(a)} \partial_\rho F^{(c)(a)}$$

$$+e_{\alpha(n)} e_{\beta(m)} e_{\gamma(l)} g^{\gamma\rho} e^\alpha_{(a);\rho} e^\beta_{(b)} F^{(a)(b)} + e_{\alpha(n)} e_{\beta(m)} e_{\gamma(l)} g^{\gamma\rho} e^\alpha_{(a)} e^\beta_{(b);\rho} F^{(a)(b)}$$

$$+e_{\alpha(n)} e_{\beta(m)} e_{\gamma(l)} g^{\gamma\rho} e^\alpha_{(a)} e^\beta_{(b)} \partial_\rho F^{(a)(b)} = 0 \, ,$$

that is

$$(e^\beta_{(b);\rho} e_{\beta(m)} e^\rho_{(n)}) \, F^{(b)}_{\ (l)} + (e^\gamma_{(c);\rho} e_{\gamma(l)} e^\rho_{(n)}) \, F^{(c)}_{\ (m)} + e^\rho_{(n)} \partial_\rho F_{(m)(l)}$$

$$+(e^\gamma_{(c);\rho} e_{\gamma(l)} e^\rho_{(m)}) \, F^{(c)}_{\ (n)} + (e^\alpha_{(a);\rho} e_{\alpha(n)} e^\rho_{(m)}) \, F^{(a)}_{\ (l)} + e^\rho_{(m)} \, \partial_\rho \, F_{(l)(n)}$$

$$+(e^\alpha_{(a);\rho} e_{\alpha(n)} e^\rho_{(l)}) \, F^{(a)}_{\ (m)} + (e^\beta_{(b);\rho} e_{\beta(m)} e^\rho_{(l)}) \, F^{(b)}_{\ (n)} + e^\rho_{(l)} \, \partial_\rho F_{(n)(m)} = 0 \, ,$$

and further

$$-\gamma_{bmn} \, F^{(b)}_{\ (l)} - \gamma_{cln} \, F^{(c)}_{\ (m)} + \partial_{(n)} F_{(m)(l)}$$

$$-\gamma_{clm} \, F^{(c)}_{\;\;(n)} - \gamma_{anm} \, F^{(a)}_{\;\;(l)} + \partial_{(m)} \, F_{(l)(n)}$$

$$-\gamma_{anl} \, F^{(a)}_{\;\;(m)} - \gamma_{bml} \, F^{(b)}_{\;\;(n)} + \partial_{(l)} F_{(n)(m)} = 0 \,.$$

The latter can be rewritten as follows:

$$\partial_{(n)} F_{(m)(l)} + \gamma_{mbn} \, F^{(b)}_{\;\;(l)} - \gamma_{lbn} \, F^{(b)}_{\;\;(m)}$$

$$+ \partial_{(m)} \, F_{(l)(n)} + \gamma_{lbm} \, F^{(b)}_{\;\;(n)} - \gamma_{nbm} \, F^{(b)}_{\;\;(l)}$$

$$+ \partial_{(l)} F_{(n)(m)} + \gamma_{nbl} \, F^{(b)}_{\;\;(m)} - \gamma_{mbl} \, F^{(b)}_{\;\;(n)} = 0 \,. \tag{2.13.4}$$

Now, let us multiply Eq. (2.13.3) by $e_{\alpha(c)}$:

$$e_{\alpha(c)} e^{\beta}_{(b);\beta} e^{\alpha}_{(a)} F^{(b)(a)} + e_{\alpha(c)} e^{\beta}_{(b)} e^{\alpha}_{(a);\beta} F^{(b)(a)} + e_{\alpha(c)} e^{\beta}_{(b)} e^{\alpha}_{(a)} \partial_{\beta} F^{(b)(a)} = e_{\alpha(c)} \frac{1}{\epsilon_0} e^{\alpha}_{(a)} j^{(a)} \,,$$

that is

$$e^{\beta}_{(b);\beta} \, F^{(b)}_{\;\;(c)} + (e^{\alpha}_{(a);\beta} e_{\alpha(c)} \, e^{\beta}_{(b)}) \, F^{(b)(a)} + e^{\beta}_{(b)} \, \partial_{\beta} F^{(b)}_{\;\;(c)} = \frac{1}{\epsilon_0} \, j_{(c)} \,,$$

or

$$\partial_{(b)} F^{(b)}_{\;\;(c)} + e^{\beta}_{(b);\beta} \, F^{(b)}_{\;\;(c)} + \gamma_{cab} \, F^{(b)(a)} = \frac{1}{\epsilon_0} \, j_{(c)} \,. \tag{2.13.5}$$

Equations (2.13.4) and (2.13.5) represent the Maxwell equations in tetrad form. Now, we can provide details for Eqs. (2.13.4) and (2.13.5):

$$n, m, l = 1, 2, 3, \qquad 0, 2, 3, \qquad 0, 3, 1, \qquad 0, 1, 2$$

$$\text{and} \quad c = 0, \quad 1, \quad 2, \quad 3 \,.$$

Let it be $n, m, l = 1, 2, 3$:

$$\partial_{(1)} F_{(2)(3)} + \gamma_{2b1} \, F^{(b)}_{\;\;(3)} - \gamma_{3b1} \, F^{(b)}_{\;\;(2)} +$$

$$+ \partial_{(2)} \, F_{(3)(1)} + \gamma_{3b2} \, F^{(b)}_{\;\;(1)} - \gamma_{1b2} \, F^{(b)}_{\;\;(3)}$$

$$+ \partial_{(3)} F_{(1)(2)} + \gamma_{1b3} \, F^{(b)}_{\;\;(2)} - \gamma_{2b3} \, F^{(b)}_{\;\;(1)} = 0 \,,$$

or

$$\partial_{(1)} F_{(2)(3)} + \gamma_{201} F^{(0)}_{\;\;(3)} + \gamma_{211} F^{(1)}_{\;\;(3)} - \gamma_{301} F^{(0)}_{\;\;(2)} - \gamma_{311} F^{(1)}_{\;\;(2)}$$

$$+ \partial_{(2)} \, F_{(3)(1)} + \gamma_{302} F^{(0)}_{\;\;(1)} + \gamma_{322} F^{(2)}_{\;\;(1)} - \gamma_{102} F^{(0)}_{\;\;(3)} - \gamma_{122} F^{(2)}_{\;\;(3)} \,)$$

$$+ \partial_{(3)} F_{(1)(2)} + \gamma_{103} F^{(0)}_{\;\;(2)} + \gamma_{133} F^{(3)}_{\;\;(2)} - \gamma_{203} F^{(0)}_{\;\;(1)} - \gamma_{233} F^{(3)}_{\;\;(1)} = 0$$

which with notation

$$(F_{(2)(3)}, F_{(3)(1)}, F_{(1)(2)}) = (cB_{(i)}) \,, \qquad (F_{(0)(1)}, F_{(0)(2)}, F_{(0)(3)}) = (E_{(i)})$$

reads

$$\partial_{(k)} cB_{(k)} - v_{21} E_{(3)} - p_{31} cB_{(2)} + v_{31} E_{(2)} + p_{21} cB_{(3)}$$

$$-v_{32}E_{(1)}-p_{12}cB_{(3)}+v_{12}E_{(3)}+p_{32}cB_{(1)}-v_{13}E_{(2)}-p_{23}cB_{(1)}+v_{23}E_{(1)}+p_{13}cB_{(2)}=0 \,,$$

$$(2.13.6)$$

or

$$-\partial_{(k)}cB_{(k)} + [\,(p_{23}-p_{32})cB_{(1)} + (p_{31}-p_{13})cB_{(2)} + (p_{12}-p_{21})cB_{(3)}\,]$$

$$-[\,(v_{23}-v_{32})E_{(1)} + (v_{31}-v_{13})E_{(2)} + (v_{12}-v_{21})E_{(3)}\,] = 0 \,.$$

This example coincides with Eq. (2.12.11):

$$\partial_{(k)}cB_k - [(p_{23}-p_{32})cB_1 + (p_{31}-p_{13})cB_2 + (p_{12}-p_{21})cB_3]$$

$$-[(v_{23}-v_{32})E_1 + (v_{31}-v_{13})E_2 + (v_{12}-v_{21})E_3 = 0 \,,$$

$$(2.13.7)$$

if

$$E_k = E_{(k)} = \mathbf{E} \,, \qquad B_k = -B_{(k)} = \mathbf{B} \,. \qquad (2.13.8)$$

<u>Let it be $n, m, l = 0, 1, 2$:</u>

$$\partial_{(0)}F_{(1)(2)} + \gamma_{1b0}\, F^{(b)}_{(2)} - \gamma_{2b0}\, F^{(b)}_{(1)}$$

$$+ \partial_{(1)}\, F_{(2)(0)} + \gamma_{2b1}\, F^{(b)}_{(0)} - \gamma_{0b1}\, F^{(b)}_{(2)}$$

$$+ \partial_{(2)}F_{(0)(1)} + \gamma_{0b2}\, F^{(b)}_{(1)} - \gamma_{1b2}F^{(b)}_{(0)} = 0 \,, \qquad (2.13.9)$$

and further

$$\partial_{(0)}F_{(1)(2)} + \gamma_{100}\, F^{(0)}_{(2)} + \gamma_{130}\, F^{(3)}_{(2)} - \gamma_{200}\, F^{(0)}_{(1)} - \gamma_{230}\, F^{(3)}_{(1)}$$

$$+ \partial_{(1)}\, F_{(2)(0)} + \gamma_{211}\, F^{(1)}_{(0)} + \gamma_{231}\, F^{(3)}_{(0)} - \gamma_{011}\, F^{(1)}_{(2)} - \gamma_{031}\, F^{(3)}_{(2)}$$

$$+ \partial_{(2)}F_{(0)(1)} + \gamma_{022}\, F^{(2)}_{(1)} + \gamma_{032}\, F^{(3)}_{(1)} - \gamma_{122}\, F^{(2)}_{(0)} - \gamma_{132}\, F^{(3)}_{(0)} = 0 \,,$$

or

$$\partial_{(0)}cB_{(3)} - v_{10}E_{(2)} - p_{20}cB_{(1)} + v_{20}E_{(1)} + p_{10}cB_{(2)}$$

$$- \partial_{(1)}\, E_{(2)} - p_{31}E_{(1)} + p_{11}E_{(3)} + v_{11}cB_{(3)} - v_{31}cB_{(1)}$$

$$+ \partial_{(2)}E_{(1)} + v_{22}cB_{(3)} - v_{32}cB_{(2)} - p_{32}E_{(2)} + p_{22}E_{(3)} = 0 \,,$$

which coincides with Eq. (2.12.16) multiplied by -1:

$$-(\partial_{(1)}E_2 - \partial_{(2)}E_1) - (v_{10}E_2 - v_{20}E_1) - [p_{31}E_1 + p_{32}E_2 - (p_{11}+p_{22})E_3]$$

$$-\partial_{(0)}cB_3 - (p_{10}cB_2 - p_{20}cB_1) + [v_{31}cB_1 + v_{32}cB_2 - (v_{11}+v_{22})cB_3] = 0 \,,$$

$$(2.13.10)$$

if

$$E_k = E_{(k)} = \mathbf{E} \,, \qquad B_k = -B_{(k)} = \mathbf{B} \,. \qquad (2.13.11)$$

Let $c = 0$ in Eq. (2.13.5):

$$\partial_{(b)} F^{(b)}_{\ (0)} + e^{\beta}_{(b);\beta} F^{(b)}_{\ (0)} + \gamma_{0ab} F^{(b)(a)} = \frac{1}{\epsilon_0} \rho \, . \qquad (2.13.12)$$

In allowing for the identity

$$e^{\beta}_{(b);\beta} F^{(b)}_{\ (0)} = -\gamma_{kc}^{\ \ c} F^{(k)}_{\ (0)} = -(\gamma_{k00} - \gamma_{k11} - \gamma_{k22} - \gamma_{k33}) F^{(k)}_{\ (0)}$$

$$= -\gamma_{k00} F^{(k)}_{\ (0)} + \gamma_{211} F^{(2)}_{\ (0)} + \gamma_{311} F^{(3)}_{\ (0)} + \gamma_{122} F^{(1)}_{\ (0)} + \gamma_{322} F^{(3)}_{\ (0)} + \gamma_{133} F^{(1)}_{\ (0)} + \gamma_{233} F^{(2)}_{\ (0)} \, ,$$

we get

$$\partial_{(k)} F^{(k)}_{\ (0)} - \gamma_{k00} F^{(k)}_{\ (0)}$$

$$+\gamma_{211} F^{(2)}_{\ (0)} + \gamma_{311} F^{(3)}_{\ (0)} + \gamma_{122} F^{(1)}_{\ (0)} + \gamma_{322} F^{(3)}_{\ (0)} + \gamma_{133} F^{(1)}_{\ (0)} + \gamma_{233} F^{(2)}_{\ (0)}$$

$$+\gamma_{010} F^{(0)(1)} + \gamma_{012} F^{(2)(1)} + \gamma_{013} F^{(3)(1)}$$

$$+\gamma_{020} F^{(0)(2)} + \gamma_{021} F^{(1)(2)} + \gamma_{023} F^{(3)(2)}$$

$$+\gamma_{030} F^{(0)(3)} + \gamma_{031} F^{(1)(3)} + \gamma_{032} F^{(2)(3)} = \frac{1}{\epsilon_0} \rho \, ,$$

or

$$\partial_{(k)} E_{(k)} + v_{k0} E_{(k)}$$

$$-p_{31} E_{(2)} + p_{21} E_{(3)} + p_{32} E_{(1)} - p_{12} E_{(3)} - p_{23} E_{(1)} + p_{13} E_{(2)}$$

$$-v_{10} E_{(1)} - v_{12} cB_{(3)} + v_{13} cB_{(2)}$$

$$-v_{20} E_{(2)} + v_{21} cB_{(3)} - v_{23} cB_{(1)}$$

$$-v_{30} E_{(3)} - v_{31} cB_{(2)} + v_{32} cB_{(1)} = \frac{1}{\epsilon_0} \rho \, ,$$

or

$$\partial_{(k)} E_{(k)} - p_{31} E_{(2)} + p_{21} E_{(3)} + p_{32} E_{(1)} - p_{12} E_{(3)} - p_{23} E_{(1)} + p_{13} E_{(2)}$$

$$-v_{12} cB_{(3)} + v_{13} cB_{(2)} + v_{21} cB_{(3)} - v_{23} cB_{(1)} - v_{31} cB_{(2)} + v_{32} cB_{(1)} = \frac{1}{\epsilon_0} \rho \, .$$

This coincides with Eq.(2.12.12):

$$\partial_{(k)} E_k - [(p_{23} - p_{32}) E_1 + (p_{31} - p_{13}) E_2 + (p_{12} - p_{21}) E_3]$$

$$-[(v_{23} - v_{32}) cB_1 + (v_{31} - v_{13}) cB_2 + (v_{12} - v_{21}) cB_3 = \frac{1}{\epsilon_0} \rho \, , \qquad (2.13.13)$$

if

$$E_k = E_{(k)} = \mathbf{E}, \qquad B_k = -B_{(k)} = \mathbf{B} \, . \qquad (2.13.14)$$

Now, let $c = 3$ in Eq. (2.13.5):

$$\partial_{(b)} F^{(b)}_{\ (3)} + e^{\beta}_{(b);\beta} F^{(b)}_{\ (3)} + \gamma_{3ab} F^{(b)(a)} = \frac{1}{\epsilon_0} j_{(3)} \, . \qquad (2.13.15)$$

Here, we will present the identities

$$e^{\beta}_{(b);\beta}\, F^{(b)}_{\ (3)} = -\gamma_{bc}\,^{c}F^{(b)}_{\ (3)} = -(\gamma_{b00} - \gamma_{b11} - \gamma_{b22} - \gamma_{b33})F^{(b)}_{\ (3)}$$

$$= -\gamma_{100}F^{(1)}_{\ (3)} - \gamma_{200}F^{(2)}_{\ (3)} + \gamma_{011}F^{(0)}_{\ (3)} + \gamma_{211}F^{(2)}_{\ (3)}$$

$$+\gamma_{022}F^{(0)}_{\ (3)} + \gamma_{122}F^{(1)}_{\ (3)} + \gamma_{033}F^{(0)}_{\ (3)} + \gamma_{133}F^{(1)}_{\ (3)} + \gamma_{233}F^{(2)}_{\ (3)} \,,$$

$$\gamma_{3ab}\, F^{(b)(a)} = -(\gamma_{30b}\, F^{(0)(b)} + \gamma_{31b}\, F^{(1)(b)} + \gamma_{32b}\, F^{(2)(b)})$$

$$= -\gamma_{301}\, F^{(0)(1)} - \gamma_{302}\, F^{(0)(2)} - \gamma_{303}\, F^{(0)(3)}$$

$$-\gamma_{310}\, F^{(1)(0)} - \gamma_{312}\, F^{(1)(2)} - \gamma_{313}\, F^{(1)(3)}$$

$$-\gamma_{320}\, F^{(2)(0)} - \gamma_{321}\, F^{(2)(1)} - \gamma_{323}\, F^{(2)(3)} \,,$$

we get

$$\partial_{(0)}F^{(0)}_{\ (3)} + \partial_{(1)}F^{(1)}_{\ (3)} + \partial_{(2)}F^{(2)}_{\ (3)}$$

$$-\gamma_{100}F^{(1)}_{\ (3)} - \gamma_{200}F^{(2)}_{\ (3)} + \gamma_{011}F^{(0)}_{\ (3)}$$

$$+\gamma_{211}F^{(2)}_{\ (3)} + \gamma_{022}F^{(0)}_{\ (3)} + \gamma_{122}F^{(1)}_{\ (3)}$$

$$+\gamma_{033}F^{(0)}_{\ (3)} + \gamma_{133}F^{(1)}_{\ (3)} + \gamma_{233}F^{(2)}_{\ (3)}$$

$$-\gamma_{301}F^{(0)(1)} - \gamma_{302}F^{(0)(2)} - \gamma_{303}F^{(0)(3)}$$

$$-\gamma_{310}F^{(1)(0)} - \gamma_{312}F^{(1)(2)} - \gamma_{313}F^{(1)(3)}$$

$$-\gamma_{320}F^{(2)(0)} - \gamma_{321}F^{(2)(1)} - \gamma_{323}F^{(2)(3)} = \frac{1}{\epsilon_0}\, j_{(3)} \,,$$

or

$$\partial_{(0)}F^{(0)}_{\ (3)} + \partial_{(1)}F^{(1)}_{\ (3)} + \partial_{(2)}F^{(2)}_{\ (3)}$$

$$-\gamma_{100}F^{(1)}_{\ (3)} - \gamma_{200}F^{(2)}_{\ (3)} + \gamma_{011}F^{(0)}_{\ (3)} + \gamma_{211}F^{(2)}_{\ (3)} + \gamma_{022}F^{(0)}_{\ (3)} + \gamma_{122}F^{(1)}_{\ (3)}$$

$$-\gamma_{301}F^{(0)(1)} - \gamma_{302}F^{(0)(2)} - \gamma_{310}F^{(1)(0)} - \gamma_{312}F^{(1)(2)} - \gamma_{320}F^{(2)(0)} - \gamma_{321}F^{(2)(1)} = \frac{1}{\epsilon_0}\, j_{(3)} \,,$$

or (additionally multiplied by -1)

$$-\partial_{(0)}E_{(3)} - \partial_{(1)}cB_{(2)} + \partial_{(2)}cB_{(1)}$$

$$-v_{10}cB_{(2)} + v_{20}cB_{(1)} - v_{11}E_{(3)} - p_{31}cB_{(1)} - v_{22}E_{(3)} - p_{32}cB_{(2)}$$

$$+v_{31}E_{(1)} + v_{32}E_{(2)} + p_{20}E_{(1)} + p_{22}cB_{(3)} - p_{10}E_{(2)} + p_{11}cB_{(3)} = -\frac{1}{\epsilon_0}\, j_{(3)} \,.$$

This coincides with Eq. (2.12.17) multiplied by (-1):

$$(\partial_{(1)}cB_2 - \partial_{(2)}cB_1) - (v_{10}cB_2 - v_{20}cB_1) + [p_{31}cB_1 + p_{32}cB_2 - (p_{11} + p_{22})cB_3]$$

$$-\partial_{(0)}E_3 - (p_{10}E_2 - p_{20}E_1) - [v_{31}E_1 + v_{32}E_2 - (v_{11} + v_{22})E_3] = \frac{1}{\epsilon_0}j^{(3)} \,, \quad (2.13.16)$$

if

$$E_k = E_{(k)} = \mathbf{E}, \qquad B_k = -B_{(k)} = \mathbf{B} \,. \qquad (2.13.17)$$

Thus, the following three forms of Maxwell equations are equivalent to each other:

matrix form

$$\alpha^{\alpha}(x) \, [\, \partial_\rho + A_\alpha(x) \,] \, \Psi = J(x) \,;$$

$$(2.13.18)$$

tensor form

$$\nabla^{\alpha}F^{\beta\gamma} + \nabla^{\beta}F^{\gamma\alpha} + \nabla^{\gamma}F^{\alpha\beta} = 0 \,,$$

$$\nabla_{\beta}F^{\beta\alpha} = \frac{1}{\epsilon_0} \, j^{\alpha} \,; \qquad (2.13.19)$$

explicit tetrad component form

$$\partial_{(k)}E_k - [(p_{23} - p_{32})E_1 + (p_{31} - p_{13})E_2 + (p_{12} - p_{21})E_3]$$

$$+ \, [(v_{23} - v_{32})cB_1 + (v_{31} - v_{13})cB_2 + (v_{12} - v_{21})cB_3] = \frac{1}{\epsilon_0}\rho \,,$$

$$\partial_{(k)}cB_k - [(p_{23} - p_{32})cB_1 + (p_{31} - p_{13})cB_2 + (p_{12} - p_{21})cB_3]$$

$$- \, [(v_{23} - v_{32})E_1 + (v_{31} - v_{13})E_2 + (v_{12} - v_{21})E_{(3)}] = 0 \,,$$

$$(\partial_{(2)}E_3 - \partial_{(3)}E_2) + (v_{20}E_3 - v_{30}E_2) + [-(p_{22} + p_{33})E_1 + p_{12}E_2 + p_{13}E_3]$$

$$+ \, \partial_{(0)}cB_1 + (p_{20}cB_3 - p_{30}cB_2) - [-(v_{22} + v_{33})cB_1 + v_{12}cB_2 + v_{13}cB_3] = 0 \,,$$

$$(\partial_{(2)}cB_3 - \partial_{(3)}cB_2) + (v_{20}cB_3 - v_{30}cB_2) + [-(p_{22} + p_{33})cB_1 + p_{12}cB_2 + p_{13}cB_3]$$

$$- \, \partial_{(0)}E_1 - (p_{20}E_3 - p_{30}E_2) + [-(v_{22} + v_{33})E_1 + v_{12}E_2 + v_{13}E_3] = \frac{1}{\epsilon_0}j^1 \,,$$

$$(\partial_{(3)}E_1 - \partial_{(1)}E_3) + (v_{30}E_1 - v_{10}E_3) + [p_{21}E_1 - (p_{11} + p_{33})E_2 + p_{23}E_3]$$

$$+ \, \partial_{(0)}cB_2 + (p_{30}cB_1 - p_{10}cB_1) - [v_{21}cB_1 - (v_{11} + v_{33})cB_2 + v_{23}cB_3] = 0 \,,$$

$$(\partial_{(3)}cB_1 - \partial_{(0)}cB_3) + (v_{30}cB_1 - v_{10}cB_3) + [+p_{31}cB_1 - (p_{11} + p_{33})cB_2 + p_{23}cB_3]$$

$$- \, \partial_{(0)}E_2 - (p_{30}E_1 - p_{10}E_3) + [v_{21}E_1 - (v_{11} + v_{33})E_2 + v_{23}E_3] = \frac{1}{\epsilon_0}j^2 \,,$$

$$(\partial_{(1)}E_2 - \partial_{(2)}E_1) + (v_{10}E_2 - v_{20}E_1) + [p_{31}E_1 + p_{32}E_2 - (p_{11} + p_{22})E_3]$$

$$+\partial_{(0)}cB_3 + (p_{10}cB_2 - p_{20}cB_1) - [v_{31}cB_1 + v_{32}cB_2 - (v_{11} + v_{22})cB_3] = 0 \ ,$$

$$(\partial_{(1)}cB_2 - \partial_{(2)}cB_1) + (v_{10}cB_2 - v_{20}cB_1) + [+p_{31}cB_1 + p_{32}cB_2 - (p_{11} + p_{22})cB_3]$$

$$- \partial_{(0)}E_3 - (p_{10}E_2 - p_{20}E_1) + [v_{31}E_1 + v_{32}E_2 - (v_{11} + v_{22})E_3] = \frac{1}{\epsilon_0}j^3 \ ,$$

$$(2.13.20)$$

where

$$\Psi = \left| \begin{matrix} 0 \\ \mathbf{E} + ic\mathbf{B} \end{matrix} \right| \ , \qquad J = \frac{1}{\epsilon_0} \left| \begin{matrix} \rho \\ i\mathbf{j} \end{matrix} \right| \ ,$$

$$e^{\rho}_{(0)}\partial_{\rho} = \partial_{(0)} \ , \qquad e^{\rho}_{(k)}\partial_{\rho} = \partial_{(k)} \ ,$$

$$(\gamma_{010}, \gamma_{020}, \gamma_{030}) = \mathbf{v}_0 \ , \qquad (\gamma_{01k}, \gamma_{02k}, \gamma_{03k}) = \mathbf{v}_k \ ,$$

$$(\gamma_{230}, \gamma_{310}, \gamma_{120}) = \mathbf{p}_0 \ , \qquad (\gamma_{23k}, \gamma_{31k}, \gamma_{12k}) = \mathbf{p}_k \ .$$

2.14 Relations Between Matrix and Tensor Equations in the Presence of Media

Let us find a detailed tetrad component form for the generally covariant Maxwell matrix equation in presence of media:

$$\alpha_\rho(x)(\partial_\rho + A_\rho) M + \beta_\rho(x)(\partial_\rho + B_\rho) N = J \ , \qquad (2.14.1)$$

where

$$M = \left| \begin{matrix} 0 \\ \mathbf{M} \end{matrix} \right| \ , \qquad N = \left| \begin{matrix} 0 \\ \mathbf{N} \end{matrix} \right| \ , \qquad J = \frac{1}{\epsilon_0} \left| \begin{matrix} \rho \\ i\,\mathbf{j} \end{matrix} \right| \ ,$$

$$\mathbf{E} + ic\mathbf{B} = \mathbf{f} \ , \qquad \frac{1}{\epsilon_0}(\mathbf{D} + i\mathbf{H}/c) = \mathbf{h} \ ,$$

$$\mathbf{M} = \frac{\mathbf{h} + \mathbf{f}}{2} = \frac{1}{2}(\frac{\mathbf{D}}{\epsilon_0} + \mathbf{E}) + \frac{i}{2}(c\mathbf{B} + \frac{\mathbf{H}}{\epsilon_0 c}) \ ,$$

$$\mathbf{N} = \frac{\mathbf{h}^* - \mathbf{f}^*}{2} = \frac{1}{2}(\frac{\mathbf{D}}{\epsilon_0} - \mathbf{E}) + \frac{i}{2}(c\mathbf{B} - \frac{\mathbf{H}}{\epsilon_0 c}) \ . \qquad (2.14.2)$$

This time, we shall use a shortened notation:

$$\frac{\mathbf{D}}{\epsilon_0} \Longrightarrow \mathbf{D} \ , \qquad c\mathbf{B} \Longrightarrow \mathbf{B} \ , \qquad \frac{\mathbf{H}}{\epsilon_0 c} \Longrightarrow \mathbf{H} \ . \qquad (2.14.3)$$

Equation (2.14.1) can be rewritten as follows:

$$\alpha^c \left(e^\rho_{(c)} \partial_\rho + \frac{1}{2} j^{ab} \gamma_{abc} \right) M + \beta^c \left(e^\rho_{(c)} \partial_\rho + \frac{1}{2} j^{*ab} \gamma_{abc} \right) N = J(x) , \qquad (2.14.4)$$

a more detailed form is

$$-i \left(e^\rho_{(0)} \partial_\rho + \frac{1}{2} j^{ab} \gamma_{ab0} \right) M + \alpha^k \left(e^\rho_{(k)} \partial_\rho + \frac{1}{2} j^{ab} \gamma_{abk} \right) M$$

$$-i \left(e^\rho_{(0)} \partial_\rho + \frac{1}{2} j^{ab} \gamma_{ab0} \right) N + \beta^k \left(e^\rho_{(k)} \partial_\rho + \frac{1}{2} j^{*ab} \gamma_{abk} \right) N = J(x) . \qquad (2.14.5)$$

By keeping in mind the notation

$$\frac{1}{2} j^{ab} \gamma_{abc} = [s^1 (\gamma_{23c} + i\gamma_{01c}) + s^2 (\gamma_{31c} + i\gamma_{02c}) + s^3 (\gamma_{12c} + i\gamma_{03c})] ,$$

$$\frac{1}{2} j^{*ab} \gamma_{abc} = [s^1 (\gamma_{23c} - i\gamma_{01c}) + s^2 (\gamma_{31c} - i\gamma_{02c}) + s^3 (\gamma_{12c} - i\gamma_{03c})] ,$$

$$e^\rho_{(0)} \partial_\rho = \partial_{(0)}, \qquad e^\rho_{(k)} \partial_\rho = \partial_{(k)} ,$$

$$(\gamma_{01a}, \gamma_{02a}, \gamma_{03a}) = \mathbf{v}_a , \qquad (\gamma_{23a}, \gamma_{31a}, \gamma_{12a}) = \mathbf{p}_a , \qquad a = 0, 1, 2, 3 ,$$

Eq. (2.14.5) can be transformed to the form

$$-i [\partial_{(0)} + \mathbf{s}(\mathbf{p}_0 + i\mathbf{v}_0)] M + \alpha^k [\partial_{(k)} + \mathbf{s}(\mathbf{p}_k + i\mathbf{v}_k)] M$$

$$-i [\partial_{(0)} + \mathbf{s}(\mathbf{p}_0 - i\mathbf{v}_0)] N + \beta^k [\partial_{(k)} + \mathbf{s}(\mathbf{p}_k - i\mathbf{v}_k)] N = J(x) . \qquad (2.14.6)$$

Let us divide Eq. (2.14.6) into real and imaginary parts (for brevity, we have omitted parameter c):

$$(\alpha^k \partial_{(k)} + \mathbf{s}\mathbf{v}_0 + \alpha^k \mathbf{s}\mathbf{p}_k) \frac{1}{2} \begin{vmatrix} 0 \\ \mathbf{D} + \mathbf{E} \end{vmatrix} + (\partial_{(0)} + \mathbf{s}\mathbf{p}_0 - \alpha^k \mathbf{s}\mathbf{v}_k) \frac{1}{2} \begin{vmatrix} 0 \\ \mathbf{B} + \mathbf{H} \end{vmatrix}$$

$$+ (\beta^k \partial_{(k)} - \mathbf{s}\mathbf{v}_0 + \beta^k \mathbf{s}\mathbf{p}_k) \frac{1}{2} \begin{vmatrix} 0 \\ \mathbf{D} - \mathbf{E} \end{vmatrix} + (\partial_{(0)} + \mathbf{s}\mathbf{p}_0 + \beta^k \mathbf{s}\mathbf{v}_k) \frac{1}{2} \begin{vmatrix} 0 \\ \mathbf{B} - \mathbf{H} \end{vmatrix} = \frac{1}{\epsilon_0} \begin{vmatrix} \rho \\ 0 \end{vmatrix} . \qquad (2.14.7)$$

$$(\alpha^k \partial_{(k)} + \mathbf{s}\mathbf{v}_0 + \alpha^k \mathbf{s}\mathbf{p}_k) \frac{1}{2} \begin{vmatrix} 0 \\ \mathbf{B} + \mathbf{H} \end{vmatrix} - (\partial_{(0)} + \mathbf{s}\mathbf{p}_0 - \alpha^k \mathbf{s}\mathbf{v}_k) \frac{1}{2} \begin{vmatrix} 0 \\ \mathbf{D} + \mathbf{E} \end{vmatrix}$$

$$+ (\beta^k \partial_{(k)} - \mathbf{s}\mathbf{v}_0 + \beta^k \mathbf{s}\mathbf{p}_k) \frac{1}{2} \begin{vmatrix} 0 \\ \mathbf{B} - \mathbf{H} \end{vmatrix} - (\partial_{(0)} + \mathbf{s}\mathbf{p}_0 + \beta^k \mathbf{s}\mathbf{v}_k) \frac{1}{2} \begin{vmatrix} 0 \\ \mathbf{D} - \mathbf{E} \end{vmatrix} = \frac{1}{\epsilon_0} \begin{vmatrix} 0 \\ \mathbf{j} \end{vmatrix} . \qquad (2.14.8)$$

By allowing for identities

$$\alpha^k \partial_{(k)} \begin{vmatrix} 0 \\ \mathbf{D} + \mathbf{E} \end{vmatrix} = \begin{vmatrix} \partial_{(k)} (D_k + E_k) \\ \partial_{(2)} (D_3 + E_3) - \partial_{(3)} (D_2 + E_2) \\ \partial_{(3)} (D_1 + E_1) - \partial_{(1)} (D_3 + E_3) \\ \partial_{(1)} (D_2 + E_2) - \partial_{(2)} (D_1 + E_1) \end{vmatrix} ,$$

$$\beta^k \, \partial_{(k)} \left| \begin{matrix} 0 \\ \mathbf{D} - \mathbf{E} \end{matrix} \right| = \left| \begin{matrix} \partial_{(k)}(D_k - E_k) \\ -\partial_{(2)}(D_3 - E_3) + \partial_{(3)}(D_2 - E_2) \\ -\partial_{(3)}(D_1 - E_1) + \partial_{(1)}(D_3 - E_3) \\ -\partial_{(1)}(D_2 - E_2) + \partial_{(2)}(D_1 - E_1) \end{matrix} \right|, \qquad (2.14.9)$$

$$\mathbf{sv}_0 \left| \begin{matrix} 0 \\ \mathbf{D} + \mathbf{E} \end{matrix} \right| = \left| \begin{matrix} 0 \\ v_{20}(D_3 + E_3) - v_{30}(D_2 + E_2) \\ v_{30}(D_1 + E_1) - v_{10}(D_3 + E_3) \\ v_{10}(D_2 + E_2) - v_{20}(D_1 + E_1) \end{matrix} \right|,$$

$$-\mathbf{sv}_0 \left| \begin{matrix} 0 \\ \mathbf{D} + \mathbf{E} \end{matrix} \right| = - \left| \begin{matrix} 0 \\ v_{20}(D_3 - E_3) - v_{30}(D_2 - E_2) \\ v_{30}(D_1 + E_1) - v_{10}(D_3 - E_3) \\ v_{10}(D_2 - E_2) - v_{20}(D_1 - E_1) \end{matrix} \right|, \qquad (2.14.10)$$

and

$$\alpha^k \, \mathbf{sp}_k = \alpha^1 (S^1 p_{11} + S^2 p_{21} + S^3 p_{31})$$

$$+\alpha^2 (S^1 p_{12} + S^2 p_{22} + S^3 p_{32})$$

$$+\alpha^3 (S^1 p_{13} + S^2 p_{23} + S^3 p_{33})$$

$$= \left| \begin{matrix} 0 & p_{32} - p_{23} & -p_{31} + p_{13} & p_{21} - p_{12} \\ 0 & -p_{22} - p_{33} & p_{12} & p_{13} \\ 0 & p_{21} & -p_{11} - p_{33} & p_{23} \\ 0 & p_{31} & p_{32} & -p_{11} - p_{22} \end{matrix} \right|,$$

$$\alpha^k \, \mathbf{sp}_k \left| \begin{matrix} 0 \\ \mathbf{D} + \mathbf{E} \end{matrix} \right|$$

$$= \left| \begin{matrix} -(p_{23} - p_{32})(D_1 + E_1) - (p_{31} - p_{13})(D_2 + E_2) - (p_{12} - p_{21})(D_3 + E_3) \\ -(p_{22} + p_{33})(D_1 + E_1) + p_{12}(D_2 + E_2) + p_{13}(D_3 + E_3) \\ p_{21}(D_1 + E_1) - (p_{11} + p_{33})(D_2 + E_2) + p_{23}(D_3 + E_3) \\ p_{31}(D_1 + E_1) + p_{32}(D_2 + E_2) - (p_{11} + p_{22})(D_3 + E_3) \end{matrix} \right|,$$

$$\beta^k \, \mathbf{sp}_k$$

$$= \beta^1 (S^1 p_{11} + S^2 p_{21} + S^3 p_{31})$$

$$+\beta^2 (S^1 p_{12} + S^2 p_{22} + S^3 p_{32})$$

$$+\beta^3 (S^1 p_{13} + S^2 p_{23} + S^3 p_{33})$$

$$= \left| \begin{matrix} 0 & p_{32} - p_{23} & -p_{31} + p_{13} & p_{21} - p_{12} \\ 0 & p_{22} + p_{33} & -p_{12} & -p_{13} \\ 0 & -p_{21} & p_{11} + p_{33} & -p_{23} \\ 0 & -p_{31} & -p_{32} & p_{11} + p_{22} \end{matrix} \right|,$$

$$\beta^k \, \mathbf{sp}_k \left| \begin{matrix} 0 \\ \mathbf{D} - \mathbf{E} \end{matrix} \right|$$

$$= \begin{vmatrix} -(p_{23} - p_{32})(D_1 - E_1) - (p_{31} - p_{13})(D_2 - E_2) - (p_{12} - p_{21})(D_3 - E_3) \\ (p_{22} + p_{33})(D_1 - E_1) - p_{12}(D_2 - E_2) - p_{13}(D_3 - E_3) \\ -p_{21}(D_1 - E_1) + (p_{11} + p_{33})(D_2 - E_2) - p_{23}(D_3 - E_3) \\ -p_{31}(D_1 - E_1) - p_{32}(D_2 - E_2) + (p_{11} + p_{22})(D_3 - E_3) \end{vmatrix} ,$$

$$(2.14.11)$$

Equation (2.12.6) (we will specify only 0-, 1- equations)

$$\partial_{(k)}(D_k + E_k) + \partial_{(k)}(D_k - E_k)$$

$$-(p_{23} - p_{32})(D_1 + E_1) - (p_{31} - p_{13})(D_2 + E_2) - (p_{12} - p_{21})(D_3 + E_3)$$

$$-(p_{23} - p_{32})(D_1 - E_1) - (p_{31} - p_{13})(D_2 - E_2) - (p_{12} - p_{21})(D_3 - E_3)$$

$$+(v_{23} - v_{32})(B_1 + H_1) + (v_{31} - v_{13})(B_2 + H_2) + (v_{12} - v_{21})(B_3 + H_3)$$

$$-(v_{23} - v_{32})(B_1 - H_1) - (v_{31} - v_{13})(B_2 - H_2) - (v_{12} - v_{21})(B_3 - H_3) = \frac{1}{\epsilon_0} 2\rho ,$$

or (recall Eq. (2.14.3))

$$\partial_{(k)} D_k - (p_{23} - p_{32})D_1 - (p_{31} - p_{13})D_2 - (p_{12} - p_{21})D_3$$

$$+(v_{23} - v_{32})\frac{H_1}{c} + (v_{31} - v_{13})\frac{H_2}{c} + (v_{12} - v_{21})\frac{H_3}{c} = \rho ;$$

$$(2.14.12)$$

and in the same manner

$$\partial_{(2)}(D_3 + E_3) - \partial_{(3)}(D_2 + E_2) - \partial_{(2)}(D_3 - E_3) + \partial_{(3)}(D_2 - E_2)$$

$$+v_{20}(D_3 + E_3) - v_{30}(D_2 + E_2) - v_{20}(D_3 - E_3) + v_{30}(D_2 - E_2)$$

$$-(p_{22} + p_{33})(D_1 + E_1) + p_{12}(D_2 + E_2) + p_{13}(D_3 + E_3)$$

$$+(p_{22} + p_{33})(D_1 - E_1) - p_{12}(D_2 - E_2) - p_{13}(D_3 - E_3)$$

$$+p_{20}(B_3 + H_3) - p_{30}(B_2 + H_2) + p_{20}(B_3 - H_3) - p_{30}(B_2 - H_2)$$

$$+(v_{22} + v_{33})(B_1 + H_1) - v_{12}(B_2 + H_2) - v_{13}(B_3 + H_3)$$

$$+(v_{22} + v_{33})(B_1 - H_1) - v_{12}(B_2 - H_2) - v_{13}(B_3 - H_3) = 0 ,$$

or (recall Eq. (2.14.3))

$$\partial_{(2)} E_3 - \partial_{(3)} E_2 + v_{20}E_3 - v_{30}E_2 - (p_{22} + p_{33})E_1 + p_{12}E_2 + p_{13}E_3$$

$$+p_{20}cB_3 - p_{30}cB_2 + (v_{22} + v_{33})cB_1 - v_{12}cB_2 - v_{13}cB_3 = 0 . \qquad (2.14.13)$$

Now, let us consider two equations from Eq. (2.14.8). The first is

$$\partial_{(k)}(B_k + H_k) + \partial_{(k)}(B_k - H_k)$$

$$-(p_{23} - p_{32})(B_1 + H_1) - (p_{31} - p_{13})(B_2 + H_2) - (p_{12} - p_{21})(B_3 + H_3)$$

$$-(p_{23} - p_{32})(B_1 - H_1) - (p_{31} - p_{13})(B_2 - H_2) - (p_{12} - p_{21})(B_3 - H_3)$$

$$-(v_{23} - v_{32})(D_1 + E_1) - (v_{31} - v_{13})(D_2 + E_2) - (v_{12} - v_{21})(D_3 + E_3)$$

$$+(v_{23} - v_{32})(D_1 - E_1) + (v_{31} - v_{13})(D_2 - E_2) + (v_{12} - v_{21})(D_3 - E_3) = 0 \;,$$

or (recall Eq. (2.14.3))

$$\partial_{(k)} cB_k - (p_{23} - p_{32}) cB_1 - (p_{31} - p_{13}) cB_2 - (p_{12} - p_{21}) cB_3$$

$$-(v_{23} - v_{32}) E_1 - (v_{31} - v_{13}) E_2 - (v_{12} - v_{21}) E_3 = 0 \;; \tag{2.14.14}$$

and in the same manner

$$\partial_{(2)}(B_3 + H_3) - \partial_{(3)}(B_2 + H_2) - \partial_{(2)}(B_3 - H_3) + \partial_{(3)}(B_2 - H_2)$$

$$+v_{20}(B_3 + H_3) - v_{30}(B_2 + H_2) - v_{20}(B_3 - H_3) + v_{30}(B_2 - H_2)$$

$$-(p_{22} + p_{33})(B_1 + H_1) + p_{12}(B_2 + H_2) + p_{13}(B_3 + H_3)$$

$$+(p_{22} + p_{33})(B_1 - H_1) - p_{12}(B_2 - H_2) - p_{13}(B_3 - H_3)$$

$$-p_{20}(D_3 + E_3) + p_{30}(D_2 + E_2) - p_{20}(D_3 - E_3) + p_{30}(D_2 - E_2)$$

$$-(v_{22} + v_{33})(D_1 + E_1) + v_{12}(D_2 + E_2) + v_{13}(D_3 + E_3)$$

$$-(v_{22} + v_{33})(D_1 - E_1) + v_{12}(D_2 - E_2) + v_{13}(D_3 - E_3) = \frac{1}{\epsilon_0} 2j^1 \;,$$

or (recall Eq. (2.14.3))

$$\partial_{(2)} \frac{H_3}{c} - \partial_{(3)} \frac{H_2}{c} + v_{20} \frac{H_3}{c}$$

$$-v_{30} \frac{H_2}{c} - (p_{22} + p_{33}) \frac{H_1}{c} + p_{12} \frac{H_2}{c} + p_{13} \frac{H_3}{c}$$

$$-p_{20} D_3 + p_{30} D_2 - (v_{22} + v_{33}) D_1 + v_{12} D_2 + v_{13} D_3 = j^1 \;.$$

$$\tag{2.14.15}$$

Evidently, these equations (and their cyclic counterparts) are equivalent to tensor generally covariant Maxwell equations

$$\nabla^\alpha F^{\beta\gamma} + \nabla^\beta F^{\gamma\alpha} + \nabla^\gamma F^{\alpha\beta} = 0 \;, \qquad \nabla_\beta H^{\beta\alpha} = j^\alpha \tag{2.14.16}$$

in tetrad representation (see Section **2.13**)

$$(F_{(2)(3)}, F_{(3)(1)}, F_{(1)(2)}) = (cB_{(i)}) \;, \qquad (F_{(0)(1)}, F_{(0)(2)}, F_{(0)(3)} = (E_{(i)}) \;,$$

$$(H_{(2)(3)}, H_{(3)(1)}, H_{(1)(2)}) = (\frac{H_{(i)}}{c}) \;, \qquad (H_{(0)(1)}, H_{(0)(2)}, H_{(0)(3)}) = (cD_{(i)}) \;,$$

$$E_k = E_{(k)} = \mathbf{E} \;, \qquad B_k = -B_{(k)} = \mathbf{B} \;,$$

$$D_k = D_{(k)} = \mathbf{D} \;, \qquad H_k = -H_{(k)} = \mathbf{H} \;. \tag{2.14.17}$$

Chapter 3

Dirac–Kähler Field Theory, A 2-Potential Approach to Electrodynamics

3.1 Introduction

In the context of studying the Maxwell equations as equations that involve a massless field with spin one and definite intrinsic parity; the well-known field of Dirac–Kähler [248] is of primary importance. Our main goal for this paper is to investigate this possibility in the context of the Maxwell theory; and, in particular, its importance when treating a 2-potential approach explored by Cabibbo-Ferrari [249]. Also, others have dealt with this concept in Vinciarelli [252], Hagen [250], Zwanzinger [251], Mignani and Recami [253], Strazhev and Kruglov [254], Kresin and Strazhev [255], Barker and Graziani [256, 257], Strazhev and Pletyukhov [258], Olive [259], Singleton [260–263], Berkovits [264], Galvão and Mignaco [265], Ferreira [266–268]. Dirac–Kähler field obeys a simple wave equation in spinor basis, this field theory is reduced to more simple theories for scalar or vector particles with different intrinsic parities when adding linear constraints on field variables; for details and references see in [270], Kruglov [269], Red'kov [40].

3.2 Dirac–Kähler Composite Boson, Wave Equation

In the Minkowski space, the Dirac–Kähler particle is described by 16-component wave function $U(x)$, a 2-bispinor, or by equivalent set of tensor fields: $\{\Phi(x), \Phi_i(x), \tilde{\Phi}(x), \tilde{\Phi}_i(x), \Phi_{mn}(x)\}$; $\Phi(x)$ is a scalar, $\Phi_i(x)$ is a vector, $\tilde{\Phi}(x)$ represents a pseudo-scalar, $\tilde{\Phi}_i(x)$ represents a pseudo-vector, $\Phi_{mn}(x)$ is an antisymmetric tensor. Connection between them is given by the formula

$$U = \left(-i\Phi + \gamma^l \Phi_l + i\sigma^{mn}\Phi_{mn} + \gamma^5\tilde{\Phi} + i\,\gamma^l\gamma^5\tilde{\Phi}_l \right) E^{-1} , \qquad (3.2.1)$$

where

$$\gamma^5 = -i\gamma^0\gamma^1\gamma^2\gamma^3 , \qquad \sigma^{ab} = \frac{1}{4}(\gamma^a\gamma^b - \gamma^b\gamma^a) ,$$

$$E^{-1} = \begin{vmatrix} \epsilon^{-1} & 0 \\ 0 & \dot{\epsilon} \end{vmatrix} = \begin{vmatrix} \epsilon^{\alpha\beta} & 0 \\ 0 & \epsilon_{\dot{\alpha}\dot{\beta}} \end{vmatrix} = \begin{vmatrix} -i\sigma^2 & 0 \\ 0 & +i\sigma^2 \end{vmatrix}. \tag{3.2.2}$$

Inverse formulas to Eq. (3.2.1) have the form

$$\Phi_l = \frac{1}{4} \, \mathrm{Sp} \, (E\gamma_l U) \,, \quad \tilde{\Phi}_l = \frac{1}{4i} \, \mathrm{Sp} \, (E\gamma^5\gamma_l U) \,,$$

$$\Phi = \frac{i}{4} \, \mathrm{Sp} \, (EU) \,, \quad \tilde{\Phi} = \frac{1}{4} \, \mathrm{Sp} \, (E\gamma^5 U) \,, \quad \Phi_{mn} = -\frac{1}{2i} \, \mathrm{Sp} \, (E\sigma_{mn} U) \,. \tag{3.2.3}$$

Next, we will use the Weyl basis (Pauli matrices are given as $\sigma^a = (I, \sigma^k)$, $\bar{\sigma}^a = (I, -\sigma^k)$)

$$U = \begin{vmatrix} \xi^{\alpha\beta} & \Delta^{\alpha}{}_{\dot{\beta}} \\ H_{\dot{\alpha}}{}^{\beta} & \eta_{\dot{\alpha}\dot{\beta}} \end{vmatrix} \,, \quad \gamma^a = \begin{vmatrix} 0 & \bar{\sigma}^a \\ \sigma^a & 0 \end{vmatrix} \,,$$

$$\Delta = (\Phi_l + i\tilde{\Phi}_l)\sigma^l\dot{\epsilon} \,, \quad H = (\Phi_l - i\tilde{\Phi}_l)\bar{\sigma}^l\epsilon^{-1} \,,$$

$$\xi = (-i\Phi - \tilde{\Phi} + i\Sigma^{mn}\Phi_{mn})\epsilon^{-1}, \quad \eta = (-i\Phi + \tilde{\Phi} + i\bar{\Sigma}^{mn}\Phi_{mn})\dot{\epsilon} \,; \tag{3.2.4}$$

and inverse relations are

$$\Phi_l + i\tilde{\Phi}_l = \frac{1}{2} \, \mathrm{Sp} \, (\dot{\epsilon}^{-1}\sigma_l\Delta) \,, \quad \Phi_l - i \, \tilde{\Phi}_l = \frac{1}{2} \, \mathrm{Sp} \, (\epsilon\bar{\sigma}_l H) \,,$$

$$-i\Phi - \tilde{\Phi} = \frac{1}{2} \, \mathrm{Sp} \, (\epsilon\xi) \,, \quad -i\Phi + \tilde{\Phi} = \frac{1}{2} \, \mathrm{Sp} \, (\dot{\epsilon}^{-1}\xi) \,,$$

$$-i\Phi^{kl} + \frac{1}{2}\epsilon^{klmn}\Phi_{mn} = \mathrm{Sp} \, (\epsilon\Sigma^{kl}\xi) \,,$$

$$-i\Phi^{kl} - \frac{1}{2}\epsilon^{klmn}\Phi_{mn} = \mathrm{Sp} \, (\dot{\epsilon}^{-1}\bar{\Sigma}^{kl}\xi) \,. \tag{3.2.5}$$

For the spinor basis, the Dirac–Kähler equation is a Dirac-like equation for 2-bispinor

$$(i\gamma^a \frac{\partial}{\partial x^a} - m) \, U(x) = 0 \,, \tag{3.2.6}$$

or in 2-spinor form

$$(A) \quad i\sigma^a \, \partial_a \, \xi(x) = m \, H(x) \,, \qquad (A') \quad i\bar{\sigma}^a \, \partial_a \, H(x) = m \, \xi(x) \,,$$

$$(B) \quad i\bar{\sigma}^a \, \partial_a \, \eta(x) = m \, \Delta(x) \,, \qquad (B') \quad i\sigma^a \, \partial_a \, \Delta(x) = m \, \eta(x) \,.$$

$$\tag{3.2.7}$$

Equations (3.2.6) and (3.32.7) are invariant under the Lorentz group.

3.3   Equations for Particles with Different Parities

Let us consider Eqs. (3.2.7) with four types of additional constraints:

$$\underline{S=0}, \qquad \tilde{\Phi}=0, \qquad \tilde{\Phi}_\alpha=0, \qquad \Phi_{\alpha\beta}=0,$$

$$\partial^l \Phi_l + m\Phi = 0, \qquad \partial_l \Phi - m\,\Phi_l = 0, \qquad \partial^d \Phi^k - \partial^k \Phi^d = 0;$$

$$\underline{S=\tilde{0}}, \qquad \Phi=0, \qquad \Phi_\alpha=0, \qquad \Phi_{\alpha\beta}=0;$$

$$\partial^l \tilde{\Phi}_l + m\,\tilde{\Phi} = 0, \qquad \partial^k \tilde{\Phi} - m\tilde{\Phi}^k = 0, \qquad \epsilon^{dkcl}\partial_c \tilde{\Phi}_l = 0;$$

$$\underline{S=1}, \qquad \Phi=0, \qquad \tilde{\Phi}=0, \qquad \tilde{\Phi}_l=0,$$

$$\partial^l \Phi_l = 0,\; \partial^l \Phi_{kl} - m\,\Phi_k = 0,\; \epsilon^{kcmn}\partial_c \Phi_{mn} = 0,\; \partial_d \Phi_k - \partial_k \Phi_d - m\Phi_{dk} = 0;$$

$$\underline{S=\tilde{1}}, \qquad \Phi=0, \qquad \tilde{\Phi}=0, \qquad \Phi_l=0,$$

$$\partial^l \tilde{\Phi}_l = 0,\; \partial^l \Phi_{kl} = 0,\; \frac{1}{2}\epsilon^{kcmn}\partial_c \Phi_{mn} + m\tilde{\Phi}^k = 0,\; \epsilon^{dkcl}\partial_c \tilde{\Phi}_l - m\Phi^{dk} = 0.$$

$$(3.3.1)$$

Let us describe some additional constraints in spinor form. For a scalar particle, we get

$$\underline{S=0} \quad \left|\begin{array}{cc} \xi & \Delta \\ H & \eta \end{array}\right| = \left|\begin{array}{cc} -\Phi\sigma^2 & +i\Phi_l\bar{\sigma}^l\sigma^2 \\ -i\Phi_l\sigma^l\sigma^2 & +\Phi\sigma^2 \end{array}\right|,$$

$$\Delta^{tr} = +H, \quad \xi = -\eta, \quad \xi^{tr} = -\xi, \quad \eta^{tr} = -\eta, \qquad (3.3.2)$$

the symbol tr stands for the matrix transposition. For a pseudo-scalar particle, we get

$$\underline{S=\tilde{0}} \quad \left|\begin{array}{cc} \xi & \Delta \\ H & \eta \end{array}\right| = \left|\begin{array}{cc} +i\tilde{\Phi}\sigma^2 & -\tilde{\Phi}_l\,\bar{\sigma}^l\sigma^2 \\ -\tilde{\Phi}_l\sigma^l\sigma^2 & +i\tilde{\Phi}\sigma^2 \end{array}\right|,$$

$$\Delta^{tr} = -H, \quad \xi = +\eta, \quad \xi^{tr} = -\xi, \quad \eta^{tr} = -\eta. \qquad (3.3.3)$$

For a vector particle, we will have

$$\underline{S=1} \quad \left|\begin{array}{cc} \xi & \Delta \\ H & \eta \end{array}\right| = \left|\begin{array}{cc} +\Sigma^{mn}\sigma^2\Phi_{mn} & +i\bar{\sigma}^l\sigma^2\Phi_l \\ -i\sigma^l\sigma^2\Phi_l & -\bar{\Sigma}^{mn}\sigma^2\Phi_{mn} \end{array}\right|,$$

$$\Delta^{tr} = +H, \quad \xi^{tr} = +\xi, \quad \eta^{tr} = +\eta. \qquad (3.3.4)$$

Here, each of symmetric spinors ξ and η depends on three independent variables:

$$\xi + \eta = -2i\,(\,\sigma^1\Phi_{23} + \sigma^2\Phi_{31} + \sigma^3\Phi_{12}\,)\,\sigma^2,$$

$$\xi - \eta = 2\,(\,\sigma^1\Phi_{01} + \sigma^2\Phi_{02} + \sigma^3\Phi_{03}\,)\,\sigma^2.$$

Finally, a pseudo-vector case is given by

$$\underline{S=\tilde{1}} \quad \left|\begin{array}{cc} \xi & \Delta \\ H & \eta \end{array}\right| = \left|\begin{array}{cc} +\Sigma^{mn}\sigma^2\Phi_{mn} & -\bar{\sigma}^l\sigma^2\tilde{\Phi}_l \\ -\sigma^l\sigma^2\tilde{\Phi}_l & -\bar{\Sigma}^{mn}\sigma^2\Phi_{mn} \end{array}\right|,$$

$$\Delta^{tr} = -H, \qquad \xi^{tr} = +\xi, \qquad \eta^{tr} = +\eta. \qquad (3.3.5)$$

3.4   Massless Vector Particle

Bearing in mind the usual photon field, it is better to use the conventional notation:

$$S = 1 \qquad \Phi_l \Longrightarrow A_l, \qquad \Phi_{kl} \Longrightarrow F_{kl} \, ,$$

$$U(x) = (+\gamma^l \, A_l + i \, \sigma^{mn} \, F_{mn} \,) \, E^{-1} \, .$$

The corresponding wave equation in spinor form reads

$$(A) \qquad i\sigma^a \, \partial_a \, \xi = m \, H \, , \qquad (A') \qquad i\bar{\sigma}^a \, \partial_a \, H = m \, \xi \, ,$$

$$(B) \qquad i\bar{\sigma}^a \, \partial_a \, \eta = m \, \Delta \, , \qquad (B') \qquad i\sigma^a \, \partial_a \, \Delta = m \, \eta \, . \qquad (3.4.1)$$

Restriction to a massless vector particle is achieved as follows

$$\begin{array}{l} (A) \\ (B) \end{array} \qquad \begin{array}{l} i\sigma^a \, \partial_a \, \xi = 0 \\ i\bar{\sigma}^a \, \partial_a \, \eta = 0 \end{array} \quad \Longleftrightarrow \quad \partial^l F_{kl} = 0 \, , \ \epsilon^{kcmn} \partial_c F_{mn} = 0 \, ; \qquad (3.4.2)$$

$$\begin{array}{l} (A') \\ (B') \end{array} \qquad \begin{array}{l} i\bar{\sigma}^a \, \partial_a \, H = \xi \\ i\sigma^a \, \partial_a \, \Delta = \eta \end{array} \quad \Longrightarrow \quad \partial^l A_l = 0 \, , \ \partial_k A_l - \partial_l A_k = F_{kl} \, . \qquad (3.4.3)$$

In a 3-vector notation the equations appear as (with $F_{01} = -E_1 = +E^1$, $F_{23} = B_1 = B^1$, etc.)

$$\underline{(A), (B)} \qquad \text{div } \mathbf{E} = 0 \, , \qquad \text{div } \mathbf{B} = 0 \, ,$$

$$\text{rot } \mathbf{E} = - \frac{\partial}{\partial t} \mathbf{B} \, , \qquad \text{rot } \mathbf{B} = + \frac{\partial}{\partial t} \mathbf{E} \, ; \qquad (3.4.4)$$

$$\underline{(A'), (B')} \qquad \frac{\partial A^0}{\partial t} + \text{div } \mathbf{A} = 0 \, , \ -\frac{\partial \mathbf{A}}{\partial t} - \text{grad } A^0 = \mathbf{E} \, , \ \text{rot } \mathbf{A} = \mathbf{B} \, . \qquad (3.4.5)$$

3.5   Massless Pseudo-Vector Particle

A pseudo-vector particle is specified by the relations:

$$S = \tilde{1} \qquad \tilde{\Phi}_l \Longrightarrow \tilde{A}_l, \qquad \Phi_{kl} \Longrightarrow \tilde{F}_{kl} \, ,$$

$$U(x) = (+i \, \sigma^{mn} \, \tilde{F}_{mn} + i \, \gamma^l \gamma^5 \, \tilde{A}_l \,) \, E^{-1} \, ;$$

$$(\tilde{A}) \qquad i\sigma^a \, \partial_a \, \xi = m \, H \, , \qquad (\tilde{A}') \qquad i\bar{\sigma}^a \, \partial_a \, H = m \, \xi \, ,$$

$$(\tilde{B}) \qquad i\bar{\sigma}^a \, \partial_a \, \eta = m \, \Delta \, , \qquad (\tilde{B}') \qquad i\sigma^a \, \partial_a \, \Delta = m \, \eta \, ;$$

$$0 = 0 \, , \qquad \partial^l \tilde{A}_l = 0 \, , \qquad \partial^l \tilde{F}_{kl} = 0 \, ,$$

$$\frac{1}{2} \, \epsilon^{kcmn} \, \partial_c \tilde{F}_{mn} + m \, \tilde{A}^k = 0 \, , \ \epsilon^{dkcl} \, \partial_c \tilde{A}_l - m \, \tilde{F}^{dk} = 0 \, . \qquad (3.5.1)$$

Transition to a massless case is realized as follows:

$$\begin{array}{l} (\tilde{A}) \\ (\tilde{B}) \end{array} \qquad \begin{array}{l} i\sigma^a \, \partial_a \, \xi = 0 \\ i\bar{\sigma}^a \, \partial_a \, \eta = 0 \end{array} \quad \Longleftrightarrow \quad \partial^l \tilde{F}_{kl} = 0 \, , \ \frac{1}{2} \, \epsilon^{kcmn} \, \partial_c \tilde{F}_{mn} = 0 \, ; \qquad (3.5.2)$$

$$
\begin{array}{ll}
(\tilde{A}') & i\bar{\sigma}^a\,\partial_a\,H = \xi \\
(\tilde{B}') & i\sigma^a\,\partial_a\,\Delta = \eta
\end{array}
\qquad\Longleftrightarrow\qquad \partial^l\tilde{A}_l = 0\,,\ \ \epsilon^{dkcl}\,\partial_c\tilde{A}_l = \tilde{F}^{dk}\,.
\qquad (3.5.3)
$$

Equations (3.5.2) coincide with Eqs. (3.4.2), so we will investigate, additionally here, only Eqs. (3.5.3) found in vector form

$$
(\tilde{A}'),(\tilde{B}') \qquad \frac{\partial \tilde{A}^0}{\partial t} + \operatorname{div}\tilde{\mathbf{A}} = 0\,,\ \ \operatorname{rot}\tilde{\mathbf{A}} = +\tilde{\mathbf{E}}\,,\ \ \tilde{\mathbf{B}} = -\frac{\partial \tilde{\mathbf{A}}}{\partial t} - \operatorname{grad}\tilde{A}^0\,. \qquad (3.5.4)
$$

3.6   Comparison of the Vector and Pseudo-Vector Fields

Let us now collect the results obtained thus far. Tensor equations for the $S = 1$ field are given as

$$
\begin{array}{lll}
(\alpha) & \partial^l F_{kl} = 0\,, & \epsilon^{kcmn}\,\partial_c F_{mn} = 0\,; \\
(\beta) & 0 = 0\,, \quad \partial^a A_a = 0\,, & \partial_b A_c - \partial_c A_b = F_{bc}\,,
\end{array}
\qquad (3.6.1)
$$

and in 3-vector form as

$$
\begin{array}{lll}
(\alpha) & \operatorname{div}\mathbf{E} = 0\,, & \operatorname{div}\mathbf{B} = 0\,, \\
 & \operatorname{rot}\mathbf{E} = -\,\partial_t\mathbf{B}\,, & \operatorname{rot}\mathbf{B} = +\,\partial_t\mathbf{E}\,, \\
(\beta) & 0 = 0\,, & \partial_t A^0 + \operatorname{div}\mathbf{A} = 0\,, \\
 & -\partial_t\mathbf{A} - \operatorname{grad}A^0 = \mathbf{E}\,, & \operatorname{rot}\mathbf{A} = +\mathbf{B}\,.
\end{array}
\qquad (3.6.2)
$$

Tensor equations for a $S = \tilde{1}$ field have the form

$$
\begin{array}{lll}
(\tilde{\alpha}) & \partial^l \tilde{F}_{kl} = 0\,, & \epsilon^{kcmn}\,\partial_c \tilde{F}_{mn} = 0\,; \\
(\tilde{\beta}) & 0 = 0\,, \quad \partial^l \tilde{A}_l = 0\,, & \epsilon_{dk}{}^{cl}\,\partial_c \tilde{A}_l = \tilde{F}_{dk}\,,
\end{array}
\qquad (3.6.3)
$$

and in 3-vector form

$$
\begin{array}{lll}
(\tilde{\alpha}) & \operatorname{div}\tilde{\mathbf{E}} = 0\,, & \operatorname{div}\tilde{\mathbf{B}} = 0\,, \\
 & \operatorname{rot}\tilde{\mathbf{E}} = -\,\partial_t\tilde{\mathbf{B}}\,, & \operatorname{rot}\tilde{\mathbf{B}} = +\,\partial_t\tilde{\mathbf{E}}\,, \\
(\tilde{\beta}) & 0 = 0\,, & \partial_t\tilde{A}^0 + \operatorname{div}\tilde{\mathbf{A}} = 0\,, \\
 & -\partial_t\tilde{\mathbf{A}} - \operatorname{grad}\tilde{A}^0 = \tilde{\mathbf{B}}\,, & \operatorname{rot}\tilde{\mathbf{A}} = +\tilde{\mathbf{E}}\,.
\end{array}
\qquad (3.6.4)
$$

3.7   Photons with Different Parities

For the case $S = 1$, spinor massless equations with sources can be obtained from the equations (A),(B) in Eq. (3.4.1)

$$
S = 1 \qquad
\begin{array}{ll}
(A) & i\sigma^a\,\partial_a\,\xi = m\,H\,, \\
(B) & i\bar{\sigma}^a\,\partial_a\,\eta = m\,\Delta
\end{array}
$$

by the formal changes

$$mH = - i\sigma^l\sigma^2\, mA_l\,, \qquad \Longrightarrow \qquad j = i\,\sigma^k\sigma^2\, j_k\,,$$

$$m\Delta = + i\bar{\sigma}^l\sigma^2\, mA_l\,, \qquad \Longrightarrow \qquad \bar{j} = -i\,\bar{\sigma}^k\sigma^2\, j_k\,.$$

Thus, we get

$$(S = 1) \qquad \begin{array}{ll} (A) & i\sigma^a\,\partial_a\,\xi = -\,j\,, \\[4pt] (B) & i\bar{\sigma}^a\,\partial_a\,\eta = -\,\bar{j}\,, \end{array}$$

$$\partial^l F_{kl} = -\,j_k\,, \qquad\qquad \epsilon^{kcmn}\partial_c F_{mn} = 0\,;$$

$$\operatorname{div}\mathbf{E} = +j^0\,, \qquad\qquad \operatorname{div}\mathbf{B} = 0\,, \qquad\qquad (3.7.1)$$

$$\operatorname{rot}\mathbf{E} = -\,\partial_t\mathbf{B}\,, \qquad\qquad \operatorname{rot}\mathbf{B} = +\,\partial_t\mathbf{E} + \mathbf{j}\,,$$

and remaining equation (A'), (B') read

$$(S = 1) \qquad \left.\begin{array}{ll} (A') & i\bar{\sigma}^a\,\partial_a\,H = \xi \\[4pt] (B') & i\sigma^a\,\partial_a\,\Delta = \eta \end{array}\right\} \qquad \Longrightarrow$$

$$\partial^l A_l = 0\,, \qquad \partial_k A_l - \partial_l A_k = F_{kl}\,,$$

$$0 = 0\,, \qquad \frac{\partial A^0}{\partial t} + \operatorname{div}\mathbf{A} = 0\,,$$

$$-\frac{\partial\mathbf{A}}{\partial t} - \operatorname{grad} A^0 = \mathbf{E}\,, \qquad \operatorname{rot}\mathbf{A} = \mathbf{B}\,. \qquad\qquad (3.7.2)$$

In the same manner, one should consider the case with opposite parity $S = \tilde{1}$. Performing in spinor equations

$$S = \tilde{1}: \qquad \begin{array}{ll} (\tilde{A}) & i\sigma^a\,\partial_a\,\xi = m\,H\,, \\[4pt] (\tilde{B}) & i\bar{\sigma}^a\,\partial_a\,\eta = m\,\Delta \end{array}$$

the formal changes

$$mH(x) = -\,\sigma^k\sigma^2\, m\tilde{A}_k \qquad \Longrightarrow \qquad j = \sigma^k\sigma^2\, \bar{\tilde{j}}_k\,,$$

$$m\Delta = -\,\bar{\sigma}^k\sigma^2\, m\tilde{A}_k \qquad \Longrightarrow \qquad \bar{\tilde{j}} = \bar{\sigma}^k\sigma^2\, \tilde{j}_k\,,$$

we obtain

$$\begin{array}{ll} (\tilde{A}) & i\sigma^a\,\partial_a\,\xi = -\,\tilde{j}\,, \\[4pt] (\tilde{B}) & i\bar{\sigma}^a\,\partial_a\,\eta = -\,\bar{\tilde{j}}\,, \end{array}$$

$$(S = \tilde{1}) \qquad \partial^l \tilde{F}_{kl} = 0\,, \qquad\qquad \tfrac{1}{2}\,\epsilon^{kcmn}\,\partial_c\tilde{F}_{mn} = +\tilde{j}^k\,,$$

$$\operatorname{div}\tilde{\mathbf{E}} = 0\,, \qquad\qquad \operatorname{div}\tilde{\mathbf{B}} = -\tilde{j}^0\,, \qquad\qquad (3.7.3)$$

$$\operatorname{rot}\tilde{\mathbf{E}} = -\,\partial_t\tilde{\mathbf{B}} + \tilde{\mathbf{j}}\,, \qquad\qquad \operatorname{rot}\tilde{\mathbf{B}} = +\,\partial_t\tilde{\mathbf{E}}\,,$$

and remaining equations $(\tilde{A}'), (\tilde{B}')$ are given as

$$(S = \tilde{1}) \qquad \begin{matrix} (\tilde{A}') & i\bar{\sigma}^a \, \partial_a \, H = \xi \\[2mm] (\tilde{B}') & i\sigma^a \, \partial_a \, \Delta = \eta \end{matrix} \left.\vphantom{\begin{matrix}1\\1\end{matrix}}\right\} \qquad \Longleftrightarrow$$

$$0 = 0 \,, \qquad \partial^l \tilde{A}_l = 0 \,, \qquad \epsilon^{dkcl} \, \partial_c \tilde{A}_l = \tilde{F}^{dk} \,,$$

$$0 = 0 \,, \qquad \partial_t \tilde{A}^0 + \text{div } \tilde{\mathbf{A}} = 0 \,,$$

$$-\partial_t \tilde{\mathbf{A}} - \text{grad } \tilde{A}^0 = \tilde{\mathbf{B}} \,, \qquad \text{rot } \tilde{\mathbf{A}} = +\tilde{\mathbf{E}} \,. \qquad (3.7.4)$$

One can notice that a simple algebraic summation of the systems related to $(A), (B)$ and $(\tilde{A}), (\tilde{B})$ in the fields of types $S = 1$ and $S = \tilde{1}$ (see Eqs. (3.7.1).)) and Eq. (3.7.3) lead to the Maxwell-like equations with electric and magnetic charges

$$\mathbf{E} + \tilde{\mathbf{E}} = \hat{\mathbf{E}} \,, \qquad \mathbf{B} + \tilde{\mathbf{B}} = \hat{\mathbf{B}} \,,$$

$$\text{div } \hat{\mathbf{E}} = j^0 \,, \qquad \text{div } \hat{\mathbf{B}} = -\tilde{j}^0 \,,$$

$$\text{rot } \hat{\mathbf{E}} = -\partial_t \hat{\mathbf{B}} + \tilde{\mathbf{j}} \,, \qquad \text{rot } \hat{\mathbf{B}} = +\partial_t \hat{\mathbf{E}} + \mathbf{j} \,. \qquad (3.7.5)$$

Description of this summation in 4-tensor representations is given by these formulas

$$(\hat{\mathbf{E}}, c\hat{\mathbf{B}}), \qquad \hat{F}_{ab} = F_{ab} + \tilde{F}_{ab} = (\partial_a A_b - \partial_b A_a) + \epsilon_{ab}{}^{kl} \partial_k \tilde{A}_l \,,$$

$$\partial^l \, [\, F_{kl} + \tilde{F}_{kl} \,] = - \, j_k(x) \,,$$

$$\frac{1}{2} \, \epsilon^{kcmn} \partial_c \, [\, F_{mn} + \tilde{F}_{mn} \,] = +\tilde{j}^k. \qquad (3.7.6)$$

The last coincide with the 2-potential approach to Maxwell electrodynamics possess two charges.

This method obtains extended Maxwell equations with two charges from two separated Maxwell theories that have different intrinsic parities

$$S = 1 \,, \qquad (A, B) \,, \qquad (A', B') \,;$$

$$S = \tilde{1} \,, \qquad (\tilde{A}, \tilde{B}) \,, \qquad (\tilde{A}', \tilde{B}') \,; \qquad (3.7.7)$$

may be sketched by the scheme

$$(S = 1) + (S = \tilde{1}): \qquad (A, B) + (\tilde{A}, \tilde{B}) \,, \quad A', B', \tilde{A}', \tilde{B}' \,. \qquad (3.7.8)$$

It is possible to introduce an alternating combination

$$(S = 1) - (S = \tilde{1}): \qquad (A, B) + (\tilde{A}, \tilde{B}) \,, \quad A', B', \tilde{A}', \tilde{B}' \,, \qquad (3.7.9)$$

which corresponds to

$$\mathbf{E} - \tilde{\mathbf{E}} = \breve{\mathbf{E}} \,, \qquad \mathbf{B} - \tilde{\mathbf{B}} = \breve{\mathbf{B}} \,,$$

$$\text{div } \breve{\mathbf{E}} = j^0 \,, \qquad \text{div } \breve{\mathbf{B}} = +\tilde{j}^0 \,,$$

$$\text{rot } \breve{\mathbf{E}} = -\partial_t \breve{\mathbf{B}} - \tilde{\mathbf{j}} \,, \qquad \text{rot } \breve{\mathbf{B}} = +\partial_t \breve{\mathbf{E}} + \mathbf{j} \,. \qquad (3.7.10)$$

In the 4-tensor description, the summation has the form

$$(\breve{\mathbf{E}}, c\breve{\mathbf{B}}), \qquad \breve{F}_{ab} = F_{ab} - \tilde{F}_{ab} = (\partial_a A_b - \partial_b A_a) - \epsilon_{ab}{}^{kl}\partial_k \tilde{A}_l \,,$$

$$\partial^l \left[F_{kl} - \tilde{F}_{kl} \right] = -\, j_k(x) \,,$$

$$\frac{1}{2}\,\epsilon^{kcmn}\partial_c \left[F_{mn} - \tilde{F}_{mn} \right] = -\tilde{j}^k \,. \tag{3.7.11}$$

Evidently, the system that consists of two independent Maxwell-like equations, $S = 1$ and $S = \tilde{1}$, is equivalent to the following pair

$$(S = 1) \pm (S = \tilde{1}): \qquad (A, B) \pm (\tilde{A}, \tilde{B}) \,, \quad A', B', \tilde{A}', \tilde{B}' \,. \tag{3.7.12}$$

Let us call the model in Eq. (3.7.12) as the extended Maxwell theory, it includes two photon fields with different parity, and two types of sources (electric and magnetic charges).

It is straightforward that we can extend the proposed approach to a more general Riemannian space-time background.

3.8   Extension of the Maxwell Theory to Riemannian Space-Time

Let us start with a generally covariant tetrad-based Dirac–Kähler equation in 4-spinor form (see Chapter 3)

$$[\, i\gamma^\alpha(x)\,(\partial/\partial x^\alpha + B_\alpha(x)) - m\,]\, U(x) = 0 \,, \tag{3.8.1}$$

where B_α is a 2-bispinor connection ($\Gamma_\alpha(x)$ is a bispinor connection)

$$B_\alpha = \frac{1}{2}J^{ab}e^\beta_{(a)}\nabla_\alpha(e_{(b)\beta}) = \Gamma_\alpha \otimes I + I \otimes \Gamma_\alpha \,,$$

and $J^{ab} = \sigma^{ab} \otimes I + I \otimes \sigma^{ab}$ stands for generators of 2-rank bispinors under the Lorentz group. From Eq. (3.8.1) written in Weyl spinor basis, one derives the following equations in 2-spinor form:

$$i\sigma^\alpha(x)\,[\,\partial/\partial x^\alpha + \Sigma_\alpha \otimes I + I \otimes \Sigma_\alpha\,]\,\zeta(x) = m\,H(x) \,,$$

$$i\bar{\sigma}^\alpha(x)\,[\,\partial/\partial x^\alpha + \bar{\Sigma}_\alpha \otimes I + I \otimes \Sigma_\alpha\,]\,H(x) = m\,\xi(x) \,,$$

$$i\bar{\sigma}^\alpha(x)\,[\,\partial/\partial x^\alpha + \bar{\Sigma}_\alpha \otimes I + I \otimes \bar{\Sigma}_\alpha\,]\,\eta(x) = m\,\Delta(x) \,,$$

$$i\sigma^\alpha(x)\,[\,\partial/\partial x^\alpha + \Sigma_\alpha \otimes I + I \otimes \bar{\Sigma}_\alpha\,]\,\Delta(x) = m\,\eta(x) \,. \tag{3.8.2}$$

Symbols $\Sigma_\alpha(x)$ and $\bar{\Sigma}_\alpha(x)$ are the Infeld–van der Vaerden connections

$$\sigma^\alpha = \sigma^a\,e^\alpha_{(a)} \,, \qquad \bar{\sigma}^\alpha = \bar{\sigma}^a\,e^\alpha_{(a)} \,,$$

$$\Sigma_\alpha = \frac{1}{2}\Sigma^{ab}\,e^\beta_{(a)}\,\nabla_\alpha(e_{(b)\beta}) \,, \qquad \bar{\Sigma}_\alpha = \frac{1}{2}\bar{\Sigma}^{ab}\,e^\beta_{(x)}\,\nabla_\alpha(e_{(b)\beta}) \,,$$

$$\Sigma^{ab} = \frac{1}{4}\,(\,\bar{\sigma}^a\sigma^b - \bar{\sigma}^b\sigma^a\,) \,, \qquad \bar{\Sigma}^{ab} = \frac{1}{4}\,(\,\sigma^a\bar{\sigma}^b - \sigma^b\bar{\sigma}^a\,) \,.$$

These spinor equations are equivalent to a generally covariant tensor system

$$\nabla^\alpha \Psi_\alpha + m\Psi = 0 , \qquad \nabla^\alpha \tilde{\Psi}_\alpha + m\tilde{\Psi} = 0 ,$$

$$\nabla_\alpha \Psi + \nabla^\beta \Psi_{\alpha\beta} - m\Psi_\alpha = 0 ,$$

$$\nabla_\alpha \tilde{\Psi} - \frac{1}{2}\epsilon_\alpha{}^{\beta\rho\sigma}(x)\nabla_\beta \Psi_{\rho\sigma} - m\tilde{\Psi}_\alpha = 0 ,$$

$$\nabla_\alpha \Psi_\beta - \nabla_\beta \Psi_\alpha + \epsilon_{\alpha\beta}{}^{\rho\sigma}(x)\nabla_\rho \tilde{\Psi}_\sigma - m\Psi_{\alpha\beta} = 0 . \qquad (3.8.3)$$

Covariant tensor field variables are connected with local tetrad tensor variables by the relations

$$\Psi_\alpha = e_\alpha^{(i)}\Psi_i , \quad \tilde{\Psi}_\alpha = e_\alpha^{(i)}\tilde{\Psi}_i , \quad \Psi_{\alpha\beta} = e_\alpha^{(m)}e_\beta^{(n)}\Psi_{mn} , \qquad (3.8.4)$$

and the Levi-Civita object is determined as

$$\epsilon^{\alpha\beta\rho\sigma}(x) = \epsilon^{abcd}e_{(a)}^\alpha e_{(b)}^\beta e_{(c)}^\rho e_{(d)}^\sigma . \qquad (3.8.5)$$

The fields Ψ, Ψ_α, $\Psi_{\alpha\beta}$ are tetrad scalars, $\tilde{\Psi}$, $\tilde{\Psi}_\alpha$ are tetrad pseudo-scalars, the Levi-Civita object $\epsilon^{\alpha\beta\rho\sigma}(x)$ is simultaneously both a generally covariant tensor and a tetrad pseudo-scalar.

Now, we have to obtain generally covariant equations for ordinary bosons with different spins and parities:

$$\underline{S = 0,} \qquad \nabla^\alpha \Psi_\alpha + m\Psi = 0 , \; \nabla_\alpha \Psi - m\Psi_\alpha = 0 , \; \nabla_\alpha \Psi_\beta - \nabla_\beta \Psi_\alpha = 0. \qquad (3.8.6)$$

The first two are the Proca equations for a scalar particle, the last equation holds identically

$$\partial_\alpha \, \partial_\beta \, \Psi \, - \, \Gamma_{\alpha\beta}^\mu \, \partial_\mu \, \Psi \, - \, \partial_\beta \, \partial_\alpha \, \Psi \, + \, \Gamma_{\beta\alpha}^\mu \, \partial_\mu \, \Psi = 0 .$$

For a pseudo-scalar field we have

$$\underline{S = \tilde{0} ,} \; \nabla^\alpha \tilde{\Psi}_\alpha + m\tilde{\Psi} = 0 , \; \nabla_\alpha \tilde{\Psi} - m\tilde{\Psi}_\alpha = 0 , \; \epsilon_{\alpha\beta}{}^{\rho\sigma}(x)\nabla_\rho \Psi_\sigma = 0. \qquad (3.8.7)$$

Here again, the last equation holds identically. Now, let $\Psi_\alpha \neq 0$, $\Psi_{\alpha\beta}(x) \neq 0$, then

$$\underline{S = 1 ,} \qquad \nabla^\alpha \Psi_\alpha = 0 , \; \nabla^\beta \Psi_{\alpha\beta} - m\Psi_\alpha = 0 ,$$

$$-\frac{1}{2}\epsilon_\alpha{}^{\beta\rho\sigma}(x)\nabla_\beta \Psi_{\rho\sigma} = 0 , \; \nabla_\alpha \Psi_\beta - \nabla_\beta \Psi_\alpha = m\Psi_{\alpha\beta} . \qquad (3.8.8)$$

Notice that the first and third equations hold identically

$$\nabla^\alpha \Psi_\alpha = \frac{1}{m} \, \nabla^\alpha \nabla^\beta \, \Psi_{\alpha\beta} = \frac{1}{2m} \, [\, \Psi_{\alpha\nu} \, R^\nu{}_\beta{}^{\beta\alpha}$$

$$-\Psi_{\beta\nu} R^\nu{}_\alpha{}^{\beta\alpha} \,] = \frac{1}{2m} \, [\, -\Psi_{\alpha\nu} \, R^{\nu\alpha} - \Psi_{\beta\nu} \, R^{\nu\beta} \,] = 0 ,$$

$$-\frac{1}{2m} \, \epsilon_\alpha{}^{\beta\rho\sigma}(x) \, \nabla_\beta \, [\, \nabla_\rho \Psi_\sigma - \nabla_\sigma \Psi_\rho \,]$$

$$= -\frac{1}{4m} \, \epsilon_\alpha^{\ \rho\beta\sigma}(x) \, [\, (\nabla_\beta \nabla_\rho - \nabla_\rho \nabla_\beta) \, \Psi_\sigma - (\nabla_\beta \nabla_\sigma - \nabla_\sigma \nabla_\beta) \, \Psi_\rho \,]$$

$$= -\frac{1}{4m} \, \epsilon_\alpha^{\ \beta\rho\sigma}(x) \, (\Psi^\nu \, R_{\nu\sigma\rho\beta} - \Psi^\nu \, R_{\nu\rho\sigma\beta}) = 0 \, . \qquad (3.8.9)$$

Now, let $\Psi(x) = \tilde{\Psi} = \Psi_\alpha = 0$, then

$$\underline{S = \tilde{1}} \, , \qquad\qquad \nabla^\alpha \tilde{\Psi}_\alpha = 0 \, , \ \nabla^\beta \tilde{\Psi}_{\alpha\beta} = 0 \, ,$$

$$\frac{1}{2} \, \epsilon_\alpha^{\ \beta\rho\sigma}(x) \, \nabla_\beta \Psi_{\rho\sigma} + m \, \tilde{\Psi}_\alpha = 0 \, , \ \epsilon_{\alpha\beta}^{\ \ \rho\sigma}(x) \, \nabla_\rho \tilde{\Psi}_\sigma - m \, \Psi_{\alpha\beta} = 0 \, . \qquad (3.8.10)$$

The first and the second equations also hold identically

$$\nabla^\alpha \tilde{\Psi}_\alpha = -\frac{1}{2m} \, \nabla^\alpha \epsilon_\alpha^{\ \beta\rho\sigma}(x) \, \nabla_\beta \, \Psi_{\rho\sigma} = -\frac{1}{2m} \, \epsilon_\alpha^{\ \beta\rho\sigma}(x) \, \nabla^\alpha \nabla_\beta \, \Psi_{\rho\sigma}$$

$$= -\frac{1}{4m} \epsilon_\alpha^{\ \beta\rho\sigma}(x) [\, \Psi_{\nu\sigma} \, R^\nu_{\ \beta\rho}{}^\alpha(x) \, + \, \Psi_{\rho\nu} \, R^\nu_{\ \sigma\beta}{}^\alpha \,] \, ,$$

$$\nabla^\beta \, \Psi_{\alpha\beta}(x) = \frac{1}{m} \, \nabla^\beta \, \epsilon_{\alpha\beta}^{\ \ \rho\sigma}(x) \, \nabla_\rho \Psi_\sigma = \frac{1}{2m} \, \epsilon_{\alpha\beta}^{\ \ \rho\sigma}(x) \, \Psi^\nu \, R_{\nu\sigma\rho}{}^\beta \, .$$

Linear constraints that distinguish four types of boson fields are the same as shown in the case of Minkowski space:

$$S = 0 \, , \qquad\qquad \tilde{\Delta} = +H \, , \ \xi = -\eta \, , \ \tilde{\xi} = -\xi \, , \ \tilde{\eta} = -\eta \, ;$$

$$S = \tilde{0} \, , \qquad\qquad \tilde{\Delta} = -H \, , \ \xi = +\eta \, , \ \tilde{\xi} = -\xi \, , \ \tilde{\eta} = -\eta \, ;$$

$$S = 1 \, , \qquad\qquad \tilde{\Delta} = +H \, , \ \tilde{\xi} = +\xi \, , \ \tilde{\eta} = +\eta \, ;$$

$$S = \tilde{1} \, , \qquad\qquad \tilde{\Delta} = -H \, , \ \tilde{\xi} = +\xi \, , \ \tilde{\eta} = +\eta \, . \qquad (3.8.11)$$

Without any additional calculation, we can write down the spinor and corresponding tensor equations for the Maxwell theories with opposite intrinsic parities:

for a vector model

$$(S = 1) \qquad \begin{array}{ll} (A) & i\sigma^\alpha(x) \, [\, \partial/\partial x^\alpha \, + \, \Sigma_\alpha \otimes I + I \otimes \Sigma_\alpha \,] \, \xi = - \, j \, , \\[2mm] (B) & i\bar{\sigma}^\alpha(x) \, [\, \partial/\partial x^\alpha \, + \, \bar{\Sigma}_\alpha \otimes I + I \otimes \bar{\Sigma}_\alpha \,] \, \eta = - \, \bar{j} \, , \end{array}$$

$$\nabla^\beta F_{\alpha\beta} = - \, j_\alpha \, , \qquad\qquad \epsilon^{\alpha\beta\rho\sigma}(x) \nabla_\beta F_{\rho\sigma} = 0 \, ;$$

$$(S = 1) \qquad \begin{array}{ll} (A') & i\bar{\sigma}^\alpha(x) \, [\, \partial/\partial x^\alpha \, + \, \bar{\Sigma}_\alpha \otimes I + I \otimes \Sigma_\alpha \,] \, H = \xi \, , \\[2mm] (B') & i\sigma^\alpha(x) \, [\, \partial/\partial x^\alpha \, + \, \Sigma_\alpha \otimes I + I \otimes \bar{\Sigma}_\alpha \,] \, \Delta = \eta \, , \end{array}$$

$$\nabla^\alpha A_\alpha = 0 \, , \qquad \nabla_\alpha A_\beta - \nabla_\beta A_\alpha = F_{\alpha\beta} \, ; \qquad (3.8.12)$$

for a pseudo-vector model

$$S = \tilde{1} \, , \qquad \begin{array}{ll} (\tilde{A}) & i\sigma^\alpha(x) \, [\, \partial/\partial x^\alpha \, + \, \Sigma_\alpha \otimes I + I \otimes \Sigma_\alpha \,] \, \xi = - \, \tilde{j} \, , \\[2mm] (\tilde{B}) & i\bar{\sigma}^\alpha(x) \, [\, \partial/\partial x^\alpha \, + \, \bar{\Sigma}_\alpha \otimes I + I \otimes \bar{\Sigma}_\alpha \,] \, \eta = - \, \tilde{\bar{j}} \, , \end{array}$$

$$(S = \tilde{1}) \qquad \nabla^\beta \tilde{F}_{\alpha\beta} = 0 \,, \qquad \tfrac{1}{2}\,\epsilon^{\alpha\beta\rho\sigma}(x)\,\nabla_\beta \tilde{F}_{\rho\sigma} = +\tilde{j}^\alpha \;;$$

$$(S = \tilde{1}) \,, \qquad \begin{matrix} (\tilde{A}') & i\bar{\sigma}^\alpha(x)\,[\,\partial/\partial x^\alpha \,+\, \bar{\Sigma}_\alpha \otimes I + I \otimes \Sigma_\alpha \,]\,H = \xi \,, \\[2mm] (\tilde{B}') & i\sigma^\alpha(x)\,[\,\partial/\partial x^\alpha \,+\, \Sigma_\alpha \otimes I + I \otimes \bar{\Sigma}_\alpha \,]\,\Delta = \eta \,, \end{matrix}$$

$$0 = 0 \,, \qquad \nabla^\alpha \tilde{A}_\alpha = 0 \,, \qquad \epsilon^{\alpha\beta\rho\sigma}(x)\,\nabla_\rho \tilde{A}_\sigma = \tilde{F}^{\alpha\beta} \,. \tag{3.8.13}$$

3.9   Dual Symmetry in the Extended Maxwell Theory

Summing and subtracting the respective tensor equations in Eqs.(3.8.12) and (3.8.13); we arrive at

$$\nabla^\beta(\, F_{\alpha\beta} + \tilde{F}_{\alpha\beta} \,) = -\,j_\alpha \,, \qquad \nabla^\beta(\, F_{\alpha\beta} - \tilde{F}_{\alpha\beta} \,) = -\,j_\alpha \,,$$

$$\tfrac{1}{2}\,\epsilon^{\alpha\beta\rho\sigma}(x)\,\nabla_\beta\,(\,F_{\rho\sigma} + \tilde{F}_{\rho\sigma}\,) = +\tilde{j}^\alpha \,, \qquad \tfrac{1}{2}\,\epsilon^{\alpha\beta\rho\sigma}(x)\,\nabla_\beta\,(\,F_{\rho\sigma} - \tilde{F}_{\rho\sigma}\,) = -\tilde{j}^\alpha \,,$$

$$F_{\alpha\beta} \pm \tilde{F}_{\alpha\beta} = \nabla_\alpha A_\beta - \nabla_\beta A_\alpha \pm \epsilon_{\alpha\beta}{}^{\rho\sigma}(x)\,\nabla_\rho \tilde{A}_\sigma \,,$$

$$\nabla^\alpha A_\alpha = 0 \,, \qquad \nabla^\alpha \tilde{A}_\alpha = 0 \,. \tag{3.9.1}$$

Here, $F_{\alpha\beta}, A_\alpha, j_\alpha$ are tetrad scalars, $\epsilon_{\alpha\beta\rho\sigma}(x), \tilde{A}_\alpha, \tilde{j}_\alpha$ are tetrad pseudo-scalars.

Let us proceed and introduce dual variables

$$F^*_{\alpha\beta} = \tfrac{1}{2}\,\epsilon_{\alpha\beta}{}^{\rho\sigma}(x)\,F_{\rho\sigma} \,, \qquad F_{\alpha\beta} = -\tfrac{1}{2}\,\epsilon_{\alpha\beta}{}^{\mu\nu}(x)\,F^*_{\mu\nu} \,,$$

$$\tilde{F}^*_{\alpha\beta} = \tfrac{1}{2}\,\epsilon_{\alpha\beta}{}^{\rho\sigma}(x)\,\tilde{F}_{\rho\sigma} \,, \qquad \tilde{F}_{\alpha\beta} = -\tfrac{1}{2}\,\epsilon_{\alpha\beta}{}^{\mu\nu}(x)\,\tilde{F}^*_{\mu\nu} \,. \tag{3.9.2}$$

One can easily prove the identity

$$\tfrac{1}{2}\,\epsilon_{\rho\sigma}{}^{\alpha\beta}(x)\,[F_{\alpha\beta} \pm \tilde{F}_{\alpha\beta}] = \tfrac{1}{2}\,\epsilon_{\rho\sigma}{}^{\alpha\beta}(x)\,[\nabla_\alpha A_\beta - \nabla_\beta A_\alpha \pm \epsilon_{\alpha\beta}{}^{\delta\gamma}(x)\,\nabla_\delta \tilde{A}_\gamma\,]$$

$$= \epsilon_{\rho\sigma}{}^{\alpha\beta}(x)\nabla_\alpha A_\beta \mp (\nabla_\rho \tilde{A}_\sigma - \nabla_\sigma \tilde{A}_\rho) \,,$$

or in new variables

$$F^*_{\rho\sigma} \pm \tilde{F}^*_{\rho\sigma} = \mp\,(\nabla_\rho \tilde{A}_\sigma - \nabla_\sigma \tilde{A}_\rho) + \epsilon_{\rho\sigma}{}^{\alpha\beta}(x)\nabla_\alpha A_\beta \,. \tag{3.9.3}$$

Thus, the system of the extended Maxwell theory can be presented as follows:

$$\nabla^\beta(\, F_{\alpha\beta} + \tilde{F}_{\alpha\beta} \,) = -\,j_\alpha \,, \qquad \nabla^\beta(\, F_{\alpha\beta} - \tilde{F}_{\alpha\beta} \,) = -\,j_\alpha \,,$$

$$\nabla^\beta\,(\, F^*_{\alpha\beta} + \tilde{F}^*_{\alpha\beta} \,) = +\tilde{j}_\alpha \,, \qquad \nabla_\beta\,(\, F^{*\alpha\beta} - \tilde{F}^{*\alpha\beta} \,) = -\tilde{j}^\alpha \,,$$

$$F_{\alpha\beta} \pm \tilde{F}_{\alpha\beta} = \nabla_\alpha A_\beta - \nabla_\beta A_\alpha \pm \epsilon_{\alpha\beta}{}^{\rho\sigma}(x)\,\nabla_\rho \tilde{A}_\sigma \,,$$

$$F^*_{\rho\sigma} \pm \tilde{F}^*_{\rho\sigma} = \mp\,(\nabla_\rho \tilde{A}_\sigma - \nabla_\sigma \tilde{A}_\rho) + \epsilon_{\rho\sigma}{}^{\alpha\beta}(x)\nabla_\alpha A_\beta \,,$$

$$\nabla^\alpha A_\alpha = 0 \,, \qquad \nabla^\alpha \tilde{A}_\alpha = 0 \,. \tag{3.9.4}$$

In these designations

$$F_{\alpha\beta} + \tilde{F}_{\alpha\beta} = F^+_{\alpha\beta} , \qquad F^*_{\alpha\beta} + \tilde{F}^*_{\alpha\beta} = F^{+*}_{\alpha\beta} ,$$

$$F_{\alpha\beta} - \tilde{F}_{\alpha\beta} = F^-_{\alpha\beta} , \qquad F^*_{\alpha\beta} - \tilde{F}^*_{\alpha\beta} = F^{-*}_{\alpha\beta} ,$$

Equations (3.9.4) then read as

$$\nabla^\beta F^+_{\alpha\beta} = -j_\alpha , \qquad\qquad \nabla^\beta F^-_{\alpha\beta} = -j_\alpha ,$$

$$\nabla^\beta F^{+*}_{\alpha\beta} = +\tilde{j}_\alpha , \qquad\qquad \nabla^\beta F^{-*}_{\alpha\beta} = -\tilde{j}_\alpha ,$$

$$F^+_{\alpha\beta} = \nabla_\alpha A_\beta - \nabla_\beta A_\alpha + \epsilon_{\alpha\beta}{}^{\rho\sigma}(x)\, \nabla_\rho \tilde{A}_\sigma ,$$

$$F^-_{\alpha\beta} = \nabla_\alpha A_\beta - \nabla_\beta A_\alpha - \epsilon_{\alpha\beta}{}^{\rho\sigma}(x)\, \nabla_\rho \tilde{A}_\sigma ,$$

$$F^{+*}_{\rho\sigma} = -(\nabla_\rho \tilde{A}_\sigma - \nabla_\sigma \tilde{A}_\rho) + \epsilon_{\rho\sigma}{}^{\alpha\beta}(x)\nabla_\alpha A_\beta ,$$

$$F^{-*}_{\rho\sigma} = +(\nabla_\rho \tilde{A}_\sigma - \nabla_\sigma \tilde{A}_\rho) + \epsilon_{\rho\sigma}{}^{\alpha\beta}(x)\nabla_\alpha A_\beta ,$$

$$\nabla^\alpha A_\alpha = 0 , \qquad \nabla^\alpha \tilde{A}_\alpha = 0 . \tag{3.9.5}$$

The obtained system is invariant with respect to the following linear transformations (extended duality operation)

$$A_\alpha \Longrightarrow -\tilde{A}'_\alpha , \qquad \tilde{A}_\alpha \Longrightarrow +A'_\alpha ,$$

$$F^+_{\alpha\beta} \Longrightarrow +F'^{+*}_{\alpha\beta} , \qquad F^{+*}_{\alpha\beta} \Longrightarrow -F'^+_{\alpha\beta} ,$$

$$F^-_{\alpha\beta} \Longrightarrow -F'^{-*}_{\alpha\beta} , \qquad F^{-*}_{\alpha\beta} \Longrightarrow +F'^-_{\alpha\beta} ,$$

$$j_\alpha \Longrightarrow -\tilde{j}'_\alpha , \qquad \tilde{j}_\alpha \Longrightarrow +j'_\alpha . \tag{3.9.6}$$

Besides, one can find one more invariance transformation

$$A_\alpha \Longrightarrow +\tilde{A}'_\alpha , \qquad \tilde{A}_\alpha \Longrightarrow -A'_\alpha ,$$

$$F^+_{\alpha\beta} \Longrightarrow -F'^{+*}_{\alpha\beta} , \qquad F^{+*}_{\alpha\beta} \Longrightarrow +F'^+_{\alpha\beta} ,$$

$$F^-_{\alpha\beta} \Longrightarrow +F'^{-*}_{\alpha\beta} , \qquad F^{-*}_{\alpha\beta} \Longrightarrow -F'^-_{\alpha\beta} ,$$

$$j_\alpha \Longrightarrow +\tilde{j}'_\alpha , \qquad \tilde{j}_\alpha \Longrightarrow -j'_\alpha . \tag{3.9.7}$$

Relations in Eqs. (3.9.6) and (3.9.7) are particular cases since the continuous extended dual transformations over 4-vector and 4-pseudo-vector:

$$\cos\chi A_\alpha + \sin\chi \tilde{A}_\alpha = A'_\alpha , \qquad \cos\chi A'_\alpha - \sin\chi \tilde{A}'_\alpha = A_\alpha ,$$

$$-\sin\chi A_\alpha + \cos\chi \tilde{A}_\alpha = \tilde{A}'_\alpha , \qquad \sin\chi A'_\alpha + \cos\chi \tilde{A}'_\alpha = \tilde{A}_\alpha , \tag{3.9.8}$$

which will generate the following transformations over field strength tensors:

$$F^+_{\alpha\beta} = \nabla_\alpha A_\beta - \nabla_\beta A_\alpha + \epsilon_{\alpha\beta}{}^{\rho\sigma}(x)\, \nabla_\rho \tilde{A}_\sigma$$

$$= \nabla_\alpha(\cos\chi\, A'_\beta - \sin\chi\, \tilde{A}'_\beta) - \nabla_\beta(\cos\chi\, A'_\alpha - \sin\chi\, \tilde{A}'_\alpha)$$

$$+\epsilon_{\alpha\beta}{}^{\rho\sigma}(x)\,\nabla_\rho(\sin\chi\, A'_\sigma + \cos\chi\, \tilde{A}'_\sigma)\,,$$

that is

$$F^+_{\alpha\beta} = \cos\chi\, F'^+_{\alpha\beta} + \sin\chi\, F'^{+*}_{\alpha\beta}. \qquad (3.9.9)$$

Analogously, we have

$$F^{+*}_{\alpha\beta} = -\sin\chi\, F'^+_{\alpha\beta} + \cos\chi\, F'^{+*}_{\alpha\beta}\,. \qquad (3.9.10)$$

In the same manner, one derives

$$F^-_{\alpha\beta} = \cos\chi\, F'^-_{\alpha\beta} - \sin\chi\, F'^{-*}_{\alpha\beta}\,,$$

$$F^{-*}_{\alpha\beta} = \sin\chi\, F'^-_{\alpha\beta} + \cos\chi\, F'^{-*}_{\alpha\beta}\,. \qquad (3.9.11)$$

Continuous dual transformations over currents are determined as

$$\cos\chi\, j_\alpha + \sin\chi\, \tilde{j}_\alpha = j'_\alpha\,, \qquad \cos\chi\, j'_\alpha - \sin\chi\, \tilde{j}'_\alpha = j_\alpha\,,$$

$$-\sin\chi\, j_\alpha + \cos\chi\, \tilde{j}_\alpha = \tilde{j}'_\alpha\,, \qquad \sin\chi\, j'_\alpha + \cos\chi\, \tilde{j}'_\alpha = \tilde{j}_\alpha\,. \qquad (3.9.12)$$

3.10   Conclusions

Let us summarize the results. Within the formalism of the 16-component Dirac–Kähler field theory, the spinor equations for two types of massless vector photon fields with different parities have been derived. The equivalent tensor equations in terms of the field strength tensor F_{ab} and corresponding 4-vector A_b and 4-pseudo-vector $\tilde{A}_b$ that depend on intrinsic photon parity are derived. The equations include additional sources, electric 4-vector j_b and magnetic 4-pseudo-vector $\tilde{j}_b$. The theories of two types of photon fields are explicitly uncoupled, their linear combinations through summation or subtraction result in Maxwell electrodynamics with electric and magnetic charges in the 2-potential approach. Our analysis has been extended to a curved space-time background.

Chapter 4

10-Dimensional Matrix Approach

4.1 Introduction

Matrix Duffin–Kemmer–Petiau formalism for boson fields has a long and rich history which is inseparably linked to the description of photons and mesons – see references in Chapter **1**. Usually the description of an interaction between a quantum mechanical particle and the external classical gravitational field looks basically different for fermions and bosons. In the case of a fermion, when starting from the Dirac equation

$$(i\gamma^a \, \partial_a \, - \, \frac{mc}{\hbar}) \, \Psi(x) \, = \, 0$$

we have to generalize through the use of the tetrad formalism [40]. For a vector boson, a totally different approach is generally used that consists of ordinary formal changing for all involved tensors and the usual derivative ∂_a into general relativity. For example, in case of a vector particle, the flat space Proca equations are

$$\partial_a \, \Psi_b \, - \, \partial_b \, \Psi_a \, = \, \frac{mc}{\hbar} \Psi_{ab} \, , \qquad \partial^b \, \Psi_{ab} = \frac{mc}{\hbar} \, \Psi_a$$

being subjected to the formal change

$$\partial_a \, \rightarrow \, \nabla_\alpha \, , \; \Psi_a \, \rightarrow \, \Psi_\alpha, \; \Psi_{ab} \, \rightarrow \, \Psi_{\alpha\beta}$$

that result in

$$\nabla_\alpha \, \Psi_\beta - \nabla_\beta \, \Psi_\alpha = \frac{mc}{\hbar} \, \Psi_{\alpha\beta} \, , \qquad \nabla^\beta \, \Psi_{\alpha\beta} = \frac{mc}{\hbar} \, \Psi_\alpha \, . \qquad (4.1.1)$$

However, the known Duffin–Kemmer–Petiau formalism in the curved space-time until recently was not used, though such a possibility has been known [30]. This situation is changing now, through other resources: Lunardi et al. [271, 272], Fainberg and Pimentel [273], Montigny et al. [274], Kanatchikov [275], Casana et al. [276, 277], Havare et al. [278, 279], Okninski [280], Bogush et al. [281–289].

4.2   Duffin–Kemmer–Petiau Equation in Gravitational Field

We start from a flat space equation in its matrix DKP-form

$$(i \, \beta^a \, \partial_a \, - \, \frac{mc}{\hbar} \,) \, \Phi(x) = 0 \; ; \qquad (4.2.1)$$

where

$$\Phi = (\Phi_0, \; \Phi_1, \; \Phi_2, \; \Phi_3; \; \Phi_{01}, \; \Phi_{02}, \; \Phi_{03}, \; \Phi_{23}, \; \Phi_{31}, \; \Phi_{12}) \; ,$$

$$\beta^a = \left| \begin{matrix} 0 & \kappa^a \\ \lambda^a & 0 \end{matrix} \right| = \kappa^a \oplus \lambda^a \; , \quad (\kappa^a)^{[mn]}_j = -i \, (\delta^m_j \, g^{na} \, - \, \delta^n_j \, g^{ma}) \; ,$$

$$(\lambda^a)^j_{[mn]} = -i \, (\delta^a_m \, \delta^j_n - \delta^a_n \, \delta^j_m) \, = \, -i \, \delta^{aj}_{mn} \; ;$$

$$(4.2.2)$$

$(g^{na}) = \mathrm{diag}(+1, -1, -1, -1)$. The basic properties of β^a are

$$\beta^c \, \beta^a \, \beta^b = \left| \begin{matrix} 0 & \kappa^c \, \lambda^a \, \kappa^b \\ \lambda^c \, \kappa^a \, \lambda^b & 0 \end{matrix} \right| , \qquad (\lambda^c \, \kappa^a \, \lambda^b)^j_{[mn]} = i \, (\, \delta^{cb}_{mn} \, g^{aj} \, - \, \delta^{cj}_{mn} \, g^{ab} \,) \; ,$$

$$(\kappa^c \, \lambda^a \, \kappa^b)^{[mn]}_j = i \, [\, \delta^a_j \, (g^{cm} \, g^{bn} \, - \, g^{cn} \, g^{bm}) \, + \, g^{ac} \, (\delta^n_j \, g^{mb} \, - \, \delta^m_j \, g^{nb}) \,] \; ,$$

$$(4.2.3)$$

and

$$\beta^c \, \beta^a \, \beta^b \, + \, \beta^b \, \beta^a \, \beta^c = \beta^c \, g^{ab} \, + \, \beta^b g^{ac} \; ,$$

$$[\beta^c, j^{ab}] = g^{ca} \, \beta^b \, - \, g^{cb} \, \beta^a \; , \qquad j^{ab} = \beta^a \, \beta^b \, - \, \beta^b \, \beta^a \; ,$$

$$[j^{mn}, j^{ab}] = (\, g^{na} \, j^{mb} \, - \, g^{nb} \, j^{ma} \,) \, - \, (\, g^{ma} \, j^{nb} \, - \, g^{mb} \, j^{na} \,) \; .$$

$$(4.2.4)$$

In accordance with this tetrad recipe, one should be able to generalize the DKP-equation as follows:

$$[\, i \, \beta^\alpha(x) \, (\partial_\alpha \, + \, B_\alpha(x)) \, - \, \frac{mc}{\hbar} \,] \, \Phi(x) = 0 \; ,$$

$$\beta^\alpha(x) = \beta^a e^\alpha_{(a)}(x) \; , \quad B_\alpha(x) = \frac{1}{2} \, j^{ab} e^\beta_{(a)} \nabla_\alpha (e_{(b)\beta}) \; . \qquad (4.2.5)$$

This equation contains the tetrad $e^\alpha_{(a)}(x)$ explicitly. Therefore, there must exist a possibility to demonstrate the equivalence of any variants of this equation associated with various tetrads:

$$e^\alpha_{(a)}(x) \; , \qquad e'^\alpha_{(b)}(x) \, = \, L_a{}^b(x) \, e^\alpha_{(b)}(x) \; , \qquad (4.2.6)$$

$L_a{}^b(x)$ is a local Lorentz transformation. We will show here, that any two such equations

$$[\, i\beta^\alpha(x) \, (\partial_\alpha + B_\alpha(x)) \, - \, mc/\hbar \,] \, \Phi(x) = 0 \; ,$$

$$[\, i\beta'^\alpha(x) \, (\partial_\alpha + B'_\alpha(x)) - mc/\hbar \,] \, \Phi'(x) = 0 \; , \qquad (4.2.7)$$

generating in tetrads $e^{\alpha}_{(a)}(x)$ and $e'^{\alpha}_{(b)}(x)$, respectively, can be converted into each other through a local gauge transformation:

$$\Phi'(x) = \left| \begin{array}{c} \phi'_a(x) \\ \phi'_{[ab]}(x) \end{array} \right| = \left| \begin{array}{cc} L_a{}^l & 0 \\ 0 & L_a{}^m L_b{}^n \end{array} \right| \left| \begin{array}{c} \phi_l(x) \\ \phi_{[mn]}(x) \end{array} \right| . \tag{4.2.8}$$

Starting from the first equation in Eq. (4.2.7), let us obtain an equation for $\Phi'(x)$. In allowing for $\Phi'(x) = S(x)\,\Phi(x)$, we get:

$$[\, i\, S\, \beta^{\alpha}\, S^{-1}(\partial_{\alpha} + S\, B_{\alpha}\, S^{-1} + S\, \partial_{\alpha}\, S^{-1}) - \frac{mc}{\hbar} \,]\, \Phi'(x) = 0 \,.$$

One should verify the relationships

$$S(x)\, \beta^{\alpha}(x)\, S^{-1}(x) = \beta'^{\alpha}(x) \,, \tag{4.2.9}$$

$$S(x)\, B_{\alpha}(x)\, S^{-1}(x) + S(x)\, \partial_{\alpha}\, S^{-1}(x) = B'_{\alpha}(x) \,. \tag{4.2.10}$$

The first one can be rewritten as

$$S(x)\, \beta^a\, e^{\alpha}_{(a)}(x)\, S^{-1}(x) = \beta^b\, e'^{\alpha}_{(b)}(x) \,;$$

from this, we come to

$$S(x)\, \beta^a\, S^{-1}(x) = \beta^b\, L_b{}^a(x) \,. \tag{4.2.11}$$

The latter condition is well-known in DKP-theory; one can verify it through the use of the block structure of β^a, which provides two relations:

$$L(x)\, \kappa^a\, [\, L^{-1}(x) \otimes L^{-1}(x)\,] = \kappa^b\, L_b{}^a(x) \,,$$

$$[\, L(x) \otimes L(x)\,]\, \lambda^a\, L^{-1}(x) = \lambda^b\, L_b{}^a(x) \,.$$

They will be satisfied identically, after we take the explicit form of κ^a and λ^a; and then allow for the pseudo-orthogonality condition: $g^{al}\, (L^{-1})_l{}^k(x) = g^{kb}\, L_b{}^a(x)$. Now, let us pass to the proof of the relationship in Eq. (4.2.10). By using the determining relation for DKP-connection, we readily produce

$$S(x)\, \partial_{\alpha}\, S^{-1}(x) = B'_{\alpha}(x) - \frac{1}{2}\, j^{mn} L_m{}^n(x)\, g_{ab}\, \partial_{\alpha}\, L_n{}^b(x)$$

therefore, Eq. (4.2.10) results in

$$S(x)\, \partial_{\alpha}\, S^{-1}(x) = \frac{1}{2}\, L_m{}^a(x)\, g_{ab}\, (\, \partial_{\alpha}\, L_n{}^b(x)\,) \,.$$

The latter condition is an identity readily verified through the use of a block structure that involves all matrices. Thus, the equations from Eq. (4.2.7) are translated into each other. In other words, they manifest a gauge symmetry under local Lorentz transformations in the complete analogy with a more familiar Dirac particle case. At the same time, the wave function from this equation represents scalar quantity relative to general coordinate transformations: that is, if $x^{\alpha} \rightarrow x'^{\alpha} = f^{\alpha}(x)$, then $\Phi'(x) = \Phi(x)$.

It remains to demonstrate that this DKP-formulation can be inverted into the Proca formalism in terms of general relativity tensors. To this end, as a first step, let us allow for the sectional structure of β^a, J^{ab} and $\Phi(x)$ in the DKP-equation; then instead of Eq. (4.2.5), we get

$$i\,[\,\lambda^c\,e^\alpha_{(c)}\,(\,\partial_\alpha\,+\,\kappa^a\,\lambda^b\,e^\beta_{(a)}\,\nabla_\alpha\,e_{(b)\beta}\,)\,]_{[mn]}{}^l\,\Phi_l = \frac{mc}{\hbar}\,\Phi_{[mn]}\,,$$

$$i\,[\kappa^c\,e^\alpha_{(c)}\,(\,\partial_\alpha\,+\,\lambda^a\,\kappa^b\,e^\beta_{(a)}\,\nabla_\alpha\,e_{(b)\beta}\,)\,]_l{}^{[mn]}\,\Phi_{[mn]} = \frac{mc}{\hbar}\Phi_l\,, \qquad (4.2.12)$$

which led to

$$(e^\alpha_{(a)}\,\partial_\alpha\,\Phi_b\,-\,e^\alpha_{(b)}\,\partial_\alpha\,\Phi_a)\,+\,(\gamma^c{}_{ab}-\gamma^c{}_{ba})\,\Phi_c = \frac{mc}{\hbar}\,\Phi_{ab}\,,$$

$$e^{(b)\alpha}\,\partial_\alpha\,\Phi_{ab}\,+\,\gamma^{nb}{}_n\Phi_{ab}+\gamma_a{}^{mn}\Phi_{mn} = \frac{mc}{\hbar}\,\Phi_a\,; \qquad (4.2.13)$$

the symbol $\gamma_{abc}(x)$ is used to designate Ricci coefficients: $\gamma_{abc}(x)\,=\,-\,e_{(a)\alpha;\beta}\,e^\alpha_{(b)}\,e^\beta_{(c)}$. In turn, Eq. (4.2.13) will look as the Proca equations in Eq. (4.1.1) after they are rewritten in terms of tetrad components

$$\Phi_a(x)\,=\,e^\alpha_{(a)}(x)\,\Phi_\alpha(x),\qquad \Phi_{ab}(x)=e^\alpha_{(a)}(x)\,e^\beta_{(b)}(x)\,\Phi_{\alpha\beta}(x)\,. \qquad (4.2.14)$$

So, as evidenced here, the manner of introducing the interaction between a spin 1 particle and external classical gravitational field can be successfully unified with the approach as it occurred with regard to a spin $1/2$ particle and was first developed by Tetrode, Weyl, Fock, and Ivanenko.

One should attach great significance to that possibility of unification. Moreover, its absence would actually be a very strange fact. We will add some further details to this interaction possiblity.

The manner of extending the flat space Dirac equation to general relativity case indicates clearly that the Lorentz group underlies equally for both these theories. In other words, the Lorentz group retains its importance and significance for changing the Minkowski space model to an arbitrary curved space-time. In contrast to this, when generalizing the Proca formulation, we automatically destroy any relations to the Lorentz group. Although the definition itself for a spin 1 particle as an elementary object was based on this group. Such a gravity sensitiveness to the fermion-boson division might appear rather strange and have an unattractive asymmetry, being subjected to the criticism. Moreover, just this feature has brought about plenty of speculation about this matter. In any case, this peculiarity of particle-gravity field interaction is recorded in almost every handbook.

4.3   Invariant Form, Conserved Currents

Let us construct tensor bilinear combinations in terms of 10-component field functions. The matrix η of bilinear invariant form should obey restriction

$$\Phi'^+\eta\Phi' = \Phi^+\eta\Phi\,, \qquad \Phi' = S\,\Phi\,. \qquad (4.3.1)$$

From Eq. (4.3.1), it follows:

$$S^+ \eta S = \eta \,. \tag{4.3.2}$$

Hence, we let c_1, c_2 stand for arbitrary numerical parameters

$$\eta = \begin{vmatrix} c_1 h & 0 \\ 0 & c_2 H \end{vmatrix} \,, \quad h^{na} = g^{na} \,, \quad H^{[ab][mn]} = g^{am} g^{bn} - g^{an} g^{bm} \,. \tag{4.3.3}$$

To ensure existence of a conserved current, we should require

$$\eta^{-1} \beta^{a+} \eta = \beta^a \implies c_1 = c_2 = 1 \,. \tag{4.3.4}$$

For simplicity, let start with the theory in Minkowski space, then an equation for conjugate field function $\bar{\Phi} = \Phi^+ \eta$ is

$$\bar{\Phi} \left(i\beta^a \overleftarrow{\partial_a} + m \right) = 0 \,.$$

Further, acting by ordinary rules, we get

$$\partial_a J^a = 0 \,, \qquad J^a = \bar{\Phi} \beta^a \Phi \,. \tag{4.3.5a}$$

When using the explicit form of η, an expression for the conserved current is written as

$$J^a = \bar{\Phi} \, \beta^a \, \Phi = -i \left(\Phi_l^* \, \Phi^{la} - \Phi_l \, \Phi^{la*} \right) \,. \tag{4.3.5b}$$

Now, let us extend analysis to the case of curved space-time. The basic Duffin–Kemmer–Petiau equation is (for brevity, m designate $mc/\hbar$)

$$\left\{ i\beta^\alpha(x) \left[(\nabla_\alpha + B_\alpha(x)) - ie A_\alpha \right] - m \right\} \Phi = 0 \,. \tag{4.3.6a}$$

Let us derive an equation for the conjugate field function – starting with

$$\Phi^+ \eta \eta^{-1} \left\{ i \left[(\overleftarrow{\nabla}_\alpha + B_\alpha^+) + ie A_\alpha \right] \eta \eta^{-1} \beta^{\alpha+} + m \right\} \eta = 0 \,,$$

then by allowing for identities

$$\eta^{-1} \beta^{a+} \eta = \beta^a \,, \qquad \eta^{-1} \beta^{\alpha+}(x)\eta = \beta^\alpha(x) \,,$$

$$\eta^{-1} j^{ab+} \eta = - j^{ab} \,, \qquad \eta^{-1} B_\alpha^+(x)\eta = -B_\alpha(x) \tag{4.3.6b}$$

we get an equation for $\bar{\Psi} = \Phi^+ \eta$:

$$\bar{\Phi} \left\{ i \left[(\overleftarrow{\nabla}_\alpha - B_\alpha) + ie A_\alpha \right] \beta^\alpha + m \right\} = 0 \,. \tag{4.3.6c}$$

Now, combining two wave equations, we readily produce

$$\bar{\Phi} \overleftarrow{\nabla}_\alpha \beta^\alpha \Phi + \bar{\Phi} \beta^\alpha \overrightarrow{\nabla}_\alpha \Phi + \bar{\Phi} \left(\beta^\alpha B_\alpha - B_\alpha \beta^\alpha \right) \Phi = 0 \,. \tag{4.3.7}$$

To proceed with relation in Eq. (4.3.7), we require a special identity to derive which can conveniently start with the known commutative relation

$$[\beta^c, j^{ab}]_- = g^{ca} \beta^b - g^{cb} \beta^a$$

and multiply this relation by a tetrad based construction $(1/2)e^{\rho}_{(c)}\, e^{\beta}_{(a)} \nabla_{\sigma} e_{(b)\beta}$, which results in

$$\beta^{\rho}(x) B_{\sigma}(x) \; - \; B_{\sigma}(x)\beta^{\rho}(x) = \nabla_{\sigma}\beta^{\rho}(x) \; . \tag{4.3.8}$$

By taking into account Eq. (4.3.8), allows Eq. (4.3.7) to read as a conservation law

$$\nabla_{\alpha}\, \bar{\Phi}\beta^{\alpha}(x)\Phi = \nabla_{\alpha}J^{\alpha}(x) = 0 \; . \tag{4.3.9a}$$

Generally covariant current 4-vector is

$$J^{\alpha}(x) = e^{\alpha}_{(b)}\, \bar{\Phi}\beta^{b}\Phi$$

that can be expressed in terms of generally covariant tensors

$$J^{\alpha} = -i\, e^{\alpha}_{(a)}(x) \; (\; \Phi^{*}_{l}\Phi^{la} - \Phi_{l}\Phi^{la*} \;) = -i \; (\; \Phi^{*}_{\beta}\Phi^{\beta\alpha} - \Phi_{\beta}\Phi^{\beta\alpha*} \;) \; . \tag{4.3.9b}$$

Note the following identities:

$$\nabla_{\alpha}(\bar{\Phi}\beta^{\alpha}\,\Phi) = \bar{\Phi}\,(\overleftarrow{\nabla}_{\alpha} - B_{\alpha})\,\beta^{\alpha}\Phi \; + \; \bar{\Phi}\,\beta^{\alpha}\,(\overrightarrow{\nabla}_{\alpha} + B_{\alpha})\,\Phi \; , \tag{4.3.10}$$

which determines the rule for calculating covariant derivative ∇_{α} of a covariant 4-vector constructed on the base of Φ and $\bar{\Phi}$.

Such a rule can be extended for more general tensor structures:

$$[\; \nabla_{\alpha}\; (\beta^{\rho_1}\beta^{\rho_2}...\beta^{\rho_n})\;]$$

$$= (\beta^{\rho_1}\beta^{\rho_2}...\beta^{\rho_n})\,B_{\alpha} - B_{\alpha}\,(\beta^{\rho_1}\beta^{\rho_2}...\beta^{\rho_n}) \; , \tag{4.3.11a}$$

which led to

$$[\; \nabla_{\alpha}\, \bar{\Phi}\,\beta^{\rho_1}\beta^{\rho_2}...\beta^{\rho_n}\,\Phi \;]$$

$$= \bar{\Phi}\,(\overleftarrow{\nabla}_{\alpha} - B_{\alpha})\,\beta^{\rho_1}\beta^{\rho_2}...\beta^{\rho_n}\Phi \; + \; \bar{\Phi}\,\beta^{\rho_1}\beta^{\rho_2}...\beta^{\rho_n}\,(\overrightarrow{\nabla}_{\alpha} + B_{\alpha})\,\Phi \; . \tag{4.3.11b}$$

Equation $(4.3.11a)$ is equivalent to the commutative relations

$$(\nabla_{\alpha} + B_{\alpha})\,\beta^{\rho_1}\beta^{\rho_2}...\beta^{\rho_n} = \beta^{\rho_1}\beta^{\rho_2}...\beta^{\rho_n}\,(\nabla_{\alpha} + B_{\alpha}) \; ,$$

$$\beta^{\rho_1}\beta^{\rho_2}...\beta^{\rho_n}\,(\overleftarrow{\nabla}_{\alpha} - B_{\alpha}) \; = (\overleftarrow{\nabla}_{\alpha} - B_{\alpha})\,\beta^{\rho_1}\beta^{\rho_2}...\beta^{\rho_n} \; . \tag{4.3.11c}$$

4.4   Energy-Momentum Tensor for a Vector Field

Let us proceed with two equations

$$(\; i\,\beta^{\alpha}(x)\; \overrightarrow{D}_{\alpha}\; -m\;)\,\Phi = 0 \; , \qquad \bar{\Phi}\,(\,i\,\beta^{\alpha}(x)\,\overleftarrow{D}_{\alpha}\; +m\,) = 0 \; , \tag{4.4.1}$$

where we can use the notation:

$$\overrightarrow{D}_{\alpha} = \overrightarrow{\nabla}_{\alpha} + B_{\alpha} - ieA_{\alpha} \; , \qquad \overleftarrow{D}_{\alpha} = \overleftarrow{\nabla}_{\alpha} - B_{\alpha} + ieA_{\alpha} \; ;$$

we remember about commutativity for D_α and $\beta^\alpha(x)$; for brevity, m and e designate, respectively $mc/\hbar$ and $e/\hbar c$.

Acting on the first equation in Eq. (4.4.1) from the left by the operator $\bar{\Phi}\,\overrightarrow{D}_\beta$, and acting on the second operator from the right by $\overrightarrow{D}_\beta\,\Phi$ and summing these results we get:

$$\bar{\Phi}\,\overrightarrow{D}_\beta\,\beta^\alpha\,\overrightarrow{D}_\alpha\,\Phi + \bar{\Phi}\,\overleftarrow{D}_\alpha\,\beta^\alpha\,\overrightarrow{D}_\beta\,\Phi = 0\,.$$

From this, after some mathematical manipulations we will obtain:

$$\bar{\Phi}\,\overrightarrow{D}_\alpha\,\beta^\alpha\,\overrightarrow{D}_\beta\,\Phi + \bar{\Phi}\,\overleftarrow{D}_\alpha\,\beta^\alpha\,\overrightarrow{D}_\beta\,\Phi = \bar{\Phi}\beta^\alpha\,[\overrightarrow{D}_\alpha\overrightarrow{D}_\beta - \overrightarrow{D}_\beta\overrightarrow{D}_\alpha]\Phi\,. \qquad (4.4.2)$$

The left-hand part of Eq. (4.4.2) can be presented as a divergence of a tensor:

$$\bar{\Phi}\,\overleftarrow{D}_\alpha\,\beta^\alpha\,\overrightarrow{D}_\beta\,\Phi + \bar{\Phi}\,\overrightarrow{D}_\alpha\,\beta^\alpha\,\overrightarrow{D}_\beta\,\Phi$$

$$= \bar{\Phi}(\overleftarrow{\nabla}_\alpha - B_\alpha)\beta^\alpha\,\overrightarrow{D}_\beta\,\Phi + \bar{\Phi}(\overrightarrow{\nabla}_\alpha + B_\alpha)\beta^\alpha\,\overrightarrow{D}_\beta\,\Phi$$

$$= (\bar{\Phi}\,\overleftarrow{\nabla}_\alpha)\beta^\alpha\,\overrightarrow{D}_\beta\,\Phi + \bar{\Phi}(\overrightarrow{\nabla}_\alpha\,\beta^\alpha\,\overrightarrow{D}_\beta\,\Phi) = \nabla_\alpha(\bar{\Phi}\beta^\alpha\,\overrightarrow{D}_\beta\,\Phi) = \nabla_\alpha(A^\alpha{}_\beta)\,, \qquad (4.4.3)$$

where we can use the notation

$$A^\alpha{}_\beta = \bar{\Phi}\,\beta^\alpha\,\overrightarrow{D}_\beta\,\Phi = \bar{\Phi}\,\beta^\alpha(\overrightarrow{\nabla}_\beta + B_\beta - ieA_\beta)\Phi\,. \qquad (4.4.4)$$

Therefore, Eq. (4.4.2) reads as

$$\nabla_\alpha(A^\alpha{}_\beta) = \bar{\Phi}\,\beta^\alpha\,[\overrightarrow{D}_\alpha, \overrightarrow{D}_\beta]_-\Phi\,. \qquad (4.4.5)$$

Let us now detail the commutator $[\overrightarrow{D}_\alpha, \overrightarrow{D}_\beta]_-$:

$$[\overrightarrow{D}_\alpha, \overrightarrow{D}_\beta]_- = -ie\,F_{\alpha\beta} + D_{\alpha\beta}\,, \qquad F_{\alpha\beta} = \partial_\alpha A_\beta - \partial_\beta A_\alpha\,,$$

$$D_{\alpha\beta} = \nabla_\alpha B_\beta - \nabla_\beta B_\alpha + B_\alpha B_\beta - B_\beta B_\alpha\,. \qquad (4.4.6)$$

Remembering that the operators act on generally covariant scalars, and allowing for the symmetry of Christoffel symbols

$$(\nabla_\alpha\nabla_\beta - \nabla_\beta\nabla_\alpha)\,\Phi = 0\,,$$

for the first part of the operator $D_{\alpha\beta}$, we get:

$$\nabla_\alpha B_\beta - \nabla_\beta B_\alpha = \frac{1}{2}j^{ab}e^\nu{}_{(a)}\,[\,e_{(b)\nu;\beta;\alpha} - e_{(b)\nu;\alpha;\beta}\,]$$

$$+ \frac{1}{2}j^{ab}\,[\,e_{(a)\nu;\alpha}e^\nu{}_{(b);\beta} - e_{(a)\nu;\beta}e^\nu{}_{(b);\alpha}\,]\,. \qquad (4.4.7)$$

The second term in expression for $D_{\alpha\beta}$ is

$$(B_\alpha B_\beta - B_\beta B_\alpha) = \frac{1}{4}\left(j^{ab}j^{kl} - j^{kl}j^{ab}\right)\,\left[\,(e^\nu{}_{(a)}e_{(b)\nu;\alpha})\;e^\mu{}_{(k)}e_{(l)\mu;\beta})\,\right],$$

with the use of commutative relations for j^{ab}, it gives

$$(B_\alpha B_\beta - B_\beta B_\alpha) = -\frac{1}{2} j^{ab} \left[e_{(a)\nu;\alpha} e^\nu_{(b);\beta} - e_{(a)\nu;\beta} e^\nu_{(b);\alpha} \right] . \qquad (4.4.8)$$

When allowing for Eq. (4.4.8), this expression for $D_{\alpha\beta}$ in Eq. (4.4.6) takes the form

$$D_{\alpha\beta} = \frac{1}{2} j^{ab} e^\nu_{(a)} \left[e_{(b)\nu;\beta;\alpha} - e_{(b)\nu;\alpha;\beta} \right]$$

$$= \frac{1}{2} j^{ab} e^\nu_{(a)} \left[e^\rho_{(b)} R_{\rho\nu\beta\alpha}(x) \right] = \frac{1}{2} j^{\nu\rho}(x) R_{\nu\rho\alpha\beta}(x) , \qquad (4.4.9)$$

where $R_{\nu\rho\alpha\beta}(x)$ stands for the Riemann curvature tensor.

Thus, Eq. (4.4.5) will read as

$$\nabla_\alpha \left(A^\alpha{}_\beta \right) = -ie\, J^\alpha\, F_{\alpha\beta} + \frac{1}{2} R_{\nu\rho\alpha\beta} \, \bar{\Phi} \, \beta^\alpha(x)\, j^{\nu\rho}(x)\, \Phi . \qquad (4.4.10)$$

By using the properties of Duffin–Kemmer–Petiau matrices, the term containing the curvature tensor can be transformed to

$$\frac{1}{2} R_{\nu\rho\alpha\beta} \, \bar{\Phi} \, \beta^\alpha(x) j^{\nu\rho}(x)\, \Phi = \frac{1}{2} R_{\nu\rho\alpha\beta}$$

$$\times \bar{\Phi} \left[\frac{g^{\alpha\nu}(x)\beta^\rho(x) - g^{\alpha\rho}(x)\beta^\nu(x)}{2} + \frac{\beta^\alpha(x)j^{\nu\rho}(x) + j^{\nu\rho}(x)\beta^\alpha(x)}{2} \right] \Phi ,$$

so that

$$\frac{1}{2} R_{\nu\rho\alpha\beta} \, \bar{\Phi} \, \beta^\alpha(x) j^{\nu\rho}(x)\, \Phi = \frac{1}{2} R_{\alpha\beta} \, J^\alpha$$

$$+ \frac{1}{4} R_{\nu\rho\alpha\beta} \, \bar{\Phi} \left(\beta^\alpha j^{\nu\rho} + j^{\nu\rho}\beta^\alpha \right) \Phi . \qquad (4.4.11)$$

Thus, Eq. (4.4.10) reduces to

$$\nabla_\alpha \left(A^\alpha{}_\beta \right) = -ie\, J^\alpha(x) F_{\alpha\beta}$$

$$+ \frac{1}{2} R_{\alpha\beta} J^\alpha + \frac{1}{4} R_{\nu\rho\alpha\beta} \, \bar{\Phi} \left(\beta^\alpha j^{\nu\rho} + j^{\nu\rho}\beta^\alpha \right) \Phi . \qquad (4.4.12)$$

By using the properties of the matrix η

$$\eta^+ = \eta , \quad \eta^{-1}\beta^{a+}\eta = \beta^a , \quad \eta^{-1}j^{ab}\eta = -j^{ab} ,$$

we readily see that in the right-hand side of Eq. (4.4.12), the first and the third terms are imaginary, and the second is real. To separate real and imaginary parts in the left-hand side as well, let us specify a conjugate tensor too:

$$A^\alpha{}_\beta = \bar{\Phi} \, \beta^\alpha \, \vec{D}_\beta \, \Phi = \bar{\Phi} \, \beta^\alpha (\vec{\nabla}_\beta + B_\beta - ieA_\beta)\Phi ,$$

$$(A^\alpha{}_\beta)^+ = \bar{\Phi}\beta^\alpha \, \overleftarrow{D}_\beta \, \Phi = \bar{\Phi}\beta^\alpha (\overleftarrow{\nabla}_\beta - B_\beta + ieA_\beta)\Phi . \qquad (4.4.13)$$

Let us introduce the special notation for real and imaginary parts:

$$U^{\alpha}_{\ \beta}(x) = \frac{1}{2} \left[A^{\alpha}_{\ \beta} + (A^{\alpha}_{\ \beta})^{+} \right]$$

$$= \frac{1}{2} \left[\bar{\Phi}\ \beta^{\alpha}(\overrightarrow{\nabla}_{\beta} + B_{\beta})\Phi + \bar{\Phi}(\overleftarrow{\nabla}_{\beta} - B_{\beta})\beta^{\alpha}\Phi \right]$$

$$= \frac{1}{2}\ \nabla_{\beta}\ \bar{\Phi}\beta^{\alpha}\Phi = \frac{1}{2}\ \nabla_{\beta}J^{\alpha}\ , \qquad (4.4.14a)$$

$$T^{\alpha}_{\ \beta}(x) = \frac{1}{2i} \left[A^{\alpha}_{\ \beta} - (A^{\alpha}_{\ \beta})^{+} \right]$$

$$= \frac{1}{2i} \left[\bar{\Phi}\ \beta^{\alpha}(\overrightarrow{\nabla}_{\beta} + B_{\beta})\Phi - \bar{\Phi}\beta^{\alpha}(\overleftarrow{\nabla}_{\beta} - B_{\beta})\Phi \right] - e\ J^{\alpha}\ A_{\beta}\ . \qquad (4.4.14b)$$

Thus, instead of Eq. (4.4.14), we can derive two real equations:

$$\nabla_{\alpha}U^{\alpha}_{\ \beta} = \frac{1}{2}R_{\alpha\beta}J^{\alpha}\ , \qquad (4.4.15)$$

$$\nabla_{\alpha}T^{\alpha}_{\ \beta} = -e\ J^{\alpha}F_{\alpha\beta} + \frac{1}{4i}\ R_{\nu\rho\alpha\beta}\ \bar{\Phi}\ (\beta^{\alpha}j^{\nu\rho} + j^{\nu\rho}\beta^{\alpha})\ \Phi\ . \qquad (4.4.16)$$

It is readily verified that Eq. (4.4.15) is a simple consequence of the current conservation law for J^{α}. Indeed, we note the identity

$$U^{\alpha}_{\ \beta}(x) = \frac{1}{2}\nabla_{\beta}\ \bar{\Phi}\beta^{\alpha}(x)\Phi = \frac{1}{2}\ \nabla_{\beta}J^{\alpha}(x)\ , \qquad (4.4.17)$$

therefore, Eq. (4.4.15) will read as

$$\nabla_{\alpha}\nabla_{\beta}J^{\alpha} = R_{\alpha\beta}J^{\alpha}$$

so that

$$(\nabla_{\alpha}\nabla_{\beta} - \nabla_{\beta}\nabla_{\alpha})J^{\alpha} + \nabla_{\beta}\nabla_{\alpha}J^{\alpha} = R_{\alpha\beta}J^{\alpha}\ .$$

Hence, in allowing for the conservation law for J^{α} we arrive at the identity

$$J^{\rho}(x)\ R_{\rho}^{\ \alpha}_{\ \beta\alpha}(x) = R_{\alpha\beta}(x)\ J^{\alpha}(x)\ . \qquad (4.4.18)$$

Thus, Eq. (4.4.15) does not contain any additional information to the current conservation law for $J^{\alpha}(x)$.

In the case of flat Minkowski space, there is known principal freedom in determination of energy-momentum tensor [320]: if $T_{b}^{\ a}(x)$ satisfies the conservation law $\partial_{a}T_{b}^{\ a} = 0$, then another tensor

$$\bar{T}_{b}^{\ a}(x) = T_{b}^{\ a}(x) + \partial_{c}\left[\Omega_{b}^{\ [ac]}(x) \right]\ , \qquad \Omega_{b}^{\ [ac]}(x) = -\Omega_{b}^{\ [ca]}(x) \qquad (4.4.19a)$$

will also satisfy the conservation law $\partial_{a}\bar{T}_{b}^{\ a} = 0$; due to an identity

$$\partial_{a}\partial_{c}\ \Omega_{b}^{\ [ca]}(x) \equiv 0\ . \qquad (4.4.19b)$$

In the case of any curved space, the situation is not so simple. Indeed, let to tensor be related according to

$$\bar{T}_\beta{}^\alpha(x) = T_\beta{}^\alpha(x) + \nabla_\rho \left[\, \Omega_\beta{}^{[\alpha\rho]}(x) \,\right] . \qquad (4.4.20)$$

Acting on both sides of this identity by covariant derivative operator ∇_α, we get

$$\nabla_\alpha \bar{T}_\beta{}^\alpha(x) = \nabla_\alpha \, T_\beta{}^\alpha(x) + \nabla_\alpha \left[\, \nabla_\rho \, \Omega_\beta{}^{[\alpha\rho]}(x) \,\right] , \qquad (4.4.21)$$

where by the use of symmetry properties for the Riemann tensor, we derive

$$\nabla_\alpha \left[\, \nabla_\rho \, \Omega_\beta{}^{[\alpha\rho]}(x) \,\right] = \frac{1}{2} \, R_{\beta\sigma\nu\rho} \, \Omega^{\sigma[\nu\rho]} . \qquad (4.4.22)$$

Thus, relationship in Eq. (4.4.21) takes the form

$$\nabla_\alpha \bar{T}_\beta{}^\alpha(x) = \nabla_\alpha T_\beta{}^\alpha(x) + \frac{1}{2} \, R_{\beta\sigma\nu\rho}(x) \, \Omega^{\sigma[\nu\rho]}(x) . \qquad (4.4.23a)$$

This equation can be rewritten (see Eq. (4.4.16)) as follows:

$$\nabla_\alpha \bar{T}_\beta{}^\alpha = -e \, J^\alpha \, F_{\alpha\beta}$$

$$+ \frac{1}{4i} \, R_{\nu\rho\sigma\beta} \, \bar{\Phi} \, (\beta^\sigma \, J^{\nu\rho} + J^{\nu\rho} \, \beta^\sigma) \, \Phi + \frac{1}{2} \, R_{\beta\sigma\nu\rho} \, \Omega^{\sigma[\nu\rho]}(x) . \qquad (4.4.23b)$$

Note, that if one takes the quantity $\Omega^{\sigma[\nu\rho]}(x)$ as

$$\Omega^{\sigma[\nu\rho]}(x) = + \frac{1}{2i} \, \bar{\Phi} \, \left[\, \beta^\sigma \, J^{\nu\rho} + J^{\nu\rho} \, \beta^\sigma \,\right] \Phi , \qquad (4.4.24)$$

then two terms in Eq. (4.4.23b), involving the curvature tensor, mutually compensate each other. In result, we arrive at

$$\nabla_\alpha \, \bar{T}_\beta{}^\alpha = e \, F_{\beta\alpha} \, J^\alpha . \qquad (4.4.25)$$

It should be noted that formally this relation coincides with one obtained by Fock for a spin 1/2 particle in Riemannian space-time in presence of electromagnetic field.

4.5 Non-Relativistic Approximation, 10-Component Formalism

Let us consider a nonrelativistic Pauli equation for a particle with spin 1 in presence of external gravitational fields. To have a nonrelativistic approximation, we must use the limitation on space-time models:

$$dS^2 = (dx^0)^2 + g_{ij}(x)dx^i dx^j , \qquad e_{(a)\alpha}(x) = \begin{vmatrix} 1 & 0 \\ 0 & e_{(i)k}(x) \end{vmatrix} . \qquad (4.5.1)$$

DKP-equation in presence of the curved space background and electromagnetic field is

$$\left[\, i\beta^0 \, D_0 + i\beta^k(x) \, D_k - \frac{mc}{\hbar} \,\right] \Psi = 0 ,$$

$$D_\alpha = \partial_\alpha + B_\alpha(x) - i\frac{e}{\hbar c}A_\alpha(x) \ . \tag{4.5.2}$$

In the metric of Eq. (4.5.1), expressions for vector connections become much simpler, indeed

$$B_0 = \frac{1}{2}J^{ik}e^m_{(i)}(\nabla_0 e_{(k)m}) \ , \tag{4.5.3}$$

$$B_l = \frac{1}{2}J^{ik}e^m_{(i)}(\nabla_l e_{(k)m}) \ , \tag{4.5.4}$$

so there is no contribution from J^{0k} generators. Due to identities

$$\beta^0\beta^0 J^{kl} = J^{kl}\beta^0\beta^0] \qquad \Longrightarrow \qquad \beta^0\beta^0 B_\alpha(x) = B_\alpha(x)\beta^0\beta^0 \tag{4.5.5}$$

the operator D_k commutes with $(\beta^0)^2$. Multiplying Eq. (4.5.2) by projective $(\beta^0)^2$ and $1 - (\beta^0)^2$, and when taking into account the relations

$$(\beta^0)^2\beta^l(x) = \beta^l(x)\,[1 - (\beta^0)^2] \ ,$$

$$[\,1 - (\beta^0)^2\,]\,\beta^l(x) = \beta^l(x)(\beta^0)^2 \ , \qquad (\beta^0)^3 = \beta^0 \ , \tag{4.5.6}$$

we get equations for χ and φ:

$$\chi = (\beta^0)^2\Psi \ , \qquad \varphi = (1 - (\beta^0)^2)\Psi \ , \qquad \Psi = \chi + \varphi \ ,$$

$$i\beta^0\,D_0\chi + i\beta^k(x)\,D_k\varphi = \frac{mc}{\hbar}\,\chi \ , \qquad i\beta^k(x)\,D_k\chi = \frac{mc}{\hbar}\,\varphi \ . \tag{4.5.7}$$

Excluding a non-dynamical part φ, we arrive at

$$i\beta^0\,D_0\chi - \frac{\hbar}{mc}\,\beta^k(x)\,\beta^l(x)\,D_k D_l\,\chi = \frac{mc}{\hbar}\,\chi \ . \tag{4.5.8}$$

Now, let us introduce two operators

$$\Pi_\pm = \frac{1}{2}\,\beta^0\,(1 \pm \beta^0) \ , \qquad \Pi_+\beta^0 = +\,\Pi_+ \ , \qquad \Pi_-\beta^0 = -\,\Pi_- \ . \tag{4.5.9}$$

From Eq. (4.5.8), it follows

$$iD_0\,\Pi_+\,\chi - \frac{\hbar}{mc}\,\Pi_+\,\beta^k(x)\beta^l(x)\,D_k D_l\,\chi - \frac{mc}{\hbar}\,\Pi_+\,\chi = 0 \ ; \tag{4.5.10}$$

with the help of

$$\Pi_+\,\beta^k(x)\beta^l(x) = \frac{1}{2}\,[\,(-\beta^l(x)\beta^k(x) + g^{lk}(x))\,\beta^0 + \beta^k(x)\beta^l(x)\,(\beta^0)^2\,] \ ,$$

Eq. (4.5.10) allows for

$$iD_0\,(\Pi_+\chi) - \frac{\hbar}{2mc}\,[\,(-\beta^l(x)\,\beta^k(x) + g^{lk}(x))\,\beta^0$$

$$+\beta^k(x)\,\beta^l(x)\,(\beta^0)^2\,]\,D_k D_l\,\chi = \frac{mc}{\hbar}\,(\Pi_+\chi) \ . \tag{4.5.11}$$

In the same manner, starting from

$$-iD_0\,\Pi_-\,\chi - \frac{\hbar}{mc}\,\Pi_-\,\beta^k(x)\,\beta^l(x)\,D_kD_l\,\chi = \frac{mc}{\hbar}\,\Pi_-\,\chi\,, \tag{4.5.12}$$

with the help of

$$\Pi_-\,\beta^k(x)\beta^l(x) = \frac{1}{2}\,[\,(-\beta^l(x)\beta^k(x) + g^{kl}(x))\,\beta^0 - \beta^k(x)\beta^l(x)\,(\beta^0)^2\,]\,, \tag{4.5.13}$$

we get

$$-iD_0\,\Pi_-\chi - \frac{\hbar}{2mc}\,[\,(-\beta^l(x)\,\beta^k(x) + g^{kl}(x))\,\beta^0$$
$$-\beta^k(x)\,\beta^l(x)\,(\beta^0)^2\,]\,D_kD_l\,\chi = \frac{mc}{\hbar}\,\Pi_-\,\chi\,. \tag{4.5.14}$$

Changing matrices β^0 and $(\beta^0)^2$ by

$$\Pi_+ + \Pi_- = \beta^0\,, \qquad \Pi_+ - \Pi_- = \beta^0\beta^0\,,$$

and using the notation (let $\Pi_-\chi = \chi_-$, $\Pi_+\chi = \chi_+$)

$$J^{[kl]}(x) = \beta^k(x)\,\beta^l(x) - \beta^l(x)\,\beta^k(x)\,, \qquad J^{(kl)}(x) = \beta^k(x)\,\beta^l(x) + \beta^l(x)\,\beta^k(x)\,,$$

we can reduce Eqs. (4.5.11) and (4.5.14) to the form

$$iD_0\chi_+ - \frac{\hbar}{2mc}\,[\,J^{[kl]}(x)D_kD_l\chi_+ - J^{(kl)}(x)D_kD_l\chi_- + D^lD_l(\chi_+ + \chi_-)\,] = \frac{mc}{\hbar}\,\chi_+\,,$$

$$-iD_0\chi_- - \frac{\hbar}{2mc}\,[\,J^{[kl]}(x)D_kD_l\chi_- - J^{(kl)}(x)D_kD_l\chi_+ + D^lD_l(\chi_+ + \chi_-)\,] = \frac{mc}{\hbar}\,\chi_-\,.$$

$$\tag{4.5.15}$$

Now, one should separate a special factor depending on the rest energy, for Eq. (4.5.15) one should make one formal change:

$$\chi \implies \exp(-i\frac{mc^2}{\hbar}t)\,\chi\,, \qquad iD_0 \implies iD_0 + \frac{mc}{\hbar}\,. \tag{4.5.16}$$

As a result, Eq. (4.5.15) gives

$$iD_0\chi_+ - \frac{\hbar}{2mc}[(J^{[kl]}(x)D_kD_l\chi_+ - J^{(kl)}(x)D_kD_l\chi_-) + D^lD_l(\chi_+ + \chi_-)] = 0\,,$$

$$-iD_0\chi_- - \frac{\hbar}{2mc}[(J^{[kl]}(x)D_kD_l\chi_- - J^{(kl)}(x)D_kD_l\chi_+) + D^lD_l(\chi_+ + \chi_-)] = 2\frac{mc}{\hbar}\chi_-\,.$$

$$\tag{4.5.17}$$

Now, when considering that χ_- is small and ignoring the term $-iD_0\chi_-$ compared with $\frac{mc}{\hbar}\chi_-$ we arrive at

$$(J^{(kl)}\,D_kD_l - D^lD_l)\,\chi_+ = \frac{4m^2c^2}{\hbar^2}\,\chi_-\,,$$

$$i\hbar D_t \, \chi_+ = \frac{\hbar^2}{2m} \, (D^l D_l + J^{[kl]} \, D_k D_l) \, \chi_+ \,. \qquad (4.5.18)$$

The second equation in Eq. (4.5.18) should be considered as a nonrelativistic Pauli equation for spin 1 particle in DKP-approach.

It is interesting to see what is the form of the nonrelativistic approximation in tensor form? At first, let us restrict ourselves to the case of the flat space. From Eq. (4.5.7), it follows:

$$\beta^0 \begin{vmatrix} \Phi_0 \\ \Phi_1 \\ \Phi_2 \\ \Phi_3 \\ \Phi_{01} \\ \Phi_{02} \\ \Phi_{03} \\ \Phi_{23} \\ \Phi_{31} \\ \Phi_{12} \end{vmatrix} = \begin{vmatrix} 0 \\ i\Phi_{01} \\ i\Phi_{02} \\ i\Phi_{03} \\ -i\Phi_1 \\ -i\Phi_2 \\ -i\Phi_3 \\ 0 \\ 0 \\ 0 \end{vmatrix}, \quad \chi = (\beta^0)^2 \, \Psi = \begin{vmatrix} 0 \\ \Phi_1 \\ \Phi_2 \\ \Phi_3 \\ \Phi_{01} \\ \Phi_{02} \\ \Phi_{03} \\ 0 \\ 0 \\ 0 \end{vmatrix}, \quad \varphi = (1 - (\beta^0)^2) \, \Psi = \begin{vmatrix} \Phi_0 \\ 0 \\ 0 \\ 0 \\ 0 \\ 0 \\ 0 \\ \Phi_{23} \\ \Phi_{31} \\ \Phi_{12} \end{vmatrix},$$

$$(4.5.19a)$$

so that

$$\chi_+ = \frac{1}{2} \begin{vmatrix} 0 \\ (\Phi_1 + i \, \Phi_{01}) \\ (\Phi_2 + i \, \Phi_{02}) \\ (\Phi_3 + i \, \Phi_{03}) \\ -i(\Phi_1 + i\Phi_{01}) \\ -i(\Phi_2 + i\Phi_{02}) \\ -i(\Phi_3 + i\Phi_{03}) \\ 0 \\ 0 \\ 0 \end{vmatrix}, \quad \chi_- = \frac{1}{2} \begin{vmatrix} 0 \\ -(\Phi_1 - i \, \Phi_{01}) \\ -(\Phi_2 - i \, \Phi_{02}) \\ -(\Phi_3 - i \, \Phi_{03}) \\ -i(\Phi_1 - i\Phi_{01}) \\ -i(\Phi_2 - i\Phi_{02}) \\ -i(\Phi_3 - i\Phi_{03}) \\ 0 \\ 0 \\ 0 \end{vmatrix}. \qquad (4.5.19b)$$

Instead of $\Phi_k \, \Phi_{0k}$, let us introduce new field variables:

$$\frac{1}{2} \, (\Phi_k \, - \, i \, \Phi_{0k}) = M_k \,, \qquad \frac{1}{2} \, (\Phi_k \, + \, i \, \Phi_{0k}) = B_k \,,$$

$$\Phi_k = B_k \, + \, M_k \,, \qquad \Phi_{0k} = -i \, (B_k \, - \, M_k) \,; \qquad (4.5.20)$$

that is

$$\chi_+ = \begin{vmatrix} 0 \\ \mathbf{B} \\ -i \, \mathbf{B} \\ 0 \\ 0 \\ 0 \end{vmatrix}, \quad \chi_- = \begin{vmatrix} 0 \\ -\mathbf{M} \\ -i \, \mathbf{M} \\ 0 \\ 0 \\ 0 \end{vmatrix}. \qquad (4.5.21)$$

Thus, in tensor representation the big and small components coincides with 3-vectors $\mathbf{B}$ and $\mathbf{M}$, respectively. Now, it is a matter of performing a simple calculation to repeat the limiting procedure in tensor basis. Indeed, starting from Proca equations (it is convenient to change the notation $mc/\hbar \Longrightarrow m$)

$$D_0 \, \Phi_k \; - \; D_k \, \Phi_0 = m \, \Phi_{0k} \; , \quad D_k \, \Phi_l \; - \; D_l \, \Phi_k = m \, \Phi_{kl} \; ,$$

$$D^l \, \Phi_{0l} = m \, \Phi_0 \; , \qquad D^0 \, \Phi_{k0} \; + \; D^l \, \Phi_{kl} = m \, \Phi_k \; , \qquad (4.5.22)$$

and excluding the non-dynamical components Φ_0, Φ_{kl},

$$D_0 \, \Phi_k \; - \; \frac{1}{m} \, D_k \, D^l \, \Phi_{0l} = m \, \Phi_{0k} \; ,$$

$$D^0 \, \Phi_{k0} \; + \; \frac{1}{m} \, D^l \, (D_k \, \Phi_l \; - \; D_l \, \Phi_k) = m \, \Phi_k \; , \qquad (4.5.23)$$

and further

$$m \, (\Phi_k \; \pm \; i \, \Phi_{0k}) = (\, D^0 \, \Phi_{k0} \; + \; \frac{1}{m} \, D^l \, D_k \, \Phi_l$$

$$- \frac{1}{m} \, D^l \, D_l \, \Phi_k \,) \; \pm i \, (\, D_0 \, \Phi_k \; - \; \frac{1}{m} \, D_k \, D^l \, \Phi_{0l} \,) \; . \qquad (4.5.24)$$

From these, with the help of Eq. (4.5.20), we get to

$$2m \, B_k = +2i \, D_0 B_k - \frac{1}{m} \, D^l D_l (B_k \; + \; M_k)$$

$$+ \frac{1}{m} \, [\, (D^l \, D_k \; - \; D_k \, D^l) \, B_l \; + \; (D^l \, D_k \; + \; D_k \, D^l) \, M_l \,] \; , \qquad (4.5.25)$$

$$2m \, M_k = -2i \, D_0 \, M_k \; - \; \frac{1}{m} \, D^l \, D_l \, (B_k \; + \; M_k)$$

$$+ \frac{1}{m} \, [\, (D^l \, D_k \; + \; D_k \, D^l) \, B_l \; + (D^l \, D_k \; - \; D_k \, D^l) \, M_l \,] \; . \qquad (4.5.26)$$

After separating the rest energy term

$$i \, D_0 \, B_k \; \Longrightarrow \; (i \, D_0 \; + \; m) \, B_k \; , \quad i \, D_0 \, M_k \; \Longrightarrow \; (i \, D_0 \; + \; m) \, M_k \; ;$$

From Eq. (4.5.26), we arrive at

$$+i \, D_0 \, B_k \; - \; \frac{1}{2m} \, \{ \, D^l \, D_l \, (B_k \; + \; M_k)$$

$$+ \, (D^l \, D_k \; - \; D_k \, D^l) \, B_l \; + (D^l \, D_k \; + \; D_k \, D^l) \, M_l \, \} = 0 \; ,$$

$$-i \, D_0 \, M_k \; - \; \frac{1}{2m} \, \{ \, D^l \, D_l \, (B_k \; + \; M_k)$$

$$+ (D^l \, D_k \; + \; D_k \, D^l) \, B_l \; + (D^l \, D_k \; - \; D_k \, D^l) \, M_l \, \} = 4m \, M_k \; . \qquad (4.5.27)$$

Therefore, a nonrelativistic Pauli equation for the big component **B** has the form (let us change the notation: $B_k(x) \Longrightarrow \psi_k(x)$)

$$+i \, D_0 \, \psi_k = \frac{1}{2m} \, [\, -D_l \, D_l \, \psi_k \; - \; (D_k \, D_l \; - \; D_l \, D_k) \, \psi_l \,] \; . \qquad (4.5.28)$$

4.6   Tetrad 3-Dimensional Nonrelativistic Equation

The nonrelativistic equation for Eq. (4.5.18) in DKP-formalism is symbolical in a sense, because it is written formally for a 10-component function though in fact the nonrelativistic function is a 3-vector. Let us turn to this limiting procedure again (for brevity, $e/\hbar c \Longrightarrow e$, $mc/\hbar \Longrightarrow m$)

$$[\, i\beta^0 \, D_0 + i\beta^l(x) \, D_l - m \,] \, \Psi = 0 \, , \qquad (4.6.1)$$

where

$$\beta^l(x) = \beta^k e^l_{(k)}(x) \, , \qquad D_l = \partial_l + B_l - ieA_l \, , \qquad D_0 = \partial_0 + B_0 - ieA_0 \, ,$$

$$B_0 = \frac{1}{2} J^{ik} e^m_{(i)} \, (\nabla_0 e_{(k)m}) \, , \qquad B_l = \frac{1}{2} J^{ik} e^m_{(i)} \, (\nabla_l e_{(k)m}) \, . \qquad (4.6.2)$$

In the use of block-form for DKP-matrices

$$\beta^0 = \begin{vmatrix} 0 & 0 & 0 & 0 \\ 0 & 0 & i_3 & 0 \\ 0 & -i_3 & 0 & 0 \\ 0 & 0 & 0 & 0 \end{vmatrix} \, , \qquad \beta^k = \begin{vmatrix} 0 & 0 & w_k & 0 \\ 0 & 0 & 0 & \tau_k \\ \tilde{w}_k & 0 & 0 & 0 \\ 0 & -\tau_k & 0 & 0 \end{vmatrix} \, , \qquad (4.6.3)$$

where

$$w_1 = (i, 0, 0) \, , \qquad w_2 = (0, i, 0) \, , \qquad w_3 = (0, 0, i) \, ,$$

$$\tilde{w}_1 = \begin{vmatrix} i \\ 0 \\ 0 \end{vmatrix} \, , \qquad \tilde{w}_2 = \begin{vmatrix} 0 \\ i \\ 0 \end{vmatrix} \, , \qquad \tilde{w}_3 = \begin{vmatrix} 0 \\ 0 \\ i \end{vmatrix} \, , \qquad i_3 = \begin{vmatrix} i & 0 & 0 \\ 0 & i & 0 \\ 0 & 0 & i \end{vmatrix} \, ,$$

$$\tau_1 = \begin{vmatrix} 0 & 0 & 0 \\ 0 & 0 & -i \\ 0 & i & 0 \end{vmatrix} \, , \qquad \tau_2 = \begin{vmatrix} 0 & 0 & i \\ 0 & 0 & 0 \\ -i & 0 & 0 \end{vmatrix} \, , \qquad \tau_3 = \begin{vmatrix} 0 & -i & 0 \\ i & 0 & 0 \\ 0 & 0 & 0 \end{vmatrix} \qquad (4.6.4)$$

when taking the explicit form of generators J^{kl}, for connections $B_0(x)$ and $B_l(x)$, we have expressions:

$$B_0(x) = \begin{vmatrix} 0 & 0 & 0 & 0 \\ 0 & b_0(x) & 0 & 0 \\ 0 & 0 & b_0(x) & 0 \\ 0 & 0 & 0 & b_0(x) \end{vmatrix} \, , \qquad B_l(x) = \begin{vmatrix} 0 & 0 & 0 & 0 \\ 0 & b_l(x) & 0 & 0 \\ 0 & 0 & b_l(x) & 0 \\ 0 & 0 & 0 & b_l(x) \end{vmatrix} \, ,$$

$$(4.6.5)$$

where

$$b_0(x) = -i \, [\, \tau_1 \, e^k_{(2)} \, \partial_0 e_{(3)k} + \tau_2 \, e^k_{(3)} \, \partial_0 e_{(1)k} + \tau_3 \, e^k_{(1)} \, \partial_0 e_{(2)k} \,] \, ,$$

$$b_l(x) = -i \, [\, \tau_1 \, e^k_{(2)} \, \nabla_l e_{(3)k} + \tau_2 \, e^k_{(3)} \, \nabla_l e_{(1)k} + \tau_3 \, e^k_{(1)} \, \nabla_l e_{(2)k} \,] \, . \qquad (4.6.6)$$

Therefore, Eq. (4.6.1) can be rewritten as a system for constituents $\Psi(x) = (\Phi_0(x), \Phi(x), E(x), H(x))$:

$$iw^l(x) \, (\nabla_l + b_l - ieA_l) \, E = m \, \Phi_0 \, ,$$

$$-(\nabla_0 + b_0 - ieA_0)\, E + i\tau^l(x)\, (\nabla_l + b_l - ieA_l)\, H = m\, \Phi \,,$$

$$i\tilde{w}^l(x)\, (\nabla_l - ieA_l)\, \Phi_0 + (\nabla_0 + b_0 - ieA_0)\, \Phi = m\, E \,,$$

$$-i\tau^l(x)\, (\nabla_l + b_l - ieA_l)\, \Phi = m\, H \,, \tag{4.6.7}$$

where

$$\tau^l(x) = e^l_{(k)}(x)\, \tau^k, \qquad w^l(x) = e^l_{(k)}(x)\, w^k, \qquad \tilde{w}^l(x) = e^l_{(k)}(x)\, \tilde{w}^k \,. \tag{4.6.8}$$

After excluding non-dynamical variables $\Phi_0(x)$ and $H(x)$

$$-(\nabla_0 + b_0 - ieA_0)\, E + i\tau^l(x)\, (\nabla_l + b_l - ieA_l)\, (-\frac{i}{m})\, \tau^k(x)\, (\nabla_k + b_k - ieA_k)\, \Phi = m\, \Phi \,,$$

$$(\nabla_0 + b_0 - ieA_0)\, \Phi + i\tilde{w}^l(x)\, (\nabla_l - ieA_l)\, \frac{i}{m} w^k(x)\, (\nabla_k + b_k - ieA_k)\, E = m\, E \,. \tag{4.6.9}$$

We shall then allow for the commutative relations

$$\tau^k(x)\, (\nabla_l + b_l - ieA_l) = (\nabla_l + b_l - ieA_l)\, \tau^k(x) \,,$$

$$(\nabla_l - ieA_l)\, w^k(x) = w^k(x)\, (\nabla_l + b_l - ieA_l) \,, \tag{4.6.10}$$

Equations (4.6.9) reduce to ($\bullet$ represents dyal multiplication of vectors)

$$-(\nabla_0 + b_0 - ieA_0)\, E + \frac{1}{m}\, \tau^l(x)\tau^k(x)\, (\nabla_l + b_l - ieA_l)\, (\nabla_k + b_k - ieA_k)\, \Phi = m\, \Phi \,,$$

$$+(\nabla_0 + b_0 - ieA_0)\, \Phi - \frac{1}{m}\, \tilde{w}^l(x) \bullet w^k(x)\, (\nabla_l + b_l - ieA_l)\, (\nabla_k + b_k - ieA_k)\, E = m\, E \,,$$

or

$$D_0 E + \frac{1}{m}\, \tau^l(x)\tau^k(x)\, D_l D_k\, \Phi = m\, \Phi \,,$$

$$D_0 \Phi - \frac{1}{m}\, \tilde{w}^l(x) \bullet w^k(x)\, D_l D_k\, E = m\, E \,. \tag{4.6.11}$$

Instead of $\Phi(x)$ and $E(x)$ let us introduce $\psi(x)$ and $\varphi(x)$:

$$\frac{1}{2}\, (\Phi + iE) = \psi \,, \qquad \frac{1}{2}\, (\Phi - iE) = \varphi \,. \tag{4.6.12}$$

Equations (4.6.11) will look

$$2m\, \psi = +2iD_0\, \psi + \frac{1}{m}\, \tau^l(x)\tau^k(x)\, D_l D_k\, (\psi + \varphi) - \frac{1}{m}\, \tilde{w}^l(x) \bullet w^k(x)\, D_l D_k\, (\psi - \varphi) \,,$$

$$2m\, \varphi = -2iD_0\, \varphi + \frac{1}{m}\, \tau^l(x)\tau^k(x)\, D_l D_k\, (\psi + \varphi) + \frac{1}{m}\, \tilde{w}^l(x) \bullet w^k(x)\, D_l D_k\, (\psi - \varphi) \,. \tag{4.6.13}$$

Making the formal change $iD_0 \Longrightarrow (iD_0 + m)$, we get to

$$0 = +2iD_0\, \psi + \frac{1}{m}\, \tau^l(x)\tau^k(x)\, D_l D_k\, (\psi + \varphi) - \frac{1}{m}\, \tilde{w}^l(x) \bullet w^k(x)\, D_l D_k\, (\psi - \varphi) \,,$$

$$4m\,\varphi = -2iD_0\,\varphi + \frac{1}{m}\,\tau^l(x)\tau^k(x)\,D_lD_k\,(\psi+\varphi) + \frac{1}{m}\,\tilde{w}^l(x)\bullet w^k(x)\,D_lD_k\,(\psi-\varphi)\,.$$

$$(4.6.14)$$

From Eq. (4.6.14), using $\psi(x)$ as a big component and $\varphi(x)$ as small, we arrive at:

$$iD_0\,\psi(x) = \frac{1}{2m}\,[\,\tilde{w}^l(x)\bullet w^k(x) - \tau^l(x)\tau^k(x)\,]\,D_lD_k\,\psi(x)\,,$$

$$4m^2\,\varphi(x) = +\,[\,\tau^l(x)\tau^k(x) + \tilde{w}^l(x)\bullet w^k(x)\,]\,D_lD_k\,\psi(x)\,. \qquad (4.6.15)$$

From Eq. (4.6.15), taking into account the identity $\tau^l(x)\tau^k(x) = -g^{lk}(x) + \tilde{w}^k(x)\bullet w^l(x)$ and using notation

$$\tilde{w}^l(x)\bullet w^k(x) + \tilde{w}^k(x)\bullet w^l(x) = w^{(lk)}(x)\,,$$

$$\tilde{w}^l(x)\bullet w^k(x) - \tilde{w}^k(x)\bullet w^l(x) = j^{lk}(x)\,,$$

$$j^{ps} = -i\epsilon_{psj}\tau_j\,, \qquad j^{lk}(x) = e^l_{(p)}e^k_{(s)}j^{ps}\,,$$

we will obtain

$$4m^2\,\varphi(x) = \left(\,-D^lD_l + w^{(lk)}(x)\,D_lD_k\,\right)\psi\,,$$

$$iD_0\,\psi(x) = \frac{1}{2m}\,\left(\,+D^lD_l + \frac{1}{2}\,j^{lk}(x)\,[D_l, D_k]_-\,\right)\psi(x)\,. \qquad (4.6.16)$$

Thus, the Pauli equation for 3-vector wave function in Riemannian space is (compare with Eq. (4.5.18))

$$iD_t\,\psi = \frac{1}{2m}\,\left(\,D^lD_l + \frac{1}{2}\,j^{lk}(x)\,[D_l, D_k]_-\,\right)\psi\,. \qquad (4.6.17)$$

One additional point should be stressed here. Take notice that the nonrelativistic wave function is constructed in terms of relativistic ones as follows:

$$\psi(x) = \frac{1}{2}\,[\,\Phi_i(x) + i\,E_i(x)\,]\,, \qquad \text{where} \qquad E_i(x) = \Phi_{0i}(x)\,; \qquad (4.6.18)$$

this function ψ is complex even if we start with real-valued relativistic components.

Let us add some details about the term $\frac{1}{2}j^{lk}(x)\,[D_l, D_k]_-$ entered Eq. (4.6.17):

$$\frac{1}{2}\,j^{lk}(x)\,[D_l, D_k]_-\,\psi = \frac{1}{2}\,j^{lk}(x)\,(\,-ieF_{lk} + \nabla_l b_k - \nabla_k b_l + b_l b_k - b_k b_l\,)\,\psi\,.$$

Now, let us take into account the relations

$$\nabla_l b_k - \nabla_k b_l = +\,j^{dc}\,(\nabla_l e_{(d)m})\,(\nabla_k e^m_{(c)}) + \frac{1}{2}\,j^{dc}\,e_{(d)m}\,\{\,\nabla_l\nabla_k - \nabla_k\nabla_l\,\}\,e_{(c)m}$$

and

$$b_l b_k - b_k b_l = \frac{1}{4}\,(j^{ps}j^{cd} - j^{cd}j^{ps})\,e^n_{(p)}\,(\nabla_l e_{(s)n})\,e^m_{(c)}\,(\nabla_k e_{(d)m})$$

$$= -\,j^{dc}\,(\nabla_l e_{(d)m})\,(\nabla_k e^m_{(c)})\,,$$

we find

$$
b_l b_k - b_k b_l + \nabla_l b_k - \nabla_k b_l = \frac{1}{2}\, j^{dc}\, e^m_{(d)}\, R_{klmn}\, e^n_{(c)} = \frac{1}{2}\, j^{mn}(x)\, R_{lkmn}\ ; \qquad (4.6.19)
$$

R_{lkmn} is a Riemann curvature tensor for 3-space, and Eq. (4.6.17) can be written as

$$
iD_t\,\psi = \frac{1}{2m}\left[\, D^l D_l - ie\,\frac{1}{2}\, j^{lk}(x)\, F_{lk} + \frac{1}{4}\, j^{lk}(x)\, j^{mn}(x)\, R_{lkmn}\,\right]\psi\ . \qquad (4.6.20)
$$

In turn, one can readily verify

$$
\frac{1}{4} j^{lk}(x)\, j^{mn}(x) R_{lkmn} = \frac{1}{4}\,(-i\epsilon_{prc}\tau_c)\, e^l_{(p)} e^k_{(r)}\,(-i\epsilon_{std}\tau_d)\, e^m_{(s)} e^n_{(t)}\, R_{lkmn}
$$

$$
= -\frac{1}{4}\,\vec{\tau}\,\left(\,\mathbf{e}^l \times \mathbf{e}^k\,\right) R_{lkmn}\,\vec{\tau}\,(\,\mathbf{e}^m \times \mathbf{e}^n\,)\ . \qquad (4.6.21)
$$

Thus, the Pauli equation for meson in a curved space looks as follows (in ordinary units)

$$
i\hbar\, D_t\,\psi = -\frac{\hbar^2}{2m}\left[\, -D^l D_l + i\,\frac{e}{\hbar c}\,\frac{1}{2}\, j^{lk}(x)\, F_{lk}\right.
$$

$$
\left. +\frac{1}{4}\,\vec{\tau}\,(\,\mathbf{e}^l \times \mathbf{e}^k\,)\, R_{lkmn}\,\vec{\tau}\,(\,\mathbf{e}^m \times \mathbf{e}^n\,)\,\right]\psi\ . \qquad (4.6.22)
$$

4.7   A Scalar Particle, Nonrelativistic Approximation

Now let us turn to the case of a scalar particle. The Klein–Fock–Gordon equation in a curved space is

$$
\left[(i\hbar\,\nabla_\alpha + \frac{e}{c}A_\alpha)\, g^{\alpha\beta}(x)\,(i\hbar\,\nabla_\beta + \frac{e}{c}A_\beta)\, - \frac{\hbar^2}{6}\, R\, - m^2 c^2\right]\Psi(x) = 0\ . \qquad (4.7.1)
$$

It is important to take notice on the additional interaction term through scalar curvature $R(x)$ (F. Gürsey [290]). This equation may be changed to the form more convenient in application. Here, the use of the known relations [91]

$$
\nabla_\alpha g^{\alpha\beta}(x)\nabla_\beta\,\Psi = \frac{1}{\sqrt{-g}}\,\frac{\partial}{\partial x^\alpha}\,\sqrt{-g}g^{\alpha\beta}\,\frac{\partial}{\partial x^\beta}\,\Psi\ ,
$$

$$
\nabla_\alpha g^{\alpha\beta} A_\beta = \frac{1}{\sqrt{-g}}\,\frac{\partial}{\partial x^\alpha}\,\sqrt{-g}g^{\alpha\beta}A_\beta\ , \qquad g = \det\,(g_{\alpha\beta})\ , \qquad (4.7.2)
$$

Eq. (4.7.1) is changed to

$$
\left[\frac{1}{\sqrt{-g}}(i\hbar\frac{\partial}{\partial x^\alpha} + \frac{e}{c}A_\alpha)\,\sqrt{-g}g^{\alpha\beta}(x)(i\hbar\frac{\partial}{\partial x^\beta} + \frac{e}{c}A_\beta) - \frac{\hbar^2}{6}\, R\, - m^2 c^2\right]\Psi(x) = 0\ .
$$
$$
(4.7.3)
$$

What is the Schrödinger's nonrelativistic equation in the curved space-time? Let us begin with a generally covariant first order equations for a scalar particle (take notice to the additional interaction term through the Ricci scalar)

$$\left(i\,\nabla_\alpha + \frac{e}{c\hbar}A_\alpha \right)\Phi = \frac{mc}{\hbar}\,\Phi_\alpha \, ,$$

$$\left(i\,\nabla_\alpha + \frac{e}{c\hbar}A_\alpha \right)\Phi^\alpha = \frac{mc}{\hbar}\left(1 + \sigma\,\frac{R(x)}{m^2c^2/\hbar^2} \right)\Phi \, . \tag{4.7.4}$$

In this notation

$$1 + \sigma\,\frac{R(x)}{m^2c^2/\hbar^2} = \Gamma(x) \, .$$

Eqs. (4.7.4) will read as

$$\left(i\,\partial_\alpha + \frac{e}{c\hbar}A_\alpha \right)\Phi(x) = \frac{mc}{\hbar}\,\Phi_\alpha \, ,$$

$$\left(\frac{i}{\sqrt{-g}}\frac{\partial}{\partial x^\alpha}\sqrt{-g} + \frac{e}{c\hbar}\,A_\alpha \right)g^{\alpha\beta}\Phi_\beta = \frac{mc}{\hbar}\,\Gamma\,\Phi \, . \tag{4.7.5}$$

In the space-time models of the considered type, one can easily separate time- and space-variables in the wave equation

$$\left(i\,\partial_0 + \frac{e}{c\hbar}A_0 \right)\Phi = \frac{mc}{\hbar}\,\Phi_0 \, , \qquad \left(i\,\partial_l + \frac{e}{c\hbar}A_l \right)\Phi = \frac{mc}{\hbar}\,\Phi_l \, ,$$

$$\left(i\frac{\partial}{\partial x^0} + \frac{i}{\sqrt{-g}}\frac{\partial\sqrt{-g}}{\partial x^0} + \frac{e}{c\hbar}\,A_0 \right)\Phi_0$$

$$+ \left(\frac{i}{\sqrt{-g}}\frac{\partial}{\partial x^k}\sqrt{-g} + \frac{e}{c\hbar}\,A_k \right)g^{kl}\Phi_l = \frac{mc}{\hbar}\,\Gamma\,\Phi \, . \tag{4.7.6}$$

Now one should separate the rest energy term by means of the substitutions:

$$\Phi \Longrightarrow \exp\left(-i\frac{mc^2t}{\hbar} \right)\Phi \, , \quad \Phi_0 \Longrightarrow \exp\left(-i\frac{mc^2t}{\hbar} \right)\Phi_0 \, , \quad \Phi_l \Longrightarrow \exp\left(-i\frac{mc^2t}{\hbar} \right)\Phi_l \, .$$

As a result, Eq. (4.7.6) will give

$$\left(i\hbar\,\partial_t + mc^2 + eA_0 \right)\Phi(x) = mc^2\,\Phi_0(x) \, , \tag{4.7.7a}$$

$$\left(i\hbar\,\partial_t + mc^2 + i\hbar\frac{1}{\sqrt{-g}}\frac{\partial\sqrt{-g}}{\partial t} + e\,A_0 \right)\Phi_0$$

$$+ c\left(\frac{i\hbar}{\sqrt{-g}}\frac{\partial}{\partial x^k}\sqrt{-g} + \frac{e}{c}\,A_k \right)g^{kl}\Phi_l = mc^2\,\Gamma\,\Phi(x) \, , \tag{4.7.7b}$$

$$\left(i\,\hbar\,\partial_l + \frac{e}{c}A_l \right)\Phi(x) = mc\,\Phi_l(x) \, . \tag{4.7.7c}$$

By using the help of Eq. (4.7.7c), the non-dynamical variable Φ_l can be readily excluded:

$$\left(i\hbar\,\partial_t + mc^2 + eA_0 \right)\Phi(x) = mc^2\,\Phi_0(x) \, , \tag{4.7.8a}$$

$$\left(i\hbar\, \partial_t + mc^2 + i\hbar \frac{1}{\sqrt{-g}}\, \frac{\partial \sqrt{-g}}{\partial t} + e\, A_0 \right) \Phi_0$$

$$+ \frac{1}{m} \left[\left(\frac{i\hbar}{\sqrt{-g}}\, \partial_k \sqrt{-g} + \frac{e}{c}\, A_k \right) g^{kl} \left(i\,\hbar\, \partial_l + \frac{e}{c} A_l \right) \right] \Phi(x) = mc^2\, \Gamma\, \Phi(x)\, . \qquad (4.7.8b)$$

Now, we are to introduce a small φ and big Ψ components:

$$\Phi - \Phi_0 = \varphi, \qquad \Phi + \Phi_0 = \Psi\, ;$$

$$\Phi = \frac{\Psi + \varphi}{2}, \qquad \Phi_0 = \frac{\Psi - \varphi}{2}\, . \qquad (4.7.9)$$

Substituting Eq. (4.7.9) into Eqs. (4.7.8a) and (4.7.8b), one gets:

$$\left(i\hbar\, \partial_t + mc^2 + e A_0 \right) \frac{\Psi + \varphi}{2} = mc^2\, \frac{\Psi - \varphi}{2}\, , \qquad (4.7.10a)$$

$$\left(i\hbar\, \partial_t + mc^2 + i\hbar \frac{1}{\sqrt{-g}}\, \frac{\partial \sqrt{-g}}{\partial t} + e\, A_0 \right) \frac{\Psi - \varphi}{2}$$

$$+ \frac{1}{m} \left[\left(\frac{i\hbar}{\sqrt{-g}}\, \partial_k \sqrt{-g} + \frac{e}{c}\, A_k \right) g^{kl} \left(i\,\hbar\, \partial_l + \frac{e}{c} A_l \right) \right] \frac{\Psi + \varphi}{2} = mc^2\, \Gamma\, \frac{\Psi + \varphi}{2}\, , \qquad (4.7.10b)$$

or after a simple calculation, we arrive at:

$$\left(i\hbar\, \partial_t + e A_0 \right) \frac{+\varphi + \Psi}{2} = -mc^2\, \varphi\, , \qquad (4.7.11a)$$

$$\left(i\hbar\, \partial_t + i\hbar \frac{1}{\sqrt{-g}}\, \frac{\partial \sqrt{-g}}{\partial t} + e\, A_0 \right) \frac{\Psi - \varphi}{2}$$

$$+ \frac{1}{m} \left[\left(\frac{i\hbar}{\sqrt{-g}}\, \partial_k \sqrt{-g} + \frac{e}{c}\, A_k \right) g^{kl} \left(i\,\hbar\, \partial_l + \frac{e}{c} A_l \right) \right] \frac{\Psi + \varphi}{2}$$

$$= mc^2\, (\Gamma + 1)\, \frac{\varphi}{2} + mc^2\, (\Gamma - 1)\, \frac{\Psi}{2}\, . \qquad (4.7.11b)$$

In this point, it is better to consider two different cases. <u>The first</u> possibility is when one poses an additional requirement $\Gamma = 1$, which means the absence of the non-minimal interaction term through R-scalar. Then at $\Gamma = 1$, from the previous equations – by ignoring the small component compared with big one – it follows:

$$\left(i\hbar\, \partial_t + e A_0 \right) \frac{\Psi}{2} = -mc^2\, \varphi\, , \qquad (4.7.12a)$$

$$\left(i\hbar\, \partial_t + i\hbar \frac{1}{\sqrt{-g}}\, \frac{\partial \sqrt{-g}}{\partial t} + e\, A_0 \right) \frac{\Psi}{2}$$

$$+ \frac{1}{m} \left[\left(\frac{i\hbar}{\sqrt{-g}}\, \partial_k \sqrt{-g} + \frac{e}{c}\, A_k \right) g^{kl} \left(i\,\hbar\, \partial_l + \frac{e}{c} A_l \right) \right] \frac{\Psi}{2} = mc^2\, \varphi\, . \qquad (4.7.12b)$$

In excluding the small constituent we arrive at:

$$\left[i\hbar\, \left(\partial_t + \frac{1}{2\sqrt{-g}}\, \frac{\partial \sqrt{-g}}{\partial t} \right) + e\, A_0 \right] \Psi$$

$$= \frac{1}{2m} \left[(\frac{i\hbar}{\sqrt{-g}} \, \partial_k \sqrt{-g} + \frac{e}{c} \, A_k) \, (-g^{kl}) \, (i \, \hbar \, \partial_l + \frac{e}{c} A_l) \right] \Psi \, . \qquad (4.7.13a)$$

Using substitution $\Psi \implies (-g)^{-1/4} \, \Psi$ the obtained equation can be simplified:

$$(\, i\hbar \, \partial_t + e \, A_0 \,) \, \Psi$$

$$= \frac{1}{2m} \left[(\frac{i\hbar}{\sqrt{-g}} \, \partial_k \sqrt{-g} + \frac{e}{c} \, A_k) \, (-g^{kl}) \, (i \, \hbar \, \partial_l + \frac{e}{c} A_l) \right] \Psi \, , \qquad (4.7.13b)$$

which is the Schrödinger equation in curved space.

The second possibility is when $\Gamma \neq 1$, then from Eq. (4.7.11b)

$$(i\hbar \, \partial_t + eA_0) \, \frac{\Psi}{2} = -mc^2 \, \varphi \, , \qquad (4.7.14a)$$

$$\left(i\hbar \, \partial_t + i\hbar \frac{1}{\sqrt{-g}} \frac{\partial \sqrt{-g}}{\partial t} + e \, A_0 \right) \frac{\Psi}{2}$$

$$+ \frac{1}{m} \left[(\frac{i\hbar}{\sqrt{-g}} \, \partial_k \sqrt{-g} + \frac{e}{c} \, A_k) \, g^{kl} (i \, \hbar \, \partial_l + \frac{e}{c} A_l) \right] \frac{\Psi}{2}$$

$$= mc^2 \, (\Gamma + 1) \, \frac{\varphi}{2} + mc^2 \, (\Gamma - 1) \, \frac{\Psi}{2} \, . \qquad (4.7.14b)$$

Through the use of Eq. (4.7.14a), we can derive the following equation for the big component Ψ:

$$\left(i\hbar \, \partial_t + i\hbar \frac{1}{\sqrt{-g}} \frac{\partial \sqrt{-g}}{\partial t} + e \, A_0 \right) \frac{\Psi}{2}$$

$$+ \frac{(\Gamma + 1)}{2} \, (i\hbar \, \partial_t + eA_0) \, \frac{\Psi}{2} \, - \, mc^2 \, (\Gamma - 1) \, \frac{\Psi}{2}$$

$$- \frac{1}{2m} \left[(\frac{i\hbar}{\sqrt{-g}} \, \partial_k \sqrt{-g} + \frac{e}{c} \, A_k) \, g^{kl} (i \, \hbar \, \partial_l + \frac{e}{c} A_l) \right] \Psi \, . \qquad (4.7.15)$$

This equation can be rewritten as follows:

$$\left[(\frac{1}{2} + \frac{1}{2} \frac{\Gamma(x) + 1}{2}) (i\hbar\partial_t + e \, A_0) + \frac{i\hbar}{2\sqrt{-g}} \frac{\partial \sqrt{-g}}{\partial t}) \right] \Psi$$

$$= \frac{1}{2m} \left[(\frac{i\hbar}{\sqrt{-g}} \, \partial_k \sqrt{-g} + \frac{e}{c} \, A_k) \, (-g^{kl}) \, (i \, \hbar \, \partial_l + \frac{e}{c} A_l) \right] \Psi + mc^2 \, \frac{(\Gamma(x) - 1)}{2} \, \Psi \, .$$

$$\qquad (4.7.16)$$

It remains to recall that

$$\Gamma(x) = 1 + \frac{1}{6} \frac{\hbar^2 R(x)}{m^2 c^2} \, ,$$

so where the previous equation will take the form

$$\left[(1 + \frac{1}{24} \frac{\hbar^2 R(x)}{m^2 c^2}) \, (i\hbar\partial_t + e \, A_0) + \frac{i\hbar}{2\sqrt{-g}} \frac{\partial \sqrt{-g}}{\partial t}) \right] \Psi$$

$$= \frac{1}{2m} \left[(\frac{i\hbar}{\sqrt{-g}} \, \partial_k \sqrt{-g} + \frac{e}{c} \, A_k) \, (-g^{kl}) \, (i \, \hbar \, \partial_l + \frac{e}{c} A_l) \right] \Psi + mc^2 \, \frac{\hbar^2 R}{12 m^2 c^2} \, \Psi \, ,$$

and finally

$$\left[(1 + \frac{1}{24} \frac{\hbar^2 R(x)}{m^2 c^2}) \, (i\hbar \partial_t + e \, A_0) + \frac{i\hbar}{2\sqrt{-g}} \frac{\partial \sqrt{-g}}{\partial t} \right] \Psi$$

$$= \frac{1}{2m} \left[(\frac{i\hbar}{\sqrt{-g}} \partial_k \sqrt{-g} + \frac{e}{c} A_k) \, (-g^{kl}) \, (i \, \hbar \, \partial_l + \frac{e}{c} A_l) \,] + \hbar^2 \frac{R}{6} \right] \Psi \, , \qquad (4.7.17)$$

which should be considered as a Schrödinger equation in a space-time with a non-vanishing scalar curvature $R(x) \neq 0$ when allowing for a non-minimal interaction term through scalar curvature $R(x)$.

In addition, several general comments may be considered here. The wave function of Schrödinger equation Ψ does not coincide with the initial scalar Klein–Fock–Gordon wave function Φ. Instead, we have the following:

$$\Psi = \Phi + \Phi_0, \qquad \Phi_0 \text{ belongs to } \{ \Phi_0, \Phi_1, \Phi_2, \Phi_3 \} . \qquad (4.7.18)$$

One may have looked at this fact as non-occasional or perhaps even a necessary one. Indeed, let one start with a neutral scalar particle theory. Such a particle cannot interact with electromagnetic field and its wave function is real. However, by general consideration, a certain nonrelativistic limit in this theory must exist. The fact in this case: for the term

$$\Phi_0 = i \frac{\hbar}{mc} \partial_0 \Phi$$

is imaginary even if $\Phi^* = +\Phi$. All the more logical in this situation, which is in accordance with the the mathematical structure of the Schrödinger equation itself, it cannot be written for real wave function at all.

This same property occurred in the theory of a vector particle: even if the wave function of the relativistic particle of spin 1 is taken as real, the corresponding wave function in the nonrelativistic approximation turned out to be complex-valued. By general consideration, one may expect an analogous result in the theory of a spin 1/2 particle: if the nonrelativistic approximation is done in the theory of Majorana neutral particle [40] with the real 4-spinor wave function then the corresponding Pauli spinor wave function must be complex-valued.

4.8 Shamaly–Capri Equation for a Spin One Particle with Anomalous Magnetic Moment in a Curved Space-Time

It has been known that when using the help of generalized relativistic wave equations involving extended sets of irreducible representations of the Lorentz group, one can treat particles with additional intrinsic electromagnetic structure: Fedorov and Pletyukhov [291–294], Capri and Shamaly [295–299], Petras [300,301], Ulehla [302], Formanek [303], Khalil [304–306], Bogush, Kisel et al. [307–311]. In particular, within this approach one describes the anomalous magnetic moment of a spin 1 particle. A natural question arises: How will this manifest such an additional structure foe? Meaning in this instance, other than the fundamental interaction; in particular, gravitational?

Let us introduce basic facts on the Shamaly–Capri equation in Minkowski space-time, starting with a free spin one particle equation in a matrix form

$$(i \, \Gamma^a \partial_a \, - \, m) \, \Psi(x) = 0 \,, \tag{4.8.1}$$

where $\Phi(x) = (\Phi, \Phi_a, \Phi_{[ab]}, \Phi_{(ab)})$ stands for a 20-component wave function, representation of the Lorentz group of the type

$$T = (0,0) \oplus (1/2, 1/2) \oplus (0,1) \oplus (1,0) \oplus (1,1) \tag{4.8.2}$$

and Γ^a stands for (20×20)-matrices [291–294]:

$$\Gamma^a = -i \, (\, \lambda_1 \, e^{4,a} \, - \, \lambda_1^* \, e^{a,4}$$

$$+ \, \lambda_2 \, g_{kn} \, e^{n,[ka]} \, - \, \lambda_2^* \, g_{kn} \, e^{[ka],n} \, - \, \lambda_3 \, g_{kn} \, e^{n,(ka)} \, - \, \lambda_3^* \, g_{kn} \, e^{(ka),n} \, ; \tag{4.8.3}$$

$(*)$ designates complex conjugation. In Eq. (4.8.3), λ_i are numerical parameters, obeying the following restrictions [291–294]:

$$\lambda_1 \lambda_1^* - \frac{3}{2} \lambda_3 \lambda_3^* = 0 \,, \qquad \lambda_2 \lambda_2^* - \lambda_3 \lambda_3^* = 1 \,. \tag{4.8.4}$$

In Eq. (4.8.3), we use elements of a complete matrix algebra $e^{A,B}$, defined according to

$$(e^{A,B})_C^{\;D} = \delta_C^{\;A} \, g^{B,D}, \qquad e^{A,B} e^{C,D} = g^{B,C} \, e^{A,D} \,,$$

$$A, B, \ldots = 0, \, a, \, [ab], \, (ab) \,, \tag{4.8.5}$$

where $\delta^B_{\;A}$ is a generalized Kronecker symbol. Symbols $g^{A,B}$ follow from $\delta_B^{\;A}$ with ordinary metric tensor of the Minkowski space. In the case under consideration, we use the following objects

$$\delta_{[cd]}^{\;[ab]} = \delta_c^a \, \delta_d^b \, - \, \delta_d^a \, \delta_c^b \,, \qquad g^{[ab],[cd]} = g^{ac} \, g^{bd} \, - \, g^{ad} \, g^{bc} \,, \tag{4.8.6}$$

$$\delta_{(cd)}^{\;(ab)} = \delta_c^a \, \delta_d^b \, + \, \delta_d^a \, \delta_c^b \, - \, \frac{1}{2} \, g^{ab} \, g_{cd} \,,$$

$$g^{(ab),(cd)} = g^{ac} \, g^{bd} + g^{ad} \, g^{bc} - \frac{1}{2} g^{ab} \, g^{cd} \,; \tag{4.8.7}$$

also we need the explicit form of the Lorentzian generators J^{ab} for Eq. (4.8.2)

$$J^{ab} = (e^{a,b} - e^{b,a}) + g_{kn}(e^{[ak],[bn]} - e^{[bk],[an]}) + g_{kn}(e^{(ak),(bn)} - e^{(bk),(an)}) \,. \tag{4.8.8}$$

In absence of external fields, Eq. (4.8.1) reduces to a corresponding Duffin–Kemmer–Petiau equation. To demonstrate this, let us transform Eq.(4.8.1) to explicit tensor form with respect to $\Phi, \Phi_a, \Phi_{[ab]}, \Phi_{(ab)}$

$$\lambda_1 \, \partial^a \Phi_a = m \, \Phi \,,$$

$$- \, \lambda_1^* \, \partial_b \Phi \, + \, \lambda_2 \, \partial^a \Phi_{[ba]} \, - \, \lambda_3 \, \partial^a \Phi_{(ba)} = m \, \Phi_b \,,$$

$$\lambda_2^* \, (\, \partial_b \Phi_a \, - \, \partial_a \Phi_b \,) = m \, \Phi_{[ba]} \,,$$

$$- \lambda_3^* \left(\partial_b \Phi_a + \partial_a \Phi_b - \frac{1}{2} g_{ab} \partial^c \Phi_c \right) = m \, \Phi_{(ab)} \, . \tag{4.8.9}$$

From the first and fourth equations in Eq. (4.8.9), by using Eq. (4.8.4), we get:

$$- \lambda_1^* \partial_b \Phi - \lambda_3 \, \partial^a \Phi_{(ba)} = \frac{1}{m} \lambda_3 \lambda_3^* \, \partial^a (\partial_a \Phi_b - \partial_b \Phi_a) \tag{4.8.10}$$

or

$$- \lambda_1^* \partial_b \Phi - \lambda_3 \, \partial^a \Phi_{(ba)} = \frac{\lambda_3 \lambda_3^*}{\lambda_2^*} \, \partial^a \Phi_{[ab]} \, . \tag{4.8.11}$$

Further, (see Eqs. (4.8.4) and (4.8.9)) we can obtain

$$\frac{1}{\lambda_2^*} \, \partial^a \Phi_{[ba]} = m \, \Phi_b \, . \tag{4.8.12}$$

By introducing new field variables

$$\Psi_a = \lambda_2^* \Phi_a \, , \ \Psi_{[ab]} = \Phi_{[ab]} \, , \tag{4.8.13}$$

we arrive at the ordinary system of tensor equation by Proca for a free spin 1 particle

$$\partial^b \Psi_{[ab]} = m \Psi_a \, , \qquad \partial_a \Psi_b - \partial_b \Psi_a = m \Psi_{[ab]} \, . \tag{4.8.14}$$

It can be presented in the matrix form

$$(\, i \, \beta^a \partial_a - m) \Psi = 0 \, , \tag{4.8.15}$$

$$\Psi = \begin{pmatrix} \Psi_a \\ \Psi_{[ab]} \end{pmatrix} \, , \ \beta^a = -i g_{bc} (e^{c,[ba]} - e^{[ba],c}) \, .$$

Now let us extend Eq. (4.8.1) to the case of arbitrary curved space-time model, with metric tensor $g_{\alpha\beta}(x)$ and corresponding tetrad $e^\mu_{(a)}(x)$, in accordance with the tetrad recipe by Tetrode–Weyl–Fock–Ivanenko. Thus, we obtain

$$[\, i\Gamma^\mu(\partial_\mu + B_\mu) - m \,] \, \Psi = 0 \, , \tag{4.8.16}$$

or

$$(\, i\Gamma^a \partial_{(a)} + \frac{i}{2} \Gamma^a J^{cd} \gamma_{cda} - m \,) \, \Psi = 0 \, , \tag{4.8.17}$$

where

$$\Gamma^\mu = \Gamma^a e^\mu_{(a)} \, , \ B_\mu = \frac{1}{2} J^{ab} e^\nu_{(a)} \nabla_\mu e_{(b)\nu} \, ,$$

$$\partial_{(a)} = e^\mu_{(a)} \partial_\mu \, , \ \gamma_{abc} = -(\nabla_\beta e_{(a)\alpha}) e^\alpha_{(b)} e^\beta_{(c)} \, ;$$

∇_μ is a covariant derivative, γ_{abc} stand for Ricci rotation coefficients.

Any two equations as in the type of Eq. (4.8.16)

$$[\, i\Gamma^\mu(\partial_\mu + B_\mu) - m \,] \, \Psi = 0 \, , \ [\, i\Gamma'^\mu(\partial_\mu + B'_\mu) - m \,] \, \Psi' = 0 \, , \tag{4.8.18}$$

based on two different tetrads, related by a local Lorentz transformation $e'^{\mu}_{(a)}(x) = L_a{}^b(x)e^{\mu}_{(b)}(x)$, can be transformed into each other by means of a special local gauge transformation $\Psi'(x) = S(x)\Psi(x)$:

$$\begin{vmatrix} \Phi' \\ \Phi'_a \\ \Phi'_{[ab]} \\ \Phi'_{(ab)} \end{vmatrix} = \begin{vmatrix} 1 & 0 & 0 & 0 \\ 0 & L_a{}^c & 0 & 0 \\ 0 & 0 & L_a{}^c L_b{}^d & 0 \\ 0 & 0 & 0 & L_a{}^c L_b{}^d \end{vmatrix} \begin{vmatrix} \Phi \\ \Phi_c \\ \Phi_{[cd]} \\ \Phi_{(cd)} \end{vmatrix} , \qquad (4.8.19)$$

two relationships must hold

$$S\Gamma^{\mu}S^{-1} = \Gamma'^{\mu} , \quad SB_{\mu}S^{-1} + S\partial_{\mu}S^{-1} = B'_{\mu} . \qquad (4.8.20)$$

The first one is equivalent to

$$S\Gamma^a e^{\mu}_{(a)}S^{-1} = \Gamma^b e'^{\mu}_{(b)} \implies S\Gamma^a S^{-1} = \Gamma^b L_b{}^a , \qquad (4.8.21)$$

which in relation can ensures the Lorentz covariance if there is a corresponding free equation in Minkowski space. Let us prove the second relation in Eq. (4.8.20). To this end, let us consider the term

$$SB_{\alpha}S^{-1} = \frac{1}{2}SJ^{ab}S^{-1}e^{\beta}_{(a)}\left(\nabla_{\alpha}e_{(b)\beta}\right) ;$$

from here, it follows

$$SB_{\alpha}S^{-1} = \frac{1}{2}(SJ^{ab}S^{-1})\,(L^{-1})_a{}^k\,[\,(\partial_{\alpha}(L^{-1})_b{}^l g_{kl} + (L^{-1})_b{}^l e'^{\beta}_{(k)}\nabla_{\alpha}e'_{(l)\beta}\,] .$$

Using explicit form of S (see Eq. (4.8.19)) and J^{ab} (see Eq. (4.8.8)), one finds identity

$$SJ^{ab}S^{-1} = J^{mn}L_m{}^a L_n{}^b , \qquad (4.8.22)$$

and

$$SB_{\alpha}S^{-1} = \frac{1}{2}J^{mn}L_n{}^b \partial_{\alpha}(L_{mb}) + B'_{\alpha} .$$

Thus, we conclude that the relations of Eqs. (4.8.20) and (4.8.18) will hold, if

$$SB_{\mu}S^{-1} = \frac{1}{2}J^{mn}L_n{}^b \partial_{\mu}(L_{mb}) . \qquad (4.8.23)$$

One can verify the last identity with the use of pseudo-orthogonality condition and explicit forms for S and J^{ab}. Thus, equations in Eq. (4.8.18) relate to each other by the gauge transformation.

Equation (4.8.17) is equivalent to the system

$$\lambda_1 \left(\partial^{(a)} + \gamma^{ba}{}_b\right) \Phi_a = m\,\Phi ,$$

$$-\lambda_1^* \partial_{(r)}\Phi + \lambda_2 \left(\partial^{(a)}\Phi_{[ra]} + \gamma_r{}^{bc}\Phi_{[bc]} + \gamma^{bc}{}_c\Phi_{[br]}\right)$$

$$-\lambda_3 \left(\partial^{(a)}\Phi_{(ra)} + \gamma_r{}^{dc}\Phi_{(dc)} + \gamma^{dc}{}_c\Phi_{(dr)}\right) = m\,\Phi_r ,$$

$$\lambda_2^* \left(\partial_{(r)} \Phi_s - \partial_{(s)} \Phi_r + \gamma^d{}_{rs} \Phi_d - \gamma^d{}_{sr} \Phi_d \right) = m \, \Phi_{[rs]} \,,$$

$$-\lambda_3^* \left[\left(\partial_{(r)} \Phi_s + \partial_{(s)} \Phi_r + \gamma_r{}^d{}_s \Phi_d + \gamma_s{}^d{}_r \Phi_d \right) \right.$$

$$\left. - \frac{1}{2} g_{rs} \left(\partial^{(a)} \Phi_a + \gamma^{ad}{}_a \Phi_d \right) \right] = m \, \Phi_{(rs)} \,. \qquad (4.8.24)$$

Let us exclude auxiliary components. To this end, expressing Φ and $\Phi_{(ad)}$ from the first and fourth equations, and substituting the results into the second equation. Then, we get

$$- \lambda_1^* \, \partial_{(r)} \Phi - \lambda_3 \left[\partial^{(a)} \Phi_{(ra)} + \gamma_r{}^{da} \Phi_{(da)} + \gamma^{ad}{}_a \Phi_{(dr)} \right]$$

$$= - \frac{\lambda_3 \lambda_3^*}{\lambda_2^*} \left[\partial^{(a)} \Phi_{[ra]} + \gamma_r{}^{bc} \Phi_{[bc]} + \gamma^{bc}{}_c \Phi_{[br]} \right]$$

$$- \frac{2\lambda_3 \lambda_3^*}{m} \left[\gamma^b{}_{ar}{}^{,(a)} \Phi_b + \gamma^{ba}{}_{b,(r)} \Phi_a \right.$$

$$\left. - \gamma^{ab}{}_r \, \partial_{(a)} \Phi_b - \gamma_r{}^{ab} \partial_{(a)} \Phi_b + \gamma_r{}^{ab} \gamma^d{}_{ba} \Phi_d - \gamma^{ab}{}_b \gamma^c{}_{ar} \Phi_c \right] \,. $$

$$(4.8.25)$$

Then

$$\frac{1}{\lambda_2^*} \left(\partial^{(a)} \Phi_{[ra]} + \gamma_r{}^{bc} \Phi_{[bc]} + \gamma^{bc}{}_c \Phi_{[br]} \right)$$

$$- \frac{2\lambda_3 \lambda_3^*}{m} \left[\gamma^b{}_{ar}{}^{,(a)} \Phi_b + \gamma^{ba}{}_{b,(r)} \Phi_a - \gamma^{ab}{}_r \partial_{(a)} \Phi_b - \gamma_r{}^{ab} \partial_{(a)} \Phi_b \right.$$

$$\left. + \gamma_r{}^{ab} \gamma^d{}_{ba} \Phi_d - \gamma^{ab}{}_b \gamma^c{}_{ar} \Phi_c \right] = m \Phi_r \,.$$

$$(4.8.26)$$

From Eqs. (4.8.24) and (4.8.26), we will obtain a generalized Proca equations

$$\partial^{(a)} \Psi_{[ra]} + \gamma_r{}^{bc} \Psi_{[bc]} + \gamma^{bc}{}_c \Psi_{[br]}$$

$$- 2\lambda_3 \lambda_3^* m^{-1} \left[\gamma^b{}_{ar}{}^{,(a)} \Psi_b + \gamma^{ba}{}_{b,(r)} \Psi_a - \gamma^{ab}{}_r \partial_{(a)} \Psi_b - \gamma_r{}^{ab} \partial_{(a)} \Psi_b \right.$$

$$\left. + \gamma_r{}^{ab} \gamma^d{}_{ba} \Psi_d - \gamma^{ab}{}_b \gamma^c{}_{ar} \Psi_c \right] = m \, \Psi_r \,, \qquad (4.8.27a)$$

$$\partial_{(a)} \Psi_b - \partial_{(b)} \Psi_a + \gamma^d{}_{ab} \Psi_d - \gamma^d{}_{ba} \Psi_d = m \, \Psi_{[ab]} \,. \qquad (4.8.27b)$$

In Eq.(4.8.27a), the term proportional to $\frac{2\lambda_3 \lambda_3^*}{m}$, determines an additional interaction with gravitational background field.

Equation (4.8.27a) can be rewritten differently if one takes into account the expression for tetrad components of the curvature tensor R_{abcd} through Ricci coefficients

$$R_{abcd} = -\gamma_{abc,(d)} + \gamma_{abd,(c)} + \gamma_{akc} \gamma^k{}_{bd}$$

$$+ \gamma_{abn} \gamma^n{}_{cd} - \gamma_{akd} \gamma^k{}_{bc} - \gamma_{abn} \gamma^n{}_{dc} \,, \qquad (4.8.28a)$$

$$R^b{}_r = R^{ab}{}_{ra} = - \gamma^{ab}{}_{r,(a)} + \gamma^{ab}{}_{a,(r)} + \gamma_{rna} \gamma^{ban} - \gamma^a{}_{ka} \gamma^{kb}{}_r \,. \qquad (4.8.28b)$$

Then

$$\partial^a \Psi_{[ra]} + \gamma_r{}^{bc} \Psi_{[bc]} + \gamma^{bc}{}_c \Psi_{[br]}$$

$$- \frac{2\lambda_3 \lambda_3^*}{m} [R^b{}_r \Psi_b - \gamma^{ab}{}_r \partial_{(a)} \Psi_b - \gamma_r{}^{ab} \partial_{(a)} \Psi_b] - m\Psi_r = 0 . \qquad (4.8.29)$$

The system in Eq.(4.8.27) can be presented in the matrix form

$$\{ i\beta^a \partial_{(a)} + \frac{i}{2} \beta^a J^{cd}{}_{(0)} \gamma_{cda}$$

$$- \frac{\lambda_3 \lambda_3^*}{m} [(\gamma^a{}_{bk} - \gamma^a{}_{kb}) (e^{b,k} - e^{k,b}) \partial_{(a)} + R_{bk} (e^{b,k} + e^{k,b})] - m \} \Psi = 0 , \qquad (4.8.30)$$

where β^a designate Duffin–Kemmer matrices as in Eq. (4.8.15), and $J^{bc}{}_{(0)} = \beta^b \beta^c - \beta^c \beta^b$.

Thus, generally covariant Shamaly–Capri equation describes a vector particle non-minimally interacting with external gravitational field through an additional term in Eq. (4.8.30)

$$- \frac{\lambda_3 \lambda_3^*}{m} \left[(\gamma^a{}_{bk} - \gamma^a{}_{kb})(e^{b,k} - e^{k,b}) \partial_{(a)} + R^a{}_{bka} (e^{b,k} + e^{k,b}) \right] \Psi .$$

Note, that previously [311] the parameter $\frac{\lambda_3 \lambda_3^*}{m}$ was interpreted as a characteristic for determining an anomalous magnetic moment of a spin 1 particle.

It is readily been verified that the system in Eq. (4.8.24) is the tetrad form of the generally covariant tensor equations (for completeness let us also take into account an external electromagnetic field by means of usual formal change $\nabla_\alpha \Longrightarrow D_\alpha = \nabla_\alpha - ieA_a(x)$)

$$\lambda_1 D_\alpha \Phi^\alpha = m \Phi ,$$

$$-\lambda_1^* D_\beta \Phi + \lambda_2 D^\alpha \Phi_{[\beta\alpha]} - \lambda_3 D^\alpha \Phi_{(\alpha\beta)} = m \Phi_\beta ,$$

$$\lambda_2^* (D_\alpha \Phi_\beta - D_\beta \Phi_\alpha) = m \Phi_{[\alpha\beta]} ,$$

$$- \lambda_3^* [D_\alpha \Phi_\beta + D_\beta \Phi_\alpha - \frac{1}{2} g_{\alpha\beta} \nabla^\rho \Phi_\rho] = m \Phi_{(\alpha\beta)} , \qquad (4.8.31)$$

where

$$\Phi_\alpha = e_\alpha^{(a)} \Phi_a , \qquad \Phi_{[\alpha\beta]} = e_\alpha^{(a)} e_\beta^{(b)} \Phi_{[ab]} , \qquad \Phi_{(\alpha\beta)} = e_\alpha^{(a)} e_\beta^{(b)} \Phi_{(ab)} .$$

As in the case of Eq. (4.8.24), the system in Eq. (4.8.31) reduces to

$$\frac{1}{\lambda_2^*} \nabla^\alpha \Phi_{[\beta\alpha]} + \frac{2\lambda_3 \lambda_3^*}{m} [D_\alpha, D_\beta]_- \Phi^\alpha = m \Phi_\beta ,$$

$$\lambda_2^* (D_\alpha \Phi_\beta - D_\beta \Phi_\alpha) = m \Phi_{[\alpha\beta]} , \qquad (4.8.32)$$

or

$$D^\alpha \Psi_{[\beta\alpha]} + \frac{2\lambda_3 \lambda_3^*}{m} [D_\alpha, D_\beta]_- \Psi^\alpha = m \Psi_\beta , \qquad (4.8.33a)$$

$$D_\alpha \Psi_\beta - D_\beta \Psi_\alpha = m \Psi_{[\alpha\beta]} . \qquad (4.8.33b)$$

Due to an identity

$$[D_\alpha, D_\beta]_- \Psi^\alpha = (-ieF_{\alpha\beta} + R_{\alpha\beta})\Psi^\alpha \ , \qquad (4.8.34)$$

we conclude that the parameter $\frac{\lambda_3\lambda_3^*}{m}$ in Eq. (4.8.33a) determines at the same time an anomalous magnetic moment and an additional term of gravitational interaction through the Ricci tensor $R_{\alpha\beta}$ (see Eqs. (4.8.28a) and (4.8.28b)).

Chapter 5

Dirac–Kähler Theory, Relation between Spinor and Tensor Formulations

5.1 Introduction

The mathematical description for the concept of elementary particles as certain relativistically invariant objects was found in the frames of 4-dimensional Minkowski space-time. It is assumed that for any particle there are given definite transformation properties for a corresponding field and a wave equation to which that field obeys; wave equation must be Lorentz (or Poincaré) invariant: Wigner [312], Pauli [313], Bhabha [314], Harish-Chandra [315], Gel'fand and Yaglom [316], Corson [317], Umezawa [318], Shirokov [319], Bogush and Moroz [320], Fedorov [246].

A common view is that the generalization of a wave equation on Riemannian space-time is substantially determined by what a particle is – boson or fermion. As a rule, they say that tensor equations for bosons are extended in a simpler way then spinor equations for fermions. This belief evidently correlates with the fact that the concepts of both flat and curved space model are based on the notion of a vector.

In that context, a very interesting problem is that the extension a wave equation for Dirac–Kähler field (there are other terms used as well: Ivanenko–Landau field, or a vector field of the general type).

The scientific literature in this field is enormous. It started with the early development of quantum mechanical wave equations theory, just after the concept of a particle with spin 1/2 arises. In particular, news objects themselves, spinors, seemed mysterious and obscure in comparison to those familiar tensors.

The main feature of the Ivanenko–Landau field [130] was that it seemingly gave possibility to perform smoothly transition from tensors to spinors, in a sense it was an attempt to eliminate spinors at all. Different aspect of that relation were investigated by many authors: Ivanenko and Landau [130], Lanczos [321, 322], Juvet [323, 324], Einstein and Mayer [325–328], Frenkel [329], Whittaker [330], Proca [331], Ruse [332], Taub [333, 334], Belinfante [155, 156], Ivanenko and Sokolov [335], Feshbach and Nikols

[336], Kähler [337, 338], Leutwyler [339], Klauder [340], Penney [341], Cereignani [342], Streater and Wilde [343], Pestov [344–346], Crumeyrolle [348], Durand [349], Strazhev et al. [350–360], Graf [361], Benn and Tucker [362–367], Banks et al. [368], Garbaczewski [369], Pletyukhov and Strazhev [370–373], Holland [374], Ivanenko et al. [375, 376], Bullinaria [377], Blau [378], Jourjine [379], Krolikowski [380, 381], Howe [382], Nikitin et al. [383], Kruglov [354, 384–386], Marchuk [387–392], Krivskij et al. [393].

The three most interesting points in connection to the general covariant extension of the wave equation for this field are; first, in flat Minkowski space there exist tensor and spinor formulations of the theory; second, in the initial tensor form there are presented tensors with different intrinsic parities; and finally, there exist different views about physical interpretation of the object: whether it is a composite boson or a set of four fermions. These three points will be of primary importance in the following treatment.

5.2 Spinor and Tensor Forms of the Wave Equation

In Minkowski space-time, the Dirac–Kähler field is described by 16-component wave function with transformation properties of 2-rank 4-bispinor $U(x)$ or by equivalent set of elementary tensor constituents

$$U(x) \qquad \text{or} \qquad \{\Psi(x), \Psi_i(x), \tilde{\Psi}(x), \tilde{\Psi}_i(x), \Psi_{mn}(x)\} \,,$$

where $\Psi(x)$ is a scalar; $\Psi_i(x)$ is a vector; $\tilde{\Psi}(x)$ is a pseudo-scalar; $\tilde{\Psi}_i(x)$ represents a pseudo-vector; $\Psi_{mn}(x)$ is an anti-symmetric tensor. Correspondingly, we have two representations for the wave equation

$$[i \, \gamma^a \partial_a - m \,] \, U(x) = 0 \,, \tag{5.2.1}$$

and

$$\partial_l \Psi + m \Psi_l = 0 \,, \qquad \partial_l \tilde{\Psi} + m \tilde{\Psi}_l = 0 \,, \qquad \partial_l \Psi + \partial^a \Psi_{la} - m \Psi_l = 0 \,,$$

$$\partial_l \tilde{\Psi} - \frac{1}{2} \, \epsilon_l{}^{amn} \, \partial_a \Psi_{mn} - m \tilde{\Psi}_l = 0 \,,$$

$$\partial_m \Psi_n - \partial_n \Psi_m + \epsilon_{mn}{}^{ab} \, \partial_a \tilde{\Psi}_b - m \Psi_{mn} = 0 \,. \tag{5.2.2}$$

Let us now describe the relation between a 2-rank bispinor $U(x)$ and corresponding tensors. It is well known that any (4×4)-matrix can be expanded on 16 Dirac matrices; and for that expanding it does not matter whether the matrix U is a 2-rank bispinor or not. However, if it is so, coefficients that arise $\{\Psi, \Psi_l, \tilde{\Psi}, \tilde{\Psi}_l, \Psi_{mn}\}$ will possess quite the definite tensorial property with respect to the Lorentz group. Let such a 2-rank bispinor $U(x)$ is parameterized according to

$$U(x) = \left[-i\Psi + \gamma^l \, \Psi_l + i\sigma^{mn} \, \Psi_{mn} + \gamma^5 \, \tilde{\Psi} + i\gamma^l \gamma^5 \, \tilde{\Psi}_l \right] E^{-1} \,; \tag{5.2.3a}$$

here E stands for a metrical bispinor matrix with simple properties

$$E = \begin{vmatrix} \epsilon & 0 \\ 0 & \dot{\epsilon}^{-1} \end{vmatrix} = \begin{vmatrix} \epsilon_{\alpha\beta} & 0 \\ 0 & \epsilon^{\dot{\alpha}\dot{\beta}} \end{vmatrix} = \begin{vmatrix} i\sigma^2 & 0 \\ 0 & -i\sigma^2 \end{vmatrix} \,,$$

$$E^2 = -I \,, \qquad \tilde{E} = -E \,, \qquad \mathrm{Sp}\, E = 0 \,, \qquad \tilde{\sigma}^{ab} E = -E\sigma^{ab} \,. \qquad (5.2.3b)$$

Inverse to Eq. $(5.2.3a)$ have relations

$$\Psi(x) = -\frac{1}{4i}\, \mathrm{Sp}\, [EU(x)] \,, \qquad \tilde{\Psi}(x) = \frac{1}{4}\, \mathrm{Sp}\, [E\gamma^5 U(x)] \,,$$

$$\Psi_l(x) = \frac{1}{4}\, \mathrm{Sp}\, [E\gamma_l U(x)] \,, \qquad \tilde{\Psi}_l(x) = \frac{1}{4i}\, \mathrm{Sp}\, [E\gamma^5 \gamma_l U(x)] \,,$$

$$\Psi_{mn}(x) = -\frac{1}{2i}\, \mathrm{Sp}\, [E\sigma_{mn} U(x)] \,. \qquad (5.2.3c)$$

We will next use also a 2-component spinor formalism; to this end, it suffices to choose Dirac matrices in spinor Weyl basis and specify additionally the notation for constituents of $U(x)$:

$$U(x) = \begin{vmatrix} \xi^{\alpha\beta}(x) & \Delta^{\alpha}{}_{\dot{\beta}}(x) \\ H_{\dot{\alpha}}{}^{\beta}(x) & \eta_{\dot{\alpha}\dot{\beta}}(x) \end{vmatrix} \,. \qquad (5.2.4a)$$

Thus, instead of Eq. $(2.3a)$, we obtain

$$\Delta(x) = [\, \Psi_l(x) + i\, \tilde{\Psi}_l(x) \,]\, \sigma^{-l} \epsilon \,, \qquad H(x) = [\, \Psi_l(x) - i\, \tilde{\Psi}_l(x)) \,]\, \bar{\sigma}^l \epsilon^{-1} \,, \quad (5.2.4b)$$

$$\xi(x) = [\, -i\, \Psi(x) - \tilde{\Psi}(x) + i\, \Sigma^{mn}\, \Psi_{mn}(x) \,]\, \epsilon^{-1} \,,$$

$$\eta(x) = [\, -i\, \Psi(x) + \tilde{\Psi}(x) + i\, \bar{\Sigma}^{mn}\, \Psi_{mn}(x) \,]\, \dot{\epsilon} \,,$$

and inverse relations

$$\Psi_l(x) + i\, \tilde{\Psi}_l(x) = \frac{1}{2}\, \mathrm{Sp}\, [\, \dot{\epsilon}^{-1}\, \sigma_l\, \Delta(x) \,] \,, \qquad \Psi_l(x) - i\, \tilde{\Psi}_l(x) = \frac{1}{2}\, \mathrm{Sp}\, [\, \epsilon\bar{\sigma}_l\, H(x) \,] \,,$$

$$-i\, \Psi(x) - \tilde{\Psi}(x) = \frac{1}{2}\, \mathrm{Sp}\, [\, \epsilon\, \xi(x) \,] \,, \qquad -i\, \Psi(x) + \tilde{\Psi}(x) = \frac{1}{2}\, \mathrm{Sp}\, [\, \dot{\epsilon}^{-1}\, \xi(x) \,] \,,$$

$$-i\, \Psi^{kl}(x) + \frac{1}{2}\, \epsilon^{klmn}\, \Psi_{mn}(x) = \mathrm{Sp}\, [\, \epsilon\, \Sigma^{kl} \xi(x) \,] \,,$$

$$-i\, \Psi^{kl}(x) - \frac{1}{2}\, \epsilon^{klmn}\, \Psi_{mn}(x) = \mathrm{Sp}\, [\, \dot{\epsilon}^{-1}\, \bar{\Sigma}^{kl}\, \xi(x) \,] \,. \qquad (5.2.4c)$$

Dirac–Kähler equation in 2-spinor form will be

$$i\sigma^a\, \partial_a\, \xi(x) = m\, H(x) \,, \qquad i\bar{\sigma}^a\, \partial_a\, H(x) = m\, \xi(x) \,,$$

$$i\bar{\sigma}^a\, \partial_a\, \eta(x) = m\, \Delta(x) \,, \qquad i\sigma^a\, \partial_a\, \Delta(x) = m\, \eta(x) \,. \qquad (5.2.5)$$

Now, let us consider a general covariant form. First, we turn to the 4-spinor approach – according to the well known recipe by Tetrode–Weyl–Fock–Ivanenko Eq. $(5.2.1)$ should be changed into

$$[\, i\gamma^{\alpha}(x)\, (\partial_{\alpha} + B_{\alpha}(x)) - m\,]\, U(x) = 0 \,; \qquad (5.2.6)$$

connection $B_{\alpha}(x)$ is defined by

$$B_{\alpha}(x) = \frac{1}{2} J^{ab}\, e^{\beta}_{(a)}(x) \nabla_{\alpha} e_{(b)\beta}(x) = \Gamma_{\alpha}(x) \otimes I + I \otimes \Gamma_{\alpha}(x) \,,$$

where $J^{ab} = [\sigma^{ab} \otimes I + I \otimes \sigma^{ab}]$ stand for generators for bispinor representation of the Lorentz group. From Eq. (5.2.6) it follows 2-spinor form equations for Dirac–Kähler field

$$i\sigma^{\alpha}(x) \left[\partial_{\alpha} + \Sigma_{\alpha}(x) \otimes I + I \otimes \Sigma_{\alpha}(x) \right] \zeta(x) = m \, H(x) \,,$$

$$i\bar{\sigma}^{\alpha}(x) \left[\partial_{\alpha} + \bar{\Sigma}_{\alpha}(x) \otimes I + I \otimes \Sigma_{\alpha}(x) \right] H(x) = m \, \xi(x) \,,$$

$$i\bar{\sigma}^{\alpha}(x) \left[\partial_{\alpha} + \bar{\Sigma}_{\alpha}(x) \otimes I + I \otimes \bar{\Sigma}_{\alpha}(x) \right] \eta(x) = m \, \Delta(x) \,,$$

$$i\sigma^{\alpha}(x) \left[\partial_{\alpha} + \Sigma_{\alpha}(x) \otimes I + I \otimes \bar{\Sigma}_{\alpha}(x) \right] \Delta(x) = m \, \eta(x) \,. \qquad (5.2.7)$$

Equations (5.2.6) and (5.2.7) possess symmetry with respect to local Lorentz group: if $U(x)$ is subject to local Lorentz transformation

$$U'(x) = \left[\, S(k(x), k^*(x)) \otimes S(k(x), k^*(x)) \, \right] U(x) \,, \qquad (5.2.8a)$$

then the new field function $U'(x)$, or set of new 2-spinors $[\, \xi'(x), \, \eta'(x), \, \Delta'(x), \, H'(x) \,]$, will obey a wave equation of the same type as before

$$\left[\, i\gamma'^{\alpha}(x) \, (\partial_{\alpha} + B'_{\alpha}(x)) \, - \, m \, \right] U'(x) = 0 \,, \qquad (5.2.8b)$$

where primed $\gamma'^{\alpha}(x)$ and $B'_{\alpha}(x)$ are constructed with the help of a primed tetrad $e'^{\alpha}_{(b)}(x)$, related to the initial one by Local Lorentz transformation

$$e'^{\alpha}_{(b)}(x) = L_b{}^a(k(x), k^*(x)) \, e^{\alpha}_{(a)}(x) \,.$$

This symmetry we have under consideration proves the correctness of this equation: the symmetry describes a gauge freedom in the case as an explicit form of the tetrad.

In addition, there exists discrete symmetry. Indeed, if $U(x)$ is subject to the following discrete operation

$$U'(x) = [i\gamma^0 \otimes i\gamma^0] \, U(x) \qquad \text{or} \qquad \begin{vmatrix} \xi'(x) & \Delta'(x) \\ H'(x) & \eta'(x) \end{vmatrix} = \begin{vmatrix} -\eta(x) & -H(x) \\ -\Delta(x) & -\xi(x) \end{vmatrix}, \qquad (5.2.9a)$$

then the new wave function $U'(x)$ (new set of 2-spinors) will obey an equation of the same form as in Eq. (5.2.6) (or Eq. (5.2.7)), but now constructed on the base of a tetrad $e'^{\alpha}_{(b)}(x)$, P-reflected to the initial

$$e'^{\alpha}_{(b)}(x) = L_b^{(p)a} \, e^{\alpha}_a(x) \,, \qquad L_b^{(p)a} = \text{diag} \, (+1, -1, -1, -1) \,. \qquad (5.2.9b)$$

For general coordinate transformations, the wave function $U(x)$ behaves as a scalar (similarly, as a wave function $\Psi(x)$ in the Dirac equation). Correspondingly, the term $\partial_{\alpha}U(x)$ represents a general covariant vector and Eq. (5.2.6) is correct in the sense of general covariance.

Now, we shall turn to extending the tensor form in Eqs. (5.2.2). In this instance, we face here a rather specific problem. Indeed, a formal change

$$\partial_l \Longrightarrow \nabla_{\alpha} \,, \qquad \Psi_i(x) \Longrightarrow \Psi_{\alpha}(x) \,,$$

$$\tilde{\Psi}_i(x) \Longrightarrow \tilde{\Psi}_{\alpha}(x) \,, \qquad \Psi_{ij}(x) \Longrightarrow \Psi_{\alpha\beta}(x)$$

This change appears to possess some vagueness: it is not clear how we should distinguish between these two functions $\Psi_\alpha(x)$ and $\tilde{\Psi}_\alpha(x)$ – because they have the same index α, the sign of covariant vector. Nevertheless, by such a formal generalization, we have the system:

$$\nabla^\alpha \Psi_\alpha(x) + m\Psi(x) = 0 , \qquad \nabla^\alpha \tilde{\Psi}_\alpha(x) + m\tilde{\Psi}(x) = 0 ,$$

$$\nabla_\alpha \Psi(x) + \nabla^\beta \Psi_{\alpha\beta}(x) - m\Psi_\alpha(x) = 0 ,$$

$$\nabla_\alpha \tilde{\Psi}(x) - \frac{1}{2}\epsilon_\alpha{}^{\beta\rho\sigma}(x) \, \nabla_\beta \Psi_{\rho\sigma}(x) - m\tilde{\Psi}_\alpha(x) = 0 ,$$

$$\nabla_\alpha \Psi_\beta(x) - \nabla_\beta \Psi_\alpha(x) + \epsilon_{\alpha\beta}{}^{\rho\sigma}(x) \, \nabla_\rho \tilde{\Psi}_\sigma(x) - m\Psi_{\alpha\beta}(x) = 0 . \qquad (5.2.10)$$

To resolve the problem of distinguishing between $\Psi_\alpha(x)$ and $\tilde{\Psi}_\alpha(x)$, along with $\Psi(x)$ and $\tilde{\Psi}(x)$; and, also determine a covariant Levi-Civita object, a comparison can be found for Eq. (5.2.10) with Eq. (5.2.6).

We will demonstrate that from Eq. (5.2.6) it will then follow onto Eqs. (5.2.10); if instead of $U(x)$ in Eq.(5.2.6) we substitute expansion of the matrix $U(x)$ in terms of tetrad tensor constituents and then translate equations to covariant tensors accordingly.

$$\Psi_\alpha(x) = e_\alpha^{(i)}(x) \, \Psi_i(x) , \qquad \tilde{\Psi}_\alpha(x) = e_\alpha^{(i)}(x) \, \tilde{\Psi}_i(x) ,$$

$$\Psi_{\alpha\beta}(x) = e_\alpha^{(m)}(x)e_\beta^{(n)}(x) \, \Psi_{mn}(x) , \qquad (5.2.11a)$$

The covariant Levi-Civita object is defined as follows:

$$\epsilon^{\alpha\beta\rho\sigma}(x) = \epsilon^{abcd} \, e_{(a)}^\alpha(x) \, e_{(b)}^\beta(x) \, e_{(c)}^\rho(x) \, e_{(d)}^\sigma(x) . \qquad (5.2.11b)$$

In this, we note that the relevant similar functions entering Eqs. (5.2.10) differs in their transformation properties with respect to tetrad P-reflection: $\Psi(x)$, $\Psi_\alpha(x)$, $\Psi_{\alpha\beta}(x)$ are tetrad scalars; $\tilde{\Psi}(x)$, $\tilde{\Psi}_\alpha(x)$ are tetrad pseudo-scalars.

Let us proceed to explain the calculations proving this. First, Eq. (5.2.6) is written in the form (the symbol $\sim$ designates matrix transposition)

$$\left[i\gamma^\alpha \, \partial_\alpha \, U + i\gamma^\alpha \, \Gamma_\alpha(x) \, U + i\gamma^\alpha \, U \, \tilde{\Gamma}_\alpha - m \, U \right] = 0 , \qquad (5.2.12a)$$

then Eq. (5.2.12a) is translated to

$$\left[i\gamma^c \, e_{(c)}^\alpha \, \partial_\alpha \, U + \frac{i}{2} \, \gamma_{abc} \, \gamma^c \, \sigma^{ab} \, U + \frac{i}{2} \, \gamma_{abc} \, \gamma^c \, U \, \tilde{\sigma}^{ab} - m \, U \right] = 0 . \qquad (5.2.12b)$$

Further, into Eq. (5.2.12b) we substitute expansion for U in terms of local tetrad tensors

$$\left\{ i\gamma^c \, e_{(c)}^\beta \, \partial_\beta \, [-i\Psi + \gamma^l \, \Psi_l + i\sigma^{mn} \, \Psi_{mn} + \gamma^5\tilde{\Psi} + i\gamma^l\gamma^5\tilde{\Psi}_l] \, E^{-1} \right.$$

$$+ \frac{i}{2}\gamma_{abc}\gamma^c\sigma^{ab} \, [-i\Psi + \gamma^l\Psi_l + i\sigma^{mn}\Psi_{mn} + \gamma^5\tilde{\Psi} + i\gamma^l\gamma^5\tilde{\Psi}_l] \, E^{-1}$$

$$+ \frac{i}{2}\gamma_{abc}\gamma^c \, [-i\Psi + \gamma^l\Psi_l + i\sigma^{mn}\Psi_{mn} + \gamma^5\tilde{\Psi} + i\gamma^l\gamma^5\tilde{\Psi}_l] \, E^{-1}\tilde{\sigma}^{ab}$$

$$-m \left[-i\Psi + \gamma^l \Psi_l + i\sigma^{mn}\Psi_{mn} + \gamma^5\tilde{\Psi} + i\gamma^l\gamma^5\tilde{\Psi}_l \right]E^{-1} \Big\} = 0 \,. \qquad (5.2.12c)$$

Now, acting subsequently from the left by operators

$$\mathrm{Sp}\,(E\times\,, \qquad \mathrm{Sp}\,(E\gamma^5\times\,, \qquad \mathrm{Sp}\,(E\gamma^k\times\,, \qquad \mathrm{Sp}\,(E\gamma^5\gamma^k\times\,, \qquad \mathrm{Sp}\,(E\sigma^{kd}\times$$

and using known formulas for traces of combinations of Dirac matrices, we arrive at

$$e^{(l)\alpha}\,\partial_\alpha\psi_l + \gamma^c{}_{lc}\,\Psi^l + m\Psi = 0\,,$$

$$e^{(l)\alpha}\,\partial_\alpha\tilde{\Psi}_l + \gamma^c{}_{lc}\,\tilde{\Psi}^l + m\tilde{\Psi} = 0\,,$$

$$e^{(k)\alpha}\,\partial_\alpha\Psi + e^\alpha_{(c)}\,\partial_\alpha\Psi^{kc} + \gamma^k{}_{mn}\,\Psi^{mn} + \gamma^c{}_{lc}\,\Psi^{kl} - m\Psi^k = 0\,,$$

$$e^{(k)\alpha}\,\partial_\alpha\tilde{\Psi} - \frac{1}{2}\epsilon^{kcmn}\,e^\alpha_{(c)}\,\partial_\alpha\Psi_{mn} + \epsilon^{kcmn}\,\gamma^n{}_{bc}\,\Psi_{mn} - m\tilde{\Psi}^k = 0\,,$$

$$e^{(d)\alpha}\,\partial_\alpha\Psi^k - e^{(k)\alpha}\,\partial_\alpha\Psi^d + (\gamma_l{}^{dk} - \gamma_l{}^{kd})\Psi^l$$
$$+\epsilon^{dkcl}\,e^\alpha_{(c)}\,\partial_\alpha\tilde{\Psi}_l + \epsilon^{acdk}\,\gamma_{bac}\tilde{\Psi}^b - m\Psi^{dk} = 0\,; \qquad (5.2.12d)$$

this represents the written tetrad components of Eq. (5.2.11a, b) for the Eqs. (5.2.10).

One important point should be specially emphasized: during calculation, a Levi-Civita object ϵ^{abcd} arose in Eq. (5.2.12d) as a direct result of the use of a trace formula for product of three Dirac matrices . This quantity ϵ^{abcd} is not a tensor with respect to the Lorentz group, but is rather just a fixed 4-index object.

We can readily demonstrate this combination

$$\epsilon^{\alpha\beta\rho\sigma}(x) = \epsilon^{abcd}\,e^\alpha_{(a)}(x)\,e^\beta_{(b)}(x)\,e^\rho_{(c)}(x)\,e^\sigma_{(d)}(x) \qquad (5.2.13a)$$

represents a tetrad pseudo-scalar. Indeed, let us compare $\epsilon^{\alpha\beta\rho\sigma}(x)$ and $\epsilon'^{\alpha\beta\rho\sigma}(x)$, constructed on the basis of tetrads $e^\alpha_{(a)}(x)$ and $e'^\alpha_{(a)}(x)$, respectively. We have

$$\epsilon'^{\alpha\beta\rho\sigma}(x) = \epsilon^{abcd}\,e'^\alpha_{(a)}(x)\,e'^\beta_{(b)}(x)\,e'^\rho_{(c)}(x)\,e'^\sigma_{(d)}(x)\,,$$

or

$$\epsilon'^{\alpha\beta\rho\sigma}(x) = \epsilon^{abcd}\,L_a{}^i(x)\,L_b{}^j(x)\,L_c{}^m(x)\,L_d{}^n(x)\,e^\alpha_{(i)}(x)\,e^\beta_{(j)}(x)\,e^\rho_{(m)}(x)\,e^\sigma_{(n)}(x)$$

$$= \left[\epsilon^{abcd}\,L_{ai}(x)\,L_{bj}(x)\,L_{cm}(x)\,L_{dn}(x)\right]\,e^{(i)\alpha}(x)\,e^{(j)\beta}(x)\,e^{(m)\rho}(x)\,e^{(n)\sigma}(x)\,.$$

By the use of the known identity [37]

$$(\epsilon^{abcd}\,A_{ai}\,A_{bj}\,A_{cm}\,A_{dn}) = -\det[A_{ab}]\times\epsilon_{ijmn}\,,$$

we get a transformation low for $\epsilon^{\alpha\beta\rho\sigma}(x)$ with respect to the local tetrad transformations:

$$\epsilon'^{\alpha\beta\rho\sigma}(x) = -\det[L_{ai}(x)]\,\epsilon^{\alpha\beta\rho\sigma}(x)\,. \qquad (5.2.13b)$$

From Eq. (5.2.13b), it follows that under tetrad P-reflection covariant Levi-Civita object Eq. (5.2.13a) behaves as a tetrad pseudo-scalar

$$\epsilon^{(p)\alpha\beta\rho\sigma}(x) = (-1)\, \epsilon^{\alpha\beta\rho\sigma}(x) \,. \tag{5.2.13c}$$

One can notice that in each equation of Eq. (5.2.10), there are combined terms with equal transformation properties with respect to the tetrad P-reflection.

The system in Eq. (5.2.10) can be translated to the form in which all the components of the wave function are tetrad scalars:

$$\Phi(x) = \{\ \Psi(x),\ \Psi_\alpha(x),\ \Psi_{\alpha\beta}(x),\ \Psi_{\alpha\beta\rho}(x)$$

$$= \epsilon_{\alpha\beta\rho\sigma}(x)\, \tilde{\Psi}^\sigma(x),\ \Psi_{\alpha\beta\rho\sigma}(x) = \epsilon_{\alpha\beta\rho\sigma}(x)\, \tilde{\Psi}(x)\ \} \,, \tag{5.2.14a}$$

then the Dirac–Kähler equation reads

$$\nabla^\rho \Psi_\rho - m\Psi = 0 \,,$$

$$\nabla^\rho \Psi_{\rho\alpha} + \nabla_\alpha \Psi + m\Psi_\alpha = 0 \,,$$

$$\nabla^\rho \Psi_{\rho\alpha\beta} + \nabla_\alpha \Psi_\beta - \nabla_\beta \Psi_\alpha - m\Psi_{\alpha\beta} = 0 \,,$$

$$\nabla^\rho \Psi_{\rho\alpha\beta\sigma} + \nabla_\alpha \Psi_{\beta\sigma} - \nabla_\beta \Psi_{\alpha\sigma} - \nabla_\sigma \Psi_{\beta\alpha} + m\Psi_{\alpha\beta\sigma}(x) = 0 \,,$$

$$\nabla_\rho \Psi_{\alpha\beta\sigma} - \nabla_\alpha \Psi_{\rho\beta\alpha} - \nabla_\beta \Psi_{\alpha\rho\sigma} - \nabla_\sigma \Psi_{\alpha\beta\rho} - m\Psi_{\rho\alpha\beta\sigma} = 0 \,. \tag{5.2.14b}$$

Deriving Eq. (5.2.14b) from Eq. (5.2.10), one should take into account that covariant derivative of the covariant tensor vanishes identically

$$\nabla_\mu \, \epsilon^{\alpha\beta\rho\sigma}(x) = 0 \,. \tag{5.2.15}$$

Now, let us prove it. By symmetry reason, it suffices to prove only one relation $\nabla_\mu \epsilon_{0123}(x) = 0$. In accordance with definition; we have:

$$\nabla_\mu \, \epsilon_{0123}(x) = [\, \partial_\mu\, \epsilon_{0123}(x)\ -\ (\Gamma^\nu_{\mu 0}\, \epsilon_{\nu 123}(x) + \Gamma^\nu_{\mu 1}\, \epsilon_{0\nu 23}(x)$$

$$+\Gamma^\nu_{\mu 2}\, \epsilon_{01\nu 3}(x) + \Gamma^\nu_{\mu 3}\, \epsilon_{012\nu}(x))\,] = \partial_\mu\, \epsilon_{0123}(x)\ -\ \Gamma^\alpha_{\mu\alpha}\, \epsilon_{0123}(x) \,.$$

Let us specify the first term $\partial_\mu\, \epsilon_{0123}(x)$, where

$$\epsilon_{0123}(x) = -\, \epsilon_{0123}\, \det{[e_{(a)\alpha}(x)]} \,;$$

with the use of the known identity [37]

$$\partial_\mu\, A = A\, (A^{-1}_{ji}\partial_\mu\, A_{ij}) \,, \qquad A = \det{[A_{ij}]}$$

and in allowing for the inverse to $e_{(a)\alpha}$ is a matrix $e^{\beta(b)}$, we get

$$e(x) = \det[e_{(a)\alpha}(x)] \,, \qquad \partial_\mu\, e(x) = e(x)\, e^{\alpha(a)}(x)\, \partial_\mu\, e_{(a)\alpha}(x) \,.$$

Therefore,

$$\partial_\mu\, \epsilon_{0123}(x) = \epsilon_{0123}(x)\, [\, e^{\alpha(a)}(x)\partial_\mu\, e_{(a)\alpha}(x)\,] \,.$$

In turn, for $\Gamma^{\alpha}_{\mu\alpha}(x)$ we have:

$$\Gamma^{\alpha}_{\mu\alpha}(x) = \frac{1}{2}g^{\alpha\rho}(x)\,\Gamma_{\rho,\mu\alpha}(x) = \frac{1}{2}g^{\alpha\rho}x)\,[\,\partial_{\mu}\,g_{\rho\alpha}(x)\,+\,\partial_{\alpha}g_{\rho\mu}(x)\,+\,\partial_{\rho}\,g_{\mu\alpha}(x)\,]$$

$$= \frac{1}{2}g^{\alpha\beta}\,[\,\partial_{\mu}\,(\,e_{(i)\rho}(x)e_{\alpha}^{(i)}(x)\,)\,+\,\partial_{\alpha}\,(\,e_{(i)\rho}(x)e_{\mu}^{(i)}(x)\,)\,+\,\partial_{\rho}\,(\,e_{(i)\mu}(x)\,e_{\alpha}^{(i)}(x)\,)\,]\,,$$

from this, after a simple calculation we derive:

$$\Gamma^{\alpha}_{\mu\alpha}(x) = e^{\alpha}_{(i)}(x)\,\partial_{\mu}\,e_{\alpha}^{(i)}(x)\,.$$

Thus, we prove the needed identity

$$\nabla_{\mu}\,\epsilon^{\alpha\beta\rho\sigma}(x) = 0\,.$$

5.3   On Two Different Covariant Levi-Civita Objects

Let us recall a standard view for a covariant Levi-Civita object – it is defined [37] as follows

$$E_{\alpha\beta\rho\sigma}(x) \equiv +\sqrt{-g(x)}\,\epsilon_{\alpha\beta\rho\sigma}\,, \qquad E^{\alpha\beta\rho\sigma}(x) \equiv \frac{1}{+\sqrt{-g(x)}}\,\epsilon^{\alpha\beta\rho\sigma}\,,$$

$$E^{\alpha\beta\rho\sigma}(x) = g^{\alpha\mu}(x)g^{\beta\nu}(x)g^{\rho\gamma}(x)g^{\sigma\delta}(x)\,E_{\mu\nu\gamma\delta}(x)\,, \qquad (5.3.1a)$$

where $g(x)$ is a determinant of a metric tensor $g_{\alpha\beta}(x)$; and $E_{0123}(x) = +\sqrt{-g(x)}$. This definition does not depend on tetrads at all, which means that $E_{\alpha\beta\rho\sigma}(x)$ is a tetrad scalar. To have the covariant Levi-Civita object invariant with respect to arbitrary coordinate changes we must assume that the object $E_{\alpha\beta\rho\sigma}(x)$ transform as a pseudo-tensor, that is we add in relevant transformation low an additional a special factor sgn $\Delta(x)$

$$\text{sgn}\,\Delta(x) = \frac{\Delta(x)}{|\,\Delta(x)\,|}\,, \qquad \Delta(x) \equiv \det\left[\frac{\partial x'^{\alpha}}{\partial x^{\alpha}}\right]\,,$$

$$E_{\alpha'\beta'\rho'\sigma'} = \frac{\Delta(x)}{|\,\Delta(x)\,|}\,\frac{\partial x^{\alpha}}{\partial x^{\alpha'}}\,\frac{\partial x^{\beta}}{\partial x^{\beta'}}\,\frac{\partial x^{\rho}}{\partial x^{\rho'}}\,\frac{\partial x^{\sigma}}{\partial x^{\sigma'}}\,E_{\alpha\beta\rho\sigma}\,. \qquad (5.3.1b)$$

In this previous case, in the frames of the tetrad formalism, the quantity $\epsilon_{\alpha\beta\rho\sigma}$ was introduced by Eq. (5.2.11b); so it is an ordinary covariant tensor with 4 indices and in the same time it is a tetrad pseudo-scalar. Two objects, $\epsilon_{\alpha\beta\rho\sigma}(x)$ and $E_{\alpha\beta\rho\sigma}$ were defined independently from each other, therefore they may not coincide. However, quite definitely a relation between them exists. Let us explore the details of this point.

First of all, let us transform the tetrad based Levi-Civita tensor $\epsilon_{\alpha\beta\rho\sigma}(x)$ to a different form similar to Eq. (5.3.1a):

$$\epsilon_{\alpha\beta\rho\sigma}(x) = \epsilon^{abcd}\,e_{(a)\alpha}(x)\,e_{(b)\beta}(x)\,e_{(c)\rho}(x)\,e_{(d)\sigma}(x) = -\,e(x)\,\epsilon_{\alpha\beta\rho\sigma}\,,$$

$$e(x) \equiv \det\,[e_{(a)\alpha}(x)]\,. \qquad (5.3.2)$$

For instance, in the case of flat Minkowski space, using a diagonal tetrad $e^\alpha_{(a)}(x) = \delta^\alpha_a$, we get $e(x) = -1$, and further derive $\epsilon_{\alpha\beta\rho\sigma}(x) = +\epsilon_{\alpha\beta\rho\sigma}$.

It is easy to obtain relation relating determinants of the tetrad and metric tensor

$$e_{(a)\alpha}(x)\, e_{(b)\beta}(x)\, g^{ab} = g_{\alpha\beta}(x) \implies -e^2(x) = g(x) ,$$

from this relation, it follows:

$$e(x) = +\sqrt{-g(x)}, \ e(x) = -\sqrt{-g(x)} .$$

By taking the solution as $e(x) = -\sqrt{-g(x)}$, we arrive at the tetrad based definition for Levi-Civita tensor given in Eq. (5.3.2); so it is equivalent by definition according to Eq. (5.3.1a). However, a tetrad determinant can be positive as well, in this case these two definitions are not equivalent because – they differ in sign.

Note this useful formula:

$$\epsilon^{\mu\nu\gamma\delta}(x) = -e(x)\, [\, g^{\mu\alpha}(x)g^{\nu\beta}(x)g^{\gamma\rho}(x)g^{\delta\sigma}(x)\, \epsilon_{\alpha\beta\rho\sigma}\,]$$

$$= +e(x)\ \det\,[g^{\alpha\beta}(x)]\ \epsilon^{\mu\nu\gamma\delta} = +e(x)\ \frac{1}{g(x)}\ \epsilon^{\mu\nu\gamma\delta} = -\frac{1}{e(x)}\ \epsilon^{\mu\nu\gamma\delta} . \tag{5.3.3}$$

Let us specify transformation properties for $e(x) = \det\,[e_{(a)\alpha}(x)]$. Under general coordinate transformations it behaves

$$x^\alpha \implies x'^\alpha , \ \ e'(x') = \det\,[\,\frac{\partial x^\alpha}{\partial x'^\alpha}\, e_{(a)\alpha}(x)\,] = \frac{1}{\Delta(x)}\ e(x) ; \tag{5.3.4a}$$

with respect to tetrad changes it is a pseudo-scalar

$$e_{(a)\alpha}(x) \implies e'_{(a)\alpha}(x) , \ \ e'(x) = \det\,[L_a{}^b e_{(b)\alpha}(x)] = \det\,[L_a{}^b]\, e(x) . \tag{5.3.4b}$$

Let us introduce special quantity

$$J(e) = -\,\frac{\det\,[e_{(a)\alpha}(x)]}{|\,\det\,[e_{(a)\alpha}(x)]\,|} = -\,\frac{e(x)}{|\,e(x)\,|} . \tag{5.3.5a}$$

which transforms as follows:

$$x^\alpha \implies x'^\alpha , \ \ J[e'(x')] = \frac{\Delta(x)}{|\,\Delta(x)\,|}\ J[e(x)] , \tag{5.3.5b}$$

$$e_{(a)\alpha}(x) \implies e'_{(a)\alpha}(x) , \ \ J[e'(x)] = \det\,(L_a{}^b)\, J[e(x)] ; \tag{5.3.5c}$$

$J[e(x)]$ is a tetrad pseudo-scalar and a coordinate pseudo-scalar.

Through collecting these results together

$$\epsilon_{\alpha\beta\rho\sigma}(x) = -e\, \epsilon_{\alpha\beta\rho\sigma} ,$$

$$E_{\alpha\beta\rho\sigma}(x) = \sqrt{-g}\, \epsilon_{\alpha\beta\rho\sigma} ,$$

$$e^2 = -g , \qquad +\sqrt{-g} = \frac{e}{|\,e\,|}\, e ,$$

we readily find relation between two Levi-Civita tensors

$$E_{\alpha\beta\rho\sigma}(x) = \frac{e}{|e|} \, e \, \epsilon_{\alpha\beta\rho\sigma} = -\frac{e}{|e|} \, \epsilon_{\alpha\beta\rho\sigma}(x) = J[e(x)] \, \epsilon_{\alpha\beta\rho\sigma}(x) \; .$$

Let us turn back to Eq. (5.2.11a). Instead of $\tilde{\Psi}(x), \tilde{\Psi}_\alpha(x)$ one can introduce new variables

$$\bar{\Psi}(x) = J(e) \, \tilde{\Psi}(x) \; , \qquad \bar{\Psi}_\alpha(x) = J(e) \, e_\alpha^{(a)}(x)\tilde{\Psi}_a(x) \; , \qquad (5.3.6a)$$

and instead of $\epsilon_{\alpha\beta\rho\sigma}(x)$ in Eq. (5.2.11b); one may determine another quantity

$$E_{\alpha\beta\rho\sigma}(x) = J[e(x)] \, \epsilon_{\alpha\beta\rho\sigma}(x) \; . \qquad (5.3.6b)$$

Correspondingly, the main system of tensor equations can be presented as follows (compare this example with Eq. (5.2.10))

$$\nabla^\alpha \Psi_\alpha + m \, \Psi = 0 \; ,$$

$$\nabla^\alpha \bar{\Psi}_\alpha + m \, \bar{\Psi} = 0 \; ,$$

$$\nabla_\alpha \Psi + \nabla^\beta \Psi_{\alpha\beta} - m \, \Psi_\alpha = 0 \; ,$$

$$\nabla_\alpha \bar{\Psi} - \frac{1}{2} E_\alpha{}^{\beta\rho\sigma}(x) \, \nabla_\beta \Psi_{\rho\sigma} - m \, \bar{\Psi}_\alpha = 0 \; ,$$

$$\nabla_\alpha \Psi_\beta - \nabla_\beta \Psi_\alpha + E_{\alpha\beta}{}^{\rho\sigma}(x) \, \nabla_\rho \bar{\Psi}_\sigma - m \, \Psi_{\alpha\beta} = 0 \; . \qquad (5.3.7)$$

Here, $\Psi(x), \Psi_\alpha(x), \Psi_{\alpha\beta}(x)$ are general covariant tensors, whereas $\bar{\Psi}(x), \bar{\Psi}_\alpha(x), E^{\rho\sigma\alpha\beta}(x)$ are general covariant pseudo-tensors; all six objects are tetrad scalars.

Thus, when describing tensor components for Dirac–Kähler field one can use alternatively both methods. Evidently, classification of the components through their tetrad properties is more preferable because it has clear Lorentzian status (such as spin and mass).

It should be noted additionally that classification for tensor quantities in the frames of the full Lorentz group within Minkowski space-time assumes four different possibilities distinguished by adding special factors in transformation low

$$1 \; , \qquad \det{(L^a{}_b)}, \qquad \text{sgn}\,(L^0{}_0) \; , \qquad \text{sgn}\,(L^0{}_0)\det{(L^a{}_b)} \; .$$

It is not clear how that Lorentz group based classification can be described in terms of a pure general covariant theory without tetrad formalism.

5.4   On Fermion Interpretation for Dirac–Kähler Field

The Dirac–Kähler equation in arbitrary curved space-time

$$[\, i\gamma^\alpha(x) \, (\, \partial_\alpha \, + \, B_\alpha(x)\,) \, - \, m \,]\, U(x) = 0 \qquad (5.4.1a)$$

does not split up into four independent equations for particles with spin 1/2:

$$\left[\, i\gamma^\beta(x) \, (\partial_\beta \, + \, \Gamma_\beta(x)) \, - \, m \,\right] \Psi^{(i)}(x) = 0 \; , \qquad (i = 1,\, 2,\, 3,\, 4) \; . \qquad (5.4.1b)$$

In other words, these two models are completely different for any curved space-time model.

Let us consider Eqs. $(5.4.1b)$ in more detail. Relevant four local bispinor fields can be developed into (4×4)-matrix $V(x)$ according to $V(x) = (\ \Psi^{(1)}, \ \Psi^{(2)}, \ \Psi^{(3)}, \ \Psi^{(4)}\ )$; then eqs. $(5.4.1b)$ read

$$[\ i\gamma^{\alpha}(x)\ (\ \partial_{\alpha}\ +\ \Gamma_{\alpha}(x)\ )\ -\ m\ ]\ V(x) = 0\ . \tag{5.4.2a}$$

Matrix $V(x)$ can be decomposed as

$$V(x) = \Big[\ -i\Psi(x)\ +\ \gamma^{l}\ \tilde{\Psi}_{l}(x)\ +\ i\sigma^{mn}\ \Psi_{mn}(x)\ +\ \gamma^{5}\ \tilde{\Psi}(x)\ +\ i\gamma^{l}\gamma^{5}\ \tilde{\Psi}_{l}(x)\ \Big]\ E^{-1}\ ; \tag{5.4.2b}$$

However, for these involved quantities $\Psi(x)$, $\tilde{\Psi}(x)$, $\Psi_{l}(x)$, $\tilde{\Psi}_{l}(x)$, $\Psi_{mn}(x)$ does not possess transformation properties of a tensor nature with respect to the local Lorentz group. At the same time, some quasi-tensor equations can be derived from Eq. $(5.4.2a)$. To this end, it is necessary repeat the same manner as previously described. For instance, turning to Eq. $(5.2.12c)$

$$\{i\gamma^{c}e_{(c)}^{\beta}\partial_{\beta}\ [\ -i\Psi + \gamma^{l}\Psi_{l} + i\sigma^{mn}\Psi_{mn} + \gamma^{5}\tilde{\Psi} + i\gamma^{l}\gamma^{5}\tilde{\Psi}_{l}\ ]\ E^{-1}$$

$$+\frac{i}{2}\gamma_{abc}\gamma^{c}\sigma^{ab}\ [\ -i\Psi + \gamma^{l}\Psi_{l} + i\sigma^{mn}\Psi_{mn} + \gamma^{5}\tilde{\Psi} + i\gamma^{l}\gamma^{5}\tilde{\Psi}_{l}\ ]\ E^{-1}$$

$$+\frac{i}{2}\gamma_{abc}\gamma^{c}\ [\ -i\Psi + \gamma^{l}\Psi_{l} + i\sigma^{mn}\Psi_{mn} + \gamma^{5}\tilde{\Psi} + i\gamma^{l}\gamma^{5}\tilde{\Psi}_{l}\ ]\ E^{-1}\tilde{\sigma}^{ab}$$

$$-m\ [\ -i\Psi + \gamma^{l}\Psi_{l} + i\sigma^{mn}\Psi_{mn} + \gamma^{5}\tilde{\Psi} + i\gamma^{l}\gamma^{5}\tilde{\Psi}_{l}\ ]\ E^{-1}\} = 0\ . \tag{5.4.3}$$

Multiplying Eq. $(5.4.3)$ from the left by E and taking the trace of the result (with the use of the rule $E\ \tilde{\sigma}^{ab} = -\sigma^{ab}\ E$), which results in

$$i\ e_{(c)}^{\beta}\ \mathrm{Sp}\ (\gamma^{c}\gamma^{l})\ \partial_{\beta}\Psi_{l} + \frac{i}{2}\ \gamma_{abc}[\ \mathrm{Sp}\ (\gamma^{c}\sigma^{ab}\gamma^{l})\ \Psi_{l} + i\ \mathrm{Sp}\ (\gamma^{c}\sigma^{ab}\gamma^{l}\gamma^{5})\ \tilde{\Psi}_{l}\ ]$$

$$+\frac{1}{2}\ \gamma_{abc}[-\ \mathrm{Sp}\ (\gamma^{c}\gamma^{l}\sigma^{ab})\ \Psi_{l} - i\ \mathrm{Sp}\ (\gamma^{c}\gamma^{l}\gamma^{5}\sigma^{ab})\ \tilde{\Psi}_{l}] + 4i\ \Psi = 0\ .$$

Further, then to allow for the known formulas

$$\frac{1}{2}\ \gamma_{abc}\ \mathrm{Sp}\ (\gamma^{c}\sigma^{ab}\gamma^{l})\ \Psi_{l} = +\frac{1}{2}\ \gamma^{cl}{}_{c}\ ,$$

$$-\frac{1}{2}\ \gamma_{abc}\ \mathrm{Sp}\ (\gamma^{c}\gamma^{l}\sigma^{ab})\ \Psi_{l} = +\frac{1}{2}\ \gamma^{cl}{}_{c}\ ,$$

$$\frac{1}{2}\ \gamma_{abc}\ \mathrm{Sp}\ (\gamma^{c}\sigma^{ab}\ \gamma^{l}\gamma^{5})\ \tilde{\Psi}_{l} = +\frac{i}{4}\ \gamma_{abc}\epsilon^{abcl}\ \tilde{\Psi}_{l}\ ,$$

$$-\frac{1}{2}\ \gamma_{abc}\ \mathrm{Sp}\ (\gamma^{c}\gamma^{l}\gamma^{5}\sigma^{ab})\ \tilde{\Psi}_{l} = -\frac{i}{4}\ \gamma_{abc}\ \epsilon^{abcl}\ \tilde{\Psi}_{l}\ ,$$

for U- and V-fields, respectively; we find

for U-field

$$e^{(l)\alpha} \partial_\alpha \Psi_l + \gamma^c_{\ lc} \Psi^l + m \Psi = 0 \,, \qquad (5.4.4a)$$

for V-field

$$e^{(l)\alpha} \partial_\alpha \Phi_l + \frac{1}{2}\gamma^c_{\ lc}(x) \Phi^l - \frac{1}{4}\gamma_{abc} \epsilon^{abcl} \tilde{\Phi}_l + m \Phi = 0 \,. \qquad (5.4.4b)$$

It should be emphasized that because Φ-fields are not of tensor nature under the local Lorentz group Eq. (5.4.4b), cannot be presented in pure covariant tensor form; whereas Eq. (5.3.4a) does.

In a similar manner, acting on Eq. (5.4.3) by the operator Sp $(E\gamma^5\times$ we get

for U-field

$$e^{(l)\alpha} \partial_\alpha \tilde{\Psi}_l + \gamma^c_{\ lc} \tilde{\Psi}^l + m \tilde{\Psi} = 0 \,; \qquad (5.4.5a)$$

for V-field

$$e^{(l)\alpha} \partial_\alpha \tilde{\Phi}_l + \frac{1}{2}\gamma^c_{\ lc} \tilde{\Phi}^l + \frac{1}{4}\gamma_{abc} \epsilon^{abcl} \Phi_l + m \tilde{\Phi} = 0 \,. \qquad (5.4.5b)$$

One more case is when multiplying Eq. (5.4.3) by Sp $(E\gamma^k$:

$$ie^{\beta}_{(c)} \partial_\beta \left[-i \, \text{Sp} \, (\gamma^k\gamma^c) \, \Psi + i \, \text{Sp} \, (\gamma^k\gamma^c\sigma^{mn})\Psi_{mn} \right] + \frac{i}{2}\gamma_{abc} \left[-i \, \text{Sp} \, (\gamma^k\gamma^c\sigma^{ab}) \, \Psi \right.$$

$$+i \, \text{Sp} \, (\gamma^k\gamma^c\sigma^{ab}\sigma^{mn}) \, \Psi_{mn} + \text{Sp} \, (\gamma^k\gamma^c\sigma^{ab}\gamma^5) \, \tilde{\Psi} \left. \right] + \frac{i}{2}\gamma_{abc} \left[+i \, \text{Sp} \, (\gamma^k\gamma^c\sigma^{ab}) \, \Psi \right.$$

$$-i \, \text{Sp} \, (\gamma^k\gamma^c\sigma^{mn}\sigma^{ab}) \, \Psi_{mn} - \text{Sp} \, (\gamma^k\gamma^c\gamma^5\sigma^{ab}) \, \tilde{\Psi} \left. \right] - m \, \text{Sp} \, (\gamma^k\gamma^l) \, \Psi_l = 0 \,,$$

which results in

for U-field

$$e^{(k)\alpha} \partial_\alpha \Psi + e^{\alpha}_{(c)} \partial_\alpha \Psi^{kc} + \gamma^k_{\ mn} \Psi^{mn} + \gamma^c_{\ lc} \Psi^{kl} - m \Psi^k = 0 \,; \qquad (5.4.6a)$$

for V-field

$$e^{(k)\alpha} \partial_\alpha \Phi + e^{\alpha}_{(c)} \partial_\alpha \Phi^{kc} + \frac{1}{2}\gamma^k_{\ mn}\Phi^{mn} + \frac{1}{2}\gamma^c_{\ lc}\Phi^{kl}$$

$$+\frac{1}{2}\gamma^{ck}_{\ \ c}(x) \Phi + \frac{1}{4}\gamma_{abc}\epsilon^{abck} \tilde{\Phi} + \frac{1}{4}\gamma_{mn}^{\ \ k} \Phi^{mn} - m \Phi^k = 0 \,. \qquad (5.4.6b)$$

Equations (5.4.6a) and (5.4.6b) substantially differ from each other, and only the first is reduced to covariant tensor form. The remaining equations can be treated similarly, where the main results are the same: only for the U-field does covariant tensor equations arise.

More importantly, there is one remark about the interpretation of the Dirac–Kähler field in flat Minkowski space as a set of four Dirac particles that should be given. The matter is that any particle as a relativistic object is determined not only by the explicitly given wave equation but also determined by a relevant operation of charge conjugation. The latter, in turn, is fixed by transformation properties of the wave function under the Lorentz group. Evidently, the Dirac–Kähler object and the system four Dirac fields assume their own and possess different charge conjugations. In particular, by introducing a definition for a particle and antiparticle in accordance with four fermions interpretation, one can immediately see that such a particle-antiparticle separating turns to be non-invariant with respect to tensor transformation rules of the Dirac–Kähler constituents. Thus, even in the flat Minkowski space, the four fermion interpretation for this field cannot be evolved successfully.

5.5   Bosons with Different Intrinsic Parities in Curved Space-Time

From the Dirac–Kähler theory, by imposing special linear restrictions, one can derive more simple equations for a particle with a single value of spin: ordinary bosons of spin 0 or 1 with different intrinsic parity.

First, let us consider tensor equations in flat Minkowski space with four different additional constraints:

$$S = 0 \,, \qquad \tilde{\Phi} = 0 \,, \qquad \tilde{\Phi}_l = 0 \,, \qquad \Phi_{lk} = 0 \,,$$

$$\partial^l \Phi_l + m\Phi = 0 \,, \qquad \partial_l \Phi - m\, \Phi_l = 0 \,, \qquad \partial^d \Phi^k - \partial^k \Phi^d = 0 \,; \qquad (5.5.1a)$$

$$S = \tilde{0} \,, \qquad \Phi = 0 \,, \qquad \Phi_l = 0 \,, \qquad \Phi_{lk} = 0 \,,$$

$$\partial^l \tilde{\Phi}_l + m\, \tilde{\Phi} = 0 \,, \qquad \partial^k \tilde{\Phi} - m\tilde{\Phi}^k = 0 \,, \qquad \epsilon^{dkcl}\partial_c \Phi_l = 0 \,; \qquad (5.5.1b)$$

$$S = 1 \,, \qquad \Phi = 0 \,, \qquad \tilde{\Phi} = 0 \,, \qquad \tilde{\Phi}_l = 0 \,,$$

$$\partial^l \Phi_l = 0 \,, \ \partial^l \Phi_{kl} - m\, \Phi_k = 0 \,, \ \epsilon^{kcmn}\partial_c \Phi_{mn} = 0 \,, \ \partial_d \Phi_k - \partial_k \Phi_d - m\Phi_{dk} = 0 \,; \qquad (5.5.2a)$$

$$S = \tilde{1} \,, \qquad \Phi = 0, \qquad \tilde{\Phi} = 0, \qquad \Phi_l = 0 \,,$$

$$\partial^l \tilde{\Phi}_l = 0 \,, \ \partial^l \Phi_{kl} = 0 \,, \ \frac{1}{2}\epsilon^{kcmn}\partial_c \Phi_{mn} + m\tilde{\Phi}^k = 0 \,, \ \epsilon^{dkcl}\partial_c \tilde{\Phi}_l - m\Phi^{dk} = 0 \,. \qquad (5.5.2b)$$

Let us describe additional constraints in spinor form. For a scalar particle, we get:

$$S = 0 \,, \qquad \begin{vmatrix} \xi & \Delta \\ H & \eta \end{vmatrix} = \begin{vmatrix} -\Phi\sigma^2 & +i\Phi_l \bar{\sigma}^l \sigma^2 \\ -i\Phi_l \sigma^l \sigma^2 & +\Phi\sigma^2 \end{vmatrix} \,,$$

$$\Delta^{tr} = +H \,, \ \xi = -\eta \,, \ \xi^{tr} = -\xi \,, \ \eta^{tr} = -\eta \,, \qquad (5.5.3)$$

the symbol of tr stands for a matrix transposition. For a pseudo-scalar particle, we get:

$$S = \tilde{0} \,, \qquad \begin{vmatrix} \xi & \Delta \\ H & \eta \end{vmatrix} = \begin{vmatrix} +i\tilde{\Phi}\sigma^2 & -\tilde{\Phi}_l \,\bar{\sigma}^l \sigma^2 \\ -\tilde{\Phi}_l \sigma^l \sigma^2 & +i\tilde{\Phi}\sigma^2 \end{vmatrix} \,,$$

$$\Delta^{tr} = -H \,, \ \xi = +\eta \,, \ \xi^{tr} = -\xi \,, \ \eta^{tr} = -\eta \,. \qquad (5.5.4)$$

For a vector particle, we will have

$$S = 1 \,, \qquad \begin{vmatrix} \xi & \Delta \\ H & \eta \end{vmatrix} = \begin{vmatrix} +\Sigma^{mn}\sigma^2 \Phi_{mn} & +i\bar{\sigma}^l \sigma^2 \Phi_l \\ -i\sigma^l \sigma^2 \Phi_l & -\bar{\Sigma}^{mn}\sigma^2 \Phi_{mn} \end{vmatrix} \,,$$

$$\Delta^{tr} = +H \,, \ \xi^{tr} = +\xi \,, \ \eta^{tr} = +\eta \,. \qquad (5.5.5a)$$

Here, each of symmetrical spinors ξ and η depends on three independent variables:

$$\xi + \eta = -2i\,(\,\sigma^1 \Phi_{23} + \sigma^2 \Phi_{31} + \sigma^3 \Phi_{12}\,)\,\sigma^2 \,,$$

$$\xi - \eta = 2 \, (\, \sigma^1 \Phi_{01} + \sigma^2 \Phi_{02} + \sigma^3 \Phi_{03} \,) \, \sigma^2 \, . \qquad (5.5.5b)$$

Finally, a pseudo-vector case is given by

$$\underline{S = \tilde{1}} \, , \qquad \left| \begin{matrix} \xi & \Delta \\ H & \eta \end{matrix} \right| = \left| \begin{matrix} +\Sigma^{mn}\sigma^2\Phi_{mn} & -\bar{\sigma}^l\sigma^2\tilde{\Phi}_l \\ -\sigma^l\sigma^2\tilde{\Phi}_l & -\bar{\Sigma}^{mn}\sigma^2\Phi_{mn} \end{matrix} \right| \, ,$$

$$\Delta^{tr} = -H \, , \quad \xi^{tr} = +\xi \, , \quad \eta^{tr} = +\eta \, . \qquad (5.5.6)$$

We plan to extend this approach to general covariant case

$$\nabla^\alpha \Psi_\alpha + m\Psi = 0 \, , \qquad \tilde{\nabla}^\alpha \Psi_\alpha + m\tilde{\Psi} = 0 \, ,$$

$$\nabla_\alpha \Psi + \nabla^\beta \Psi_{\alpha\beta}(x) - m\Psi_\alpha = 0 \, ,$$

$$\tilde{\nabla}_\alpha \Psi - \frac{1}{2}\epsilon_\alpha{}^{\beta\rho\sigma}(x)\nabla_\beta \Psi_{\rho\sigma} - m\tilde{\Psi}_\alpha = 0 \, ,$$

$$\nabla_\alpha \Psi_\beta - \nabla_\beta \Psi_\alpha + \epsilon_{\alpha\beta}{}^{\rho\sigma}(x)\nabla_\rho \tilde{\Psi}_\sigma - m\Psi_{\alpha\beta} = 0 \, . \qquad (5.5.7)$$

First, let it be

$$\underline{S = 0} \, , \qquad \nabla^\alpha \Psi_\alpha + m\Psi = 0 \, ,$$

$$\nabla_\alpha \Psi - m\Psi_\alpha = 0 \, , \quad \nabla_\alpha \Psi_\beta - \nabla_\beta \Psi_\alpha = 0 \, . \qquad (5.5.8)$$

The first two are the Proca equations for scalar particle. The last equation holds identically

$$\partial_\alpha \, \partial_\beta \, \Psi \, - \, \Gamma^\mu_{\alpha\beta} \, \partial_\mu \, \Psi \, - \, \partial_\beta \, \partial_\alpha \, \Psi \, + \, \Gamma^\mu_{\beta\alpha} \, \partial_\mu \, \Psi = 0 \, .$$

For a pseudo-scalar field, we have

$$\underline{S = \tilde{0}} \, , \quad \nabla^\alpha \tilde{\Psi}_\alpha + m\tilde{\Psi} = 0 \, ,$$

$$\nabla_\alpha \tilde{\Psi} - m\tilde{\Psi}_\alpha = 0 \, , \quad \epsilon_{\alpha\beta}{}^{\rho\sigma}(x)\nabla_\rho \Psi_\sigma = 0 \, ; \qquad (5.5.9)$$

Here, the last equation holds identically. Now, let only $\Psi_\alpha \neq 0$, $\Psi_{\alpha\beta}(x) \neq 0$, then

$$\underline{S = 1} \, , \qquad \nabla^\alpha \Psi_\alpha = 0 \, , \quad \nabla^\beta \Psi_{\alpha\beta} - m\Psi_\alpha = 0 \, ,$$

$$-\frac{1}{2}\epsilon_\alpha{}^{\beta\rho\sigma}(x)\nabla_\beta \Psi_{\rho\sigma} = 0 \, , \quad \nabla_\alpha \Psi_\beta - \nabla_\beta \Psi_\alpha = m\Psi_{\alpha\beta} \, . \qquad (5.5.10)$$

Here, the first and third equations hold identically:

$$\nabla^\alpha \Psi_\alpha = \frac{1}{m} \, \nabla^\alpha \nabla^\beta \, \Psi_{\alpha\beta} = \frac{1}{2m} \, [\, \Psi_{\alpha\nu} \, R^\nu{}_\beta{}^{\beta\alpha}$$

$$-\Psi_{\beta\nu} R^\nu{}_\alpha{}^{\beta\alpha} \,] = \frac{1}{2m} \, [\, -\Psi_{\alpha\nu} \, R^{\nu\alpha} - \Psi_{\beta\nu} \, R^{\nu\beta} \,] = 0 \, ,$$

$$-\frac{1}{2m} \, \epsilon_\alpha{}^{\beta\rho\sigma}(x) \, \nabla_\beta \, [\, \nabla_\rho \Psi_\sigma - \nabla_\sigma \Psi_\rho \,]$$

$$= -\frac{1}{4m} \, \epsilon_\alpha{}^{\rho\beta\sigma}(x) \, [\, (\nabla_\beta\nabla_\rho - \nabla_\rho\nabla_\beta) \, \Psi_\sigma - (\nabla_\beta\nabla_\sigma - \nabla_\sigma\nabla_\beta) \, \Psi_\rho \,]$$

$$= -\frac{1}{4m} \, \epsilon_\alpha{}^{\beta\rho\sigma}(x) \, (\Psi^\nu \, R_{\nu\sigma\rho\beta} - \Psi^\nu \, R_{\nu\rho\sigma\beta}) \; = 0 \; .$$

Now, let $\Psi(x) = \tilde{\Psi} = \Psi_\alpha = 0$, then

$$\underline{S = \tilde{1}} \, , \qquad\qquad \nabla^\alpha \tilde{\Psi}_\alpha = 0 \, , \; \nabla^\beta \Psi_{\alpha\beta} = 0 \, ,$$

$$\frac{1}{2} \, \epsilon_\alpha{}^{\beta\rho\sigma}(x) \, \nabla_\beta \Psi_{\rho\sigma} + m \, \tilde{\Psi}_\alpha = 0 \, ,$$

$$\epsilon_{\alpha\beta}{}^{\rho\sigma}(x) \, \nabla_\rho \tilde{\Psi}_\sigma - m \, \Psi_{\alpha\beta} = 0 \; . \qquad\qquad (5.5.11)$$

The first and the second equations hold identically:

$$\nabla^\alpha \tilde{\Psi}_\alpha = -\frac{1}{2m} \, \nabla^\alpha \epsilon_\alpha{}^{\beta\rho\sigma}(x) \, \nabla_\beta \, \Psi_{\rho\sigma} = -\frac{1}{2m} \, \epsilon_\alpha{}^{\beta\rho\sigma}(x) \, \nabla^\alpha \nabla_\beta \, \Psi_{\rho\sigma}$$

$$= -\frac{1}{4m} \epsilon_\alpha{}^{\beta\rho\sigma}(x) [\, \Psi_{\nu\sigma} \, R^\nu{}_{\beta\rho}{}^\alpha(x) \; + \; \Psi_{\rho\nu} \, R^\nu{}_{\sigma\beta}{}^\alpha \,] \, ,$$

$$\nabla^\beta \, \Psi_{\alpha\beta}(x) = \frac{1}{m} \, \nabla^\beta \, \epsilon_{\alpha\beta}{}^{\rho\sigma}(x) \, \nabla_\rho \Psi_\sigma = \frac{1}{2m} \, \epsilon_{\alpha\beta}{}^{\rho\sigma}(x) \, \Psi^\nu \, R_{\nu\sigma\rho}{}^\beta \; .$$

Constraints separating four boson fields are the same as in the case of Minkowski space:

$$S = 0 \, , \qquad\qquad \Delta^{tr} = +H \, , \; \xi = -\eta \, , \; \xi^{tr} = -\xi \, , \; \eta^{tr} = -\eta \; ;$$

$$S = \tilde{0} \, , \qquad\qquad \Delta^{tr} = -H \, , \; \xi = +\eta \, , \; \xi^{tr} = -\xi \, , \; \eta^{tr} = -\eta \; ;$$

$$S = 1 \, , \qquad\qquad \Delta^{tr} = +H \, , \; \xi^{tr} = +\xi \, , \; \eta^{tr} = +\eta \; ;$$

$$S = \tilde{1} \, , \qquad\qquad \Delta^{tr} = -H \, , \; \xi^{tr} = +\xi \, , \; \eta^{tr} = +\eta \; . \qquad (5.5.12)$$

We may conclude that the use of tetrad formalism permit us to apply results on classification of the particles with respect to discrete Lorentzian transformations (including intrinsic parity of bosons) when treating relevant particle fields on the background of arbitrary curved space-time model.

5.6 Conserved Quantities in the Theory of the Dirac–Kähler Field

It is known that the wave equation of the Dirac–Kähler field allows for sixteen conservation laws and that there exists a number of studies where such conserved Dirac–Kähler currents have been examined. However, previously as a rule, the main attention was given to the existence of those conservation laws and through the description of its underlying symmetry instead of elucidating its physical meaning. There is one significant point to be outlined here, and that is, the Dirac–Kähler theory gives a possibility to obtain many facts of a more simple theory for ordinary boson particles of spin 0 or 1. This is achieved by imposing special linear conditions on the 16-component Dirac–Kähler wave function. The same trick may be done with the Dirac–Kähler conserved currents, and it is the object of our next section.

Before proceeding to the problem of conserved quantities in the Dirac–Kähler theory, with the subsequent analysis for the most interesting cases of ordinary bosons particles of spin 0 or 1 and different intrinsic parity; let us consider in the beginning the simple example of the Dirac particle – in this there exists the possibility to introduce the notation required.

The Dirac equation and its conjugate for $\bar{\Psi}(x) = \Psi^+(x)\gamma^0$ have the form

$$[\, i\gamma^a \overrightarrow{\partial}_a - m\,]\, \Psi(x) = 0 \,, \qquad \bar{\Psi}(x)\,[\, i\gamma^a \overleftarrow{\partial}_a + m\,] = 0 \,. \qquad (5.6.1)$$

Here, we take the formulas (the use of spinor basis is presupposed)

$$(\gamma^0)^+ = +\,\gamma^0, \qquad (\gamma^i)^+ = -\,\gamma^i\,, \qquad \gamma^0(\gamma^a)^+\gamma^0 = +\gamma^a \,. \qquad (5.6.2)$$

From Eq. (5.6.1), through this known procedure it follows the charge conservation law

$$\bar{\Psi}\,[\, i\gamma^a \overrightarrow{\partial}_a - m\,]\, \Psi + \bar{\Psi}\,[\, i\gamma^a \overleftarrow{\partial}_a + m\,]\, \Psi = 0 \,,$$

$$\partial_a J^a(x) = 0 \,, \qquad \text{where} \qquad J^a(x) = \bar{\Psi}(x)\gamma^a\Psi(x) \,. \qquad (5.6.3)$$

The symmetry properties of the vector $J^a(x)$ under double spin covering of the full Lorentz group G_L^{spin} [40]:

$$S(k,\bar{k}^*) \,, \qquad M = i\gamma^0 \,, \qquad N = \gamma^0\gamma^5$$

can be easily found. Indeed, in the first place consider continuous transformations

$$S(k,\bar{k}^*): \qquad J^{\,'a}(x') = \Psi^+(x)\, S^+(k,\bar{k}^*)\, \gamma^0\gamma^a\, S(k,\bar{k}^*)\Psi(x) \,;$$

from where, with the use of the formula

$$S^+(k,\bar{k}^*)\, \gamma^0 = +\, \gamma^0\, S^{-1}(k,\bar{k}^*) \,, \qquad (5.6.4a)$$

we get

$$J^{\,'a}(x') = \Psi^+(x)\, \gamma^0\, [\, S^{-1}(k,\bar{k}^*)\gamma^a S(k,\bar{k}^*)\,]\, \Psi(x) \,.$$

Further, by taking in mind the known relation

$$S^{-1}(k,\bar{k}^*)\gamma^a S(k,\bar{k}^*) = \gamma^b\, L_b{}^a(\bar{k},\bar{k}^*) = L^a{}_b(k,\,,k^*)\, \gamma^b \,, \qquad (5.6.4b)$$

we arrive at the law

$$S(k,\bar{k}^*): \qquad J^{\,'a}(x') = L^a{}_b(k,k^*)\, J^b(x) \,. \qquad (5.6.4c)$$

For the discrete transformation M

$$M = i\gamma^0: \qquad J^{\,'a}(x') = \Psi^+(x)\, M^+\, \gamma^0\gamma^a\, M\, \Psi(x) \,;$$

in using the formula

$$M^+\gamma^0 = +\, \gamma^0 M^{-1} \,, \qquad (5.6.5a)$$

we have

$$M: \qquad J^{\,'a}(x') = \Psi^+(x)\, \gamma^0 M^{-1}\, \gamma^a\, M\, \Psi(x) \,.$$

And further, to allow for

$$M^{-1}\,\gamma^a\,M = \bar{\delta}_b^a\,\gamma^b = L^{(P)a}{}_b\,\gamma^b\,,\tag{5.6.5b}$$

we can arrive at

$$M:\qquad J^{\prime a}(x') = L^{(P)a}{}_b\,J^b(x)\,.\tag{5.6.5c}$$

Analogously, for the second discrete element N

$$N = \gamma^0\gamma^5:\qquad J^{\prime a}(x') = \Psi^+(x)\,N^+\,\gamma^0\gamma^a\,N\,\Psi(x)\,;$$

with the formula

$$N^+\gamma^0 = -\,\gamma^0 N^{-1}\,,\tag{5.6.6a}$$

we get

$$N:\qquad J^{\prime a}(x') = -\,\Psi^+(x)\,\gamma^0 N^{-1}\,\gamma^a\,N\,\Psi(x)$$

and then with the formula

$$N^{-1}\,\gamma^a\,N = -\,\bar{\delta}_b^a\,\gamma^b = L^{(T)a}{}_b\,\gamma^b\,,\tag{5.6.6b}$$

arrive at

$$N:\qquad J^{\prime a}(x') = -L^{(T)a}{}_b\,J^b(x)\,.\tag{5.6.6c}$$

Next, we collect the previous results together:

$$S(k,\bar{k}^*):\qquad J^{\prime a}(x') = L^a{}_b(k,k^*)\,J^b(x)\,,$$
$$M:\qquad J^{\prime a}(x') = +\,L^{(P)a}{}_b\,J^b(x)\,,\tag{5.6.7}$$
$$N:\qquad J^{\prime a}(x') = -\,L^{(T)a}{}_b\,J^b(x)\,.$$

The problem of conserved currents in the Dirac–Kähler theory may be considered in much the same manner. The main equation and its conjugate are

$$[\,i\gamma^a\,\overrightarrow{\partial}_a - m\,]\,U(x) = 0\,,\qquad \bar{U}(x)\,[\,i\gamma^a\,\overleftarrow{\partial}_a + m\,] = 0\,,\tag{5.6.8a}$$

where

$$\bar{U}(x) = U^+(x)\,(\gamma^0 \otimes \gamma^0) = \tilde{\gamma}^0 U^+(x)\gamma^0\,.\tag{5.6.8b}$$

The Dirac–Kähler functions behave under the group G_L^{spin} as follows:

$$S(k,\bar{k}^*)\ :\qquad U'(x') = (S \otimes S)\,U(x) = S\,U(x)\,\tilde{S}\,,$$
$$\bar{U}'(x') = \bar{U}(x)\,(S^{-1} \otimes S^{-1}) = \tilde{S}^{-1}\,\bar{U}(x)\,S^{-1}\,,\tag{5.6.9a}$$

$$M = i\gamma^0\ :\qquad U'(x') = (M \otimes M)\,U(x) = M\,U(x)\,\tilde{M}\,,$$
$$\bar{U}'(x') = \bar{U}(x)\,(M^{-1} \otimes M^{-1}) = \tilde{M}^{-1}\,\bar{U}(x)\,M^{-1}\,,\tag{5.6.9b}$$

and

$$N = \gamma^0\gamma^5\ :\qquad U'(x') = (N \otimes N)\,U(x) = N\,U(x)\,\tilde{N}\,,$$
$$\bar{U}'(x') = \bar{U}(x)\,(N^{-1} \otimes N^{-1}) = \tilde{N}^{-1}\,\bar{U}(x)\,N^{-1}\,.\tag{5.6.9c}$$

When combining Eqs. (5.6.8a) and (5.6.8b), we get to sixteen conserved currents:

$$\bar{U}[\,i\gamma^a\,\overrightarrow{\partial}_a\,-m\,]\,U(x)+\bar{U}(x)\,[\,i\gamma^a\,\overleftarrow{\partial}_a\,+m\,]U(x)=0 \qquad \Longrightarrow$$

$$\partial_a\,J^a(x)=0\,,\qquad \text{where}\qquad J^a(x)=\bar{U}(x)\gamma^aU(x)\,. \qquad (5.6.10a)$$

Take notice that $J^a(x)$ is a (4×4)-matrix; that is it represents 16 conserved currents. On the base of these one may construct special linear combinations which will be of simple symmetry properties under the group G_L^{spin}:

$$j^{(a)}(x)=\text{Sp}\,J^a(x)\,,\qquad \partial_a j^{(a)}(x)=0\,;$$

$$\nu^{(a)}(x)=\text{Sp}\,\gamma^5\,J^a(x)\,,\qquad \partial_a\nu^{(a)}(x)=0\,;$$

$$T^{(a)l}(x)=\text{Sp}\,\tilde{\gamma}^l\,J^a(x)\,,\qquad \partial_a T^{(a)l}(x)=0\,; \qquad (5.6.10b)$$

$$\nu^{(a)l}(x)=\text{Sp}\,\gamma^5\tilde{\gamma}^l\,J^a(x)\,,\qquad \partial_a\nu^{(a)l}(x)=0\,;$$

$$L^{(a)kl}(x)=\text{Sp}\,\tilde{\sigma}^{kl}\,J^a(x)\,,\qquad \partial_a\varphi^{(a)kl}(x)=0\,.$$

Now the task is to follow the symmetry properties of these separate currents under the group G_L^{spin}. First, let us consider the current $j^{(a)}(x)$:

$$j^{('a)}(x')=\text{Sp}\,\bar{U}'(x')\gamma^aU'(x')\,;$$

$$S(k,\bar{k}^*),\qquad j^{('a)}(x')=\text{Sp}\,\tilde{S}^{-1}\,\bar{U}(x)\,[S^{-1}\gamma^aS]\,U(x)\,\tilde{S}$$
$$=L^a{}_b(k,\,,k^*)\,\text{Sp}\,\bar{U}(x)\gamma^bU(x)=L^a{}_b(k,\,,k^*)\,j^{(b)}(x)\,; \qquad (5.6.11a)$$

$$M,\qquad j^{('a)}(x')=\text{Sp}\,\tilde{M}^{-1}\,\bar{U}(x)\,[M^{-1}\gamma^aM]\,U(x)\,\tilde{M}$$
$$=L^{(P)a}{}_b\,\text{Sp}\,\bar{U}(x)\gamma^bU(x)=L^{(P)a}{}_b\,j^{(b)}(x)\,; \qquad (5.6.11b)$$

$$N,\qquad j^{('a)}(x')=\text{Sp}\,\tilde{N}^{-1}\,\bar{U}(x)\,[N^{-1}\gamma^aN]\,U(x)\,\tilde{N}$$
$$=L^{(T)a}{}_b\,\text{Sp}\,\bar{U}(x)\gamma^bU(x)=L^{(T)a}{}_b\,j^{(b)}(x)\,. \qquad (5.6.11c)$$

For the current $\nu^{(a)}(x)$:

$$S(k,\bar{k}^*),\qquad \nu^{('a)}(x')=+L^a{}_b(k,\,,k^*)\,\nu^{(b)}(x)\,, \qquad (5.6.12a)$$

$$M,\qquad \nu^{('a)}(x')=-\,L^{(P)a}{}_b\,\nu^{(b)}(x)\,, \qquad (5.6.12b)$$

$$N,\qquad \nu^{('a)}(x')=-\,L^{(T)a}{}_b\,\nu^{(b)}(x)\,. \qquad (5.6.12c)$$

For the current $T^{(a)l}(x)$:

$$S(k,\bar{k}^*),\qquad T^{('a)l}(x')=L^l{}_k\,L^a{}_b\,T^{(b)k}(x)\,, \qquad (5.6.13a)$$

$$M,\qquad T^{('a)l}(x')=L^{(P)l}{}_k\,L^{(P)a}{}_b\,T^{(b)k}(x)\,, \qquad (5.6.13b)$$

$$N,\qquad T^{('a)l}(x')=L^{(T)l}{}_k\,L^{(T)a}{}_b\,T^{(b)k}(x)\,. \qquad (5.6.13c)$$

For the current $\nu^{(a)l}(x)$:

$$S(k, \bar{k}^*): \qquad T^{('a)l}(x') = L^l{}_k L^a{}_b \nu^{(b)k}(x), \qquad (5.6.14a)$$

$$M, \qquad \nu^{('a)l}(x') = - L^{(P)l}{}_k L^{(P)a}{}_b \nu^{(b)k}(x), \qquad (5.6.14b)$$

$$N, \qquad \nu^{('a)l}(x') = - L^{(T)l}{}_k L^{(T)a}{}_b \nu^{(b)k}(x). \qquad (5.6.14c)$$

And, for the current $L'^{(a)bc}(x')$ we have

$$S(k, \bar{k}^*): \qquad L'^{(a)bc}(x') = L^a{}_b L^b{}_k L^c{}_l L^{(b)kl}(x), \qquad (5.6.15a)$$

$$M, \qquad L'^{(a)bc}(x') = L^{(P)a}{}_b L^{(P)b}{}_k L^{(P)c}{}_l L^{(b)kl}(x), \qquad (5.6.15b)$$

$$N, \qquad L'^{(a)bc}(x') = L^{(T)a}{}_b L^{(T)b}{}_k L^{(T)c}{}_l L^{(b)kl}(x). \qquad (5.6.15c)$$

Here, we collect the results together as a table:

	S	M	N
$j^{(a)}$:	L	$+ L^{(P)}$	$+ L^{(T)}$
$\nu^{(a)}$:	L	$- L^{(P)}$	$- L^{(T)}$
$T^{(a)l}$:	$L \otimes L$	$+ L^{(P)} \otimes L^{(P)}$	$+ L^{(T)} \otimes L^{(T)}$
$\nu^{(a)l}$:	$L \otimes L$	$- L^{(P)} \otimes L^{(P)}$	$- L^{(T)} \otimes L^{(T)}$
$L^{(a)bc}$:	$L \otimes L \otimes L$	$+ L^{(P)} \otimes L^{(P)} \otimes L^{(P)}$	$+ L^{(T)} \otimes L^{(T)} \otimes L^{(T)}$

$$(5.6.16)$$

5.7   Conserved Currents for Boson Fields in Tensor Form

The next task is to obtain explicit tensor form for these currents in Eq. (5.6.10b). Then use of the formula for the bispinor metric matrix

$$E = \begin{vmatrix} i\sigma^2 & 0 \\ 0 & -i\sigma^2 \end{vmatrix} = i\,\gamma^0\gamma^2, \qquad E^{-1} = -\,i\,\gamma^0\gamma^2,$$

the Dirac–Kähler function looks as

$$U(x) = \left[-i\,A + \gamma^5 B + \gamma^l A_l + i\,\gamma^l\gamma^5 B_l + i\,\sigma^{mn} F_{mn} \right] (-i\,\gamma^0\gamma^2), \qquad (5.7.1a)$$

also

$$U^+(x) = (i\,\gamma^0\gamma^2)\left[i\,A^* + \gamma^5 B^* + (\gamma^l)^+ A_l^* - i\,\gamma^5(\gamma^l)^+ B_l^* - i\,(\sigma^{mn})^+ F_{mn}^* \right],$$

and further

$$\bar{U}(x) = \gamma^0 U^+ \gamma^0$$

$$= (-i\,\gamma^0\gamma^2)\left[i\,A^* - \gamma^5 B^* + \gamma^l A_l^* + i\,\gamma^5\gamma^l B_l^* + i\,\sigma^{mn} F_{mn}^* \right]. \qquad (5.7.1b)$$

Therefore, the conserved current matrix may be given as

$$J^{(a)}(x) = \bar{U}(x)\gamma^a U(x)$$

$$= (-i\,\gamma^0\gamma^2)\left(+i\,A^* - \gamma^5 B^* + \gamma^p A_p^* + i\,\gamma^5\gamma^p B_p^* + i\,\sigma^{bc} F_{bc}^*\right)$$

$$\times\gamma^a\left(-i\,A + \gamma^5 B + \gamma^s A_s + i\,\gamma^s\gamma^5 B_s + i\,\sigma^{mn} F_{mn}\right)(-i\,\gamma^0\gamma^2)\,. \tag{5.7.2}$$

Also, we shall need two relations:

$$\tilde{\gamma}^l\left(-i\gamma^0\gamma^l\right) = -\gamma^l\left(-i\gamma^0\gamma^l\right)\,,$$

$$\tilde{\sigma}^{ab}\left(-i\gamma^0\gamma^l\right) = -\left(-i\gamma^0\gamma^l\right)\sigma^{ab}\,. \tag{5.7.3}$$

So that, from Eq. (5.6.10b) it follows the formulas for conserved currents of the Dirac–Kähler field:

$$j^{(a)}(x) = \mathrm{Sp}\,J^a(x)$$

$$= -\mathrm{Sp}\left[\left(i\,A^* - \gamma^5 B^* + \gamma^p A_p^* + i\,\gamma^5\gamma^p B_p^* + i\,\sigma^{bc} F_{bc}^*\right)\right.$$

$$\left.\times\gamma^a\left(-i\,A + \gamma^5 B + \gamma^s A_s + i\,\gamma^s\gamma^5 B_s + i\,\sigma^{mn} F_{mn}\right)\right]\,, \tag{5.7.4a}$$

$$\nu^{(a)}(x) = \mathrm{Sp}\,\gamma^5 J^a(x)$$

$$= -\mathrm{Sp}\left[\gamma^5\left(i\,A^* - \gamma^5 B^* + \gamma^p A_p^* + i\,\gamma^5\gamma^p B_p^* + i\,\sigma^{bc} F_{bc}^*\right)\right.$$

$$\left.\times\gamma^a\left(-i\,A + \gamma^5 B + \gamma^s A_s + i\,\gamma^s\gamma^5 B_s + i\,\sigma^{mn} F_{mn}\right)\right]\,, \tag{5.7.4b}$$

$$T^{(a)l}(x) = \mathrm{Sp}\,\tilde{\gamma}^l J^a(x)$$

$$= +\mathrm{Sp}\left[\gamma^l\left(+i\,A^* - \gamma^5 B^* + \gamma^p A_p^* + i\,\gamma^5\gamma^p B_p^* + i\,\sigma^{bc} F_{bc}^*\right)\right.$$

$$\left.\times\gamma^a\left(-i\,A + \gamma^5 B + \gamma^s A_s + i\,\gamma^s\gamma^5 B_s + i\,\sigma^{mn} F_{mn}\right)\right]\,, \tag{5.7.4c}$$

$$\nu^{(a)l}(x) = \mathrm{Sp}\,\gamma^5\tilde{\gamma}^l\,J^a(x)$$

$$= +\mathrm{Sp}\left[\gamma^5\gamma^l\left(i\,A^* - \gamma^5 B^* + \gamma^p A_p^* + i\,\gamma^5\gamma^p B_p^* + i\,\sigma^{bc} F_{bc}^*\right)\right.$$

$$\left.\times\gamma^a\left(-i\,A + \gamma^5 B + \gamma^s A_s + i\,\gamma^s\gamma^5 B_s + i\,\sigma^{mn} F_{mn}\right)\right]\,, \tag{5.7.4d}$$

$$L^{(a)kl}(x) = \mathrm{Sp}\,\tilde{\sigma}^{kl} J^a(x)$$

$$= +\mathrm{Sp}\left[\sigma^{kl}\left(+i\,A^* - \gamma^5 B^* + \gamma^p A_p^* + i\,\gamma^5\gamma^p B_p^* + i\,\sigma^{bc} F_{bc}^*\right)\right.$$

$$\left.\times\gamma^a\left(-i\,A + \gamma^5 B + \gamma^s A_s + i\,\gamma^s\gamma^5 B_s + i\,\sigma^{mn} F_{mn}\right)\right]\,. \tag{5.7.4e}$$

The relationships presuppose here possess rather unwieldy calculation. For us, the most interesting concept are the ordinary particles with one value of spin, 0 or 1, and certain intrinsic parity. So, instead of Eq. (5.7.4) we will get the four expansions for more simple cases $S = 0, \tilde{0}, 1, \tilde{1}$.

Particle $S = 0$

$$j^{(a)}(x) = - \; \mathrm{Sp} \; \left[(\; +i \, A^* + \gamma^p \, A^*_p \;) \, \gamma^a \; (\; -i \, A + \gamma^s \, A_s \;) \right] \; ,$$

$$\nu^{(a)}(x) = - \; \mathrm{Sp} \; \left[\gamma^5 \; (\; +i \, A^* + \gamma^p \, A^*_p \;) \, \gamma^a \; (\; -i \, A + \gamma^s \, A_s +) \right] \; ,$$

$$T^{(a)l}(x) = + \; \mathrm{Sp} \; \left[\gamma^l \; (\; +i \, A^* + \gamma^p \, A^*_p \;) \, \gamma^a \; (\; -i \, A + \gamma^s \, A_s \;) \right] \; , \qquad (5.7.5)$$

$$\nu^{(a)l}(x) = + \; \mathrm{Sp} \; \left[\gamma^5 \, \gamma^l \; (+i \, A^* + \gamma^p \, A^*_p) \, \gamma^a \; (\; -i \, A + \gamma^s \, A_s \;) \right] \; ,$$

$$L^{(a)kl}(x) = + \; \mathrm{Sp} \; \left[\sigma^{kl} \; (\; +i \, A^* + \gamma^p \, A^*_p \;) \, \gamma^a \; (\; -i \, A + \gamma^s \, A_s \;) \right] \; .$$

Particle $S = \tilde{0}$

$$j^{(a)}(x) = - \; \mathrm{Sp} \; \left[(\; -\gamma^5 \, B^* + i \, \gamma^5 \gamma^p \, B^*_p \;) \, \gamma^a \; (\; +\gamma^5 \, B + i \, \gamma^s \gamma^5 \, B_s \;) \right] \; ,$$

$$\nu^{(a)}(x) = - \; \mathrm{Sp} \; \left[\gamma^5 \; (\; -\gamma^5 \, B^* + i \, \gamma^5 \gamma^p \, B^*_p \;) \, \gamma^a \; (\; +\gamma^5 \, B + i \, \gamma^s \gamma^5 \, B_s \;) \right] \; ,$$

$$T^{(a)l}(x) = + \; \mathrm{Sp} \; \left[\gamma^l \; (\; -\gamma^5 \, B^* + i \, \gamma^5 \gamma^p \, B^*_p \;) \, \gamma^a \; (\; +\gamma^5 \, B + i \, \gamma^s \gamma^5 \, B_s \;) \right] \; ,$$

$$\nu^{(a)l}(x) = + \; \mathrm{Sp} \; \left[\gamma^5 \, \gamma^l \; (\; -\gamma^5 \, B^* + i \, \gamma^5 \gamma^p \, B^*_p \;) \, \gamma^a \; (\; +\gamma^5 \, B + i \, \gamma^s \gamma^5 \, B_s \;) \right] \; ,$$

$$L^{(a)kl}(x) = + \; \mathrm{Sp} \; \left[\sigma^{kl} \; (\; -\gamma^5 \, B^* + i \, \gamma^5 \gamma^p \, B^*_p \;) \, \gamma^a \; (\; +\gamma^5 \, B + i \, \gamma^s \gamma^5 \, B_s \;) \right] \; .$$
$$(5.7.6)$$

Particle $S = 1$

$$j^{(a)}(x) = - \; \mathrm{Sp} \; \left[(\; +\gamma^p \, A^*_p + i \, \sigma^{bc} \, F^*_{bc} \;) \, \gamma^a \; (\; +\gamma^s \, A_s + i \, \sigma^{mn} \, F_{mn} \;) \right] \; ,$$

$$\nu^{(a)}(x) = - \; \mathrm{Sp} \; \left[\gamma^5 \; (\; +\gamma^p \, A^*_p + i \, \sigma^{bc} \, F^*_{bc} \;) \, \gamma^a \; (\; +\gamma^s \, A_s + i \, \sigma^{mn} \, F_{mn} \;) \right] \; ,$$

$$T^{(a)l}(x) = + \; \mathrm{Sp} \; \left[\gamma^l \; (\; +\gamma^p \, A^*_p + i \, \sigma^{bc} \, F^*_{bc} \;) \, \gamma^a \; (\; +\gamma^s \, A_s + i \, \sigma^{mn} \, F_{mn} \;) \right] \; ,$$

$$\nu^{(a)l}(x) = + \; \mathrm{Sp} \; \left[\gamma^5 \, \gamma^l \; (\; +\gamma^p \, A^*_p + i \, \sigma^{bc} \, F^*_{bc} \;) \, \gamma^a \; (\; +\gamma^s \, A_s + i \, \sigma^{mn} \, F_{mn} \;) \right] \; ,$$

$$L^{(a)kl}(x) = + \; \mathrm{Sp} \; \left[\sigma^{kl} \; (\; +\gamma^p \, A^*_p + i \, \sigma^{bc} \, F^*_{bc} \;) \, \gamma^a \; (\; +\gamma^s \, A_s + i \, \sigma^{mn} \, F_{mn} \;) \right] \; .$$
$$(5.7.7)$$

Particle $S = \tilde{1}$

$$j^{(a)}(x) = -\mathrm{Sp} \left[(\; +i \, \gamma^5 \gamma^p B^*_p + i \, \sigma^{bc} F^*_{bc} \;) \, \gamma^a \; (\; +i \, \gamma^s \gamma^5 B_s + i \, \sigma^{mn} F_{mn} \;) \right],$$

$$\nu^{(a)}(x) = -\mathrm{Sp} \left[\gamma^5 \; (\; +i \, \gamma^5 \gamma^p B^*_p + i \, \sigma^{bc} F^*_{bc} \;) \, \gamma^a \; (\; +i \, \gamma^s \gamma^5 B_s + i \, \sigma^{mn} F_{mn} \;) \right],$$

$$T^{(a)l}(x) = +\mathrm{Sp} \left[\gamma^l \; (\; +i \, \gamma^5 \gamma^p B^*_p + i \, \sigma^{bc} F^*_{bc} \;) \, \gamma^a \; (\; +i \, \gamma^s \gamma^5 B_s + i \, \sigma^{mn} F_{mn} \;) \right],$$

$$\nu^{(a)l}(x) = + \; \mathrm{Sp} \; \left[\gamma^5 \gamma^l \; (\; +i \, \gamma^5 \gamma^p B^*_p + i \, \sigma^{bc} F^*_{bc} \;) \, \gamma^a (\; +i \, \gamma^s \gamma^5 B_s + i \, \sigma^{mn} F_{mn} \;) \right],$$

$$L^{(a)kl}(x) = +\mathrm{Sp} \left[\sigma^{kl} \; (\; +i \, \gamma^5 \gamma^p B^*_p + i \, \sigma^{bc} F^*_{bc} \;) \, \gamma^a \; (\; +i \, \gamma^s \gamma^5 B_s + i \, \sigma^{mn} F_{mn} \;) \right].$$
$$(5.7.8)$$

First, let us consider the case of scalar particle $S = 0$, starting from the current $j^{(a)}(x)$ (take notice on factor $1/4$)

$$S = 0, \qquad j^{(a)}(x) = - \frac{1}{4} \; \mathrm{Sp} \; \left[(\; +i \, A^* + \gamma^p \, A^*_p \;) \, \gamma^a \; (\; -i \, A + \gamma^s \, A_s \;) \right], \quad (5.7.9a)$$

an auxiliary table (the all non-zero traces of matrix combinations)

$$
\begin{array}{ccc}
 & \gamma^a & \gamma^a\gamma^s \\
I & 0 & \mathrm{Sp}\,(\gamma^a\gamma^s) \\
\gamma^p & \mathrm{Sp}\,(\gamma^p\gamma^a) & 0
\end{array}
\qquad (5.7.9b)
$$

from this it follows

$$
S = 0, \qquad j^{(a)}(x) = \frac{1}{i}\,(\,A^* A^a - A\,A^{a*}\,). \qquad (5.7.9c)
$$

The current vanishes identically for a real-valued scalar field. Now, in the same manner consider the current

$$
S = 0, \qquad \nu^{(a)}(x) = -\frac{1}{4}\,\mathrm{Sp}\,\left[\gamma^5\,(\,+i\,A^* + \gamma^p\,A_p^*\,)\,\gamma^a\,(\,-i\,A + \gamma^s\,A_s\,)\,\right]; \qquad (5.7.10a)
$$

an auxiliary table

$$
\begin{array}{ccc}
 & \gamma^a & \gamma^a\gamma^s \\
\gamma^5 & 0 & 0 \\
\gamma^5\gamma^p & 0 & 0
\end{array}
\qquad (5.7.10b)
$$

so that

$$
S = 0, \qquad \nu^{(a)}(x) \equiv 0. \qquad (5.7.10c)
$$

Consider the current

$$
S = 0, \qquad T^{(a)l}(x) = +\frac{1}{4}\,\mathrm{Sp}\,\left[\gamma^l\,(\,+i\,A^* + \gamma^p\,A_p^*\,)\,\gamma^a\,(\,-i\,A + \gamma^s\,A_s\,)\,\right]; \qquad (5.7.11a)
$$

an auxiliary table

$$
\begin{array}{ccc}
 & \gamma^a & \gamma^a\gamma^s \\
\gamma^l & \mathrm{Sp}\,(\gamma^l\gamma^a) & 0 \\
\gamma^l\gamma^p & 0 & \mathrm{Sp}\,(\gamma^l\gamma^p\gamma^a\gamma^s)
\end{array}
\qquad (5.7.11b)
$$

non-zero traces

$$
\frac{1}{4}\,\mathrm{Sp}\,(\gamma^l\gamma^a) = g^{la}, \qquad \frac{1}{4}\,\mathrm{Sp}\,(\gamma^l\gamma^p\gamma^a\gamma^s) = g^{lp}g^{as} - g^{la}g^{ps} + g^{ls}g^{pa},
$$

the conserved current is

$$
S = 0, \qquad T^{(a)l}(x) = A^* A\,g^{al} - g^{al}\,(A_s^* A^s) + (A^{a*}\,A^l + A^a\,A^{l*}). \qquad (5.7.11c)
$$

This quantity differs from zero both for complex- and real-valued scalar field; it is symmetrical with respect to indices al, and evidently should be considered as the energy-momentum tensor for a scalar particle. For the current

$$
S = 0, \qquad \nu^{(a)l}(x) = +\,\mathrm{Sp}\,\left[\gamma^5\,\gamma^l\,(+i\,A^* + \gamma^p\,A_p^*)\,\gamma^a\,(\,-i\,A + \gamma^s\,A_s\,)\,\right], \qquad (5.7.12a)
$$

an auxiliary table

$$
\begin{array}{ccc}
 & \gamma^a & \gamma^a\gamma^s \\
\gamma^5\gamma^l & 0 & 0 \\
\gamma^5\gamma^l\gamma^p & 0 & \mathrm{Sp}\ (\gamma^5\gamma^l\gamma^p\gamma^a\gamma^s)
\end{array}
\qquad , \qquad (5.7.12b)
$$

the non-zero trace

$$
\frac{1}{4}\ \mathrm{Sp}\ (\gamma^5\gamma^l\gamma^p\gamma^a\gamma^s) = i\ \epsilon^{lpas}\ ,
$$

and the current expression is

$$
S = 0, \qquad \nu^{(a)l}(x) = i\ \epsilon^{alps}\ A_p^*\ A_s\ . \qquad (5.7.12c)
$$

The current vanishes for all real-valued fields. And now consider the last current

$$
S = 0, \qquad L^{(a)kl}(x) = +\ \mathrm{Sp}\ \left[\sigma^{kl}\ (\ +i\ A^* + \gamma^p\ A_p^*\ )\ \gamma^a\ (\ -i\ A + \gamma^s\ A_s\ )\right],
$$
$$
(5.7.13a)
$$

an auxiliary table

$$
\begin{array}{ccc}
 & \gamma^a & \gamma^a\gamma^s \\
\sigma^{kl} & 0 & \mathrm{Sp}\ (\sigma^{kl}\gamma^a\gamma^s) \\
\sigma^{kl}\gamma^p & \mathrm{Sp}\ (\sigma^{kl}\gamma^p\gamma^a) & 0
\end{array}
\qquad , \qquad (5.7.13b)
$$

the non-zero traces

$$
\frac{1}{4}\ \mathrm{Sp}\ (\sigma^{kl}\gamma^a\gamma^s) = \frac{1}{2}\ (-g^{ka}g^{ls} + g^{ks}g^{la})\ ,
$$

$$
\frac{1}{4}\ \mathrm{Sp}\ (\sigma^{kl}\gamma^s\gamma^a) = \frac{1}{2}\ (-g^{ks}g^{la} + g^{ka}g^{ls})\ ,
$$

an expression for the current is

$$
S = 0, \qquad L^{(a)kl}(x) = \frac{i}{2}\ \left[\ (A^*\ A^k + A\ A^{k*})\ g^{la} - (A^*\ A^l + A\ A^{l*})\ g^{ka}\ \right].
$$
$$
(5.7.13c)
$$

The current is non-zero both for complex- and real-valued field, it should be called the angular momentum. By collecting all results together (take notice that we have no interpretation for the current $\nu^{(a)l}(x)$):

$S = 0,$

$$
j^{(a)}(x) = -i\ (\ A^*A^a - A\ A^{a*}\ )\ ,
$$
$$
\nu^{(a)}(x) \equiv 0\ ,
$$
$$
T^{(a)l}(x) = A^*A\ g^{al} - g^{al}\ (A_s^*\ A^s) + (A^{a*}\ A^l + A^a\ A^{l*})\ , \qquad (5.7.14)
$$
$$
\nu^{(a)l}(x) = i\ \epsilon^{alps}\ A_p^*\ A_s\ ,
$$
$$
L^{(a)kl}(x) = (i/2)\ \left[\ (A^*\ A^k + A\ A^{k*})\ g^{la} - (A^*\ A^l + A\ A^{l*})\ g^{ka}\ \right]\ .
$$

For the pseudo-scalar particle it will look the same (and again, we have no interpretation for the current $\nu^{(a)l}(x)$)

$$\underline{S = \tilde{0},}$$

$$j^{(a)}(x) = +i \ (\ B^* B^a - B \ B^{a*} \ ) \ ,$$

$$\nu^{(a)}(x) \equiv 0 \ ,$$

$$T^{(a)l}(x) = B^* B \ g^{al} - g^{al} \ (B^*_s \ B^s) + (B^{a*} \ B^l + B^a \ B^{l*}) \ , \tag{5.7.15}$$

$$\nu^{(a)l}(x) = -i \ \epsilon^{alps} \ B^*_p \ B_s \ ,$$

$$L^{(a)kl}(x) = (i/2) \ \left[\ (B^* \ B^k + B \ B^{k*}) \ g^{la} - (B^* \ B^l + B \ B^{l*}) \ g^{ka} \ \right] \ .$$

Now, let us consider the currents for a vector particle.

$$S = 1, \qquad j^{(a)}(x) = -\frac{1}{4} \ \mathrm{Sp} \left[(\ +\gamma^p A^*_p + i \ \sigma^{bc} F^*_{bc} \ ) \ \gamma^a \ (\ +\gamma^s A_s + i \ \sigma^{mn} F_{mn} \ ) \right] \tag{5.7.16a}$$

An auxiliary table looks as

	$\gamma^a \gamma^s$	$\gamma^a \sigma^{mn}$	
γ^p	0	$\mathrm{Sp} \ (\gamma^p \gamma^a \sigma^{mn})$	
σ^{bc}	$\mathrm{Sp} \ (\sigma^{bc} \gamma^a \gamma^s)$	0	$(5.7.16b)$

non-zero traces are

$$\frac{1}{4} \ \mathrm{Sp} \ (\gamma^s \gamma^a \sigma^{mn}) = \frac{1}{2} \ (-g^{ms} g^{na} + g^{ma} g^{ns}) \ ,$$

$$\frac{1}{4} \ \mathrm{Sp} \ (\gamma^a \gamma^s \sigma^{mn}) = \frac{1}{2} \ (-g^{ma} g^{ns} + g^{ms} g^{na}) \ ,$$

so that

$$S = 1, \qquad j^{(a)} = +i \ (A^*_n F^{na} - A_n F^{na*}) \ , \tag{5.7.16c}$$

which vanishes for a real-valued (neutral) field. For the current

$$S = 1, \qquad \nu^{(a)}(x) = -\frac{1}{4} \ \mathrm{Sp} \left[\gamma^5 (\ +\gamma^p A^*_p + i \ \sigma^{bc} F^*_{bc} \ ) \ \gamma^a (\ +\gamma^s A_s + i \ \sigma^{mn} F_{mn} \ ) \right] , \tag{5.7.17a}$$

an auxiliary table

	$\gamma^a \gamma^s$	$\gamma^a \sigma^{mn}$	
$\gamma^5 \gamma^p$	0	$\mathrm{Sp} \ (\gamma^5 \gamma^p \gamma^a \sigma^{mn})$	
$\gamma^5 \sigma^{bc}$	$\mathrm{Sp} \ (\gamma^5 \sigma^{bc} \gamma^a \gamma^s)$	0	$(5.7.17b)$

non-zero traces

$$\frac{1}{4} \ \mathrm{Sp} \ (\gamma^5 \gamma^p \gamma^a \sigma^{mn}) = \frac{i}{2} \ \epsilon^{pamn} \ , \qquad \mathrm{Sp} \ (\gamma^5 \sigma^{bc} \gamma^a \gamma^s) = \frac{i}{2} \ \epsilon^{asbc} \ ,$$

so that the current is

$$S = 1, \qquad \nu^{(a)} = -\frac{1}{2} \ (A^*_s F_{mn} - A_s F^*_{mn}) \ \epsilon^{asmn} \ . \tag{5.7.17c}$$

It differs from zero only for complex-valued fields.

Now for

$$S = 1, \qquad T^{(a)l}(x) = +\frac{1}{4} \operatorname{Sp} \left[\gamma^l (+\gamma^p A_p^* + i\,\sigma^{bc} F_{bc}^*)\, \gamma^a (+\gamma^s A_s + i\,\sigma^{mn} F_{mn}) \right],$$
$$(5.7.18a)$$

an auxiliary table looks

	$\gamma^a \gamma^s$	$\gamma^a \sigma^{mn}$
$\gamma^l \gamma^p$	$\operatorname{Sp}(\gamma^l \gamma^p \gamma^a \gamma^s)$	0
$\gamma^l \sigma^{bc}$	0	$\operatorname{Sp}(\gamma^l \sigma^{bc} \gamma^a \sigma^{mn})$

$$(5.7.18b)$$

In order to obtain the formula for a six-matrix product let us write down the formulas

$$\gamma^l \sigma^{bc} = \frac{1}{2}\left(-\gamma^b g^{lc} + \gamma^c g^{lb} + i\gamma^5 \epsilon^{lbcd} \gamma_d\right),$$

$$\gamma^a \sigma^{mn} = \frac{1}{2}\left(-\gamma^m g^{an} + \gamma^n g^{am} + i\gamma^5 \epsilon^{amne} \gamma_e\right),$$

and then

$$(\gamma^l \sigma^{bc})(\gamma^a \sigma^{mn}) = \frac{1}{4}\left(-\gamma^b g^{lc} + \gamma^c g^{lb} + i\gamma^5 \epsilon^{lbcd} \gamma_d\right)\left(-\gamma^m g^{an} + \gamma^n g^{am} + i\gamma^5 \epsilon^{amne} \gamma_e\right)$$

$$= \frac{1}{4}\Big[-g^{lc}(-\gamma^b \gamma^m g^{an} + \gamma^b \gamma^n g^{am} + i\gamma^b \gamma^5 \epsilon^{amne} \gamma_e)$$

$$+ g^{lb}(-\gamma^c \gamma^m g^{an} + \gamma^c \gamma^n g^{am} + i\gamma^c \gamma^5 \epsilon^{amne} \gamma_e)$$

$$+ i\epsilon^{lbcd}(-\gamma^5 \gamma_d \gamma^m g^{an} + \gamma^5 \gamma_d \gamma^n g^{am} + i\gamma^5 \gamma_d \gamma^5 \epsilon^{amne} \gamma_e)\Big].$$
$$(5.7.19a)$$

In taking the trace of such a matrix expression, we will arrive at the formula we need

$$\frac{1}{4}\operatorname{Sp}\left[(\gamma^l \sigma^{bc})(\gamma^a \sigma^{mn})\right]$$

$$= \frac{1}{4}\left[-g^{lc}(-g^{bm}g^{an} + g^{bn}g^{am}) + g^{lb}(-g^{cm}g^{an} + g^{cn}g^{am}) + \epsilon^{lbcd}\,\epsilon^{amne} g_{de}\right].$$
$$(5.7.19b)$$

Now, from Eq. (5.7.18a) we get for the current

$$T^{(a)l} = A_p^* A_s \frac{1}{4}\operatorname{Sp}(\gamma^l \gamma^p \gamma^a \gamma^s) - F_{bc}^* F_{mn} \frac{1}{4}\operatorname{Sp}\left[(\gamma^l \sigma^{bc})(\gamma^a \sigma^{mn})\right]$$

$$= A_p^* A_s \left(g^{lp}g^{as} - g^{la}g^{ps} + g^{ls}g^{pa}\right)$$

$$-\frac{1}{4} F_{bc}^* F_{mn} \left[-g^{lc}(-g^{bm}g^{an} + g^{bn}g^{am}) + g^{lb}(-g^{cm}g^{an} + g^{cn}g^{am}) + \epsilon^{lbcd}\,\epsilon^{amne} g_{de}\right]$$

and further

$$T^{(a)l} = -g^{al} A_p^* A^p + (A^{a*} A^l + A^a A^{l*})$$

$$+ F^{*ln} F_n{}^a - \frac{1}{4} F_{bc}^* F_{mn} \epsilon^{lbcd}\,\epsilon^{amne} g_{de}.$$
$$(5.7.20a)$$

Here, the last term may be changed to

$$-\frac{1}{4} F_{bc}^* F_{mn} \, \epsilon^{lbcd} \, \epsilon^{amne} g_{de} = +\frac{1}{4} F_{bc}^* F_{mn} \begin{vmatrix} g^{la} & g^{lm} & g^{ln} \\ g^{ba} & g^{bm} & g^{bn} \\ g^{ca} & g^{cm} & g^{cn} \end{vmatrix}$$

$$= +\frac{1}{4} F_{bc}^* F_{mn}[g^{la}(g^{bm} g^{cn} - g^{bn} g^{cm}) - g^{ba}(g^{lm} g^{cn} - g^{ln} g^{cm}) + g^{ca}(g^{lm} g^{bn} - g^{ln} g^{bm})]$$

$$= \frac{1}{2} g^{la} F^{*mn} F_{mn} + F^{ln} F_n^{*\ a} \,. \tag{5.7.20b}$$

When we use the help of Eq. (5.7.20b), from Eq. (5.7.20a) it follows

$$T^{(a)l} = -g^{al} \, A_p^* A^p + (A^{a*} A^l + A^a A^{l*})$$

$$+\frac{1}{2} g^{la} F^{*mn} F_{mn} + (F^{*ln} F_n^{\ a} + F^{ln} F_n^{*\ a}) \,. \tag{5.7.20c}$$

The tensor is symmetrical under indices al, it differs from zero for both the complex- and real-valued fields, and may be considered as the energy-momentum tensor for a vector $S = 1$ particle.

Now, consider another current

$$S = 1: \quad \nu^{(a)l}(x) = + \text{Sp} \left[\gamma^5 \gamma^l (\, +\gamma^p A_p^* + i \, \sigma^{bc} F_{bc}^* \,) \, \gamma^a (+\gamma^s A_s + i \, \sigma^{mn} F_{mn} \,) \right] \,, \tag{5.7.21a}$$

an auxiliary table is

	$\gamma^a \gamma^s$	$\gamma^a \sigma^{mn}$
$\gamma^5 \gamma^l \gamma^p$	$\text{Sp} \, (\gamma^5 \gamma^l \gamma^p \gamma^a \gamma^s)$	0
$\gamma^5 \gamma^l \sigma^{bc}$	0	$\text{Sp} \, (\gamma^5 \gamma^l \sigma^{bc} \gamma^a \sigma^{mn})$

$$\tag{5.7.21b}$$

We will need the trace-formula for the second term:

$$\frac{1}{4} \text{Sp} \, [\gamma^5 \, (\gamma^l \sigma^{bc})(\gamma^a \sigma^{mn})]$$

$$= \frac{1}{16} \text{Sp} \left\{ \gamma^5 \left[-g^{lc}(-\gamma^b \gamma^m g^{an} + \gamma^b \gamma^n g^{am} + i\gamma^b \gamma^5 \epsilon^{amne} \gamma_e) \right. \right.$$

$$+ g^{lb} \, (-\gamma^c \gamma^m g^{an} + \gamma^c \gamma^n g^{am} + i\gamma^c \gamma^5 \epsilon^{amne} \gamma_e)$$

$$\left. \left. + i\epsilon^{lbcd} \, (-\gamma^5 \gamma_d \gamma^m g^{an} + \gamma^5 \gamma_d \gamma^n g^{am} + i\gamma^5 \gamma_d \gamma^5 \epsilon^{amne} \gamma_e) \right] \right\}$$

$$= \frac{i}{4} \left[(g^{lc} \epsilon^{amnb} - g^{lb} \epsilon^{amnc}) + \epsilon^{lbcd} \, (\delta_d^n \, g^{am} - \delta_d^m \, g^{an}) \right] \,. \tag{5.7.21c}$$

When using the help of Eq. (5.7.21c) for the current, we will find

$$\nu^{(a)l}(x) = i \, A_p^* A_s \, \epsilon^{alps} + \frac{i}{2} \, F_{bc}^* F_{mn} \left[g^{an} \, \epsilon^{lbcm} + g^{lb} \, \epsilon^{amnc} \right] \,. \tag{5.7.21d}$$

The conservation laws for the quantity $\nu^{(a)l}(x)$ may be written down in detailed form:

$$\frac{\partial}{\partial x^a} \, \nu^{(a)0}(x) = 0 \,,$$

$\nu^{(a)0}$:

$$\nu^{(0)0} = \quad -i\,[\,+E^{*1}\,B^1 + E^{*2}\,B^2 + E^{*3}\,B^3 - \text{conj.}\,]\,,$$
$$\nu^{(1)0} = \quad +i\,[\,-A_2^*\,A_3 + (E_2^*\,E_3 + B_2^*\,B_3) - \text{conj.}\,]\,,$$
$$\nu^{(2)0} = \quad +i\,[\,-A_3^*\,A_1 + (E_3^*\,E_1 + B_3^*\,B_1) - \text{conj.}\,]\,,$$
$$\nu^{(3)0} = \quad +i\,[\,-A_1^*\,A_2 + (E_1^*\,E_2 + B_1^*\,B_2) - \text{conj.}\,]\,,$$

$$\text{(5.7.22a)}$$

$$\frac{\partial}{\partial x^a}\,\nu^{(a)1}(x) = 0\,,$$

$\nu^{(a)1}$:

$$\nu^{(0)1}(x) = \quad +i\,[\,+A_2^*\,A_3 + (E_2^*\,E_3 + B_2^*\,B_3) - \text{conj.}\,]\,,$$
$$\nu^{(1)1}(x) = \quad -i\,[\,-E^{*1}\,B^1 + E^{*2}\,B^2 + E^{*3}\,B^3 - \text{conj.}\,]\,,$$
$$\nu^{(2)1}(x) = \quad i\,[\,-A_0^*\,A_3 + (E_2^*\,B_1 + E_1^*\,B_2) - \text{conj.}\,]\,,$$
$$\nu^{(3)1}(x) = \quad i\,[\,+A_0^*\,A_2 + (E_3^*\,B_1 + E_1^*\,B_3) - \text{conj.}\,]\,,$$

$$\text{(5.7.22b)}$$

$$\frac{\partial}{\partial x^a}\,\nu^{(a)2}(x) = 0\,,$$

$\nu^{(a)2}$:

$$\nu^{(0)2}(x) = \quad +i\,[\,+A_3^*\,A_1 + (E_3^*\,E_1 + B_3^*\,B_1) - \text{conj.}\,]\,,$$
$$\nu^{(1)2}(x) = \quad i\,[\,+A_0^*\,A_3 + (E_1^*\,B_2 + E_2^*\,B_1) - \text{conj.}\,]\,,$$
$$\nu^{(2)2}(x) = \quad -i\,[\,+E^{*1}\,B^1 - E^{*2}\,B^2 + E^{*3}\,B^3 - \text{conj.}\,]\,,$$
$$\nu^{(3)2}(x) = \quad i\,[\,-A_0^*\,A_1 + (E_3^*\,B_2 + E_2^*\,B_3) - \text{conj.}\,]\,,$$

$$\text{(5.7.22c)}$$

$$\frac{\partial}{\partial x^a}\,\nu^{(a)3}(x) = 0\,,$$

$\nu^{(a)3}$:

$$\nu^{(0)3}(x) = \quad +i\,[\,+A_1^*\,A_2 + (E_1^*\,E_2 + B_1^*\,B_2) - \text{conj.}\,]\,,$$
$$\nu^{(1)3}(x) = \quad i\,[\,-A_0^*\,A_2 + (E_1^*\,B_3 + E_3^*\,B_1) - \text{conj.}\,]\,,$$
$$\nu^{(2)3}(x) = \quad i\,[\,+A_0^*\,A_1 + (E_2^*\,B_3 + E_3^*\,B_2) - \text{conj.}\,]\,,$$
$$\nu^{(3)3}(x) = \quad -i\,[\,+E^{*1}\,B^1 + E^{*2}\,B^2 - E^{*3}\,B^3 - \text{conj.}\,]\,.$$

$$\text{(5.7.22d)}$$

And finally, let us obtain an explicit form of the current

$$S = 1: \qquad L^{(a)kl}(x) = +\frac{1}{4}\,\text{Sp}\left[\sigma^{kl}(+\gamma^p A_p^* + i\sigma^{bc}F_{bc}^*)\,\gamma^a(+\gamma^s A_s + i\sigma^{mn}F_{mn})\right],$$

$$\text{(5.7.23a)}$$

an auxiliary table is

	$\gamma^a\gamma^s$	$\gamma^a\sigma^{mn}$	
$\sigma^{kl}\gamma^p$	0	$\text{Sp}\,(\sigma^{kl}\gamma^p\,\gamma^a\sigma^{mn})$	(5.7.23b)
$\sigma^{kl}\sigma^{bc}$	$\text{Sp}\,(\sigma^{kl}\sigma^{bc}\gamma^a\gamma^s)$	0	

non-zero traces are

$$\frac{1}{4}\,\text{Sp}\,(\sigma^{kl}\sigma^{bc}\gamma^a\gamma^s) = \frac{1}{4}\,\text{Sp}\,(\sigma^{bc}\gamma^a\gamma^s\sigma^{kl})$$

$$= \frac{1}{16}\,\text{Sp}\,\left[(\gamma^b g^{ca} - \gamma^c g^{ba} + i\gamma^5\,\epsilon^{bcad}\gamma_d)(-\gamma^k g^{sl} + \gamma^l g^{sk} + i\gamma^5\epsilon^{skle}\gamma_e)\right]$$

$$= \frac{1}{4}\,\left[g^{ca}(-g^{bk}g^{sl} + g^{bl}g^{sk}) - g^{ba}(-g^{ck}g^{sl} + g^{cl}g^{sk}) - \epsilon^{bcad}\epsilon^{skle}g_{de}\right]$$

$$= \frac{1}{4} \, \text{Sp} \, (\sigma^{kl} \gamma^p \gamma^a \sigma^{mn})$$

$$= \frac{1}{16} \, \text{Sp} \, [(\gamma^k g^{lp} - \gamma^l g^{kp} + i\gamma^5 \epsilon^{klpd} \gamma_d)(-\gamma^m g^{an} + \gamma^n g^{am} + i\gamma^5 \epsilon^{amne} \gamma_e)]$$

$$= \frac{1}{4} \, \left[g^{lp}(-g^{km} g^{an} + g^{kn} g^{am}) - g^{kp}(-g^{lm} g^{an} + g^{ln} g^{am}) + \epsilon^{klpd} \epsilon^{amne} g_{de}) \right] \, .$$

In using the two relations

$$i \, F^*_{bc} A_s \, \frac{1}{4} \, \text{Sp} \, (\sigma^{kl} \sigma^{bc} \gamma^a \gamma^s)$$

$$= \frac{i}{2} \, \left[-A^a F^{*kl} + (F^{*ak} A^l - F^{*al} A^k) + (-g^{ak} F^{*ls} A_s + g^{al} F^{*ks} A_s) \right] \, ,$$

$$i A^*_p F_{mn} \, \frac{1}{4} \, \text{Sp} \, (\sigma^{kl} \gamma^p \gamma^a \sigma^{mn})$$

$$= \frac{i}{2} \, \left[-A^{*a} F^{kl} + (F^{ak} A^{*l} - F^{al} A^{*k}) + (-g^{ak} F^{ls} A^*_s + g^{al} F^{ks} A^*_s) \right]$$

one readily produces the expression for this current

$$L^{(a)kl} = \frac{i}{2} \, \Big\{ -(A^a F^{*kl} + A^{*a} F^{kl}) + [\, (F^{*ak} A^l - F^{*al} A^k) + (F^{ak} A^{*l} - F^{al} A^{*k}) \,]$$

$$+ [\, (-g^{ak} F^{*ls} A_s + g^{al} F^{*ks} A_s) + (-g^{ak} F^{ls} A^*_s + g^{al} F^{ks} A^*_s) \,] \Big\} \, . \qquad (5.7.24c)$$

Here, we collect together all the results for the particle of $S = 1$:

$$S = 1: \qquad j^{(a)} = +i \, (A^*_n F^{na} - A_n F^{na*}) \, ,$$

$$\nu^{(a)} = -(1/2) \, (A^*_s F_{mn} - A_s F^*_{mn}) \, \epsilon^{asmn} \, ,$$

$$T^{(a)l} = -g^{al} A^*_p A^p + (A^{a*} A^l + A^a A^{l*})$$

$$+ (1/2) g^{al} F^{*mn} F_{mn} + (F^{*ln} F_n{}^a + F^{ln} F^*_n{}^a) \, ,$$

$$\nu^{(a)l}(x) = i \, A^*_p A_s \, \epsilon^{alps}$$

$$+ (i/2) \, F^*_{bc} F_{mn} \, (\, g^{an} \, \epsilon^{lbcm} + g^{lb} \, \epsilon^{amnc} \,) \, ,$$

$$L^{(a)kl} = (i/2) \, \Big\{ -(A^a F^{*kl} + A^{*a} F^{kl})$$

$$+ [\, (F^{*ak} A^l - F^{*al} A^k) + (F^{ak} A^{*l} - F^{al} A^{*k}) \,]$$

$$+ [\, (-g^{ak} F^{*ls} A_s + g^{al} F^{*ks} A_s) + (-g^{ak} F^{ls} A^*_s + g^{al} F^{ks} A^*_s) \,] \Big\} \, .$$

Take notice that we have no interpretation for two currents, $\nu^{(a)}$ and $\nu^{(a)l}(x)$, which are non-zero for complex-valued fields only.

Here, we shall not consider in detail the case of a vector particle with another intrinsic parity. For the particle type of $S = \tilde{1}$, any additional ideas will not be addressed.

Chapter 6

Graviton in a Curved Space-Time Background and Gauge Symmetry

6.1 Introduction

Theory of massive and massless fields of spin 2, starting from Pauli and Fierz investigations [394, 395], always attracted attention: de Broglie [396], Pauli [313], Gel'fand and Yaglom [316], Fradkin [397], Fedorov et al. [398–400, 402, 403], Fainberg [404], Regge [405], Buchdahl [406, 407], Velo and Zwanziger [408, 409], Cox [413], Barut [414], Loide [416], Vasiliev [418], Buchbinder et al. [419, 420], Bogush et al. [421–424]). Most of investigations were performed within the second order wave equation approach. However, it is known that many problems arose for fields with a higher spin which may be avoided if one starts with the theory of first order equations. One of the earliest considerations of the theory for spin 2 particle was given by Fedorov [398]. It turned out that such a description requires a 30-component wave function. Afterwards, Fedorov proposed another description for spin 2 particle, a 50-component model, [401, 412]. The primary question is about the relation between the two models. Bogush and Kisel [417] demonstrated (within spinor formalism) that a 50-component model describes a spin 2 particle with additional electromagnetic characteristics, anomalous magnetic moment[1]. A more detailed analysis of this spinor description was given in [309–311, 415, 417]. In [423, 424], that theory was transformed to a more simple tensor technique.

In the present chapter, the 30-component first order theory is investigated in the case of a vanishing mass for the particle and external curved space-time background.

6.2 Spin 2 Particle in the Flat Space-Time

A system of first order wave equations that describe a massless spin 2 particle in a flat space-time has the form

$$\partial^a \Phi_a = 0 , \qquad (6.2.1a)$$

---

[1]Here one may see an analogy with the known Petras [300, 301, 303, 310] theory for a spin 1/2 particle, or Shamaly – Capri [295–299, 427] theory for a particle with spin 1, when an increase in the number of field variables permits us to introduce an additional parameter for a particle, anomalous magnetic moment.

$$\frac{1}{2}\partial_a\Phi - \frac{1}{3}\partial^b\Phi_{ab} = \Phi_a \,, \qquad\qquad (6.2.1b)$$

$$\frac{1}{2}(\partial^k\Phi_{kab} + \partial^k\Phi_{kba} - \frac{1}{2}g_{ab}\partial^k\Phi_{kn}{}^n) + \partial_a\Phi_b + \partial_b\Phi_a - \frac{1}{2}g_{ab}\partial^k\Phi_k = 0 \,, \qquad (6.2.1c)$$

$$\partial_a\Phi_{bc} - \partial_b\Phi_{ac} + \frac{1}{3}(g_{bc}\partial^k\Phi_{ak} - g_{ac}\partial^k\Phi_{bk}) = \Phi_{abc} \,. \qquad (6.2.1d)$$

A 30-component wave function consists of a scalar Φ, vector Φ_a, symmetric 2-rank tensor Φ_{ab}, and 3-rank tensor Φ_{abc} is antisymmetric in the first two indices. From Eq. (6.2.1d), it follows four conditions that are satisfied by the 3-index field:

$$\Phi_{abc} + \Phi_{bca} + \Phi_{cab} = 0 \quad \text{or} \quad \epsilon^{kabc}\Phi_{abc} = 0 \,. \qquad (6.2.2a)$$

Simplifying Eq. (6.2.1d) in indices b and c, one produces

$$\partial_a\Phi_b{}^b = \Phi_{ac}{}^c \,. \qquad\qquad (6.2.2b)$$

Thus, a total number of independent components entering the theory equals 31 (instead of 30 in massive case):

$$\Phi(x) \Longrightarrow 1 \,, \qquad \Phi_a \Longrightarrow 4 \,, \qquad \Phi_{ab} \Longrightarrow 10 \,,$$

$$\Phi_{abc} \Longrightarrow 6 \times 4 - 4 - 4 = 16 \,.$$

After excluding fields Φ_a and Φ_{kab} from Eqs. (6.2.1a) – (6.2.1d) one gets to a pair of second order equations on fields $\Phi(x)$ and $\Phi_{ab}(x)$[2]:

$$\frac{1}{2}\nabla^2\Phi - \frac{1}{3}\partial^k\partial^l\Phi_{kl} = 0 \,, \qquad\qquad (6.2.3a)$$

$$(\partial_a\partial_b - \frac{1}{4}g_{ab}\nabla^2)\Phi - \frac{1}{4}g_{ab}\nabla^2\Phi_c^c$$

$$+\nabla^2\Phi_{ab} - \partial_a\partial^l\Phi_{bl} - \partial_b\partial^l\Phi_{al} + \frac{1}{2}g_{ab}\partial^k\partial^l\Phi_{kl} = 0 \,. \qquad (6.2.3b)$$

Then we allow for Eq. (6.2.3a), where Eq. (6.2.3b) can be rewritten as

$$(\partial_a\partial_b + \frac{1}{2}g_{ab}\nabla^2)\Phi - \frac{1}{4}g_{ab}\nabla^2\Phi_c^c + \nabla^2\Phi_{ab} - \partial_a\partial^l\Phi_{bl} - \partial_b\partial^l\Phi_{al} = 0 \,. \qquad (6.2.3c)$$

The fact of prime significance in the theory under consideration is that these equations permit specific gauge symmetry[3]. That means the following: this second order system for Eq. (6.2.3c) is satisfied by the following substitution (class of trivial or gradient-like solutions)

$$\Phi^{(0)} = \partial^l\Lambda_l \,, \Phi_{ab}^{(0)} = \partial_a\Lambda_b + \partial_b\Lambda_a - \frac{1}{2}g_{ab}\partial^l\Lambda_l \,, \qquad (6.2.4)$$

at any 4-vector function $\Lambda_a(x)$. Indeed,

$$-\frac{1}{3}\,\partial^a\partial^b\Phi_{ab}^{(0)} = -\frac{1}{2}\nabla^2\partial^l\Lambda_l = -\frac{1}{2}\nabla^2\Phi^{(0)} \,, \qquad (6.2.5)$$

---

[2]The notation $\nabla^2 = \partial^a\partial_a$ is used.

[3]The fact was first established by Pauli and Fierz

and therefore, the set in Eq. (6.2.4) turns Eq. (6.2.3a) into an identity. Further, let us take into account

$$\frac{1}{2}(\partial^k \Phi_{kab} + \partial^k \Phi_{kba} - \frac{1}{2}g_{ab}\partial^k \Phi_{kn}{}^n) = +\frac{1}{3}\partial^l \partial_l(\partial_b \Lambda_a + \partial_a \Lambda_b) - \frac{2}{3}\partial_a \partial_b \partial^l \Lambda_l ,$$

$$\partial_a \Phi_b + \partial_b \Phi_a - \frac{1}{2}g_{ab}\partial^k \Phi_k = -\frac{1}{3}\partial^l \partial_l(\partial_b \Lambda_a + \partial_a \Lambda_b) + \frac{2}{3}\,\partial_a \partial_b\,\partial^l \Lambda_l ,$$

one can verify that the set in Eq. (6.2.4) satisfies Eq. (6.2.3b) as well.

So, a massless spin-2 field in Minkowski space-time can be described by the first order system, or by the second order system. By this aspect, their solutions are not determined uniquely; in general, to any chosen one we may add an arbitrary Λ_a-dependent term.

6.3   Spin 2 Particle in Curved Space-Time

In using the principle for minimal coupling to a curved space-time background (external gravitational field), it is expected that generally covariant equations for a spin-2 particle are to be taken in the form

$$\nabla^\alpha \Phi_\alpha = 0 , \qquad (6.3.1a)$$

$$\frac{1}{2}\nabla_\alpha \Phi - \frac{1}{3}\nabla^\beta \Phi_{\alpha\beta} = \Phi_\alpha , \qquad (6.3.1b)$$

$$\frac{1}{2}\left(\nabla^\rho \Phi_{\rho\alpha\beta} + \nabla^\rho \Phi_{\rho\beta\alpha} - \frac{1}{2}g_{\alpha\beta}(x)\nabla^\rho \Phi_{\rho\sigma}{}^\sigma\right)$$

$$+\left(\nabla_\alpha \Phi_\beta + \nabla_\beta \Phi_\alpha - \frac{1}{2}g_{\alpha\beta}(x)\nabla^\rho \Phi_\rho\right) = 0 , \qquad (6.3.1c)$$

$$\nabla_\alpha \Phi_{\beta\sigma} - \nabla_\beta \Phi_{\alpha\sigma} + \frac{1}{3}\left(g_{\beta\sigma}(x)\nabla^\rho \Phi_{\alpha\rho} - g_{\alpha\sigma}(x)\nabla^\rho \Phi_{\beta\rho}\right) = \Phi_{\alpha\beta\sigma} . \qquad (6.3.1d)$$

Here, ∇_α designates a generally covariant derivative. As in the flat space-time, the system exhibits the property

$$\nabla_\alpha\,\Phi_\beta{}^\beta = \Phi_{\alpha\beta}{}^\beta . \qquad (6.3.2)$$

Now, we are to investigate the question of possible gauge symmetry in the system. To this end, we will try to satisfy these equations by a substitution

$$\Phi^{(0)} = \nabla^\beta \Lambda_\beta ,$$

$$\Phi^{(0)}_{\alpha\beta} = \nabla_\alpha \Lambda_\beta + \nabla_\beta \Lambda_\alpha - \frac{1}{2}g_{\alpha\beta}(x)\nabla^\sigma \Lambda_\sigma , \qquad (6.3.3)$$

where $\Lambda_\alpha(x)$ is an arbitrary 4-vector function. By using Eq. (6.3.1b), a vector field corresponding to the set in Eq. (6.3.3) takes the form

$$\Phi^{(0)}_\alpha = \frac{2}{3}\nabla_\alpha \nabla^\beta \Lambda_\beta - \frac{1}{3}\nabla^\beta \nabla_\alpha \Lambda_\beta - \frac{1}{3}(\nabla^\beta \nabla_\beta)\Lambda_\alpha . \qquad (6.3.4)$$

After we substitute this into Eq. (6.3.1a), one produces

$$0 = \frac{2}{3}(\nabla^\alpha \nabla_\alpha)\nabla^\beta \Lambda_\beta - \frac{1}{3}\nabla^\alpha \nabla^\beta \nabla_\alpha \Lambda_\beta - \frac{1}{3}\nabla^\beta(\nabla^\alpha \nabla_\alpha)\Lambda_\beta . \qquad (6.3.5a)$$

To employ conventionally the Riemann and Ricci tensors

$$(\nabla_\beta \nabla_\alpha - \nabla_\alpha \nabla_\beta)\Lambda_\rho = R_{\beta\alpha\rho\sigma}\Lambda^\sigma, \qquad R_{\beta\alpha}{}^\beta{}_\sigma = R_{\alpha\sigma},$$

the second term in Eq. $(6.3.5a)$ can be rewritten as

$$-\frac{1}{3}\nabla^\alpha\nabla_\beta\nabla_\alpha\Lambda^\beta = -\frac{1}{3}\nabla^\alpha(\nabla_\alpha\nabla_\beta\Lambda^\beta + R_{\alpha\beta}\Lambda^\beta),$$

with the use of Eq. $(6.3.5a)$, will take the form

$$0 = \frac{1}{3}\ [\nabla^\alpha\nabla_\alpha, \nabla^\beta]_-\Lambda_\beta - \frac{1}{3}\nabla^\alpha(R_{\alpha\beta}\Lambda^\beta). \qquad (6.3.5b)$$

The latter, with the commutator

$$[\nabla^\alpha\nabla_\alpha, \nabla^\beta]_-\Lambda_\beta = -\nabla^\alpha(R_{\alpha\sigma}\Lambda^\sigma), \qquad (6.3.6)$$

will read as

$$0 = -\frac{2}{3}\nabla^\alpha(R_{\alpha\beta}\Lambda^\beta). \qquad (6.3.7)$$

This equation means: if $R_{\alpha\beta} \neq 0$, the present spin-2 particle equation does not have any trivial Λ_α-based solution. In other terms, a gauge principle in accordance with Einstein gravitational equations the equality $R_{\alpha\beta} \neq 0$ relates that at those x^α-points any material fields will vanish. However, in $(R_{\alpha\beta} = 0)$-region, the wave equation under consideration includes such Λ_α-based solutions and correspondingly a gauge principle.

Now, analogously, we should consider Eq. $(6.3.1c)$: what we will have is on substituting Λ_α-set into it. We must exclude all auxiliary fields from Eq. $(6.3.1c)$:

$$\frac{1}{2}\left(\nabla^\rho\Phi^{(0)}_{\rho\alpha\beta} + \nabla^\rho\Phi^{(0)}_{\rho\beta\alpha} - \frac{1}{2}g_{\alpha\beta}(x)\nabla^\rho\Phi^{(0)}_{\rho\sigma}{}^\sigma\right)$$

$$+\nabla_\alpha\Phi^{(0)}_\beta + \nabla_\beta\Phi^{(0)}_\alpha - \frac{1}{2}g_{\alpha\beta}(x)\nabla^\rho\Phi^{(0)}_\rho = 0. \qquad (6.3.8)$$

Let us step by step calculate all terms entering Eq. $(6.3.8)$. For the first (1) term, we have that

$$(1) \overset{def}{=} \frac{1}{2}\nabla^\rho\Phi^{(0)}_{\rho\alpha\beta} = \frac{1}{2}(\nabla^\rho\nabla_\rho)(\nabla_\alpha\Lambda_\beta)$$

$$+\frac{1}{2}(\nabla^\rho\nabla_\rho)(\nabla_\beta\Lambda_\alpha) - \frac{1}{4}g_{\alpha\beta}(\nabla^\rho\nabla_\rho)(\nabla^\gamma\Lambda_\gamma)$$

$$-\frac{1}{2}(\nabla^\rho\nabla_\rho)\nabla_\alpha\Lambda_\beta - \frac{1}{2}\nabla^\rho[\nabla_\alpha, \nabla_\rho]_-\Lambda_\beta$$

$$-\frac{1}{2}\nabla_\alpha\nabla_\beta(\nabla^\rho\Lambda_\rho) - \frac{1}{2}[\nabla^\rho, \nabla_\alpha\nabla_\beta]_-\Lambda_\rho$$

$$+\frac{1}{4}(\nabla_\beta\nabla_\alpha)(\nabla^\gamma\Lambda_\gamma) + \frac{1}{6}g_{\alpha\beta}(\nabla^\rho\nabla_\rho)(\nabla^\sigma\Lambda_\sigma)$$

$$+\frac{1}{6}g_{\alpha\beta}\nabla^\rho[\nabla^\sigma, \nabla_\rho]_-\Lambda_\sigma + \frac{1}{6}g_{\alpha\beta}(\nabla^\sigma\nabla_\sigma)(\nabla^\rho\Lambda_\rho)$$

$$+\frac{1}{6}g_{\alpha\beta}[\nabla^\rho, \nabla^\sigma\nabla_\sigma]_-\Lambda_\rho - \frac{1}{12}g_{\alpha\beta}(\nabla^\rho\nabla_\rho)(\nabla^\gamma\Lambda_\gamma)$$

$$-\frac{1}{6}\nabla_\beta\nabla_\alpha(\nabla^\sigma\Lambda_\sigma) - \frac{1}{6}\nabla_\beta[\nabla^\sigma, \nabla_\alpha]_-\Lambda_\sigma$$

$$-\frac{1}{6}(\nabla^\sigma\nabla_\sigma)\nabla_\beta\Lambda_\alpha - \frac{1}{6}[\nabla_\beta, \nabla^\sigma\nabla_\sigma]_-\Lambda_\alpha + \frac{1}{12}\nabla_\beta\nabla_\alpha(\nabla^\gamma\Lambda_\gamma) \ .$$

The second term in Eq. (6.3.8) can be produced by straightforward symmetry considerations from Eq. (6.3.8). Where, in this case, the third term in Eq. (6.3.8) turns out to vanish

$$(3) \overset{def}{=\!=\!=} -\frac{1}{4}g_{\alpha\beta}\nabla^\rho\Phi^{(0)}_{\rho\gamma}{}^\gamma = -\frac{1}{2}g_{\alpha\beta}\nabla^\rho\nabla_\rho\Phi^{(0)\beta}_\beta$$

$$= -\frac{1}{4}g_{\alpha\beta}\nabla^\rho\nabla_\rho \left(\nabla_\beta\Lambda^\beta + \nabla_\beta\Lambda^\beta - \frac{1}{2}\delta^\beta_\beta \nabla^\gamma\Lambda_\gamma\right) = 0 \ .$$

Then for the fourth and fifth terms, we have

$$(4) \overset{def}{=\!=\!=} \nabla_\alpha \Phi^{(0)}_\beta = \frac{1}{2}\nabla_\alpha\nabla_\beta \nabla^\gamma\Lambda_\gamma - \frac{1}{3}\nabla_\alpha\nabla_\beta(\nabla^\rho \Lambda_\rho)$$

$$-\frac{1}{3}\nabla_\alpha[\nabla^\rho, \nabla_\beta]_-\Lambda_\rho - \frac{1}{3}(\nabla^\rho \nabla_\rho)\nabla_\alpha\Lambda_\beta$$

$$-\frac{1}{3}[\nabla_\alpha, \nabla^\rho \nabla_\rho]_-\Lambda_\beta + \frac{1}{6}\nabla_\alpha\nabla_\beta \nabla^\gamma\Lambda_\gamma \ ,$$

$$(5) \overset{def}{=\!=\!=} \nabla_\beta \Phi^{(0)}_\alpha = \frac{1}{2}\nabla_\beta\nabla_\alpha \nabla^\gamma\Lambda_\gamma$$

$$-\frac{1}{3}\nabla_\beta\nabla_\alpha(\nabla^\rho \Lambda_\rho) - \frac{1}{3}\nabla_\beta[\nabla^\rho, \nabla_\alpha]_-\Lambda_\rho - \frac{1}{3}(\nabla^\rho \nabla_\rho)\nabla_\beta\Lambda_\alpha$$

$$-\frac{1}{3}[\nabla_\beta, \nabla^\rho \nabla_\rho]_-\Lambda_\alpha + \frac{1}{6}\nabla_\beta\nabla_\alpha \nabla^\gamma\Lambda_\gamma \ ;$$

and term (6) is

$$(6) \overset{def}{=\!=\!=} -\frac{1}{2}g_{\alpha\beta}\nabla^\rho\Phi^{(0)}_\rho - \frac{1}{2}g_{\alpha\beta}\nabla^\rho\Phi^{(0)}_\rho$$

$$= -\frac{1}{4}g_{\alpha\beta}(\nabla^\rho\nabla_\rho)(\nabla^\gamma\Lambda_\gamma) + \frac{1}{6}g_{\alpha\beta}(\nabla^\rho\nabla_\rho)(\nabla^\sigma\Lambda_\sigma)$$

$$+\frac{1}{6}g_{\alpha\beta}\nabla^\rho[\nabla^\sigma, \nabla_\rho]_-\Lambda_\sigma + \frac{1}{6}g_{\alpha\beta}(\nabla^\sigma \nabla_\sigma)(\nabla^\rho\Lambda_\rho)$$

$$+\frac{1}{6}g_{\alpha\beta}[\nabla^\rho, \nabla^\sigma\nabla_\sigma]_-\Lambda_\rho - \frac{1}{12}g_{\alpha\beta}(\nabla^\rho\nabla_\rho)(\nabla^\gamma\Lambda_\gamma) \ .$$

Summing up all six expressions and by taking into account similar terms (factors at all terms without commutators turn out to be equal zero as should be expected):

$$0 = (\nabla^\rho\nabla_\rho) (\nabla_\alpha\Lambda_\beta)\left[\left(\frac{1}{2}-\frac{1}{2}\right) + \left(\frac{1}{2}-\frac{1}{6}\right) - \frac{1}{3}\right]$$

$$+(\nabla^\rho\nabla_\rho)(\nabla_\beta\Lambda_\alpha)\left[\left(\frac{1}{2}-\frac{1}{6}\right) + \left(\frac{1}{2}-\frac{1}{2}\right) - \frac{1}{3}\right]$$

$$+g_{\alpha\beta}(\nabla^\rho\nabla_\rho)(\nabla^\gamma\Lambda_\gamma)\left[\left(-\frac{1}{4}+\frac{1}{6}+\frac{1}{6}-\frac{1}{12}\right)+\left(-\frac{1}{4}+\frac{1}{6}+\frac{1}{6}-\frac{1}{12}\right)\right]$$

$$+\nabla_\alpha\nabla_\beta\,(\nabla^\rho\Lambda_\rho)\left[\left(-\frac{1}{2}+\frac{1}{4}-\frac{1}{6}+\frac{1}{12}\right)+\left(-\frac{1}{2}+\frac{1}{4}-\frac{1}{6}+\frac{1}{12}\right)+\left(\frac{1}{2}-\frac{1}{3}+\frac{1}{6}\right)+\left(\frac{1}{2}-\frac{1}{3}+\frac{1}{6}\right)\right]$$

$$+\left\{-\frac{1}{2}\nabla^\rho[\nabla_\alpha,\nabla_\rho]_-\Lambda_\beta-\frac{1}{2}[\nabla^\rho,\nabla_\alpha\nabla_\beta]_-\Lambda_\rho\right.$$

$$+\frac{1}{6}g_{\alpha\beta}\nabla^\rho[\nabla^\sigma,\nabla_\rho]_-\Lambda_\sigma+\frac{1}{6}g_{\alpha\beta}[\nabla^\rho,\nabla^\sigma\nabla_\sigma]_-\Lambda_\rho-$$

$$\left.-\frac{1}{6}\nabla_\beta[\nabla^\sigma,\nabla_\alpha]_-\Lambda_\sigma-\frac{1}{6}[\nabla_\beta,\nabla^\sigma\nabla_\sigma]_-\Lambda_\alpha\right\}$$

$$+\left\{-\frac{1}{2}\nabla^\rho[\nabla_\beta,\nabla_\rho]_-\Lambda_\alpha-\frac{1}{2}[\nabla^\rho,\nabla_\beta\nabla_\alpha]_-\Lambda_\rho\right.$$

$$+\frac{1}{6}g_{\beta\alpha}\nabla^\rho[\nabla^\sigma,\nabla_\rho]_-\Lambda_\sigma+\frac{1}{6}g_{\beta\alpha}[\nabla^\rho,\nabla^\sigma\nabla_\sigma]_-\Lambda_\rho$$

$$\left.-\frac{1}{6}\nabla_\alpha[\nabla^\sigma,\nabla_\beta]_-\Lambda_\sigma-\frac{1}{6}[\nabla_\alpha,\nabla^\sigma\nabla_\sigma]_-\Lambda_\beta\right\}+$$

$$+\left\{-\frac{1}{3}\nabla_\alpha[\nabla^\rho,\nabla_\beta]_-\Lambda_\rho-\frac{1}{3}[\nabla_\alpha,\nabla^\rho\nabla_\rho]_-\Lambda_\beta\right\}$$

$$+\left\{-\frac{1}{3}\nabla_\beta[\nabla^\rho,\nabla_\alpha]_-\Lambda_\rho-\frac{1}{3}[\nabla_\beta,\nabla^\rho\nabla_\rho]_-\Lambda_\alpha\right\}$$

$$+\frac{1}{6}g_{\alpha\beta}(\nabla^\rho[\nabla^\sigma,\nabla_\rho]_-\Lambda_\sigma+[\nabla^\rho,\nabla^\sigma\nabla_\sigma]_-\Lambda_\rho)\,.$$

By calculating in series all the commutators; after simple calculation, we will produce

$$0=g_{\alpha\beta}\,\nabla_\rho\,(R^{\rho\sigma}\,\Lambda_\sigma)+\,\Lambda^\sigma\left[\,\nabla_\rho R^\rho{}_{\alpha\beta\sigma}+\,\nabla_\rho R^\rho{}_{\beta\alpha\sigma}\right]$$

$$+\,(\nabla_\rho\Lambda_\sigma)\left[\,R^\rho{}_{\alpha\beta}{}^\sigma+\,R^\rho{}_{\beta\alpha}{}^\sigma\right]$$

$$-\,\Lambda^\rho\left[\,\nabla_\alpha R_{\beta\rho}+\nabla_\beta R_{\alpha\rho}\right]-\frac{3}{2}\left[\,R_\beta{}^\rho\,(\nabla_\alpha\Lambda_\rho)+\,R_\alpha{}^\rho\,(\nabla_\beta\Lambda_\rho)\right]$$

$$+\frac{1}{2}\left[R_\beta^\rho(\nabla_\rho\Lambda_\alpha)+R_\alpha{}^\rho\,(\nabla_\rho\Lambda_\beta)\right].\tag{6.3.9}$$

It must be noted that contrary to the expectations the equation obtained contains explicitly the curvature Riemann tensor. It enters into Eq. (6.3.9) in two combinations:

$$\Lambda^\sigma\,(\nabla_\rho R^\rho{}_{\alpha\beta\sigma}+\,\nabla_\rho R^\rho{}_{\beta\alpha\sigma})\,,\tag{6.3.10a}$$

$$(\nabla_\rho\Lambda_\sigma)\,(R^\rho{}_{\alpha\beta}{}^\sigma+\,R^\rho{}_{\beta\alpha}{}^\sigma)\,.\tag{6.3.10b}$$

The curvature tensor in combination with Eq. (6.3.10a) can be readily escaped. To this end, it suffices for the Bianchi identity

$$\nabla_\gamma R^\rho{}_{\alpha\,\beta\sigma}+\nabla_\sigma R^\rho{}_{\alpha\,\gamma\beta}+\nabla_\beta R^\rho{}_{\alpha\,\sigma\gamma}=0\,,$$

$$\nabla_\rho R^\rho{}_{\alpha\,\beta\sigma}+\nabla_\sigma R_{\beta\alpha}-\nabla_\beta R_{\alpha\sigma}=0\,.$$

Thus,

$$\nabla_\rho R^\rho{}_{\alpha\,\beta\sigma} + \nabla_\rho R^\rho{}_{\beta\alpha\sigma} = (\nabla_\alpha R_{\beta\sigma} + \nabla_\beta R_{\alpha\sigma}) - 2\nabla_\sigma R_{\beta\alpha}\,. \tag{6.3.11}$$

Then by Eq. (6.3.11), Eq. (6.3.9) takes the form

$$0 = g_{\alpha\beta}\,\nabla_\rho\,(R^{\rho\sigma}\,\Lambda_\sigma) - 2\Lambda^\sigma\nabla_\sigma R_{\alpha\beta} + (\nabla_\rho\Lambda_\sigma)\,[\,R^\rho{}_{\alpha\beta}{}^\sigma + R^\rho{}_{\beta\alpha}{}^\sigma]$$

$$-\frac{3}{2}\,[\,R_\beta{}^\rho\,(\nabla_\alpha\Lambda_\rho) + R_\alpha{}^\rho(\nabla_\beta\Lambda_\rho)\,] + \frac{1}{2}\,[\,R_\beta{}^\rho\,(\nabla_\rho\Lambda_\alpha) + R_\alpha{}^\rho\,(\nabla_\rho\Lambda_\beta)\,]\,. \tag{6.3.12}$$

However, the curvature tensor still remains to enter Eq. (6.3.12). And this means that in regions that involve curvature this massless spin-2 equation does not allow any gauge principle.

Now, we will show that in order to overcome such a difficulty these starting equations should be slightly altered. To this end, let us add a special term (not a minimal gravitational interaction term) into Eq. (6.3.1c):

$$\frac{1}{2}\Big(\nabla^\rho\Phi_{\rho\alpha\beta} + \nabla^\rho\Phi_{\rho\beta\alpha} - \frac{1}{2}g_{\alpha\beta}(x)\nabla^\rho\Phi_{\rho\sigma}{}^\sigma\Big)$$

$$+\Big(\nabla_\alpha\Phi_\beta + \nabla_\beta\Phi_\alpha - \frac{1}{2}g_{\alpha\beta}(x)\nabla^\rho\Phi_\rho\Big) = A\,(\,R^\rho{}_{\alpha\beta}{}^\sigma + R^\rho{}_{\beta\alpha}{}^\sigma\,)\,\Phi_{\rho\sigma}\,. \tag{6.3.13}$$

Let us show that at special parameter A, the theory of massless spin-2 particle can be done satisfactory in this gauge principle sense. Indeed,

$$AR^\rho{}_{\alpha\beta}{}^\sigma\Phi^{(0)}_{\rho\sigma} = AR^\rho{}_{\alpha\beta}{}^\sigma\Big(\nabla_\rho\Lambda_\sigma + \nabla_\sigma\Lambda_\rho - \frac{1}{2}g_{\rho\sigma}\nabla^\gamma\Lambda_\gamma\Big)$$

$$= A\Big(R^\rho{}_{\alpha\beta}{}^\sigma\nabla_\rho\Lambda_\sigma + R^\rho{}_{\beta\alpha}{}^\sigma\nabla_\rho\Lambda_\sigma + \frac{1}{2}R_{\alpha\beta}\nabla^\gamma\Lambda_\gamma\Big)\,,$$

and therefore a contribution of that additional term into Eq. (6.3.12) is equal to

$$A\,(\,R^\rho{}_{\alpha\beta}{}^\sigma + R^\rho{}_{\beta\alpha}{}^\sigma\,)\,\Phi^{(0)}_{\rho\sigma}$$

$$= 2A\,(\,R^\rho{}_{\alpha\beta}{}^\sigma + R^\rho{}_{\beta\alpha}{}^\sigma\,)\,(\nabla_\rho\Lambda_\sigma) + A\,R_{\alpha\beta}\,(\nabla^\gamma\Lambda_\gamma)\,. \tag{6.3.14}$$

So, instead of Eq. (6.3.12); we have

$$2A(\nabla_\rho\Lambda_\sigma)(R^\rho{}_{\alpha\beta}{}^\sigma + R^\rho{}_{\beta\alpha}{}^\sigma) + AR_{\alpha\beta}(\nabla^\gamma\Lambda_\gamma)$$

$$= g_{\alpha\beta}\nabla_\rho(R^{\rho\sigma}\Lambda_\sigma) - 2\Lambda^\sigma\nabla_\sigma R_{\alpha\beta} + (\nabla_\rho\Lambda_\sigma)[R^\rho{}_{\alpha\beta}{}^\sigma + R^\rho{}_{\beta\alpha}{}^\sigma]$$

$$-\frac{3}{2}\Big[R_\beta{}^\rho(\nabla_\alpha\Lambda_\rho) + R_\alpha{}^\rho(\nabla_\beta\Lambda_\rho)\Big] + \frac{1}{2}\Big[R_\beta{}^\rho(\nabla_\rho\Lambda_\alpha) + R_\alpha{}^\rho(\nabla_\rho\Lambda_\beta)\Big]\,.$$

$$\tag{6.3.15a}$$

Setting $A = \frac{1}{2}$, both terms with curvature tensor will be canceled by each other:

$$\frac{1}{2}R_{\alpha\beta}(\nabla^\gamma\Lambda_\gamma) = g_{\alpha\beta}\nabla_\rho(R^{\rho\sigma}\Lambda_\sigma) - 2\Lambda^\sigma\nabla_\sigma R_{\alpha\beta}$$

$$-\frac{3}{2}\left[R_\beta{}^\rho(\nabla_\alpha\Lambda_\rho) + R_\alpha{}^\rho(\nabla_\beta\Lambda_\rho)\right] + \frac{1}{2}\left[R_\beta{}^\rho(\nabla_\rho\Lambda_\alpha) + R_\alpha{}^\rho(\nabla_\rho\Lambda_\beta)\right].$$

$$(6.3.15b)$$

Finally the obtained relationship does not contain the curvature tensor and will turn into an identity at $R_{\alpha\beta}(x) = 0$ which was required. So, the required system is one which changes Eq. $(6.3.1c)$ by

$$\frac{1}{2}\left(\nabla^\rho\Phi_{\rho\alpha\beta} + \nabla^\rho\Phi_{\rho\beta\alpha} - \frac{1}{2}g_{\alpha\beta}(x)\nabla^\rho\Phi_{\rho\sigma}{}^\sigma\right)$$

$$+\left(\nabla_\alpha\Phi_\beta + \nabla_\beta\Phi_\alpha - \frac{1}{2}g_{\alpha\beta}(x)\nabla^\rho\Phi_\rho\right) = \frac{1}{2}(R^\rho{}_{\alpha\beta}{}^\sigma + R^\rho{}_{\beta\alpha}{}^\sigma)\Phi_{\rho\sigma}.$$

$$(6.3.16)$$

Chapter 7

Particle with Spin 2 and Anomalous Magnetic Moment

7.1 50-Component Model in Curved Space-Time

In this present Chapter, we consider a more complicated 50-component model for a massive spin 2 particle in presence of external electromagnetic and gravitational fields. The primary question is: how an additional intrinsic structure of the particle (anomalous magnetic moment) manifests itself in any curved space-time background?

We start with tensor equations given in [423, 424] for flat Minkowski space-time, and extend them by changing the ordinary derivative into one that is a covariant ($\partial_b \implies \nabla_b$), so we arrive at

$$2\,\lambda_1 D^a \Psi_a^{(1)} + 2\,\lambda_2 D^a \Psi_a^{(2)} + iM\,\Psi = 0\,, \tag{7.1.1a}$$

$$\lambda_3\,D_a\Psi + 2\lambda_4\,D^b\Psi_{(ba)} + iM\,\Psi_a^{(1)} = 0\,, \tag{7.1.1b}$$

$$\lambda_5\,D_a\Psi + 2\lambda_6\,D^b\Psi_{(ba)} + iM\,\Psi_a^{(2)} = 0\,, \tag{7.1.1c}$$

$$\frac{\lambda_7}{2}\left(D_a\Psi_b^{(1)} + D_b\Psi_a^{(1)} - \frac{1}{2}\,g_{ab}D^c\Psi_c^{(1)}\right)$$
$$+\frac{\lambda_8}{2}\left(D_a\Psi_b^{(2)} + D_b\Psi_a^{(2)} - \frac{1}{2}\,g_{ab}D^c\Psi_c^{(2)}\right)$$
$$+2\lambda_9\,D^c\Psi_{(abc)} - 2\lambda_{10}\left(D^c\Psi_{a[bc]} + D^c\Psi_{b[ac]}\right) + iM\,\Psi_{(ab)} = 0\,, \tag{7.1.1d}$$

$$\frac{\lambda_{11}}{2}\left(D_c\Psi_{(ab)} - D_b\Psi_{(ac)} - \frac{1}{3}g_{ca}D^m\Psi_{(mb)} + \frac{1}{3}g_{ba}D^m\Psi_{(mc)}\right) + iM\,\Psi_{a[bc]} = 0\,, \tag{7.1.1e}$$

$$\frac{\lambda_{12}}{3}\left(D_a\Psi_{(bc)} + D_b\Psi_{(ca)} + D_c\Psi_{(ab)}\right.$$
$$\left. -\frac{1}{3}\,g_{ac}D^m\Psi_{(mb)} - \frac{1}{3}\,g_{cb}D^m\Psi_{(ma)} - \frac{1}{3}\,g_{ba}D^m\Psi_{(mc)}\right) + iM\Psi_{(abc)} = 0\,.$$

$$(7.1.1f)$$

Here, $D_a = \nabla_a + ieA_a$, where ∇_a is a covariant derivative, A_a stands for electromagnetic potential; $\lambda_1, ..., \lambda_{12}$ are 12 numerical constants obeying additional restrictions[1]

$$2\lambda_{10}\lambda_{11} - \frac{2}{3}\lambda_9\lambda_{12} = 1 , \qquad \lambda_4\lambda_7 + \lambda_6\lambda_8 + \frac{8}{9}\lambda_9\lambda_{12} = \frac{1}{3} ,$$

$$\lambda_1\lambda_3 + \lambda_2\lambda_5 = -\frac{1}{4} , \qquad (\lambda_1\lambda_4 + \lambda_2\lambda_6)(\lambda_3\lambda_7 + \lambda_5\lambda_8) = -\frac{1}{12} .$$

$$(7.1.2a)$$

In the 50-component model for a spin 2 particle, we employ one scalar, two vectors, and three tensors:

$$\Psi , \qquad \Psi_a^{(1)} , \Psi_a^{(2)} , \qquad \Psi_{(ab)} , \Psi_{a[bc]} , \Psi_{(abc)} . \qquad (7.1.2b)$$

Recall that in a 30-component model, there involved a scalar, vector, and two tensors

$$\Phi , \qquad \Phi_a , \qquad \Phi_{(ab)} , \Phi_{[ab]c} ; \qquad (7.1.3a)$$

with 30 independent variables

$$\Phi(x) \implies 1 , \qquad \Phi_a \implies 4 ,$$

$$\Phi_{(ab)} \implies (10-1) = 9 , \qquad \Phi_{[ab]c} \implies 6 \times 4 - 4 - 4 = 16 ;$$

and equations (compare with Eqs. (7.1.1a)–(7.1.1f))

$$D^a\Phi_a - M \Phi = 0 , \qquad \frac{1}{2} D_a\Phi - \frac{1}{3} D^b\Phi_{(ab)} - M \Phi_a = 0 ,$$

$$D_a\Phi_b + D_b\Phi_a - \frac{1}{2}g_{ab} D^k\Phi_k + \frac{1}{2} (D^k\Phi_{[ka]b}$$

$$+D^k\Phi_{[kb]a} - \frac{1}{2}g_{ab}D^k\Phi_{[kn]}{}^n) - M \Phi_{(ab)} = 0 ,$$

$$D_a\Phi_{(bc)} - D_b\Phi_{(ac)} + \frac{1}{3} (g_{bc}D^k\Phi_{(ak)} - g_{ac} D^k\Phi_{(bk)}) - M \Phi_{[ab]c} = 0 . \qquad (7.1.3b)$$

7.2 Excluding Superfluous Variables

Here, we will show that excluding from the 50-component models superfluous variables (formally they consist a 4-vector and 3-rank tensor) and by introducing new field variables, one can get a 30-component model modified by the presence of additional interaction terms with electromagnetic and gravitational fields.

To this end, first, instead of $\Psi_a^{(1)}, \Psi_a^{(2)}$ let us introduce new variables

$$\begin{vmatrix} B_a \\ C_a \end{vmatrix} = \begin{vmatrix} \lambda_1 & \lambda_2 \\ \lambda_7 & \lambda_8 \end{vmatrix} \begin{vmatrix} \Psi_a^{(1)} \\ \Psi_a^{(2)} \end{vmatrix} , \qquad (7.2.1a)$$

---

[1]Here, it will be clear that only one parameter has physical sense, referring to anomalous magnetic moment, all other can be eliminated from the model.

and inverse one

$$\left| \begin{array}{c} \Psi_a^{(1)} \\ \Psi_a^{(2)} \end{array} \right| = \frac{1}{\lambda_1\lambda_8 - \lambda_2\lambda_7} \left| \begin{array}{cc} \lambda_8 & -\lambda_2 \\ -\lambda_7 & \lambda_1 \end{array} \right| \left| \begin{array}{c} B_a \\ C_a \end{array} \right| . \qquad (7.2.1b)$$

The system in Eq. (7.1.1) can be presented as follows:

$$2\, D^a B_a + im\, \Psi = 0 \,, \qquad (7.2.2a)$$

$$-\frac{1}{4}\, D_a\Psi + 2\,(\lambda_1\lambda_4 + \lambda_2\lambda_6)\, D^b\Psi_{(ba)} + iM\, B_a = 0 \,, \qquad (7.2.2b)$$

$$(\lambda_7\lambda_3 + \lambda_8\lambda_5)\, D_a\Psi + 2\,(\lambda_7\lambda_4 + \lambda_8\lambda_6)\, D^b\Psi_{(ba)} + iM\, C_a = 0 \,, \qquad (7.2.2c)$$

$$\frac{1}{2}\left(D_a C_b + D_b C_a - \frac{1}{2}\, g_{ab} D^c C_c \right)$$
$$+ 2\lambda_9\, D^c\Psi_{(abc)} - 2\lambda_{10}\left(D^c\Psi_{a[bc]} + D^c\Psi_{b[ac]} \right) + iM\, \Psi_{(ab)} = 0 \,,$$

$$(7.2.2d)$$

$$\frac{\lambda_{11}}{2}\left(D_c\Psi_{(ab)} - D_b\Psi_{(ac)} - \frac{1}{3} g_{ca} D^m\Psi_{(mb)} + \frac{1}{3} g_{ba} D^m\Psi_{(mc)} \right) + iM\, \Psi_{a[bc]} = 0 \,,$$

$$(7.2.2e)$$

$$\frac{\lambda_{12}}{3}\Big[D_a\Psi_{(bc)} + D_b\Psi_{(ca)} + D_c\Psi_{(ab)}$$
$$-\frac{1}{3} g_{ac} D^m\Psi_{(mb)} - \frac{1}{3}\, g_{cb} D^m\Psi_{(ma)} - \frac{1}{3} g_{ba} D^m\Psi_{(mc)} \Big] + iM\Psi_{(abc)} = 0 \,.$$

$$(7.2.2f)$$

Then by multiplying Eq. (7.2.2c) by $(\lambda_1\lambda_4 + \lambda_2\lambda_6)$ and taking into account Eq. (7.2.2a), we get

$$-\frac{1}{12} D_a\Psi + 2\,(\lambda_1\lambda_4 + \lambda_2\lambda_6)(\lambda_7\lambda_4 + \lambda_8\lambda_6)\, D^b\Psi_{(ba)} + iM\,(\lambda_1\lambda_4 + \lambda_2\lambda_6)\, C_a = 0 \,.$$

By substituting the expression for $D_a\Psi$ from Eq. (7.2.2b), we arrive at

$$-\frac{2}{3}(\lambda_1\lambda_4 + \lambda_2\lambda_6)\, D^b\Psi_{(ba)} - \frac{iM}{3}\, B_a$$

$$+ 2(\lambda_1\lambda_4 + \lambda_2\lambda_6)(\lambda_7\lambda_4 + \lambda_8\lambda_6)\, D^b\Psi_{(ba)} + iM\,(\lambda_1\lambda_4 + \lambda_2\lambda_6) C_a = 0 \,,$$

from this, it follows:

$$C_a = \frac{1}{3(\lambda_1\lambda_4 + \lambda_2\lambda_6)}\, B_a - \frac{2}{iM}\left[(\lambda_7\lambda_4 + \lambda_8\lambda_6) - \frac{1}{3} \right] D^n\Psi_{(na)} \,. \qquad (7.2.3a)$$

This identity permits to exclude a superfluous vector C_a. In particular, then Eq. (7.2.2d) gives

$$\frac{1}{6(\lambda_1\lambda_4 + \lambda_2\lambda_6)} \left(D_a B_b + D_b B_a - \frac{1}{2} g_{ab} D^c B_c \right)$$

$$-\frac{1}{iM} \left[(\lambda_7\lambda_4 + \lambda_8\lambda_6) - \frac{1}{3} \right] \left(D_a D^n \Psi_{(nb)} + D_b D^n \Psi_{(na)} - \frac{1}{2} g_{ab} D^c D^n \Psi_{(nc)} \right)$$

$$+2\lambda_9 \, D^c \Psi_{(abc)} - 2\lambda_{10} \, (\, D^c \Psi_{a[bc]} + D^c \Psi_{b[ac]}) + iM \, \Psi_{(ab)} = 0 \, .$$

$$(7.2.3b)$$

Therefore, instead of Eqs. (7.2.2); we can use an equivalent one

$$2 \, D^a B_a + iM \, \Psi = 0 \, , \qquad (7.2.4a)$$

$$-\frac{1}{4} \, D_a \Psi + 2 \, (\lambda_1\lambda_4 + \lambda_2\lambda_6) \, D^b \Psi_{(ba)} + iM \, B_a = 0 \, , \qquad (7.2.4b)$$

$$C_a = \frac{1}{3(\lambda_1\lambda_4 + \lambda_2\lambda_6)} \, B_a - \frac{2}{iM} \left[(\lambda_7\lambda_4 + \lambda_8\lambda_6) - \frac{1}{3} \right] D^n \Psi_{(na)} \, ,$$

$$(7.2.4c)$$

$$\frac{1}{6(\lambda_1\lambda_4 + \lambda_2\lambda_6)} \, (D_a B_b + D_b B_a - \frac{1}{2} \, g_{ab} D^c B_c)$$

$$-\frac{1}{iM} \left[(\lambda_7\lambda_4 + \lambda_8\lambda_6) - \frac{1}{3} \right] \left(D_a D^n \Psi_{(nb)} + D_b D^n \Psi_{(na)} - \frac{1}{2} g_{ab} D^c D^n \Psi_{(nc)} \right)$$

$$+2\lambda_9 \, D^c \Psi_{(abc)} - 2\lambda_{10} \, (\, D^c \Psi_{a[bc]} + D^c \Psi_{b[ac]}) + iM \, \Psi_{(ab)} = 0 \, ,$$

$$(7.2.4d)$$

$$\frac{\lambda_{11}}{2} \, (D_c \Psi_{(ab)} - D_b \Psi_{(ac)} - \frac{1}{3} g_{ca} D^m \Psi_{(mb)} + \frac{1}{3} g_{ba} D^m \Psi_{(mc)}) + iM \, \Psi_{a[bc]} = 0 \, ,$$

$$(7.2.4e)$$

$$\frac{\lambda_{12}}{3} \left[D_a \Psi_{(bc)} + D_b \Psi_{(ca)} + D_c \Psi_{(ab)} \right.$$

$$\left. -\frac{1}{3} \, g_{ac} D^m \Psi_{(mb)} - \frac{1}{3} \, g_{cb} D^m \Psi_{(ma)} - \frac{1}{3} \, g_{ba} D^m \Psi_{(mc)} \right] + iM\Psi_{(abc)} = 0 \, .$$

$$(7.2.4f)$$

By the help of Eqs. (7.2.4e) and (7.2.4f), let us express tensors $\Psi_{a[bc]}$ and $\Psi_{(abc)}$ through the 2-rank tensor:

$$\Psi_{a[bc]} = \frac{i\lambda_{11}}{2M} \left(D_c \Psi_{(ab)} - D_b \Psi_{(ac)} - \frac{1}{3} g_{ca} D^m \Psi_{(mb)} + \frac{1}{3} g_{ba} D^m \Psi_{(mc)} \right) \, ,$$

$$(7.2.5a)$$

$$\Psi_{(abc)} = \frac{i\lambda_{12}}{3M} \left(D_a \Psi_{(bc)} + D_b \Psi_{(ca)} + D_c \Psi_{(ab)} \right.$$

$$-\frac{1}{3}\,g_{ac}D^m\Psi_{(mb)} - \frac{1}{3}\,g_{cb}D^m\Psi_{(ma)} - \frac{1}{3}\,g_{ba}D^m\Psi_{(mc)}\Bigg).$$

$$(7.2.5b)$$

Through substitution into Eq. (7.2.4d), we get

$$\frac{1}{6(\lambda_1\lambda_4 + \lambda_2\lambda_6)}\,(D_aB_b + D_bB_a - \frac{1}{2}\,g_{ab}D^cB_c)$$

$$+\frac{i}{M}\left[(\lambda_7\lambda_4 + \lambda_8\lambda_6) - \frac{1}{3}\right]\left(D_aD^c\Psi_{(cb)} + D_bD^c\Psi_{(ca)} - \frac{1}{2}\,g_{ab}D^cD^n\Psi_{(nc)}\right)$$

$$+i\frac{2\lambda_9\lambda_{12}}{3M}D^c\left(D_a\Psi_{(bc)} + D_b\Psi_{(ca)} + D_c\Psi_{(ab)}\right.$$

$$-\frac{1}{3}g_{ac}D^m\Psi_{(mb)} - \frac{1}{3}g_{cb}D^m\Psi_{(ma)} - \frac{1}{3}g_{ba}D^m\Psi_{(mc)}\Bigg)$$

$$-i\frac{\lambda_{10}\lambda_{11}}{M}\left[D^c\left(D_c\Psi_{(ab)} - D_b\Psi_{(ac)} - \frac{1}{3}g_{ca}D^m\Psi_{(mb)} + \frac{1}{3}g_{ba}D^m\Psi_{(mc)}\right)\right.$$

$$+D^c\left(D_c\Psi_{(ba)} - D_a\Psi_{(bc)} - \frac{1}{3}g_{cb}D^m\Psi_{(ma)} + \frac{1}{3}g_{ab}D^m\Psi_{(mc)}\right)\Bigg] + iM\,\Psi_{(ab)} = 0\,.$$

Now, we allow for

$$\lambda_{10}\lambda_{11} = \frac{1}{2} + \frac{1}{3}\lambda_9\lambda_{12}\,, \qquad \lambda_4\lambda_7 + \lambda_6\lambda_8 - \frac{1}{3} = -\frac{8}{9}\lambda_9\lambda_{12}\,,$$

and by using the notation $\lambda_9\lambda_{12} = \mu$, we obtain

$$\frac{M}{6i(\lambda_1\lambda_4 + \lambda_2\lambda_6)}\left(D_aB_b + D_bB_a - \frac{1}{2}\,g_{ab}D^cB_c\right)$$

$$- \mu\frac{8}{9}D_aD^c\Psi_{(cb)} - \mu\frac{8}{9}D_bD^c\Psi_{(ca)} + \mu\frac{4}{9}\,g_{ab}D^cD^n\Psi_{(nc)}$$

$$+\mu\frac{2}{3}D^cD_a\Psi_{(bc)} + \mu\frac{2}{3}D^cD_b\Psi_{(ca)} + \mu\frac{2}{3}D^cD_c\Psi_{(ab)} - \mu\frac{2}{9}g_{ac}D^cD^m\Psi_{(mb)}$$

$$- \mu\frac{2}{9}g_{cb}D^cD^m\Psi_{(ma)} - \mu\frac{2}{9}g_{ba}D^cD^m\Psi_{(mc)}$$

$$- \frac{1}{2}D^cD_c\Psi_{(ab)} + \frac{1}{2}D^cD_b\Psi_{(ac)} + \frac{1}{6}g_{ca}D^cD^m\Psi_{(mb)} - \frac{1}{6}g_{ba}D^cD^m\Psi_{(mc)}$$

$$- \frac{1}{2}D^cD_c\Psi_{(ba)} + \frac{1}{2}D^cD_a\Psi_{(bc)} + \frac{1}{6}g_{cb}D^cD^m\Psi_{(ma)} - \frac{1}{6}g_{ab}D^cD^m\Psi_{(mc)}$$

$$- \mu\frac{1}{3}D^cD_c\Psi_{(ab)} + \mu\frac{1}{3}D^cD_b\Psi_{(ac)} + \mu\frac{1}{9}g_{ca}D^cD^m\Psi_{(mb)} - \mu\frac{1}{9}g_{ba}D^cD^m\Psi_{(mc)}$$

$$- \mu\frac{1}{3}D^cD_c\Psi_{(ba)} + \mu\frac{1}{3}D^cD_a\Psi_{(bc)} + \mu\frac{1}{9}g_{cb}D^cD^m\Psi_{(ma)} - \mu\frac{1}{9}g_{ab}D^cD^m\Psi_{(mc)}$$

$$+ M^2\,\Psi_{(ab)} = 0\,.$$

By this, after simple manipulations, we arrive at (commutator will be noted as $[..., ...]_-$)

$$\frac{1}{6i\,(\lambda_1\lambda_4 + \lambda_2\lambda_6)}\,\left(D_a B_b + D_b B_a - \frac{1}{2}\,g_{ab}D^c B_c\right)$$

$$-\frac{1}{M}\left[D^c D_c \Psi_{(ba)} - \frac{1}{2}\,(\,D^c D_b \Psi_{(ac)} + D^c D_a \Psi_{(bc)}\,)\right.$$

$$\left.+\frac{1}{3}\,g_{ab}\,D^n D^m \Psi_{(nm)} - \frac{1}{6}\,(\,D_a D^m \Psi_{(mb)} + D_b D^m \Psi_{(ma)}\,)\right]$$

$$+\frac{\mu}{M}\,([D^c, D_a]_-\Psi_{(bc)} + [D^c, D_b]_-\Psi_{(ac)}) + M\,\Psi_{(ab)} = 0\,. \qquad (7.2.6)$$

Now, let us introduce a new variable (numerical parameter γ will be specified here)

$$\Phi_{[bc]a} = -\frac{1}{M}\frac{\gamma}{2}\left(D_c\Psi_{(ab)} - D_b\Psi_{(ac)} + \frac{1}{3}g_{ab}D^m\Psi_{(mc)} - \frac{1}{3}g_{ac}D^m\Psi_{(mb)}\right),$$

then, we now derive an identity

$$\frac{1}{\gamma}\,\left(D^c\Phi_{[bc]a} + D^c\Phi_{[ac]b}\right) = -\frac{1}{M}$$

$$\times\left[\frac{1}{2}\left(D^c D_c\Psi_{(ab)} - D^c D_b\Psi_{(ac)} + \frac{g_{ab}}{3}D^c D^m\Psi_{(mc)} - \frac{g_{ac}}{3}D^c D^m\Psi_{(mb)}\right)\right.$$

$$\left.+\frac{1}{2}\left(D^c D_c\Psi_{(ba)} - D^c D_a\Psi_{(bc)} + \frac{g_{ba}}{3}D^c D^m\Psi_{(mc)} - \frac{g_{bc}}{3}D^c D^m\Psi_{(ma)}\right)\right]$$

$$= -\frac{1}{M}\left(D^c D_c\Psi_{(ab)} - \frac{1}{2}D^c D_b\Psi_{(ac)} - \frac{1}{3}D^c D_a\Psi_{(bc)} + \frac{g_{ab}}{3}D^c D^m\Psi_{(mc)}\right.$$

$$\left.-\frac{g_{ac}}{6}D^c D^m\Psi_{(mb)} - \frac{g_{bc}}{6}D^c D^m\Psi_{(ma)}\right),$$

which coincides with the expression in brackets in Eq. (7.2.6). Therefore, Eq. (7.2.6) may be presented as (let it be $\gamma = \sqrt{2}$)

$$\frac{1}{6i\,(\lambda_1\lambda_4 + \lambda_2\lambda_6)}\,\left(D_a B_b + D_b B_a - \frac{1}{2}\,g_{ab}D^c B_c\right) + \frac{1}{\sqrt{2}}\,\left(D^c\Phi_{[bc]a} + D^c\Phi_{[ac]b}\right)$$

$$+\frac{\mu}{M}\,(\,[D^c, D_a]_-\Psi_{(bc)} + [D^c, D_b]_-\Psi_{(ac)}\,) + M\,\Psi_{(ab)} = 0\,.$$

$$(7.2.7)$$

In the following procedure, it will be convenient to use two variables

$$\Phi = -\,\frac{1}{4\sqrt{3}(\lambda_1\lambda_4 + \lambda_2\lambda_6)}\,\Psi\,, \qquad \Phi_a = \frac{i}{\sqrt{6}(\lambda_1\lambda_4 + \lambda_2\lambda_6)}\,B_a\,. \qquad (7.2.8)$$

Thus, from the 50-component system, we have arrived at a modified 30-component model

$$\frac{1}{\sqrt{2}}D^a\Phi_a + M\,\Phi = 0\,,$$

$$\frac{1}{\sqrt{2}} \, D_a \Phi + \sqrt{\frac{2}{3}} \, D^b \Psi_{(ba)} + M \, \Phi_a = 0 \, ,$$

$$-\frac{1}{\sqrt{6}} \, (\, D_a \Phi_b + D_b \Phi_a - \frac{1}{2} \, g_{ab} D^c \Phi_c \,) \, + \frac{1}{\sqrt{2}} \, (\, D^c \Phi_{[bc]a} + D^c \Phi_{[ac]b} \,)$$

$$+ \frac{\mu}{M} \, (\, [D^c, D_a]_- \Psi_{(bc)} + [D^c, D_b]_- \Psi_{(ac)} \,) + M \, \Psi_{(ab)} = 0 \, ,$$

$$\frac{1}{\sqrt{2}} \, (\, D_c \Psi_{(ab)} - D_b \Psi_{(ac)} + \frac{1}{3} g_{ab} D^m \Psi_{(lc)} - \frac{1}{3} g_{ac} D^m \Psi_{(lb)} \,) + M \, \Phi_{a[bc]} = 0 \, .$$

$$(7.2.9)$$

By the linear transformations

$$\Phi = -\tilde{\Phi} \, , \qquad \Psi_a = \sqrt{2} \, \tilde{\Phi}_a \, ,$$

$$\Phi_{(ab)} = \frac{1}{\sqrt{3}} \, \tilde{\Phi}_{(ab)}, \qquad \Phi_{[bc]a} = \frac{1}{\sqrt{6}} \, \tilde{\Phi}_{[bc]a} \qquad\qquad (7.2.10)$$

it becomes simpler

$$D^a \tilde{\Phi}_a - M \, \tilde{\Phi} = 0 \, ,$$

$$\frac{1}{2} \, D_a \tilde{\Phi} - \frac{1}{3} \, D^b \tilde{\Psi}_{(ba)} - M \, \tilde{\Phi}_a = 0 \, ,$$

$$(\, D_a \tilde{\Phi}_b + D_b \tilde{\Phi}_a - \frac{1}{2} \, g_{ab} D^c \tilde{\Phi}_c \,) \, + \frac{1}{2} \, (\, D^c \tilde{\Phi}_{[ca]b} + D^c \tilde{\Phi}_{[cb]a} \,)$$

$$- \frac{\mu}{M} \, \left(\, [D^c, D_a]_- \, \tilde{\Phi}_{(bc)} + [D^c, D_b]_- \, \tilde{\Phi}_{(ac)} \, \right) - M \, \tilde{\Phi}_{(ab)} = 0 \, ,$$

$$D_c \tilde{\Phi}_{(ba)} - D_b \tilde{\Phi}_{(ca)} + \frac{1}{3} g_{ba} D^m \tilde{\Psi}_{(mc)} - \frac{1}{3} g_{ca} D^m \tilde{\Psi}_{(mb)} - M \, \tilde{\Phi}_{[cb]a} = 0 \, .$$

$$(7.2.11)$$

If $\mu = 0$, we obtain a (massive particle) 30-component theory (the sign of $\sim$ is taken away):

$$D^a \Phi_a - M \, \Phi = 0 \, ,$$

$$\frac{1}{2} \, D_a \Phi - \frac{1}{3} \, D^b \Phi_{(ab)} - M \, \Phi_a = 0 \, ,$$

$$D_a \Phi_b + D_b \Phi_a - \frac{1}{2} g_{ab} \, D^c \Phi_c + \frac{1}{2} \, (D^c \Phi_{[ca]b} + D^c \Phi_{[cb]a} \,) - M \, \Phi_{(ab)} = 0 \, ,$$

$$D_c \Phi_{(ba)} - D_b \Phi_{(ca)} + \frac{1}{3} g_{ba} D^m \Phi_{(mc)} - \frac{1}{3} g_{ca} D^m \Phi_{(mb)} - M \, \Phi_{[cb]a} = 0 \, .$$

$$(7.2.12)$$

Let us find an explicit form for (see 4-th equation in Eq. (7.2.11))

$$\mu M^{-1} \left([D^c, D_a]_- \Phi_{(bc)} + [D^c, D_b]_- \Phi_{(ac)} \right). \qquad (7.2.13)$$

It suffices to consider the first term

$$[D^c, D_a]_- \tilde{\Phi}_{(bc)} = [\nabla_c + ieA_c, \nabla_a + ieA_a]_- \Phi_b{}^c$$

$$= (\nabla_c \nabla_a - \nabla_a \nabla_c) \Phi_b{}^c + ieF_{ca} \Phi_b{}^c ;$$

from this point, with the help of known rules

$$(\nabla_c \nabla_a - \nabla_a \nabla_c) A_{bk} = -A_{nk} R^n{}_{b\,ca} - A_{bn} R^n{}_{k\,ca}$$

it follows

$$(\nabla_c \nabla_a - \nabla_a \nabla_c) A_b{}^c = -A_n{}^c R^n{}_{b\,ca} - A_{bn} R^{nc}{}_{ca} .$$

Further, we then allow for symmetry of curvature tensor to find

$$(\nabla^c \nabla_a - \nabla_a \nabla^c) A_{bc} = R_{ca\,bn} A^{nc} + A_b{}^n R_{na} ,$$

we derive

$$(\nabla^c \nabla_a - \nabla_a \nabla^c) \Phi_{bc} = R_{ca\,bn} \Phi^{cn} + R_{ac} \Phi^c{}_b . \qquad (7.2.14)$$

Therefore,

$$[D^c, D_a]_- \tilde{\Phi}_{(bc)} = ieF_{ca}\Phi_b{}^c + R_{ca\,bn} \Phi^{cn} + R_{ac} \Phi^c{}_b , \qquad (7.2.15)$$

and through an additional interaction term (in Eq. (7.2.11)) is specified by

$$\frac{\mu}{M} \left([D^c, D_a]_- \Phi_{(bc)} + [D^c, D_b]_- \Phi_{(ac)} \right)$$

$$= \frac{\mu}{M} ie (\Phi_a{}^c F_{cb} + \Phi_b{}^c F_{ca})$$

$$+ \frac{\mu}{M} (R_{ca\,bn} \Phi^{cn} + R_{cb\,an} \Phi^{cn})$$

$$+ \frac{\mu}{M} (R_{ac} \Phi^c{}_b + R_{bc} \Phi^c{}_a) . \qquad (7.2.16)$$

The relation in Eq. (7.2.16) means that the parameter μ, initially interpreted as defining anomalous magnetic moment, also determines the additional interaction with geometrical background, through Ricci R_{kl} and Riemann curvature tensor R_{klmn}.

It should be noted that in the case of spin 1/2 particle with anomalous magnetic moment [300, 301, 303, 310] there arises an additional interaction through Ricci scalar [425, 426]; in the case of spin 1 particle [295–299, 427] there arises an additional interaction through Ricci tensor [428]. In other words, sensitiveness of the anomalous magnetic moment to the space-time geometry substantially depends on spin of the particle.

Chapter 8

Spherical Solutions for Dirac–Kähler and Dirac Particles

8.1 Spherical Solutions of the Dirac–Kähler Field

The Dirac–Kähler field (other terms are Ivanenko–Landau field or vector field of general type) was investigated by many authors – see bibliography in the chapter **5**. The most intriguing question is: what does this field describe? Is it a boson or a composite fermion with internal degree of freedom? The goal of this chapter is to construct spherical solutions for the Dirac–Kähler field for both boson and fermion type. Then, we describe the relations between them. The problem is solved in the flat Minkowski space. Additionally, we specify the case of a curved space-time background (3-space with constant positive curvature) where any fermion solutions cannot be constructed.

The Dirac–Kähler theory is written in a diagonal spherical tetrad in flat Minkowski space-time, and has the form

$$\left[i\gamma^0 \, \partial_t \; + \; i \; (\gamma^3 \, \partial_r \; + \; \frac{\gamma^1 J^{31} \; + \; \gamma^2 \, J^{32}}{r}) \; + \; \frac{1}{r} \, \Sigma_{\theta,\phi} \; - \; m \right] U(x) = 0 \;, \quad (8.1.1a)$$

$$\Sigma_{\theta,\phi} = i\gamma^1 \, \partial_\theta \; + \; \gamma^2 \, \frac{i\partial_\phi \; + \; iJ^{12} \, \cos\theta}{\sin\theta} \;, \quad J^{12} = \sigma^{12} \otimes I \; + \; I \otimes \sigma^{12} \;. \quad (8.1.1b)$$

By diagonalizing operators $\mathbf{J}^2$, J_3 of the total angular momentum (first constructing solutions for boson type)

$$J_1 = l_1 \; + \; \frac{iJ^{12} \cos\phi}{\sin\theta} \;, \quad J_2 = l_2 \; + \; \frac{iJ^{12} \sin\phi}{\sin\theta} \;, \quad J_3 = l_3 \;, \quad (8.1.2a)$$

for the wave function, we obtain substitution (details of the relevant general formalism can be seen in [41])

$$U_{\epsilon JM}(t,r,\theta,\phi) = \frac{e^{-i\epsilon t}}{r} \begin{vmatrix} f_{11} \, D_{-1} & f_{12} \, D_0 & f_{13} \, D_{-1} & f_{14} \, D_0 \\ f_{21} \, D_0 & f_{22} \, D_{+1} & f_{23} \, D_0 & f_{24} \, D_{+1} \\ f_{31} \, D_{-1} & f_{32} \, D_0 & f_{33} \, D_{-1} & f_{34} \, D_0 \\ f_{41} \, D_0 & f_{42} \, D_{+1} & f_{43} \, D_0 & f_{44} \, D_{+1} \end{vmatrix} \;, \quad (8.1.2b)$$

$f_{ab} = f_{ab}(r)$, Wigner functions $D_\sigma = D^J_{-M,\sigma}(\phi, \theta, 0)$, a quantum number J takes on the values $0, 1, 2, \ldots.$. When calculating the action of angular operator, $\Sigma_{\theta,\phi} U_{\epsilon JM}$, we need to employ the known formulas [429]

$$\partial_\theta D_{-1} = \frac{1}{2}(b D_{-2} - a D_0), \qquad [(-M + \cos\theta)/\sin\theta] D_{-1} = \frac{1}{2}(-b D_{-2} - a D_0),$$

$$\partial_\theta D_{+1} = \frac{1}{2}(a D_0 - b D_{+2}), \qquad [(-M - \cos\theta)/\sin\theta] D_{+1} = \frac{1}{2}(-a D_0 - b D_{+2}),$$

$$\partial_\theta D_0 = \frac{1}{2}(a D_{-1} - a D_{+1}), \qquad [-M/\sin\theta] D_0 = \frac{1}{2}(-a D_{-1} - a D_{+1}),$$

$$a = \sqrt{J(J+1)}, \qquad b = \sqrt{(J-1)(J+1)}. \tag{8.1.3a}$$

So, for $\Sigma_{\theta,\phi} U_{\epsilon JM}$ we get

$$\Sigma_{\theta,\phi} U = i\sqrt{J(J+1)} \begin{vmatrix} -f_{41} D_{-1} & -f_{42} D_0 & -f_{43} D_{-1} & -f_{44} D_0 \\ f_{31} D_0 & f_{32} D_{+1} & f_{33} D_0 & f_{34} D_{+1} \\ f_{21} D_{-1} & f_{22} D_0 & f_{23} D_{-1} & f_{24} D_0 \\ -f_{11} D_0 & -f_{12} D_{+1} & -f_{13} D_0 & -f_{14} D_{+1} \end{vmatrix}. \tag{8.1.3b}$$

To simplify the problem, for the functions $U_{\epsilon JM}$; let us diagonalize additionally an operator of P-reflection for the Dirac–Kähler field, in the basis of spherical tetrad it has the form [41]

$$\hat{\Pi}_{sph} = \begin{vmatrix} 0 & 0 & 0 & -1 \\ 0 & 0 & -1 & 0 \\ 0 & -1 & 0 & 0 \\ -1 & 0 & 0 & 0 \end{vmatrix} \otimes \begin{vmatrix} 0 & 0 & 0 & -1 \\ 0 & 0 & -1 & 0 \\ 0 & -1 & 0 & 0 \\ -1 & 0 & 0 & 0 \end{vmatrix} \hat{P}. \tag{8.1.4a}$$

From eigenvalue equation $\hat{\Pi}_{sph} U_{\epsilon JM} = \Pi U_{\epsilon JM}$ we derive the following restrictions

$$f_{31} = \pm f_{24}, \quad f_{32} = \pm f_{23}, \quad f_{33} = \pm f_{22}, \quad f_{34} = \pm f_{21},$$

$$f_{41} = \pm f_{14}, \quad f_{42} = \pm f_{13}, \quad f_{43} = \pm f_{12}, \quad f_{44} = \pm f_{11}; \tag{8.1.4b}$$

upper sign refers to the value $\Pi = (-1)^{J+1}$, lower – the values $\Pi = (-1)^J$. Next, to distinguish between these two cases that have different parity, we will associate $\Pi = (-1)^{J+1}$ with the symbol $\Delta = -1$, and $\Pi = (-1)^J$ with $\Delta = +1$.

Correspondingly, the substitution will read

$$U_{\epsilon JM\Delta}(t, r, \theta, \phi) = \frac{e^{-i\epsilon t}}{r} \begin{vmatrix} f_{11} D_{-1} & f_{12} D_0 & f_{13} D_{-1} & f_{14} D_0 \\ f_{21} D_0 & f_{22} D_{+1} & f_{23} D_0 & f_{24} D_{+1} \\ \Delta f_{24} D_{-1} & \Delta f_{23} D_0 & \Delta f_{22} D_{-1} & \Delta f_{21} D_0 \\ \Delta f_{14} D_0 & \Delta f_{13} D_{+1} & \Delta f_{12} D_0 & \Delta f_{11} D_{+1} \end{vmatrix}. \tag{8.1.5a}$$

The system of radial equations at $\Delta = +1$ is

$$\epsilon f_{24} - i\frac{d}{dr} f_{24} - i\frac{a}{r} f_{14} - m f_{11} = 0,$$

$$\epsilon f_{23} - i\frac{d}{dr}f_{23} + i\frac{1}{r}f_{14} - i\frac{a}{r}f_{13} - mf_{12} = 0 \,,$$

$$\epsilon f_{22} - i\frac{d}{dr}f_{22} - i\frac{a}{r}f_{12} - mf_{13} = 0 \,,$$

$$\epsilon f_{21} - i\frac{d}{dr}f_{21} + i\frac{1}{r}f_{12} - i\frac{a}{r}f_{11} - mf_{14} = 0 \,,$$

$$\epsilon f_{14} + i\frac{d}{dr}f_{14} + i\frac{1}{r}f_{23} + i\frac{a}{r}f_{24} - mf_{21} = 0 \,,$$

$$\epsilon f_{13} + i\frac{d}{dr}f_{13} + i\frac{a}{r}f_{23} - mf_{22} = 0 \,,$$

$$\epsilon f_{12} + i\frac{d}{dr}f_{12} + i\frac{1}{r}f_{21} + i\frac{a}{r}f_{22} - mf_{23} = 0 \,,$$

$$\epsilon f_{11} + i\frac{d}{dr}f_{11} + i\frac{1}{r}f_{21} + i\frac{a}{r}f_{21} - mf_{24} = 0 \,. \qquad (8.1.5b)$$

Thus, in having changed m into $-m$, we produce analogous equations at $\Delta = -1$.

Now, let us translate equations to new field variables

$$A = (f_{11} + f_{22})/\sqrt{2} \,, \qquad B = (f_{11} - f_{22})/i\sqrt{2} \,,$$
$$C = (f_{12} + f_{21})/\sqrt{2} \,, \qquad D = (f_{11} - f_{22})/i\sqrt{2} \,,$$
$$K = (f_{13} + f_{24})/\sqrt{2} \,, \qquad L = (f_{13} - f_{24})/i\sqrt{2} \,,$$
$$M = (f_{14} + f_{23})/\sqrt{2} \,, \qquad N = (f_{14} + f_{23})/i\sqrt{2} \,. \qquad (8.1.6a)$$

As result, we obtain equations without imaginary i

$$\epsilon K - \frac{dL}{dr} + \frac{a}{r}N - mA = 0 \,,$$

$$\epsilon L + \frac{dK}{dr} + \frac{a}{r}N + mB = 0 \,,$$

$$\epsilon A - \frac{dB}{dr} + \frac{a}{r}D - mK = 0 \,,$$

$$\epsilon B + \frac{dA}{dr} + \frac{a}{r}C + mL = 0 \,,$$

$$\epsilon M - \frac{dN}{dr} + \frac{1}{r}N + \frac{a}{r}L - mC = 0 \,,$$

$$\epsilon N + \frac{dM}{dr} + \frac{1}{r}M + \frac{a}{r}K + mD = 0 \,,$$

$$\epsilon C - \frac{dD}{dr} + \frac{1}{r}D + \frac{a}{r}B - mM = 0 \,,$$

$$\epsilon D + \frac{dC}{dr} + \frac{1}{r}C + \frac{a}{r}A + mN = 0 \,. \qquad (8.1.6b)$$

Note: Eqs. (8.1.6b) permit the following linear constraints

$$A = \lambda K \,, \quad B = \lambda L \,, \quad C = \lambda M \,, \quad D = \lambda N \,, \qquad (8.1.7a)$$

where $\lambda = \pm 1$. In particular, at $\lambda = +1$ we get a system of four equations

$$\frac{dK}{dr} + \frac{a}{r}M + (\epsilon + m)L = 0 \,,$$

$$\frac{dL}{dr} - \frac{a}{r}N - (\epsilon - m)K = 0 \,,$$

$$(\frac{d}{dr} + \frac{1}{r})M + \frac{a}{r}K + (\epsilon + m)N = 0 \,,$$

$$(\frac{d}{dr} - \frac{1}{r})N - \frac{a}{r}L - (\epsilon - m)M = 0 \,. \tag{8.1.7b}$$

By formally changing m into $-m$, we will obtain equations for the case $\lambda = -1$.

When taking these restrictions into account, the substitution for solution $U^\lambda_{\epsilon JM\Delta}$ can be written in a simpler form

$$U^\lambda_{\epsilon JM\Delta}(t,r,\theta,\phi) = \frac{e^{-i\epsilon t}}{r\sqrt{2}}$$

$$\times \begin{vmatrix} \lambda\,(K+iL)\,D_{-1} & \lambda\,(M+iN)\,D_{\,0} & (K+iL)\,D_{-1} & (M+iN)\,D_{\,0} \\ \lambda\,(M-iN)\,D_{\,0} & \lambda\,(K-iL)\,D_{+1} & (M-iN)\,D_{\,0} & (K-iL)\,D_{+1} \\ \Delta\,(K-iL)\,D_{-1} & \Delta\,(M-iN)\,D_{\,0} & \Delta\,\lambda\,(K-iL)\,D_{-1} & \Delta\,\lambda\,(M-iN)\,D_{\,0} \\ \Delta\,(M+iN)\,D_{\,0} & \Delta\,(K+iL)\,D_{+1} & \Delta\,\lambda\,(M+iN)\,D_{\,0} & \Delta\,\lambda\,(K+iL)\,D_{+1} \end{vmatrix} \,.$$

$$\tag{8.1.7c}$$

Equations (8.1.7b) can be solved with the use of two different substitutions:

$$I. \quad \sqrt{J+1}\,K(r) = f(r)\,, \quad \sqrt{J+1}\,L(r) = g(r)\,,$$

$$\sqrt{J}\,M(r) = f(r)\,, \quad \sqrt{J}\,N(r) = g(r)\,; \tag{8.1.8a}$$

$$II. \quad \sqrt{J}\,K(r) = f(r)\,, \quad \sqrt{J}\,L(r) = g(r)\,,$$

$$\sqrt{J+1}\,M(r) = -f(r)\,, \quad \sqrt{J+1}\,N(r) = -g(r)\,. \tag{8.1.8b}$$

In the case of Eq. (8.1.8a), we get

$$I. \quad (\frac{d}{dr} + \frac{J+1}{r})f + (\epsilon + m)g = 0\,,$$

$$(\frac{d}{dr} - \frac{J+1}{r})g - (\epsilon - m)f = 0\,, \tag{8.1.9a}$$

and similarly for Eq. (8.1.8b), we obtain

$$II. \quad (\frac{d}{dr} - \frac{J}{r})f + (\epsilon + m)g = 0\,,$$

$$(\frac{d}{dr} + \frac{J}{r})g - (\epsilon - m)f = 0\,. \tag{8.1.9b}$$

Remember that Eqs. $(8.1.9a, b)$ refers to the case of $\Delta = +1$ and $\lambda = +1$.

Thus, at fixed quantum numbers $(\epsilon, \ J, \ M, \ \Delta)$, there exists four types of solutions: due to the two numbers for $\lambda = \pm 1$ and due to the existence of two substitutions I and II (see Eq. $(8.1.8a, b)$). Solutions of the type I are described by

$$U^{I,\lambda}_{\epsilon J M \Delta}(x) = \frac{e^{i\epsilon t}}{r}$$

$$\times \begin{vmatrix} \lambda D_{-1}/\sqrt{J+1} & \lambda\, D_0/\sqrt{J} & D_{-1}/\sqrt{J+1} & D_0/\sqrt{J} \\ \lambda D_0/\sqrt{J} & \lambda D_{+1}/\sqrt{J+1} & D_0/\sqrt{J} & D_{+1}/\sqrt{J+1} \\ D_{-1}/\sqrt{J+1} & D_0/\sqrt{J} & \lambda D_{-1}/\sqrt{J+1} & \lambda D_0/\sqrt{J} \\ D_0/\sqrt{J} & D_{+1}/\sqrt{J+1} & \lambda D_0/\sqrt{J} & \lambda D_{+1}/\sqrt{J+1} \end{vmatrix} \begin{matrix} \leftarrow (f+ig) \\ \leftarrow (f-ig) \\ \leftarrow \Delta(f-ig) \\ \leftarrow \Delta(f+ig) \end{matrix} ;$$

$$(8.1.10a)$$

where all elements of each line should by multiplied by a function from the right; at $\Delta = +1, \lambda = +1$, the functions f and g obey Eqs. $(8.1.9a)$, whereas for the three remaining cases we should perform in Eq. $(8.1.9a)$ formal changes in accordance with the rules

$$(\Delta = -1, \ \lambda = +1) \qquad m \to -m \ ;$$

$$(\Delta = +1, \ \lambda = -1) \qquad m \to -m \ ;$$

$$(\Delta = -1, \ \lambda = -1) \qquad m \to +m \ . \qquad (8.1.10b)$$

Analogously, solutions for the second type; we then have

$$U^{II,\lambda}_{\epsilon J M \Delta}(x) = \frac{e^{-i\epsilon t}}{r}$$

$$\times \begin{vmatrix} -\lambda D_{-1}/\sqrt{J} & \lambda D_0/\sqrt{J+1} & D_{-1}/\sqrt{J} & -D_0/\sqrt{J+1} \\ \lambda D_0/\sqrt{J+1} & -\lambda; D_{+1}/\sqrt{J} & -D_0/\sqrt{J+1} & D_{+1}/\sqrt{J} \\ D_{-1}/\sqrt{J} & -D_0/\sqrt{J+1} & -\lambda D_{-1}/\sqrt{J} & \lambda D_0/\sqrt{J+1} \\ -D_0/\sqrt{J+1} & D_{+1}/\sqrt{J} & \lambda\, D_0/\sqrt{J+1} & -\lambda D_{+1}/\sqrt{J} \end{vmatrix} \begin{matrix} \leftarrow (f+ig) \\ \leftarrow (f-ig) \\ \leftarrow \Delta(f-ig) \\ \leftarrow \Delta(f+ig) \end{matrix} .$$

$$(8.1.10c)$$

At $\Delta = +1, \ \lambda = +1$, the functions f and g obey Eqs. $(8.1.9b)$; in the three remaining cases, one should use the rules in Eq. $(8.1.10b)$.

In the case of a minimal value for $J = 0$, this needs a separate consideration. Indeed, initial substitution for the wave function $U_{\epsilon 00}(x)$ turns to be independent on angular variables

$$U_{\epsilon 00}(t, r) = \frac{e^{-i\epsilon t}}{r} \begin{vmatrix} 0 & f_{12} & 0 & f_{14} \\ f_{21} & 0 & f_{23} & 0 \\ 0 & f_{32} & 0 & f_{34} \\ f_{41} & 0 & f_{43} & 0 \end{vmatrix} . \qquad (8.1.11)$$

The operator of spacial inversion is only a matrix operation, and permits separate functions of Eq. $(8.1.11)$ in two classes – eigenvalue equation $\hat{\Pi}\, U_{\epsilon 00} = \Pi\, U_{\epsilon 00}$ gives:

$\Pi = +1 \ (\Delta = +1)$,

$$f_{32} = +f_{23} \ , \quad f_{34} = +f_{21} \ , \quad f_{41} = +f_{14} \ , \quad f_{43} = +f_{12} \ ; \qquad (8.1.12a)$$

$\Pi = -1 (\Delta = -1)$,

$$f_{32} = -f_{23} \ , \quad f_{34} = -f_{21} \ , \quad f_{41} = -f_{14} \ , \quad f_{43} = -f_{12} \ . \qquad (8.1.12b)$$

Then, we allow for the relation $\Sigma_{\theta,\phi} \, U_{\epsilon 00} = 0$, to derive the radial system (for states with $J = 0$, functions $A, \ B, \ K, \ L$ in Eq. $(8.1.6a)$) vanish identically)

$$\epsilon \, M - \frac{dN}{dr} + \frac{N}{r} - m \, C = 0 \ ,$$

$$\epsilon \, N + \frac{dM}{dr} + \frac{M}{r} + m \, D = 0 \ ,$$

$$\epsilon \, C - \frac{dD}{dr} + \frac{D}{r} - m \, M = 0 \ ,$$

$$\epsilon \, D + \frac{dC}{dr} + \frac{C}{r} - m \, N = 0 \ . \qquad (8.1.12c)$$

To obtain equations when $\Delta = -1$, one should change m into $-m$.

The system in Eq. (8.1.12) can be simplified by two substitutions:

$C = +M \ , \ D = +N \ (\lambda = +1)$

$$(\frac{d}{dr} + \frac{1}{r})M + (\epsilon + m)N = 0 \ ,$$

$$(\frac{d}{dr} - \frac{1}{r})N - (\epsilon - m)M = 0 \ ; \qquad (8.1.13a)$$

$C = -M, \ D = -N \ (\lambda = -1)$

$$(\frac{d}{dr} + \frac{1}{r})M + (\epsilon - m)N = 0 \ ,$$

$$(\frac{d}{dr} - \frac{1}{r})N - (\epsilon + m)M = 0 \ . \qquad (8.1.13b)$$

Thus, at $J = 0$ and a fixed parity, there exist two different solutions (doubling by $\lambda = \pm 1$):

$$U_{\epsilon 00 \Delta}^{\lambda}(t, r) = \frac{e^{-i\epsilon t}}{r} \begin{vmatrix} 0 & \lambda & 0 & 1 \\ \lambda & 0 & 1 & 0 \\ 0 & 1 & 0 & \lambda \\ 1 & 0 & \lambda & 0 \end{vmatrix} \begin{matrix} \leftarrow & (M + iN) \\ \leftarrow & (M - iN) \\ \leftarrow & \Delta(M - iN) \\ \leftarrow & \Delta(M + iN) \end{matrix} \ . \qquad (8.1.14)$$

8.2   On Relations Between Boson and Fermion Solutions

Here, we will now relate our earlier spherical solutions of the boson type with spherical solutions of the ordinary Dirac equation [41]

$$
\Psi_{\epsilon j m \delta}(x) = \frac{e^{-i\epsilon t}}{r}
\begin{vmatrix}
D_{-1/2} & \leftarrow & (F + iG) \\
D_{+1/2} & \leftarrow & (F - iG) \\
D_{-1/2} & \leftarrow & \delta(F - iG) \\
D_{+1/2} & \leftarrow & \delta(F + iG)
\end{vmatrix} ,
\tag{8.2.1a}
$$

where $\delta = +1$ refers to the parity $P = (-1)^{j+1}$, and $\delta = -1$ refers to the parity $P = (-1)^j$. Radial equations for F and G at $\delta = +1$ are

$$
\left(\frac{d}{dr} + \frac{j + 1/2}{r} \right) F + (\epsilon + m)\, G = 0 \, ,
$$

$$
\left(\frac{d}{dr} - \frac{j + 1/2}{r} \right) G - (\epsilon - m)\, F = 0 \; ;
\tag{8.2.1b}
$$

Through changing the sign of m in Eq. (8.2.1b), we obtain the equation for states with a different parity (the case $\delta = -1$).

In order to connect explicitly this boson solution of the Dirac–Kähler field with spherical solutions of the (four) Dirac equations, one must perform over the matrix $U(x)$ a special transformation $U(x) \;\rightarrow\; V(x)$. In choosing this method, a new representation of the Dirac–Kähler equation is split into four separate Dirac-like equation. This presents the possibility to decompose four rows of the (4×4)-matrix $V(x)$, related to the Dirac–Kähler equation, in terms of solutions to four Dirac equation. The transformation we shall need has the form

$$
V(x) = (I \otimes S(x))\, U(x) \, , \qquad
S(x) =
\begin{vmatrix}
B(x) & 0 \\
0 & B(x)
\end{vmatrix} ,
$$

$$
B(x) =
\begin{vmatrix}
\cos \frac{\theta}{2}\, e^{-i\phi/2} & \sin \frac{\theta}{2}\, e^{-i\phi/2} \\
- \sin \frac{\theta}{2}\, e^{+i\phi/2} & \cos \frac{\theta}{2}\, e^{+i\phi/2}
\end{vmatrix} .
\tag{8.2.2a}
$$

Spherical bispinor connection Γ_α

$$
\Gamma_t = 0 \, , \;\; \Gamma_r = 0 \, , \;\; \Gamma_\theta = \sigma^{12} \, , \;\; \Gamma_\phi = \sin \theta \, \sigma^{32} + \cos \theta \, \sigma^{12}
$$

entering the Dirac–Kähler equation translated to $V(x)$-representation

$$
\{\, [\, i\gamma^\alpha(x)\, (\partial_\alpha + \Gamma_\alpha(x))\, -\, m\,]\, V(x)
$$

$$
+ i\,\gamma^\alpha(x)\, V(x)\, [\, S(x)\, \Gamma_\alpha(x)\, S^{-1}(x)\, +\, S(x)\, \partial_\alpha\, S^{-1}(x)\,]\,\} = 0
\tag{8.2.2b}
$$

this will make to zero for the following term

$$
S(x)\, \Gamma_\alpha(x)\, S^{-1}(x)\, +\, S(x)\, \partial_\alpha\, S^{-1}(x) = 0 \, ,
$$

and we obtain what is needed

$$[\, i\gamma^\alpha(x)\,(\partial_\alpha \,+\, \Gamma_\alpha(x))\, -\, m\,]\, V(x) = 0 \,. \qquad (8.2.2c)$$

Thus, the task consists in the following:

1) First, we should translate the earlier spherical solutions in U-form to corresponding V-form;

2) Second, we should expand four rows of the matrix V in terms of Dirac spherical waves.

When using Eq. (8.2.2a), the matrix $U_{\epsilon JM}$ in Eq. (8.1.2b) will assume the form (we have written V_{ij} by rows)

$$(V_{i1}) = \begin{vmatrix} f_{11}D_{-1}\cos\frac{\theta}{2}e^{-i\phi/2} \,-\, f_{12}D_0\sin\frac{\theta}{2}e^{-i\phi/2} \\ f_{21}D_0\cos\frac{\theta}{2}e^{-i\phi/2} \,-\, f_{22}D_{+1}\sin\frac{\theta}{2}e^{-i\phi/2} \\ f_{31}D_{-1}\cos\frac{\theta}{2}e^{-i\phi/2} \,-\, f_{32}D_0\sin\frac{\theta}{2}e^{-i\phi/2} \\ f_{41}D_0\cos\frac{\theta}{2}e^{-i\phi/2} \,-\, f_{42}D_{+1}\sin\frac{\theta}{2}e^{-i\phi/2} \end{vmatrix} \,,$$

$$(V_{i2}) = \begin{vmatrix} f_{11}D_{-1}\sin\frac{\theta}{2}e^{+i\phi/2} \,+\, f_{12}D_0\cos\frac{\theta}{2}e^{+i\phi/2} \\ f_{21}D_0\sin\frac{\theta}{2}e^{+i\phi/2} \,+\, f_{22}D_{+1}\cos\frac{\theta}{2}e^{+i\phi/2} \\ f_{31}D_{-1}\sin\frac{\theta}{2}e^{+i\phi/2} \,+\, f_{32}D_0\cos\frac{\theta}{2}e^{+i\phi/2} \\ f_{41}D_0\sin\frac{\theta}{2}e^{+i\phi/2} \,+\, f_{42}D_{+1}\cos\frac{\theta}{2}e^{+i\phi/2} \end{vmatrix} \,,$$

$$(V_{i3}) = \begin{vmatrix} f_{13}D_{-1}\cos\frac{\theta}{2}e^{-i\phi/2} \,-\, f_{14}D_0\sin\frac{\theta}{2}e^{-i\phi/2} \\ f_{23}D_0\cos\frac{\theta}{2}e^{-i\phi/2} \,-\, f_{24}D_{+1}\sin\frac{\theta}{2}e^{-i\phi/2} \\ f_{33}D_{-1}\cos\frac{\theta}{2}e^{-i\phi/2} \,-\, f_{34}D_0\sin\frac{\theta}{2}e^{-i\phi/2} \\ f_{43}D_0\cos\frac{\theta}{2}e^{-i\phi/2} \,-\, f_{44}D_{+1}\sin\frac{\theta}{2}e^{-i\phi/2} \end{vmatrix} \,,$$

$$(V_{i4}) = \begin{vmatrix} f_{13}D_{-1}\sin\frac{\theta}{2}e^{+i\phi/2} \,+\, f_{14}D_0\cos\frac{\theta}{2}e^{+i\phi/2} \\ f_{23}D_0\sin\frac{\theta}{2}e^{+i\phi/2} \,+\, f_{24}D_{+1}\cos\frac{\theta}{2}e^{+i\phi/2} \\ f_{33}D_{-1}\sin\frac{\theta}{2}e^{+i\phi/2} \,+\, f_{34}D_0\cos\frac{\theta}{2}e^{+i\phi/2} \\ f_{43}D_0\sin\frac{\theta}{2}e^{+i\phi/2} \,+\, f_{44}D_{+1}\cos\frac{\theta}{2}e^{+i\phi/2} \end{vmatrix} \,.$$

We will apply 8 formulas relating to D-functions for integer and half-integer j [429]; two of the formulas have been written here:

$$\cos\frac{\theta}{2}e^{i\phi/2}\, D^J_{-M,0} = \sqrt{\frac{J(J-M)}{2J+1}}\; D^{J-1/2}_{-M-1/2,-1/2}$$

$$+\, \sqrt{\frac{(J+1)(J+M+1)}{2J+1}}\; D^{J+1/2}_{-M-1/2,-1/2}\,,$$

$$\cos\frac{\theta}{2}e^{i\phi/2}\, D^J_{-M,+1} = \sqrt{\frac{(J+1)(J-M)}{2J+1}}\; D^{J-1/2}_{-M-1/2,+1/2}$$

$$+ \sqrt{\frac{J(J+M+1)}{2J+1}} \, D^{J+1/2}_{-M-1/2,+1/2} \, .$$

Utilizing these relations, for (V_{ij}) we obtain (the factor $e^{-i\epsilon t}/r$ is omitted)

$$V_{\epsilon jm} = V^{(J-1/2)}_{\epsilon JM} + V^{(J+1/2)}_{\epsilon JM} \, , \qquad (8.2.3)$$

where

$$(V^{(J-1/2)}_{i1}) = \sqrt{\frac{J+M}{2J+1}} \begin{vmatrix} (\sqrt{J+1}f_{11} - \sqrt{J}f_{12}) \, D^{J-1/2}_{-M+1/2,-1/2} \\ (\sqrt{J}f_{21} - \sqrt{J+1}f_{22}) \, D^{J-1/2}_{-M+1/2,+1/2} \\ (\sqrt{J+1}f_{31} - \sqrt{J}f_{32}) \, D^{J-1/2}_{-M+1/2,-1/2} \\ (\sqrt{J}f_{41} - \sqrt{J+1}f_{42}) \, D^{J-1/2}_{-M+1/2,+1/2} \end{vmatrix} \, ,$$

$$(V^{(J-1/2)}_{i2}) = \sqrt{\frac{J-M}{2J+1}} \begin{vmatrix} -(\sqrt{J+1}f_{11} - \sqrt{J}f_{12}) \, D^{J-1/2}_{-M-1/2,-1/2} \\ -(\sqrt{J}f_{11} - \sqrt{J+1}f_{22}) \, D^{J-1/2}_{-M-1/2,+1/2} \\ -(\sqrt{J+1}f_{21} - \sqrt{J}f_{22}) \, D^{J-1/2}_{-M-1/2,-1/2} \\ -(\sqrt{J}f_{41} - \sqrt{J+1}f_{42}) \, D^{J-1/2}_{-M-1/2,+1/2} \end{vmatrix} \, ,$$

$$(V^{(J-1/2)}_{i3}) = \sqrt{\frac{J+M}{2J+1}} \begin{vmatrix} (\sqrt{J+1}f_{13} - \sqrt{J}f_{14}) \, D^{J-1/2}_{-M+1/2,-1/2} \\ (\sqrt{J}f_{23} - \sqrt{J+1}f_{24}) \, D^{J-1/2}_{-M+1/2,+1/2} \\ (\sqrt{J+1}f_{33} - \sqrt{J}f_{34}) \, D^{J-1/2}_{-M+1/2,-1/2} \\ (\sqrt{J}f_{43} - \sqrt{J+1}f_{44}) \, D^{J-1/2}_{-M+1/2,+1/2} \end{vmatrix} \, ,$$

$$(V^{(J-1/2)}_{i4}) = \sqrt{\frac{J-M)}{2J+1}} \begin{vmatrix} -(\sqrt{J+1}f_{13} - \sqrt{J}f_{14})D^{J-1/2}_{-M-1/2,-1/2} \\ -(\sqrt{J}f_{23} - \sqrt{J+1}f_{24})D^{J-1/2}_{-M-1/2,+1/2} \\ -(\sqrt{J+1}f_{33} - \sqrt{J}f_{34})D^{J-1/2}_{-M-1/2,-1/2} \\ -(\sqrt{J}f_{44} - \sqrt{J+1}f_{44})D^{J-1/2}_{-M-1/2,+1/2} \end{vmatrix} \, ,$$

and

$$(V^{(J+1/2)}_{i2}) = \sqrt{\frac{J-M+1}{2J+1}} \begin{vmatrix} (\sqrt{J}f_{11} + \sqrt{J+1}f_{12}) \, D^{J+1/2}_{-M+1/2,-1/2} \\ (\sqrt{J+1}f_{21} + \sqrt{J}f_{22}) \, D^{J+1/2}_{-M+1/2,+1/2} \\ (\sqrt{J}f_{31} + \sqrt{J+1}f_{32}) \, D^{J+1/2}_{-M+1/2,-1/2} \\ (\sqrt{J+1}f_{41} + \sqrt{J}f_{42}) \, D^{J+1/2}_{-M+1/2,+1/2} \end{vmatrix} \, ,$$

$$(V^{(J+1/2)}_{i2}) = \sqrt{\frac{J+M+1}{2J+1}} \begin{vmatrix} (\sqrt{J}f_{11} + \sqrt{J+1}f_{12}) \, D^{J+1/2}_{-M-1/2,-1/2} \\ (\sqrt{J+1}f_{11} + \sqrt{J}f_{22}) \, D^{J+1/2}_{-M-1/2,+1/2} \\ (\sqrt{J}f_{21} + \sqrt{J+1}f_{22}) \, D^{J+1/2}_{-M-1/2,-1/2} \\ (\sqrt{J+1}f_{41} + \sqrt{J}f_{42}) \, D^{J+1/2}_{-M-1/2,+1/2} \end{vmatrix} \, ,$$

$$(V_{i3}^{(J+1/2)}) = \sqrt{\frac{J-M+1}{2J+1}} \begin{vmatrix} (\sqrt{J}f_{13} + \sqrt{J+1}f_{14}) \, D_{-M+1/2,-1/2}^{J+1/2} \\ (\sqrt{J+1}f_{23} + \sqrt{J}f_{24}) \, D_{-M+1/2,+1/2}^{J+1/2} \\ (\sqrt{J}f_{33} + \sqrt{J+1}f_{34}) \, D_{-M+1/2,-1/2}^{J+1/2} \\ (\sqrt{J+1}f_{43} + \sqrt{J}f_{44}) \, D_{-M+1/2,+1/2}^{J+1/2} \end{vmatrix},$$

$$(V_{i4}^{(J+1/2)}) = \sqrt{\frac{J+M+1}{2J+1}} \begin{vmatrix} (\sqrt{J}f_{13} + \sqrt{J+1}f_{14}) \, D_{-M-1/2,-1/2}^{J+1/2} \\ (\sqrt{J+1}f_{23} + \sqrt{J}f_{24}) \, D_{-M-1/2,+1/2}^{J+1/2} \\ (\sqrt{J}f_{33} + \sqrt{J+1}f_{34}) \, D_{-M-1/2,-1/2}^{J+1/2} \\ (\sqrt{J+1}f_{43} + \sqrt{J}f_{44}) \, D_{-M-1/2,+1/2}^{J+1/2} \end{vmatrix}.$$

Now, the functions for f_{ab} in Eq. (8.1.2b) should be taken in accordance with the next substitutions given

$$f_{ab} = \begin{vmatrix} \lambda \, (K+iL) & \lambda \, (M+iN) & (K+iL) & (M+iN) \\ \lambda \, (M-iN) & \lambda \, (K-iL) & (M-iN) & (K-iL) \\ \Delta \, (K-iL) & \Delta \, (M-iN) & \Delta\lambda \, (K-iL) & \Delta\lambda (M-iN) \\ \Delta \, (M+iN) & \Delta \, (K+iL) & \Delta\lambda (M+iN) & \Delta\lambda (K+iL) \end{vmatrix},$$

where $\lambda = \pm 1$ and $\Delta = \pm 1$. Thus, from Eq. (8.1.2b) it follows:

$$V_{\epsilon JM\Delta\lambda}^{(J-1/2)} = \begin{vmatrix} \lambda \, \Omega & -\lambda \, \Xi & \Omega & \Xi \\ -\lambda \, \Upsilon & \lambda \, Z & \Upsilon & Z \\ \Omega & \Xi & \lambda \, \Omega & -\lambda \, \Xi \\ \Upsilon & Z & -\lambda \, \Upsilon & \lambda \, Z \end{vmatrix} \begin{matrix} \leftarrow H^+ \\ \leftarrow H^- \\ \leftarrow \Delta H^+ \\ \leftarrow \Delta H^- \end{matrix}, \qquad (8.2.4a)$$

where symbols Ω, Ξ, Υ, Z stand for expressions

$$\Omega = \sqrt{\frac{J+M}{J(J+1)}} \, D_{-M+1/2,-1/2}^{J-1/2} , \quad Y = \sqrt{\frac{J+M}{J(J+1)}} \, D_{-M+1/2,+1/2}^{J-1/2} ,$$

$$\Xi = \sqrt{\frac{J-M}{J(J+1)}} \, D_{-M-1/2,-1/2}^{J-1/2} , \quad Z = \sqrt{\frac{J-M}{J(J+1)}} \, D_{-M-1/2,-1/2}^{J-1/2} ,$$

$$(8.2.4b)$$

and $H^\pm(r)$ represent

$$H^\pm(r) = \sqrt{\frac{J(J+1)}{2J+1}} \left[\sqrt{J+1} \, (K \pm iL) - \sqrt{J} \, (M \pm iN) \right] . \qquad (8.2.4c)$$

In the same manner for $V_{\epsilon JM}^{(J+1/2)}$, we get

$$V_{\epsilon JM\Delta\lambda}^{(J+1/2)} = \begin{vmatrix} \lambda \, \Omega & \lambda \, \Xi & \Omega & \Xi \\ \lambda \, \Upsilon & \lambda \, Z & \Upsilon & Z \\ \Omega & \Xi & \lambda \, \Omega & \lambda \, \Xi \\ \Upsilon & Z & \lambda \, \Upsilon & \lambda \, Z \end{vmatrix} \begin{matrix} \leftarrow H^+ \\ \leftarrow H^- \\ \leftarrow \Delta \, H^+ \\ \leftarrow \Delta \, H^- \end{matrix}, \qquad (8.2.5a)$$

where now for the symbols Ω, Ξ, Υ, Z; we will note the following expressions (compare with Eq. (8.2.4b)):

$$\Omega = \sqrt{\frac{J-M+1}{J(J+1)}}\, D^{J+1/2}_{-M+1/2,-1/2} \;, \quad \Xi = \sqrt{\frac{J+M+1}{J(J+1)}}\, D^{J+1/2}_{-M-1/2,-1/2} \;,$$

$$\Upsilon = \sqrt{\frac{J-M+1}{J(J+1)}}\, D^{J+1/2}_{-M+1/2,+1/2} \;, \quad Z = \sqrt{\frac{J+M+1}{J(J+1)}}\, D^{J+1/2}_{-M-1/2,-1/2} \;,$$

$$(8.2.5b)$$

and

$$H^{\pm}(r) = \sqrt{\frac{J}{2J+1}} \left[\sqrt{J+1}\,(K \pm iL) + \sqrt{J+1}\,(M \pm iN) \right] . \qquad (8.2.5c)$$

Now, we should take into account the substitution in Eq. (8.1.8a, b), which results in

$$U^{I}_{\epsilon JM\Delta\lambda} \;\rightarrow\; \{\, V^{(J+1/2)}_{\epsilon JM\Delta\lambda}\,, \; V^{(J-1/2)}_{\epsilon JM\Delta\lambda} = 0 \,\}\,,$$

$$U^{II}_{\epsilon JM\Delta\lambda} \;\rightarrow\; \{\, V^{(J+1/2)}_{\epsilon JM\Delta\lambda} = 0\,, \; V^{(J-1/2)}_{\epsilon JM\Delta\lambda} \,\}\,. \qquad (8.2.6)$$

Let us expand the rows of (4×4)-matrices $V^{(J\pm1/2)}_{\epsilon JM\Delta\lambda}(x)$ in terms of Dirac solutions $\Psi_{\epsilon jm\delta}(x)$. First, let us consider $V^{(J+1/2)}(x)$. We should take $(\Delta,\,\lambda)$ subsequently as

$$+1, +1; \;\; +1, -1; \;\; -1, +1; \;\; -1, -1$$

and we should also note that the expressions in Eqs. (8.2.4a) and (8.2.4b) have only signs at Ω, Ξ, Y, Z. Thus, we get

$$V^{(J+1/2)}_{JM,+1,+1} = \begin{vmatrix} + & + & + & + \\ + & + & + & + \\ + & + & + & + \\ + & + & + & + \end{vmatrix}, \quad V^{(J+1/2)}_{JM,+1,-1} = \begin{vmatrix} - & - & + & + \\ - & - & + & + \\ + & + & - & - \\ + & + & - & - \end{vmatrix},$$

$$V^{J+1/2}_{JM,-1,+1} = \begin{vmatrix} + & + & + & + \\ + & + & + & + \\ - & - & - & - \\ - & - & - & - \end{vmatrix}, \quad V^{J+1/2}_{JM,-1,-1} = \begin{vmatrix} - & - & + & + \\ - & - & + & + \\ - & - & + & + \\ - & - & + & + \end{vmatrix}.$$

$$(8.2.7)$$

The functions f and g that enter the matrix $V^{(J+1/2)}_{\epsilon JM,+1,+1}(x)$ obey Eqs. (8.1.9a) (see also Eq. (8.1.10b)). Comparing these equations for f and g with those for F and G in Eq. (8.2.1), and by noting the relevant Wigner functions, we conclude that the row from the matrix in Eq. (8.2.7) satisfy the ordinary Dirac equation. Therefore, at this we find the corresponding quantum number an explicit form of linear expansions.

$$V_{JM,+1,+1}^{(J+1/2)} = \left\{ \sqrt{\frac{J-M-1}{J(J+1)}} \left[\Psi_{J+1/2,M-1/2,+1}^{(1)} + \Psi_{J+1/2,M-1/2,+1}^{(3)} \right] \right.$$

$$\left. + \sqrt{\frac{J+M+1}{J(J+1)}} \left[\Psi_{J+1/2,M+1/2,+1}^{(2)} + \Psi_{J+1/2,M+1/2,+1}^{(4)} \right] \right\}. \qquad (8.2.8)$$

In this same manner, we will consider the three remaining cases.

Let us introduce the notation

$$j = (J+1/2) \,, \quad m = (M+1/2) \,, \quad m' = M - 1/2 \,,$$

$$\alpha = \sqrt{\frac{J-M+1}{J(J+1)}} \,, \quad \beta = \sqrt{\frac{J+M+1}{J(J+1)}} \,, \qquad (8.2.9a)$$

the expansions of the Dirac–Kähler boson solutions in terns of fermion Dirac solutions can be presented as follows:

$$V_{JM,+1,+1}^{(J+1/2)} = \alpha \left(\Psi_{jm',+1}^{(1)} + \Psi_{jm',+1}^{(3)} \right) + \beta \left(\Psi_{jm,+1}^{(2)} + \Psi_{jm,+1}^{(4)} \right),$$

$$V_{JM,-1,-1}^{(J+1/2)} = \alpha \left(-\Psi_{jm',+1}^{(1)} + \Psi_{jm',+1}^{(3)} \right) + \beta \left(-\Psi_{jm,+1}^{(2)} + \Psi_{jm,+1}^{(4)} \right),$$

$$V_{JM,+1,-1}^{(J+1/2)} = \alpha \left(-\Psi_{jm',-1}^{(1)} + \Psi_{jm',-1}^{(3)} \right) + \beta \left(-\Psi_{jm,-1}^{(2)} + \Psi_{jm,-1}^{(4)} \right),$$

$$V_{JM,-1,+1}^{(J+1/2)} = \alpha \left(\Psi_{jm',-1}^{(1)} + \Psi_{jm',-1}^{(3)} \right) + \beta \left(\Psi_{jm,-1}^{(2)} + \Psi_{jm,-1}^{(4)} \right).$$

$$(8.2.9b)$$

Not, let us consider solutions $V_{JM\Delta\lambda}^{(J-1/2)}$ with the structure (only the first row is given)

$$V_{JM,+1,+1}^{(J-1/2)} = \begin{vmatrix} (f+ig)\,\Omega & . & . & . \\ -(f-ig)\,\Upsilon & . & . & . \\ (f-ig)\,\Omega & . & . & . \\ -(f+ig)\,\Upsilon & . & . & . \end{vmatrix} ; \qquad (8.2.10a)$$

where f and g obey

$$(\frac{d}{dr} - \frac{J}{r})f + (\epsilon + m)g = 0 \,,$$

$$(\frac{d}{dr} + \frac{J}{r})g - (\epsilon - m)f = 0 \,. \qquad (8.2.10b)$$

Note that Eqs. (8.2.10b) do not coincide with relevant equation in the Dirac case at $\delta = \pm 1$; besides, in the rows of the matrix $V_{JM,+1,+1}^{(J-1/2)}$ (in Eq. (8.2.10b)) only the first row is written). We do not have the structure required to relate these with the Dirac solutions

$$\Psi_{jm,\delta=+1} \sim \begin{vmatrix} + \\ + \\ + \\ + \end{vmatrix} \,, \quad \Psi_{jm,\delta=-1} \sim \begin{vmatrix} + \\ + \\ - \\ - \end{vmatrix} .$$

However, both impediments can be removed by a simple change in notation $f = -G, g = +F$. Then

$$(f + ig) = i(F + iG) \qquad (f - ig) = -i(F - iG) \, ,$$

and Eqs. $(8.2.10b)$ is read as

$$(\frac{d}{dr} + \frac{J}{r})F + (\epsilon - m)G = 0 \, ,$$

$$(\frac{d}{dr} - \frac{J}{r})G - (\epsilon + m)F = 0 \, ; \qquad (8.2.11a)$$

and Eq. $(8.2.10a)$ reduces to

$$V_{JM,+1,+1}^{(J-1/2)} = \begin{vmatrix} i(F + iG)\,\Omega & . & . & . \\ i(F - iG)\,\Upsilon & . & . & . \\ -i(F - iG)\,\Omega & . & . & . \\ -i(F + iG)\,\Upsilon & . & . & . \end{vmatrix} . \qquad (8.2.11b)$$

Further, we can allow for Eq. (8.2.11a,b) and structure the matrix into four cases

$$V_{JM,+1,+1}^{(J-1/2)} = \begin{vmatrix} + & - & + & - \\ + & - & + & - \\ - & + & - & + \\ - & + & - & + \end{vmatrix}, \quad V_{JM,+1,-1}^{(J-1/2} = \begin{vmatrix} - & + & + & - \\ - & + & + & - \\ - & + & + & - \\ - & + & + & - \end{vmatrix},$$

$$V_{JM,-1,+1}^{(J-1/2} = \begin{vmatrix} + & - & + & - \\ + & - & + & - \\ + & - & + & - \\ + & - & + & - \end{vmatrix}, \quad V_{JM,-1,-1}^{(J-1/2)} = \begin{vmatrix} - & + & + & - \\ - & + & + & - \\ + & - & - & + \\ + & - & - & + \end{vmatrix},$$

$$(8.2.12a)$$

we arrive at the needed expansions for $V_{JM\Delta\lambda}^{(J-1/2)}$

$$V_{JM,+1,-1}^{(J-1/2)} = \rho \left(i\Psi_{jm',-1}^{(1)} + i\Psi_{jm',-1}^{(3)} \right) - \sigma \left(i\Psi_{jm,-1}^{(2)} + i\Psi_{jm,-1}^{(4)} \right),$$

$$V_{JM,-1,-1}^{(J-1/2)} = \rho \left(-i\Psi_{jm',-1}^{(1)} + i\Psi_{jm',-1}^{(3)} \right) + \sigma \left(i\Psi_{jm,-1}^{(2)} - i\Psi_{jm,-1}^{(4)} \right),$$

$$V_{JM,+1,-1}^{(J-1/2} = \rho \left(-i\Psi_{jm',+1}^{(1)} + i\Psi_{jm',+1}^{(3)} \right) + \sigma \left(i\Psi_{jm,+1}^{(2)} - i\Psi_{jm,+1}^{(4)} \right),$$

$$V_{JM,-1,+1}^{(J-1/2)} = \rho \left(i\Psi_{jm',+1}^{(1)} + i\Psi_{jm',+1}^{(3)} \right) - \sigma \left(i\Psi_{jm,+1}^{(2)} + i\Psi_{jm,+1}^{(4)} \right),$$

$$(8.2.12b)$$

where

$$j = (J - 1/2) \, , \quad \rho = \sqrt{\frac{J + M}{J(J+1)}} \, , \quad \sigma = \sqrt{\frac{J - M}{J(J+1)}} \, .$$

Expansions in the case of the minimal value for $J = 0$ will be much more simple

$$V_{00,+1,+1} = \Psi^{(1)}_{1/2,-1/2,+1} + \Psi^{(3)}_{1/2,-1/2,+1} + \Psi^{(2)}_{1/2,+1/2,+1} + \Psi^{(4)}_{1/2,+1/2,+1} \, ,$$

$$V_{00,-1,-1} = -\Psi^{(1)}_{1/2,-1/2,+1} + \Psi^{(3)}_{1/2,-1/2,+1} + -\Psi^{(2)}_{1/2,+1/2,+1} + \Psi^{(4)}_{1/2,+1/2,+1} \, ,$$

$$V_{00,+1,+1} = -\Psi^{(1)}_{1/2,-1/2,-1} + \Psi^{(3)}_{1/2,-1/2,-1} + -\Psi^{(2)}_{1/2,+1/2,-1} + \Psi^{(4)}_{1/2,+1/2,-1} \, ,$$

$$V_{00,+1,+1} = \Psi^{(1)}_{1/2,-1/2,-1} + \Psi^{(3)}_{1/2,-1/2,-1} + \Psi^{(2)}_{1/2,+1/2,-1} + \Psi^{(4)}_{1/2,+1/2,-1} \, .$$

$$(8.2.13)$$

8.3 Discussion

By performing a special transformation over (4×4)-matrix $U(x) \implies V(x)$ for a spherical boson solution of the Dirac–Kähler equation with simple linear expansions to four rows; a new representative that pertains to the Dirac–Kähler field $V(x)$, in terms of spherical fermion solutions for $\Psi_i(x)$ on four ordinary Dirac equations have been derived. However, this fact cannot be interpreted as the possibility to ignore the differences of being able to distinguish between the Dirac–Kähler field and the system with four Dirac fermions. The main formal argument is that the special transformation $(S(x) \otimes I)$ involved does not belong to the group of tetrad local gauge transformations for a Dirac–Kähler field, 2-rank bispinor under the Lorentz group. Therefore, the linear expansions between boson and fermion functions are not gauge invariant under the group of local tetrad rotations.

The formal possibility to produce such expansions exists only for the case of flat Minkowski space-time, and cannot be extended to any other space-time with curvature. For instance, let us specify the situation for spherical space with constant positive curvature. In spherical coordinates and tetrad

$$dS^2 = dt^2 - d\chi^2 - \sin^2 \chi (\, d\theta^2 + \sin^2 \theta d\phi^2 \,) \, ,$$

$$e^\alpha_{(0)} = (1, \, 0, \, 0, \, 0) \, , \quad e^\alpha_{(1)} = (0, \, 0, \, \sin^{-1} \chi, \, 0) \, ,$$

$$e^\alpha_{(2)} = (0, \, 0, \, 0, \, \sin^{-1} \chi \sin^{-1} \theta) \, , \quad e^\alpha_{(3)} = (0, 1, 0, 0) \, , \qquad (8.3.1)$$

the Dirac–Kähler equation takes the form

$$\left[i\gamma^0 \, \partial_t + i \left(\gamma^3 \, \partial_\chi + \frac{\gamma^1 J^{31} + \gamma^2 J^{32}}{\tan \chi} \right) + \frac{1}{\sin \chi} \Sigma_{\theta,\phi} - m \right] U(x) = 0 \, . \qquad (8.3.2)$$

Most of calculations performed are valid here; with small formal changes. In particular, instead of Eq. $(8.1.7b)$ we have $\Delta = +1$, $\lambda = +1$

$$\frac{dK}{d\chi} + \frac{a}{\sin \chi} M + (\epsilon + m)L = 0 \, ,$$

$$\frac{dL}{d\chi} - \frac{a}{\sin \chi} N - (\epsilon - m)K = 0 \, ,$$

$$\left(\frac{d}{d\chi} + \frac{1}{\tan\chi}\right)M + \frac{a}{\sin\chi}K + (\epsilon + m)N = 0 \,,$$

$$\left(\frac{d}{d\chi} - \frac{1}{\tan\chi}\right)N - \frac{a}{\sin\chi}L - (\epsilon - m)M = 0 \,. \tag{8.3.3}$$

However, using these substitutions in the form of Eq. $(8.1.8a, b)$, cannnot be imposed because they are not consistent with Eqs. (8.3.3). This means that no solutions for radial functions of (formally) fermion type exist in spherical space. The latter, is due to the fact that one cannot find any transformation like $I \otimes S(x)$, which would divide the Dirac–Kähler equation into four separate Dirac equations.

Chapter 9

Electromagnetic Spherical Solutions in Models S_3 and H_3

9.1 Matrix Maxwell Equation in Riemannian Space

A matrix Maxwell equation in the form of Riemann–Silberstein–Majorana–Oppenheimer [5, 128, 139, 140] is extended to the case of an arbitrary Riemannian space-time. In this instance, in accordance with the tetrad approach of Tetrode–Weyl–Fock–Ivanenko [133–135]

$$\alpha^\rho(x)\,[\,\partial_\rho + A_\rho(x)\,]\,\psi(x) = J(x)\,, \tag{9.1.1}$$

$$\alpha^\rho(x) = \alpha^c\,e^\rho_{(c)}(x)\,, \qquad A_\rho(x) = \frac{1}{2}j^{ab}\,e^\beta_{(a)}\,\nabla_\rho e_{(n)\beta}\,, \tag{9.1.2}$$

where $e^\rho_{(c)}(x)$ stands for the tetrad, j^{ab} stands for – generators of the complex vector representation of complex orthogonal group $SO(3.C)$. Equation (9.1.2) can be rewritten in terms of rotational Ricci coefficients

$$\alpha^c\,(\,e^\rho_{(c)}\partial_\rho + \frac{1}{2}j^{ab}\gamma_{abc}\,)\,\psi = J(x)\,, \tag{9.1.3}$$

where

$$j^{23} = s_1,\ j^{01} = is_1,\ j^{31} = s_2,\ j^{02} = is_2,\ j^{12} = s_3,\ j^{03} = is_3\,,$$

$$s_1 = \begin{vmatrix} 0 & 0 & 0 & 0 \\ 0 & 0 & 0 & 0 \\ 0 & 0 & 0 & -1 \\ 0 & 0 & 1 & 0 \end{vmatrix}, \qquad s_2 = \begin{vmatrix} 0 & 0 & 0 & 0 \\ 0 & 0 & 0 & 1 \\ 0 & 0 & 0 & 0 \\ 0 & -1 & 0 & 0 \end{vmatrix},$$

$$s_3 = \begin{vmatrix} 0 & 0 & 0 & 0 \\ 0 & 0 & -1 & 0 \\ 0 & 1 & 0 & 0 \\ 0 & 0 & 0 & 0 \end{vmatrix}, \qquad s_i = \begin{vmatrix} 0 & 0 \\ 0 & \tau_i \end{vmatrix}.$$

9.2   Spherical Coordinates and Tetrad in the Space S_3

Consider the spherical coordinates in the Riemann spherical space S_3

$$dS^2 = c^2 dt^2 - d\chi^2 - \sin^2\chi \left(d\theta^2 + \sin^2\theta d\phi^2\right), \quad x^\alpha = (ct, \chi, \theta, \phi),$$

$$g_{\alpha\beta} = \begin{vmatrix} 1 & 0 & 0 & 0 \\ 0 & -1 & 0 & 0 \\ 0 & 0 & -\sin^2\chi & 0 \\ 0 & 0 & 0 & -\sin^2\chi\sin^2\theta \end{vmatrix}, \tag{9.2.1}$$

let us use the following tetrad

$$e^\alpha_{(0)} = (1, 0, 0, 0), \qquad e^\alpha_{(3)} = (0, 1, 0, 0),$$

$$e^\alpha_{(1)} = \left(0, 0, \frac{1}{\sin\chi}, 0\right), \qquad e^\alpha_{(2)} = \left(1, 0, 0, \frac{1}{\sin\chi\sin\theta}\right). \tag{9.2.2}$$

Christoffel symbols are given by

$$\Gamma^\chi_{\phi\phi} = -\sin\chi\cos\chi\sin^2\theta, \qquad \Gamma^\chi_{\theta\theta} = -\sin\chi\cos\chi,$$

$$\Gamma^\theta_{\phi\phi} = -\sin\theta\cos\theta, \ \Gamma^\theta_{\theta\chi} = \frac{\cos\chi}{\sin\chi}, \qquad \Gamma^\phi_{\phi\theta} = \cot\theta, \ \Gamma^\phi_{\chi\phi} = \frac{\cos\chi}{\sin\chi}. \tag{9.2.3}$$

The Ricci rotation coefficients are $\gamma_{ab0} = 0$, $\gamma_{ab3} = 0$ and

$$\gamma_{ab1} = \begin{vmatrix} 0 & 0 & 0 & 0 \\ 0 & 0 & 0 & -\frac{1}{\tan\chi} \\ 0 & 0 & 0 & 0 \\ 0 & \frac{1}{\tan\chi} & 0 & 0 \end{vmatrix}, \ \gamma_{ab2} = \begin{vmatrix} 0 & 0 & 0 & 0 \\ 0 & 0 & \frac{1}{\tan\theta\sin\chi} & 0 \\ 0 & -\frac{1}{\tan\theta\sin\chi} & 0 & -\frac{1}{\tan\chi} \\ 0 & 0 & \frac{1}{\tan\chi} & 0 \end{vmatrix}.$$

$$\tag{9.2.4}$$

Correspondingly, for $\alpha^\alpha(x)$ and $A_\alpha(x)$ we get

$$\alpha^\alpha(x) = \left(\alpha^0, \ \alpha^3, \ \frac{\alpha^1}{\sin\chi}, \ \frac{\alpha^2}{\sin\chi\sin\theta}\right), \qquad A_0(x) = 0,$$

$$A_\chi(x) = 0, \ A_\theta(x) = j^{31}, \ A_\phi(x) = \sin\theta\, j^{32} + \cos\theta\, j^{12}. \tag{9.2.5}$$

Therefore, Eq. (9.1.3) takes the form

$$\left[-i\partial_0 + \alpha^3\partial_r + \frac{\alpha^1 j^{31} + \alpha^2 j^{32}}{\tan\chi} + \frac{1}{\sin\chi}\Sigma_{\theta,\phi} \right]\psi(x) = 0,$$

$$\Sigma_{\theta,\phi} = \alpha^1\,\partial_\theta + \alpha^2\,\frac{\partial_\phi + \cos\theta j^{12}}{\sin\theta}. \tag{9.2.6}$$

It is more convenient to have the matrix j^{12} as diagonal one. To this end, we needed to use a cyclic basis

$$\psi' = U_4\psi, \qquad U_4 = \begin{vmatrix} 1 & 0 \\ 0 & U \end{vmatrix}, \qquad U = \begin{vmatrix} -1/\sqrt{2} & i/\sqrt{2} & 0 \\ 0 & 0 & 1 \\ 1/\sqrt{2} & i/\sqrt{2} & 0 \end{vmatrix}. \tag{9.2.7}$$

It is a matter of simple calculation to find

$$
U\tau_1 U^{-1} = \frac{1}{\sqrt{2}} \begin{vmatrix} 0 & -i & 0 \\ -i & 0 & -i \\ 0 & -i & 0 \end{vmatrix} = \tau_1' , \qquad j'^{23} = s_1' = \begin{vmatrix} 0 & 0 \\ 0 & \tau_1' \end{vmatrix} ,
$$

$$
U\tau_2 U^{-1} = \frac{1}{\sqrt{2}} \begin{vmatrix} 0 & -1 & 0 \\ 1 & 0 & -1 \\ 0 & 1 & 0 \end{vmatrix} = \tau_2' , \qquad j'^{31} = s_2' = \begin{vmatrix} 0 & 0 \\ 0 & \tau_2' \end{vmatrix} ,
$$

$$
U\tau_3 U^{-1} = -i \begin{vmatrix} +1 & 0 & 0 \\ 0 & 0 & 0 \\ 0 & 0 & -1 \end{vmatrix} = \tau_3' , \qquad j'^{12} = s_3' = \begin{vmatrix} 0 & 0 \\ 0 & \tau_3' \end{vmatrix} ,
$$

and the matrices involved for the wave equation show this new basis as

$$
\alpha'^1 = \frac{1}{\sqrt{2}} \begin{vmatrix} 0 & -1 & 0 & 1 \\ 1 & 0 & -i & 0 \\ 0 & -i & 0 & -i \\ -1 & 0 & -i & 0 \end{vmatrix} , \quad
\alpha'^2 = \frac{1}{\sqrt{2}} \begin{vmatrix} 0 & -i & 0 & -i \\ -i & 0 & -1 & 0 \\ 0 & 1 & 0 & -1 \\ -i & 0 & 1 & 0 \end{vmatrix} ,
$$

$$
\alpha'^3 = \begin{vmatrix} 0 & 0 & 1 & 0 \\ 0 & -i & 0 & 0 \\ -1 & 0 & 0 & 0 \\ 0 & 0 & 0 & +i \end{vmatrix} .
$$

In a cyclic basis, Eq. (9.2.6) is read as

$$
\left[-i\frac{\partial}{\partial t} + \alpha'^3 \frac{\partial}{\partial r} + \frac{\alpha'^1 s_2' - \alpha'^2 s_1'}{\tan \chi} + \frac{1}{\sin \chi} \Sigma'_{\theta,\phi} \right] \psi'(x) = 0 ,
$$

$$
\Sigma'_{\theta,\phi} = \alpha'^1 \, \partial_\theta + \alpha'^2 \frac{\partial_\phi + \cos\theta \, s_3'}{\sin\theta} . \tag{9.2.8}
$$

9.3   Separation of Variables and Wigner Functions

In the spherical tetrad basis, expression for components of the total angular momentum re

$$
J_1 = l_1 + \frac{\cos\phi}{\sin\theta} s_3 , \qquad J_2 = l_2 + \frac{\sin\phi}{\sin\theta} s_3 , \qquad J_3 = l_3 .
$$

Many years ago such a tetrad basis was used by Schrödinger [430] and Pauli [431] when looking at the problem of single-valuedness of wave functions in quantum theory (then the case of spin $S = 1/2$ particle was specified).

Let us now construct electromagnetic spherical waves, then the field function should be taken in the form

$$
\psi = e^{-i\omega t} \begin{vmatrix} 0 \\ f_1(r)D_{-1} \\ f_2(r)D_0 \\ f_3(r)D_{+1} \end{vmatrix} , \tag{9.3.1}
$$

where the Wigner D-function are used $D_\sigma = D^j_{-m,\sigma}(\phi, \theta, 0)$, $\sigma = -1, 0, +1; j, m$ determine $\mathbf{J}^2$ and J_3 eigenvalues. When separating the variables we will need the recurrent relations for Wigner's functions [429]

$$\partial_\theta D_{-1} = \frac{1}{2}\left(a\, D_{-2} - \nu\, D_0 \right), \qquad \frac{m - \cos\theta}{\sin\theta} D_{-1} = \frac{1}{2}\left(a\, D_{-2} + \nu\, D_0 \right),$$

$$\partial_\theta D_0 = \frac{1}{2}\left(\nu\, D_{-1} - \nu\, D_{+1} \right), \qquad \frac{m}{\sin\theta} D_0 = \frac{1}{2}\left(\nu\, D_{-1} + \nu\, D_{+1} \right),$$

$$\partial_\theta D_{+1} = \frac{1}{2}\left(\nu\, D_0 - a\, D_{+2} \right), \qquad \frac{m + \cos\theta}{\sin\theta} D_{+1} = \frac{1}{2}\left(\nu\, D_0 + a\, D_{+2} \right),$$

$$(9.3.2)$$

where $\nu = \sqrt{j(j+1)}$, $a = \sqrt{(j-1)(j+2)}$. Let us find the action of the angular operator (the factor $e^{-i\omega t}$ will be omitted for brevity)

$$\Sigma_{\theta\phi}\psi = \frac{\nu}{\sqrt{2}} \begin{vmatrix} (f_1 + f_3)D_0 \\ -i\, f_2 D_{-1} \\ i\,(f_1 - f_3)D_0 \\ +i\, f_2 D_{+1} \end{vmatrix}. \qquad (9.3.3)$$

From this Maxwell equation, and by taking into account Eq. (9.3.3) and identities

$$-i\partial_0\psi = -\omega\, e^{-i\omega t} \begin{vmatrix} 0 \\ f_1(r)D_{-1} \\ f_2(r)D_0 \\ f_3(r)D_{+1} \end{vmatrix}, \qquad \alpha^3\partial_r\psi = e^{-i\omega t} \begin{vmatrix} f_2'D_0 \\ -i\, f_1'D_{-1} \\ 0 \\ +i\, f_3'D_{+1} \end{vmatrix},$$

$$\frac{\alpha^1 s_2 - \alpha^2 s_1}{\tan\chi}\, \psi = \frac{e^{-i\omega t}}{\tan\chi} \begin{vmatrix} 2f_2(r)D_0 \\ -i\, f_1(r)D_{-1} \\ 0 \\ +i\, f_3(r)D_{+1} \end{vmatrix}, \qquad (9.3.4)$$

we get the radial equations

$$f_2' + \frac{2}{\tan\chi}f_2 + \frac{1}{\sin\chi}\,\frac{\nu}{\sqrt{2}}\,(f_1 + f_3) = 0\,,$$

$$-\omega f_1 - i\, f_1' - \frac{i}{\tan\chi}f_1 - \frac{i}{\sin\chi}\,\frac{\nu}{\sqrt{2}}\, f_2 = 0\,,$$

$$-\omega f_2 + \frac{i}{\sin\chi}\,\frac{\nu}{\sqrt{2}}(f_1 - f_3) = 0\,,$$

$$-\omega f_3 + i\, f_3' + \frac{i}{\tan\chi}f_3 + \frac{i}{\sin\chi}\,\frac{\nu}{\sqrt{2}}\, f_2 = 0\,. \qquad (9.3.5)$$

After using substitutions

$$f_1 = \frac{1}{\sin\chi}\, F_1\,, \qquad f_2 = \frac{1}{\sin\chi}\, F_2\,, \qquad f_3 = \frac{1}{\sin\chi}\, F_3\,,$$

the systems reads simpler

$$(1) \quad \left(\frac{d}{d\chi} + \frac{1}{\tan\chi}\right)\omega F_2 + \frac{\omega\,\nu}{\sqrt{2}\sin\chi}\,(F_1 + F_3) = 0 \,,$$

$$(2) \quad -\omega^2 F_1 - i\omega\,F_1' - \frac{i\,\nu}{\sqrt{2}\sin\chi}\,\omega F_2 = 0 \,,$$

$$(3) \quad \omega F_2 = \frac{i\nu}{\sqrt{2}\sin\chi}(F_1 - F_3) \,,$$

$$(4) \quad -\omega^2 F_3 + i\,\omega F_3' + \frac{i\nu}{\sqrt{2}\sin\chi}\,\omega F_2 = 0 \,. \qquad (9.3.6)$$

By combining Eqs. (2) and (4), we get:

$(4) + (2)$,

$$-\omega(F_1 + F_3) - i(F_1' - F_3') = 0 \,;$$

$(4) - (2)$,

$$\omega^2(F_1 - F_3) + i\omega(F_1' + F_3') + \frac{2i\nu}{\sqrt{2}\sin\chi}\,\frac{i\nu}{\sqrt{2}\sin\chi}(F_1 - F_3) = 0 \,. \qquad (9.3.7)$$

Here, we arrive at an identity

$$\left(\frac{d}{d\chi} + \frac{1}{\tan\chi}\right)\frac{i\nu}{\sqrt{2}\sin\chi}(F_1 - F_3) + \frac{\omega\,\nu}{\sqrt{2}\sin\chi}\,(-i)(F_1' - F_3') = 0 \,. \qquad (9.3.8)$$

Therefore, the Maxwell equations reduce to only three independent equations

$$\omega F_2 = \frac{i\nu}{\sqrt{2}\sin\chi}(F_1 - F_3) \,, \qquad -\omega(F_1 + F_3) - i(F_1' - F_3') = 0 \,,$$

$$\omega^2(F_1 - F_3) + i\omega(F_1' + F_3') + \frac{2i\nu}{\sqrt{2}\sin\chi}\,\frac{i\nu}{\sqrt{2}\sin\chi}(F_1 - F_3) = 0 \,. \qquad (9.3.9)$$

Let us introduce new variables

$$F = \frac{F_1 + F_3}{\sqrt{2}} \,, \qquad G = \frac{F_1 - F_3}{\sqrt{2}} \,,$$

then Eq. (9.3.9) will read as

$$F_2 = \frac{i\nu}{\omega\,\sin\chi}\,G \,, \qquad F = -\frac{i}{\omega}\frac{d}{d\chi}G \,,$$

$$\left(\frac{d^2}{d\chi^2} + \omega^2 - \frac{\nu^2}{\sin^2\chi}\right)G = 0 \,. \qquad (9.3.10)$$

9.4   Solutions of the Radial Equation in S_3

Let us now solve for Eq. (9.3.10). To this end, we shall need to introduce a new variable

$$z = 1 - e^{-2i\chi} , \qquad z = 2\sin\chi \; e^{i(-\chi+\pi/2)} ; \tag{9.4.1}$$

z runs along closed path in the complex plane. We then allow for identities

$$\frac{d}{d\chi} = 2i(1-z)\frac{d}{dz} , \qquad \frac{\cos\chi}{\sin\chi} = i\frac{2-z}{z} , \qquad \frac{1}{\sin^2\chi} = -\frac{4(1-z)}{z^2} ,$$

Equation (9.3.10) reduces to

$$4(1-z)^2\frac{d^2G}{dz^2} - 4(1-z)\frac{dG}{dz} - \omega^2 G - \frac{4(1-z)\nu^2}{z^2} G = 0 . \tag{9.4.2}$$

Let us use the substitution $G = z^a(1-z)^b g(z)$ from Eq. (9.4.2) to arrive at

$$z(1-z)\frac{d^2g}{dz^2} + [2a - (2a+2b+1)z]\frac{dg}{dz}$$

$$+ \left[\frac{\omega^2}{4} - (a+b)^2 + (a(a-1) - \nu^2)\frac{1}{z} + (b^2 - \frac{\omega^2}{4})\frac{1}{1-z}\right] g = 0 . \tag{9.4.3}$$

Requiring $a = j+1, -j$, $b = \pm(\omega/2)$ for g we obtain a simpler equation

$$z(1-z)\frac{d^2g}{dz^2} + [2a - (2a+2b+1)z]\frac{dg}{dz} - \left[(a+b)^2 - \frac{\omega^2}{4}\right] g = 0 , \tag{9.4.4}$$

which is known as the hypergeometric type

$$z(1-z)\,F'' + [\gamma - (\alpha+\beta+1)z]\,F' - \alpha\beta\,F = 0 ,$$

$$\gamma = 2a , \qquad \alpha = a+b-\frac{\omega}{2} , \qquad \beta = a+b+\frac{\omega}{2} . \tag{9.4.5}$$

The function G is given by

$$G = z^a(1-z)^b g(z) = \left[\, 2i\sin\chi e^{-i\chi} \,\right]^a \left[\, 1 - 2i\sin\chi e^{-i\chi} \,\right]^b g(z) ; \tag{9.4.6}$$

it is finite at the points $\chi = 0$ and $\chi = \pi$, only if a is positive

$$a = j+1 . \tag{9.4.7}$$

Also, we must require $b = -\omega/2$, when hypergeometric series can be reduced to a polynomial (we take $\omega > 0$):

$$\alpha = j+1-\omega = -n = \{0, -1, -2, ...\} \qquad \Longrightarrow \qquad \omega = n+1+j . \tag{9.4.8}$$

Thus, physical solutions of the Maxwell equations in the Riemann spherical space S_3 are given by relations

$$G = \left[\, 2i\sin\chi e^{-i\chi} \,\right]^a \left[\, 1 - 2i\sin\chi e^{-i\chi} \,\right]^b g(z) ,$$

$$g(z) = F(-n, , j+1, 2j+2; 2i \sin \chi e^{-i\chi}) \,, \qquad (9.4.9a)$$

where

$$\omega = n + 1 + j \,, \qquad j = 0, 1, 2, ..., \qquad n = 0, 1, 2, ...; \qquad (9.4.9b)$$

or in usual units

$$\omega = \frac{c}{\rho} (n + 1 + j) \,; \qquad (9.4.9c)$$

ρ stands for the curvature radius of the space, c is velocity of the light.

9.5   Spherical Tetrad in Lobachevsky Space H_3

Let us consider Maxwell equation in spherical coordinates of the Lobachevsky model

$$dS^2 = c^2 dt^2 - d\chi^2 - \sinh^2 \chi \, (d\theta^2 + \sin^2 \theta d\phi^2) \,, \qquad x^\alpha = (ct, \chi, \theta, \phi) \,,$$

$$g_{\alpha\beta} = \begin{vmatrix} 1 & 0 & 0 & 0 \\ 0 & -1 & 0 & 0 \\ 0 & 0 & -\sinh^2 \chi & 0 \\ 0 & 0 & 0 & -\sinh^2 \chi \sin^2 \theta \end{vmatrix} \,, \qquad (9.5.1)$$

in the following tetrad

$$e_{(0)}^\alpha = (1, 0, 0, 0) \,, \qquad e_{(3)}^\alpha = (0, 1, 0, 0) \,,$$

$$e_{(1)}^\alpha = (0, 0, \frac{1}{\sinh \chi}, 0) \,, \qquad e_{(2)}^\alpha = (1, 0, 0, \frac{1}{\sinh \chi \sin \theta}) \,. \qquad (9.5.2)$$

The Christoffel symbols are

$$\Gamma_{\phi\phi}^\chi = -\sinh \chi \cosh \chi \sin^2 \theta \,, \qquad \Gamma_{\theta\theta}^\chi = -\sinh \chi \cosh \chi \,,$$

$$\Gamma_{\phi\phi}^\theta = -\sin \theta \cos \theta \,, \quad \Gamma_{\theta\chi}^\theta = \frac{\cosh \chi}{\sinh \chi} \,, \qquad \Gamma_{\phi\theta}^\phi = \cot \theta \,, \quad \Gamma_{\chi\phi}^\phi = \frac{\cosh \chi}{\sinh \chi} \,, \qquad (9.5.3)$$

and the Ricci rotation coefficients are $\gamma_{ab0} = 0$, $\gamma_{ab3} = 0$ and

$$\gamma_{ab1} = \begin{vmatrix} 0 & 0 & 0 & 0 \\ 0 & 0 & 0 & -\frac{1}{\tanh \chi} \\ 0 & 0 & 0 & 0 \\ 0 & \frac{1}{\tanh \chi} & 0 & 0 \end{vmatrix} , \gamma_{ab2} = \begin{vmatrix} 0 & 0 & 0 & 0 \\ 0 & 0 & +\frac{1}{\tan \theta \sinh \chi} & 0 \\ 0 & -\frac{1}{\tan \theta \sinh \chi} & 0 & -\frac{1}{\tanh \chi} \\ 0 & 0 & \frac{1}{\tanh \chi} & 0 \end{vmatrix} .$$

$$(9.5.4)$$

For $\alpha^\alpha(x)$ and $A_\alpha(x)$, we get

$$\alpha^\alpha(x) = (\alpha^0, \alpha^3, \frac{\alpha^1}{\sinh \chi}, \frac{\alpha^2}{\sinh \chi \sin \theta}) \,, \qquad A_0(x) = 0 \,,$$

$$A_\chi(x) = 0 \,, \qquad A_\theta(x) = j^{31} \,, \qquad A_\phi(x) = \sin \theta \, j^{32} + \cos \theta \, j^{12} \,. \qquad (9.5.5)$$

Therefore, Maxwell equation will read as (here, the cyclic basis will be used)

$$[-i\frac{\partial}{\partial t} + \alpha^3 \frac{\partial}{\partial r} + \frac{\alpha^1 s_2 - \alpha^2 s_1}{\tanh \chi} + \frac{1}{\sinh \chi} \Sigma_{\theta,\phi}] \, \psi(x) = 0 \,,$$

$$\Sigma_{\theta,\phi} = \alpha^1 \, \partial_\theta + \alpha^2 \frac{\partial_\phi + \cos \theta \, s_3}{\sin \theta} \,. \qquad (9.5.6)$$

9.6   Separation of the Variables

We start with spherical substitution

$$\psi = e^{-i\omega t} \begin{vmatrix} 0 \\ f_1(r)D_{-1} \\ f_2(r)D_0 \\ f_3(r)D_{+1} \end{vmatrix}. \tag{9.6.1}$$

Further calculations would be completely identical. These were the same procedures that we used in our previous case. Therefore, we can go just to the final results.

$$f_2' + \frac{2}{\tanh \chi} f_2 + \frac{1}{\sinh \chi} \frac{\nu}{\sqrt{2}} (f_1 + f_3) = 0 \,,$$

$$-\omega f_1 - i\, f_1' - \frac{i}{\tanh \chi} f_1 - \frac{i}{\sinh \chi} \frac{\nu}{\sqrt{2}} f_2 = 0 \,,$$

$$-\omega f_2 + \frac{i}{\sinh \chi} \frac{\nu}{\sqrt{2}} (f_1 - f_3) = 0 \,,$$

$$-\omega f_3 + i\, f_3' + \frac{i}{\tanh \chi} f_3 + \frac{i}{\sinh \chi} \frac{\nu}{\sqrt{2}} f_2 = 0 \,. \tag{9.6.2}$$

The system becomes simpler in variables F_i

$$f_1 = \frac{1}{\sinh \chi} F_1 \,, \qquad f_2 = \frac{1}{\sinh \chi} F_2 \,, \qquad f_3 = \frac{1}{\sinh \chi} F_3 \,,$$

so we get

$$(1) \qquad (\frac{d}{d\chi} + \frac{1}{\tanh \chi})\, \omega F_2 + \frac{\omega\, \nu}{\sqrt{2}\sinh \chi} (F_1 + F_3) = 0 \,,$$

$$(2) \qquad - \omega^2 F_1 - i\omega\, F_1' - \frac{i\, \nu}{\sqrt{2}\sinh \chi}\, \omega F_2 = 0 \,,$$

$$(3) \qquad \omega F_2 = \frac{i\nu}{\sqrt{2}\sinh \chi}(F_1 - F_3) \,,$$

$$(4) \qquad - \omega^2 F_3 + i\, \omega F_3' + \frac{i\nu}{\sqrt{2}\sinh \chi}\, \omega F_2 = 0 \,. \tag{9.6.3}$$

By combining (2) and (4), we obtain

$(4) + (2) \,,$

$$-\omega(F_1 + F_3) - i(F_1' - F_3') = 0 \,; \tag{9.6.4a}$$

$(4) - (2) \,,$

$$\omega^2(F_1 - F_3) + i\omega(F_1' + F_3') + \frac{2i\nu}{\sqrt{2}\sinh \chi} \frac{i\nu}{\sqrt{2}\sinh \chi}(F_1 - F_3) = 0 \,; \tag{9.6.4b}$$

Equation (1) in Eq. (9.6.3) reduces to identity $0 \equiv 0$ (when allowing for (3) and Eq. (9.6.4a)). So we have only three independent equations

$$\omega F_2 = \frac{i\nu}{\sqrt{2}\sinh \chi}(F_1 - F_3) \,, \qquad -\omega(F_1 + F_3) - i(F_1' - F_3') = 0 \,,$$

$$\omega^2(F_1 - F_3) + i\omega(F_1' + F_3') + \frac{2i\nu}{\sqrt{2}\sinh \chi}\frac{i\nu}{\sqrt{2}\sinh \chi}(F_1 - F_3) = 0 \,. \qquad (9.6.5)$$

In new field variables

$$F = \frac{F_1 + F_3}{\sqrt{2}} \,, \qquad G = \frac{F_1 - F_3}{\sqrt{2}} \,,$$

Equations (9.6.5) will read as

$$F_2 = \frac{i\nu}{\omega \sinh \chi} G \,, \qquad F = -\frac{i}{\omega}\frac{d}{d\chi}G \,,$$

$$\left(\frac{d^2}{d\chi^2} + \omega^2 - \frac{\nu^2}{\sinh^2 \chi} \right) G = 0 \,. \qquad (9.6.6)$$

9.7   Solutions of the Radial Equations in H_3

In Eq. (9.6.6), we will need to introduce new variable

$$z = 1 - e^{-2\chi} \,, \qquad z = 2\sinh \chi \, e^{-\chi} \,; \qquad (9.7.1)$$

Equation (9.6.6) reduces to

$$4(1-z)^2\frac{d^2G}{dz^2} - 4(1-z)\frac{dG}{dz} + \omega^2 G - \frac{4(1-z)\nu^2}{z^2} G = 0 \,. \qquad (9.7.2)$$

Then with the use of substitution $G = z^a(1-z)^b g(z)$, we arrive at

$$z(1-z)\frac{d^2g}{dz^2} + [2a - (2a + 2b + 1)z]\frac{dg}{dz}$$

$$+ \left[-\frac{\omega^2}{4} - (a+b)^2 + (a(a-1) - \nu^2)\frac{1}{z} + (b^2 + \frac{\omega^2}{4})\frac{1}{1-z} \right] g = 0 \,. \qquad (9.7.3)$$

By the additional restrictions $a = j + 1, -j$, $b = \pm\frac{i\omega}{2}$, for g we obtain equation

$$z(1-z)\frac{d^2g}{dz^2} + [2a - (2a + 2b + 1)z]\frac{dg}{dz} - \left[(a+b)^2 + \frac{\omega^2}{4} \right] g = 0 \,, \qquad (9.7.4)$$

of hypergeometric type with parameters

$$\gamma = 2a \,, \qquad \alpha = a + b - \frac{i\omega}{2} \,, \qquad \beta = a + b + \frac{i\omega}{2} \,. \qquad (9.7.5)$$

Therefore, the function G is given by

$$G = z^a(1-z)^b g(z) = \left[\, 2\sinh \chi e^{-\chi}\,\right]^a \left[\, 1 - 2\sinh \chi e^{-\chi}\,\right]^b g(z) \,. \qquad (9.7.6)$$

9.8   Conclusions

Complex formalism of Riemann–Silberstein–Majorana–Oppenheimer in Maxwell electrodynamics is extended to the case of arbitrary pseudo-Riemannian space-time; where, in
accordance with the tetrad recipe has been applied to solve Maxwell equations on the background of the simplest static cosmological models for spaces with constant curvature in
Riemann and Lobachevsky parameterized by spherical coordinates.

Separation of variables is realized in the basis of a Schrödinger–Pauli type. The description of angular dependence in electromagnetic complex 3-vectors is given in terms
of Wigner D-functions. In the case of a compact Riemann model, a discrete frequency
spectrum for electromagnetic modes that depend on the curvature radius of space. Our
solution determined that three discrete parameters were found. In the case of hyperbolic
Lobachevsky model, there does not exist discrete spectrum for frequencies in electromagnetic modes.

Chapter 10

10-Dimensional Spherical Solutions in Riemann Space S_3

10.1 Separation of Variables

The task of this chapter is to obtain in explicit form spherical waves solutions to Maxwell equations in space of positive curvature, spherical Riemann S_3 model, when it is parameterized by spherical coordinates. We will use the known Duffin–Kemmer approach to Maxwell theory that extends to arbitrary curved space-times; in accordance with general tetrad formalism – see chapter **4**.

The Duffin–Kemmer equation for massless vector field is

$$[\, i\, \beta^\alpha(x)\, (\partial_\alpha \; + \; B_\alpha(x)) \; - \; P_6 \,]\, \Phi(x) = 0 \,,$$

$$\beta^\alpha(x) = \beta^a e^\alpha_{(a)}(x) \,, \qquad B_\alpha(x) = \frac{1}{2}\, J^{ab} e^\beta_{(a)} \nabla_\alpha(e_{(b)\beta}) \,,$$

$$J^{ab} = \beta^a \beta^b - \beta^b \beta^a \,, \qquad P_6 = \begin{vmatrix} 0 & 0 & 0 & 0 \\ 0 & 0 & 0 & 0 \\ 0 & 0 & I & 0 \\ 0 & 0 & 0 & I \end{vmatrix} \,, \tag{10.1.1}$$

or with the use of notation for Ricci rotation coefficients

$$\{i\, \beta^c \, (\, e^\alpha_{(c)} \partial_\alpha + \frac{1}{2} J^{ab} \gamma_{abc} \,) \; - \; P_6 \} \Psi = 0 \,,$$

$$\gamma_{bac} = -\gamma_{abc} = -e_{(b)\beta;\alpha}\, e^\beta_{(a)} e^\alpha_{(c)} \,. \tag{10.1.2}$$

In spherical coordinates for Riemann space S_3

$$dS^2 = c^2 dt^2 - d\chi^2 - \sin^2\chi\, (d\theta^2 + \sin^2\theta d\phi^2)$$

let us take the following tetrad

$$e^\alpha_{(0)} = (1, 0, 0, 0) \,, \qquad e^\alpha_{(3)} = (0, 1, 0, 0) \,,$$

$$e^{\alpha}_{(1)} = (0,0,\frac{1}{\sin\chi},0)\,, \qquad e^{\alpha}_{(2)} = (1,0,0,\frac{1}{\sin\chi\sin\theta})\,.$$

Christoffel symbols are

$$\Gamma^{\chi}_{\phi\phi} = -\sin\chi\cos\chi\sin^2\theta\,, \qquad \Gamma^{\chi}_{\theta\theta} = -\sin\chi\cos\chi\,,$$

$$\Gamma^{\theta}_{\phi\phi} = -\sin\theta\cos\theta\,, \ \Gamma^{\theta}_{\theta\chi} = \frac{\cos\chi}{\sin\chi}\,, \ \Gamma^{\phi}_{\phi\theta} = \cot\theta\,, \ \Gamma^{\phi}_{\chi\phi} = \frac{\cos\chi}{\sin\chi}\,.$$

Ricci rotation coefficients are $\gamma_{ab0} = 0$, $\gamma_{ab3} = 0$ and

$$\gamma_{ab1} = \begin{vmatrix} 0 & 0 & 0 & 0 \\ 0 & 0 & 0 & -\frac{1}{\tan\chi} \\ 0 & 0 & 0 & 0 \\ 0 & +\frac{1}{\tan\chi} & 0 & 0 \end{vmatrix}, \ \gamma_{ab2} = \begin{vmatrix} 0 & 0 & 0 & 0 \\ 0 & 0 & +\frac{1}{\tan\theta\sin\chi} & 0 \\ 0 & -\frac{1}{\tan\theta\sin\chi} & 0 & -\frac{1}{\tan\chi} \\ 0 & 0 & +\frac{1}{\tan\chi} & 0 \end{vmatrix}.$$

Therefore, Eq. (10.1.2) takes the form

$$\left[i\,\beta^0\frac{\partial}{\partial t} + i\,\beta^3\frac{\partial}{\partial\chi} + \frac{i}{\sin\chi}(\beta^1 J^{31} + \beta^2 J^{32}) + \frac{1}{\sin\chi}\Sigma^{\kappa}_{\theta,\phi} - P_6\right]\Phi(x) = 0\,,$$

where

$$\Sigma^{\kappa}_{\theta,\phi} = i\,\beta^1\partial_\theta + \beta^2\frac{i\partial + iJ^{12}\cos\theta}{\sin\theta}\,. \tag{10.1.3}$$

Here, we will use Duffin–Kemmer matrices in cyclic basis

$$\beta^0 = \begin{vmatrix} 0 & 0 & 0 & 0 & 0 & 0 & 0 & 0 & 0 & 0 \\ 0 & 0 & 0 & 0 & +i & 0 & 0 & 0 & 0 & 0 \\ 0 & 0 & 0 & 0 & 0 & +i & 0 & 0 & 0 & 0 \\ 0 & 0 & 0 & 0 & 0 & 0 & +i & 0 & 0 & 0 \\ 0 & -i & 0 & 0 & 0 & 0 & 0 & 0 & 0 & 0 \\ 0 & 0 & -i & 0 & 0 & 0 & 0 & 0 & 0 & 0 \\ i & 0 & 0 & -i & 0 & 0 & 0 & 0 & 0 & 0 \\ 0 & 0 & 0 & 0 & 0 & 0 & 0 & 0 & 0 & 0 \\ 0 & 0 & 0 & 0 & 0 & 0 & 0 & 0 & 0 & 0 \\ 0 & 0 & 0 & 0 & 0 & 0 & 0 & 0 & 0 & 0 \end{vmatrix},$$

$$\beta^3 = \begin{vmatrix} 0 & 0 & 0 & 0 & 0 & i & 0 & 0 & 0 & 0 \\ 0 & 0 & 0 & 0 & 0 & 0 & 0 & +1 & 0 & 0 \\ 0 & 0 & 0 & 0 & 0 & 0 & 0 & 0 & 0 & 0 \\ 0 & 0 & 0 & 0 & 0 & 0 & 0 & 0 & 0 & -1 \\ 0 & 0 & 0 & 0 & 0 & 0 & 0 & 0 & 0 & 0 \\ 0 & 0 & 0 & 0 & 0 & 0 & 0 & 0 & 0 & 0 \\ i & 0 & 0 & 0 & 0 & 0 & 0 & 0 & 0 & 0 \\ 0 & -1 & 0 & 0 & 0 & 0 & 0 & 0 & 0 & 0 \\ 0 & 0 & 0 & 0 & 0 & 0 & 0 & 0 & 0 & 0 \\ 0 & 0 & +1 & 0 & 0 & i & 0 & 0 & 0 & 0 \end{vmatrix},$$

$$\beta^1 = \frac{1}{\sqrt{2}} \begin{vmatrix} 0 & 0 & 0 & 0 & -i & 0 & +i & 0 & 0 & 0 \\ 0 & 0 & 0 & 0 & 0 & 0 & 0 & 0 & +1 & 0 \\ 0 & 0 & 0 & 0 & 0 & 0 & 0 & +1 & 0 & +1 \\ 0 & 0 & 0 & 0 & 0 & 0 & 0 & 0 & +1 & 0 \\ -i & 0 & 0 & 0 & 0 & 0 & 0 & 0 & 0 & 0 \\ 0 & 0 & 0 & 0 & 0 & 0 & 0 & 0 & 0 & 0 \\ +i & 0 & 0 & 0 & 0 & 0 & 0 & 0 & 0 & 0 \\ 0 & 0 & -1 & 0 & 0 & 0 & 0 & 0 & 0 & 0 \\ 0 & -1 & 0 & -1 & 0 & 0 & 0 & 0 & 0 & 0 \\ 0 & 0 & -1 & 0 & 0 & 0 & 0 & 0 & 0 & 0 \end{vmatrix} ,$$

$$\beta^2 = \frac{1}{\sqrt{2}} \begin{vmatrix} 0 & 0 & 0 & 0 & 1 & 0 & 1 & 0 & 0 & 0 \\ 0 & 0 & 0 & 0 & 0 & 0 & 0 & 0 & -i & 0 \\ 0 & 0 & 0 & 0 & 0 & 0 & 0 & +i & 0 & -i \\ 0 & 0 & 0 & 0 & 0 & 0 & 0 & 0 & +i & 0 \\ -1 & 0 & 0 & 0 & 0 & 0 & 0 & 0 & 0 & 0 \\ 0 & 0 & 0 & 0 & 0 & 0 & 0 & 0 & 0 & 0 \\ -1 & 0 & 0 & 0 & 0 & 0 & 0 & 0 & 0 & 0 \\ 0 & 0 & +i & 0 & 0 & 0 & 0 & 0 & 0 & 0 \\ 0 & -i & 0 & +i & 0 & 0 & 0 & 0 & 0 & 0 \\ 0 & 0 & -i & 0 & 0 & 0 & 0 & 0 & 0 & 0 \end{vmatrix} ,$$

when the matrix iJ^{12} has a diagonal structure

$$iJ^{12} = \begin{vmatrix} 0 & 0 & 0 & 0 \\ 0 & t_3 & 0 & 0 \\ 0 & 0 & t_3 & 0 \\ 0 & 0 & 0 & t_3 \end{vmatrix} , \qquad t_3 = \begin{vmatrix} +1 & 0 & 0 \\ 0 & 0 & 0 \\ 0 & 0 & -1 \end{vmatrix} .$$

The components for a total conserved angular momentum will read as

$$J_1 = l_1 + \frac{\cos\phi}{\sin\theta} iJ^{12} , \qquad J_2 = l_2 + \frac{\sin\phi}{\sin\theta} iJ^{12} , \qquad J_3 = l_3 . \tag{10.1.4}$$

The field function of a vector particle with parameters (ϵ, j, m) should be constructed in terms of Wigner functions; this is shown as

$$\Phi_{\omega jm}(x) = e^{-i\omega t} [f_1(r) D_0, \ f_2(r) D_{-1}, \ f_3(r) D_0, \ f_4(r) D_{+1},$$

$$f_5(r) D_{-1}, f_6(r) D_0, \ f_7(r) D_{+1}, f_8(r) D_{-1}, \ f_9(r) D_0, \ f_{10}(r) D_{+1}] .$$

$$\tag{10.1.5}$$

We then need the following recurrent relations [429]:

$$\partial_\theta D_{-1} = (1/2) (a D_{-2} - \sqrt{j(j+1)} D_0) ,$$

$$\frac{-m + \cos\theta}{\sin\theta} D_{-1} = (1/2) (-a D_{-2} - \sqrt{j(j+1)} D_0) ,$$

$$\partial_\theta D_0 = (1/2) \sqrt{j(j+1)} (D_{-1} - D_{+1}) ,$$

$$\frac{-m}{\sin\theta}\, D_0 = (1/2)\, \sqrt{j(j+1)}\, (-D_{-1} - D_{+1})\,,$$

$$\partial_\theta\, D_{+1} = (1/2)\, (\,\sqrt{j(j+1)}\, D_0 - a\, D_{+2}\,)\,,$$

$$\frac{-m - \cos\theta}{\sin\theta}\, D_{+1} = (1/2)\, (\,-\sqrt{j(j+1)}\, D_0 - a\, D_{+2}\,)\,,$$

$$a = \sqrt{(j-1)(j+2)}\,, \tag{10.1.6}$$

through the use of these relations, we readily find

$$\Sigma^\kappa_{\theta,\phi}\, \Phi \;=\; \exp(-i\omega t)\, \sqrt{j(j+1)}\, \begin{vmatrix} (-f_5 - f_7)\, D_0 \\ -\, i\, f_9\, D_{-1} \\ (\,-if_8 + if_{10})\, D_0 \\ -if_9\, D_{+1} \\ f_1\, D_{-1} \\ 0 \\ f_1\, D_{+1} \\ -if_3\, D_{-1} \\ (+if_2 - if_4)\, D_0 \\ +if_3\, D_{+1} \end{vmatrix}\,. \tag{10.1.7}$$

Then to allow for Eq. (10.1.7) and identities

$$i\, \beta^0\, \partial_t\, \Phi = \omega\, \exp(-i\omega t)\, \begin{vmatrix} 0 \\ if_5\, D_{-1} \\ if_6\, D_0 \\ if_7\, D_{+1} \\ -if_2\, D_{-1} \\ -if_3\, D_0 \\ -if_4\, D_{+1} \\ 0 \\ 0 \\ 0 \end{vmatrix}\,,$$

$$i\, (\, \beta^3\, \frac{\partial}{\partial\chi} + \frac{\beta^1\beta^{31} + \beta^2\beta^{32}}{\sin\chi}\,)\Phi_{\omega jm} = \exp(-i\omega t);\ \begin{vmatrix} (-d/d\chi - 2/\sin\chi)f_6\, D_0 \\ (id/d\chi + i/\sin\chi)f_8\, D_{-1} \\ 0 \\ (-id/d\chi - i/\sin\chi)f_{10}\, D_{+1} \\ 0 \\ 0 \\ 0 \\ (-id/d\chi - i/\sin\chi)f_2\, D_{-1} \\ 0 \\ (id/d\chi + i/\chi)f_4\, D_{+1} \end{vmatrix}\,,$$

we arrive at 10 radial equations ($\nu = \sqrt{j(j+1)/2}$):

$$-(\frac{d}{d\chi} + \frac{2}{\tan\chi})\, f_6 - \frac{\nu}{\chi}\, (f_5 + f_7) = 0\,,$$

$$i\omega f_5 + i(\frac{d}{d\chi} + \frac{1}{\tan \chi}) \, f_8 + i\frac{\nu}{\sin \chi} \, f_9 = 0 \, ,$$

$$i\epsilon f_6 + i\frac{\nu}{\sin \chi}(-f_8 + f_{10}) = 0 \, ,$$

$$i\omega f_7 - i(\frac{d}{d\chi} + \frac{1}{\tan \chi}) \, f_{10} - i\frac{\nu}{\sin \chi} \, f_9 = 0 \, ,$$

$$-i\omega f_2 + \frac{\nu}{\sin \chi} \, f_1 - f_5 = 0 \, ,$$

$$-i\omega f_3 - \frac{d}{d\chi} \, f_1 - f_6 = 0 \, ,$$

$$-i\omega f_4 + \frac{\nu}{\sin \chi} \, f_1 - f_7 = 0 \, ,$$

$$-i(\frac{d}{d\chi} + \frac{1}{\tan \chi}) \, f_2 - i\frac{\nu}{\sin \chi} \, f_3 - f_8 = 0 \, ,$$

$$i\frac{\nu}{r} \, (f_2 - f_4) - f_9 = 0 \, ,$$

$$i(\frac{d}{d\chi} + \frac{1}{\tan \chi}) \, f_4 + i\frac{\nu}{\sin \chi} \, f_3 - f_{10} = 0 \, . \tag{10.1.8}$$

Here, let us diagonalize additionally the P-inversion operator (its form is given in cyclic basis)

$$\hat{P}^{cycl.}_{sph.} = \begin{vmatrix} 1 & 0 & 0 & 0 & 0 & 0 & 0 & 0 & 0 & 0 \\ 0 & 0 & 0 & 1 & 0 & 0 & 0 & 0 & 0 & 0 \\ 0 & 0 & 1 & 0 & 0 & 0 & 0 & 0 & 0 & 0 \\ 0 & 1 & 0 & 0 & 0 & 0 & 0 & 0 & 0 & 0 \\ 0 & 0 & 0 & 0 & 0 & 0 & 1 & 0 & 0 & 0 \\ 0 & 0 & 0 & 0 & 0 & 1 & 0 & 0 & 0 & 0 \\ 0 & 0 & 0 & 0 & 1 & 0 & 0 & 0 & 0 & 0 \\ 0 & 0 & 0 & 0 & 0 & 0 & 0 & 0 & 0 & -1 \\ 0 & 0 & 0 & 0 & 0 & 0 & 0 & 0 & -1 & 0 \\ 0 & 0 & 0 & 0 & 0 & 0 & 0 & -1 & 0 & 0 \end{vmatrix} \hat{P} \, ,$$

the eigenvalues equation $\hat{P}^{cycl.}_{sph.} \, \Phi_{jm} = P \, \Phi_{jm}$ leads us to

$$P = (-1)^{j+1} \, ,$$

$$f_1 = f_3 = f_6 = 0 \, , \quad f_4 = -f_2 \, , \quad f_7 = -f_5 \, , \quad f_{10} = +f_8 \, ; \tag{10.1.9}$$

$$P = (-1)^j \, ,$$

$$f_9 = 0 \, , \quad f_4 = +f_2 \, , \quad f_7 = +f_5 \, , \quad f_{10} = -f_8 \, . \tag{10.1.10}$$

Correspondingly, we have two different systems:

$$P = (-1)^{j+1} \, ,$$

$$i\omega \, f_5 + i(\frac{d}{d\chi} + \frac{1}{\tan\chi}) \, f_8 + i\frac{\nu}{\sin \chi} \, f_9 = 0 \, , \qquad -i\omega \, f_2 - f_5 = 0 \, ,$$

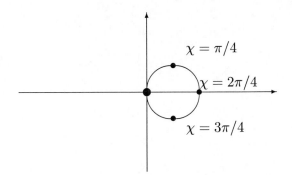

Figure 10.1: Domain $\tilde{G}_1$

$$-i(\frac{d}{d\chi} + \frac{1}{\tan \chi})\, f_2 - f_8 = 0\,, \qquad 2i\frac{\nu}{\sin\chi}\, f_2 - f_9 = 0\,; \qquad (10.1.11)$$

$P = (-1)^j\,,$

$$(\frac{d}{d\chi} + \frac{2}{\tan \chi})\, f_6 + \frac{2\nu}{\sin\chi}\, f_5 = 0\,, \qquad i\omega\, f_5 + i(\frac{d}{d\chi} + \frac{1}{\tan \chi})\, f_8 = 0\,,$$

$$i\omega\, f_6 - \frac{2i\nu}{\sin\chi}\, f_8 = 0\,, \qquad -i\omega\, f_2 + \frac{\nu}{\sin\chi}\, f_1 - f_5 = 0\,,$$

$$i\omega\, f_3 + \frac{d}{d\chi}f_1 + f_6 = 0\,, \qquad i(\frac{d}{d\chi} + \frac{1}{\tan\chi})\, f_2 + i\frac{\nu}{\sin\chi}\, f_3 + f_8 = 0\,. \qquad (10.1.12)$$

10.2 Waves with Parity $P = (-1)^{j+1}$

Next, consider Eqs. (10.1.11). Expressing f_5, f_8, f_9 through f_2 and then substituting them into the first equation we get

$$(\frac{d}{d\chi} + \frac{1}{\tan\chi})(\frac{d}{d\chi} + \frac{1}{\tan\chi})f_2 + (\, \omega^2 - \frac{j(j+1)}{\sin^2\chi}\,)\, f_2 = 0\,; \qquad (10.2.1)$$

it is simplified by $f_2 = \sin^{-1}\chi\ f(\chi)$

$$\frac{d^2}{d\chi^2}\, f + (\, \omega^2 - \frac{j(j+1)}{\sin^2\chi}\,)\, f = 0\,. \qquad (10.2.2)$$

In a new variable

$$z = 1 - e^{-2i\chi}\,, \qquad z = 2\sin\chi\ e^{i(-\chi+\pi/2)}\,,$$

which can be clarified by figure

$$\frac{d}{d\chi} = 2i(1-z)\frac{d}{dz}\,, \qquad \frac{\cos\chi}{\sin\chi} = i\frac{2-z}{z}\,, \qquad \frac{1}{\sin^2\chi} = -\frac{4(1-z)}{z^2}\,,$$

Equation (10.2.2) transforms into

$$4(1-z)^2\frac{d^2f}{dz^2} - 4(1-z)\frac{df}{dz} - \omega^2 f - \frac{4(1-z)\nu^2}{z^2}\, f = 0 \;. \qquad (10.2.3)$$

By the substitution

$$f = z^a(1-z)^b F(z) \;,$$

$$f' = az^{a-1}(1-z)^b F(z) - bz^a(1-z)^{b-1}F(z) + z^a(1-z)^b\frac{dF(z)}{dz} \;,$$

$$f'' = a(a-1)z^{a-2}(1-z)^b F(z) - abz^{a-1}(1-z)^{b-1}F(z) + az^{a-1}(1-z)^b\frac{dF(z)}{dz}$$

$$-abz^{a-1}(1-z)^{b-1}F(z) + b(b-1)z^a(1-z)^{b-2}F(z) - bz^a(1-z)^{b-1}\frac{dF(z)}{dz}$$

$$+az^{a-1}(1-z)^b\frac{dF(z)}{dz} - bz^a(1-z)^{b-1}\frac{dF(z)}{dz} + z^a(1-z)^b\frac{d^2F(z)}{dz^2}$$

Equation (10.2.3) gives

$$z(1-z)\frac{d^2F}{dz^2} + [2a - (2a+2b+1)z]\frac{dF}{dz} +$$

$$\left[\frac{\omega^2}{4} - (a+b)^2 + (a(a-1)-\nu^2)\frac{1}{z} + (b^2 - \frac{\omega^2}{4})\frac{1}{1-z}\right] F = 0 \;.$$

Requiring

$$a(a-1) - \nu^2 = 0 \;, \qquad b^2 - \frac{\omega^2}{4} = 0$$

or

$$a = j+1, -j \;, \qquad b = \pm\frac{\omega}{2} \;, \qquad (10.2.4)$$

we get

$$z(1-z)\frac{d^2F}{dz^2} + [2a - (2a+2b+1)z]\frac{dF}{dz} - \left[(a+b)^2 - \frac{\omega^2}{4}\right] F = 0 \;. \qquad (10.2.5)$$

It is of hypergeometric type equation

$$z(1-z)\, F'' + [\gamma - (\alpha+\beta+1)z]\, F' - \alpha\beta\, F = 0$$

with parameters

$$\gamma = 2a \;, \qquad \alpha + \beta = 2a + 2b \;, \qquad \alpha\beta = (a+b)^2 - \frac{\omega^2}{4} \;,$$

or

$$\alpha = a + b - \frac{\omega}{2} \;, \qquad \beta = a + b + \frac{\omega}{2} \;. \qquad (10.2.6)$$

The function $f(z)$ is

$$f = z^a(1-z)^b F(z) = \left[\, 2i\sin\chi e^{-i\chi}\,\right]^a \left[\, 1 - 2i\sin\chi e^{-i\chi}\,\right]^b F(z) \;; \qquad (10.2.7a)$$

it is finite at the points $\chi = 0$ and $\chi = \pi$ only if $a = j + 1$; and if $b = -\omega/2$ – then one can reduce the hypergeometric series to a polynomial (supposing $\omega > 0$):

$$\alpha = j + 1 - \omega = -n = \{0, -1, -2, \dots\} \implies \omega = n + 1 + j . \tag{10.2.7b}$$

Thus, solution with parity $P = (-1)^{j+1}$ is given by (it is a spherical wave of magnetic type)

$$P = (-1)^{j+1}$$

$$f_2 = \frac{1}{\sin \chi} f(\chi) , \quad f = z^a (1 - z)^b F(z)$$

$$= \left[2i \sin \chi e^{-i\chi} \right]^a \left[1 - 2i \sin \chi e^{-i\chi} \right]^b F(z) ,$$

$$F(z) = F(-n, j + 1, 2j + 2; z) = F(-n, j + 1, 2j + 2; 2i \sin \chi e^{-i\chi}) ;$$

$$\tag{10.2.8}$$

for frequency ω only special values are permitted:

$$\omega = n + 1 + j , \qquad j = 0, 1, 2, \dots, \qquad n = 0, 1, 2, \dots; \tag{10.2.9a}$$

or in usual units

$$\omega = \frac{c}{\rho} (n + 1 + j) , \tag{10.2.9b}$$

ρ stands for the curvature radius.

Let us write down (tetrad) 4-potential of these waves

$$\begin{vmatrix} A_{(0)} \\ A_{(1)} \\ A_{(2)} \\ A_{(3)} \end{vmatrix} = \begin{vmatrix} 0 \\ +f_2 D_{-1} \\ 0 \\ -f_2 D_{+1} \end{vmatrix} . \tag{10.2.10}$$

10.3   The Lorentz Gauge in Spherical Space

Let us detail the Lorentz gauge

$$\nabla_\beta \Phi^\beta(x) = 0 . \tag{10.3.1}$$

In tetrad components, this will read as

$$\Phi_{(a)} e^{(a)\alpha}{}_{;\alpha} + e^{(a)\alpha} \partial_a \Phi_{(a)} = 0 , \tag{10.3.2}$$

where $\Phi_{(a)}$ are components of a 4-vector in Cartesian basis

$$\begin{vmatrix} \Phi_{(0)} \\ \Phi_{(1)} \\ \Phi_{(2)} \\ \Phi_{(3)} \end{vmatrix} = \begin{vmatrix} 1 & 0 & 0 & 0 \\ 0 & -1/\sqrt{2} & 0 & 1/\sqrt{2} \\ 0 & -i/\sqrt{2} & 0 & -i/\sqrt{2} \\ 0 & 0 & 1 & 0 \end{vmatrix} \begin{vmatrix} f_1 D_0 \\ f_2 D_{-1} \\ f_3 D_0 \\ f_4 D_{+1} \end{vmatrix} . \tag{10.3.3}$$

When we use the known formula

$$e^{(a)\alpha}{}_{;\alpha} = \frac{1}{\sqrt{-g}} \frac{\partial}{\partial x^\alpha} \sqrt{-g} \, e^{(a)\alpha} \, ,$$

we get

$$e^{(0)\alpha}{}_{;\alpha} = 0 \, , \qquad e^{(3)\alpha}{}_{;\alpha} = -\frac{2}{\tan \chi} \, ,$$

$$e^{(1)\alpha}{}_{;\alpha} = -\frac{1}{\sin \chi} \frac{\cos \theta}{\sin \theta} \, , \qquad e^{(2)\alpha}{}_{;\alpha} = 0 \, , \qquad (10.3.4)$$

and therefore Eq. (10.3.2) gives

$$\Phi_1 \left(-\frac{1}{\sin \chi} \frac{\cos \theta}{\sin \theta}\right) + \Phi_3 \left(-\frac{2}{\tan \chi}\right) + \partial_t \Phi_0 - \partial_\chi \Phi_3$$

$$-\frac{1}{\sin \chi} \partial_\theta \Phi_1 - \frac{1}{\sin \chi \sin \theta} \partial_\phi \Phi_2 = 0 \, ,$$

or

$$\frac{1}{\sqrt{2}} \left(-f_2 D_{-1} + f_4 D_{+1}\right) \left(-\frac{1}{\sin \chi} \frac{\cos \theta}{\sin \theta}\right) + f_3 D_0 \left(-\frac{2}{\tan \chi}\right) - i\omega f_1 D_0 - \partial_\chi f_3 D_0$$

$$-\frac{1}{\sin \chi} \partial_\theta \frac{1}{\sqrt{2}} \left(-f_2 D_{-1} + f_4 D_{+1}\right) - \frac{1}{\sin \chi \sin \theta} \partial_\phi \frac{1}{\sqrt{2}} \left(-i f_2 D_{-1} - i f_4 D_{+1}\right) = 0 \, ,$$

and further

$$-i\omega f_1 D_0 - \partial_\chi f_3 D_0 - \frac{2}{\tan \chi} f_3 D_0$$

$$+\frac{1}{\sqrt{2} \sin \chi} \left[f_2 \left(\frac{-m + \cos \theta}{\sin \theta} D_{-1} + \partial_\theta D_{-1}\right) - f_4 \left(\frac{m + \cos \theta}{\sin \theta} D_{+1} + \partial_\theta D_{+1}\right) \right] = 0 \, .$$

$$(10.3.5)$$

Now, we are to use the known recurrent formulas [429]

$$\frac{-m + \cos \theta}{\sin \theta} D_{-1} + \partial_\theta D_{-1} = -\sqrt{j(j+1)} \, D_0 \, ,$$

$$\frac{+m + \cos \theta}{\sin \theta} D_{+1} + \partial_\theta D_{+1} = +\sqrt{j(j+1)} \, D_0 \, , \qquad (10.3.6)$$

then the Lorentz condition gives us

$$-i\omega f_1 - \left(\frac{\partial}{\partial \chi} + \frac{2}{\tan \chi}\right) f_3 - \frac{\sqrt{j(j+1)}}{\sqrt{2} \sin \chi} \left(f_2 + f_4\right) = 0 \, ; \qquad (10.3.7)$$

This is the Lorentz gauge in radial form.

There exist restrictions due to P-parity. For waves with $P = (-1)^{j+1}$ the Lorentz gauge Eq. (10.3.7) holds identically. When $P = (-1)^j$, the Lorentz gauge in Eq. (10.3.7) looks simpler

$$P = (-1)^j \, , \qquad i\omega f_1 + \left(\frac{\partial}{\partial \chi} + \frac{2}{\tan \chi}\right) f_3 + 2\frac{\nu}{\sin \chi} f_2 = 0 \, . \qquad (10.3.8)$$

The Lorentz condition being imposed substantially confines the gauge freedom. However, the freedon gauge still remains. The known way to exclude the freedom is to impose Landau gauge:

$$\Phi_0 = 0 , \qquad \nabla^j \Phi_j = 0 ; \qquad (10.3.9)$$

and instead of Eq. (10.3.9), we have

$$P = (-1)^j , \quad f_1 = 0 , \qquad (\frac{\partial}{\partial \chi} + \frac{2}{\tan \chi}) f_3 + 2 \frac{\nu}{\sin \chi} f_2 = 0 . \qquad (10.3.10)$$

10.4   Waves of Electric Type

Now, let us turn to radial equations in Eq. (10.1.12) for waves with $P = (-1)^j$ in Landau gauge:

$$(\frac{d}{d\chi} + \frac{2}{\tan \chi}) f_6 + \frac{2\nu}{\sin \chi} f_5 = 0 , \qquad i\omega f_5 + i(\frac{d}{d\chi} + \frac{1}{\tan \chi}) f_8 = 0 ,$$

$$i\omega f_6 - \frac{2i\nu}{\sin \chi} f_8 = 0 , \qquad -i\omega f_2 - f_5 = 0 ,$$

$$i\omega f_3 + f_6 = 0 , \qquad i(\frac{d}{d\chi} + \frac{1}{\tan \chi}) f_2 + i\frac{\nu}{\sin \chi} f_3 + f_8 = 0 . \qquad (10.4.1)$$

After substitutions

$$f_2 = \frac{F_2}{\sin \chi} , \quad f_3 = F_3 , \quad f_5 = \frac{F_5}{\sin \chi} ,$$

$$f_6 = \frac{F_6}{\sin^2 \chi} , \qquad f_8 = \frac{F_8}{\sin \chi} , \qquad (10.4.2)$$

Equations (10.4.1) become simpler

$$\frac{d}{d\chi} F_6 + 2\nu F_5 = 0 , \qquad \omega F_5 + \frac{d}{d\chi} F_8 = 0 , \qquad \omega F_6 - 2\nu F_8 = 0 ,$$

$$-i\omega F_2 - F_5 = 0 , \qquad i\omega F_3 + \frac{F_6}{\sin^2 \chi} = 0 , \qquad i\frac{d}{d\chi} F_2 + i\nu F_3 + F_8 = 0 .$$

$$(10.4.3)$$

The system in Eq. (10.4.3) gives

$$F_5 = -\frac{1}{\omega} \frac{d}{d\chi} F_8 , \qquad F_6 = \frac{2\nu}{\omega} F_8 ,$$

$$F_2 = -\frac{i}{\omega^2} \frac{d}{d\chi} F_8 , \qquad F_3 = \frac{2i\nu}{\omega^2} \frac{1}{\sin^2 \chi} F_8 ,$$

$$(\frac{d^2}{d\chi^2} + \omega^2 - \frac{j(j+1)}{\sin^2 \chi}) F_8 = 0 ; \qquad (10.4.4)$$

the second order differential equation for F_8 has been solved in section **10.2**.

Thus, the waves of electric type have been constructed:

$$
\begin{vmatrix} A_{(0)} \\ A_{(1)} \\ A_{(2)} \\ A_{(3)} \end{vmatrix} = \begin{vmatrix} 0 \\ +f_2 D_{-1} \\ f_3 D_0 \\ f_2 D_{+1} \end{vmatrix}.
\tag{10.4.5}
$$

Chapter 11

Solutions with Cylindric Symmetry in Spherical Space

In this chapter, we use the complex formalism of Riemann–Silberstein–Majorana–Oppenheimer to treat the Maxwell equations in Riemann and Lobachevsky spaces for constant curvature parameterized by cylindric coordinates.

11.1 Cylindric Coordinates and Tetrad in Spherical Space S_3

Let us consider the Maxwell equation in the cylindrical coordinates and tetrad in spherical space S_3

$$n_1 = \sin r \cos \phi \, , \ n_2 = \sin r \sin \phi \, , \ n_3 = \cos r \sin z \, , \ n_4 = \cos r \cos z \, ,$$

$$dS^2 = dt^2 - dr^2 - \sin^2 r \, d\phi^2 - \cos^2 r \, dz^2 \, , \qquad x^\alpha = (t, r, \phi, z) \, ,$$

$$e^{\beta}_{(a)}(y) = \begin{vmatrix} 1 & 0 & 0 & 0 \\ 0 & 1 & 0 & 0 \\ 0 & 0 & \sin^{-1} r & 0 \\ 0 & 0 & 0 & \cos^{-1} r \end{vmatrix} , \qquad (11.1.1)$$

where (r, ϕ, z) run within

$$\rho \in [0, \ +\pi/2] \, , \ \phi \in [-\pi, \ +\pi] \, , \ z \in [-\pi, \ +\pi] \, .$$

Christoffel symbols are given by $\Gamma^0_{\beta\sigma} = 0$, $\Gamma^i_{00} = 0$, $\Gamma^i_{0j} = 0$ and

$$\Gamma^r_{\ jk} = \begin{vmatrix} 0 & 0 & 0 \\ 0 & -\sin r \cos r & 0 \\ 0 & 0 & \sin r \cos r \end{vmatrix} ,$$

$$\Gamma^\phi_{\ jk} = \begin{vmatrix} 0 & \frac{\cos r}{\sin r} & 0 \\ \frac{\cos r}{\sin r} & 0 & 0 \\ 0 & 0 & 0 \end{vmatrix} , \qquad \Gamma^z_{\ jk} = \begin{vmatrix} 0 & 0 & -\frac{\sin r}{\cos r} \\ 0 & 0 & 0 \\ -\frac{\sin r}{\cos r} & 0 & 0 \end{vmatrix} . \qquad (11.1.2)$$

For covariant derivatives of tetrad vectors, we get

$$A_{\beta;\alpha} = \frac{\partial A_\beta}{\partial x^\alpha} - \Gamma^\sigma_{\alpha\beta} A_\sigma$$

and therefore

$$e_{(0)\beta;\alpha} = 0 \, , \qquad e_{(1)\beta;\alpha} = \begin{vmatrix} 0 & 0 & 0 & 0 \\ 0 & 0 & 0 & 0 \\ 0 & 0 & -\sin r \cos r & 0 \\ 0 & 0 & 0 & \sin r \cos r \end{vmatrix} ,$$

$$e_{(2)\beta;\alpha} = \begin{vmatrix} 0 & 0 & 0 & 0 \\ 0 & 0 & \cos r & 0 \\ 0 & 0 & 0 & 0 \\ 0 & 0 & 0 & 0 \end{vmatrix} , \qquad e_{(3)\beta;\alpha} = \begin{vmatrix} 0 & 0 & 0 & 0 \\ 0 & 0 & 0 & -\sin r \\ 0 & 0 & 0 & 0 \\ 0 & 0 & 0 & 0 \end{vmatrix} .$$

What remains is finding the Ricci rotation coefficients:

$$\gamma_{ab0} = e_{(a)}{}^\beta e_{(b)\beta;t} e^t_{(0)} = 0 \, , \qquad \gamma_{ab1} = e_{(a)}{}^a e_{(b)a;r} e^r_{(1)} \, ,$$

$$\gamma_{ab2} = e_{(a)}{}^\beta e_{(b)\beta;\phi} e^\phi_{(2)} \, , \qquad \gamma_{ab3} = e_{(a)}{}^\beta e_{(b)\beta;z} e^z_{(3)} \, ;$$

from which it follows

$$\gamma_{011} = \gamma_{021} = \gamma_{031} = 0 \, , \; \gamma_{012} = \gamma_{022} = \gamma_{032} = 0 \, , \; \gamma_{013} = \gamma_{023} = \gamma_{033} = 0 \, ,$$

$$\gamma_{231} = 0 \, , \qquad \gamma_{311} = 0 \, , \qquad \gamma_{121} = 0 \, , \gamma_{232} = 0 \, , \; \gamma_{312} = 0 \, ,$$

$$\gamma_{122} = \frac{\cos r}{\sin r} \, , \; \gamma_{233} = 0 \, , \qquad \gamma_{313} = \frac{\sin r}{\cos r} \, , \qquad \gamma_{123} = 0 \, .$$

In taking into account the identities

$$e^\rho_{(0)} \partial_\rho = \partial_{(0)} = \partial_t \, , \qquad e^\rho_{(1)} \partial_\rho = \partial_{(1)} = \partial_r \, ,$$

$$e^\rho_{(2)} \partial_\rho = \partial_{(2)} = \frac{1}{\sin r} \partial_\phi \, , \qquad e^\rho_{(3)} \partial_\rho = \partial_{(3)} = \frac{1}{\cos r} \partial_z \, ,$$

$$\mathbf{v}_0 = (\gamma_{010}, \gamma_{020}, \gamma_{030}) \equiv 0 \, , \qquad \mathbf{v}_1 = (\gamma_{011}, \gamma_{021}, \gamma_{031}) \equiv 0 \, ,$$

$$\mathbf{v}_2 = (\gamma_{0120}, \gamma_{022}, \gamma_{032}) \equiv 0 \, , \qquad \mathbf{v}_3 = (\gamma_{013}, \gamma_{023}, \gamma_{033}) \equiv 0 \, ,$$

$$\mathbf{p}_0 = (\gamma_{230}, \gamma_{310}, \gamma_{120}) = 0 \, , \qquad \mathbf{p}_1 = (\gamma_{231}, \gamma_{311}, \gamma_{121}) = 0 \, ,$$

$$\mathbf{p}_2 = (\gamma_{232}, \gamma_{312}, \gamma_{122}) = (0, 0, \frac{\cos r}{\sin r}) \, , \qquad \mathbf{p}_3 = (\gamma_{233}, \gamma_{313}, \gamma_{123}) = (0, \frac{\sin r}{\cos r}, 0) \, ,$$

in the absence of an external source, the Maxwell equation in a complex form by Riemann–Silberstein–Majorana–Oppenheimer reads

$$\left(-i\partial_t + \alpha^1 \partial_r + \alpha^2 \frac{1}{\sin r} \partial_\phi + \alpha^3 \frac{1}{\cos r} \partial_z + \alpha^2 s_3 \frac{\cos r}{\sin r} + \alpha^3 s_2 \frac{\sin r}{\cos r} \right) \begin{vmatrix} 0 \\ \mathbf{E} + ic\mathbf{B} \end{vmatrix} = 0.$$

$$(11.1.3)$$

11.2   Separation of Variables in S_3, Solutions at $m = 0$

Wave Maxwell operator from Eq. (11.1.3) commutes with $i\partial_t$, $-i\partial_\phi$, $-i\partial_z$. Therefore, for a field function we get a substitution

$$\Psi = \begin{vmatrix} 0 \\ \mathbf{E} + ic\mathbf{B} \end{vmatrix} = e^{-i\omega t}\, e^{im\phi}\, e^{ikz} \begin{vmatrix} 0 \\ f_1(r) \\ f_2(r) \\ f_3(r) \end{vmatrix}. \tag{11.2.1}$$

Correspondingly, Eq. (11.1.3) will read as

$$\left(-\omega + \alpha^1 \frac{d}{dr} + \frac{im}{\sin r}\alpha^2 + \frac{ik}{\cos r}\alpha^3 + \frac{\cos r}{\sin r}\alpha^2 s_3 + \frac{\sin r}{\cos r}\alpha^3 s_2 \right) \begin{vmatrix} 0 \\ f_1 \\ f_2 \\ f_3 \end{vmatrix} = 0 . \tag{11.2.2}$$

After a simple calculation, we get the following radial system

$$(\frac{d}{dr} + \frac{\cos r}{\sin r} - \frac{\sin r}{\cos r})f_1 + \frac{im}{\sin r}\, f_2 + \frac{ik}{\cos r}\, f_3 = 0 ,$$

$$-\omega f_1 - \frac{ik}{\cos r}\, f_2 + \frac{im}{\sin r}\, f_3 = 0 ,$$

$$-\omega f_2 - (\frac{d}{dr} - \frac{\sin r}{\cos r})\, f_3 + \frac{ik}{\cos r} f_1 = 0 ,$$

$$-\omega f_3 + (\frac{d}{dr} + \frac{\cos r}{\sin r})\, f_2 - \frac{im}{\sin r} f_1 = 0 . \tag{11.2.3}$$

First, let us consider the case $m = 0$, when the radial equations become more simple

$$(\frac{d}{dr} + \frac{\cos r}{\sin r} - \frac{\sin r}{\cos r})f_1 + \frac{ik}{\cos r}\, f_3 = 0 ,$$

$$f_1 = -\frac{ik}{\omega \cos r}\, f_2 ,$$

$$-\omega f_2 - (\frac{d}{dr} - \frac{\sin r}{\cos r})\, f_3 + \frac{ik}{\cos r} f_1 = 0 ,$$

$$f_3 = \frac{1}{\omega}\, (\frac{d}{dr} + \frac{\cos r}{\sin r})\, f_2 . \tag{11.2.4}$$

Using 2nd and 4th equations, from the first equation it follows

$$(\frac{d}{dr} + \frac{\cos r}{\sin r} - \frac{\sin r}{\cos r})\, \frac{-ik}{\omega \cos r}\, f_2 + \frac{ik}{\cos r}\, \frac{1}{\omega}\, (\frac{d}{dr} + \frac{\cos r}{\sin r})\, f_2 = 0 ,$$

$$-\omega f_2 - (\frac{d}{dr} - \frac{\sin r}{\cos r})\, \frac{1}{\omega}\, (\frac{d}{dr} + \frac{\cos r}{\sin r})\, f_2 + \frac{ik}{\cos r}\, \frac{-ik}{\omega \cos r}\, f_2 = 0 ,$$

that is equivalent to the identity $0 \equiv 0$; and the equation for f_2 is

$$\frac{d^2}{dr^2}\, f_2 + (\frac{\cos r}{\sin r} - \frac{\sin r}{\cos r})\, \frac{d}{dr}\, f_2 + (\omega^2 - 1 - \frac{1}{\sin^2 r} - \frac{k^2}{\cos^2 r})\, f_2 = 0 . \tag{11.2.5}$$

The latter can be simplified

$$f_2(r) = \frac{1}{\sin r} E(r) , \quad \frac{d^2 E}{dr^2} - \frac{1}{\sin r \cos r} \frac{dE}{dr} + (\omega^2 - \frac{k^2}{\cos^2 r}) E = 0 ; \quad (11.2.6)$$

two continuous functions are given by

$$f_1(r) = \frac{-ik}{\omega} \frac{1}{\cos r \sin r} E(r) , \quad f_3 = \frac{1}{\omega} \frac{1}{\sin r} \frac{d}{dr} E(r) . \quad (11.2.7)$$

Turning back to Eq. (11.2.6), first let us consider a particular case when $k^2 = \omega^2$

$$\frac{\sin r}{\cos r} \frac{d}{dr} \frac{\cos r}{\sin r} \frac{d}{dr} E + k^2 (1 - \frac{1}{\cos^2 r}) E = 0 ,$$

from this, it follows

$$(\frac{\cos r}{\sin r} \frac{d}{dr}) (\frac{\cos r}{\sin r} \frac{d}{dr}) E = k^2 E ,$$

so that

$$\frac{\cos r}{\sin r} \frac{d}{dr} = \frac{d}{dx} \quad \Longrightarrow \quad \frac{dr}{dx} = \frac{\cos r}{\sin r} ,$$

$$-dx = d \ln \cos r , \quad x = \ln(C \cos^{-1} r) , \quad C = \text{const}$$

and

$$\frac{d^2}{dx^2} E = k^2 E ,$$

which has two solutions

$$k^2 = \omega^2 , \quad E_\pm = e^{\mp kx} = E_0 (\cos r)^{\pm k} .$$

In these solutions

$$(\cos r)^{+k} , \quad \frac{1}{(\cos r)^k} , \quad r \in [\, 0, \frac{\pi}{2} \,]$$

at $k > 0$ the second one must be rejected because it trends to infinity as $r \to \pi/2$; when for $k < 0$ the first solution must be rejected by analogous reason. Thus, physical solutions are

$$k^2 = \omega^2 , \; k > 0 , \quad E = E_0 \cos^k r \; e^{-i(\omega t - kz)} ;$$

$$k^2 = \omega^2 , \; k < 0 , \quad E = E_0 \cos^{-k} r \; e^{-i(\omega t - kz)} . \quad (11.2.8)$$

Because at $r \neq 0, \pi/2$, the values $z = -\pi$ and $z = +\pi$ determine the same point in spherical space S_3, solutions for Eq. (11.2.5) represent a continuous function in S_3 only if k takes on integer values:

$$k = \pm n , \quad n = 1, 2, 3, ... ; \quad (11.2.9)$$

or in usual units

$$k = \pm \frac{\omega \rho}{c} = n , \quad \omega = \frac{c}{\rho} n , \quad n = 1, 2, 3, ... \quad (11.2.10)$$

Turning again to Eq. (11.2.6), let us construct one more special solution. To this end, consider an approximate solution in the vicinity of the point $r = 0$: $E = \sin^A r$, we get $A = 0, +2$. In the same manner, the nearby point $r = \pi/2$, an approximate solution is $E = \cos^B r$, $B^2 = k^2$. Let us demonstrate that there exist values k such that an exact solution can be constructed as follows (the case $B = 0$ was previously considered earlier):

$$E = \sin^2 r \cos^B r . \tag{11.2.11}$$

In this substitution, Eq. (11.2.6) gives

$$2\cos^4 r - 2(B+1)\sin^2 r \, \cos^2 r - 3B\sin^2 r \, \cos^2 r + B(B-1)\sin^4 r$$

$$-2\cos^2 r + B\sin^2 r - k^2 \sin^2 r + \omega^2 \sin^2 r \cos^2 r = 0 .$$

Using the notation $\cos^2 r = x$, it can be written as

$$x^2 \left(4 + 4B - \omega^2 + B^2\right) + x \left(-4 - 4B + \omega^2 - B^2\right) + x^0 \left(B^2 - k^2\right) = 0 .$$

The latter, is satisfied, if $B^2 = k^2$, $(B+2)^2 - \omega^2 = 0$, that is

$$B = -2 + \omega, \; -2 - \omega , \qquad k = \pm B , \quad E = \sin^2 r \; \cos^B r . \tag{11.2.12}$$

Thus, the corresponding solutions of this type are

$$E = \sin^2 r \; \cos^B r \, e^{-i(\omega t - kz)} . \tag{11.2.13}$$

Solutions with negative B must be rejected because they give infinite electromagnetic field at the point $r = \pi/2$. Besides, periodicity requirement for z led to $k = \pm 1, \pm 2, \pm 3, ...$

Therefore, the wave propagating in the positive direction is given by

$$B = +k = +1, +2, +3, ...,$$

$$k = -2 + \omega , \qquad \omega = 2 + k = 3, 4, 5, ... ,$$
$$E = \sin^2 r \; \cos^k r \, e^{-i(\omega t - kz)} . \tag{11.2.14}$$

In turn, the wave propagating in the negative direction is given by

$$B = -k = +1, +2, +3, ... ,$$

$$-k = -2 + \omega , \qquad \omega = 2 - k = +3, +4, +5, ... ,$$
$$E = \sin^2 r \; \cos^{-k} r \, e^{-i(\omega t - kz)} . \tag{11.2.15}$$

Turning to the general equation in Eq. (11.2.6), one may try to construct all the other solutions of that type $m = 0$ on the basis of the following substitution:

$$E(r) = \sin^2 r \; \cos^B r \, F(r) ; \tag{11.2.16}$$

Equation (11.2.6) gives (let $\cos^2 r = x$)

$$4x(1 - x)\frac{d^2}{dx^2}F + 4[1 - 3x + B(1 - x)]\frac{d}{dx}F$$

$$+ \left[-(2+5B) + \frac{2x}{1-x} + B(B-1)\frac{1-x}{x} - \frac{2}{1-x} + \frac{B}{x} - \frac{k^2}{x} + \omega^2 \right] F = 0 \,.$$

Requiring $k^2 = B^2$, for F we get the equation

$$4x(1-x)\frac{d^2}{dx^2}F + 4[(1+B)-(3+B)x]\,\frac{d}{dx} - [\,(B+2)^2 - \omega^2\,]\,F = 0 \,,$$

which is of hypergeometric type

$$z(1-z)\,F + [\gamma - (\alpha + \beta + 1)z]\,F' - \alpha\beta\,F = 0 \,,$$

$$k = \pm B \,, \qquad \gamma = 1 + B \,, \quad \alpha = \frac{B+2-\omega}{2} \,, \quad \beta = \frac{B+2+\omega}{2} \,.$$

Thus, the general solution of the type $m = 0$ takes the form

$$E = \sin^2 r \cos^B r \; F(\alpha, \beta, \gamma, \cos^2 r) \; e^{-i(\omega t - kz)} \,. \tag{11.2.17}$$

We are to separate single-valued and continuous functions in S_3. For a wave propagating in the positive direction z:

$$k > 0 \,, \qquad k = +1, +2, +3, \dots$$

the function $E(r)$ is finite at $r = \pi/2$ only if $B = +k$, besides polynomial solutions can arise only if

$$\alpha = \frac{k+2-\omega}{2} = -n = 0, -1, -2, \dots \implies \omega = k + 2(n+1) = N \,. \tag{11.2.18}$$

For a wave propagating in the positive direction z:

$$k < 0 \,, \qquad -k = 1, 2, 3, \dots$$

the function $E(r)$ is finite at $r = \pi/2$ only if $B = -k$; additionally, one must obtain polynomials to get

$$\alpha = \frac{-k+2-\omega}{2} = -n = 0, -1, -2, \dots \implies \omega = -k + 2(n+1) = N \,. \tag{11.2.19}$$

All the constructed solutions of the Maxwell equations are finite, single-valued, and continuous functions in spherical Riemann space S_3.

11.3   Maxwell Solutions at $k = 0$

Now, we can turn to Eqs. (11.2.3) at $k = 0$

$$(\frac{d}{dr} + \frac{\cos r}{\sin r} - \frac{\sin r}{\cos r})f_1 + \frac{im}{\sin r}\,f_2 = 0 \,,$$

$$f_1 = \frac{im}{\omega \sin r}\,f_3 \,,$$

$$f_2 = -\frac{1}{\omega}(\frac{d}{dr} - \frac{\sin r}{\cos r})\, f_3\,,$$

$$-\omega f_3 + (\frac{d}{dr} + \frac{\cos r}{\sin r})\, f_2 - \frac{im}{\sin r} f_1 = 0\,. \qquad (11.3.1)$$

Here, we use the 2nd and 3rd from the 1st and 4th, to get an identity $0 \equiv 0$ and the following equation for f_3:

$$\frac{d^2}{dr^2}\, f_3 + (\frac{\cos r}{\sin r} - \frac{\sin r}{\cos r})\, \frac{d}{dr}\, f_3 + (\omega^2 - 1 - \frac{1}{\cos^2 r} - \frac{k^2}{\sin^2 r})\, f_3 = 0\,, \qquad (11.3.2)$$

which gives

$$f_3(r) = \frac{1}{\cos r}\, E(r)\,, \qquad \frac{d^2 E}{dr^2} + \frac{1}{\sin r \cos r}\, \frac{dE}{dr} + (\omega^2 - \frac{m^2}{\sin^2 r})E = 0\,. \qquad (11.3.3)$$

First, consider a particular case $m^2 = \omega^2$

$$\frac{d^2 E}{dr^2} + \frac{1}{\sin r \cos r}\, \frac{dE}{dr} + m^2(1 - \frac{1}{\sin^2 r})\, E = 0\,, \qquad (11.3.4)$$

with the solutions

$$(\sin r)^m\,, \qquad \frac{1}{(\sin r)^m}\,, \qquad r \in [\,0, \frac{\pi}{2}\,]\,. \qquad (11.3.5)$$

Physical solutions are

$$m^2 = \omega^2\,,\ m > 0\,, \qquad E = E_0\, \sin^m r\, e^{-i(\omega t - m\phi)}\,;$$

$$m^2 = \omega^2\,,\ m < 0\,, \qquad E = E_0\, \sin^{-m} r\, e^{-i(\omega t - m\phi)}\,. \qquad (11.3.6)$$

Performing analysis like in previous Section, we easily construct solutions:

$$B = +m = +1, +2, +3, \dots\,,$$

$$m = -2 + \omega\,, \qquad \omega = 2 + m = 3, 4, 5, \dots\,,$$

$$F_{02} = \cos^2 r\ \sin^m r\, e^{-i(\omega t - m\phi)}\,, \qquad (11.3.7)$$

and

$$B = -m = +1, +2, +3, \dots\,,$$

$$-m = -2 + \omega\,, \qquad \omega = 2 - m = +3, +4, +5, \dots\,,$$

$$F_{02} = \cos^2 r\ \sin^{-m} r\, e^{-i(\omega t - mz)}\,. \qquad (11.3.8)$$

All possible solutions of Eq. (11.3.3) can be constructed as

$$E(r) = \cos^2 r\ \sin^B r\, F(r)\,, \qquad (11.3.9)$$

and further (let $\sin^2 r = x$) we get

$$m^2 = B^2\,, \qquad 4x(1-x)\frac{d^2}{dx^2}F + 4[(1+B) - (3+B)x]\,\frac{d}{dx}F - [\,(B+2)^2 - \omega^2\,]\,F = 0\,,$$

what is of hypergeometric type

$$z(1-z)\,F + [\gamma - (\alpha+\beta+1)z]\,F' - \alpha\beta\,F = 0\,,$$

$$k = \pm B\,, \qquad \gamma = 1+B\,, \quad \alpha = \frac{B+2-\omega}{2}\,, \quad \beta = \frac{B+2+\omega}{2}\,.$$

Thus, the Maxwell equations solutions of the type $k = 0$ is given by

$$E = \cos^2 r \sin^B r\, F(\alpha,\beta,\gamma,\sin^2 r)\, e^{-i(\omega t - m\phi)}\,. \tag{11.3.10}$$

We can separate physical waves (expressed in polynomials):

$$m > 0\,, \qquad B = +m\,,$$

$$\alpha = \frac{m+2-\omega}{2} = -n \qquad \Longrightarrow \qquad \omega = m + 2(n+1) = N\,; \tag{11.3.11}$$

$$m < 0\,, \qquad B = -m,$$

$$\alpha = \frac{-m+2-\omega}{2} = -n \qquad \Longrightarrow \qquad \omega = -m + 2(n+1) = N\,. \tag{11.3.12}$$

11.4 Radial System at Arbitrary m, k

Now, let us solve radial equations in the general case of Eq. (11.2.3). When taking into account the remaining three:

$$-\omega f_1 = \frac{ik}{\cos r}\,f_2 - \frac{im}{\sin r}\,f_3\,,$$

$$-\omega f_2 = (\frac{d}{dr} - \frac{\sin r}{\cos r})\,f_3 - \frac{ik}{\cos r}f_1\,,$$

$$-\omega f_3 = -(\frac{d}{dr} + \frac{\cos r}{\sin r})\,f_2 + \frac{im}{\sin r}f_1\,, \tag{11.4.1}$$

the first equation reduces to

$$(\frac{d}{dr} + \frac{\cos r}{\sin r} - \frac{\sin r}{\cos r})(\frac{ik}{\cos r}\,f_2 - \frac{im}{\sin r}\,f_3) + \frac{im}{\sin r}\left[(\frac{d}{dr} - \frac{\sin r}{\cos r})\,f_3 - \frac{ik}{\cos r}f_1\right]$$

$$+ \frac{ik}{\cos r}\left[-(\frac{d}{dr} + \frac{\cos r}{\sin r})\,f_2 + \frac{im}{\sin r}f_1\right] = 0\,,$$

what is the identity $0 = 0$. The system in Eq. (11.4.1) is simplified with substitutions

$$f_2 = \frac{1}{\sin r}\,F_2\,, \qquad f_3 = \frac{1}{\cos r}\,F_3\,,$$

so that

$$-\omega\,f_1 = i\,\frac{k\,F_2 - m\,F_3}{\sin r \cos r}\,,$$

$$-\omega\,\frac{F_2}{\sin r} = \frac{1}{\cos r}\frac{dF_3}{dr} - \frac{ik}{\cos r}f_1\,,$$

$$-\omega \, \frac{F_3}{\cos r} = -\frac{1}{\sin r} \, \frac{dF_2}{dr} + \frac{im}{\sin r} \, f_1 \; . \tag{11.4.2}$$

Excluding f_1, we arrive at

$$(\frac{\omega}{\cos r} \frac{d}{dr} + \frac{km}{\sin r \cos^2 r}) \, F_3 + \frac{1}{\sin r} \, (\omega^2 - \frac{k^2}{\cos^2 r}) \, F_2 = 0 \; ,$$

$$(\frac{\omega}{\sin r} \frac{d}{dr} - \frac{km}{\cos r \sin^2 r}) \, F_2 + \frac{1}{\cos r} \, (-\omega^2 + \frac{m^2}{\sin^2 r}) \, F_3 = 0 \; . \tag{11.4.3}$$

By using a new variable $y = (1 - \cos 2r)/2$, the system reads

$$(2\omega \frac{d}{dy} - \frac{km}{y(1-y)}) \, F_2 + (-\frac{\omega^2}{1-y} + \frac{m^2}{y(1-y)}) \, F_3 = 0 \; ,$$

$$(2\omega \frac{d}{dy} + \frac{km}{y(1-y)}) \, F_3 + (+\frac{\omega^2}{y} - \frac{k^2}{y(1-y)}) \, F_2 = 0 \; . \tag{11.4.4}$$

Instead of F_2, F_3, let us introduce new functions by means of a linear transformation with unit determinant $\alpha N - \beta M = 1$:

$$F_2 = \alpha(y) \, G_2 + \beta(y) \, G_3 \; , \qquad F_3 = M(y) \, G_2 + N(y) \, G_3 \; , \tag{11.4.5}$$

and the inverse is given by

$$G_2 = N(y) \, F_2 - \beta(y) \, F_3 \; , \qquad G_3 = -M(y) \, F_2 + \alpha(y) \, F_3 \; . \tag{11.4.6}$$

In combining Eqs. (11.4.4), we get

$$N \, (2\omega \frac{d}{dy} - \frac{km}{y(1-y)}) \, F_2 + 2\omega \frac{dN}{dy} F_2 - 2\omega \frac{dN}{dy} F_2 + N \, (-\frac{\omega^2}{1-y} + \frac{m^2}{y(1-y)}) \, F_3$$

$$-\beta \, (2\omega \frac{d}{dy} + \frac{km}{y(1-y)}) \, F_3 - 2\omega \frac{d\beta}{dy} F_3 + 2\omega \frac{d\beta}{dy} F_3 - \beta \, (+\frac{\omega^2}{y} - \frac{k^2}{y(1-y)}) \, F_2 = 0 \; ,$$

$$-M \, (2\omega \frac{d}{dy} - \frac{km}{y(1-y)}) \, F_2 - 2\omega \frac{dM}{dy} F_2 + 2\omega \frac{dM}{dy} F_2 - M(-\frac{\omega^2}{1-y} + \frac{m^2}{y(1-y)}) \, F_3$$

$$+\alpha \, (2\omega \frac{d}{dy} + \frac{km}{y(1-y)}) \, F_3 + 2\omega \frac{d\alpha}{dy} F_3 - 2\omega \frac{d\alpha}{dy} F_3 + \alpha \, (+\frac{\omega^2}{y} - \frac{k^2}{y(1-y)}) \, F_2 = 0 \; , \tag{11.4.7}$$

from hence, it follows that

$$2\omega \frac{d}{dy} \, G_2 - N \, \frac{km}{y(1-y)} \, F_2 - 2\omega \frac{dN}{dy} F_2 + N \, (-\frac{\omega^2}{1-y} + \frac{m^2}{y(1-y)}) \, F_3$$

$$-\beta \, \frac{km}{y(1-y)} \, F_3 + 2\omega \frac{d\beta}{dy} F_3 - \beta \, (+\frac{\omega^2}{y} - \frac{k^2}{y(1-y)}) \, F_2 = 0 \; ,$$

$$2\omega \frac{d}{dy} \, G_3 + M \, \frac{km}{y(1-y)} \, F_2 + 2\omega \frac{dM}{dy} F_2 - M \, (-\frac{\omega^2}{1-y} + \frac{m^2}{y(1-y)}) \, F_3$$

$$+\alpha\,\frac{km}{y(1-y)}\,F_3 - 2\omega\frac{d\alpha}{dy}F_3 + \alpha(+\frac{\omega^2}{y} - \frac{k^2}{y(1-y)})\,F_2 = 0\ .$$

Instead of F_2, F_3, we substitute their expression through G_2, G_3 according to Eq. (11.4.5):

$$2\omega\frac{dG_2}{dy} + \left[-(N\alpha + \beta M)\,\frac{km}{y(1-y)} - 2\omega\frac{dN}{dy}\,\alpha + NM\,\frac{-\omega^2 y + m^2}{y(1-y)}\right.$$

$$\left. +2\omega\frac{d\beta}{dy}M - \beta\alpha\,\frac{\omega^2(1-y)-k^2}{y(1-y)}\right]\,G_2$$

$$+ \left[-2N\beta\,\frac{km}{y(1-y)} - 2\omega\frac{dN}{dy}\,\beta + N^2\frac{-\omega^2 y + m^2}{y(1-y)}\right.$$

$$\left. +2\omega\frac{d\beta}{dy}N - \beta^2\,\frac{\omega^2(1-y)-k^2}{y(1-y)}\right]\,G_3 = 0\ , \tag{11.4.8}$$

$$2\omega\frac{dG_3}{dy} + \left[(M\beta + \alpha N)\,\frac{km}{y(1-y)} + 2\omega\frac{dM}{dy}\,\beta - NM\,\frac{-\omega^2 y + m^2}{y(1-y)}\right.$$

$$\left. -2\omega\frac{d\alpha}{dy}N + \beta\alpha\,\frac{\omega^2(1-y)-k^2}{y(1-y)}\right]\,G_3$$

$$+ \left[2M\alpha\,\frac{km}{y(1-y)} + 2\omega\frac{dM}{dy}\,\alpha - M^2\,\frac{-\omega^2 y + m^2}{y(1-y)}\right.$$

$$\left. -2\omega\frac{d\alpha}{dy}M + \alpha^2\,\frac{\omega^2(1-y)-k^2}{y(1-y)}\right]\,G_2 = 0\ . \tag{11.4.9}$$

Let us assume that the transformation used is an orthogonal one:

$$\alpha G_2 + \beta G_3 = \cos A\,G_2 + \sin A\,G_3\ ,$$

$$M G_2 + N G_3 = -\sin A\,G_2 + \cos A\,G_3\ ,$$

then

$$-2\omega\frac{dN}{dy}\,\alpha + 2\omega\frac{d\beta}{dy}M = -2\omega[(\cos A)'\cos A + (\sin A)'\sin A] = 0\ ,$$

$$-2\omega\frac{dN}{dy}\,\beta + 2\omega\frac{d\beta}{dy}N = 2\omega[-(\cos A)'\sin A + (\sin A)'\cos A] = +\,2\omega\,A'\ ,$$

$$2\omega\frac{dM}{dy}\,\beta - 2\omega\frac{d\alpha}{dy}N = 2\omega[-(\sin A)'\sin A - (\cos A)'\cos A] = 0\ ,$$

$$2\omega\frac{dM}{dy}\,\alpha - 2\omega\frac{d\alpha}{dy}M = 2\omega[-(\sin A)'\cos A + (\cos A)'\sin A] = -\,2\omega\,A'\ ,$$

and

$$N\alpha + \beta M = \cos 2A\ , \qquad 2N\beta = \sin 2A\ , \qquad 2M\alpha = -\sin 2A\ ,$$

$$\alpha\beta = \sin A\cos A = \frac{1}{2}\sin 2A\ , \qquad NM = -\sin A\cos A = -\frac{1}{2}\sin 2A\ ,$$

$$N^2 = \cos^2 A\ , \qquad \beta^2 = \sin^2 A\ , \qquad M^2 = \sin^2 A\ , \qquad \alpha^2 = \cos^2 A.$$

Therefore, Eqs. (11.4.8) and (11.4.9) take the form

$$2\omega\frac{dG_2}{dy} - \left[\cos 2A\frac{km}{y(1-y)} + \frac{1}{2}\sin 2A\frac{-\omega^2 y + m^2 + \omega^2(1-y) - k^2}{y(1-y)}\right]G_2$$

$$+ \left[2\omega A' - \sin 2A\frac{km}{y(1-y)} + \cos^2 A\frac{-\omega^2 y + m^2}{y(1-y)}\right.$$

$$\left. - \sin^2 A\frac{\omega^2(1-y) - k^2}{y(1-y)}\right]G_3 = 0\;,$$

$$2\omega\frac{dG_3}{dy} + \left[\cos 2A\frac{km}{y(1-y)} + \frac{1}{2}\sin 2A\frac{-\omega^2 y + m^2 + \omega^2(1-y) - k^2}{y(1-y)}\right]G_3$$

$$+ \left[-2\omega A' - \sin 2A\frac{km}{y(1-y)} - \sin^2 A\frac{-\omega^2 y + m^2}{y(1-y)}\right.$$

$$\left. + \cos^2 A\frac{\omega^2(1-y) - k^2}{y(1-y)}\right]G_2 = 0\;. \tag{11.4.10}$$

In supposing that the used linear transformation does not depend on the coordinate y, we get more simple expressions

$$2\omega\frac{dG_2}{dy} - \left[\cos 2A\frac{km}{y(1-y)} + \frac{1}{2}\sin 2A\frac{-\omega^2 y + m^2 + \omega^2(1-y) - k^2}{y(1-y)}\right]G_2$$

$$+\frac{-2km\sin 2A + (1 + \cos 2A)[-\omega^2 y + m^2] - (1 - \cos 2A)[\omega^2(1-y) - k^2]}{2y(1-y)}G_3 = 0\;,$$

$$2\omega\frac{dG_3}{dy} + \left[\cos 2A\frac{km}{y(1-y)} + \frac{1}{2}\sin 2A\frac{-\omega^2 y + m^2 + \omega^2(1-y) - k^2}{y(1-y)}\right]G_3$$

$$+\frac{-2km\sin 2A - (1 - \cos 2A)[-\omega^2 y + m^2] + (1 + \cos 2A)[\omega^2(1-y) - k^2]}{2y(1-y)}G_2 = 0\;.$$

$$\tag{11.4.11}$$

Let

$$\cos 2A = 0\;, \qquad 2A = \frac{\pi}{2} \qquad \sin 2A = 1\;,$$

then the previous equations will read as

$$\left(2\omega\frac{d}{dy} - \frac{-\omega^2 y + \omega^2(1-y) + m^2 - k^2}{2y(1-y)}\right)G_2 + \frac{-\omega^2 + (m-k)^2}{2y(1-y)}\,G_3 = 0\;,$$

$$\left(2\omega\frac{d}{dy} + \frac{-\omega^2 y + \omega^2(1-y) + m^2 - k^2}{2y(1-y)}\right)G_3 + \frac{\omega^2 - (m+k)^2}{2y(1-y)}G_2 = 0\;. \tag{11.4.12}$$

It is remarkable that in the system produced, all the singularities are located at the points $y = 0, 1, \infty$ only. From Eq. (11.4.12) after excluding G_3 one straightforwardly gets an equation for G_2

$$G_3 = -2\omega\frac{2y(1-y)}{-\omega^2 + (m-k)^2}\frac{dG_2}{dy} + \frac{-\omega^2 y + \omega^2(1-y) + m^2 - k^2}{-\omega^2 + (m-k)^2}G_2\;,$$

$$4y(1-y)\frac{d^2G_2}{dy^2} + 4(1-2y)\frac{dG_2}{dy} + (2\omega + \omega^2 - \frac{m^2}{y(1-y)} + \frac{m^2-k^2}{1-y})G_2 = 0 \ .$$

$$(11.4.13)$$

Through the substitution $G_2 = y^A(1-y)^B G(y)$, Eq. (11.4.13) takes the form

$$4y(1-y)\,G'' + 4\left[A(1-y) - By + A(1-y) - By + (1-2y)\right]G'$$

$$+\left[4A(A-1)\frac{1}{y} + 4B(B-1)\frac{1}{1-y} - 4A(A-1) - 4B(B-1) - 8AB\right.$$

$$\left.+4(-2A - 2B + \frac{A}{y} + \frac{B}{1-y}) + 2\omega + \omega^2 - m^2(\frac{1}{y} + \frac{1}{1-y}) + \frac{m^2-k^2}{1-y}\right]G = 0 \ .$$

$$(11.4.14)$$

Requiring

$$4A(A-1) + 4A - m^2 = 0 \implies A = \pm\frac{1}{2}\mid m\mid ,$$

$$4B(B-1) + 4B - k^2 = 0 \implies B = \pm\frac{1}{2}\mid k\mid , \qquad (11.4.15)$$

we arrive at

$$y(1-y)G'' + [2A + 1 - 2(A + B + 1)y]G'$$

$$-[(A + B)(A + B + 1) - \frac{\omega}{2}(\frac{\omega}{2} + 1)]G = 0 \ ,$$

this is of the hypergeometric type

$$\gamma = 2A + 1 , \qquad \alpha + \beta = 2A + 2B + 1 ,$$

$$\alpha\beta = (A + B)(A + B + 1) - \frac{\omega}{2}(\frac{\omega}{2} + 1) ,$$

that is

$$\alpha = A + B - \frac{\omega}{2} , \qquad \beta = A + B + 1 + \frac{\omega}{2} , \qquad \gamma = 2A + 1 \ . \qquad (11.4.16)$$

The functions are finite on the sphere S_3 only if

$$A = +\frac{1}{2}\mid m\mid , \qquad B = +\frac{1}{2}\mid k\mid , \qquad \alpha = -n = 0, -1, -2, \ldots \qquad (11.4.17)$$

which led to the frequency spectrum in the form

$$\omega = 2(n + A + B) = 2n + \mid m\mid + \mid k\mid ; \qquad (11.4.17)$$

the parameters m and k are to be as an integer only, m , $k \in \{0, \pm 1, \pm 2, \ldots\}$. The function $G_2(y)$ is

$$G_2(y) = M_2 y^{\mid m\mid/2}(1-y)^{\mid k\mid/2}F(-n, n + 1 + \mid m\mid + \mid k\mid, \mid m\mid + 1; y) \ . \qquad (11.4.18)$$

In Eqs. (11.4.12), we might exclude G_2

$$G_2(y) = -2\omega\frac{2y(1-y)}{\omega^2 - (m+k)^2}\frac{dG_3}{dy} - \frac{-\omega^2 y + \omega^2(1-y) + m^2 - k^2}{\omega^2 - (m+k)^2}G_3 \ ,$$

$$4y(1-y)\frac{d^2G_3}{dy^2} + 4(1-2y)\frac{dG_3}{dy} + (-2\omega + \omega^2 - \frac{m^2}{y(1-y)} + \frac{m^2-k^2}{1-y})G_3 = 0 \ .$$

$$(11.4.19)$$

In using the substitution $G_3 = y^A(1-y)^B F(y)$, the equation for G_3 reduces to

$$4y(1-y)\ F'' + 4\ [A(1-y)\ -By+A(1-y)-By\ +(1-2y)\ ]\ \ F'$$

$$+\left[\ 4A(A-1)\frac{1}{y} + 4B(B-1)\frac{1}{1-y} - 4A(A-1) - 4B(B-1) - 8AB\right.$$

$$+4(-2A-2B+\frac{A}{y}+\frac{B}{1-y}) - 2\omega+\omega^2-m^2(\frac{1}{y}+\frac{1}{1-y})+\frac{m^2-k^2}{1-y}\Bigg]\ F = 0\ .$$

$$(11.4.20)$$

Requiring

$$4A(A-1)+4A-m^2=0\ \implies\ A=+\frac{1}{2}\mid m\mid\ ,$$

$$4B(B-1)+4B-k^2=0\ \implies\ B=+\frac{1}{2}\mid k\mid\ ,$$

we arrive at a hypergeometric type equation

$$y(1-y)\ F'' + [2A+1-2(A+B+1)y\ ]\ F'$$

$$-\left[(A+B)(A+B+1)-\frac{\omega}{2}(\frac{\omega}{2}-1)\right]\ F=0\ ,$$

$$a=A+B+1-\frac{\omega}{2}\ ,\qquad b=A+B+\frac{\omega}{2}\ ,\qquad c=2A+1\ .$$

Further, we get

$$a=A+B+1-\frac{\omega}{2}=-N,\qquad N=0,1,2,...,$$

$$\omega = 2(A+B+1+N) = \mid m \mid + \mid k \mid +2(1+N)\ ,\qquad \underline{N+1=n}\ ,$$

$$G_3 = M_3\ y^{\mid m\mid/2}\ (1-y)^{\mid k\mid/2}\ F(-n+1,\ n+\mid m\mid+\mid k\mid,\ \mid m\mid+1;\ y)\ .$$

$$(11.4.21)$$

It remains to find a relative factor in two functions G_2 and G_3:

$$G_2 = M_2 y^{\mid m\mid/2}(1-y)^{\mid k\mid/2}F(-n, n+1+\mid m\mid+\mid k\mid, \mid m\mid+1; y)\ ,$$

$$G_3 = M_3 y^{\mid m\mid/2}(1-y)^{\mid k\mid/2}F(-n+1, n+\mid m\mid+\mid k\mid, \mid m\mid+1; y)\ ,$$

$$(11.4.22)$$

and the relationship

$$G_3\ [(m-k)^2-\omega^2] = -4\omega\ y(1-y)\ \frac{dG_2}{dy} + [m^2-k^2+\omega^2(1-2y)]\ G_2$$

must hold. Using the expressions for G_2 G_3, we get

$$(m - k - \omega)\,(m - k + \omega)\,\frac{M_3}{M_2}\,F_3 = -4\omega[\frac{\mid m \mid}{2}(1 - y)F_2 - \frac{\mid k \mid}{2}y\,F_2$$

$$+y(1 - y)\frac{d}{dy}\,F_2] + [m^2 - k^2 + \omega^2(1 - 2y)]\,F_2\,.$$

It is sufficient to consider this equation for the point $y = 0$ only; this results in

$$-(\omega + m - k)\,(\omega - m + k)\,\frac{M_3}{M_2} = (\omega - \mid m \mid - k)\,(\omega - \mid m \mid + k)\,,$$

and therefore (M stands for a numerical constant)

$$M_2 = M\,(\omega + m - k)(\omega - m + k)\,,$$

$$M_3 = -M\,(\omega - \mid m \mid - k)(\omega - \mid m \mid + k)\,. \tag{11.4.23}$$

11.5   Maxwell Solutions in Elliptical Model

Let us consider the problem of Maxwell solutions in elliptical space S_3'. This space S_3' is a space of constant positive curvature also this differs from the spherical model in topological properties only: S_3 is 1-connected, S_3' is a 2-connected. The question is on the role of these differences for electromagnetic field solutions. To obtain explicit realizations for two models, it is convenient to use relations known in the theory of unitary and orthogonal groups. To each point in S_3, there exists a corresponding element in the unitary group $SU(2)$: $B = \sigma^0 n_0 - i\sigma^k n_k$. In turn, for each point in elliptic space S_3' there exists corresponding a element in $SO(3)$ parameterized by Gibbs 3-vector (see Fedorov book [246])

$$0(\mathbf{c}) = I + 2\,\frac{\mathbf{c}^\times + (\mathbf{c}^\times)^2}{(1 + \mathbf{c}^2)}\,, \quad (\mathbf{c}^\times)_{kl} = -\,\epsilon_{klj}\,c_j\,; \tag{11.5.1}$$

note that two infinite length vectors represent one; the same point in S_3'

$$\mathbf{c}_\infty^+ = +\infty\,\mathbf{c}_0\,, \quad \mathbf{c}_\infty^- = -\infty\,\mathbf{c}_0\,, \quad \mathbf{c}_0^2 = 1\,, \quad 0(\mathbf{c}^{\pm\infty}) = I + 2\,(\mathbf{c}_0^\times)^2\,. \tag{11.5.2}$$

Mapping $2 \to 1$ from $SU(2)$ to $SO(3)$ is given by $\mathbf{c} = \mathbf{n}/n_0$. Cylindrical coordinates (ρ, ϕ, z) in elliptic space can be defined by

$$c_1 = \frac{\tan \rho}{\cos \rho}\,\cos z\,, \quad c_2 = \frac{\tan \rho}{\cos z}\,\sin z\,, \quad c_3 = \tan\,z\,,$$

$$\tilde{G}\,, \qquad \rho \in [0, \pi/2]\,, \quad \phi \in [-\pi,\,+\pi]\,, \quad z \in [-\pi/2, +\pi/2]\,. \tag{11.5.3}$$

Additionally, we must define such an identification rule on the boundary of the region $\tilde{G}$, which agrees with the identification rule for vectors $\mathbf{c}_\infty^+$ and $\mathbf{c}_\infty^-$.

To this end, it is convenient to divide the region $\tilde{G}$ into three parts. For the region $\tilde{G}_1 = \tilde{G}(\rho \neq 0, \pi/2)$ identification is given in Fig. 11.1 (for more detail, see [41]): each

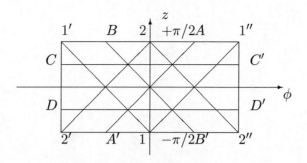

Figure 11.1: Domain $\tilde{G}_1$

pair (A, A') , (B, B'); and so on, represents the same point in elliptical model S'_3. Also $(1, 1', 1'')$ and $(2, 2', 2'')$ correspond to one respective point in S'_3.

Now, we should find what constructed Maxwell solutions in case of spherical model S_3 will be single-valued ones when considering elliptical model S'_3 (here, we examine only points parameterized by the region $\tilde{G}_1$). Evidently, it is sufficient to examine the behavior of the vector $f = e^{im\phi}\, e^{ikz}$. From the relations

$$f(C) = f(C') , \qquad f(D) = f(D') , \quad \ldots$$

No additional restrictions arise besides both M and K should be integers. Equations $f(1) = f(1') = f(1'')$ give

$$e^{-ik(\pi/2)} = e^{-im\pi}\, e^{+ik(\pi/2)} = e^{+im\pi}\, e^{+ik(\pi/2)} ,$$

Hence, it follows that $e^{i2m\pi} = 1$, $e^{i(k-m)\pi} = 1$, $e^{i(k+m)\pi} = 1$; and therefore, $(k - m)$ and $(k + m)$ must be even. The same results follows from the consideration of points $(2, 2', 2'')$. Therefore, Maxwell solutions constructed in spherical space will be single-valued solutions in elliptical space (in region $\tilde{G}_1$) only if m and k are both even, or both odd. Correspondingly, ω parameter N in the expression for the frequency spectrum given by

$$\omega = 2n + \mid m \mid + \mid k \mid = N \tag{11.5.4}$$

takes on even values: $N = 0, 2,\ 4,\ 6,\ \ldots$ Behavior of the wave solutions in two other domains, at the values $\rho = 0, \rho = \pi/2$, can be considered in the same manner. The result is that all functions $\Phi_{\omega mk}(\rho, \phi, z)$ at $\omega = N = 0,\ 2,\ 4,\ \ldots$ are single-valued and continuous in elliptic space, and they represent physical solutions for the Maxwell equation in this space, whereas all remaining functions $\Phi_{\omega mk}(\rho, \phi, z)$, $N = 1,\ 3,\ \ldots$ should be rejected as non-single-valued and discontinuous in S'_3 space.

Chapter 12

Solutions with Cylindric Symmetry in Lobachevsky Space

12.1 Cylindrical Coordinate and Tetrad in Lobachevsky Space H_3

Let us consider Maxwell equations in cylindric coordinates of hyperbolic Lobachevsky space H_3:

$$n_1 = \sinh r \cos \phi \,, \qquad n_2 = \sinh r \sin \phi \,,$$

$$n_3 = \cosh r \sinh z \,, \qquad n_4 = \cosh r \cosh z \,,$$

$$dS^2 = dt^2 - dr^2 - \sinh^2 r \, d\phi^2 - \cosh^2 r \, dz^2 \,, \qquad x^\alpha = (t, r, \phi, z) \,,$$

$$e_{(a)}^{\beta}(y) = \begin{vmatrix} 1 & 0 & 0 & 0 \\ 0 & 1 & 0 & 0 \\ 0 & 0 & \sinh^{-1} r & 0 \\ 0 & 0 & 0 & \cosh^{-1} r \end{vmatrix} \,, \qquad (12.1.1)$$

where $(r, \ \phi, \ z)$ run within $r \in [0, +\infty) \,$, $\phi \in [0, 2\pi] \,$, $z \in (-\infty, +\infty)$. Christoffel symbols are

$$\Gamma^0_{\beta\sigma} = 0 \,, \qquad \Gamma^i_{00} = 0 \,, \qquad \Gamma^i_{0j} = 0 \,, \qquad \Gamma^z_{\ jk} = \begin{vmatrix} 0 & 0 & \frac{\sinh r}{\cosh r} \\ 0 & 0 & 0 \\ \frac{\sinh r}{\cosh r} & 0 & 0 \end{vmatrix} \,,$$

$$\Gamma^r_{\ jk} = \begin{vmatrix} 0 & 0 & 0 \\ 0 & -\sinh r \cosh r & 0 \\ 0 & 0 & -\sinh r \cosh r \end{vmatrix} , \Gamma^\phi_{\ jk} = \begin{vmatrix} 0 & \frac{\cosh r}{\sinh r} & 0 \\ \frac{\cosh r}{\sinh r} & 0 & 0 \\ 0 & 0 & 0 \end{vmatrix} .$$

Derivatives of tetrad vectors read

$$e_{(0)\beta;\alpha} = 0 \,, \qquad e_{(1)\beta;\alpha} = \begin{vmatrix} 0 & 0 & 0 & 0 \\ 0 & 0 & 0 & \\ 0 & 0 & -\sinh r \cosh r & 0 \\ 0 & 0 & 0 & -\sinh r \cosh r \end{vmatrix} \,,$$

$$e_{(2)\beta;\alpha} = \begin{vmatrix} 0 & 0 & 0 & 0 \\ 0 & 0 & \cosh r & 0 \\ 0 & 0 & 0 & 0 \\ 0 & 0 & 0 & 0 \end{vmatrix}, \quad e_{(3)\beta;\alpha} = \begin{vmatrix} 0 & 0 & 0 & 0 \\ 0 & 0 & 0 & \sinh r \\ 0 & 0 & 0 & 0 \\ 0 & 0 & 0 & 0 \end{vmatrix}.$$

Ricci rotation coefficients are

$$\gamma_{011} = \gamma_{021} = \gamma_{031} = 0 \ , \ \ \gamma_{012} = \gamma_{022} = \gamma_{032} = 0 \ ,$$

$$\gamma_{013} = \gamma_{023} = \gamma_{033} = 0 \ , \ \ \gamma_{231} = 0 \ , \ \ \gamma_{311} = 0 \ , \ \ \gamma_{121} = 0 \ ,$$

$$\gamma_{232} = 0 \ , \ \ \gamma_{312} = 0 \ , \ \ \gamma_{122} = \frac{\cosh r}{\sinh r} \ ,$$

$$\gamma_{233} = 0 \ , \ \ \gamma_{313} = -\frac{\sinh r}{\cosh r} \ , \ \ \gamma_{123} = 0 \ . \tag{12.1.2}$$

For the relations

$$e^{\rho}_{(0)}\partial_\rho = \partial_{(0)} = \partial_t \ , \qquad e^{\rho}_{(1)}\partial_\rho = \partial_{(1)} = \partial_r \ ,$$

$$e^{\rho}_{(2)}\partial_\rho = \partial_{(2)} = \frac{1}{\sinh r}\partial_\phi \ , \qquad e^{\rho}_{(3)}\partial_\rho = \partial_{(3)} = \frac{1}{\cosh r}\partial_z \ ,$$

$$\mathbf{v}_0 = (\gamma_{010}, \gamma_{020}, \gamma_{030}) \equiv 0 \ , \qquad \mathbf{v}_1 = (\gamma_{011}, \gamma_{021}, \gamma_{031}) \equiv 0 \ ,$$

$$\mathbf{v}_2 = (\gamma_{010}, \gamma_{022}, \gamma_{032}) \equiv 0 \ , \qquad \mathbf{v}_3 = (\gamma_{013}, \gamma_{023}, \gamma_{033}) \equiv 0 \ ,$$

$$\mathbf{p}_0 = (\gamma_{230}, \gamma_{310}, \gamma_{120}) = 0 \ , \qquad \mathbf{p}_1 = (\gamma_{231}, \gamma_{311}, \gamma_{121}) = 0 \ ,$$

$$\mathbf{p}_2 = (\gamma_{232}, \gamma_{312}, \gamma_{122}) = (0, 0, \frac{\cosh r}{\sinh r}) \ ,$$

$$\mathbf{p}_3 = (\gamma_{233}, \gamma_{313}, \gamma_{123}) = (0, -\frac{\sinh r}{\cosh r}, 0) \ ,$$

Maxwell matrix equation in complex form will read as

$$\left(-i\partial_t + \alpha^1 \partial_r + \alpha^2 \frac{1}{\sinh r}\partial_\phi + \alpha^3 \frac{1}{\cosh r}\partial_z \right.$$

$$\left. +\alpha^2 s_3 \frac{\cosh r}{\sinh r} - \alpha^3 s_2 \frac{\sinh r}{\cosh r} \right) \begin{vmatrix} 0 \\ \mathbf{E} + ic\mathbf{B} \end{vmatrix} = 0 \ . \tag{12.1.3}$$

12.2   Separation of Variables in Space H_3

Maxwell matrix operator in Eq. (12.1.3) commutes with three operators: $i\partial_t, \ i\partial_\phi, \ i\partial_z$; therefore solutions can be constructed on the base of the following substitution

$$\Psi = \begin{vmatrix} 0 \\ \mathbf{E} + ic\mathbf{B} \end{vmatrix} = e^{-i\omega t} \ e^{im\phi} \ e^{ikz} \begin{vmatrix} 0 \\ f_1(r) \\ f_2(r) \\ f_3(r) \end{vmatrix} \ , \tag{12.2.1}$$

and Eq. (12.1.3) takes the form

$$
\left(-\omega + \alpha^1 \frac{d}{dr} + \frac{im}{\sinh r}\,\alpha^2 + \frac{ik}{\cosh r}\,\alpha^3 + \frac{\cosh r}{\sinh r}\,\alpha^2 s_3 - \frac{\sinh r}{\cosh r}\,\alpha^3 s_2 \right)
\begin{vmatrix} 0 \\ f_1 \\ f_2 \\ f_3 \end{vmatrix} = 0 \, .
$$

(12.2.2)

After these simple calculations, we get the radial system in the form

$$
(\frac{d}{dr} + \frac{\cosh r}{\sinh r} + \frac{\sinh r}{\cosh r})\, f_1 + \frac{im}{\sinh r}\, f_2 + \frac{ik}{\cosh r}\, f_3 = 0 \, ,
$$

$$
-\omega f_1 - \frac{ik}{\cosh r}\, f_2 + \frac{im}{\sinh r}\, f_3 = 0 \, ,
$$

$$
-\omega f_2 - (\frac{d}{dr} + \frac{\sinh r}{\cosh r})\, f_3 + \frac{ik}{\cosh r} f_1 = 0 \, ,
$$

$$
-\omega f_3 + (\frac{d}{dr} + \frac{\cosh r}{\sinh r})\, f_2 - \frac{im}{\sinh r} f_1 = 0 \, .
$$

(12.2.3)

12.3   Solutions at $m = 0$

First let $m = 0$, then

$$
(\frac{d}{dr} + \frac{\cosh r}{\sinh r} + \frac{\sinh r}{\cosh r})\, f_1 + \frac{ik}{\cosh r}\, f_3 = 0 \, ,
$$

$$
f_1 = \frac{-ik}{\omega\,\cosh r}\, f_2 \, ,
$$

$$
-\omega f_2 - (\frac{d}{dr} + \frac{\sinh r}{\cosh r})\, f_3 + \frac{ik}{\cosh r} f_1 = 0 \, ,
$$

$$
f_3 = \frac{1}{\omega}\,(\frac{d}{dr} + \frac{\cosh r}{\sinh r})\, f_2 \, .
$$

(12.3.1)

By the help of the 2nd and 4th equations from the 1st and the 3rd ones it follows the identity $0 = 0$ and equation for f_2:

$$
f_2(r) = \frac{1}{\sinh r} E(r) \, , \quad \frac{d^2 E}{dr^2} - \frac{1}{\sinh r \cosh r} \frac{dE}{dr} + (\omega^2 - \frac{k^2}{\cosh^2 r}) E = 0 \, ;
$$

(12.3.2)

besides

$$
f_1(r) = \frac{-ik}{\omega}\, \frac{1}{\cosh r \sinh r}\, E(r) \, , \qquad f_3 = \frac{1}{\omega}\, \frac{1}{\sinh r}\, \frac{d}{dr} E(r) \, .
$$

Eq. (12.3.2) can be resolved easily when $k^2 = \omega^2$

$$
\frac{\sinh r}{\cosh r}\, \frac{d}{dr}\, \frac{\cosh r}{\sinh r}\, \frac{d}{dr}\, E + k^2(1 - \frac{1}{\cosh^2 r})\, E = 0 \, ,
$$

from this, it follows

$$
(\frac{\cosh r}{\sinh r}\, \frac{d}{dr})\, (\frac{\cosh r}{\sinh r}\, \frac{d}{dr})\, E = -k^2\, E \, .
$$

(12.3.3)

Using the help of a new variable

$$\frac{\cosh r}{\sinh r}\frac{d}{dr} = \frac{d}{dx} \qquad \Longrightarrow \qquad \frac{dr}{dx} = \frac{\cosh r}{\sinh r} \, ,$$

$$dx = d\ln\cosh r \, , \qquad x = \ln(C \, \cosh r) \, , \qquad C = \text{const}$$

we arrive at

$$\frac{d^2}{dx^2}E = -k^2 \, E \, , \qquad E = e^{ikx} = \text{const} \, (\cosh r)^{ik} \, , \qquad k = \pm\,\omega \, . \qquad (12.3.4)$$

Two constructed solutions are conjugated

$$(\cosh r)^{+ik} = (e^{\ln\cosh r})^{+ik} = \cos(\,k\ln\cosh r\,) + i \, \sin(\,k\ln\cosh r\,) \, ,$$

$$(\cosh r)^{-ik} = [e^{\ln\cosh r}]^{-ik} = \cos(\,k\ln\cosh r\,] - i \, \sin(\,k\ln\cosh r\,) \, ,$$

$$(12.3.5)$$

therefore, this can now separate into two independent real ones

$$E_+(r) = \cos\,(k_0\ln\cosh r) \, , \qquad E_-(r) = \sin\,(k_0\ln\cosh r) \, . \qquad (12.3.6)$$

In the limit of a vanishing curvature, they reduce to those that are known

$$r \to 0 \, , \qquad E_+(r) = \cos\,(\,k_0\ln\cosh r\,) \quad \longrightarrow \quad +1 \, ,$$

$$r \to 0 \, , \qquad E_-(r) = \sin\,(\,k_0\ln\cosh r\,) \quad \longrightarrow \quad \text{const}\,r^2 \, ; \qquad (12.3.7)$$

these satisfy the equation in the flat space-time

$$\frac{d}{dr}\frac{1}{r}\frac{d}{dr}\,E(r) = 0 \qquad \Longrightarrow \qquad E(r) \sim 1,\,r^2 \, . \qquad (12.3.8)$$

In contrast to the flat space, in Lobachevsky model the waves are both oscillating at infinity. Let us show that Eq. (12.3.2) has another simple and more exact solution in the form

$$E(r) = \sinh^2 r \, \cosh^B r \, . \qquad (12.3.9)$$

Indeed, through the substitution of Eq. (12.3.9) into Eq. (12.3.2) results in

$$-2\,\cosh^2 r - B \, \sinh^2 r \, - k^2 \, \sinh^2 r + \omega^2 \, \sinh^2 r \, \cosh^2 r = 0 \, .$$

By the notation $\cosh^2 r = x$, this then is read as

$$x^2 \, (4 + 4B + \omega^2 + B^2) + x \, (-4 - 4B - \omega^2 - B^2) + x^0 \, (B^2 + k^2) = 0 \, .$$

The latter is only satisfied if

$$B^2 = -k^2 \, , \qquad (B + 2)^2 + \omega^2 = 0 \, , \qquad (12.3.10)$$

that is

$$B = -2 + i\,\omega,\ -2 - i\,\omega\,, \qquad k = \pm i\,B = \begin{cases} \mp\,(2i + \omega)\,, \\ \mp\,(2i - \omega)\,. \end{cases} \qquad (12.3.11)$$

Corresponding solutions appear as

$$E(t, r, z) = E_0\ \sinh^2 r\ \cosh^B r\ e^{-i(\omega t - kz)}\,, \qquad (12.3.12)$$

their real and imaginary parts are given by

$$B = -2 + i\omega\,,\ k = iB = -2i - \omega\,,$$

$$E(t, r, z) = E_0\ \sinh^2 r\ \cosh^{-2 + i\,\omega} r\ e^{i(-2zi - \omega z - \omega t)}$$

$$= E_0 \tanh^2 r e^{2z} \cos(\omega \ln \cosh r - \omega z - \omega t)$$

$$+ i E_0\ \tanh^2 r e^{2z} \sin(\omega \ln \cosh r - \omega z - \omega t)\,,$$

$$B = -2 + i\omega\,,\ k = -iB = 2i + \omega\,,$$

$$E(t, r.z) = E_0 \sinh^2 r \cosh^{-2 + i\,\omega} r e^{i(2zi + \omega z - \omega t)}$$

$$= E_0 \tanh^2 r e^{-2z} \cos(\omega \ln \cosh r + \omega z - \omega t)$$

$$+ i E_0 \tanh^2 r e^{-2z} \sin(\omega \ln \cosh r + \omega z - \omega t)\,,$$

$$B = -2 - i\omega\,,\ k = iB = -2i + \omega\,,$$

$$E(t, r.z) = E_0\ \sinh^2 r\ \cosh^{-2 - i\,\omega} r\ e^{i(-2zi + \omega z - \omega t)}$$

$$= E_0\ \tanh^2 r\ e^{2z} \cos(-\omega \ln \cosh r + \omega z - \omega t)$$

$$+ i E_0\ \tanh^2 r\ e^{2z} \sin(-\omega \ln \cosh r + \omega z - \omega t)\,,$$

$$B = -2 - i\omega\,,\ k = -iB = 2i - \omega\,,$$

$$E(t, r.z) = E_0\ \sinh^2 r\ \cosh^{-2 - i\,\omega} r\ e^{i(2zi - \omega z - \omega t)}$$

$$= E_0 \tanh^2 r e^{-2z} \cos(-\omega \ln \cosh r - \omega z - \omega t)$$

$$+ i E_0 \tanh^2 r e^{-2z} \sin(-\omega \ln \cosh r - \omega z - \omega t)\,. \qquad (12.3.13)$$

A physical sense of these waves is not clear. More insight can be reached when constructing all the possible solutions of Eq. (12.3.2) through the substitution

$$E(r) = \sinh^2 r\ \cosh^B r\ F(r)\,.$$

In that way, from Eq. (12.3.2) we can arrive at the equation of hypergeometric type (if $x = \sinh r$, $k^2 + B^2 = 0$)

$$4x(1 - x)\frac{d^2}{dx^2}F + 4[(1 + B) - (3 + B)x]\frac{d}{dx}F - [\,(B + 2)^2 + \omega^2\,]\,F = 0\,,$$

with

$$\gamma = 1 + B , \qquad \alpha + \beta = 2 + B , \qquad \alpha\beta = \frac{(B+2)^2 + \omega^2}{4} ,$$

that is

$$B = \pm i k , \quad \gamma = 1 + B , \quad \alpha = \frac{B + 2 - i\omega}{2} , \quad \beta = \frac{B + 2 + i\omega}{2} , \qquad (12.3.14a)$$

and

$$E(t, r, z) = \sinh^2 r \ \cosh^B r \ F(\alpha, \beta, \gamma, \cosh^2 r) \ e^{-i(\omega t - kz)} . \qquad (12.3.14b)$$

Evidently, these previously constructed solutions in Eq. (12.3.13) can be obtained from the general relations in Eqs. (12.3.14a) and (12.3.14b), if one demands $\alpha = 0$ or $\beta = 0$. However, no physical ground exists to impose such (polynomial) restrictions in the case of Lobachevsky space. Instead, the complete electromagnetic basis should include waves spreading along z, with real parameters k.

12.4   Solutions at $k = 0$

Now, let us turn to Eqs. (12.2.3) with $k = 0$

$$(\frac{d}{dr} + \frac{\cosh r}{\sinh r} + \frac{\sinh r}{\cosh r}) f_1 + \frac{im}{\sinh r} \ f_2 = 0 ,$$

$$f_1 = \frac{im}{\omega \sinh r} \ f_3 , \qquad f_2 = -\frac{1}{\omega}(\frac{d}{dr} + \frac{\sinh r}{\cosh r}) \ f_3 ,$$

$$-\omega f_3 + (\frac{d}{dr} + \frac{\cosh r}{\sinh r}) \ f_2 - \frac{im}{\sinh r} f_1 = 0 . \qquad (12.4.1)$$

The first and fourth equations gives

$$(\frac{d}{dr} + \frac{\cosh r}{\sinh r} + \frac{\sinh r}{\cosh r}) \frac{im}{\omega \sinh r} \ f_3 + \frac{im}{\sinh r} \ (-\frac{1}{\omega}(\frac{d}{dr} + \frac{\sinh r}{\cosh r}) \ f_3) = 0 ,$$

$$-\omega f_3 + (\frac{d}{dr} + \frac{\cosh r}{\sinh r}) \ (-\frac{1}{\omega}(\frac{d}{dr} + \frac{\sinh r}{\cosh r}) \ f_3) - \frac{im}{\sinh r} \frac{im}{\omega \sinh r} \ f_3 = 0 ;$$

they reduce, respectively, to the identity $0 \equiv 0$ and

$$f_3(r) = \frac{1}{\cosh r} \ E(r) , \quad \frac{d^2 E}{dr^2} + \frac{1}{\sinh r \cosh r} \frac{dE}{dr} + (\omega^2 - \frac{m^2}{\sinh^2 r}) E = 0 . \quad (12.4.2)$$

In variable $y = -\sinh^2 r$, the last is rewritten as

$$-4y(1 - y)\frac{d^2}{dy^2} \ E - 4(1 - y)\frac{d}{dy} \ E + (\omega^2 + \frac{m^2}{y}) \ E = 0 , \qquad (12.4.3)$$

that is solved in hypergeometric functions $E = y^a(1 - y)^b Y(y)$

$$4y(1 - y)\frac{d^2}{dy^2}Y + 4[1 + 2a - (2a + 2b + 1)y] \frac{d}{dy}Y$$

$$+ \left[-4(a+b)^2 - \omega^2 + (4a^2 - m^2)\frac{1}{y} + 4b(b-1)\frac{1}{1-y} \right] Y = 0 \, . \qquad (12.4.4)$$

$$m = \pm 2a \, , \qquad b = 1 \, , \qquad b = 0 \, , \qquad (12.4.5)$$

we get an equation of hypergeometric type with

$$\gamma = 1 + 2a \, , \qquad \alpha + \beta = 2a + 2b \, , \qquad \alpha\beta = \frac{4(a+b)^2 + \omega^2}{4} \, ,$$

that is

$$\gamma = 1 + 2a \, , \qquad \alpha = a + b \mp \frac{i\omega}{2} \, , \qquad \beta = a + b \pm \frac{i\omega}{2} \, . \qquad (12.4.6)$$

12.5 Solutions with Arbitrary m, k

Now, let us consider radial equations in the general case of Eq. (12.2.3); the first equation in Eq. (12.2.3) turns to be the identity $0 = 0$ when the three remaining hold

$$-\omega f_1 = \frac{ik}{\cosh r} \, f_2 - \frac{im}{\sinh r} \, f_3 \, ,$$

$$-\omega f_2 = \left(\frac{d}{dr} + \frac{\sinh r}{\cosh r} \right) f_3 - \frac{ik}{\cosh r} f_1 \, ,$$

$$-\omega f_3 = -\left(\frac{d}{dr} + \frac{\cosh r}{\sinh r} \right) f_2 + \frac{im}{\sinh r} f_1 \, ;$$

with substitutions

$$f_2 = \frac{1}{\sinh r} \, F_2 \, , \qquad f_3 = \frac{1}{\cosh r} \, F_3 \, ,$$

one obtains

$$-\omega \, f_1 = i \, \frac{k \, F_2 - m \, F_3}{\sinh r \, \cosh r} \, ,$$

$$-\omega \, \frac{F_2}{\sinh r} = \frac{1}{\cosh r} \frac{dF_3}{dr} - \frac{ik}{\cosh r} \, f_1 \, ,$$

$$-\omega \, \frac{F_3}{\cosh r} = -\frac{1}{\sinh r} \frac{dF_2}{dr} + \frac{im}{\sinh r} \, f_1 \, . \qquad (12.5.1)$$

After excluding f_1, we get

$$\left(\frac{\omega}{\cosh r} \frac{d}{dr} + \frac{km}{\sinh r \, \cosh^2 r} \right) F_3 + \left(\frac{\omega^2}{\sinh r} - \frac{k^2}{\sinh r \, \cosh^2 r} \right) F_2 = 0 \, ,$$

$$\left(\frac{\omega}{\sinh r} \frac{d}{dr} - \frac{km}{\cosh r \, \sinh^2 r} \right) F_2 - \left(\frac{\omega^2}{\cosh r} - \frac{m^2}{\cosh r \, \sinh^2 r} \right) F_3 = 0 \, .$$

$$(12.5.2)$$

In variable $\cosh \, 2r - 1 = 2y$, Eqs. (12.5.2) takes the form

$$\left(2\omega \frac{d}{dy} - \frac{km}{y(1+y)} \right) F_2 + \left(-\frac{\omega^2}{1+y} + \frac{m^2}{y(1+y)} \right) F_3 = 0 \, ,$$

$$(2\omega \frac{d}{dy} + \frac{km}{y(1+y)}) \, F_3 + (+\frac{\omega^2}{y} - \frac{k^2}{y(1+y)}) \, F_2 = 0 \, . \qquad (12.5.3)$$

Let us introduce new functions by means of a linear transformation (with unit determinant $\alpha N - \beta M = 1$ and numerical parameters)

$$F_2 = \alpha \, G_2 + \beta \, G_3 \, , \qquad F_3 = M \, G_2 + N \, G_3 \, ; \qquad (12.5.4a)$$

and inverse one given by

$$G_2 = N \, F_2 - \beta \, F_3 \, , \qquad G_3 = -M \, F_2 + \alpha \, F_3 \, . \qquad (12.5.4b)$$

By combining equations in Eq. (12.5.3), we get

$$2\omega \frac{d}{dy} G_2 - N \frac{km}{y(1+y)} F_2 + N(-\frac{\omega^2}{1+y} + \frac{m^2}{y(1+y)}) F_3$$

$$-\beta \frac{km}{y(1+y)} F_3 - \beta(+\frac{\omega^2}{y} - \frac{k^2}{y(1+y)}) F_2 = 0 \, ,$$

$$2\omega \frac{d}{dy} G_3 + M \frac{km}{y(1+y)} F_2 - M(-\frac{\omega^2}{1+y} + \frac{m^2}{y(1+y)}) F_3$$

$$+\alpha \frac{km}{y(1+y)} F_3 + \alpha(+\frac{\omega^2}{y} - \frac{k^2}{y(1+y)}) F_2 = 0 \, .$$

Now, by expressing F_2, F_3 through G_2, G_3, according to Eq. (12.5.3)

$$2\omega \frac{dG_2}{dy} + \left[-(N\alpha + \beta M) \frac{km}{y(1+y)} + NM \frac{-\omega^2 y + m^2}{y(1+y)} - \beta\alpha \frac{\omega^2(1+y) - k^2}{y(1+y)} \right] G_2$$

$$+ \left[-2N\beta \frac{km}{y(1+y)} + N^2 \frac{-\omega^2 y + m^2}{y(1+y)} - \beta^2 \frac{\omega^2(1+y) - k^2}{y(1-y)} \right] G_3 = 0 \, ,$$

$$(12.5.5a)$$

$$2\omega \frac{dG_3}{dy} + \left[(M\beta + \alpha N) \frac{km}{y(1+y)} - NM \frac{-\omega^2 y + m^2}{y(1+y)} + \beta\alpha \frac{\omega^2(1+y) - k^2}{y(1+y)} \right] G_3$$

$$+ \left[2M\alpha \frac{km}{y(1+y)} - M^2 \frac{-\omega^2 y + m^2}{y(1+y)} + \alpha^2 \frac{\omega^2(1+y) - k^2}{y(1+y)} \right] G_2 = 0 \, .$$

$$(12.5.5b)$$

Let us detail the coefficients at G_3 and G_2:

$$\left[-2N\beta \frac{km}{y(1+y)} + N^2 \frac{-\omega^2 y + m^2}{y(1+y)} - \beta^2 \frac{\omega^2(1+y) - k^2}{y(1-y)} \right] G_3$$

$$= \frac{1}{y(1+y)} \left[-2km \, N\beta - \omega^2(N^2 + \beta^2) \, y + N^2 \, m^2 + \beta^2 \, k^2 \right] G_3 \, ,$$

$$\left[2M\alpha \frac{km}{y(1+y)} - M^2 \frac{-\omega^2 y + m^2}{y(1+y)} + \alpha^2 \frac{\omega^2(1+y) - k^2}{y(1+y)} \right] G_2$$

$$= \frac{1}{y(1+y)} \left[2km\, M\alpha + \omega^2(M^2 + \alpha^2)\, y - M^2\, m^2 - \alpha^2\, k^2 \right] G_2 \; .$$

Redundant singularities will be excluded if

$$N^2 + \beta^2 = 0 \; , \qquad M^2 + \alpha^2 = 0 \; . \tag{12.5.6}$$

Through an additional assumption of unitarity for the transformations in Eq. (12.5.4a,b)

$$\alpha\, G_2 + \beta\, G_3 = \cos A\; G_2 + i \sin A\; G_3 \; ,$$

$$M\, G_2 + N\, G_3 = i \sin A\; G_2 + \cos A\; G_3 \; , \tag{12.5.7}$$

we get

$$N\alpha + \beta M = \cos 2A \; , \qquad 2N\beta = i\,\sin 2A \; , \qquad 2M\alpha = i\,\sin 2A \; ,$$

$$\alpha\beta = i\,\sin A \cos A = \frac{i}{2}\sin 2A \; , \qquad NM = i\,\sin A \cos A = \frac{i}{2}\sin 2A \; ,$$

$$N^2 = \cos^2 A \; , \qquad \beta^2 = -\sin^2 A \; , \qquad M^2 = -\sin^2 A \; , \qquad \alpha^2 = \cos^2 A \; ;$$

Correspondingly, as in the previous equations, then give

$$2\omega \frac{dG_2}{dy} + \left[-\cos 2A \frac{km}{y(1+y)} + \frac{i}{2}\sin 2A \frac{-\omega^2 y + m^2 - \omega^2(1+y) + k^2}{y(1+y)} \right] G_2$$

$$+ \left[-i\sin 2A \frac{km}{y(1+y)} + \cos^2 A \frac{-\omega^2 y + m^2}{y(1+y)} + \sin^2 A \frac{\omega^2(1+y) - k^2}{y(1+y)} \right] G_3 = 0 \; ,$$

$$2\omega \frac{dG_3}{dy} + \left[\cos 2A \frac{km}{y(1+y)} - \frac{i}{2}\sin 2A \frac{-\omega^2 y + m^2 - \omega^2(1+y) + k^2}{y(1+y)} \right] G_3$$

$$+ \left[i\sin 2A \frac{km}{y(1+y)} + \sin^2 A \frac{-\omega^2 y + m^2}{y(1+y)} + \cos^2 A \frac{\omega^2(1+y) - k^2}{y(1+y)} \right] G_2 = 0 \; . \tag{12.5.8}$$

By an additional requirement

$$A = \pi/4 \; , \qquad \cos^2 A = \sin^2 A = \frac{1}{2} \; , \qquad \sin 2A = 1 \; , \qquad \cos 2A = 0 \; ,$$

the system becomes much more simple

$$\left(2\omega \frac{d}{dy} - i\frac{\omega^2 y + \omega^2(1+y) - m^2 - k^2}{2y(1+y)} \right) G_2 + \frac{\omega^2 + (m - ik)^2}{2y(1+y)}\, G_3 = 0 \; ,$$

$$\left(2\omega \frac{d}{dy} + i\frac{\omega^2 y + \omega^2(1+y) - m^2 - k^2}{2y(1+y)} \right) G_3 + \frac{\omega^2 + (m + ik)^2}{2y(1+y)}\, G_2 = 0 \; . \tag{12.5.9}$$

From this, it follows

$$G_3 = -2\omega \frac{2y(1+y)}{\omega^2 + (m-ik)^2} \frac{dG_2}{dy} + i\frac{\omega^2 y + \omega^2(1+y) - m^2 - k^2}{\omega^2 + (m-ik)^2} G_2 \,,$$

$$4y(1+y)\frac{d^2G_2}{dy^2} + 4(1+2y)\frac{dG_2}{dy} + \left(-2i\omega + \omega^2 - \frac{m^2}{y(1+y)} - \frac{m^2+k^2}{1+y}\right) G_2 = 0 \,.$$

$$(12.5.10)$$

After changing the variable y to $-y$, the equation for G_2 will read as

$$4y(1-y)\frac{d^2G_2}{dy^2} + 4(1-2y)\frac{dG_2}{dy} - \left(-2i\omega + \omega^2 + \frac{m^2}{y(1-y)} - \frac{m^2+k^2}{1-y}\right) G_2 = 0 \,.$$

$$(12.5.11)$$

Through substitution $G_2 = y^A(1-y)^B G(y)$, we get

$$4y(1-y)\frac{d^2G}{dy^2} + 4[1 + 2A - (2A + 2B + 1 + 1)y]\frac{dG}{dy}$$

$$- \left[-\omega(2i - \omega) + 4(A+B)(A+B+1) - \frac{4A^2 - m^2}{y} - \frac{4B^2 + k^2}{1-y}\right] G = 0 \,.$$

By additional requirements $m = \pm 2A$, $k = \pm 2iB$ it becomes an equation of hypergeometric type

$$4y(1-y)\frac{d^2G}{dy^2} + 4[1 + 2A - (2A + 2B + 1 + 1)y]\frac{dG}{dy}$$

$$- [-\omega(2i - \omega) + 4(A+B)(A+B+1)] G = 0 \,,$$

$$\gamma = 1 + 2A \,, \qquad \alpha + \beta = 2A + 2B + 1 \,,$$

$$\alpha\beta = \frac{-\omega(2i - \omega) + 4(A+B)(A+B+1)}{4} \,, \qquad (12.5.12)$$

that is

$$\alpha = A + B - \frac{i\omega}{2} \,, \qquad \beta = A + B + 1 + \frac{i\omega}{2} \,. \qquad (12.5.13)$$

In the same manner, from Eqs. (12.5.9) it follows

$$G_2 = -2\omega \frac{2y(1+y)}{\omega^2 + (m+ik)^2} \frac{dG_3}{dy} - i\frac{\omega^2 y + \omega^2(1+y) - m^2 - k^2}{\omega^2 + (m+ik)^2} G_3 \,,$$

$$2\omega \frac{d}{dy} \left(-2\omega \frac{2y(1+y)}{\omega^2 + (m+ik)^2} \frac{dG_3}{dy} - i\frac{\omega^2 y + \omega^2(1+y) - m^2 - k^2}{\omega^2 + (m+ik)^2} G_3\right)$$

$$-i\frac{\omega^2 y + \omega^2(1+y) - m^2 - k^2}{2y(1+y)} \left(-2\omega \frac{2y(1+y)}{\omega^2 + (m+ik)^2} \frac{dG_3}{dy}\right.$$

$$\left.-i\frac{\omega^2 y + \omega^2(1+y) - m^2 - k^2}{\omega^2 + (m+ik)^2} G_3\right) + \frac{\omega^2 + (m-ik)^2}{2y(1+y)} G_3 = 0 \,,$$

$$(12.5.14)$$

or

$$4y(1+y)\frac{d^2G_3}{dy^2} + 4(1+2y)\frac{dG_3}{dy}$$

$$+ \left(2i\omega + \omega^2 - \frac{m^2}{y(1+y)} - \frac{m^2+k^2}{1+y} \right) G_3 = 0 . \qquad (12.5.15)$$

After changing the variable y to $-y$

$$4y(1-y)\frac{d^2G_3}{dy^2} + 4(1-2y)\frac{dG_3}{dy} - \left(2i\omega + \omega^2 + \frac{m^2}{y(1-y)} - \frac{m^2+k^2}{1-y} \right) G_3 = 0 ,$$

and with the substitution $G_3 = y^A(1-y)^B G(y)$, we get

$$4y(1-y)\frac{d^2G}{dy^2} + 4[1+2A-(2A+2B+1+1)y]\frac{dG}{dy}$$

$$- \left[\omega(2i+\omega) + 4(A+B)(A+B+1) - \frac{4A^2-m^2}{y} - \frac{4B^2+k^2}{1-y} \right] G = 0 . \quad (12.5.16)$$

Introducing the help of the additional restriction $m = \pm 2A$, $k = \pm 2iB$ the latter reads as an equation of hypergeometric type

$$4y(1-y)\frac{d^2G}{dy^2} + 4[1+2A-(2A+2B+1+1)y]\frac{dG}{dy}$$

$$- [\omega(2i+\omega) + 4(A+B)(A+B+1)] G = 0 ,$$

$$c = 1+2A , \qquad a+b = 2A+2B+1 ,$$

$$ab = \frac{\omega(2i+\omega) + 4(A+B)(A+B+1)}{4} , \qquad (12.5.17)$$

that is

$$a = A+B+1-\frac{i\omega}{2} , \qquad b = A+B+\frac{i\omega}{2} . \qquad (12.5.18)$$

Let us find a relative factor in two functions G_2 and G_3 starting from

$$G_2 = M_2\, y^{|m|/2}\, (1-y)^{|k|/2i}\, F\left(-n,\ n+1+|m|+\frac{|k|}{i},\ |m|+1;\ y\right)$$

$$= M_2\, y^{(c-1)/2}(1-y)^{(a+b-c)/2}F(a,b,c,y) ,$$

$$G_3 = M_3\, y^{|m|/2i}\, (1-y)^{|k|/2i}\, F\left(-n+1,\ n+|m|+\frac{|k|}{i},\ |m|+1;\ y\right)$$

$$= M_3\, y^{(c-1)/2}(1-y)^{(a+b-c)/2}F(a+1,b-1,c,y) ,$$

and by allowing for

$$G_3\,[(m-ik)^2+\omega^2] = -4\omega\, y(1-y)\frac{dG_2}{dy} + i[-m^2-k^2+\omega^2(1-2y)]\, G_2 ,$$

we arrive at

$$(m - ik - i\omega)\,(m - ik + i\omega)\,\frac{M_3}{M_2}\,F_3(y)$$

$$= -4\omega\,[\,\frac{|\,m\,|}{2}\,(1 - y)\,F_2(y) - \frac{|\,k\,|}{2i}\,y\,F_2(y)$$

$$+ y(1 - y)\,\frac{d}{dy}\,F_2(y)\,] + i[-m^2 - k^2 + \omega^2(1 - 2y)]\,F_2(y)\,. \qquad (12.5.19)$$

To find the relative factor it is sufficient to consider the latter at the point $y = 0$ which results in

$$i(-i\omega + m - ik)\,(i\omega + m - ik)\,\frac{M_3}{M_2} = (-i\omega + |\,m\,| - ik)\,(i\omega + |\,m\,| + ik)\,,$$

that is

$$M_2 = iM\,(-i\omega + m - ik)(i\omega + m - ik)\,,$$

$$M_3 = M\,(-i\omega + |\,m\,| - ik)(i\omega + |\,m\,| + ik)\,. \qquad (12.5.20)$$

Chapter 13

On Simulating a Medium with Special Properties by Lobachevsky Geometry

13.1 Introduction

The aim of this chapter is to obtain exact solutions for the Maxwell equations in 3-dimensional Lobachevsky space H_3. A coordinate system used is one from the list given by Olevsky [42], which generalizes Cartesian coordinate in flat Euclidean space.

   For the treatment of Maxwell equations, we make use of a complex representation according to the known approach by Riemann–Silberstein–Oppenheimer–Majorana. On the basis of this technique, new exact solutions for the extended plane wave in Lobachevsky space have been constructed explicitly. These may be interesting in the cosmological sense; besides, of interest in the context of geometric simulating electromagnetic field in a special medium.

13.2 Cartesian Coordinates in Lobachevsky Space

We will use the following (horospherical) coordinate system in Lobachevsky space H_3 as specified

$$x^a = (t, x, y, z), \qquad dS^2 = dt^2 - e^{-2z}(dx^2 + dy^2) - dz^2 , \qquad (13.2.1)$$

the element of volume is given by

$$dV = \sqrt{-g}\, dxdydz = e^{-2z} dxdydz , \qquad x, y, z \in (-\infty, +\infty) ;$$

the magnitude and sign of the z are substantial, in particular, when dealing with localization. For example, as in the energy of the electromagnetic field

$$dW = \frac{1}{2}(\mathbf{E}^2 + \mathbf{B}^2)dV = \frac{1}{2}(\mathbf{E}^2 + \mathbf{B}^2)\, e^{-2z}\, dxdydz . \qquad (13.2.2)$$

It is helpful to have at hand some details of the parametrization of the model H_3 by (x, y, z). It is known that this model can be identified with a branch of hyperboloid in 4-dimension flat space

$$u_0^2 - u_1^2 - u_2^2 - u_3^2 = \rho^2 , \qquad u_0 = +\sqrt{\rho^2 + \mathbf{u}^2} .$$

Coordinate by using, (x, y, z), as referred to u_a by relations

$$u_1 = xe^{-z} , \quad u_2 = ye^{-z} ,$$

$$u_3 = \frac{1}{2}[(e^z - e^{-z}) + (x^2 + y^2)e^{-z}] ,$$

$$u_0 = \frac{1}{2}[(e^z + e^{-z}) + (x^2 + y^2)e^{-z}] . \tag{13.2.3}$$

It is convenient to employ 3-dimensional Poincaré realization for Lobachevsky space for the inside part of 3-sphere

$$q_i = \frac{u_i}{u_0} = \frac{u_i}{\sqrt{\rho^2 + u_1^2 + u_2^2 + u_3^2}} , \qquad q_i q_i < +1 . \tag{13.2.4}$$

Quasi-Cartesian coordinates (x, y, z) are referred to q_i as follows

$$q_1 = \frac{2x}{x^2 + y^2 + e^{2z} + 1} , \quad q_2 = \frac{2y}{x^2 + y^2 + e^{2z} + 1} , \quad q_3 = \frac{x^2 + y^2 + e^{2z} - 1}{z^2 + y^2 + e^{2z} + 1} . \tag{13.2.5}$$

Inverses to Eq. (13.2.5) relations are

$$x = \frac{q_1}{1 - q_3} , \qquad y = \frac{q_2}{1 - q_3} , \qquad e^z = \frac{\sqrt{1 - q^2}}{1 - q_3} . \tag{13.2.6}$$

In particular, note that on the axis $q_1 = 0, q_2 = 0, q \in (-1, +1)$ relations in Eq. (13.2.6) assume the form

$$x = 0 , \qquad y = 0 , \qquad e^z = \sqrt{\frac{1 + q_3}{1 - q_3}} ,$$

that is

$$q_3 \longrightarrow +1 , \qquad e^z \longrightarrow +\infty , \qquad z \longrightarrow +\infty ;$$

$$q_3 \longrightarrow -1 , \qquad e^z \longrightarrow +0 , \qquad z \longrightarrow -\infty . \tag{13.2.7}$$

Solutions of the Maxwell equation, constructed here, can be of interest in the context for the description of electromagnetic waves in special media, because the Lobachevsky geometry simulates effectively a definite special medium (see chapter **1**), inhomogeneous along the axis z. Effective electric permittivity tensor $\epsilon^{ik}(x)$ is given by

$$\epsilon^{ik}(x) = -\sqrt{-g} \, g^{00}(x)g^{ik}(x) = \begin{vmatrix} 1 & 0 & 0 \\ 0 & 1 & 0 \\ 0 & 0 & e^{-2z} \end{vmatrix} , \tag{13.2.8}$$

whereas the corresponding effective magnetic permittivity tensor is

$$(\mu^{-1})^{ik}(x) = \sqrt{-g} \begin{vmatrix} g^{22}g^{33} & 0 & 0 \\ 0 & g^{33}g^{11} & 0 \\ 0 & 0 & g^{11}g^{22} \end{vmatrix} = \begin{vmatrix} 1 & 0 & 0 \\ 0 & 1 & 0 \\ 0 & 0 & e^{2z} \end{vmatrix} . \tag{13.2.9}$$

In the explicit form, effective constitutive relations (the system *SI* is used) are

$$D^i = \epsilon_0 \epsilon^{ik} E_k , \qquad B_i = \mu_0 \mu^{ik} H^k , \tag{13.2.10}$$

note that two matrices coincide: $\epsilon^{ik}(x) = \mu^{ik}(x)$.

13.3   Tetrads and Maxwell Equations in Complex Form

In the coordinate of Eq. (13.2.1), let us introduce a tetrad

$$e^{\beta}_{(a)} = \begin{vmatrix} 1 & 0 & 0 & 0 \\ 0 & e^z & 0 & 0 \\ 0 & 0 & e^z & 0 \\ 0 & 0 & 0 & 1 \end{vmatrix} . \tag{13.3.1}$$

One should find Christoffel symbols; some of them evidently vanish: $\Gamma^0_{\beta\sigma} = 0$, $\Gamma^i_{00} = 0$, $\Gamma^i_{0j} = 0$, remaining ones are determined by relations

$$\Gamma^x_{jk} = \begin{vmatrix} 0 & 0 & -1 \\ 0 & 0 & 0 \\ -1 & 0 & 0 \end{vmatrix} , \Gamma^y_{jk} = \begin{vmatrix} 0 & 0 & 0 \\ 0 & 0 & -1 \\ 0 & -1 & 0 \end{vmatrix} , \Gamma^z_{jk} = \begin{vmatrix} e^{-2z} & 0 & 0 \\ 0 & e^{-2z} & 0 \\ 0 & 0 & 0 \end{vmatrix} .$$

Ricci rotation coefficients are (only the non-vanishing are written down)

$$\gamma_{311} = -1 , \qquad \gamma_{232} = 1 .$$

Using the notation

$$e^{\rho}_{(0)}\partial_\rho = \partial_{(0)} = \partial_t , \qquad e^{\rho}_{(1)}\partial_\rho = \partial_{(1)} = e^z \partial_x ,$$

$$e^{\rho}_{(2)}\partial_\rho = \partial_{(2)} = e^z \partial_y , \qquad e^{\rho}_{(3)}\partial_\rho = \partial_{(3)} = \partial_z ,$$

$$\mathbf{v}_0 = (\gamma_{010}, \gamma_{020}, \gamma_{030}) \equiv 0 , \qquad \mathbf{v}_1 = (\gamma_{011}, \gamma_{021}, \gamma_{031}) \equiv 0 ,$$

$$\mathbf{v}_2 = (\gamma_{0120}, \gamma_{022}, \gamma_{032}) \equiv 0 , \qquad \mathbf{v}_3 = (\gamma_{013}, \gamma_{02\,3}, \gamma_{033}) \equiv 0 ,$$

$$\mathbf{p}_0 = (\gamma_{230}, \gamma_{310}, \gamma_{120}) = 0 , \qquad \mathbf{p}_1 = (\gamma_{231}, \gamma_{311}, \gamma_{121}) = (0, -1, 0) ,$$

$$\mathbf{p}_2 = (\gamma_{232}, \gamma_{312}, \gamma_{122}) = (1, 0, 0) , \qquad \mathbf{p}_3 = (\gamma_{233}, \gamma_{313}, \gamma_{123}) = 0 ,$$

the Maxwell equations in the complex matrix form read

$$\left[\alpha^k \partial_{(k)} + \mathbf{s}\mathbf{v}_0 + \alpha^k \mathbf{s}\mathbf{p}_k - i \left(\partial_{(0)} + \mathbf{s}\mathbf{p}_0 - \alpha^k \mathbf{s}\mathbf{v}_k \right) \right] \begin{vmatrix} 0 \\ \mathbf{E} + i\mathbf{B} \end{vmatrix} = 0 ; \tag{13.3.2}$$

in the used retrad it assumes the form

$$\left(-i\partial_t + \alpha^1 e^z \partial_x + \alpha^2 e^z \partial_y + \alpha^3 \partial_z - \alpha^1 s_2 + \alpha^2 s_1 \right) \begin{vmatrix} 0 \\ \mathbf{E} + i\mathbf{B} \end{vmatrix} = 0 . \tag{13.3.3}$$

13.4   Separation of the Variables

Let us use the substitution

$$\begin{vmatrix} 0 \\ \mathbf{E} + i\mathbf{B} \end{vmatrix} = e^{-i\omega t}\, e^{ik_1 x}\, e^{ik_2 y} \begin{vmatrix} 0 \\ \mathbf{f}(z) \end{vmatrix} , \qquad (13.4.1)$$

Correspondingly, Eq. (13.4.1) gives

$$\left(-\omega + \alpha^1 e^z i k_1 + \alpha^2 e^z i k_2 + \alpha^3 \frac{d}{dz} - \alpha^1 s_2 + \alpha^2 s_1 \right) \begin{vmatrix} 0 \\ f_1(z) \\ f_2(z) \\ f_3(z) \end{vmatrix} = 0 . \qquad (13.4.2)$$

After the simple calculation, we derive a first order system for f_i:

$$ik_1\, e^z f_1 + ik_2\, e^z\, f_2 + (\frac{d}{dz} - 2) f_3 = 0 ,$$

$$-\omega f_1 - (\frac{d}{dz} - 1) f_2 + ik_2\, e^z\, f_3 = 0 ,$$

$$-\omega f_2 + (\frac{d}{dz} - 1) f_1 - ik_1\, e^z\, f_3 = 0 ,$$

$$-\omega f_3 - e^z\, ik_2 f_1 + ik_1\, e^z\, f_2 = 0 . \qquad (13.4.3)$$

In allowing the three last equations in the first one, we get an identity $0 = 0$. So, there exist only three independent equations (here, the notation $k_1 = a,\, k_2 = b$ is used):

$$\omega f_3 = -ib\, e^z\, f_1 + ia\, e^z\, f_2 ,$$

$$\omega f_1 = -(\frac{d}{dz} - 1) f_2 + ib\, e^z\, f_3 , \qquad \omega f_2 = +(\frac{d}{dz} - 1) f_1 - ia\, e^z\, f_3 . \qquad (13.4.4)$$

By the substitutions $f_1 = e^z F_1(z)$, $f_2 = e^z F_2(z)$, Eqs. (13.4.4) give

$$\omega f_3 = -ib\, e^{2z}\, F_1 + ia\, e^{2z}\, F_2 ,$$

$$\omega F_1 = -\frac{d}{dz} F_2 + ib\, f_3 , \qquad \omega F_2 = +\frac{d}{dz} F_1 - ia\, f_3 . \qquad (13.4.5)$$

There exist a particular case readily treatable, when $a = 0,\ b = 0,\ f_3 = 0$:

$$\omega F_1 = -\frac{d}{dz} F_2 , \qquad \omega F_2 = +\frac{d}{dz} F_1 \implies$$

$$F_1(z) = e^{\pm i\omega z} , \qquad F_2 = \pm i\, e^{\pm i\omega z} , \qquad (13.4.6)$$

which gives

$$\Phi^{\pm} = \begin{vmatrix} 0 \\ \mathbf{E} + i\mathbf{B} \end{vmatrix} = e^{-i\omega t} e^z \begin{vmatrix} 0 \\ e^{\pm i\omega z} \\ \pm i\, e^{\pm i\omega z} \\ 0 \end{vmatrix} ,$$

or

$$E_1^{(\pm)} + iB_1^{(\pm)} = \cos(\omega t \mp \omega z) - i\sin(\omega t \mp \omega z) ,$$

$$E_2^{(\pm)} + iB_2^{(\pm)} = \pm\sin(\omega t \mp \omega z) \pm i\,\cos(\omega t \mp \omega z) . \qquad (13.4.7)$$

This is easily checked by the known presupposed property

$$\mathbf{E}^{(\pm)} \times \mathbf{B}^{(\pm)} = \pm\mathbf{e}_z . \qquad (13.4.8)$$

Let us turn back to the general case of Eq. (13.4.5), from the first equation it follows

$$f_3 = \frac{-ib}{\omega} e^{2z} F_1 + \frac{ia}{\omega} e^{2z} F_2 , \qquad (13.4.9)$$

and further, we get a system for F_1 and F_2

$$(\frac{d}{dz} + \frac{ab\,e^{2z}}{\omega})\,F_2 = \frac{b^2 e^{2z} - \omega^2}{\omega}\,F_1 ,$$

$$(\frac{d}{dz} - \frac{ab\,e^{2z}}{\omega})\,F_1 = \frac{\omega^2 - a^2 e^{2z}}{\omega}\,F_2 . \qquad (13.4.10)$$

By introducing a new variable $e^z = \sqrt{\omega}\,Z$, two last are written as

$$Z\,(\frac{d}{dZ} + ab\,Z)\,F_2 = +(b^2 Z^2 - \omega)\,F_1 ,$$

$$Z\,(\frac{d}{dZ} - ab\,Z)\,F_1 = -(a^2 Z^2 - \omega)\,F_2 . \qquad (13.4.11)$$

This system can be solved straightforwardly in terms of the Heun confluent functions. Indeed, from Eq. (13.4.11) it follows a second order differential equation for F_1

$$\frac{d^2 F_1}{dZ^2} - \frac{a^2 Z^2 + \omega}{Z\,(a^2 Z^2 - \omega)}\,\frac{dF_1}{dZ} + \left[\frac{\omega^2}{Z^2} + \frac{2\,ab\,\omega}{a^2 Z^2 - \omega} - (a^2 + b^2)\,\omega\right] F_1 = 0, \qquad (13.4.12)$$

Here, we note an additional singular point $Z = \pm\sqrt{\omega}/a$. By the new variable, we get

$$y = \frac{a^2 Z^2}{\omega} , \qquad \frac{d^2 F_1}{dy^2} + \left[\frac{1}{y} - \frac{1}{y-1}\right]\frac{dF_1}{dy}$$

$$+ \left[\frac{\omega^2}{4\,y^2} - \frac{2\,ab\,\omega + (a^2 + b^2)\,\omega^2}{4\,a^2 y} + \frac{b\omega}{2a\,(y-1)}\right] F_1 = 0, \qquad (13.4.13)$$

from this, with the substitution $F_1(y) = y^c\,g_1(y)$; we arrive at

$$\frac{d^2 g_1}{dy^2} + \left[\frac{2c+1}{y} - \frac{1}{y-1}\right]\frac{dg_1}{dy} + \left[\frac{\omega^2/4 + c^2}{y^2}\right.$$

$$+ \frac{2c - \omega^2/2 - b\omega/a - b^2\omega^2/(2a^2)}{2\,y} + \left.\frac{-2c + b\omega/a}{2\,(y-1)}\right] g_1 = 0 . \qquad (13.4.14)$$

When $c = \pm i\omega/2$, Eq. (13.4.14) becomes simpler

$$\frac{d^2 g_1}{dy^2} + \left[\frac{2c+1}{y} - \frac{1}{y-1} \right] \frac{dg_1}{dy}$$

$$+ \left[\frac{2c - \omega^2/2 - b\,\omega/a - b^2\omega^2/(2a^2)}{2\,y} + \frac{-2c + b\omega/a}{2\,(y-1)} \right] g_1 = 0\,,$$

which can be identified with the confluent Heun function

$$H(\alpha,\,\beta,\,\gamma,\,\delta,\,\eta,\,z)\,, \qquad \frac{d^2 H}{dz^2} + \left[\alpha + \frac{1+\beta}{z} + \frac{1+\gamma}{z-1} \right] \frac{dH}{dz}$$

$$+ \left[\frac{1}{2} \frac{\alpha + \alpha\beta - \beta\gamma - \beta - \gamma - 2\eta}{z} + \frac{1}{2} \frac{\alpha\gamma + \beta + \alpha + 2\eta + 2\delta + \beta\gamma + \gamma}{z-1} \right] H = 0$$

$$(13.4.15)$$

with parameters

$$\alpha = 0\,, \qquad \beta = 2c\,, \qquad \gamma = -2\,, \qquad \delta = -\frac{1}{4} \frac{(a^2 + b^2)\,\omega^2}{a^2}\,,$$

$$\eta = \frac{1}{4} \frac{2\,ab\,\omega + (a^2 + b^2)\,\omega^2 + 4\,a^2}{a^2}\,, \qquad F_1 = y^{\pm i\omega/2}\,H(\alpha,\,\beta,\,\gamma,\,\delta,\,\eta,\,y)\,. \quad (13.4.16)$$

Next, is to develop a method that makes it possible to construct solutions for the system in Eq. (13.4.11) in terms of the Bessel functions.

13.5   Additional Study on the System

Let us perform a special linear transformation in Eq. (13.4.11) (suppose $(\alpha n - \beta m) = 1$)

$$F_1 = \alpha\,G_1 + \beta\,G_2\,, \qquad F_2 = m\,G_1 + n\,G_2\,;$$

$$G_1 = n\,F_1 - \beta\,F_2\,, \qquad G_2 = -m\,F_1 + \alpha\,F_2\,. \qquad (13.5.1)$$

Through combining equations from Eq. (13.4.11), we get

$$n\,Z\,(\frac{d}{dZ} - abZ)F_1 - \beta Z(\frac{d}{dZ} + abZ)F_2 = -n(a^2 Z^2 - \omega)F_2 - \beta(b^2 Z^2 - \omega)F_1\,,$$

$$-mZ(\frac{d}{dZ} - abZ)F_1 + \alpha Z(\frac{d}{dZ} + abZ)F_2 = m(a^2 Z^2 - \omega)F_2 + \alpha(b^2 Z^2 - \omega)F_1\,,$$

from this, it follows

$$Z\,\frac{d}{dZ}\,G_1 - Z^2\,ab\,(nF_1 + \beta F_2) = -Z^2\,(na^2 F_2 + \beta b^2 F_1) + \omega\,(nF_2 + \beta F_1)\,,$$

$$Z\,\frac{d}{dZ}\,G_2 + Z^2\,ab\,(mF_1 + \alpha F_2) = Z^2\,(ma^2 F_2 + \alpha b^2 F_1) - \omega\,(mF_2 + \alpha F_1)\,.$$

$$(13.5.2)$$

We take into account Eq. (13.5.1), to allow Eqs. (13.5.2) reduce to

$$\left[Z\frac{d}{dZ} - Z^2 ab(n\alpha + \beta m) + Z^2(a^2 mn + b^2 \alpha\beta) - \omega(nm + \alpha\beta) \right] G_1$$

$$= \left[-Z^2(an - b\beta)^2 + \omega(n^2 + \beta^2) \right] G_2 ,$$

$$\left[Z\frac{d}{dZ} + Z^2 ab\,(m\beta + n\alpha) - Z^2(a^2 mn + b^2 \alpha\beta) + \omega(nm + \alpha\beta) \right] G_2$$

$$= \left[Z^2(am - b\alpha)^2 - \omega(m^2 + \alpha^2) \right] G_1 . \tag{13.5.3}$$

Let us impose an additional restriction

$$an - b\beta = 0 \qquad \Longrightarrow \qquad \frac{\beta}{n} = \frac{a}{b} ,$$

$$\left[Z\frac{d}{dZ} - Z^2 ab(n\alpha + \beta m) + Z^2(a^2 mn + b^2 \alpha\beta) - \omega(nm + \alpha\beta) \right] G_1$$

$$= +\omega(n^2 + \beta^2)G_2 ,$$

$$\left[Z\frac{d}{dZ} + Z^2 ab(m\beta + n\alpha) - Z^2(a^2 mn + b^2 \alpha\beta) + \omega(nm + \alpha\beta) \right] G_2$$

$$= \left[Z^2(am - b\alpha)^2 - \omega\,(m^2 + \alpha^2) \right] G_1 ; \tag{13.5.4}$$

note that there exists an alternative possibility or

$$am - b\alpha = 0 \qquad \Longrightarrow \qquad \frac{\alpha}{m} = \frac{a}{b} ,$$

$$\left[Z\frac{d}{dZ} - Z^2 ab(n\alpha + \beta m) + Z^2(a^2 mn + b^2 \alpha\beta) - \omega(nm + \alpha\beta) \right] G_1$$

$$= \left[-Z^2(an - b\beta)^2 + \omega(n^2 + \beta^2) \right] G_2 ,$$

$$\left[Z\frac{d}{dZ} + Z^2 ab(m\beta + n\alpha) - Z^2(a^2 mn + b^2 \alpha\beta) + \omega(nm + \alpha\beta) \right] G_2$$

$$= -\omega(m^2 + \alpha^2)G_1 . \tag{13.5.5}$$

The two variants are equivalent to each other, for definiteness we will use the variant Eq. (13.5.4). It can be presented in a more symmetrical form

$$F_1 = \alpha\,G_1 + \beta\,G_2 = +\frac{b}{\sqrt{a^2 + b^2}}\,G_1 + \frac{a}{\sqrt{a^2 + b^2}}\,G_2 ,$$

$$F_2 = m\,G_1 + n\,G_2 = -\frac{a}{\sqrt{a^2 + b^2}}\,G_1 + \frac{b}{\sqrt{a^2 + b^2}}\,G_2 ; \tag{13.5.6}$$

by this, Eqs. (13.5.4) assumes the form

$$\left[Z\frac{d}{dZ} - Z^2 ab\frac{b^2 - a^2}{b^2 + a^2} + Z^2 ab\frac{b^2 - a^2}{b^2 + a^2} - \omega\left(-\frac{ab}{a^2 + b^2} + \frac{ab}{a^2 + b^2}\right) \right] G_1$$

$$= +\omega \left(\frac{b^2}{a^2 + b^2} + \frac{a^2}{a^2 + b^2} \right) G_2 \,,$$

$$\left[Z \frac{d}{dZ} + Z^2 ab \frac{b^2 - a^2}{a^2 + b^2} - Z^2 ab \frac{b^2 - a^2}{a^2 + b^2} + \omega(-\frac{ab}{a^2 + b^2} + \frac{ab}{a^2 + b^2}) \right] G_2$$

$$= \left[Z^2(-\frac{a^2}{\sqrt{a^2 + b^2}} - \frac{b^2}{\sqrt{a^2 + b^2}})^2 - \omega(\frac{a^2}{a^2 + b^2} + \frac{b^2}{a^2 + b^2}) \right] G_1 \,,$$

that is

$$Z \frac{d}{dZ} G_1 = \omega \, G_2 \,, \qquad Z \frac{d}{dZ} G_2 = [Z^2(a^2 + b^2) - \omega] \, G_1 \,. \tag{13.5.7}$$

From Eq. (13.5.7), we derive the Bessel equation for G_1

$$\left(Z^2 \frac{d^2}{dZ^2} + Z \frac{d}{dZ} + \omega^2 - \omega(a^2 + b^2)Z^2 \right) G_1 = 0 \,. \tag{13.5.8}$$

To understand better the physical meaning of Eq. (13.5.8), it is convenient to translate the equation to the variable z, then it will read as

$$e^z = \sqrt{\omega} \, Z \,, \qquad \left(\frac{d^2}{dz^2} + \omega^2 - (a^2 + b^2)e^{2z} \right) G_1 = 0 \,. \tag{13.5.9}$$

It can be associated with the Schrödinger equation

$$\left(\frac{d^2}{dz^2} + \epsilon - U(z) \right) \varphi(z) = 0 \tag{13.5.10}$$

with potential function $U(z) = (a^2 + b^2)e^{2z}$, and an effective force acting on the left $F_z = -2(a^2 + b^2)e^{2z}$. Note, that when $a = k_1 = 0$, $b = k_2 = 0$, the effective force vanishes. The corresponding quantum-mechanical system can be illustrated by Fig. (13.1).

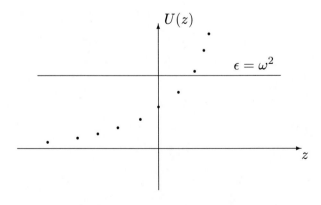

Figure 13.1: Effective potential curve

Therefore, we should expect properties of the electromagnetic solutions similar to those existing in the associated quantum-mechanical problem.

Let us turn back to Eq. (13.5.8) – in the variable

$$x = i \sqrt{\omega(a^2 + b^2)} \, Z = i \sqrt{a^2 + b^2} \, e^z$$

it assumes the form of the Bessel equation

$$\left(\frac{d^2}{dx^2} + \frac{1}{x}\frac{d}{dx} + 1 + \frac{\omega^2}{x^2} \right) G_1 = 0 \, . \tag{13.5.11}$$

The first order equation for the system in Eq. (13.5.7), is that variable x takes the form

$$x \frac{d}{dx} G_1 = \omega \, G_2 \, , \qquad x \frac{d}{dx} G_2 = -\frac{\omega^2 + x^2}{\omega} G_1 \, . \tag{13.5.12}$$

A second order equation allows for G_2 to read as

$$\left[\frac{d^2}{dx^2} + (\frac{1}{x} - \frac{2x}{\omega^2 + x^2}) \frac{d}{dx} + \frac{x^2 + \omega^2}{x^2} \right] G_2 = 0 \, . \tag{13.5.13}$$

Note that by substituting Eq. (13.5.6)

$$F_1 = \frac{b}{\sqrt{a^2 + b^2}} G_1 + \frac{a}{\sqrt{a^2 + b^2}} G_2 \, , \qquad F_2 = -\frac{a}{\sqrt{a^2 + b^2}} G_1 + \frac{b}{\sqrt{a^2 + b^2}} G_2$$

into Eq. (13.4.9), we get

$$f_3 = \frac{e^{2z}}{\omega} (-ib \, F_1 + ia \, F_2) = \frac{\sqrt{a^2 + b^2}}{i \, \omega} G_1 \, .$$

13.6   Asymptotic Behavior of Solutions

Mostly used for Bessel equation are solutions

<u>in Bessel functions</u>

$$G_1^I(x) = J_{+i\omega}(x) \, , \qquad\qquad G_1^{II}(x) = J_{-i\omega}(x) \, ; \tag{13.6.1}$$

<u>in Hankel functions</u>

$$G_1^I(x) = H_{+i\omega}^{(1)}(x) \, , \qquad\qquad G_1^{II}(x) = H_{+i\omega}^{(2)}(x) \, ,$$

$$I' \qquad G_1(x) = H_{-i\omega}^{(1)}(x) \, , \qquad II' \qquad G_1(x) = H_{-i\omega}^{(2)}(x) \, ; \tag{13.6.2}$$

Note that for $H_{-i\omega}^{(1)}(x) = e^{-\omega\pi} H_{i\omega}^{(2)}(x)$, where the primed cases I', II' coincide, respectively, with II, I; and, for this reason will not be considered in the following example.

<u>in Neyman functions</u>

$$G_1^I(x) = N_{+i\omega}(x) \, , \qquad G_1^{II}(x) = N_{-i\omega}(x) \, . \tag{13.6.3}$$

For brevity, the notation $+\sqrt{a^2 + b^2} = 2\sigma$ is used. First, let us consider the solutions in Bessel's functions when

$$z \to -\infty, \ x = i\sigma e^z \to i0 \,,$$

$$G_1^I(x) = J_{+i\omega}(x) = \frac{1}{\Gamma(1+i\omega)} \left(\frac{x}{2}\right)^{+i\omega} = \frac{(i\,\sigma)^{i\omega}}{\Gamma(1+i\omega)} \ e^{+i\omega z} \,,$$

$$G_1^{II}(x) = J_{-i\omega}(x) = \frac{1}{\Gamma(1-i\omega)} \left(\frac{x}{2}\right)^{-i\omega} = \frac{(i\,\sigma)^{-i\omega}}{\Gamma(1-i\omega)} \ e^{-i\omega z} \,. \qquad (13.6.4)$$

In the region $z \to +\infty$, $(x = i\sigma e^z = iX \to i\infty)$, using the established asymptotic formula

$$J_{i\omega}(x) \sim \sqrt{\frac{2}{\pi x}} \ \cos\left(x - (i\omega + \frac{1}{2})\frac{\pi}{2}\right),$$

we get

$$G_1^I(z \to \infty) = J_{+i\omega}(z \to \infty) \sim e^{i\pi/4} \sqrt{\frac{1}{2\pi iX}} \ e^{-\omega\pi/2} \ e^{+X} \,,$$

$$G_1^{II}(z \to \infty) = J_{-i\omega}(z \to \infty) \sim e^{i\pi/4} \sqrt{\frac{1}{2\pi iX}} \ e^{+\omega\pi/2} \ e^{+X} \,. \qquad (13.6.5)$$

Let us consider solutions in Hankel's functions determined in terms of $J_{\pm i\omega}(x)$ as follows

$$H_{i\omega}^{(1)}(x) = +\frac{i}{\sin(i\omega\pi)} \left(e^{\omega\pi} J_{+i\omega}(x) - J_{-i\omega}(x)\right) \,,$$

$$H_{i\omega}^{(2)}(x) = -\frac{i}{\sin(i\omega\pi)} \left(e^{-\omega\pi} J_{+i\omega}(x) - J_{-i\omega}(x)\right) \,, \qquad (13.6.6)$$

so that $z \to -\infty$, $x \to i0$,

$$G_1^I(x) = H_{i\omega}^{(1)}(x) = +\frac{i}{\sin(i\omega\pi)} \left(e^{+\omega\pi} \frac{(i\,\sigma)^{i\omega}}{\Gamma(1+i\omega)} \ e^{+i\omega z} - \frac{(i\,\sigma)^{-i\omega}}{\Gamma(1-i\omega)} \ e^{-i\omega z}\right),$$

$$G_1^{II}(x) = H_{i\omega}^{(2)}(x) = -\frac{i}{\sin(i\omega\pi)} \left(e^{-\omega\pi} \frac{(i\,\sigma)^{i\omega}}{\Gamma(1+i\omega)} \ e^{+i\omega z} - \frac{(i\,\sigma)^{-i\omega}}{\Gamma(1-i\omega)} \ e^{+i\omega z}\right).$$

$$(13.6.7)$$

Their behavior, when $z \to +\infty$, in this case is governed by this known relation

$$H_{i\omega}^{(1)}(x) \sim \sqrt{\frac{2}{\pi x}} \ \exp\left[+i\left(x - \frac{\pi}{2}(i\omega + \frac{1}{2})\right)\right],$$

$$H_{i\omega}^{(2)}(x) \sim \sqrt{\frac{2}{\pi x}} \ \exp\left[-i\left(x - \frac{\pi}{2}(i\omega + \frac{1}{2})\right)\right];$$

from this, it follows

$z \to +\infty, \; x = iX \to i\infty$,

$$G_1^I(x) = H_{i\omega}^{(1)}(x) \sim e^{-i\pi/4} \sqrt{\frac{2}{i\pi X}} \; e^{+\omega\pi/2} \; e^{-X} \,,$$

$$G_1^{II}(x) = H_{i\omega}^{(2)}(x) \sim e^{+i\pi/4} \sqrt{\frac{2}{i\pi X}} \; e^{-\omega\pi/2} \; e^{+X} \,. \qquad (13.6.8)$$

Let us consider the interpretation for the solution of the first type: this wave goes from the left, then it is partly reflected and partly goes forward through an effective potential barrier but gradually damping as z rises. The corresponding reflection coefficient is determined as follows:

$$G(z) \sim M_+^I e^{+i\omega z} + M_-^I e^{-i\omega z} \,, \qquad R = \frac{\mid M_-^I \mid^2}{\mid M_+^I \mid^2} \,. \qquad (13.6.9)$$

By taking into account the identities

$$(i\sigma)^{+i\omega} = (e^{i\pi/2} e^{\ln \sigma})^{+i\omega} = e^{-\omega\pi/2} e^{+i\omega \ln \sigma} \,,$$

$$(i\sigma)^{-i\omega} = (e^{i\pi/2} e^{\ln \sigma})^{-i\omega} = e^{+\omega\pi/2} e^{-i\omega \ln \sigma} \,; \qquad (13.6.10)$$

we derive

$$\mid M_+^I \mid^2 = \frac{1}{\sin(+i\omega\pi)\sin(-i\omega\pi)} \frac{e^{+\omega\pi}}{\Gamma(1-i\omega)\Gamma(1+i\omega)} \,,$$

$$\mid M_-^I \mid^2 = \frac{1}{\sin(+i\omega\pi)\sin(-i\omega\pi)} \frac{e^{+\omega\pi}}{\Gamma(1-i\omega)\Gamma(1+i\omega)} \,. \qquad (13.6.11)$$

This means that for all solutions of that type the reflection coefficient always equals to 1:

$$R = 1 \,. \qquad (13.6.12)$$

Solutions of the second type, rise to infinity as $z \to +\infty$, are characterized by

$$M_+^{II} e^{+i\omega z} + M_-^{II} e^{-i\omega z} \,, \qquad R = \frac{\mid M_-^{II} \mid^2}{\mid M_+^{II} \mid^2} = e^{4\omega\pi} > 1 \,. \qquad (13.6.13)$$

Finally, let us specify asymptotic behavior for these solutions in terms of Neyman functions. These functions are defined by

$$N_{i\omega}(x) = \frac{\cos(i\omega\pi) \, J_{i\omega}(x) - J_{-i\omega}(x)}{\sin(i\omega\pi)} \,,$$

$$N_{-i\omega}(x) = \frac{J_{i\omega}(x) - \cos(i\omega\pi) \, J_{-i\omega}(x)}{\sin(i\omega\pi)} \,. \qquad (13.6.14)$$

In the region $z \to +\infty, \; (x = iX \to i\infty)$, with the use of the known relation

$$N_{i\omega}(x) \sim \sqrt{\frac{2}{i\pi X}} \; \sin\left(iX - (i\omega + \frac{1}{2})\frac{\pi}{2} \right) \,,$$

we get

$$G_1^I(x) = N_{+i\omega(x)} \sim ie^{+i\pi/4}\sqrt{\frac{1}{2i\pi X}}e^{-\omega\pi/2}\,e^X\,,$$

$$G_1^{II}(x) = N_{-i\omega(x)} \sim +ie^{+i\pi/4}\sqrt{\frac{1}{2i\pi X}}e^{+\omega\pi/2}\,e^X\,. \tag{13.6.15}$$

In the region $z \to -\infty$ their behavior is given by

$$G^I(z) = \frac{\cos(i\omega\pi)}{\sin(i\omega\pi)}\frac{(i\,\sigma)^{i\omega}}{\Gamma(1+i\omega)}\,e^{+i\omega z} - \frac{1}{\sin(i\omega\pi)}\frac{(i\,\sigma)^{-i\omega}}{\Gamma(1-i\omega)}\,e^{-i\omega z}\,,$$

$$G^{II}(z) = \frac{1}{\sin(i\omega\pi)}\frac{(i\,\sigma)^{i\omega}}{\Gamma(1+i\omega)}\,e^{+i\omega z} - \frac{\cos(i\omega\pi)}{\sin(i\omega\pi)}\frac{(i\,\sigma)^{-i\omega}}{\Gamma(1-i\omega)}\,e^{-i\omega z}\,. \tag{13.6.16}$$

For these solutions, we have, respectively;

$$R^I = \frac{e^{2\omega\pi}}{(e^{2\omega\pi}+e^{-2\omega\pi})/4} = \frac{4}{1+e^{-4\omega\pi}}\,,$$

$$R^{II} = e^{2\omega\pi}\,(e^{2\omega\pi}+e^{-2\omega\pi})/4 = \frac{1+e^{4\omega\pi}}{4}\,. \tag{13.6.17}$$

13.7 Explicit Form of the Function G_2

The function $G_1(x)$ satisfies the Bessel equation

$$\left(\frac{d^2}{dx^2} + \frac{1}{x}\frac{d}{dx} + 1 + \frac{\omega^2}{x^2}\right)G_1 = 0\,; \tag{13.7.1}$$

the second function $G_2(x)$ is determined by

$$G_2 = \frac{x}{\omega}\frac{d}{dx}\,G_1\,. \tag{13.7.2}$$

Solutions of the Bessel equation obey the following recurrent formulas

$$x\frac{d}{dx}\,F_{i\omega} = i\omega\,F_{i\omega} - xF_{i\omega+1}\,,$$

$$x\frac{d}{dx}\,F_{-i\omega} = +i\omega\,F_{-i\omega}(x) + xF_{-i\omega-1}\,, \tag{13.7.3}$$

where $F_{\pm\nu}$ stands for

$$J_{\pm\nu}(x)\,, \qquad H_{\pm\nu}^{(1)}(x)\,, \qquad H_{\pm\nu}^{(2)}(x)\,, \qquad N_{\pm\nu}(x)\,.$$

Therefore, with the help of Eq. (13.7.3), one can express G_2 in terms of the known G_1. For instance,

$$G_1^I(x) = H_{+i\omega}^{(1)}(x)\,, \qquad G_2^I(x) = i\,H_{+i\omega}^{(1)}(x) - \frac{x}{\omega}\,H_{i\omega+1}^{(1)}(x)\,,$$

$$G_1^{II}(x) = H_{+i\omega}^{(2)}(x) , \qquad G_2^{II} = i\, H_{+i\omega}^{(2)}(x) - \frac{x}{\omega}\, H_{i\omega+1}^{(2)}(x) . \qquad (13.7.4)$$

Remember that

$$F_1^I = \frac{b}{\sqrt{a^2 + b^2}} G_1 + \frac{a}{\sqrt{a^2 + b^2}} G_2 ,$$

$$F_2^I = -\frac{a}{\sqrt{a^2 + b^2}} G_1 + \frac{b}{\sqrt{a^2 + b^2}} G_2 ,$$

$$f_3^I = \frac{e^{2z}}{\omega}(-ib\, F_1^I + ia\, F_2^I) = \frac{\sqrt{a^2 + b^2}}{i\,\omega}\, G_1 . \qquad (13.7.5)$$

Let us examine asymptotic behavior of G_2. Starting with

$$H_{i\omega}^{(1)}(x) = +\frac{i}{\sin(i\omega\pi)}\left(e^{\omega\pi} J_{+i\omega}(x) - J_{-i\omega}(x)\right) ,$$

$$H_{i\omega}^{(2)}(x) = -\frac{i}{\sin(i\omega\pi)}\left(e^{-\omega\pi} J_{+i\omega}(x) - J_{-i\omega}(x)\right) ,$$

$$H_{i\omega+1}^{(1)}(x) = +\frac{i}{\sin(i\omega+1)\pi}\left(e^{-i(i\omega+1)\pi} J_{+i\omega+1}(x) - J_{-(i\omega+1)}(x)\right) ,$$

$$H_{i\omega+1}^{(2)}(x) = -\frac{i}{\sin(i\omega+1)\pi}\left(e^{i(i\omega+1)\pi} J_{+i\omega+1}(x) - J_{-(i\omega+1)}(x)\right) ,$$

$$(13.7.6)$$

using the help of these established relations

$$z \to -\infty, \ x \to i0 ,$$

$$J_{+i\omega}(x) \sim \frac{(i\,\sigma)^{i\omega}}{\Gamma(1 + i\omega)}\, e^{+i\omega z} , \qquad J_{-i\omega}(x) \sim \frac{(i\,\sigma)^{-i\omega}}{\Gamma(1 - i\omega)}\, e^{-i\omega z} ,$$

we get

$$z \to -\infty, \ x \to i0 ,$$

$$H_{i\omega}^{(1)} \sim +\frac{i}{\sin(i\omega\pi)}\left(e^{+\omega\pi}\frac{(i\,\sigma)^{i\omega}}{\Gamma(1 + i\omega)}\, e^{+i\omega z} - \frac{(i\,\sigma)^{-i\omega}}{\Gamma(1 - i\omega)}\, e^{-i\omega z}\right) ,$$

$$H_{i\omega}^{(2)} \sim -\frac{i}{\sin(i\omega\pi)}\left(e^{-\omega\pi}\frac{(i\,\sigma)^{i\omega}}{\Gamma(1 + i\omega)}\, e^{+i\omega z} - \frac{(i\,\sigma)^{-i\omega}}{\Gamma(1 - i\omega)}\, e^{-i\omega z}\right) ,$$

$$H_{i\omega+1}^{(1)} \sim \frac{i}{\sin(i\omega+1)\pi}\left(e^{-i(i\omega+1)\pi}\frac{(i\sigma)^{i\omega+1}}{\Gamma(2 + i\omega)}e^{i\omega z}e^{z} - \frac{(i\sigma)^{-i\omega-1}}{\Gamma(-i\omega)}e^{-i\omega z}e^{-z}\right)$$

$$\sim -\frac{i}{\sin(i\omega+1)\pi}\frac{(i\sigma)^{-i\omega-1}}{\Gamma(-i\omega)}\, e^{-i\omega z}e^{-z} ,$$

$$H_{i\omega+1}^{(2)}(x) \sim \frac{-i}{\sin(i\omega+1)\pi}\left(e^{i(i\omega+1)\pi}\frac{(i\sigma)^{i\omega+1}}{\Gamma(2 + i\omega)}\, e^{i\omega z}e^{z} - \frac{(i\sigma)^{-i\omega-1}}{\Gamma(-i\omega)}e^{-i\omega z}e^{-z}\right)$$

$$\sim \frac{i}{\sin(i\omega + 1)\pi} \frac{(i\sigma)^{-i\omega - 1}}{\Gamma(-i\omega)} e^{-i\omega z} e^{-z} .$$

$$(13.7.7)$$

So, we get

$$G_2^I(x) = -\frac{1}{\sin(i\omega\pi)} \left(e^{+\omega\pi} \frac{(i\,\sigma)^{i\omega}}{\Gamma(1 + i\omega)} e^{+i\omega z} - \frac{(i\,\sigma)^{-i\omega}}{\Gamma(1 - i\omega)} e^{-i\omega z} \right)$$

$$-\frac{2\sigma}{\omega} \frac{1}{\sin(i\omega + 1)\pi} \frac{(i\sigma)^{-i\omega - 1}}{\Gamma(-i\omega)} e^{-i\omega z} .$$

$$(13.7.8)$$

Here, through the consideration of an identity

$$-\frac{2\sigma}{\omega} \frac{1}{\sin(i\omega + 1)\pi} \frac{(i\sigma)^{-i\omega - 1}}{\Gamma(-i\omega)} e^{-i\omega z}$$

$$= +\frac{2\sigma}{\omega} \frac{1}{\sin(i\omega\pi)} \frac{(i\sigma)^{-i\omega}(-i\omega)}{(i\sigma)\Gamma(1 - i\omega)} e^{-i\omega z} = -2 \frac{1}{\sin(i\omega\pi)} \frac{(i\sigma)^{-i\omega}}{\Gamma(1 - i\omega)} e^{-i\omega z}$$

$$(13.7.9)$$

one reduces this relation in Eq. (13.7.8) to the form

$$G_2^I(x) = -\frac{1}{\sin(i\omega\pi)} \left(e^{+\omega\pi} \frac{(i\,\sigma)^{i\omega}}{\Gamma(1 + i\omega)} e^{+i\omega z} + \frac{(i\,\sigma)^{-i\omega}}{\Gamma(1 - i\omega)} e^{-i\omega z} \right) .$$

$$(13.7.10)$$

In a similar manner, can then treat this case as:

$$G_2^{II} = \frac{1}{\sin(i\omega\pi)} \left(e^{-\omega\pi} \frac{(i\,\sigma)^{i\omega}}{\Gamma(1 + i\omega)} e^{+i\omega z} - \frac{(i\,\sigma)^{-i\omega}}{\Gamma(1 - i\omega)} e^{-i\omega z} \right)$$

$$+\frac{2\sigma}{\omega} \frac{1}{\sin(i\omega + 1)\pi} \frac{(i\sigma)^{-i\omega - 1}}{\Gamma(-i\omega)} e^{-i\omega z}$$

$$= \frac{1}{\sin(i\omega\pi)} \left(e^{-\omega\pi} \frac{(i\,\sigma)^{i\omega}}{\Gamma(1 + i\omega)} e^{+i\omega z} + \frac{(i\,\sigma)^{-i\omega}}{\Gamma(1 - i\omega)} e^{-i\omega z} \right) .$$

$$(13.7.11)$$

Behavior of these solutions when $z \to +\infty$ is the governed relation

$$H_{i\omega}^{(1)}(x) \sim \sqrt{\frac{2}{\pi x}} \exp\left[+i\left(x - \frac{\pi}{2}(i\omega + \frac{1}{2}) \right) \right] ,$$

$$H_{i\omega}^{(2)}(x) \sim \sqrt{\frac{2}{\pi x}} \exp\left[-i\left(x - \frac{\pi}{2}(i\omega + \frac{1}{2}) \right) \right] ;$$

from this, it follows

$$H_{i\omega}^{(1)}(x) \sim e^{-i\pi/4} \sqrt{\frac{2}{i\pi X}} e^{+\omega\pi/2} e^{-X} ,$$

$$H_{i\omega}^{(2)}(x) \sim e^{+i\pi/4} \sqrt{\frac{2}{i\pi X}} e^{-\omega\pi/2} e^{+X} ,$$

$$H^{(1)}_{i\omega+1}(x) \sim \sqrt{\frac{2}{i\pi X}} \; \exp\left[+i\left(iX - \frac{\pi}{2}(i\omega+1+\frac{1}{2})\right)\right]$$

$$\sim -i \, e^{-i\pi/4} \sqrt{\frac{2}{i\pi X}} \; e^{+\omega\pi/2} \, e^{-X} \; ,$$

$$H^{(2)}_{i\omega+1}(x) \sim \sqrt{\frac{2}{i\pi X}} \; \exp\left[-i\left(iX - \frac{\pi}{2}(i\omega+1+\frac{1}{2})\right)\right]$$

$$\sim i \, e^{+i\pi/4} \sqrt{\frac{2}{i\pi X}} \; e^{-\omega\pi/2} \, e^{+X} \; . \tag{13.7.12}$$

Therefore, we arrive at the formulas

$$G_2^I(x) = i \, H^{(1)}_{i\omega}(x) - \frac{x}{\omega} \, H^{(1)}_{i\omega+1}(x)$$

$$\sim ie^{-i\pi/4} \sqrt{\frac{2}{i\pi X}} \; e^{+\omega\pi/2} \, e^{-X} - \frac{X}{\omega} e^{-i\pi/4} \sqrt{\frac{2}{i\pi X}} \; e^{+\omega\pi/2} \, e^{-X} \; ,$$

$$G_2^{II} = i \, H^{(2)}_{i\omega}(x) - \frac{x}{\omega} \, H^{(2)}_{i\omega+1}(x)$$

$$\sim ie^{+i\pi/4} \sqrt{\frac{2}{i\pi X}} \; e^{-\omega\pi/2} \, e^{+X} + \frac{X}{\omega} e^{+i\pi/4} \sqrt{\frac{2}{i\pi X}} \; e^{-\omega\pi/2} \, e^{+X} \; .$$

$$\tag{13.7.13}$$

Evidently, to find asymptotic for G_2, it is sufficient to make use of the known asymptotic for G_1. For instance,

$$G_2^I \sim \frac{1}{\omega} \frac{d}{dz} \frac{i}{\sin(i\omega\pi)} \left(e^{+\omega\pi} \frac{(i\,\sigma)^{i\omega}}{\Gamma(1+i\omega)} e^{+i\omega z} - \frac{(i\,\sigma)^{-i\omega}}{\Gamma(1-i\omega)} e^{-i\omega z} \right)$$

$$= -\frac{1}{\sin(i\omega\pi)} \left(e^{+\omega\pi} \frac{(i\,\sigma)^{i\omega}}{\Gamma(1+i\omega)} e^{+i\omega z} + \frac{(i\,\sigma)^{-i\omega}}{\Gamma(1-i\omega)} e^{-i\omega z} \right) \; ;$$

$$\tag{13.7.14}$$

which coincides with Eq. (13.7.10). It is a superposition of two plane waves with reflection coefficient $R = 1$.

13.8   Concluding Remarks

In accordance with Eq. (13.5.10), this equation

$$\omega^2 = U(z) \qquad \omega^2 = (a^2 + b^2)e^{2z_0} \tag{13.8.1}$$

determines a critical point z_0 in which behavior of the function $G_1(x)$ must change dramatically. To such a point z_0, where it corresponds with

$$x_0 = i\sqrt{a^2 + b^2}e^{z_0} = i\omega \; . \tag{13.8.2}$$

In order to examine this behavior for the solutions in the vicinity of x_0, it is convenient to introduce a new coordinate

$$x = x_0 + i\omega\, u = i\omega(1+u)\,, \qquad \frac{d}{dx} = \frac{1}{i\omega}\frac{d}{du}\,; \qquad (13.8.3)$$

Eq. (13.5.11) for $G_1(x)$ assumes the form

$$\left(\frac{d^2}{du^2} + \frac{1}{1+u}\frac{d}{du} - \omega^2 + \frac{\omega^2}{(1+u)^2} \right) G_1 = 0\,. \qquad (13.8.4)$$

Close to $u = 0$, we have

$$\left(\frac{d^2}{du^2} + \frac{d}{du} \right) G_1 = 0\,.$$

that is

$$G_1 = e^{Bu}, \qquad B^2 + B = 0, \qquad B = 0, -1\,; \qquad (13.8.5)$$

physically interesting is the choice $B = -1$. To such a critical value $x_0 = i\omega$, there corresponds

$$\omega = \sqrt{k_1^2 + k_2^2}\; e^{z_0} \qquad \Longrightarrow \qquad z_0 = \ln \frac{\omega}{\sqrt{k_1^2 + k_2^2}}\,; \qquad (13.8.6a)$$

in usual units, this relation will read as

$$z_0 = \rho\, \ln \frac{\omega}{c\,\sqrt{k_1^2 + k_2^2}}\,, \qquad (13.8.6b)$$

where ρ is a curvature radius of the Lobachevsky space.

Penetration of the electromagnetic field into the effective medium, depends on the parameters of an electromagnetic wave: frequency ω, $k_1^2 + k_2^2$, and the curvature radius ρ – see Eq. (13.8.6b). See the illustration in Fig. (13.2).

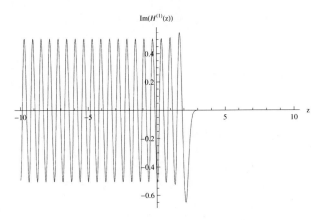

Figure 13.2: Im $H^{(1)}_{+i\omega}$, $\omega = 10$

Chapter 14

Maxwell Equations, Squaring Procedure, and Separation of the Variables

14.1   Introduction

Our treatment here, possesses quite a definite accent: the main attention is given to the possibilities given by the matrix approach for explicit constructing solutions to the Maxwell equations. In vacuum case, the matrix form includes three real (4×4)-matrices α^j:

$$(-i\partial_0 + \alpha^j \partial_j)\Psi(x) = 0 ,$$

where $\Psi = (0, \mathbf{E} + ic\mathbf{B})$. In the use of a squaring procedure, one may construct four formal solutions of the Maxwell equations on the basis of the scalar solution of the Klein–Fock–Gordon equation:

$$\{\Psi^0, \Psi^1, \Psi^2, \Psi^3\} = (i\partial_0 + \alpha^1\partial_1 + \alpha^2\partial_2 + \alpha^3\partial_3)\,\Phi(x) ,$$

$$\partial^a \partial_a \Phi(x) = 0 .$$

The problem of separating physical electromagnetic solutions in the linear space $\{\lambda_0 \Psi^0 + \lambda_1 \Psi^1 + \lambda_2 \Psi^2 + \lambda_3 \Psi^3\}$ is investigated. The Maxwell equations reduce to a algebraic system for numerical parameters λ_a. Several particular cases, such as plane waves and cylindrical waves are considered in detail. A possible extension of this technique to a curved space-time models is also discussed.

14.2   Complex Matrix Form of the Maxwell Theory

Let us start with Maxwell equations in vacuum (with the use of usual notation for current 4-vector $j^a = (\rho, \mathbf{J}/c)$, $c^2 = 1/\epsilon_0\mu_0$)

$$\operatorname{div} c\mathbf{B} = 0 , \qquad \operatorname{rot}\mathbf{E} = -\frac{\partial c\mathbf{B}}{\partial ct} ,$$

$$\text{div } \mathbf{E} = \frac{\rho}{\epsilon_0}, \qquad \text{rot } c\mathbf{B} = \frac{\mathbf{j}}{\epsilon_0} + \frac{\partial \mathbf{E}}{\partial ct}. \qquad (14.2.1)$$

Let us introduce 3-dimensional complex vector $\psi^k = E^k + icB^k$, through the help of the previous equations, can be combined into

$$\partial_1 \Psi^1 + \partial_2 \Psi^0 + \partial_3 \Psi^3 = j^0/\epsilon_0, \quad -i\partial_0 \psi^1 + (\partial_2 \psi^3 - \partial_3 \psi^2) = i \, j^1/\epsilon_0,$$

$$-i\partial_0 \psi^2 + (\partial_3 \psi^1 - \partial_1 \psi^3) = i \, j^2/\epsilon_0, \quad -i\partial_0 \psi^3 + (\partial_1 \psi^2 - \partial_2 \psi^1) = i \, j^3/\epsilon_0;$$

let $x_0 = ct$, $\partial_0 = c\partial_t$. These four relations can be rewritten in a matrix form using a 4-dimensional column Ψ with one additional zero-element

$$(-i\partial_0 + \alpha^j \partial_j)\Psi = J, \qquad \Psi = \begin{vmatrix} 0 \\ \psi^1 \\ \psi^2 \\ \psi^3 \end{vmatrix}, \qquad J = \frac{1}{\epsilon_0} \begin{vmatrix} j^0 \\ i \, j^1 \\ i \, j^2 \\ i \, j^3 \end{vmatrix}, \qquad (14.2.2)$$

where

$$\alpha^1 = \begin{vmatrix} 0 & 1 & 0 & 0 \\ -1 & 0 & 0 & 0 \\ 0 & 0 & 0 & -1 \\ 0 & 0 & 1 & 0 \end{vmatrix}, \alpha^2 = \begin{vmatrix} 0 & 0 & 1 & 0 \\ 0 & 0 & 0 & 1 \\ -1 & 0 & 0 & 0 \\ 0 & -1 & 0 & 0 \end{vmatrix}, \alpha^3 = \begin{vmatrix} 0 & 0 & 0 & 1 \\ 0 & 0 & -1 & 0 \\ 0 & 1 & 0 & 0 \\ -1 & 0 & 0 & 0 \end{vmatrix}. \qquad (14.2.3)$$

14.3   Constructing Electromagnetic Vector Solutions from Scalar Ones

This matrix form for the Maxwell theory

$$(-i\partial_0 + \alpha^j \partial_j)\Psi = 0, \qquad \Psi = \begin{vmatrix} 0 \\ \psi^1 \\ \psi^2 \\ \psi^3 \end{vmatrix} \qquad (14.3.1)$$

permits us to develop a simple method of finding solutions for the Maxwell equations on the basis of already established solutions for the scalar massless equation by d'Alembert. Indeed, in virtue of these commutative relations, we have this operator identity:

$$(-i\partial_0 + \alpha^1 \partial_1 + \alpha^2 \partial_2 + \alpha^3 \partial_3)(-i\partial_0 - \alpha^1 \partial_1 - \alpha^2 \partial_2 - \alpha^3 \partial_3)$$

$$= (-\partial_0^2 + \partial_1^2 + \partial_2^2 + \partial_3^2).$$

Therefore, with any special scalar solution (in arbitrary system of coordinates) for d'Alembert equation

$$(-\partial_0^2 + \partial_1^2 + \partial_2^2 + \partial_3^2) \, \Phi(x) = 0,$$

one can immediately construct four (formal) solutions of the Maxwell equation

$$(-i\partial_0 + \alpha^1 \partial_1 + \alpha^2 \partial_2 + \alpha^3 \partial_3) \, \Psi^a = 0,$$

where Ψ^a are columns in the matrix

$$\left(i\partial_0 + \alpha^1\partial_1 + \alpha^2\partial_2 + \alpha^3\partial_3\right)\Phi(x)$$

$$= \begin{vmatrix} i\partial_0\Phi & \partial_1\Phi & \partial_2\Phi & \partial_3\Phi \\ -\partial_1\Phi & i\partial_0\Phi & -\partial_3\Phi & \partial_2\Phi \\ -\partial_2\Phi & \partial_3\Phi & i\partial_0\Phi & -\partial_1\Phi \\ -\partial_3\Phi & -\partial_2\Phi & \partial_1\Phi & i\partial_0\Phi \end{vmatrix} = \{\Psi^0, \Psi^1, \Psi^2, \Psi^3\}. \qquad (14.3.2)$$

Thus, we have four formal solutions of the free Maxwell equations (let $\partial_a\Phi = F_a$)

$$\{\Psi^0, \Psi^1, \Psi^2, \Psi^3\} = \begin{vmatrix} iF_0 & F_1 & F_2 & F_3 \\ -F_1 & iF_0 & -F_3 & F_2 \\ -F_2 & F_3 & iF_0 & -F_1 \\ -F_3 & -F_2 & F_1 & iF_0 \end{vmatrix}. \qquad (14.3.3)$$

14.4 Electromagnetic Plane Waves from Scalar Ones

Let us specify this method for an elementary example, with starting from a scalar plane wave propagating along the axis z

$$\Phi = A\sin(\omega t - kz) = A\sin(k_0 x_0 - k_3 x_3) = A\cos\varphi.$$

The recipe of Eq. (14.3.2) gives

$$\{\Psi^a\} = A \begin{vmatrix} ik_0 & 0 & 0 & -k_3 \\ 0 & ik_0 & k_3 & 0 \\ 0 & -k_3 & ik_0 & 0 \\ k_3 & 0 & 0 & ik_0 \end{vmatrix} \cos\varphi. \qquad (14.4.1)$$

Because the left and the right columns have non-vanishing zero-components, they cannot represent any real solutions of the Maxwell equations. However, the following two that remain seem to be suitable (the factor $A\cos\varphi$ is omitted)

$$\Psi^I = \begin{vmatrix} 0 \\ ik_0 \\ -k_3 \\ 0 \end{vmatrix} = \begin{vmatrix} 0 \\ E^1 + icB^1 \\ E^2 + icB^2 \\ E^3 + ciB^3 \end{vmatrix}, \qquad \Psi^{II} = A \begin{vmatrix} 0 \\ k_3 \\ ik_0 \\ 0 \end{vmatrix} = \begin{vmatrix} 0 \\ E^1 + icB^1 \\ E^2 + icB^2 \\ E^3 + ciB^3 \end{vmatrix}.$$

Thus, we have two wave solutions

$$I \qquad E^2 = -k_3 A\cos\varphi, \quad B^1 = \frac{k_0}{c}A\cos\varphi,$$

$$II \qquad E^1 = k_3 A\cos\varphi, \quad B^2 = \frac{k_0}{c}A\cos\varphi. \qquad (14.4.2)$$

The waves obey relations

$$I \qquad (\mathbf{E} \times \mathbf{B}) = +\mathbf{e}_3 \frac{k_3 k_0}{c}A^2\cos^2\varphi,$$

$$II \qquad (\mathbf{E} \times \mathbf{B}) = +\mathbf{e}_3 \, \frac{k_3 k_0}{c} A^2 \cos^2 \varphi \,. \tag{14.4.3}$$

Two waves are orthogonal to each other: $\mathbf{E}^I \mathbf{E}^{II} = 0$, $\mathbf{B}^I \mathbf{B}^{II} = 0$.

In the same manner we can solve a more general problem of constructing plane wave solutions with arbitrary wave vector $\mathbf{k}$. Let us start with a scalar wave $\Phi = A \, \sin(k_0 x_0 - k_i x_i) = A \, \cos \varphi$. The matrix of solutions will take the form

$$\{\Psi^0, \Psi^1, \Psi^2, \Psi^3\} = A \begin{vmatrix} ik_0 & -k_1 & -k_2 & -k_3 \\ k_1 & ik_0 & +k_3 & -k_2 \\ k_2 & -k_3 & ik_0 & +k_1 \\ k_3 & +k_2 & -k_1 & ik_0 \end{vmatrix} \cos \varphi \,. \tag{14.4.4}$$

We have four formal solutions Ψ^a, but they cannot be regarded as physical because each of them has a non-vanishing zero-component. However, we can use linearity of the Maxwell equation and combine elementary columns with any coefficients. In this way, we are able to construct physical solutions. For brevity, let us omit the factor $A \, \sin \varphi$ and operate only with the columns of the matrix

$$\{\Psi^0, \Psi^1, \Psi^2, \Psi^3\} = \begin{vmatrix} ik_0 & -k_1 & -k_2 & -k_3 \\ k_1 & ik_0 & +k_3 & -k_2 \\ k_2 & -k_3 & ik_0 & +k_1 \\ k_3 & +k_2 & -k_1 & ik_0 \end{vmatrix} \,.$$

When taking into account properties of the wave vector

$$k_0^2 - \mathbf{k}^2 = 0 \qquad \Longrightarrow \qquad \mathbf{k} = k_0 \mathbf{n} \,, \quad \mathbf{n}^2 = 1 \,, \tag{14.4.5}$$

previous matrix can be rewritten as follows (the common factor k_0 is omitted)

$$\{\Psi^0, \Psi^1, \Psi^2, \Psi^3\} \sim \begin{vmatrix} i & -n_1 & -n_2 & -n_3 \\ n_1 & i & +n_3 & -n_2 \\ n_2 & -n_3 & i & +n_1 \\ n_3 & +n_2 & -n_1 & i \end{vmatrix} \,.$$

First, with the help of the column (0) let us produce a zero at the first component of the columns $(1) - (2) - (3)$

$$\{\Psi^0, \Psi^1, \Psi^2, \Psi^3\} \sim \begin{vmatrix} i & 0 & 0 & 0 \\ n_1 & -in_1 n_1 + i & -in_2 n_1 + n_3 & -in_3 n_1 - n_2 \\ n_2 & -in_1 n_2 - n_3 & -in_2 n_2 + i & -in_3 n_2 + n_1 \\ n_3 & -in_1 n_3 + n_2 & -in_2 n_3 - n_1 & -in_3 n_3 + i \end{vmatrix} \,.$$

Noting that the column (3) is a linear combination of both columns (1) and (2)

$$i \, n_2 \, (1) - i \, n_1 \, (2) = (3) \,;$$

Now, solution (3) is a linear combination for both columns (1) and (2). Therefore, from this it follows: (multiplying the columns (1) and (2) by imaginary i)

$$\{\Psi^0, \Psi^1, \Psi^2, \Psi^3\} \sim \begin{vmatrix} i & 0 & 0 & 0 \\ n_1 & n_1^2 - 1 & n_2 n_1 + in_3 & 0 \\ n_2 & n_1 n_2 - in_3 & n_2 n_2 - 1 & 0 \\ n_3 & n_1 n_3 + in_2 & n_2 n_3 - in_1 & 0 \end{vmatrix} \,. \tag{14.4.6}$$

Thus, the two physical Maxwell solutions are

$$
\Psi^I = \begin{vmatrix} 0 \\ n_1^2 - 1 \\ n_1 n_2 - i n_3 \\ n_1 n_3 + i n_2 \end{vmatrix} = \begin{vmatrix} 0 \\ E^1 + i c B^1 \\ E^2 + i c B^2 \\ E^3 + c i B^3 \end{vmatrix} ,
$$

$$
\Psi^{II} = \begin{vmatrix} 0 \\ n_2 n_1 + i n_3 \\ n_2 n_2 - 1 \\ n_2 n_3 - i n_1 \end{vmatrix} = \begin{vmatrix} 0 \\ E^1 + i c B^1 \\ E^2 + i c B^2 \\ E^3 + c i B^3 \end{vmatrix} ,
\tag{14.4.7}
$$

or differently (the factor $A \cos \varphi$ is omitted)

$$
I \qquad \mathbf{E} = (n_1^2 - 1, \ n_1 n_2, \ n_1 n_3) , \qquad c\mathbf{B} = (0, \ -n_3, \ n_2) ;
$$

$$
II \qquad \mathbf{E} = (n_2 n_1, \ n_2 n_2 - 1, \ n_2 n_3) , \qquad c\mathbf{B} = (n_3, \ 0, \ -n_1) .
$$

$$\tag{14.4.8}$$

It is the matter of simple calculation to verify the identity for amplitudes $cB = E$. Also, both of these waves behave according to:

$$
\mathbf{E}^I \times \mathbf{B}^I = \begin{vmatrix} \mathbf{e}_1 & \mathbf{e}_2 & \mathbf{e}_3 \\ n_1^2 - 1 & n_1 n_2 & n_1 n_3 \\ 0 & -n_3 & n_2 \end{vmatrix} = (1 - n_1^2)(n_1 \mathbf{e}_1 + n_2 \mathbf{e}_2 + n_3 \mathbf{e}_1) ,
$$

$$
\mathbf{E}^{II} \times \mathbf{B}^{II} = \begin{vmatrix} \mathbf{e}_1 & \mathbf{e}_2 & \mathbf{e}_3 \\ n_2 n_1 & n_2 n_2 - 1 & n_2 n_3 \\ n_3 & 0 & -n_1 \end{vmatrix} = (1 - n_2^2)(n_1 \mathbf{e}_1 + n_2 \mathbf{e}_2 + n_3 \mathbf{e}_1) .
$$

Besides, they are independent solutions (but not orthogonal ones)

$$
\mathbf{E}^I \mathbf{E}^{II} = -n_1 n_2 , \qquad \mathbf{B}^I \mathbf{B}^{II} = -n_1 n_2 .
\tag{14.4.9}
$$

14.5   On Separating Physical Electromagnetic Solutions at a Real-Valued Scalar Function Φ

Let us consider four formal solutions of the Maxwell equations

$$
\{\Psi^0, \Psi^1, \Psi^2, \Psi^3\} = \begin{vmatrix} i F_0 & F_1 & F_2 & F_3 \\ -F_1 & i F_0 & -F_3 & F_2 \\ -F_2 & F_3 & i F_0 & -F_1 \\ -F_3 & -F_2 & F_1 & i F_0 \end{vmatrix} , \quad F_a(x) = \partial_a \Phi(x) .
$$

Physical solutions should be associated with the following structure:

$$
\begin{vmatrix} 0 \\ \mathbf{E} + i c \mathbf{B} \end{vmatrix} ,
$$

when zero-component of the (4×1)-column vanishes. Let the function $\Phi(x)$ be taken as real-valued. We should find all possible solutions to the following equation

$$\lambda_0 \Psi^0 + \lambda_1 \Psi^1 + \lambda_2 \Psi^2 + \lambda_3 \Psi^3 = \begin{vmatrix} 0 \\ \mathbf{E} + ic\mathbf{B} \end{vmatrix} . \qquad (14.5.1)$$

Let us separate real and imaginary parts in $\lambda_c : \lambda_c = a_c + ib_c$. The relation in Eq. (14.5.1) takes the form

$$(a_0 + ib_0)iF_0 + (a_1 + ib_1)F_1 + (a_2 + ib_2)F_2 + (a_3 + ib_3)F_3 = 0 \; ,$$

$$-(a_0 + ib_0)F_1 + (a_1 + ib_1)iF_0 - (a_2 + ib_2)F_3 + (a_3 + ib_3)F_2 = E_1 + icB_1 \; ,$$

$$-(a_0 + ib_0)F_2 + (a_1 + ib_1)F_3 + (a_2 + ib_2)iF_0 - (a_3 + ib_3)F_1 = E_2 + icB_2 \; ,$$

$$-(a_0 + ib_0)F_3 - (a_1 + ib_1)F_2 + (a_2 + ib_2)F_1 + (a_3 + ib_3)iF_0 = E_3 + icB_3 \; ,$$

from this, it follows

$$-b_0 F_0 + a_1 F_1 + a_2 F_2 + a_3 F_3 = 0 \; , \quad a_0 F_0 + b_1 F_1 + b_2 F_2 + b_3 F_3 = 0 \; ,$$

$$-a_0 F_1 - b_1 F_0 - a_2 F_3 + a_3 F_2 = E_1 \; , \quad -b_0 F_1 + a_1 F_0 - b_2 F_3 + b_3 F_2 = cB_1 \; ,$$

$$-a_0 F_2 + a_1 F_3 - b_2 F_0 - a_3 F_1 = E_2 \; , \quad -b_0 F_2 + b_1 F_3 + a_2 F_0 - b_3 F_1 = cB_2 \; ,$$

$$-a_0 F_3 - a_1 F_2 + a_2 F_1 - b_3 F_0 = E_3 \; , \quad -b_0 F_3 - b_1 F_2 + b_2 F_1 + a_3 F_0 = cB_3 \; .$$

Let us now consider the equation rot $\mathbf{E} = -\partial_0 c\mathbf{B}$. Here, for brevity, everywhere instead $c\mathbf{B}$ we shall write $\mathbf{B}$. By taking these two identities

$$\partial_1 E_2 - \partial_2 E_1 = \partial_1[-a_0 F_2 + a_1 F_3 - b_2 F_0 - a_3 F_1]$$

$$-\partial_2[-a_0 F_1 - b_1 F_0 - a_2 F_3 + a_3 F_2]$$

$$= a_1 \partial_1 F_3 - b_2 \partial_1 F_0 - a_3 \partial_1 F_1 + b_1 \partial_2 F_0 + a_2 \partial_2 F_3 - a_3 \partial_2 F_2 \; ,$$

$$-\partial_0 cB_3 = b_0 \partial_0 F_3 + b_1 \partial_0 F_2 - b_2 \partial_0 F_1 - a_3 \partial_0 F_0 \; ;$$

we produce an equation

$$a_1 \partial_1 F_3 - b_2 \partial_1 F_0 - a_3 \partial_1 F_1 + b_1 \partial_2 F_0 + a_2 \partial_2 F_3 - a_3 \partial_2 F_2$$

$$= b_0 \partial_0 F_3 + b_1 \partial_0 F_2 - b_2 \partial_0 F_1 - a_3 \partial_0 F_0 \; ,$$

from substituting the identity $\partial_0 F_0 = \partial_1 F_1 + \partial_2 F_2 + \partial_3 F_3$, we obtain

$$[\, b_0 \partial_0 - (a_1 \partial_1 + a_2 \partial_2 + a_3 \partial_3) \,] \, F_3 = 0 \; . \qquad (14.5.2)$$

In the same manner, we get

$$[\, b_0 \partial_0 - (a_1 \partial_1 + a_2 \partial_2 + a_3 \partial_3) \,] \, F_1 = 0 \; ,$$

$$[\, b_0 \partial_0 - (a_1 \partial_1 + a_2 \partial_2 + a_3 \partial_3) \,] \, F_2 = 0 \; . \qquad (14.5.3)$$

Now, consider the equation rot $c\mathbf{B} = \partial_0\mathbf{E}$. By calculating two terms

$$\partial_1 cB_2 - \partial_2 cB_1$$

$$= b_1\partial_1 F_3 + a_2\partial_1 F_0 - b_3\partial_1 F_1 - a_1\partial_2 F_0 + b_2\partial_2 F_3 - b_3\partial_2 F_2 \ ,$$

$$\partial_0 E_3 = -a_0\partial_0 F_3 - a_1\partial_0 F_2 + a_2\partial_0 F_1 - b_3\partial_0 F_0$$

$$= -a_0\partial_0 F_3 - a_1\partial_0 F_2 + a_2\partial_0 F_1 - b_3\partial_1 F_1 - b_3\partial_2 F_2 - b_3\partial_3 F_3 \ ,$$

we arrive at

$$b_1\partial_1 F_3 + a_2\partial_1 F_0 - b_3\partial_1 F_1 - a_1\partial_2 F_0 + b_2\partial_2 F_3 - b_3\partial_2 F_2$$

$$= -a_0\partial_0 F_3 - a_1\partial_0 F_2 + a_2\partial_0 F_1 - b_3\partial_1 F_1 - b_3\partial_2 F_2 - b_3\partial_3 F_3 \ ,$$

or

$$[\, a_0\partial_0 + (b_1\partial_1 + b_2\partial_2 + b_3\partial_3)\,]\, F_3 = 0 \ . \tag{14.5.4}$$

Analogously, we get

$$[\, a_0\partial_0 + (b_1\partial_1 + b_2\partial_2 + b_3\partial_3)\,]\, F_1 = 0 \ ,$$

$$[\, a_0\partial_0 + (b_1\partial_1 + b_2\partial_2 + b_3\partial_3)\,]\, F_2 = 0 \ . \tag{14.5.5}$$

Now, let us consider equation div $\mathbf{E} = 0$

$$0 = \partial_1 E_1 + \partial_2 E_2 + \partial_3 E_3$$

$$= -a_0\partial_1 F_1 - b_1\partial_1 F_0 - a_2\partial_1 F_3 + a_3\partial_1 F_2$$

$$-a_0\partial_2 F_2 + a_1\partial_2 F_3 - b_2\partial_2 F_0 - a_3\partial_2 F_1$$

$$-a_0\partial_3 F_3 - a_1\partial_3 F_2 + a_2\partial_3 F_1 - b_3\partial_3 F_0 \ ,$$

that is

$$-a_0(\partial_1 F_1 + \partial_2 F_2 + \partial_3 F_3) - (b_1\partial_1 + b_2\partial_2 + b_3\partial_3)F_0 = 0 \ ,$$

which is equivalent to

$$[\, a_0\partial_0 + (b_1\partial_1 + b_2\partial_2 + b_3\partial_3)\,]\, F_0 = 0 \ . \tag{14.5.6}$$

It remains to consider equation div $\mathbf{B} = 0$

$$0 = -b_0\partial_1 F_1 + a_1\partial_1 F_0 - b_2\partial_1 F_3 + b_3\partial_1 F_2$$

$$-b_0\partial_2 F_2 + b_1\partial_2 F_3 + a_2\partial_2 F_0 - b_3\partial_2 F_1$$

$$-b_0\partial_3 F_3 - b_1\partial_3 F_2 + b_2\partial_3 F_1 + a_3\partial_3 F_0 \ ,$$

or

$$-b_0(\partial_1 F_1 + \partial_2 F_2 + \partial_3 F_3) + (a_1\partial_1 + a_2\partial_2 + a_3\partial_3)F_0 = 0 \ ,$$

which is equivalent to

$$[b_0\partial_0 - (a_1\partial_1 + a_2\partial_2 + a_3\partial_3)]\, F_0 = 0 \ . \tag{14.5.7}$$

Thus, to construct physical solutions of the Maxwell equations as linear combinations from non-physical

$$(a_0 + ib_0)\Psi^0 + (a_1 + ib_1)\Psi^1 + (a_2 + ib_2)\Psi^2 + (a_3 + ib_3)\Psi^3 = \begin{vmatrix} 0 \\ \mathbf{E} + ic\mathbf{B} \end{vmatrix}$$

one must satisfy the following 8 equations

$$[\, b_0\partial_0 - (a_1\partial_1 + a_2\partial_2 + a_3\partial_3)\,]\, F_c = 0\,,$$

$$[\, a_0\partial_0 + (b_1\partial_1 + b_2\partial_2 + b_3\partial_3)\,]\, F_c = 0\,, \qquad (14.5.8)$$

where $c = 0, 1, 2, 3$; $F_c = \partial_c\Phi$.

14.6   On Separating Physical Electromagnetic Solutions at a Complex-Valued Scalar Function Φ

Let us consider four formal solutions of the Maxwell equations

$$\{\Psi^0, \Psi^1, \Psi^2, \Psi^3\} = \begin{vmatrix} iF_0 & F_1 & F_2 & F_3 \\ -F_1 & iF_0 & -F_3 & F_2 \\ -F_2 & F_3 & iF_0 & -F_1 \\ -F_3 & -F_2 & F_1 & iF_0 \end{vmatrix}, \quad F_a = \partial_a\Phi\,.$$

Let the function $\Phi(x)$ be taken as complex-valued. We should examine relationship for λ_a defining all possible solutions to the following equation:

$$\lambda_0\Psi^0 + \lambda_1\Psi^1 + \lambda_2\Psi^2 + \lambda_3\Psi^3 = \begin{vmatrix} 0 \\ \mathbf{E} + ic\mathbf{B} \end{vmatrix}\cdot \qquad (14.6.1)$$

Let us separate real and imaginary parts in λ_c, $\Phi(x)$, and $F_c(x)$

$$\lambda_c = a_c + i\, b_c\,, \qquad \Phi(x) = L(x) + i\, K(x)\,, \qquad F_c(x) = L_c(x) + i\, K_c(x)\,.$$

The relation Eq. (14.6.1) takes the form

$$(a_0 + ib_0)i(L_0 + iK_0) + (a_1 + ib_1)(L_1 + iK_1)$$

$$+(a_2 + ib_2)(L_2 + iK_2) + (a_3 + ib_3)(L_3 + iK_3) = 0\,,$$

$$-(a_0 + ib_0)(L_1 + iK_1) + (a_1 + ib_1)i(L_0 + iK_0)$$

$$-(a_2 + ib_2)(L_3 + iK_3) + (a_3 + ib_3)(L_2 + iK_2) = E_1 + icB_1\,,$$

$$-(a_0 + ib_0)(L_2 + iK_2) + (a_1 + ib_1)(L_3 + iK_3)$$

$$+(a_2 + ib_2)i(L_0 + iK_0) - (a_3 + ib_3)(L_1 + iK_1) = E_2 + icB_2\,,$$

$$-(a_0 + ib_0)(L_3 + iK_3) - (a_1 + ib_1)(L_2 + iK_2)$$

$$+(a_2 + ib_2)(L_1 + iK_1) + (a_3 + ib_3)i(L_0 + iK_0) = E_3 + icB_3\,,$$

from this, it follows

$$0 = (-b_0 L_0 + a_1 L_1 + a_2 L_2 + a_3 L_3) + (-a_0 K_0 - b_1 K_1 - b_2 K_2 - b_3 K_3) \,,$$

$$0 = (+a_0 L_0 + b_1 L_1 + b_2 L_2 + b_3 L_3) + (-b_0 K_0 + a_1 K_1 + a_2 K_2 + a_3 K_3) \,,$$

$$E_1 = (-a_0 L_1 - b_1 L_0 - a_2 L_3 + a_3 L_2) + (+b_0 K_1 - a_1 K_0 + b_2 K_3 - b_3 K_2) \,,$$

$$B_1 = (-b_0 L_1 + a_1 L_0 - b_2 L_3 + b_3 L_2) + (-a_0 K_1 - b_1 K_0 - a_2 K_3 + a_3 K_2) \,,$$

$$E_2 = (-a_0 L_2 - b_2 L_0 + a_1 L_3 - a_3 L_1) + (+b_0 K_2 - a_2 K_0 - b_1 K_3 + b_3 K_1) \,,$$

$$B_2 = (-b_0 L_2 + a_2 L_0 + b_1 L_3 - b_3 L_1) + (-a_0 K_2 - b_2 K_0 + a_1 K_3 - a_3 K_1) \,,$$

$$E_3 = (-a_0 L_3 - b_3 L_0 - a_1 L_2 + a_2 L_1) + (+b_0 K_3 - a_3 K_0 + b_1 K_2 - b_2 K_1) \,,$$

$$B_3 = (-b_0 L_3 + a_3 L_0 - b_1 L_2 + b_2 L_1) + (-a_0 K_3 - b_3 K_0 - a_1 K_2 + a_2 K_1) \,.$$

Substituting these expressions into Maxwell equations and performing the same calculation as in previous section; we get

$$\mathrm{div}\,\mathbf{E} = 0 \implies -(a_0 \partial_0 + b_j \partial_j)\,L_0 + (b_0 \partial_0 - a_j \partial_j)\,K_0 = 0 \,,$$

$$\mathrm{div}\,\mathbf{B} = 0 \implies +(b_0 \partial_0 - a_j \partial_j)\,L_0 + (a_0 \partial_0 + b_j \partial_j)\,K_0 = 0 \,,$$

$$\partial_2 B_3 - \partial_3 B_2 = +\partial_0 E_1 \implies -(a_0 \partial_0 + b_j \partial_j)\,L_1 + (b_0 \partial_0 - a_j \partial_j)\,K_1 = 0 \,,$$

$$\partial_2 E_3 - \partial_3 E_2 = -\partial_0 B_1 \implies +(b_0 \partial_0 - a_j \partial_j)\,L_1 + (a_0 \partial_0 + b_j \partial_j)\,K_1 = 0 \,,$$

$$\partial_3 B_1 - \partial_1 B_3 = +\partial_0 E_2 \implies -(a_0 \partial_0 + b_j \partial_j)\,L_2 + (b_0 \partial_0 - a_j \partial_j)\,K_2 = 0 \,,$$

$$\partial_3 E_1 - \partial_1 E_3 = -\partial_0 B_2 \implies +(b_0 \partial_0 - a_j \partial_j)\,L_2 + (a_0 \partial_0 + b_j \partial_j)\,K_2 = 0 \,,$$

$$\partial_1 B_2 - \partial_2 B_1 = +\partial_0 E_3 \implies -(a_0 \partial_0 + b_j \partial_j)\,L_3 + (b_0 \partial_0 - a_j \partial_j)\,K_3 = 0 \,,$$

$$\partial_1 E_2 - \partial_2 E_1 = -\partial_0 B_3 \implies +(b_0 \partial_0 - a_j \partial_j)\,L_3 + (a_0 \partial_0 + b_j \partial_j)\,K_3 = 0 \,.$$

Thus, for the physical Maxwell solutions the following equations must hold:

$$-(a_0 \partial_0 + b_j \partial_j)\,L_c + (b_0 \partial_0 - a_j \partial_j)\,K_c = 0 \,,$$

$$+(b_0 \partial_0 - a_j \partial_j)\,L_c + (a_0 \partial_0 + b_j \partial_j)\,K_c = 0 \,. \tag{14.6.2}$$

In particular, we have two more simple equations when taking real or imaginary scalar functions

$$\Phi = L + i\,0 \,, \qquad (a_0\,\partial_0 + b_j\,\partial_j)\,L_c = 0 \,, \quad (b_0\,\partial_0 - a_j\,\partial_j)\,L_c = 0 \,;$$

$$\Phi = 0 + i\,K \,, \qquad (a_0'\,\partial_0 + b_j'\,\partial_j)\,K_c = 0 \,, \quad (b_0'\,\partial_0 - a_j'\,\partial_j)\,K_c = 0 \,. \tag{14.6.3}$$

In the simplest case of a plane scalar wave

$$\Phi = e^{i(k_0 x^0 - k^j x^j)} \,, \quad \varphi = k_0\,x^0 - k^j\,x^j = k_c\,x^c \,,$$

$$L_c + i K_c = -k_c \sin\varphi + i\,k_c \cos\varphi \tag{14.6.4}$$

previous equations take the form

$$\Phi = L + i\, 0\,, \qquad (a_0\, k_0 + b_j\, k_j)\, k_c = 0\,, \quad (b_0\, k_0 - a_j\, k_j)\, k_c = 0\,;$$

$$\Phi = 0 + i\, K\,, \qquad (a_0'\, k_0 + b_j'\, k_j)\, k_c = 0\,, \quad (b_0'\, k_0 - a_j'\, k_j)\, k_c = 0\,,$$

$$(14.6.5)$$

that is

$$a_c' = a_c\,, \qquad b_c' = b_c\,,$$

$$(a_0\, k_0 + b_j\, k_j)\, k_c = 0\,, \quad (b_0\, k_0 - a_j\, k_j)\, k_c = 0\,. \qquad (14.6.6)$$

In the general case Eq. (14.6.2), equations for a_c, b_c and a_c', b_c' may not coincide.

Let us turn again to Eqs. (14.6.2) and translate them to variables

$$L_c = \frac{F_c^* + F_c}{2}\,, \qquad K_c = +i\,\frac{F_c^* - F_c}{2}\,,$$

$$-(a_0\partial_0 + b_j\partial_j)\,\frac{F_c^* + F_c}{2} + (b_0\partial_0 - a_j\partial_j)\,i\,\frac{F_c^* - F_c}{2} = 0\,;$$

$$+(b_0\partial_0 - a_j\partial_j)\,\frac{F_c^* + F_c}{2} + (a_0\partial_0 + b_j\partial_j)\,i\,\frac{F_c^* - F_c}{2} = 0\,,$$

or

$$[\,-(a_0\partial_0 + b_j\partial_j) - i\,(b_0\partial_0 - a_j\partial_j)\,]\, F_c$$

$$+[\,-(a_0\partial_0 + b_j\partial_j) + i\,(b_0\partial_0 - a_j\partial_j)\,]\, F_c^* = 0\,,$$

$$[\,+i\,(b_0\partial_0 - a_j\partial_j) + (a_0\partial_0 + b_j\partial_j)\,]\, F_c$$

$$+[\,+i\,(b_0\partial_0 - a_j\partial_j) - (a_0\partial_0 + b_j\partial_j)\,]\, F_c^* = 0\,.$$

Through summing and subtracting two relations, we arrive at

$$[\,-(a_0\partial_0 + b_j\partial_j) + i\,(b_0\partial_0 - a_j\partial_j)\,]\, F_c^* = 0\,,$$

$$[\,-(a_0\partial_0 + b_j\partial_j) - i\,(b_0\partial_0 - a_j\partial_j)\,]\, F_c = 0\,. \qquad (14.6.7)$$

These can be rewritten, as follows

$$[\,- (a_0 + ib_0)\,\partial_0 + i(a_j + ib_j)\,\partial_j\,]\, F_c = 0\,,$$

$$[\,- (a_0 - ib_0)\,\partial_0 - i(a_j - ib_j)\,\partial_j\,]\, F_c^* = 0\,,$$

or shorter

$$(\,- \lambda_0\,\partial_0 + i\,\lambda_j\,\partial_j\,)\, F_c = 0\,,$$

$$(\,- \lambda_0\,\partial_0^* - i\,\lambda_j^*\,\partial_j\,)\, F_c^* = 0\,. \qquad (14.6.8)$$

These relationships play the role of the Maxwell equations.

14.7   Separating Physical Electromagnetic Solutions of the Plane Wave Type

Let us apply the general relations in Eq. (14.5.8) to the case when

$$\Phi = A \, \sin\varphi \, , \qquad \varphi = k_0 x_0 - k_3 x_3 \, ,$$

$$F_0 = k_0 A \cos\varphi \, , \quad F_1 = 0 \, , \quad F_2 = 0 \, , \quad F_3 = -k_3 A \cos\varphi \, .$$

Equations (14.5.8), then give

$$b_0 k_0 + a_3 k_3 = 0 \, , \qquad a_0 k_0 - b_3 k_3 = 0 \, . \tag{14.7.1}$$

For a wave spreading in the positive direction $k_3 = +k_0 > 0$, and Eqs. (14.7.1) give

$$b_0 = -a_3 \, , \qquad b_3 = a_0 \, ,$$

and correspondingly the main linear relationship looks

$$(a_0 - ia_3)\Psi^0 + (a_3 + ia_0)\Psi^3 + (a_1 + ib_1)\Psi^1$$

$$+(a_2 + ib_2)\Psi^2 = \begin{vmatrix} 0 \\ \mathbf{E} + ic\mathbf{B} \end{vmatrix} \, ; \tag{14.7.2}$$

coefficients at Ψ^1 and Ψ^2 are arbitrary. To understand this fact let us recall the explicit form of Ψ^a

$$\{\Psi^0, \Psi^1, \Psi^2, \Psi^3\} = A \begin{vmatrix} ik_0 & 0 & 0 & -k_0 \\ 0 & ik_0 & k_0 & 0 \\ 0 & -k_0 & ik_0 & 0 \\ k_0 & 0 & 0 & ik_0 \end{vmatrix} \cos\varphi \, .$$

One may separate two subsets of non-physical solutions

$$(a_0 - ia_3)\Psi^0 + (a_3 + ia_0)\Psi^3$$

$$= (a_0 - ia_3)k_0 \left\{ \begin{vmatrix} i \\ 0 \\ 0 \\ +1 \end{vmatrix} + i \begin{vmatrix} -1 \\ 0 \\ 0 \\ i \end{vmatrix} \right\} \equiv \begin{vmatrix} 0 \\ 0 \\ 0 \\ 0 \end{vmatrix} \, , \tag{14.7.3}$$

so the relationship in Eq. (14.7.2) reduces to

$$(a_1 + ib_1)\Psi^1 + (a_2 + ib_2)\Psi^2 = (a_1 + ib_1) \begin{vmatrix} 0 \\ ik_0 \\ -k_3 \\ 0 \end{vmatrix} + (a_2 + ib_2) \begin{vmatrix} 0 \\ k_3 \\ ik_3 \\ 0 \end{vmatrix} \, . \tag{14.7.4}$$

Now, let us consider a more general case when

$$\Phi(x) = A \sin(k_0 x_0 - \mathbf{k} \mathbf{x}) = A \, \sin\varphi \, ,$$

$$F_0 = k_0 A \cos\varphi\,, \qquad F_1 = -k_1 A \cos\varphi\,,$$

$$F_2 = -k_2 A \cos\varphi\,, \qquad F_3 = -k_3 A \cos\varphi\,. \tag{14.7.5}$$

Equations (14.5.8) give

$$b_0 k_0 + a_1 k_1 + a_2 k_2 + a_3 k_3 = 0\,, \quad a_0 k_0 - b_1 k_1 - b_2 k_2 - b_3 k_3 = 0\,, \tag{14.7.6}$$

and additionally the identity $k_0 = +\sqrt{k_1^2 + k_2^2 + k_3^2}$ holds. One may introduce parametrization $k_j = k_0 n_j,\ n_j n_j = 1$, then Eqs. (14.7.6) will read

$$b_0 = -(a_1 n_1 + a_2 n_2 + a_3 n_3)\,, \quad a_0 = (b_1 n_1 + b_2 n_2 + b_3 n_3)\,. \tag{14.7.7}$$

Now, turning to Eq. (14.7.6) and excluding variables a_0, b_0; one gets

$$[(b_1 n_1 + b_2 n_2 + b_3 n_3) - i(a_1 n_1 + a_2 n_2 + a_3 n_3)]\Psi^0$$

$$+(a_1 + ib_1)\Psi^1 + (a_2 + ib_2)\Psi^2 + (a_3 + ib_3)\Psi^3 = \Psi\,. \tag{14.7.8}$$

By taking into account

$$\{\Psi^0, \Psi^1, \Psi^2, \Psi^3\} = \begin{vmatrix} i & -n_1 & -n_2 & -n_3 \\ n_1 & i & n_3 & -n_2 \\ n_2 & -n_3 & i & n_1 \\ n_3 & n_2 & -n_1 & i \end{vmatrix} k_0 A \cos\varphi\,,$$

from Eq. (14.7.8), we get $\Psi_0 = 0$ and (factor $k_0 A \cos\varphi$ is omitted)

$$\Psi_1 = [(b_1 - ia_1)n_1 + (b_2 - ia_2)n_2 + (b_3 - ia_3)n_3]\,n_1$$

$$+(a_1 + ib_1)i + (a_2 + ib_2)n_3 - (a_3 + ib_3)n_2\,,$$

$$\Psi_2 = [(b_1 - ia_1)n_1 + (b_2 - ia_2)n_2 + (b_3 - ia_3)n_3]\,n_2$$

$$-(a_1 + ib_1)n_3 + (a_2 + ib_2)i + (a_3 + ib_3)n_1\,,$$

$$\Psi_3 = [(b_1 - ia_1)n_1 + (b_2 - ia_2)n_2 + (b_3 - ia_3)n_3]\,n_3$$

$$+(a_1 + ib_1)n_2 - (a_2 + ib_2)n_1 + (a_3 + ib_3)i\,. \tag{14.7.9}$$

Let us introduce elementary solutions, associated with coefficients $a_j + ib_j$

$$\Psi_{(1)} = \begin{vmatrix} 0 \\ (b_1 - ia_1)\,n_1 n_1 + (a_1 + ib_1)i \\ (b_1 - ia_1)\,n_1 n_2 - (a_1 + ib_1)n_3 \\ (b_1 - ia_1)\,n_1 n_3 + (a_1 + ib_1)n_2 \end{vmatrix}\,,$$

$$\mathbf{E}_{(1)} = \begin{vmatrix} b_1 n_1 n_1 - b_1 \\ b_1 n_1 n_2 - a_1 n_3 \\ b_1 n_1 n_3 + a_1 n_2 \end{vmatrix}\,, \qquad c\mathbf{B}_{(1)} = \begin{vmatrix} -a_1 n_1 n_1 + a_1 \\ -a_1 n_1 n_2 - b_1 n_3 \\ -a_1 n_1 n_3 + b_1 n_2 \end{vmatrix}\,,$$

$$\Psi_{(2)} = \begin{vmatrix} 0 \\ (b_2 - ia_2)\, n_2 n_1 + (a_2 + ib_2) n_3 \\ (b_2 - ia_2)\, n_2 n_2 + (a_2 + ib_2) i \\ (b_2 - ia_2)\, n_2 n_3 - (a_2 + ib_2) n_1 \end{vmatrix} ,$$

$$\mathbf{E}_{(2)} = \begin{vmatrix} b_2 n_2 n_1 + a_2 n_3 \\ b_2 n_2 n_2 - b_2 \\ b_2 n_2 n_3 - a_2 n_1 \end{vmatrix} , \qquad c\mathbf{B}_{(2)} = \begin{vmatrix} -a_2 n_2 n_1 + b_2 n_3 \\ -a_2 n_2 n_2 + a_2 \\ -a_2 n_2 n_3 - b_2 n_1 \end{vmatrix} ,$$

$$\Psi_{(3)} = \begin{vmatrix} 0 \\ (b_3 - ia_3)\, n_3 n_1 - (a_3 + ib_3) n_2 \\ (b_3 - ia_3)\, n_3 n_2 + (a_3 + ib_3) n_1 \\ (b_3 - ia_3)\, n_3\, n_3 + (a_3 + ib_3) i \end{vmatrix} ,$$

$$\mathbf{E}_{(3)} = \begin{vmatrix} b_3 n_3 n_1 - a_3 n_2 \\ b_3 n_3 n_2 + a_3 n_1 \\ b_3 n_3\, n_3 - b_3 \end{vmatrix} , \qquad c\mathbf{B}_{(3)} = \begin{vmatrix} -a_3 n_3 n_1 - b_3 n_2 \\ -a_3 n_3 n_2 + b_3 n_1 \\ -a_3 n_3\, n_3 + a_3 \end{vmatrix} . \qquad (14.7.10)$$

Let us show that three types of solutions are linearly dependent. It suffices to examine their linear combinations (for definite consider electric field)

$$A_1\, \mathbf{E}_{(1)} + A_2\, \mathbf{E}_{(2)} + A_3\, \mathbf{E}_{(3)} = 0 ,$$

that is

$$A_1 \begin{vmatrix} b_1 n_1 n_1 - b_1 \\ b_1 n_1 n_2 - a_1 n_3 \\ b_1 n_1 n_3 + a_1 n_2 \end{vmatrix} + A_2 \begin{vmatrix} b_2 n_2 n_1 + a_2 n_3 \\ b_2 n_2 n_2 - b_2 \\ b_2 n_2 n_3 - a_2 n_1 \end{vmatrix} + A_3 \begin{vmatrix} b_3 n_3 n_1 - a_3 n_2 \\ b_3 n_3 n_2 + a_3 n_1 \\ b_3 n_3 n_3 - b_3 \end{vmatrix} = \begin{vmatrix} 0 \\ 0 \\ 0 \end{vmatrix} .$$

It remains to show that the determinant of the (3×3)-matrix vanishes

$$\det \begin{vmatrix} b_1 n_1 n_1 - b_1 & b_2 n_2 n_1 + a_2 n_3 & b_3 n_3 n_1 - a_3 n_2 \\ b_1 n_1 n_2 - a_1 n_3 & b_2 n_2 n_2 - b_2 & b_3 n_3 n_2 + a_3 n_1 \\ b_1 n_1 n_3 + a_1 n_2 & b_2 n_2 n_3 - a_2 n_1 & b_3 n_3\, n_3 - b_3 \end{vmatrix} = 0 . \qquad (14.7.11)$$

Let $b_1 n_1 = s_1, b_2 n_2 = s_2, b_3 n_3 = s_3$, then

$$0 = \begin{vmatrix} s_1 n_1 - b_1 & s_2 n_1 + a_2 n_3 & s_3 n_1 - a_3 n_2 \\ s_1 n_2 - a_1 n_3 & s_2 n_2 - b_2 & s_3 n_2 + a_3 n_1 \\ s_1 n_3 + a_1 n_2 & s_2 n_3 - a_2 n_1 & s_3 n_3 - b_3 \end{vmatrix}$$

$$= (s_1 n_1 - b_1)[(s_2 n_2 - b_2)(s_3 n_3 - b_3) - (s_3 n_2 + a_3 n_1)(s_2 n_3 - a_2 n_1)]$$

$$-(s_1 n_2 - a_1 n_3)[(s_2 n_1 + a_2 n_3)(s_3 n_3 - b_3) - (s_3 n_1 - a_3 n_2)(s_2 n_3 - a_2 n_1)]$$

$$+(s_1 n_3 + a_1 n_2)[(s_2 n_1 + a_2 n_3)(s_3 n_2 + a_3 n_1) - (s_3 n_1 - a_3 n_2)(s_2 n_2 - b_2)] .$$

It is easily verified that all the terms cancel out each other indeed.

Let us consider one other example: and, start with a complex scalar plane wave

$$\Phi(x) = e^{i(k_0 x_0 - k_j x_j)} , \qquad L_c = k_c \cos \varphi , \qquad K_c = k_c \sin \varphi ;$$

Equations (14.6.2) take the form

$$-(a_0 k_0 + b_j k_j)\, k_c \cos\varphi + (b_0 k_0 - a_j k_j)\, k_c \sin\varphi = 0 \ ,$$

$$+(b_0 k_0 - a_j k_j)\, k_c \cos\varphi + (a_0 k_0 + b_j k_j)\, k_c \sin\varphi = 0 \ ,$$

from this, it follow

$$b_0 k_0 + (a_1 k_1 + a_2 k_2 + a_3 k_3) = 0 \ , \qquad a_0 k_0 - (b_1 k_1 + b_2 k_2 + b_3 k_3 = 0 \ ,$$

or

$$b_0 = -(a_1 n_1 + a_2 n_2 + a_3 n_3) \ , \qquad a_0 = +(b_1 n_1 + b_2 n_2 + b_3 n_3) \ . \tag{14.7.12}$$

Correspondingly, the main relation gives

$$\Psi(x) = \left[+(b_1 n_1 + b_2 n_2 + b_3 n_3) - i(a_1 n_1 + a_2 n_2 + a_3 n_3) \right] \Psi^0$$

$$+(a_1 + ib_1)\Psi^1 + (a_2 + ib_2)\Psi^2 + (a_3 + ib_3)\Psi^3 = \left| \begin{array}{c} 0 \\ \mathbf{E} + ic\mathbf{B} \end{array} \right| ,$$

where

$$\{\Psi^0, \Psi^1, \Psi^2, \Psi^3\} = \left| \begin{array}{cccc} i & -n_1 & -n_2 & -n_3 \\ n_1 & i & n_3 & -n_2 \\ n_2 & -n_3 & i & n_1 \\ n_3 & n_2 & -n_1 & i \end{array} \right| \, i k_0 A \, (\cos\varphi + i\sin\varphi) \ .$$

Further, we get

$$\Psi = k_0 A \, (\cos\varphi + i\sin\varphi)$$

$$\times \left| \begin{array}{c} 0 \\ (\mathbf{a}+i\mathbf{b})\mathbf{n}n_1 - (a_1 + ib_1) + i(a_2 + ib_2)n_3 - i(a_3 + ib_3)n_2 \\ (\mathbf{a}+i\mathbf{b})\mathbf{n}n_2 - i(a_1 + ib_1)n_3 - (a_2 + ib_2) + i(a_3 + ib_3)n_1 \\ (\mathbf{a}+i\mathbf{b})\mathbf{n}n_3 + i(a_1 + ib_1)n_2 - i(a_2 + ib_2)n_1 - (a_3 + ib_3) \end{array} \right| ,$$

which is equivalent to

$$\Psi = k_0 A \, (\cos\varphi + i\sin\varphi)$$

$$\times \left| \begin{array}{c} 0 \\ (\mathbf{a}\mathbf{n}n_1 - a_1 - b_2 n_3 + b_3 n_2) + i(\mathbf{b}\mathbf{n}n_1 - b_1 - n_2 a_3 + n_3 a_2) \\ (\mathbf{a}\mathbf{n}n_2 - a_2 - b_3 n_1 + b_1 n_3) + i(\mathbf{b}\mathbf{n}n_2 - b_2 - n_3 a_1 + n_1 a_2) \\ (\mathbf{a}\mathbf{n}n_3 - a_3 - b_1 n_2 + b_2 n_1) + i(\mathbf{b}\mathbf{n}n_3 - b_3 - n_1 a_2 + n_2 a_1) \end{array} \right| . \tag{14.7.13}$$

Through the use of notation

$$\mathbf{L} = \mathbf{n}\,(\mathbf{n}\mathbf{a}) - \mathbf{a} - \mathbf{b} \times \mathbf{n} \ , \qquad \mathbf{C} = \mathbf{n}\,(\mathbf{n}\mathbf{b}) - \mathbf{b} + \mathbf{a} \times \mathbf{n} \ ,$$

relationship in Eq. (14.7.13) can be written shorter

$$\mathbf{E} + ic\mathbf{B} = (\mathbf{L} + i\mathbf{C})\, k_0 A \, (\cos\varphi + i\sin\varphi) \ , \tag{14.7.14}$$

from this, it follow

$$\mathbf{E} = k_0 A \left(\cos\varphi \mathbf{L} - \sin\varphi \mathbf{C}\right) , \quad c\mathbf{B} = k_0 A \left(\sin\varphi \mathbf{L} + \cos\varphi \mathbf{C}\right) . \qquad (14.7.15)$$

One can readily prove the identities

$$\mathbf{L}^2 = \mathbf{C}^2 = \mathbf{a}^2 + \mathbf{b}^2 - (\mathbf{na})^2 - (\mathbf{nb})^2 + 2\mathbf{n}\left(\mathbf{a} \times \mathbf{b}\right) ,$$

$$\mathbf{L}\,\mathbf{C} = 0 , \qquad \mathbf{E}\mathbf{B} = 0 , \qquad \mathbf{E}^2 = c^2 \mathbf{B}^2 ,$$

$$\mathbf{Ln} = 0 , \qquad \mathbf{Cn} = 0 , \qquad \mathbf{En} = 0 , \qquad \mathbf{Bn} = 0 . \qquad (14.7.16)$$

General expressions for $\mathbf{L}, \mathbf{C}$ may be decomposed into the sum

$$\mathbf{L} = \mathbf{L}_1 + \mathbf{L}_2 , \qquad \mathbf{C} = \mathbf{C}_1 + \mathbf{C}_2 ,$$

$$I \quad \underline{\mathbf{b} = 0}: \quad \mathbf{L}_1 = \mathbf{n}\,(\mathbf{na}) - \mathbf{a} , \qquad \mathbf{C}_1 = \mathbf{a} \times \mathbf{n} ,$$

$$\mathbf{E}_1 \times c\mathbf{B}_1 = k_0^2 A^2 \left(\mathbf{L_1} \times \mathbf{C}_1\right) = k_0^2 A^2 \left[\, a^2 - (\mathbf{na})^2\,\right] \mathbf{n} ;$$

$$II \quad \underline{\mathbf{a} = 0}: \quad \mathbf{L}_2 = -\mathbf{b} \times \mathbf{n} , \qquad \mathbf{C}_2 = \mathbf{n}\,(\mathbf{nb}) - \mathbf{b} ,$$

$$\mathbf{E}_2 \times c\mathbf{B}_2 = k_0^2 A^2 \left(\mathbf{L_2} \times \mathbf{C}_2\right) = k_0^2 A^2 \left[\, b^2 - (\mathbf{nb})^2\,\right] \mathbf{n} . \qquad (14.7.17)$$

In other words, these two electromagnetic wave are clockwise polarized.

For a particular case when $\mathbf{b} = \mathbf{a}$, we get

$$I \quad \mathbf{L}_1 = \mathbf{n}\,(\mathbf{na}) - \mathbf{a} , \qquad \mathbf{C}_1 = \mathbf{a} \times \mathbf{n} ,$$

$$\mathbf{E}_1 = k_0 A(\cos\varphi \, \mathbf{L}_1 - \sin\varphi \, \mathbf{C}_1) ,$$

$$c\mathbf{B}_1 = k_0 A \left(\sin\varphi \mathbf{L}_1 + \cos\varphi \mathbf{C}_1\right) ;$$

$$II \quad \mathbf{L}_2 = -\mathbf{a} \times \mathbf{n} = -\mathbf{C}_1 , \qquad \mathbf{C}_2 = \mathbf{n}\,(\mathbf{na}) - \mathbf{a} = \mathbf{L}_1 ,$$

$$\mathbf{E}_2 = k_0 A(-\sin\varphi \, \mathbf{L}_1 - \cos\varphi \, \mathbf{C}_1) ,$$

$$c\mathbf{B}_2 = k_0 A \left(-\sin\varphi \mathbf{C}_1 + \cos\varphi \mathbf{L}_1\right) ; \qquad (14.7.18)$$

so constructed waves are linearly independent and orthogonal:

$$\mathbf{E}_1 \mathbf{E}_2 = 0 , \qquad \mathbf{B}_1 \mathbf{B}_2 = 0 . \qquad (14.7.19)$$

In view of linearity of the Maxwell equations any linear combination of the type is a solution as well:

$$\mathbf{E} = c_1 \, \mathbf{E}_1 + c_2 \, \mathbf{E}_2 , \qquad \mathbf{B} = c_1 \, \mathbf{B}_1 + c_2 \, \mathbf{B}_2 . \qquad (14.7.20)$$

14.8   Cylindrical Waves

Let us start with a cylindrical scalar wave (for brevity $k_0 = E$):

$$\Phi = e^{iEx_0}\, e^{ikz}\, e^{im\phi}\, R(\rho)\,, \quad x_1 = \rho\,\cos\phi\,, \quad x_2 = \rho\,\sin\phi\,, \quad x_3 = z\,,$$

$$\frac{\partial}{\partial x_1} = \cos\phi\,\frac{\partial}{\partial \rho} - \frac{\sin\phi}{\rho}\,\frac{\partial}{\partial \phi}\,, \quad \frac{\partial}{\partial x_2} = \sin\phi\,\frac{\partial}{\partial \rho} + \frac{\cos\phi}{\rho}\,\frac{\partial}{\partial \phi}\,. \qquad (14.8.1)$$

Corresponding electromagnetic solutions are to be constructed on the base of relations

$$\{\Psi^0, \Psi^1, \Psi^2, \Psi^3\} = \lambda_0\Psi^0 + \lambda_1\Psi^1 + \lambda_2\Psi^2 + \lambda_3\Psi^3 = \left|\begin{array}{c} 0 \\ \mathbf{E} + ic\mathbf{B} \end{array}\right|\,. \qquad (14.8.2)$$

Let us find F_a

$$F_0 = iE\, e^{iEx_0}\, e^{ikz}\, e^{im\phi}\, R(\rho)\,, \qquad F_3 = ik\, e^{iEx_0}\, e^{ikz}\, e^{im\phi}\, R(\rho)\,,$$

$$F_1 = e^{iEx_0}\, e^{ikz}\, e^{im\phi}\, (\cos\phi\,\frac{d}{d\rho} - im\,\frac{\sin\phi}{\rho})\, R\,,$$

$$F_2 = e^{iEx_0}\, e^{ikz}\, e^{im\phi}\, (\sin\phi\,\frac{d}{d\rho} + im\,\frac{\cos\phi}{\rho})\, R\,. \qquad (14.8.3)$$

The main equations to solve are

$$(-\lambda_0\partial_0 + i\lambda_j\partial_j)\, F_c = 0\,, \qquad (14.8.4)$$

or

$$\left[\, -i\lambda_0 E - \lambda_3 k + i(\lambda_1\cos\phi + \lambda_2\sin\phi)\,\frac{\partial}{\partial \rho}\right.$$

$$\left. +i\,\frac{-\lambda_1\sin\phi + \lambda_2\cos\phi}{\rho}\,\frac{\partial}{\partial \phi}\,\right] F_c = 0\,. \qquad (14.8.5)$$

Let $c = 1$

$$\left[-i\lambda_0 E - \lambda_3 k + i(\lambda_1\cos\phi + \lambda_2\sin\phi)\,\frac{\partial}{\partial \rho}\right.$$

$$\left. +i\,\frac{-\lambda_1\sin\phi + \lambda_2\cos\phi}{\rho}\,\frac{\partial}{\partial \phi}\right]e^{im\phi}\, (\cos\phi\frac{d}{d\rho} - im\frac{\sin\phi}{\rho})R = 0\,.$$

Let $c = 2$

$$\left[-i\lambda_0 E - \lambda_3 k + i(\lambda_1\cos\phi + \lambda_2\sin\phi)\,\frac{\partial}{\partial \rho}\right.$$

$$\left. +i\,\frac{-\lambda_1\sin\phi + \lambda_2\cos\phi}{\rho}\,\frac{\partial}{\partial \phi}\right]e^{im\phi}\, (\sin\phi\frac{d}{d\rho} + im\frac{\cos\phi}{\rho})R = 0\,.$$

Noting that

$$F_1 + iF_2 = e^{iEx_0}\, e^{ikz}\, e^{i(m+1)\phi}(\frac{d}{d\rho} - \frac{m}{\rho})R\,,$$

$$F_1 - iF_2 = e^{iEx_0}\, e^{ikz}\, e^{i(m-1)\phi}(\frac{d}{d\rho} + \frac{m}{\rho})R\,, \qquad (14.8.6)$$

from previous two equations, we get

$$\left[-i\lambda_0 E - \lambda_3 k + i(\lambda_1 \cos\phi + \lambda_2 \sin\phi)\,\frac{\partial}{\partial\rho}\right.$$

$$\left.+\frac{\lambda_1 \sin\phi - \lambda_2 \cos\phi}{\rho}\,(m+1)\right](\frac{d}{d\rho} - \frac{m}{\rho})R = 0\,,$$

$$\left[-i\lambda_0 E - \lambda_3 k + i(\lambda_1 \cos\phi + \lambda_2 \sin\phi)\,\frac{\partial}{\partial\rho}\right.$$

$$\left.+\frac{\lambda_1 \sin\phi - \lambda_2 \cos\phi}{\rho}\,(m-1)\right](\frac{d}{d\rho} + \frac{m}{\rho})R = 0\,. \tag{14.8.7}$$

In turn, when $c = 0, 3$, we get one the same equation

$$\left[-i\lambda_0 E - \lambda_3 k + i(\lambda_1 \cos\phi + \lambda_2 \sin\phi)\,\frac{\partial}{\partial\rho} + \frac{\lambda_1 \sin\phi - \lambda_2 \cos\phi}{\rho}\,m\right]R(\rho) = 0\,. \tag{14.8.8}$$

We readily note that there exists very simple way to satisfy these three equations on parameters λ_c

$$-i\lambda_0 E - \lambda_3 k = 0\,, \qquad \lambda_1 = 0\,, \qquad \lambda_2 = 0\,. \tag{14.8.9}$$

Let us demonstrate that no other solutions exist. To this end, with notation

$$C = -i\lambda_0 E - \lambda_3 k\,, \quad A = \lambda_1 \cos\phi + \lambda_2 \sin\phi\,, \quad B = \lambda_1 \sin\phi - \lambda_2 \cos\phi\,,$$

let us rewrite Eqs. (14.8.7) and (14.8.8) in the form

$$\left[C + iA\frac{d}{d\rho} + \frac{B}{\rho}\,(m+1)\right](\frac{dR}{d\rho} - \frac{m}{\rho}\,R) = 0\,,$$

$$\left[C + iA\frac{d}{d\rho} + \frac{B}{\rho}\,(m-1)\right](\frac{dR}{d\rho} + \frac{m}{\rho}\,R) = 0\,,$$

$$\left[C + iA\frac{d}{d\rho} + \frac{B}{\rho}m\right]R = 0\,. \tag{14.8.10}$$

Combining two first equations in Eq. (14.8.10) (summing and subtracting), we get

$$C\frac{dR}{d\rho} + iA\,\frac{d^2 R}{d\rho^2} + \frac{mB}{\rho}\frac{dR}{d\rho} - \frac{mB}{\rho^2}R = 0\,,$$

$$-\frac{mC}{\rho}R - imA\,\frac{d}{d\rho}\frac{R}{\rho} + \frac{B}{\rho}\frac{dR}{d\rho} - m^2\frac{B}{\rho^2}R = 0\,,$$

$$C\,R + iA\,\frac{dR}{d\rho} + m\,\frac{B}{\rho}\,R = 0\,. \tag{14.8.11}$$

After differentiating the third equation will look

$$C\,\frac{dR}{d\rho} + iA\,\frac{d^2 R}{d\rho^2} + \frac{mB}{\rho}\frac{dR}{d\rho} - \frac{mB}{\rho^2}\,R = 0\,,$$

which coincides with the first equation in Eq. (14.8.11). Therefore, the system in Eq. (14.8.11) is equivalent to

$$-\frac{mC}{\rho}R - imA\,\frac{d}{d\rho}\frac{R}{\rho} + \frac{B}{\rho}\frac{dR}{d\rho} - m^2\frac{B}{\rho^2}R = 0\,,$$

$$C\,R + iA\,\frac{dR}{d\rho} + m\,\frac{B}{\rho}\,R = 0\,. \qquad (14.8.12)$$

The system in Eq. (14.8.12) can be rewritten as follows:

$$-m\,\frac{C}{\rho}\,R + \frac{B}{\rho}\frac{dR}{d\rho} + i\,\frac{mA}{\rho^2}\,R - \frac{m}{\rho}\,(iA\,\frac{dR}{d\rho} + m\,\frac{B}{\rho}\,R) = 0\,,$$

$$C\,R + iA\,\frac{dR}{d\rho} + m\,\frac{B}{\rho}\,R = 0\,.$$

Here, the first equation with the use of the second equation gives

$$\frac{B}{\rho}\frac{dR}{d\rho} + i\,\frac{mA}{\rho^2}\,R = 0\,.$$

Therefore, system in Eq. (14.8.12) is equivalent to

$$B\,\frac{dR}{d\rho} + i\,\frac{mA}{\rho}\,R = 0\,, \qquad iA\,\frac{dR}{d\rho} + m\,\frac{B}{\rho}\,R + C\,R = 0\,,$$

which in turn is equivalent to

$$(B + iA)(\frac{d}{d\rho} + \frac{m}{\rho})R + C\,R = 0\,,$$

$$(B - iA)(\frac{d}{d\rho} - \frac{m}{\rho})R - C\,R = 0\,. \qquad (14.8.13)$$

Noting the identity $(A + iB)(A - iB) = \lambda_1^2 + \lambda_2^2$, this reduces Eqs. (14.8.13) to the form

$$(\lambda_1^2 + \lambda_2^2)(\frac{d}{d\rho} + \frac{m}{\rho})R + (B - iA)C\,R = 0\,,$$

$$(\lambda_1^2 + \lambda_2^2)(\frac{d}{d\rho} - \frac{m}{\rho})R + (B + iA)C\,R = 0\,. \qquad (14.8.14)$$

Remembering that

$$C = -i\lambda_0 E - \lambda_3 k\,, \quad A = \lambda_1\cos\phi + \lambda_2\sin\phi\,, \quad B = \lambda_1\sin\phi - \lambda_2\cos\phi\,,$$

we immediately conclude that Eqs. (14.8.14) can be satisfied only by the following way

$$(\lambda_1^2 + \lambda_2^2)\,, \qquad C = 0\,. \qquad (14.8.15)$$

In other words, this means that relations in Eq. (14.8.9) provides us with the only possible solution in terms of two (1×4)-columns

$$\Psi = \lambda_0\Psi^0 + \lambda_3\Psi^3\,, \qquad -\lambda_0\,E + i\,\lambda_3 k = 0\,,$$

$$\Psi^0 = \begin{vmatrix} iF_0(x) \\ -F_1(x) \\ -F_2(x) \\ -F_3(x) \end{vmatrix}, \qquad \Psi^3 = \begin{vmatrix} F_3(x) \\ F_2(x) \\ -F_1(x) \\ iF_0(x) \end{vmatrix},$$

$$F_0 = iE \; \Phi, \qquad F_1 = (\cos\phi \, \frac{d}{d\rho} - im \, \frac{\sin\phi}{\rho}) \; \Phi,$$

$$F_2 = (\sin\phi \, \frac{d}{d\rho} + im \, \frac{\cos\phi}{\rho}) \; \Phi, \qquad F_3 = ik \; \Phi. \qquad (14.8.16)$$

Solution explicitly reads by components:

$$(\Psi)_0 = (-\lambda_0 \, E + i \, \lambda_3 k) \; \Phi = 0,$$

$$E_3 + iB_3 = (\Psi)_3 = (-\lambda_0 \, ik - \lambda_3 E) \; \Phi = -\lambda_3 \, \frac{E^2 - k^2}{E} \; \Phi,$$

$$E_1 + iB_1 = (\Psi)_1$$

$$= \left[-\lambda_0 \, (\cos\phi \, \frac{d}{d\rho} - im \, \frac{\sin\phi}{\rho}) + \lambda_3 \, (\sin\phi \, \frac{d}{d\rho} + im \, \frac{\cos\phi}{\rho}) \right] \Phi,$$

$$E_2 + iB_2 = (\Psi)_2$$

$$= \left[(-\lambda_0 \, (\sin\phi \, \frac{d}{d\rho} + im \, \frac{\cos\phi}{\rho}) - \lambda_3 \, (\cos\phi \, \frac{d}{d\rho} - im \, \frac{\sin\phi}{\rho}) \right] \Phi.$$

After a simple rewriting, we get

$$E_3 + iB_3 = (\Psi)_3 = -\lambda_3 \frac{E^2 - k^2}{E} \; \Phi,$$

$$E_1 + iB_1 = \frac{\lambda_3}{E} \left[-ik(\cos\phi \frac{d}{d\rho} - im\frac{\sin\phi}{\rho}) + E(\sin\phi \frac{d}{d\rho} + im\frac{\cos\phi}{\rho}) \right] \Phi,$$

$$E_2 + iB_2 = \frac{\lambda_3}{E} \left[(-ik(\sin\phi \frac{d}{d\rho} + im\frac{\cos\phi}{\rho}) - E(\cos\phi \frac{d}{d\rho} - im\frac{\sin\phi}{\rho}) \right] \Phi.$$

$$(14.8.17)$$

In these two special cases we get the simplicity

$$\underline{k = +E}, \qquad E_3 + iB_3 = 0,$$

$$E_1 + iB_1 = -i \, \lambda_3 \; e^{i\phi}(\frac{d}{d\rho} - \frac{m}{\rho})\Phi, \qquad E_2 + iB_2 = -\lambda_3 \; e^{i\phi}(\frac{d}{d\rho} - \frac{m}{\rho})\Phi;$$

$$\underline{k = -E}, \qquad E_3 + iB_3 = 0,$$

$$E_1 + iB_1 = i \, \lambda_3 \; e^{-i\phi}(\frac{d}{d\rho} + \frac{m}{\rho}) \; \Phi, \qquad E_2 + iB_2 = -\lambda_3 \; e^{-i\phi}(\frac{d}{d\rho} + \frac{m}{\rho}) \; \Phi.$$

$$(14.8.18)$$

It seems reasonable to expect further developments in this matrix based approach to Maxwell theory in the frame of complex vector formalism, as a possible base to explore general method to separate the variables for Maxwell equations in different coordinates. Also it would be desirable to extent this approach to Maxwell theory in curved space-time models, when using Riemann–Silberstein–Majorana–Oppenheimer formalism in tetrad form.

Chapter 15

Helicity Operator and Spin 1 Field in Lobachevsky and Rieman Models

15.1 Helicity Operator and Separation of the Variables

In Lobachevsky space-time, let us use quasi-Cartesian coordinates and corresponding tetrad

$$x^a = (t, x, y, z), \qquad dS^2 = dt^2 - e^{-2z}(dx^2 + dy^2) - dz^2,$$

$$e^{\beta}_{(a)} = \begin{vmatrix} 1 & 0 & 0 & 0 \\ 0 & e^z & 0 & 0 \\ 0 & 0 & e^z & 0 \\ 0 & 0 & 0 & 1 \end{vmatrix}. \tag{15.1.1}$$

Christoffel symbols are $\Gamma^0_{\beta\sigma} = 0$, $\Gamma^i_{00} = 0$, $\Gamma^i_{0j} = 0$, and

$$x^i = (x, y, z), \qquad \Gamma^i{}_{jk} = \frac{1}{2} g^{il} \left(-\partial_l g_{jk} + \partial_j g_{lk} + \partial_k g_{lj} \right),$$

$$\Gamma^x{}_{jk} = \begin{vmatrix} 0 & 0 & -1 \\ 0 & 0 & 0 \\ -1 & 0 & 0 \end{vmatrix}, \Gamma^y{}_{jk} = \begin{vmatrix} 0 & 0 & 0 \\ 0 & 0 & -1 \\ 0 & -1 & 0 \end{vmatrix}, \Gamma^z{}_{jk} = \begin{vmatrix} e^{-2z} & 0 & 0 \\ 0 & e^{-2z} & 0 \\ 0 & 0 & 0 \end{vmatrix}.$$

The Ricci rotations coefficients are

$$\gamma_{011} = \gamma_{021} = \gamma_{031} = 0, \; \gamma_{012} = \gamma_{022} = \gamma_{032} = 0, \; \gamma_{013} = \gamma_{023} = \gamma_{033} = 0,$$

$$\gamma_{231} = 0, \qquad \gamma_{311} = -1, \; \gamma_{121} = 0, \; \gamma_{232} = 1, \qquad \gamma_{312} = 0, \; \gamma_{122} = 0,$$

$$\gamma_{233} = 0, \; \gamma_{313} = 0, \; \gamma_{123} = 0.$$

Duffin–Kemmer equation

$$\left[i\beta^c \left(e^{\beta}_{(c)} \partial_\beta + \frac{1}{2} J^{ab} \gamma_{abc} \right) - M \right] \Psi = 0 \tag{15.1.2}$$

here takes the form

$$\left[+i\beta^0 \frac{\partial}{\partial t} + i\beta^1 e^z \frac{\partial}{\partial x} + i\beta^2 e^z \frac{\partial}{\partial y} + i \left(\beta^3 \frac{\partial}{\partial z} - \beta^1 J^{31} + \beta^2 J^{23} \right) - M \right] \Psi = 0 \,.$$

(15.1.3)

We will search solutions in the form of quasi-plane waves

$$\Psi = e^{-i\epsilon t} \, e^{iax} \, e^{iby} \begin{vmatrix} \Phi_0(z) \\ \Phi_j(z) \\ E_j(z) \\ H_j(z) \end{vmatrix} .$$

(15.1.4)

Equations (15.1.3) gives

$$\left[\epsilon \beta^0 - a\beta^1 e^z - b\beta^2 e^z + i \left(\beta^3 \frac{\partial}{\partial z} - \beta^1 J^{31} + \beta^2 J^{23} \right) - M \right] \Psi = 0 \,.$$

(15.1.5)

Next, the cyclic basis for Duffin–Kemmer matrices is used, when a third projection of the spin is a diagonal matrix

$$\beta^0 = \begin{vmatrix} 0 & 0 & 0 & 0 \\ 0 & 0 & i & 0 \\ 0 & -i & 0 & 0 \\ 0 & 0 & 0 & 0 \end{vmatrix}, \beta^i = \begin{vmatrix} 0 & 0 & e_i & 0 \\ 0 & 0 & 0 & \tau_i \\ -e_i^+ & 0 & 0 & 0 \\ 0 & -\tau_i & 0 & 0 \end{vmatrix},$$

$$e_1 = \frac{1}{\sqrt{2}}(-i,\, 0,\, i)\,, \qquad e_2 = \frac{1}{\sqrt{2}}(1,\, 0,\, 1)\,, \qquad e_3 = (0, i, 0)\,,$$

$$\tau_1 = \frac{1}{\sqrt{2}} \begin{vmatrix} 0 & 1 & 0 \\ 1 & 0 & 1 \\ 0 & 1 & 0 \end{vmatrix}, \tau_2 = \frac{1}{\sqrt{2}} \begin{vmatrix} 0 & -i & 0 \\ i & 0 & -i \\ 0 & i & 0 \end{vmatrix}, \tau_3 = \begin{vmatrix} 1 & 0 & 0 \\ 0 & 0 & 0 \\ 0 & 0 & -1, \end{vmatrix} = s_3 \,,$$

$$J^{12} = \beta^1 \beta^2 - \beta^2 \beta^1 = -i \begin{vmatrix} 0 & 0 & 0 & 0 \\ 0 & \tau_3 & 0 & 0 \\ 0 & 0 & \tau_3 & 0 \\ 0 & 0 & 0 & \tau_3 \end{vmatrix} = -iS_3 \,,$$

$$J^{13} = \beta^1 \beta^3 - \beta^3 \beta^1 = i \begin{vmatrix} 0 & 0 & 0 & 0 \\ 0 & \tau_2 & 0 & 0 \\ 0 & 0 & \tau_2 & 0 \\ 0 & 0 & 0 & \tau_2 \end{vmatrix} = iS_2 \,,$$

$$J^{23} = \beta^2 \beta^3 - \beta^3 \beta^2 = -i \begin{vmatrix} 0 & 0 & 0 & 0 \\ 0 & \tau_1 & 0 & 0 \\ 0 & 0 & \tau_1 & 0 \\ 0 & 0 & 0 & \tau_1 \end{vmatrix} = -iS_1 \,,$$

$$-\beta^1 J^{31} + \beta^2 J^{23} = i \begin{vmatrix} 0 & 0 & (e_1\tau_2 - e_2\tau_1) & 0 \\ 0 & 0 & 0 & (\tau_1\tau_2 - \tau_2\tau_1) \\ 0 & 0 & 0 & 0 \\ 0 & -(\tau_1\tau_2 - \tau_2\tau_1) & 0 & 0 \end{vmatrix} .$$

Equation (15.1.5) in block form reads

$$
\left[\epsilon \begin{vmatrix} 0 & 0 & 0 & 0 \\ 0 & 0 & i & 0 \\ 0 & -i & 0 & 0 \\ 0 & 0 & 0 & 0 \end{vmatrix} - ae^z \begin{vmatrix} 0 & 0 & e_1 & 0 \\ 0 & 0 & 0 & \tau_1 \\ -e_1^+ & 0 & 0 & 0 \\ 0 & -\tau_1 & 0 & 0 \end{vmatrix} - be^z \begin{vmatrix} 0 & 0 & e_2 & 0 \\ 0 & 0 & 0 & \tau_2 \\ -e_2^+ & 0 & 0 & 0 \\ 0 & -\tau_2 & 0 & 0 \end{vmatrix} \right.
$$

$$
+i \left(\begin{vmatrix} 0 & 0 & e_3 & 0 \\ 0 & 0 & 0 & \tau_3 \\ -e_3^+ & 0 & 0 & 0 \\ 0 & -\tau_3 & 0 & 0 \end{vmatrix} \frac{\partial}{\partial z} \right.
$$

$$
\left. \left. +i \begin{vmatrix} 0 & 0 & (e_1\tau_2 - e_2\tau_1) & 0 \\ 0 & 0 & 0 & (\tau_1\tau_2 - \tau_2\tau_1) \\ 0 & 0 & 0 & 0 \\ 0 & -(\tau_1\tau_2 - \tau_2\tau_1) & 0 & 0 \end{vmatrix} \right) - M \right] \begin{vmatrix} \Phi_0 \\ \Phi_j \\ E_j \\ H_j \end{vmatrix} = 0 \,,
$$

$$(15.1.6a)$$

which is equivalent to

$$
-a\,e^z\,e_1\,\mathbf{E} - b\,e^z\,e_2\,\mathbf{E} + i\,e_3\,\frac{\partial \mathbf{E}}{\partial z} - (e_1\tau_2 - e_2\,\tau_1)\,\mathbf{E} - M\,\Phi_0 = 0 \,,
$$

$$
\epsilon\,I\,\mathbf{E} - a\,e^z\,\tau_1\,\mathbf{H} - b\,e^z\,\tau_2\,\mathbf{H} + i\,\tau_3\,\frac{\partial \mathbf{H}}{\partial z} - (\tau_1\tau_2 - \tau_2\,\tau_1)\,\mathbf{H} - M\,\vec{\Phi} = 0 \,,
$$

$$
-\epsilon\,I\,\vec{\Phi} + a\,e^z\,e_1^+\,\Phi_0 + b\,e^z\,e_2^+\,\Phi_0 - i\,e_3^+\,\frac{\partial \Phi_0}{\partial z} - M\,\mathbf{E} = 0 \,,
$$

$$
a\,e^z\,\tau_1\,\vec{\Phi} + b\,e^z\,\tau_2\,\vec{\Phi} - i\,\tau_3\,\frac{\partial \vec{\Phi}}{\partial z} + (\tau_1\tau_2 - \tau_2\,\tau_1)\,\vec{\Phi} - M\,\mathbf{H} = 0 \,.
$$

$$(15.1.6b)$$

In explicit form, 10 equations look (let $\gamma = 1/\sqrt{2}$)

$$
i\gamma\,a\,e^z\,(E_1 - E_3) - \gamma\,b\,e^z\,(E_1 + E_3) - \frac{dE_2}{dz} + 2\,E_2 - M\,\Phi_0 = 0 \,, \qquad (15.1.7a)
$$

$$
i\,\epsilon\,E_1 - \gamma\,a\,e^z\,H_2 + i\gamma\,b\,e^z\,H_2 + i\,\frac{dH_1}{dz} - i\,H_1 - M\,\Phi_1 = 0 \,,
$$

$$
i\,\epsilon\,E_2 - \gamma a\,e^z\,(H_1 + H_3) - i\gamma\,b\,e^z\,(H_1 - H_3) - M\,\Phi_2 = 0 \,,
$$

$$
i\,\epsilon\,E_3 - \gamma a\,e^z\,H_2 - i\gamma\,b\,e^z\,H_2 - i\,\frac{dH_3}{dz} + i\,H_3 - M\,\Phi_3 = 0 \,, \qquad (15.1.7b)
$$

$$
-i\,\epsilon\,\Phi_1 + i\gamma\,a\,e^z\,\Phi_0 + \gamma\,b\,e^z\,\Phi_0 - M\,E_1 = 0 \,,
$$

$$
-i\,\epsilon\,\Phi_2 - \frac{d\Phi_0}{dz} - M\,E_2 = 0 \,,
$$

$$
-i\,\epsilon\,\Phi_3 - i\gamma\,a\,e^z\,\Phi_0 + \gamma\,b\,e^z\,\Phi_0 - M\,E_3 = 0 \,, \qquad (15.1.7c)
$$

$$\gamma a\, e^z\, \Phi_2 - i\gamma\, b\, e^z\, \Phi_2 - i\,\frac{d\Phi_1}{dz} + i\,\Phi_1 - M\,H_1 = 0\,,$$

$$\gamma a\, e^z\,(\Phi_1 + \Phi_3) + i\gamma\, b\, e^z\,(\Phi_1 - \Phi_3) - M\,H_2 = 0\,,$$

$$\gamma a\, e^z\, \Phi_2 + i\gamma\, b\, e^z\, \Phi_2 + i\,\frac{d\Phi_3}{dz} - i\,\Phi_3 - M\,H_3 = 0\,.$$

$$(15.1.7d)$$

After evident regrouping, we have

$$\gamma(ia - b)e^z\, E_1 - \gamma(ia + b)e^z E_3 - (\frac{d}{dz} - 2)\, E_2 - M\,\Phi_0 = 0\,, \qquad (15.1.8a)$$

$$i\,\epsilon\, E_1 - \gamma(a - ib)\, e^z\, H_2 + i\,(\frac{d}{dz} - 1)\, H_1 - M\,\Phi_1 = 0\,,$$

$$i\,\epsilon\, E_2 - \gamma(a + ib)\, e^z\, H_1 - \gamma(a - ib)\, e^z\, H_3 - M\,\Phi_2 = 0\,,$$

$$i\,\epsilon\, E_3 - \gamma(a + ib)\, e^z\, H_2 - i(\frac{d}{dz} - 1)\, H_3 - M\,\Phi_3 = 0\,,$$

$$(15.1.8b)$$

$$-i\,\epsilon\,\Phi_1 + \gamma\,(b + ia)\, e^z\,\Phi_0 - M\,E_1 = 0\,,$$

$$-i\,\epsilon\,\Phi_2 - \frac{d\Phi_0}{dz} - M\,E_2 = 0\,,$$

$$-i\,\epsilon\,\Phi_3 + \gamma(b - ia)\, e^z\,\Phi_0 - M\,E_3 = 0\,, \qquad (15.1.8c)$$

$$\gamma(a - ib)\, e^z\,\Phi_2 - i(\frac{d}{dz} - 1)\,\Phi_1 - M\,H_1 = 0\,,$$

$$\gamma(a + ib)\, e^z\,\Phi_1 + \gamma(a - ib)\, e^z\,\Phi_3 - M\,H_2 = 0\,,$$

$$\gamma(a + ib)\, e^z\,\Phi_2 + i\,(\frac{d}{dz} - 1)\,\Phi_3 - M\,H_3 = 0\,. \qquad (15.1.8d)$$

To find explicit solutions for this system, it is convenient to diagonalize additionally a generalized helicity operator (its explicit form can be proposed by analogy with the case of similar operator in Minkowski space; it commutes with the wave operator[1]):

$$\Psi_0(z) = (\Phi_0(z),\,\Phi_j(z),\,E_j(z),\,H_j(z))\,,$$

$$\left[ae^z S^1 + be^z S^2 - i(S^3\frac{d}{dz} - S^1 J^{31} + S^2 J^{23})\right]\Psi_z = \sigma\Psi(z)\,,$$

$$-S^1 J^{31} + S^2 J^{23} = -S^3\,; \qquad (15.1.9a)$$

in block form, it reads

$$\left[ae^z \begin{vmatrix} 0 & 0 & 0 & 0 \\ 0 & \tau_1 & 0 & 0 \\ 0 & 0 & \tau_1 & 0 \\ 0 & 0 & 0 & \tau_1 \end{vmatrix} + be^z \begin{vmatrix} 0 & 0 & 0 & 0 \\ 0 & \tau_2 & 0 & 0 \\ 0 & 0 & \tau_2 & 0 \\ 0 & 0 & 0 & \tau_2 \end{vmatrix} \right.$$

[1]To avoid misunderstanding, it should be noted that next we use an helicity operator which differs in the factor i from usual, so its eigenvalues are imaginary or zero.

$$-i \begin{vmatrix} 0 & 0 & 0 & 0 \\ 0 & \tau_3 & 0 & 0 \\ 0 & 0 & \tau_3 & 0 \\ 0 & 0 & 0 & \tau_3 \end{vmatrix} (\frac{d}{dz} - 1) \Bigg] \begin{vmatrix} \Phi_0(z) \\ \Phi_j(z) \\ E_j(z) \\ H_j(z) \end{vmatrix} = \sigma \begin{vmatrix} \Phi_0(z) \\ \Phi_j(z) \\ E_j(z) \\ H_j(z) \end{vmatrix}, \tag{15.1.9b}$$

so that

$$0 = \sigma \, \Phi_0 \,,$$

$$a \, e^z \, \tau_1 \, \vec{\Phi} + b \, e^z \, \tau_2 \, \vec{\Phi} - i \, \tau_3 \, (\frac{d}{dz} - 1) \vec{\Phi} = \sigma \, \vec{\Phi} \,,$$

$$a \, e^z \, \tau_1 \, \mathbf{E} + b \, e^z \, \tau_2 \, \mathbf{E} - i \, \tau_3 \, \frac{d\mathbf{E}}{dz} = \sigma \, \mathbf{E} \,,$$

$$a \, e^z \, \tau_1 \, \mathbf{H} + b \, e^z \, \tau_2 \, \mathbf{H} - i \, \tau_3 \, (\frac{d}{dz} - 1) \mathbf{H} = \sigma \, \mathbf{H} \,. \tag{15.1.9c}$$

After simple calculations, we get 10 equations

$$0 = \sigma \, \Phi_0 \,,$$

$$\gamma \, (a - ib) \, e^z \, \Phi_2 - i \, (\frac{d}{dz} - 1) \Phi_1 = \sigma \, \Phi_1 \,,$$

$$\gamma (a + ib) \, e^z \, \Phi_1 + i \gamma \, (a - ib) \, e^z \, \Phi_3 = \sigma \, \Phi_2 \,,$$

$$\gamma (a + ib) \, e^z \, \Phi_2 + i \, (\frac{d}{dz} - 1) \Phi_3 = \sigma \, \Phi_3 \,,$$

$$\gamma \, (a - ib) \, e^z \, E_2 - i \, \frac{dE_1}{dz} = \sigma \, E_1 \,,$$

$$\gamma (a + ib) \, e^z \, E_1 + \gamma \, (a - ib) \, e^z \, E_3 = \sigma \, E_2 \,,$$

$$\gamma (a + ib) \, e^z \, E_2 + i \, \frac{dE_3}{dz} = \sigma \, E_3 \,,$$

$$\gamma \, (a - ib) \, e^z \, H_2 - i \, (\frac{d}{dz} - 1) H_1 = \sigma \, H_1 \,,$$

$$\gamma (a + ib) \, e^z \, H_1 + \gamma \, (a - ib) \, e^z \, H_3 = \sigma \, H_2 \,,$$

$$\gamma (a + ib) \, e^z \, H_2 + i \, (\frac{d}{dz} - 1) H_3 = \sigma \, H_3 \,.$$

It is more convenient to rewrite them in the form

$$0 = \sigma \, \Phi_0 \,, \tag{15.1.10a}$$

$$-i \, (\frac{d}{dz} - 1) \Phi_1 = \sigma \, \Phi_1 - \gamma \, (a - ib) \, e^z \, \Phi_2 \,,$$

$$\gamma (a + ib) \, e^z \, \Phi_1 + \gamma \, (a - ib) \, e^z \, \Phi_3 = \sigma \, \Phi_2 \,,$$

$$+i \, (\frac{d}{dz} - 1) \Phi_3 = \sigma \, \Phi_3 - \gamma (a + ib) \, e^z \, \Phi_2 \,, \tag{15.1.10b}$$

$$-i \, \frac{dE_1}{dz} = \sigma \, E_1 - \gamma \, (a - ib) \, e^z \, E_2 \,,$$

$$\gamma(a + ib) e^z E_1 + \gamma (a - ib) e^z E_3 = \sigma E_2 \,,$$

$$+i \frac{dE_3}{dz} = \sigma E_3 - \gamma(a + ib) e^z E_2 \,, \qquad (15.1.10c)$$

$$-i \left(\frac{d}{dz} - 1\right) H_1 = \sigma H_1 - \gamma (a - ib) e^z H_2 \,,$$

$$\gamma(a + ib) e^z H_1 + \gamma (a - ib) e^z H_3 = \sigma H_2 \,,$$

$$+i \left(\frac{d}{dz} - 1\right) H_3 = \sigma H_3 - \gamma(a + ib) e^z H_2 \,. \qquad (15.1.10d)$$

Note, that we see three similar groups of equations for Φ_j, E_j, H_i, respectively. Let us examine one of them, for instance

$$\gamma (a - ib) e^z H_2 - i \left(\frac{d}{dz} - 1\right) H_1 = \sigma H_1 \,,$$

$$\gamma(a + ib) e^z H_1 + \gamma (a - ib) e^z H_3 = \sigma H_2 \,,$$

$$\gamma(a + ib) e^z H_2 + i \left(\frac{d}{dz} - 1\right) H_3 = \sigma H_3 \,. \qquad (15.1.11)$$

The case $\sigma = 0$ reduces to (functions H_1 and H_3 turn to be proportional to each other)

$$(a + ib)H_1 = -(a - ib)H_3 \,,$$

$$H_2 = +\frac{ie^{-z}}{\gamma(a - ib)} \left(\frac{d}{dz} - 1\right) H_1 = -\frac{ie^{-z}}{\gamma(a + ib)} \left(\frac{d}{dz} - 1\right) H_3 \,. \qquad (15.1.12)$$

When $\sigma \neq 0$, one can exclude H_2 – then the first and the third equations provide us with the linear differential system for H_1 and H_3:

$$(a - ib) e^{2z} \left[(a + ib)H_1 + (a - ib)H_3\right] - i\sigma \left(\frac{d}{dz} - 1\right) H_1 = 2\sigma^2 H_1 \,,$$

$$(a + ib) e^{2z} \left[(a + ib)H_1 + (a - ib)H_3\right] + i\sigma \left(\frac{d}{dz} - 1\right) H_3 = 2\sigma^2 H_3 \,.$$

$$(15.1.13)$$

One may observe the symmetry in Eq. (15.1.13):

$$H_1 \Longrightarrow H_3 \,, \qquad \sigma \Longrightarrow -\sigma \,, \qquad b \Longrightarrow -b \,.$$

From Eq. (15.1.13), we get

$$H_3 = \frac{2i\,\sigma}{e^{2z} (a - i\,b)^2} \left(\frac{d}{dz} - 1 - i\,\sigma\right) H_1 - \frac{a^2 + b^2}{(a - i\,b)^2} H_1 \,,$$

$$\frac{d^2 H_1}{dz^2} - 4\frac{dH_1}{dz} - \left[e^{2z} (a^2 + b^2) - 4 - (i + \sigma)^2\right] H_1 = 0 \,;$$

$$(15.1.14a)$$

and the second symmetrical variant

$$H_1 = \frac{-2i\,\sigma}{e^{2z}\,(a+i\,b)^2}\left(\frac{d}{dz} - 1 + i\,\sigma\right)H_3 - \frac{a^2+b^2}{(a+i\,b)^2}\,H_3\,,$$

$$\frac{d^2H_3}{dz^2} - 4\frac{dH_3}{dz} - \left[e^{2z}\,(a^2+b^2) - 4 - (i-\sigma)^2\right]H_3 = 0\,.$$

$$(15.1.14b)$$

Let us translate Eqs. (15.1.14a) to another variable

$$i\sqrt{a^2+b^2}\,e^z = Z\,, \qquad \frac{d}{dz} = Z\frac{d}{dZ}\,,$$

then the second order equations read as

$$\left[Z^2\frac{d^2}{dZ^2} - 3Z\frac{d}{dZ} + Z^2 + 4 - (1-i\sigma)^2\right]H_1 = 0\,, \qquad (15.1.15a)$$

or differently, $H_1(Z) = Z^2h_1(Z)$:

$$\left[\frac{d^2}{dZ^2} + \frac{1}{Z}\frac{d}{dZ} + 1 - \frac{(1-i\sigma)^2}{Z^2}\right]h_1 = 0\,, \qquad (15.1.15b)$$

which is the Bessel equations with solutions

$$h_1 = J_{\pm\mu}(Z)\,, \qquad \mu = 1 - i\sigma\,. \qquad (15.1.15c)$$

Expression h_3 of Eq. (15.1.14a) is defined by

$$Z^2h_3 = \frac{a+ib}{a-ib}\left[-2i\sigma(Z\frac{d}{dZ} + 1 - i\sigma) - Z^2\right]h_1\,. \qquad (15.1.15d)$$

Similar formulas can be produced for Eq. (15.1.14b):

$$H_3 = e^{2Z}h_3\,,$$

and

$$\left[\frac{d^2}{dZ^2} + \frac{1}{Z}\frac{d}{dZ} + 1 - \frac{(1+i\sigma)^2}{Z^2}\right]h_3 = 0\,, \qquad (15.1.16a)$$

$$h_3 = J_{\pm\nu}(Z)\,, \qquad \nu = 1 + i\sigma\,, \qquad (15.1.16b)$$

$$Z^2h_1 = \frac{a-ib}{a+ib}\left[+2i\sigma(Z\frac{d}{dZ} + 1 + i\sigma) - Z^2\right]h_3\,. \qquad (15.1.16c)$$

Now, we joint together the main equations of Eq. (15.1.8) and Eqs. (15.1.10a)–(15.1.10d). First, let us consider the case of non-zero σ. At this, from the very beginning, one must assume (see Eq. (15.1.10a)) $\Phi_0 = 0$. Equations (15.1.8) give

$$\gamma(ia-b)e^z E_1 - \gamma(ia+b)e^z E_3 - (\frac{d}{dz} - 2)\,E_2 = 0\,, \qquad (15.1.17a)$$

$$i \, \epsilon \, E_1 - \sigma \, H_1 - M \, \Phi_1 = 0 \,,$$

$$i \, \epsilon \, E_2 - \sigma \, H_2 - M \, \Phi_2 = 0 \,,$$

$$i \, \epsilon \, E_3 - \sigma \, H_3 - M \, \Phi_3 = 0 \,, \qquad (15.1.17b)$$

$$-i \, \epsilon \, \Phi_1 - M \, E_1 = 0 \,, \quad -i \, \epsilon \, \Phi_2 - M \, E_2 = 0 \,, \quad -i \, \epsilon \, \Phi_3 - M \, E_3 = 0 \,, \qquad (15.1.17c)$$

$$\sigma \, \Phi_1 - M \, H_1 = 0 \,, \quad \sigma \, \Phi_2 - M \, H_2 = 0 \,, \quad \sigma \, \Phi_3 - M \, H_3 = 0 \,. \qquad (15.1.17d)$$

On can exclude E_j from this equation

$$\gamma(ia - b)e^z \, \Phi_1 - \gamma(ia + b)e^z \Phi_3 - (\frac{d}{dz} - 2) \, \Phi_2 = 0 \,. \qquad (15.1.18)$$

This relation coincides with the Lorentz condition (when $\Phi_0 = 0$). Indeed, let us start with the Lorentz gauge in tensor form

$$\nabla_\beta \Phi^{\beta(cart)} = 0 \qquad \Longrightarrow \qquad \nabla_\beta (e^{(b)\beta} \Phi^{cart}_{(b)} = 0 \qquad \Longrightarrow$$

$$\frac{\partial \Phi_{(b)cart}}{\partial x^\beta} \, e^{(b)\beta} + \Phi^{cart}_{(b)} \nabla_\beta e^{(b)\beta} = 0 \,, \qquad (15.1.19a)$$

or

$$\frac{\partial \Phi^{cart}_{(b)}}{\partial x^\beta} \, e^{(b)\beta} + \Phi^{cart}_{(b)} \frac{1}{\sqrt{-g}} \frac{\partial}{\partial x^\beta} \sqrt{-g} e^{(b)\beta} = 0 \,. \qquad (15.1.19b)$$

In allowing for the relations in Eq. (15.1.1), Eq. (15.1.19b) reduces to

$$\frac{\partial \Phi^{cart}_{(0)}}{\partial t} - e^z \frac{\partial \Phi^{cart}_{(1)}}{\partial x} - e^z \frac{\partial \Phi^{cart}_{(2)}}{\partial y} - (\frac{\partial}{\partial z} - 2)\Phi^{cart}_{(3)} = 0 \,, \qquad (15.1.19c)$$

or with the substitution

$$(\Phi^{cart}_a) = e^{-i\epsilon t} e^{iax} e^{iby} \begin{vmatrix} \Phi^{cart}_0(z) \\ \Phi^{cart}_1(z) \\ \Phi^{cart}_2(z) \\ \Phi^{cart}_3(z) \end{vmatrix}$$

we arrive at

$$-i\epsilon\Phi^{cart}_{(0)} - iae^z \Phi^{cart}_{(1)} - ibe^z \Phi^{cart}_{(2)} - (\frac{d}{dz} - 2)\Phi^{cart}_{(3)} = 0 \,. \qquad (15.1.20a)$$

Translating the last relation to the variables of a cyclic basis

$$\Phi_2 = \Phi^{cart}_{(3)} \,, \qquad \Phi_3 - \Phi_1 = \sqrt{2}\Phi^{cart}_{(1)} \,, \qquad \Phi_3 + \Phi_1 = \sqrt{2}i \, \Phi^{cart}_{(2)} \,,$$

we get

$$-i\epsilon\Phi_0 - iae^z \frac{\Phi_3 - \Phi_1}{\sqrt{2}} - ibe^z \frac{\Phi_3 + \Phi_1}{\sqrt{2}i} - (\frac{d}{dz} - 2)\Phi_2 = 0 \,,$$

that is

$$-i\epsilon\Phi_0 + \frac{ia - b}{\sqrt{2}}e^z\Phi_1 - \frac{ia + b}{\sqrt{2}}e^z\Phi_3 - (\frac{d}{dz} - 2)\Phi_2 = 0 \,. \qquad (15.1.20b)$$

Equation (15.1.20b) when $\Phi_0 = 0$ coincides with Eq. (15.1.18)

$$\gamma(ia - b)e^z\,\Phi_1 - \gamma(ia + b)e^z\Phi_3 - (\frac{d}{dz} - 2)\,\Phi_2 = 0\,.$$

Therefore, Eq. (15.1.18) (or Eq. (15.1.17a)) is the identity that is valid automatically due to the structure of the wave equation for spin 1 field. For the remaining 9 equations for Eq. (15.1.17b,c,d) lead us to linear constrains between 9 nontrivial constituents:

$$i\,\epsilon\,E_j - \sigma\,H_j - M\,\Phi_j = 0\,,$$

$$-i\,\epsilon\,\Phi_j - M\,E_j = 0\,,$$

$$\sigma\,\Phi_j - M\,H_j = 0\,, \tag{15.1.21a}$$

that is

$$\sigma = \pm\sqrt{\epsilon^2 - M^2}\,, \qquad \Phi_0 = 0\,,$$

$$E_j = -\frac{i\epsilon}{M}\Phi_j\,, \qquad H_j = \frac{\sigma}{M}\Phi_j\,. \tag{15.1.21b}$$

Now, we are to proceed with the case $\sigma = 0$. Eigenvalue helicity equations here are

$$0 = 0\,, \tag{15.1.22a}$$

$$-i\,(\frac{d}{dz} - 1)\Phi_1 = -\gamma\,(a - ib)\,e^z\,\Phi_2\,,$$

$$\gamma(a + ib)\,e^z\,\Phi_1 + \gamma\,(a - ib)\,e^z\,\Phi_3 = 0\,,$$

$$+i\,(\frac{d}{dz} - 1)\Phi_3 = -\gamma(a + ib)\,e^z\,\Phi_2\,, \tag{15.1.22b}$$

$$-i\,\frac{dE_1}{dz} = -\gamma\,(a - ib)\,e^z\,E_2\,,$$

$$\gamma(a + ib)\,e^z\,E_1 + \gamma\,(a - ib)\,e^z\,E_3 = 0\,,$$

$$+i\,\frac{dE_3}{dz} = -\gamma(a + ib)\,e^z\,E_2\,, \tag{15.1.22c}$$

$$-i\,(\frac{d}{dz} - 1)H_1 = -\gamma\,(a - ib)\,e^z\,H_2\,,$$

$$\gamma(a + ib)\,e^z\,H_1 + \gamma\,(a - ib)\,e^z\,H_3 = 0\,,$$

$$+i\,(\frac{d}{dz} - 1)H_3 = -\gamma(a + ib)\,e^z\,H_2\,. \tag{15.1.22d}$$

When taking into consideration Eq. (15.1.22) from Eq. (15.1.8), we get

$$\gamma(ia - b)e^z\,E_1 - \gamma(ia + b)e^z E_3 - (\frac{d}{dz} - 2)\,E_2 - M\,\Phi_0 = 0\,, \tag{15.1.23a}$$

$$i\,\epsilon\,E_1 - M\,\Phi_1 = 0\,,$$

$$i\,\epsilon\,E_2 - M\,\Phi_2 = 0\,,$$

$$i\,\epsilon\,E_3 - M\,\Phi_3 = 0\,, \tag{15.1.23b}$$

$$-i\,\epsilon\,\Phi_1 + \gamma\,(b+ia)\,e^z\,\Phi_0 - M\,E_1 = 0\,,$$

$$-i\,\epsilon\,\Phi_2 - \frac{d\Phi_0}{dz} - M\,E_2 = 0\,,$$

$$-i\,\epsilon\,\Phi_3 + \gamma(b-ia)\,e^z\,\Phi_0 - M\,E_3 = 0\,, \qquad (15.1.23c)$$

$$-M\,H_1 = 0\,, \qquad -M\,H_2 = 0\,, \qquad -M\,H_3 = 0\,. \qquad (15.1.23d)$$

Excluding E_j in Eq. (15.1.23c), we derive

$$(\epsilon^2 - M^2)\Phi_1 - \gamma\epsilon(a-ib)\,e^z\,\Phi_0 = 0\,,$$

$$(\epsilon^2 - M^2)\Phi_2 - i\epsilon\frac{d}{dz}\Phi_0 = 0\,,$$

$$(\epsilon^2 - M^2)\Phi_3 + \gamma\epsilon(a+ib)\,e^z\,\Phi_0 = 0\,. \qquad (15.1.24a)$$

Now, let us recall Eqs. (15.1.12)

$$(a+ib)\Phi_1 + (a-ib)\Phi_3 = 0\,,$$

$$\Phi_2 = +\frac{ie^{-z}}{\gamma(a-ib)}\Big(\frac{d}{dz}-1\Big)\Phi_1 = -\frac{ie^{-z}}{\gamma(a+ib)}\Big(\frac{d}{dz}-1\Big)\Phi_3\,. \qquad (15.1.24b)$$

Thus, we obtain equation for the independent variables

$$\Phi_1 = +\frac{\gamma\epsilon(a-ib)}{(\epsilon^2 - M^2)}\,e^z\,\Phi_0\,,$$

$$\Phi_3 = -\frac{\gamma\epsilon(a+ib)}{\epsilon^2 - M^2}e^z\Phi_0\,,$$

$$\Phi_2 = \frac{i\epsilon}{(\epsilon^2 - M^2)}\frac{d}{dz}\Phi_0\,. \qquad (15.1.24c)$$

Substituting them into Eq. (15.1.23a)

$$\gamma(ia-b)e^z\,\Phi_1 - \gamma(ia+b)e^z\Phi_3 - \Big(\frac{d}{dz}-2\Big)\,\Phi_2 - i\epsilon\Phi_0 = 0\,,$$

we arrive at

$$\gamma(ia-b)e^z\,\frac{\gamma\epsilon(a-ib)}{(\epsilon^2 - M^2)}\,e^z\,\Phi_0 + \gamma(ia+b)\frac{\gamma\epsilon(a+ib)}{\epsilon^2 - M^2}e^z\Phi_0 e^z$$

$$-\Big(\frac{d}{dz}-2\Big)\,\frac{i\epsilon}{(\epsilon^2 - M^2)}\frac{d}{dz}\Phi_0 - i\epsilon\Phi_0 = 0\,,$$

that is

$$\left[\Big(\frac{d}{dz}-2\Big)\frac{d}{dz} + (\epsilon^2 - M^2) - e^{2z}(a^2+b^2)\right]\Phi_0 = 0\,. \qquad (15.1.25)$$

This type for Eq. (15.1.14a) and it can be solved in Bessel functions.

In the end, one may mention that in the massless case, instead of Eq. (15.1.23), we would have

$$0 = 0\,,$$

$$i \epsilon E_1 = 0, \qquad i \epsilon E_2 = 0, \qquad i \epsilon E_3 = 0,$$

$$-i \epsilon \Phi_1 + \gamma (b + ia) e^z \Phi_0 = 0,$$

$$-i \epsilon \Phi_2 - \frac{d\Phi_0}{dz} = 0, \qquad -i \epsilon \Phi_3 + \gamma(b - ia) e^z \Phi_0 = 0,$$

$$H_1 = 0, \qquad H_2 = 0, \qquad H_3 = 0. \qquad (15.1.26)$$

These equations describe gauge solution of gradient type, with vanishing electromagnetic tensor.

15.2   Nonrelativistic Approximation for Quasi-Planes Waves

Let us start with the system of equations obtained after the separation of variables in the Duffin–Kemmer equation for a spin 1 field in quasi-Cartesian coordinates of the Lobachevsky space

$$\gamma(ia - b)e^z E_1 - \gamma(ia + b)e^z E_3 - (\frac{d}{dz} - 2) E_2 - M \Phi_0 = 0, \qquad (15.2.1a)$$

$$i \epsilon E_1 - \gamma(a - ib) e^z H_2 + i (\frac{d}{dz} - 1) H_1 - M \Phi_1 = 0,$$

$$i \epsilon E_2 - \gamma(a + ib) e^z H_1 - \gamma(a - ib) e^z H_3 - M \Phi_2 = 0,$$

$$i \epsilon E_3 - \gamma(a + ib) e^z H_2 - i(\frac{d}{dz} - 1) H_3 - M \Phi_3 = 0, \qquad (15.2.1b)$$

$$-i \epsilon \Phi_1 + \gamma (b + ia) e^z \Phi_0 - M E_1 = 0,$$

$$-i \epsilon \Phi_2 - \frac{d\Phi_0}{dz} - M E_2 = 0,$$

$$-i \epsilon \Phi_3 + \gamma(b - ia) e^z \Phi_0 - M E_3 = 0, \qquad (15.2.1c)$$

$$\gamma(a - ib) e^z \Phi_2 - i(\frac{d}{dz} - 1) \Phi_1 - M H_1 = 0,$$

$$\gamma(a + ib) e^z \Phi_1 + \gamma(a - ib) e^z \Phi_3 - M H_2 = 0,$$

$$\gamma(a + ib) e^z \Phi_2 + i (\frac{d}{dz} - 1) \Phi_3 - M H_3 = 0. \qquad (15.2.1d)$$

When performing nonrelativistic approximation in Eq. (15.2.1a,b,c,d) we will adhere the method elaborated in chapter **4**. First, with the help of Eqs. (15.2.1a) and (15.2.1d) let us exclude non-dynamical variables Φ_0, H_1, H_2, H_3 in eqs. (15.2.1b), (15.2.1c) (components that are not differentiated in time)

$$\Phi_0 = \frac{1}{M} \left[\gamma(ia - b)e^z E_1 - \gamma(ia + b)e^z E_3 - (\frac{d}{dz} - 2) E_2 \right],$$

$$H_1 = \frac{1}{M} \left[\gamma(a - ib) e^z \Phi_2 - i(\frac{d}{dz} - 1) \Phi_1 \right],$$

$$H_2 = \frac{1}{M} \left[\gamma(a+ib)\, e^z\, \Phi_1 + \gamma(a-ib)\, e^z\, \Phi_3 \right],$$

$$H_3 = \frac{1}{M} \left[\gamma(a+ib)\, e^z\, \Phi_2 + i\,(\frac{d}{dz} - 1)\, \Phi_3 \right],$$

which results in

$$i\,\epsilon\, E_1 - \gamma(a-ib)\, e^z\, \frac{1}{M} \left[\gamma(a+ib)\, e^z\, \Phi_1 + \gamma(a-ib)\, e^z\, \Phi_3 \right]$$

$$+i\,(\frac{d}{dz} - 1)\, \frac{1}{M} \left[\gamma(a-ib)\, e^z\, \Phi_2 - i(\frac{d}{dz} - 1)\, \Phi_1 \right] - M\, \Phi_1 = 0 \,,$$

$$-i\epsilon\Phi_1 + \gamma(b+ia)e^z\, \frac{1}{M} \left[\gamma(ia-b)e^z E_1 - \gamma(ia+b)e^z E_3 - (\frac{d}{dz} - 2)E_2 \right] - M E_1 = 0 \,,$$

$$i\,\epsilon\, E_2 - \gamma(a+ib)\, e^z\, \frac{1}{M} \left[\gamma(a-ib)\, e^z\, \Phi_2 - i(\frac{d}{dz} - 1)\, \Phi_1 \right]$$

$$-\gamma(a-ib)\, e^z\, \frac{1}{M} \left[\gamma(a+ib)\, e^z\, \Phi_2 + i\,(\frac{d}{dz} - 1)\, \Phi_3 \right] - M\, \Phi_2 = 0 \,,$$

$$-i\epsilon\Phi_2 - \frac{d}{dz}\, \frac{1}{M} \left[\gamma(ia-b)e^z E_1 - \gamma(ia+b)e^z E_3 - (\frac{d}{dz} - 2)E_2 \right] - M E_2 = 0 \,,$$

$$i\epsilon E_3 - \gamma(a+ib)e^z\, \frac{1}{M} \left[\gamma(a+ib)e^z\Phi_1 + \gamma(a-ib)e^z\Phi_3 \right]$$

$$-i(\frac{d}{dz} - 1)\, \frac{1}{M} \left[\gamma(a+ib)e^z\Phi_2 + i(\frac{d}{dz} - 1)\Phi_3 \right] - M\Phi_3 = 0 \,,$$

$$-i\epsilon\Phi_3 + \gamma(b-ia)e^z\, \frac{1}{M} \left[\gamma(ia-b)e^z E_1 - \gamma(ia+b)e^z E_3 - (\frac{d}{dz} - 2)E_2 \right] - M E_3 = 0 \,.$$

$$(15.2.2)$$

For the system in Eqs. (15.2.2), let us introduce big and small components, Ψ_j and ψ_j:

$$\Phi_j = \Psi_j + \psi_j \,, \qquad E_j = -i(\Psi_j - \psi_j) \,,$$

Equations (15.2.2) take the form

$$\epsilon(\Psi_1 - \psi_1) - \gamma(a-ib)e^z\, \frac{1}{M} \left[\gamma(a+ib)e^z(\Psi_1 + \psi_1) + \gamma(a-ib)e^z(\Psi_3 + \psi_3) \right]$$

$$+i(\frac{d}{dz} - 1)\, \frac{1}{M} \left[\gamma(a-ib)e^z(\Psi_2 + \psi_2) - i(\frac{d}{dz} - 1)(\Psi_1 + \psi_1) \right] - M(\Psi_1 + \psi_1) = 0 \,,$$

$$\epsilon(\Psi_1 + \psi_1) + \gamma(b+ia)e^z\, \frac{1}{M} \left[\gamma(ia-b)e^z(\Psi_1 - \psi_1) - \gamma(ia+b)e^z(\Psi_3 - \psi_3) \right.$$

$$\left. -(\frac{d}{dz} - 2)(\Psi_2 - \psi_2) \right] - M(\Psi_1 - \psi_1) = 0 \,,$$

$$\epsilon(\Psi_2 - \psi_2) - \gamma(a + ib)e^z \frac{1}{M}\left[\gamma(a - ib)e^z(\Psi_2 + \psi_2) - i(\frac{d}{dz} - 1)(\Psi_1 + \psi_1)\right]$$

$$-\gamma(a - ib)e^z \frac{1}{M}\left[\gamma(a + ib)e^z(\Psi_2 + \psi_2) + i(\frac{d}{dz} - 1)(\Psi_3 + \psi_3)\right] - M(\Psi_2 + \psi_2) = 0\,,$$

$$\epsilon(\Psi_2 + \psi_2) - \frac{d}{dz}\frac{1}{M}\left[\gamma(ia - b)e^z(\Psi_1 - \psi_1) - \gamma(ia + b)e^z(\Psi_3 - \psi_3)\right.$$

$$\left.-(\frac{d}{dz} - 2)(\Psi_2 - \psi_2)\right] - M(\Psi_2 - \psi_2) = 0\,,$$

$$\epsilon(\Psi_3 - \psi_3) - \gamma(a + ib)e^z \frac{1}{M}\left[\gamma(a + ib)\,e^z(\Psi_1 + \psi_1) + \gamma(a - ib)e^z(\Psi_3 + \psi_3)\right]$$

$$-i(\frac{d}{dz} - 1)\frac{1}{M}\left[\gamma(a + ib)e^z(\Psi_2 + \psi_2) + i(\frac{d}{dz} - 1)(\Psi_3 + \psi_3)\right] - M(\Psi_3 + \psi_3) = 0\,,$$

$$\epsilon(\Psi_3 + \psi_3) + \gamma(b - ia)e^z \frac{1}{M}\left[\gamma(ia - b)e^z(\Psi_1 - \psi_1) - \gamma(ia + b)e^z(\Psi_3 - \psi_3)\right.$$

$$\left.-(\frac{d}{dz} - 2)(\Psi_2 - \psi_2)\right] - M(\Psi_3 - \psi_3) = 0\,.$$

$$(15.2.3)$$

Within each pair of equations, let us sum and subtract two equations

$$2\epsilon\Psi_1 - 2M\Psi_1 + \frac{1}{M}\left(\frac{d^2}{dz^2} - 2\frac{d}{dz} + 1\right)(\Psi_1 + \psi_1) - \frac{2\gamma^2 e^{2z}(a^2 + b^2)}{M}\Psi_1$$

$$-\frac{2i\gamma e^z(ib - a)}{M}(\Psi_2 - \psi_2) - \frac{2i\gamma e^z(ib - a)}{M}\frac{d\psi_2}{dz} - \frac{2\gamma^2 e^{2z}(ib - a)^2}{M}\psi_3 = 0\,,$$

$$-2\epsilon\psi_1 - 2M\psi_1 + \frac{1}{M}\left(\frac{d^2}{dz^2} - 2\frac{d}{dz} + 1\right)(\Psi_1 + \psi_1) - \frac{2\gamma^2 e^{2z}(a^2 + b^2)}{M}\psi_1$$

$$+\frac{2i\gamma e^z(ib - a)}{M}(\Psi_2 - \psi_2) - \frac{2i\gamma e^z(ib - a)}{M}\frac{d\Psi_2}{dz} - \frac{2\gamma^2 e^{2z}(ib - a)^2}{M}\Psi_3 = 0\,,$$

$$2\,\epsilon\Psi_2 - 2M\Psi_2 - \frac{2i\gamma e^z(a + ib)}{M}\Psi_1$$

$$+\frac{2i\gamma e^z(a + ib)}{M}\frac{d\psi_1}{dz} - \frac{2\gamma^2 e^{2z}(a^2 + b^2)}{M}(\Psi_2 + \psi_2)$$

$$+\frac{1}{M}\left(\frac{d^2}{dz^2}-2\frac{d}{dz}\right)(\Psi_2-\psi_2)-\frac{2i\gamma e^z(-a+ib)}{M}\Psi_3+\frac{2i\gamma e^z(-a+ib)}{M}\frac{d\psi_3}{dz}=0\,,$$

$$-2\epsilon\psi_2-2M\psi_2-\frac{2i\gamma e^z(a+ib)}{M}\psi_1$$

$$+\frac{2i\gamma e^z(a+ib)}{M}\frac{d\Psi_1}{dz}-\frac{2\gamma^2 e^{2z}(a^2+b^2)}{M}(\Psi_2+\psi_2)$$

$$-\frac{1}{M}\left(\frac{d^2}{dz^2}-2\frac{d}{dz}\right)(\Psi_2-\psi_2)-\frac{2i\gamma e^z(-a+ib)}{M}\psi_3+\frac{2i\gamma e^z(-a+ib)}{M}\frac{d\Psi_3}{dz}=0\,,$$

$$2\epsilon\Psi_3-2M\Psi_3-\frac{2\gamma^2 e^{2z}(a+ib)^2}{M}\psi_1$$

$$-\frac{2i\gamma e^z(a+ib)}{M}\frac{d\psi_2}{dz}-\frac{2i\gamma e^z(a+ib)}{M}(\Psi_2-\psi_2)$$

$$+\frac{1}{M}\left(\frac{d^2}{dz^2}-2\frac{d}{dz}+1\right)(\Psi_3+\psi_3)-\frac{2\gamma^2 e^{2z}(a^2+b^2)}{M}\Psi_3=0\,,$$

$$-2\epsilon\psi_3-2M\psi_3-\frac{2\gamma^2 e^{2z}(a+ib)^2}{M}\Psi_1$$

$$-\frac{2i\gamma e^z(a+ib)}{M}\frac{d\Psi_2}{dz}+\frac{2i\gamma e^z(a+ib)}{M}(\Psi_2-\psi_2)$$

$$+\frac{1}{M}\left(\frac{d^2}{dz^2}-2\frac{d}{dz}+1\right)(\Psi_3+\psi_3)-\frac{2\gamma^2 e^{2z}(a^2+b^2)}{M}\psi_3=0\,.$$

$$(15.2.4)$$

Next step, separate the rest energy

$$e^{-i\epsilon t}=e^{-i(M+E)t}\qquad\Longrightarrow\qquad i\frac{\partial}{\partial t}e^{-i\epsilon t}=(M+E)e^{-i\epsilon t}\,,$$

to this end, there is enough in Eqs. (15.2.4) to perform one formal change ϵ $(M+E)$:

$$2E\Psi_1+\frac{1}{M}\left(\frac{d^2}{dz^2}-2\frac{d}{dz}+1\right)(\Psi_1+\psi_1)-\frac{2\gamma^2 e^{2z}(a^2+b^2)}{M}\Psi_1$$

$$-\frac{2i\gamma e^z(ib-a)}{M}(\Psi_2-\psi_2)-\frac{2i\gamma e^z(ib-a)}{M}\frac{d\psi_2}{dz}-\frac{2\gamma^2 e^{2z}(ib-a)^2}{M}\psi_3=0\,,$$

$$-2E\psi_1+\frac{1}{M}\left(\frac{d^2}{dz^2}-2\frac{d}{dz}+1\right)(\Psi_1+\psi_1)-\frac{2\gamma^2 e^{2z}(a^2+b^2)}{M}\psi_1$$

$$+\frac{2i\gamma e^z(ib-a)}{M}(\Psi_2-\psi_2)-\frac{2i\gamma e^z(ib-a)}{M}\frac{d\Psi_2}{dz}-\frac{2\gamma^2 e^{2z}(ib-a)^2}{M}\Psi_3=4M\psi_1\,,$$

$$2E\Psi_2 - \frac{2i\gamma e^z(a+ib)}{M}\Psi_1 + \frac{2i\gamma e^z(a+ib)}{M}\frac{d\psi_1}{dz} - \frac{2\gamma^2 e^{2z}(a^2+b^2)}{M}(\Psi_2+\psi_2)$$

$$+\frac{1}{M}\left(\frac{d^2}{dz^2}-2\frac{d}{dz}\right)(\Psi_2-\psi_2) - \frac{2i\gamma e^z(-a+ib)}{M}\Psi_3 + \frac{2i\gamma e^z(-a+ib)}{M}\frac{d\psi_3}{dz} = 0\,,$$

$$-2E\psi_2 - \frac{2i\gamma e^z(a+ib)}{M}\psi_1 + \frac{2i\gamma e^z(a+ib)}{M}\frac{d\Psi_1}{dz} - \frac{2\gamma^2 e^{2z}(a^2+b^2)}{M}(\Psi_2+\psi_2)$$

$$-\frac{1}{M}\left(\frac{d^2}{dz^2}-2\frac{d}{dz}\right)(\Psi_2-\psi_2) - \frac{2i\gamma e^z(-a+ib)}{M}\psi_3 + \frac{2i\gamma e^z(-a+ib)}{M}\frac{d\Psi_3}{dz} = 4M\psi_2\,,$$

$$2E\Psi_3 - \frac{2\gamma^2 e^{2z}(a+ib)^2}{M}\psi_1 - \frac{2i\gamma e^z(a+ib)}{M}\frac{d\psi_2}{dz} - \frac{2i\gamma e^z(a+ib)}{M}(\Psi_2-\psi_2)$$

$$+\frac{1}{M}\left(\frac{d^2}{dz^2}-2\frac{d}{dz}+1\right)(\Psi_3+\psi_3) - \frac{2\gamma^2 e^{2z}(a^2+b^2)}{M}\Psi_3 = 0\,,$$

$$-2E\psi_3 - \frac{2\gamma^2 e^{2z}(a+ib)^2}{M}\Psi_1 - \frac{2i\gamma e^z(a+ib)}{M}\frac{d\Psi_2}{dz} + \frac{2i\gamma e^z(a+ib)}{M}(\Psi_2-\psi_2)$$

$$+\frac{1}{M}\left(\frac{d^2}{dz^2}-2\frac{d}{dz}+1\right)(\Psi_3+\psi_3) - \frac{2\gamma^2 e^{2z}(a^2+b^2)}{M}\psi_3 = 4M\psi_3\,.$$

$$(15.2.5)$$

Conditions of non-relativity reduce to two restrictions:

$$E << M\,, \qquad \psi_j << \Psi_j\,. \tag{15.2.6}$$

By taking into account Eq. (15.2.6), from Eq. (15.2.5) one can derive three equations for big components Ψ_j (also there arises expressions for small components in terms of big ones). Next, we shall write down only three equations for big components

$$\left(\frac{d^2}{dz^2}-2\frac{d}{dz}+2EM+1-e^{2z}(a^2+b^2)\right)\Psi_1 + 2i\gamma e^z(a-ib)\Psi_2 = 0\,, \quad (15.2.7a)$$

$$\left(\frac{d^2}{dz^2}-2\frac{d}{dz}+2EM+1-e^{2z}(a^2+b^2)\right)\Psi_3 - 2i\gamma e^z(a+ib)\Psi_2 = 0\,, \quad (15.2.7b)$$

$$\left(\frac{d^2}{dz^2}-2\frac{d}{dz}+2EM-e^{2z}(a^2+b^2)\right)\Psi_2$$

$$-2i\gamma e^z(a+ib)\Psi_1 + 2i\gamma e^z(a-ib)\Psi_3 = 0\,. \tag{15.2.7c}$$

These will be necessary for the system of equations in a Pauli approximation.

Let us consider how these equations behave for a vanishing curvature. To this end, it is better to translate Eq. (15.2.7) to usual units

$$E = \frac{\epsilon \rho}{\hbar c}, \qquad M = \frac{mc\rho}{\hbar}, \qquad 2EM = 2m\epsilon \frac{\rho^2}{\hbar^2},$$

$$a = \frac{P_1 \rho}{\hbar}, \qquad b = \frac{P_2 \rho}{\hbar}, \qquad z = \frac{Z}{\rho}.$$

For instance, the first equation takes the form

$$\frac{2\,\epsilon\,m}{\hbar^2}\,\Psi_1 + \left(\frac{d^2}{dZ^2} - \frac{2}{\rho}\frac{d}{dZ} + \frac{1}{\rho^2} \right)\Psi_1$$

$$- e^{2Z/\rho}\frac{P_1^2 + P_2^2}{\hbar^2}\,\Psi_1 - 2\,i\,\gamma\,e^{Z/\rho}\frac{iP_2 - P_1}{\hbar\rho}\,\Psi_2 = 0 ;$$

at vanishing curvature, $\rho \to \infty$, it reduces to

$$\frac{2\,\epsilon\,m}{\hbar^2}\,\Psi_1 + \frac{d^2}{dZ^2}\,\Psi_1 - \frac{P_1^2 + P_2^2}{\hbar^2}\,\Psi_1 = 0 . \qquad (15.2.8a)$$

Similarly, one can consider two remaining equations – so we get

$$\frac{2\,\epsilon\,m}{\hbar^2}\,\Psi_1 + \frac{d^2}{dZ^2}\,\Psi_1 - \frac{P_1^2 + P_2^2}{\hbar^2}\,\Psi_2 = 0 , \qquad (15.2.8b)$$

$$\frac{2\,\epsilon\,m}{\hbar^2}\,\Psi_1 + \frac{d^2}{dZ^2}\,\Psi_1 - \frac{P_1^2 + P_2^2}{\hbar^2}\,\Psi_3 = 0 . \qquad (15.2.8c)$$

Equations (15.2.8) coincide with those which arise when we examine the same problem in Minkowski space

$$(2\,EM - a^2 - b^2 - p_3^2)\,\Psi_1 = 0 ,$$

$$(2\,EM - a^2 - b^2 - p_3^2)\,\Psi_2 = 0 ,$$

$$(2\,EM - a^2 - b^2 - p_3^2)\,\Psi_3 = 0 . \qquad (15.2.9)$$

Note, that the most simple solutions for Eqs. (15.2.9) in Minkowski space are taken according to

$$\Psi_1 \neq 0 , \qquad \Psi_2 = 0 , \qquad \Psi_3 = 0 ,$$

$$\Psi_1 = 0 , \qquad \Psi_2 \neq 0 , \qquad \Psi_3 = 0 ,$$

$$\Psi_1 = 0 , \qquad \Psi_2 = 0 , \qquad \Psi_3 \neq 0 ,$$

this does not provide us with eigenstates for the helicity operator with a spin 1 field.

To construct solutions for a system in Eqs. (15.2.7a,b,c) in Lobachevsky space, we must use an additional operator, the generalized helicity operator Σ. In a nonrelativistic limit, the eigenvalue equation $\Sigma\Psi = \sigma\Psi$ leads to three relations

$$\gamma\,(a - ib)\,e^z\,\Psi_2 = \left(+i\,(\frac{d}{dz} - 1) + \sigma \right)\Psi_1 ,$$

$$\gamma(a + ib)\, e^z\, \Psi_1 + \gamma\, (a - ib)\, e^z\, \Psi_3 = \sigma\, \Psi_2\,,$$

$$\gamma(a + ib)\, e^z\, \Psi_2 = \left(-i\,(\frac{d}{dz} - 1) + \sigma\right) \Psi_3\,. \qquad (15.2.10)$$

Depending on the values of σ (zero or non-zero) we have substantially different constrains in Eq. (15.2.10).

First, let us examine the case $\sigma \neq 0$ (from general consideration it is evident that in the nonrelativistic limit the following relation must hold: $\sigma = \pm\sqrt{2ME}$)

$$\gamma\, (a - ib)\, e^z\, \Psi_2 = \left(+i\,(\frac{d}{dz} - 1) + \sigma\right) \Psi_1\,,$$

$$\gamma(a + ib)\, e^z\, \Psi_1 + \gamma\, (a - ib)\, e^z\, \Psi_3 = \sigma\, \Psi_2\,,$$

$$\gamma(a + ib)\, e^z\, \Psi_2 = \left(-i\,(\frac{d}{dz} - 1) + \sigma\right) \Psi_3\,. \qquad (15.2.11a)$$

Because $\sigma \neq 0$, with the help of the second relation one can exclude the function Ψ_2

$$\Psi_2 = \frac{\gamma}{\sigma}e^z\,[(a + ib)\, \Psi_1 + (a - ib)\, \Psi_3] \qquad (15.2.11b)$$

in the first and the third equations, thus obtaining the system for only two functions

$$(a - ib)^2 e^{2z}\Psi_3 = \left[+2i\sigma(\frac{d}{dz} - 1) + 2\sigma^2 - (a^2 + b^2)e^{2z}\right] \Psi_1\,,$$

$$(a + ib)^2 e^{2z}\Psi_1 = \left[-2i\sigma(\frac{d}{dz} - 1) + 2\sigma^2 - (a^2 + b^2)e^{2z}\right] \Psi_3\,.$$

$$(15.2.11c)$$

Let us find a second order equation for Ψ_1:

$$(a + ib)^2 e^{2z}\Psi_1 = \left[-2i\sigma(\frac{d}{dz} - 1) + 2\sigma^2 - (a^2 + b^2)e^{2z}\right]$$

$$\times \frac{e^{-2z}}{(a - ib)^2} \left[+2i\sigma(\frac{d}{dz} - 1) + 2\sigma^2 - (a^2 + b^2)e^{2z}\right] \Psi_1\,,$$

or

$$(a^2 + b^2)^2 e^{4z}\Psi_1$$

$$= \left[-2i\sigma(\frac{d}{dz} - 3) + 2\sigma^2 - (a^2 + b^2)e^{2z}\right]$$

$$\times \left[+2i\sigma(\frac{d}{dz} - 1) + 2\sigma^2 - (a^2 + b^2)e^{2z}\right] \Psi_1\,,$$

and further

$$(a^2 + b^2)^2 e^{4z}\Psi_1$$

$$= \left[\, 4\sigma^2(\frac{d}{dz} - 3)(\frac{d}{dz} - 1) - i\sigma 4\sigma^2(\frac{d}{dz} - 3) + 2i\sigma(a^2 + b^2)e^{2z}(\frac{d}{dz} - 1)\right.$$

$$+\sigma^2 4i\sigma(\frac{d}{dz}-1)+4\sigma^2\sigma^2-2\sigma^2(a^2+b^2)e^{2z}$$

$$-(a^2+b^2)e^{2z}2i\sigma(\frac{d}{dz}-1)-(a^2+b^2)e^{2z}2\sigma^2+(a^2+b^2)^2e^{4z}\Bigg]\Psi_1\,.$$

From here, after evident simplifying we can arrive at

$$0=4\sigma^2\left[\,(\frac{d^2}{dz^2}-4\frac{d}{dz}+3)+2i\sigma+\sigma^2-(a^2+b^2)e^{2z}\right]\Psi_1$$

that is

$$\left[\frac{d^2}{dz^2}-4\frac{d}{dz}+\sigma^2+3+2i\sigma-(a^2+b^2)e^{2z}\right]\Psi_1=0\,. \tag{15.2.12a}$$

Similarly, equation for Ψ_3 is

$$\left[\frac{d^2}{dz^2}-4\frac{d}{dz}+\sigma^2+3-2i\sigma-e^{2z}\,(a^2+b^2)\right]\Psi_3=0\,. \tag{15.2.12b}$$

Remembering, that Ψ_1, Ψ_3 Eqs. (15.2.12a,b) are not independent, instead they are connected by the first order differential system of Eq. (15.2.11c). Moreover, the relation of Eq. (15.2.11b) permits us to construct explicitly Ψ_2 in terms of the known Ψ_1, Ψ_3, which provides us with the complete solution to the problem.

In order to detail the problem, let us translate Eqs. (15.2.12a) and (15.2.12b) to a new variable

$$x=i\sqrt{a^2+b^2}\,e^z\,,$$

$$\left[x\frac{d^2}{dx^2}-3\frac{d}{dx}+x+\frac{(\sigma-i)\,(\sigma+3i)}{x}\right]\Psi_1=0\,,$$

$$\left[x\frac{d^2}{dx^2}-3\frac{d}{dx}+x+\frac{(\sigma+i)\,(\sigma-3i)}{x}\right]\Psi_3=0\,.$$

$$\tag{15.2.13a}$$

Through the substitutions

$$\Psi_1=x^2f_1(x)\,,\qquad\Psi_3=x^2f_3(x)\,,$$

Equations (15.2.13a) reduce to the Bessel equations

$$\frac{d^2}{dx^2}f_1+\frac{1}{x}\frac{d}{dx}f_1+\left(1-\frac{(1-i\sigma)^2}{x^2}\right)f_1=0\,,$$

$$\frac{d^2}{dx^2}f_3+\frac{1}{x}\frac{d}{dx}f_3+\left(1-\frac{(1+i\sigma)^2}{x^2}\right)f_3=0\,, \tag{15.2.13b}$$

with the following solutions (let it be: $\nu=1-i\sigma$, $\mu=1+i\sigma$)

$$f_1^+(x)=A_1^+J_{+\nu}(x)\,,\qquad f_1^-(x)=A_1^-J_{-\nu}(x)\,,$$

$$f_3^+(x)=A_3^+J_{+\mu}(x)\,,\qquad f_3^-(x)=A_3^-J_{-\mu}(x)\,. \tag{15.2.13c}$$

We wish to know the pairs in the solution of Eq. (15.2.13c) that are linked by the first order relations in Eq. (15.2.11c). To clarify that let us translate Eq. (15.2.11c) to the variable x:

$$x = i\sqrt{a^2 + b^2}\, e^z,$$

$$\left(+2\,i\,\sigma\,x\frac{d}{dx} - 2\,i\,\sigma + 2\,\sigma^2 + x^2\right)\Psi_1 + \frac{(a - i\,b)}{a + i\,b}\, x^2\Psi_3 = 0\,,$$

$$\left(-2\,i\,\sigma\,x\frac{d}{dx} + 2\,i\,\sigma + 2\,\sigma^2 + x^2\right)\Psi_3 + \frac{(a + i\,b)}{a - i\,b}\, x^2\Psi_1 = 0\,;$$

$$(15.2.14a)$$

these reduce to

$$\Psi_1 = x^2 f_1(x)\,, \qquad \Psi_3 = x^2 f_3(x)\,,$$

$$\left(+2\,i\,\sigma\,x\frac{d}{dx} + 2\,i\,\sigma + 2\,\sigma^2 + x^2\right) f_1 + \frac{(a - i\,b)}{a + i\,b}\, x^2 f_3 = 0\,,$$

$$\left(-2\,i\,\sigma\,x\frac{d}{dx} - 2\,i\,\sigma + 2\,\sigma^2 + x^2\right) f_3 + \frac{(a + i\,b)}{a - i\,b}\, x^2 f_1 = 0\,.$$

$$(15.2.14b)$$

It is easily to simplify the formulas by the following notation

$$(a - ib)f_3 = \bar{f}_3, \qquad (a + ib)f_1 = \bar{f}_1\,,$$

$$\left(+2\,i\,\sigma\,(x\frac{d}{dx} + \nu) + x^2\right)\bar{f}_1 + x^2\bar{f}_3 = 0\,,$$

$$\left(-2\,i\,\sigma\,(x\frac{d}{dx} + \mu) + x^2\right)\bar{f}_3 + x^2\bar{f}_1 = 0\,. \qquad (15.2.14c)$$

Let us demonstrate that one possible solution is realized on the functions

$$\bar{f}_1 = J_{+\nu}(x), \qquad \bar{f}_3 = J_{-\mu}(x)\,,$$

$$(+\nu = 1 - i\sigma, -\mu = -1 - i\sigma)\,. \qquad (15.2.15a)$$

To this end, it suffices to apply the first equation

$$\left(+2\,i\,\sigma\,(x\frac{d}{dx} + \nu) + x^2\right) J_\nu(x) + x^2 J_{-\mu}(x) = 0\,.$$

By the help of the known relation for the Bessel functions

$$(x\frac{d}{dx} + \nu)J_\nu = +x J_{\nu-1}$$

it can be translated to the form

$$2i\sigma\, J_{-i\sigma}(x) + [\, x J_{-i\sigma+1}(x) + x J_{-i\sigma-1}(x)\,] = 0\,. \qquad (15.2.15b)$$

From this, in allowing for the known recursive relation

$$x J_{a+1} + x J_{a-1} = 2a J_a , \qquad \text{when} \qquad a = -i\sigma ,$$

we get an identity $0 = 0$.

Thus, we have constructed the first class of solutions.

$\sigma = \pm\sqrt{2ME}$,

$$\Psi_1^I = \frac{x^2}{a+ib} J_{+\nu}(x) , \qquad \Psi_3^I = \frac{x^2}{a-ib} J_{-\mu}(x) ,$$

$$\Psi_2^I = \frac{\gamma}{\sigma} e^z \left[(a+ib)\Psi_1^I + (a-ib)\Psi_3^I \right]$$

$$= \frac{\gamma}{\sigma} e^z \left[x^2 J_{+\nu}(x) + x^2 J_{-\mu}(x) \right] . \tag{15.2.15c}$$

Similarly, we can prove existent of the second class of solutions

$\sigma = \pm\sqrt{2ME}$,

$$\Psi_1^{II} = \frac{x^2}{a+ib} J_{-\nu}(x) , \qquad \Psi_3^{II} = \frac{x^2}{a-ib} J_{+\mu}(x) ,$$

$$\Psi_2^{II} = \frac{\gamma}{\sigma} e^z \left[(a+ib)\Psi_1^{II} + (a-ib)\Psi_3^{II} \right]$$

$$= \frac{\gamma}{\sigma} e^z \left[x^2 J_{-\nu}(x) + x^2 J_{+\mu}(x) \right] . \tag{15.2.15d}$$

Now, let us turn to the case of $\sigma = 0$, it is specified by the following relations

$$\Psi_3 = -\frac{a+ib}{a-ib} \Psi_1 , \qquad \Psi_2 = +\frac{ie^{-z}}{\gamma(a-ib)} (\frac{d}{dz} - 1)\Psi_1 . \tag{15.2.16}$$

In allowing for Eq. (15.2.16) from Eq. (15.2.7), we obtain

$$\left(\frac{d^2}{dz^2} - 4\frac{d}{dz} + 2EM + 3 - e^{2z}(a^2+b^2) \right) \Psi_1 = 0 , \tag{15.2.17a}$$

$$\left(\frac{d^2}{dz^2} - 4\frac{d}{dz} + 2EM + 3 - e^{2z}(a^2+b^2) \right) \Psi_3 = 0 , \tag{15.2.17b}$$

$$\left(\frac{d^2}{dz^2} - 2\frac{d}{dz} + 2EM - e^{2z}(a^2+b^2) \right) \Psi_2$$

$$+2i\gamma e^z (a-ib)\Psi_3 - 2i\gamma e^z (a+ib)\Psi_1 = 0 . \tag{15.2.17c}$$

The second relation in Eq. (15.2.16) says that Ψ_2 can be found on the basis of the known Ψ_1. In fact, this gives the complete solution to the problem when $\sigma = 0$.

Note that equations for Ψ_1 and Ψ_3 are the same. Let us translate Eqs. (15.2.17a) and (15.2.17b) to the new variable

$$x = i\sqrt{a^2+b^2}\, e^z ,$$

$$\left(x \frac{d^2}{dx^2} - 3 \frac{d}{dx} + x + \frac{3 + 2\,E\,M}{x} \right) \Psi_1 = 0 \,,$$

$$\left(x \frac{d^2}{dx^2} - 3 \frac{d}{dx} + x + \frac{3 + 2\,E\,M}{x} \right) \Psi_3 = 0 \,. \qquad (15.2.18a)$$

By substitution $\Psi_1 = x^2 g_1(x)$ reduces it to the Bessel type

$$\frac{d^2}{dx^2} g_1 + \frac{1}{x} \frac{d}{dx} g_1 + \left(1 - \frac{1 - 2\,E\,M}{x^2} \right) g_1 = 0 \,. \qquad (15.2.18b)$$

Thus, at $\sigma = 0$ there are constructed solutions of two types:

$\sigma = 0$,

$$\Psi_1^I(x) = + \frac{x^2}{a + ib} J_{+\nu}(x) \,, \qquad \Psi_3^I(x) = - \frac{x^2}{a - ib} J_{+\nu}(x) \,,$$

$$\Psi_1^{II}(x) = + \frac{x^2}{a + ib} J_{-\nu}(x) \,, \qquad \Psi_1^{II}(x) = - \frac{x^2}{a - ib} J_{-\nu}(x) \,.$$

$$(15.2.18c)$$

It remains to prove that Eq. (15.2.17c) is consistent with Eq. (15.2.16) and Eqs. (15.2.17a, b). To this end, it is convenient to introduce new functions

$$(a + ib)\Psi_1 = e^z \bar{\Psi}_1 \,, \qquad (a - ib)\Psi_3 = e^z \bar{\Psi}_3 \,, \qquad (15.2.19a)$$

then

$$\bar{\Psi}_3 = -\bar{\Psi}_1 \,, \qquad \bar{\Psi}_2 = + \frac{i}{\gamma(a^2 + b^2)} \frac{d}{dz} \bar{\Psi}_1 \,, \qquad (15.2.19b)$$

$$\left(\frac{d^2}{dz^2} - 2 \frac{d}{dz} + 2\,E\,M - e^{2z}(a^2 + b^2) \right) \bar{\Psi}_1 = 0 \,, \qquad (15.2.19c)$$

$$\left(\frac{d^2}{dz^2} - 2 \frac{d}{dz} + 2\,E\,M - e^{2z}(a^2 + b^2) \right) \bar{\Psi}_2 - 4\,i\,\gamma\,e^{2z} \bar{\Psi}_1 = 0 \,.$$

$$(15.2.19d)$$

Let us substitute the expression for $\bar{\Psi}_2$ (15.2.19b) into the Eq. (15.2.19d):

$$\left(\frac{d^2}{dz^2} - 2 \frac{d}{dz} + 2\,E\,M - e^{2z}(a^2 + b^2) \right) \frac{d}{dz} \bar{\Psi}_1 - 2(a^2 + b^2)e^{2z} \bar{\Psi}_1 = 0 \,.$$

Now, allow for the following relation

$$\left(\frac{d^2}{dz^2} - 2 \frac{d}{dz} + 2EM \right) \frac{d}{dz} \bar{\Psi}_1 = \frac{d}{dz} e^{2z}(a^2 + b^2)\bar{\Psi}_1 \,,$$

we translate it to the form

$$\frac{d}{dz} e^{2z}(a^2 + b^2)\bar{\Psi}_1 - e^{2z}(a^2 + b^2)\frac{d}{dz} \bar{\Psi}_1 - 2(a^2 + b^2)e^{2z} \bar{\Psi}_1 = 0 \,,$$

which is an identity.

Thus, this problem of solving the Duffin–Kemmer equation for spin 1 particle on the background of Lobachevsky space in a Pauli approximation is completed. Ultimately, the problem ends with a second order differential equation of the Schrödinger type with a barrier in a special form that generates a reflection coefficient $R = 1$ for all states – see chapter **13**.

15.3   Spin 1 Field in the Lobachevsky Space H_3: Horospherical Coordinates

The task is to construct a complete system of solutions for a field with spin 1 (massive and massless cases) in the space of constant negative curvature, Lobachevsky space H_3. Treatment is based on 10-dimensional Duffin–Kemmer formalism extended to a curved model according to the tetrad method. The main equation will read as

$$\left\{ \beta^c [i\hbar \, (\, e^{\beta}_{(c)} \partial_\beta + \frac{1}{2} J^{ab}\gamma_{abc} \,) \, - \, \frac{e}{c}A_c \,] \, - \, mc \right\} \Psi = 0 \, . \qquad (15.3.1)$$

We will use horospherical coordinates and corresponding diagonal tetrad

$$x^a = (t, r, \phi, z) \, , \qquad dS^2 = dt^2 - e^{-2z}(dr^2 + r^2 d\phi^2) - dz^2 \, ,$$

$$e^{\beta}_{(a)} = \begin{vmatrix} 1 & 0 & 0 & 0 \\ 0 & e^z & 0 & 0 \\ 0 & 0 & \frac{e^z}{r} & 0 \\ 0 & 0 & 0 & 1 \end{vmatrix} \, , \qquad e_{(a)\beta} = \begin{vmatrix} 1 & 0 & 0 & 0 \\ 0 & -e^{-z} & 0 & 0 \\ 0 & 0 & -re^{-z} & 0 \\ 0 & 0 & 0 & -1 \end{vmatrix} \, . \qquad (15.3.2a)$$

For the Christoffel symbol, we have $\Gamma^0_{\beta\sigma} = 0$, $\Gamma^i_{00} = 0$, $\Gamma^i_{0j} = 0$, and

$$x^i = (r, \phi, z) \, , \qquad \Gamma^i_{jk} = g^{il} \, \Gamma_{l,jk} = \frac{1}{2} \, g^{il} \, (-\partial_l \, g_{jk} + \partial_j \, g_{lk} + \partial_k \, g_{lj}) \, ,$$

$$\Gamma^r_{jk} = \begin{vmatrix} 0 & 0 & -1 \\ 0 & -r & 0 \\ -1 & 0 & 0 \end{vmatrix} \, , \; \Gamma^\phi_{jk} = \begin{vmatrix} 0 & \frac{1}{r} & 0 \\ \frac{1}{r} & 0 & -1 \\ 0 & -1 & 0 \end{vmatrix} \, , \; \Gamma^z_{jk} = \begin{vmatrix} e^{-2z} & 0 & 0 \\ 0 & r^2 e^{-2z} & 0 \\ 0 & 0 & 0 \end{vmatrix} \, .$$

$$(15.3.2b)$$

Let us calculate covariant derivative for tetrad 4-vectors:

$$A_{\beta;\alpha} = \frac{\partial A_\beta}{\partial x^\alpha} - \Gamma^\sigma_{\alpha\beta} A_\sigma \implies$$

$$e_{(0)\beta;\alpha} = \frac{\partial e_{(0)\beta}}{\partial x^\alpha} - \Gamma^\sigma_{\alpha\beta} e_{(0)\sigma} = \frac{\partial e_{(0)\beta}}{\partial x^\alpha} - \Gamma^0_{\alpha\beta} \, e_{(0)0} = 0 \, ,$$

$$e_{(1)\beta;\alpha} = \frac{\partial e_{(1)\beta}}{\partial x^\alpha} - \Gamma^r_{\alpha\beta} \, e_{(1)r} = \frac{\partial e_{(1)\beta}}{\partial x^\alpha} + \Gamma^r_{\alpha\beta} \, e^{-z}$$

$$= \begin{vmatrix} 0 & 0 & 0 & 0 \\ 0 & 0 & 0 & e^{-z} \\ 0 & 0 & 0 & 0 \\ 0 & 0 & 0 & 0 \end{vmatrix} + \begin{vmatrix} 0 & 0 & 0 & 0 \\ 0 & 0 & 0 & -e^{-z} \\ 0 & 0 & -re^{-z} & 0 \\ 0 & -e^{-z} & 0 & 0 \end{vmatrix} = \begin{vmatrix} 0 & 0 & 0 & 0 \\ 0 & 0 & 0 & 0 \\ 0 & 0 & -re^{-z} & 0 \\ 0 & -e^{-z} & 0 & 0 \end{vmatrix},$$

$$e_{(2)\beta;\alpha} = \frac{\partial e_{(2)\beta}}{\partial x^\alpha} - \Gamma^\phi_{\alpha\beta}\, e_{(2)\phi} = \frac{\partial e_{(2)\beta}}{\partial x^\alpha} + \Gamma^\phi_{\alpha\beta} re^{-z}$$

$$= \begin{vmatrix} 0 & 0 & 0 & 0 \\ 0 & 0 & 0 & 0 \\ 0 & -e^{-z} & 0 & re^{-z} \\ 0 & 0 & 0 & 0 \end{vmatrix} + \begin{vmatrix} 0 & 0 & 0 & 0 \\ 0 & 0 & e^{-z} & 0 \\ 0 & e^{-z} & 0 & -re^{-z} \\ 0 & 0 & -re^{-z} & 0 \end{vmatrix} = \begin{vmatrix} 0 & 0 & 0 & 0 \\ 0 & 0 & e^{-z} & 0 \\ 0 & 0 & 0 & 0 \\ 0 & 0 & -re^{-z} & 0 \end{vmatrix},$$

$$e_{(3)\beta;\alpha} = \frac{\partial e_{(3)\beta}}{\partial x^\alpha} - \Gamma^z_{\alpha\beta}\, e_{(3)z} = 0 + \Gamma^z_{\alpha\beta} \implies e_{(3)\beta;\alpha} = \begin{vmatrix} 0 & 0 & 0 & 0 \\ 0 & e^{-2z} & 0 & 0 \\ 0 & 0 & r^2 e^{-2z} & 0 \\ 0 & 0 & 0 & 0 \end{vmatrix}.$$

Now, we are ready to find the Ricci rotation coefficients $\gamma_{abc} = e_{(a)}^{\ \beta}\, e_{(b)\beta;\alpha}\, e_{(c)}^{\ \alpha}$, so we get

$$\gamma_{ab0} = e_{(a)}^{\ \beta}\, e_{(b)\beta;t}\, e_{(0)}^{\ t} = 0 , \qquad \gamma_{ab1} = e_{(a)}^{\ a}\, e_{(b)a;r}\, e_{(1)}^{\ r} ,$$

$$\gamma_{ab2} = e_{(a)}^{\ \beta}\, e_{(b)\beta;\phi}\, e_{(2)}^{\ \phi} , \qquad \gamma_{ab3} = e_{(a)}^{\ \beta}\, e_{(b)\beta;z}\, e_{(3)}^{\ z} .$$

Thus we obtain

$$\gamma_{011} = \gamma_{021} = \gamma_{031} = 0 , \qquad \gamma_{012} = \gamma_{022} = \gamma_{032} = 0 , \qquad \gamma_{013} = \gamma_{023} = \gamma_{033} = 0 ,$$

and

$$\gamma_{231} = e_{(2)}^{\ \phi}\, e_{(3)\phi;r}\, e_{(1)}^{\ r} = 0 , \qquad \gamma_{311} = e_{(3)}^{\ z}\, e_{(1)z;r}\, e_{(1)}^{\ r} = -1 ,$$

$$\gamma_{121} = e_{(1)}^{\ r}\, e_{(2)r;r}\, e_{(1)}^{\ r} = 0 , \qquad \gamma_{232} = e_{(2)}^{\ \phi}\, e_{(3)\phi;\phi}\, e_{(2)}^{\ \phi} = 1 ,$$

$$\gamma_{312} = e_{(3)}^{\ z}\, e_{(1)z;\phi}\, e_{(2)}^{\ \phi} = 0 , \qquad \gamma_{122} = e_{(1)}^{\ r}\, e_{(2)r;\phi}\, e_{(2)}^{\ \phi} = \frac{e^z}{r} ,$$

$$\gamma_{233} = e_{(2)}^{\ \phi}\, e_{(3)\phi;z}\, e_{(3)}^{\ z} = 0 , \qquad \gamma_{313} = e_{(3)}^{\ z}\, e_{(1)z;z}\, e_{(3)}^{\ z} = 0 ,$$

$$\gamma_{123} = e_{(1)}^{\ r}\, e_{(2)r;z}\, e_{(3)}^{\ z} = 0 . \tag{15.3.3}$$

When taking into account these tetrad and Ricci coefficients, we obtain an explicit representation for Duffin–Kemmer equation in Eq. (15.3.1)

$$\left[i\beta^0 \partial_t + e^z [i\beta^1\, \partial_r + \frac{i\beta^2}{r}(\, \partial_\phi + J^{12})] + i\beta^3 \partial_z + i\beta^1 J^{13} + i\beta^2 J^{23} - M \right] \Psi = 0 . \tag{15.3.4}$$

To separate the variable, we use the following substitution

$$\Psi = e^{-i\epsilon t} e^{im\phi} \begin{vmatrix} \Phi_0(r,z) \\ \vec{\Phi}(r,z) \\ \mathbf{E}(r,z) \\ \mathbf{H}(r,z) \end{vmatrix} , \tag{15.3.5}$$

then present Eq. (15.3.4) in a block form

$$
\left[\epsilon \begin{vmatrix} 0 & 0 & 0 & 0 \\ 0 & 0 & i & 0 \\ 0 & -i & 0 & 0 \\ 0 & 0 & 0 & 0 \end{vmatrix} + i\,e^z \begin{vmatrix} 0 & 0 & e_1 & 0 \\ 0 & 0 & 0 & \tau_1 \\ -e_1^+ & 0 & 0 & 0 \\ 0 & -\tau_1 & 0 & 0 \end{vmatrix} \frac{\partial}{\partial r} \right.
$$

$$
-\frac{e^z}{r} \begin{vmatrix} 0 & 0 & e_2 & 0 \\ 0 & 0 & 0 & \tau_2 \\ -e_2^+ & 0 & 0 & 0 \\ 0 & -\tau_2 & 0 & 0 \end{vmatrix} (m - S_3) + i \begin{vmatrix} 0 & 0 & e_3 & 0 \\ 0 & 0 & 0 & \tau_3 \\ -e_3^+ & 0 & 0 & 0 \\ 0 & -\tau_3 & 0 & 0 \end{vmatrix} \frac{\partial}{\partial z}
$$

$$
\left. +i \begin{vmatrix} 0 & 0 & -2e_3 & 0 \\ 0 & 0 & 0 & -\tau_3 \\ 0 & 0 & 0 & 0 \\ 0 & +\tau_3 & 0 & 0 \end{vmatrix} - M \right] \begin{vmatrix} \Phi_0 \\ \vec{\Phi} \\ \mathbf{E} \\ \mathbf{H} \end{vmatrix} = 0 \,,
$$

$$(15.3.6a)$$

or

$$
i\,e^z\,e_1 \partial_r \mathbf{E} - \frac{e^z}{r} e_2(m - S_3)\mathbf{E} + i(\partial_z - 2)e_3 \mathbf{E} = M\,\Phi_0 \,,
$$

$$
i\epsilon\,\mathbf{E} + i\,e^z\,\tau_1 \partial_r \mathbf{H} - e^z \frac{\tau_2}{r}(\,m - S_3\,)\mathbf{H} + i(\partial_z - 1)\tau_3 \mathbf{H} = M\,\vec{\Phi} \,,
$$

$$
-i\epsilon\,\vec{\Phi} - i\,e^z\,e_1^+ \partial_r \Phi_0 + e^z \frac{m}{r} e_2^+ \Phi_0 - i\,e_3^+ \partial_z \Phi_0 = M\,\mathbf{E} \,,
$$

$$
-i\,e^z\,\tau_1 \partial_r \vec{\Phi} + e^z \frac{(m - S_3)}{r} \tau_2 \vec{\Phi} - i(\partial_z - i)\tau_3 \vec{\Phi} = M\,\mathbf{H} \,.
$$

$$(15.3.6b)$$

In performing the needed calculation, we get the system of 10 equations

$$
\gamma\,e^z \left(\frac{\partial E_1}{\partial r} - \frac{\partial E_3}{\partial r} \right) - e^z \frac{\gamma}{r}[(m-1)E_1 + (m+1)E_3]
$$

$$
-(\frac{\partial}{\partial z} - 2)E_2 = M\,\Phi_0 \,, \qquad (15.3.7a)
$$

$$
+i\epsilon\,E_1 + i\gamma\,e^z \frac{\partial H_2}{\partial r} + ie^z \gamma \frac{m}{r} H_2 + i(\frac{\partial}{\partial z} - 1)H_1 = M\,\Phi_1 \,,
$$

$$
+i\epsilon\,E_2 + i\gamma\,e^z \left(\frac{\partial H_1}{\partial r} + \frac{\partial H_3}{\partial r} \right) - e^z \frac{i\gamma}{r}[(m-1)H_1 - (m+1)H_3] = M\,\Phi_2 \,,
$$

$$
+i\epsilon\,E_3 + i\gamma\,e^z \frac{\partial H_2}{\partial r} - ie^z \gamma \frac{m}{r} H_2 - i(\frac{\partial}{\partial z} - 1)H_3 = M\,\Phi_3 \,, \qquad (15.3.7b)
$$

$$
-i\epsilon\,\Phi_1 + \gamma\,e^z \frac{\partial \Phi_0}{dr} + e^z \gamma \frac{m}{r} \Phi_0 = M\,E_1 \,,
$$

$$
-i\epsilon\Phi_2 - \frac{\partial \Phi_0}{\partial z} = ME_2 \,,
$$

$$-i\epsilon \ \Phi_3 - \gamma \, e^z \frac{\partial \Phi_0}{\partial r} + e^z \, \gamma \frac{m}{r} \Phi_0 = M \ E_3 \ , \qquad (15.3.7c)$$

$$-i\gamma \, e^z \frac{\partial \Phi_2}{\partial r} - ie^z \, \gamma \frac{m}{r} \Phi_2 - i(\frac{\partial}{\partial z} - 1)\Phi_1 = M \ H_1 \ ,$$

$$-i\gamma \, e^z \left(\frac{\partial \Phi_1}{\partial r} + \frac{\partial \Phi_3}{\partial r} \right) + \frac{ie^z \, \gamma}{r} [(m-1)\Phi_1 - (m+1)\Phi_3] = M \ H_2 \ ,$$

$$-i\gamma \, e^z \frac{\partial \Phi_2}{\partial r} + i \, e^z \, \gamma \frac{m}{r} \Phi_2 + i(\frac{\partial}{\partial z} - 1)\Phi_3 = M \ H_3 \ . \qquad (15.3.7d)$$

Using the notation

$$\gamma \, (\frac{\partial}{\partial r} + \frac{m-1}{r}) = a_- , \qquad \gamma \, (\frac{\partial}{\partial r} + \frac{m+1}{r}) = a_+ \ , \qquad \gamma \, (\frac{\partial}{\partial r} + \frac{m}{r}) = a \ ,$$

$$\gamma \, (-\frac{\partial}{\partial r} + \frac{m-1}{r}) = b_- , \qquad \gamma \, (-\frac{\partial}{\partial r} + \frac{m+1}{r}) = b_+ \ , \qquad \gamma \, (-\frac{\partial}{\partial r} + \frac{m}{r}) = b \ ,$$

$$(15.3.8)$$

they can be written shorter

$$-e^z \, b_- \ E_1 - e^z \, a_+ \ E_3 - (\frac{\partial}{\partial z} - 2)E_2 = M \ \Phi_0 \ , \qquad (15.3.9a)$$

$$i \, e^z \, a \ H_2 + i\epsilon \, E_1 + i \, (\frac{\partial}{\partial z} - 1)H_1 = M \ \Phi_1 \ ,$$

$$-i \, e^z \, b_- \ H_1 + i \, e^z \, a_+ \ H_3 + i\epsilon \, E_2 = M \ \Phi_2 \ ,$$

$$-i \, e^z \, b \ H_2 + i\epsilon \, E_3 - i(\frac{\partial}{\partial z} - 1)H_3 = M \ \Phi_3 \ , \qquad (15.3.9b)$$

$$e^z \, a \ \Phi_0 - i\epsilon \ \Phi_1 = M \ E_1 \ ,$$

$$-i\epsilon \Phi_2 - \frac{\partial}{\partial z}\Phi_0 = M \ E_2 \ ,$$

$$e^z \, b \ \Phi_0 - i\epsilon \ \Phi_3 = M \ E_3 \ , \qquad (15.3.9c)$$

$$-i \, e^z \, a \ \Phi_2 \ - i \, (\frac{\partial}{\partial z} - 1)\Phi_1 = M \ H_1 \ ,$$

$$i \, e^z \, b_- \ \Phi_1 - i \, e^z \, a_+ \ \Phi_3 = M \ H_2 \ ,$$

$$i \, e^z \, b \ \Phi_2 + i \, (\frac{\partial}{\partial z} - 1)\Phi_3 = M \ H_3 \ . \qquad (15.3.9d)$$

To solve the problem let us consider an additional operator, an extended helicity operator in the Lobachevsky space

$$\Sigma \Psi = \sigma \Psi, \qquad \Psi = e^{-i\epsilon t} e^{im\phi} \begin{vmatrix} \Phi_0(r, z) \\ \vec{\Phi}(r, z) \\ \mathbf{E}(r, z) \\ \mathbf{H}(r, z) \end{vmatrix} ,$$

$$\left[e^z \left(S_1 \frac{\partial}{\partial r} + i S_2 \frac{m - S_3}{r} \right) + \left(\frac{\partial}{\partial z} - 1 \right) S_3 \right] \begin{vmatrix} \Phi_0 \\ \vec{\Phi} \\ \mathbf{E} \\ \mathbf{H} \end{vmatrix} = \sigma \begin{vmatrix} \Phi_0 \\ \vec{\Phi} \\ \mathbf{E} \\ \mathbf{H} \end{vmatrix} . \tag{15.3.10}$$

From Eq. (15.3.10), it follows a system of 10 equations ($\gamma = 1/\sqrt{2}$):

$$0 = \sigma \, \Phi_0 , \tag{15.3.11a}$$

$$\gamma \, e^z \frac{\partial}{\partial r} \Phi_2 + \gamma \, e^z \frac{m}{r} \Phi_2 + \left(\frac{\partial}{\partial z} - 1 \right) \Phi_1 = \sigma \, \Phi_1 ,$$

$$\gamma \, e^z \left(\frac{\partial}{\partial r} \Phi_1 + \frac{\partial}{\partial r} \Phi_3 \right) - e^z \frac{\gamma}{r} [(m-1)\Phi_1 - (m+1)\Phi_3] = \sigma \, \Phi_2 ,$$

$$\gamma \, e^z \frac{\partial}{\partial r} \Phi_2 - \gamma \, e^z \frac{m}{r} \Phi_2 - \left(\frac{\partial}{\partial z} - 1 \right) \Phi_3 = \sigma \, \Phi_3 , \tag{15.3.11b}$$

$$\gamma \, e^z \frac{\partial}{\partial r} E_2 + \gamma \, e^z \frac{m}{r} E_2 + \left(\frac{\partial}{\partial z} - 1 \right) E_1 = \sigma \, E_1 ,$$

$$\gamma \, e^z \left(\frac{\partial}{\partial r} E_1 + \frac{\partial}{\partial r} E_3 \right) - e^z \frac{\gamma}{r} [(m-1)E_1 - (m+1)E_3] = \sigma \, E_2 ,$$

$$\gamma \, e^z \frac{\partial}{\partial r} E_2 - \gamma \, e^z \frac{m}{r} E_2 - \left(\frac{\partial}{\partial z} - 1 \right) E_3 = \sigma \, E_3 , \tag{15.3.11c}$$

$$\gamma \, e^z \frac{\partial}{\partial r} H_2 + \gamma \, e^z \frac{m}{r} H_2 + \left(\frac{\partial}{\partial z} - 1 \right) H_1 = \sigma \, H_1 ,$$

$$\gamma \, e^z \left(\frac{\partial}{\partial r} H_1 + \frac{\partial}{\partial r} H_3 \right) - e^z \frac{\gamma}{r} [(m-1)H_1 - (m+1)H_3] = \sigma \, H_2 ,$$

$$\gamma \, e^z \frac{\partial}{\partial r} H_2 - \gamma \, e^z \frac{m}{r} H_2 - \left(\frac{\partial}{\partial z} - 1 \right) H_3 = \sigma \, H_3 . \tag{15.3.11d}$$

It can be presented shorter by:

$$0 = \sigma \, \Phi_0 , \tag{15.3.12a}$$

$$\left(\frac{\partial}{\partial z} - 1 \right) \Phi_1 = \sigma \, \Phi_1 - a \, e^z \, \Phi_2 ,$$

$$- e^z \, b_- \Phi_1 + e^z \, a_+ \Phi_3 = \sigma \, \Phi_2 ,$$

$$-\left(\frac{\partial}{\partial z} - 1 \right) \Phi_3 = \sigma \, \Phi_3 + b \, e^z \, \Phi_2 , \tag{15.3.12b}$$

$$\left(\frac{\partial}{\partial z} - 1 \right) E_1 = \sigma \, E_1 - a \, e^z \, E_2 ,$$

$$- e^z \, b_- E_1 + e^z \, a_+ E_3 = \sigma \, E_2 ,$$

$$-\left(\frac{\partial}{\partial z} - 1 \right) E_3 = \sigma \, E_3 + b \, e^z \, E_2 , \tag{15.3.12c}$$

$$\left(\frac{\partial}{\partial z} - 1 \right) H_1 = \sigma \, H_1 - a \, e^z \, H_2 ,$$

$$- e^z \, b_- H_1 + e^z \, a_+ H_3 = \sigma \, H_2 ,$$

$$-(\frac{\partial}{\partial z} - 1)H_3 = \sigma\, H_3 + b\ e^z\, H_2 \,. \tag{15.3.12d}$$

Substituting Eq. (15.3.12) into Eq. (15.3.9), we arrive at much simpler set of equations (note that Eqs. (15.3.13b) and (15.3.13d) provide us with linear restrictions)

$$-e^z\, b_-\, E_1 - e^z\, a_+\, E_3 - (\frac{\partial}{\partial z} - 2)E_2 = M\,\Phi_0 \,, \tag{15.3.13a}$$

$$i\epsilon\, E_1 + i\sigma\, H_1 = M\,\Phi_1 \,,$$

$$i\sigma\, H_2 + i\epsilon\, E_2 = M\,\Phi_2 \,,$$

$$i\epsilon\, E_3 + i\sigma\, H_3 = M\,\Phi_3 \,, \tag{15.3.13b}$$

$$e^z\, a\,\Phi_0 - i\epsilon\,\Phi_1 = M\, E_1 \,,$$

$$-i\epsilon\Phi_2 - \frac{\partial}{\partial z}\Phi_0 = M\, E_2 \,,$$

$$e^z\, b\,\Phi_0 - i\epsilon\,\Phi_3 = M\, E_3 \,, \tag{15.3.13c}$$

$$-\sigma\,\Phi_1 = M\, H_1 \,, \qquad -i\sigma\,\Phi_2 = M\, H_2 \,, \qquad -i\sigma\,\Phi_3 = M\, H_3 \,. \tag{15.3.13d}$$

In the system for Eq. (15.3.12), one can note three groups of similar equations (in essence, these are much the same). Let us detail the equations for Eq. (15.3.12a) for $\Phi_j(z, r)$:

$$(\frac{\partial}{\partial z} - 1)\Phi_1 = \sigma\,\Phi_1 - a\, e^z\,\Phi_2 \,,$$

$$-e^z\, b_-\Phi_1 + e^z\, a_+\Phi_3 = \sigma\,\Phi_2 \,,$$

$$-(\frac{\partial}{\partial z} - 1)\Phi_3 = \sigma\,\Phi_3 + b\ e^z\,\Phi_2 \,. \tag{15.3.14a}$$

By substitutions

$$\Phi_1 = e^z\varphi_1(r, z) \,, \qquad \Phi_2 = e^{2z}\varphi_2(r, z) \,, \qquad \Phi_3 = e^z\varphi_3(r, z) \tag{15.3.14b}$$

the system in Eq. (15.3.14a) is simplified to

$$a\,\varphi_2 = e^{-2z}(+\sigma - \frac{\partial}{\partial z})\,\varphi_1 \,,$$

$$-b_-\varphi_1 + a_+\varphi_3 = \sigma\,\varphi_2 \,,$$

$$b\,\varphi_2 = e^{-2z}(-\sigma - \frac{\partial}{\partial z})\,\varphi_3 \,. \tag{15.3.14c}$$

In using r-dependent differential operators b_-, a_+ (see Eq. (15.3.8)), let us introduce new variables

$$b_-\varphi_1 = \bar{\varphi}_1 \,, \qquad a_+\varphi_3 = \bar{\varphi}_3 \,; \tag{15.3.15a}$$

the from Eq. (15.3.14c), it follows

$$b_- a\,\varphi_2 = e^{-2z}(+\sigma - \frac{\partial}{\partial z})\bar{\varphi}_1 \,,$$

$$\bar{\varphi}_3 - \bar{\varphi}_1 = \sigma\,\varphi_2 \,,$$

$$a_+ b \; \varphi_2 = e^{-2z}(-\sigma - \frac{\partial}{\partial z}) \; \bar{\varphi}_3 \; . \tag{15.3.15b}$$

Note that the first and the third equations involve the same operator

$$b_- a = a_+ b = \frac{1}{2}\left(-\frac{\partial^2}{\partial r^2} - \frac{1}{r}\frac{\partial}{\partial r} + \frac{m^2}{r^2}\right) = \Delta \; . \tag{15.3.15c}$$

First, let us assume $\sigma \neq 0$. From the first and the third equations in Eq. (15.3.15b) it follows

$$\sigma(\bar{\varphi}_1 + \bar{\varphi}_3) = -\frac{\partial}{\partial z}(\bar{\varphi}_3 - \bar{\varphi}_1) = -\sigma\frac{\partial}{\partial z}\varphi_2 \; . \tag{15.3.16a}$$

Thus, we derive two simple relations

$$\bar{\varphi}_3 + \bar{\varphi}_1 = -\frac{\partial}{\partial z}\varphi_2 \; , \qquad \bar{\varphi}_3 - \bar{\varphi}_1 = \sigma \; \varphi_2 \; ;$$

this will provide us with expressions for $\bar{\varphi}_1$ and $\bar{\varphi}_3$ through φ_2:

$$\bar{\varphi}_3 = \frac{1}{2}(+\sigma - \frac{\partial}{\partial z})\varphi_2 \; , \qquad \bar{\varphi}_1 = \frac{1}{2}(-\sigma - \frac{\partial}{\partial z})\varphi_2 \; . \tag{15.3.16b}$$

In turn, through substituting them into the first and the third equations in Eq. (15.3.15b), we get the same equation for φ_2:

$$b_- a \; \varphi_2 = e^{-2z}(\sigma - \frac{\partial}{\partial z})\frac{1}{2}(-\sigma - \frac{\partial}{\partial z})\varphi_2 \; ,$$

$$a_+ b \; \varphi_2 = e^{-2z}(-\sigma - \frac{\partial}{\partial z}) \; \frac{1}{2}(+\sigma - \frac{\partial}{\partial z})\varphi_2 \; . \tag{15.3.17a}$$

In Eq. (15.3.17a), one can separate the variables:

$$\varphi_2(r, z) = \varphi_2(r) \; \varphi_2(z) \; ,$$

$$\frac{1}{h_2(r)} \; (2\Delta) \; h_2(r) = \frac{1}{h_2(z)}e^{-2z}(\frac{d^2}{dz^2} - \sigma^2)\varphi_2(z) = \Lambda \; ;$$

We get two ordinary differential equations; in r and z variables, respectively:

$$2\Delta \; \varphi_2(r) = \Lambda \; \varphi_2(r) \; , \tag{15.3.17b}$$

$$(\frac{d^2}{dz^2} - \sigma^2)\varphi_2(z) = \Lambda e^{+2z} \; \varphi_2(z) \; . \tag{15.3.17c}$$

By what is known in $\varphi_2 = \varphi_2(r)\varphi_2(z)$, one can calculate the two remaining components.

$$\bar{\varphi}_1 = \frac{1}{2}(-\sigma - \frac{\partial}{\partial z})\varphi_2 \; , \qquad \bar{\varphi}_3 = \frac{1}{2}(+\sigma - \frac{\partial}{\partial z})\varphi_2 \; . \tag{15.3.17d}$$

Similar analysis gives

$$2\Delta \; e_2(r) = \Lambda \; e_2(r) \; ,$$

$$(\frac{d^2}{dz^2} - \sigma^2)e_2(z) = \Lambda e^{+2z} \; e_2(z) \; ,$$

$$\bar{e}_1 = \frac{1}{2}(-\sigma - \frac{\partial}{\partial z})e_2 \,, \qquad \bar{e}_3 = \frac{1}{2}(+\sigma - \frac{\partial}{\partial z})e_2 \,; \qquad (15.3.18a)$$

$$2\Delta\, \varphi_2(r) = \Lambda\, h_2(r) \,,$$

$$(\frac{d^2}{dz^2} - \sigma^2)\varphi_2(z) = \Lambda e^{+2z}\, h_2(z) \,,$$

$$\bar{h}_1 = \frac{1}{2}(-\sigma - \frac{\partial}{\partial z})h_2 \,, \qquad \bar{h}_3 = \frac{1}{2}(+\sigma - \frac{\partial}{\partial z})h_2 \,. \qquad (15.3.18b)$$

The explicit form of Eq. $(15.3.17b)$ is

$$\left[\frac{d^2}{dr^2} + \frac{1}{r}\frac{d}{dr} - \frac{m^2}{r^2} + \Lambda\right]\varphi_2 = 0 \,;$$

in the new variable $y = \sqrt{\Lambda}\, r$ it is the Bessel equation

$$\left(\frac{d^2}{dy^2} + \frac{1}{y}\frac{d}{dy} + 1 - \frac{m^2}{y^2}\right)\varphi_2 = 0 \,,$$

with solutions

$$\varphi_2 = A\, J_m(y) + B\, Y_m(y) \,. \qquad (15.3.19a)$$

In turn, Eq. $(15.3.17c)$ with the variable $x = i\,\sqrt{\Lambda}\, e^{+z}$ is the Besell equation, as well.

$$\left(\frac{d^2}{dx^2} + \frac{1}{x}\frac{d}{dx} + 1 - \frac{\sigma^2}{x^2}\right)\varphi_2 = 0 \,,$$

Here, solution is

$$\varphi_2 = C\, J_\sigma(x) + D\, Y_\sigma(x) \,. \qquad (15.3.19b)$$

Now, let us turn to the system in Eqs. (15.3.15b) for the case $\sigma = 0$:

$$\bar{\varphi}_3 = \bar{\varphi}_1 = \bar{\varphi} \,,$$

$$\Delta\, \varphi_2 = -e^{-2z}\frac{\partial}{\partial z}\bar{\varphi} \,, \qquad \Delta\, \varphi_2 = -e^{-2z}\frac{\partial}{\partial z}\,\bar{\varphi} \,. \qquad (15.3.20)$$

Similar equations take place for E_j, H_j. The system in Eq.(15.3.13) becomes simpler as well

$$-e^z\, b_-\, E_1 - e^z\, a_+\, E_3 - (\frac{\partial}{\partial z} - 2)E_2 = M\, \Phi_0 \,, \qquad (15.3.21a)$$

$$i\epsilon\, E_j = M\, \Phi_j \,, \qquad (15.3.21b)$$

$$e^z\, a\, \Phi_0 - i\epsilon\, \Phi_1 = M\, E_1 \,,$$

$$-i\epsilon\Phi_2 - \frac{\partial}{\partial z}\Phi_0 = M\, E_2 \,,$$

$$e^z\, b\, \Phi_0 - i\epsilon\, \Phi_3 = M\, E_3 \,, \qquad (15.3.21c)$$

$$H_j = 0 \,, \qquad (15.3.21d)$$

or in other variables

$$-2\bar{e} - \frac{\partial}{\partial z}e_2 = M\ e^{-2z}\Phi_0\ ,\qquad (15.3.22a)$$

$$i\epsilon\ \bar{e} = M\ \bar{\varphi}\ ,\qquad i\epsilon\ e_2 = M\ \varphi_2\ ,\qquad (15.3.22b)$$

$$-i\epsilon e^{2z}\varphi_2 - \frac{\partial}{\partial z}\Phi_0 = M\ e^{2z}e_2\ ,$$

$$\Delta\ \Phi_0 - i\epsilon\ \bar{\varphi} = M\ \bar{e}\ ,\qquad (15.3.22c)$$

$$H_j = 0\ .\qquad (15.3.22d)$$

Let us exclude the variables $\bar{e}, e_2$:

$$H_j = 0,\qquad \bar{e} = \frac{M}{i\epsilon}\ \bar{\varphi}\ ,\qquad e_2 = \frac{M}{i\epsilon}\ \varphi_2\ ,\qquad (15.3.23a)$$

$$-2\frac{M}{i\epsilon}\ \bar{\varphi} - \frac{\partial}{\partial z}\frac{M}{i\epsilon}\ \varphi_2 = M\ e^{-2z}\Phi_0\ ,$$

$$-i\epsilon e^{2z}\varphi_2 - \frac{\partial}{\partial z}\Phi_0 = M\ e^{2z}\frac{M}{i\epsilon}\ \varphi_2\ ,$$

$$\Delta\ \Phi_0 - i\epsilon\ \bar{\varphi} = M\ \frac{M}{i\epsilon}\ \bar{\varphi}\ ,\qquad (15.3.23b)$$

or

$$-2\ \bar{\varphi} - \frac{\partial}{\partial z}\varphi_2 = i\epsilon e^{-2z}\Phi_0\ ,$$

$$\varphi_2 = \frac{i\epsilon}{(\epsilon^2 - M^2)}e^{-2z}\frac{\partial}{\partial z}\Phi_0\ ,$$

$$\bar{\varphi} = \frac{i\epsilon}{(M^2 - \epsilon^2)}\Delta\ \Phi_0\ .\qquad (15.3.23c)$$

From Eq. (15.3.23c), that allow for the second and the third equations; from the first equation we can derive

$$-2\ \frac{i\epsilon}{(M^2 - \epsilon^2)}\Delta\ \Phi_0 - \frac{\partial}{\partial z}\frac{i\epsilon}{(\epsilon^2 - M^2)}e^{-2z}\frac{\partial}{\partial z}\Phi_0 = i\epsilon e^{-2z}\Phi_0\ ,$$

that is

$$\left[2\Delta - \frac{\partial}{\partial z}e^{-2z}\frac{\partial}{\partial z} - (\epsilon^2 - M^2)e^{-2z}\right]\Phi_0 = 0\ .\qquad (15.3.24a)$$

Let it be $\Phi_0(r,z) = \Phi_0(r)\Phi_0(z)$, then the variables are separated

$$\frac{1}{\Phi_0(r)}2\Delta\Phi_0(r) = \frac{1}{\Phi_0(r)}\left(\frac{\partial}{\partial z}e^{-2z}\frac{\partial}{\partial z} + (\epsilon^2 - M^2)e^{-2z}\right)\Phi_0(z) = \Lambda\ ,\qquad (15.3.24b)$$

and further, we get

$$\frac{1}{\Phi_0(r)}2\Delta\Phi_0(r) = \Lambda\ ,$$

$$\frac{1}{\Phi_0(z)}\left(\frac{\partial}{\partial z}e^{-2z}\frac{\partial}{\partial z} + (\epsilon^2 - M^2)e^{-2z}\right)\Phi_0(z) = \Lambda\ .\qquad (15.3.24c)$$

In conclusion, let us specify the massless case. Instead of Eq. (15.3.13), now we have:

$$-e^z b_- \ E_1 - e^z a_+ \ E_3 - (\frac{\partial}{\partial z} - 2)E_2 = 0 \, ,$$

$$i\epsilon \ E_1 + i\sigma \ H_1 = 0 \, , \quad i\sigma \ H_2 + i\epsilon \ E_2 = 0 \, , \quad i\epsilon \ E_3 + i\sigma \ H_3 = 0 \, ,$$

$$e^z a \ \Phi_0 - i\epsilon \ \Phi_1 = E_1 \, ,$$

$$-i\epsilon\Phi_2 - \frac{\partial}{\partial z}\Phi_0 = E_2 \, ,$$

$$e^z b \ \Phi_0 - i\epsilon \ \Phi_3 = E_3 \, ,$$

$$-\sigma \ \Phi_1 = H_1 \, , \qquad -i\sigma \ \Phi_2 = H_2 \, , \qquad -i\sigma \ \Phi_3 = H_3 \, . \qquad (15.3.25)$$

The most interesting case is for $\sigma = 0$, the system Eq. (15.3.25) gives

$$E_j = 0 \, , \qquad H_j = 0 \, ,$$

$$a \ \Phi_0 - i\epsilon \ \varphi_1 = 0 \, ,$$

$$-i\epsilon\Phi_2 - \frac{\partial}{\partial z}\Phi_0 = 0 \, ,$$

$$b \ \Phi_0 - i\epsilon \ \varphi_3 = 0 \, ; \qquad (15.3.26)$$

from this, it follows:

$$E_j = 0 \, , \qquad H_j = 0 \, , \qquad (15.3.27a)$$

$$\bar{\varphi} = \frac{1}{i\epsilon}\Delta \ \Phi_0 \, , \qquad \varphi_2 = -\frac{1}{i\epsilon}e^{-2z}\frac{\partial}{\partial z}\Phi_0 = 0 \, , \qquad (15.3.27b)$$

where Φ_0 is an arbitrary function. This class describes the solutions of a gradient type with a vanishing electromagnetic tensor.

15.4   Spin 1 Particle on 4-Dimensional Sphere

In the case of a 3-dimensional spherical Riemann space S_3 that is parameterized by the system of quasi-cylindric coordinates (see chapter **13**).

$$dS^2 = dt^2 - \cos^2 z(dr^2 + \sin^2 r d\phi^2) - dz^2,$$

$$z \in [-\pi/2, +\pi/2] \, , \ r \in [0, +\pi], \ \phi \in [0, 2\pi] \, ; \qquad (15.4.1)$$

we will use a diagonal tetrad (let $x^\alpha = (t, r, \phi, z)$)

$$e^{\beta}_{(a)}(x) = \begin{vmatrix} 1 & 0 & 0 & 0 \\ 0 & \cos^{-1} z & 0 & 0 \\ 0 & 0 & \cos^{-1} z \ \sin^{-1} r & 0 \\ 0 & 0 & 0 & 1 \end{vmatrix} . \qquad (15.4.2)$$

The corresponding Ricci coefficients are

$$\gamma_{122} = \frac{1}{\cos z \tan r} \, , \ \gamma_{311} = -\tan z \, , \ \gamma_{322} = -\tan z \, . \qquad (15.4.3)$$

Tetrad-based Duffin–Kemmer equation takes the form

$$\left\{ i\beta^0 \frac{\partial}{\partial t} + \frac{1}{\cos z}\left(i\beta^1 \frac{\partial}{\partial r} + \beta^2 \frac{i\partial_\phi + iJ^{12}\cos r}{\sin r} \right) \right.$$

$$\left. + i\beta^3 \frac{\partial}{\partial z} + i\frac{\sin z}{\cos z}(\beta^1 J^{13} + \beta^2 J^{23}) - M \right\} \Psi = 0 . \tag{15.4.4}$$

To separate the variables, we take the substitution

$$\Psi = e^{-i\epsilon t} e^{im\phi} \begin{vmatrix} \Phi_0(r,z) \\ \vec{\Phi}(r,z) \\ \mathbf{E}(r,z) \\ \mathbf{H}(r,z) \end{vmatrix} , \tag{15.4.5a}$$

and use a block-representation (in cyclic basis for (10×10)-matrices)

$$\left[\epsilon \cos z \begin{vmatrix} 0 & 0 & 0 & 0 \\ 0 & 0 & i & 0 \\ 0 & -i & 0 & 0 \\ 0 & 0 & 0 & 0 \end{vmatrix} + i \begin{vmatrix} 0 & 0 & e_1 & 0 \\ 0 & 0 & 0 & \tau_1 \\ -e_1^+ & 0 & 0 & 0 \\ 0 & -\tau_1 & 0 & 0 \end{vmatrix} \frac{\partial}{\partial r} \right.$$

$$-\frac{1}{\sin r} \begin{vmatrix} 0 & 0 & e_2 & 0 \\ 0 & 0 & 0 & \tau_2 \\ -e_2^+ & 0 & 0 & 0 \\ 0 & -\tau_2 & 0 & 0 \end{vmatrix} (\nu - \cos r\, S_3) + i\cos z \begin{vmatrix} 0 & 0 & e_3 & 0 \\ 0 & 0 & 0 & \tau_3 \\ -e_3^+ & 0 & 0 & 0 \\ 0 & -\tau_3 & 0 & 0 \end{vmatrix} \frac{\partial}{\partial z}$$

$$\left. + i\sin z \begin{vmatrix} 0 & 0 & -2e_3 & 0 \\ 0 & 0 & 0 & -\tau_3 \\ 0 & 0 & 0 & 0 \\ 0 & +\tau_3 & 0 & 0 \end{vmatrix} - M\cos z \right] \begin{vmatrix} \Phi_0 \\ \vec{\Phi} \\ \mathbf{E} \\ \mathbf{H} \end{vmatrix} = 0 . \tag{15.4.5b}$$

After performing the needed calculations, we arrive at the system

$$\gamma\left(\frac{\partial E_1}{\partial r} - \frac{\partial E_3}{\partial r}\right) - \frac{\gamma}{\sin r}[(m - \cos r)E_1 + (m + \cos r)E_3]$$

$$-(\cos z\frac{\partial}{\partial z} - 2\sin z)E_2 = M\cos z\Phi_0 , \tag{15.4.6a}$$

$$i\epsilon\cos z E_1 + i\gamma\frac{\partial H_2}{\partial r} + i\gamma\frac{m}{\sin r}H_2 + i(\cos z\frac{\partial}{\partial z} - \sin z)H_1 = M\cos z\Phi_1 ,$$

$$+i\epsilon\cos z E_2 + i\gamma\left(\frac{\partial H_1}{\partial r} + \frac{\partial H_3}{\partial r}\right)$$

$$-\frac{i\gamma}{\sin r}[(m - \cos r)H_1 - (m + \cos r)H_3] = M\cos z\Phi_2 ,$$

$$+i\epsilon\cos z E_3 + i\gamma\frac{\partial H_2}{\partial r} - i\gamma\frac{m}{\sin r}H_2 - i(\cos z\frac{\partial}{\partial z} - \sin z)H_3 = M\cos z\Phi_3 ,$$

$$\tag{15.4.6b}$$

$$-i\epsilon \cos z \Phi_1 + \gamma \frac{\partial \Phi_0}{dr} + \gamma \frac{m}{\sin r} \Phi_0 = M \cos z E_1 \,,$$

$$-i\epsilon \Phi_2 - \frac{\partial \Phi_0}{\partial z} = M E_2 \,,$$

$$-i\epsilon \cos z \Phi_3 - \gamma \frac{\partial \Phi_0}{\partial r} + \gamma \frac{m}{\sin r} \Phi_0 = M \cos z E_3 \,,$$

$$(15.4.6c)$$

$$-i\gamma \frac{\partial \Phi_2}{\partial r} - i\gamma \frac{m}{\sin r} \Phi_2 - i(\cos z \frac{\partial}{\partial z} - \sin z) \Phi_1 = M \cos z H_1 \,,$$

$$-i\gamma (\frac{\partial \Phi_1}{\partial r} + \frac{\partial \Phi_3}{\partial r}) + \frac{i\gamma}{\sin r} [(m - \cos r) \Phi_1 - (m + \cos r) \Phi_3] = M \cos z H_2 \,,$$

$$-i\gamma \frac{\partial \Phi_2}{\partial r} + i\gamma \frac{m}{\sin r} \Phi_2 + i(\cos z \frac{\partial}{\partial z} - \sin z) \Phi_3 = M \cos z H_3 \,.$$

$$(15.4.6d)$$

When using the notation

$$\gamma \left(\frac{\partial}{\partial r} + \frac{m - \cos r}{\sin r}\right) = a_- \,, \ \gamma \left(\frac{\partial}{\partial r} + \frac{m + \cos r}{\sin r}\right) = a_+ \,, \ \gamma \left(\frac{\partial}{\partial r} + \frac{m}{\sin r}\right) = a \,,$$

$$\gamma \left(-\frac{\partial}{\partial r} + \frac{m - \cos r}{\sin r}\right) = b_- \,, \ \gamma \left(-\frac{\partial}{\partial r} + \frac{m + \cos r}{\sin r}\right) = b_+ \,, \ \gamma \left(-\frac{\partial}{\partial r} + \frac{m}{\sin r}\right) = b \,,$$

$$(15.4.7)$$

it will read simply as

$$-b_- E_1 - a_+ E_3 - \cos z (\frac{\partial}{\partial z} - 2 \tan z) E_2 = M \cos z \Phi_0 \,, \qquad (15.4.8a)$$

$$iaH_2 + i\epsilon \cos z E_1 + i \cos z (\frac{\partial}{\partial z} - \tan z) H_1 = M \cos z \Phi_1 \,,$$

$$-ib_- H_1 + ia_+ H_3 + i\epsilon \cos z E_2 = M \cos z \Phi_2 \,,$$

$$-ibH_2 + i\epsilon \cos z E_3 - i(\frac{\partial}{\partial z} - \tan z) H_3 = M \cos z \ \Phi_3 \,, \qquad (15.4.8b)$$

$$a\Phi_0 - i\epsilon \cos z \Phi_1 = M \cos z E_1 \,,$$

$$-i\epsilon \Phi_2 - \frac{\partial}{\partial z} \Phi_0 = M E_2 \,,$$

$$b \ \Phi_0 - i\epsilon \cos z \Phi_3 = M \cos z E_3 \,, \qquad (15.4.8c)$$

$$-ia\Phi_2 - i \cos z (\frac{\partial}{\partial z} - \tan z) \Phi_1 = M \cos z H_1 \,,$$

$$ib_- \Phi_1 - ia_+ \Phi_3 = M \cos z H_2 \,,$$

$$ib\Phi_2 + i \cos(\frac{\partial}{\partial z} - \tan z) \Phi_3 = M \cos H_3 \,. \qquad (15.4.8d)$$

Let us employ an additional operator, a generalized helicity operator – such that (note that the usual factor $-i$ at the operator Σ is absent, so the eigenvalues are multiplied by $-i$)

$$\Sigma\Psi = \sigma\Psi, \qquad \left[\frac{1}{\cos z}\left(S_1\frac{\partial}{\partial r} + iS_2\frac{m - S_3\cos r}{\sin r}\right)\right.$$

$$\left.+(\frac{\partial}{\partial z} - \tan z)\,S_3\right]\begin{vmatrix}\Phi_0\\ \vec{\Phi}\\ \mathbf{E}\\ \mathbf{H}\end{vmatrix} = \sigma\begin{vmatrix}\Phi_0\\ \vec{\Phi}\\ \mathbf{E}\\ \mathbf{H}\end{vmatrix}. \qquad (15.4.9)$$

From Eq. (15.4.9), it follows the system of 10 equations (let $\gamma = 1/\sqrt{2}$):

$$0 = \sigma\,\Phi_0, \qquad (15.4.10a)$$

$$\gamma\frac{\partial}{\partial r}\Phi_2 + \gamma\frac{m}{\sin r}\Phi_2 + \cos z(\frac{\partial}{\partial z} - \tan z)\Phi_1 = \sigma\,\cos z\,\Phi_1,$$

$$\gamma(\frac{\partial}{\partial r}\Phi_1 + \frac{\partial}{\partial r}\Phi_3) - \frac{\gamma}{\sin r}[(m - \cos r)\Phi_1 - (m + \cos r)\Phi_3] = \sigma\,\cos z\,\Phi_2,$$

$$\gamma\frac{\partial}{\partial r}\Phi_2 - \gamma\frac{m}{\sin r}\Phi_2 - \cos z(\frac{\partial}{\partial z} - \tan z)\Phi_3 = \sigma\,\cos z\,\Phi_3,$$

$$(15.4.10b)$$

$$\gamma\frac{\partial}{\partial r}E_2 + \gamma\frac{m}{\sin r}E_2 + \cos z(\frac{\partial}{\partial z} - \tan z)E_1 = \sigma\,\cos z\,E_1,$$

$$\gamma(\frac{\partial}{\partial r}E_1 + \frac{\partial}{\partial r}E_3) - \frac{\gamma}{\sin r}[(m - \cos r)E_1 - (m + \cos r)E_3] = \sigma\,\cos z\,E_2,$$

$$\gamma\frac{\partial}{\partial r}E_2 - \gamma\frac{m}{\sin r}E_2 - \cos z(\frac{\partial}{\partial z} - \tan z)E_3 = \sigma\,\cos z\,E_3,$$

$$(15.4.10c)$$

$$\gamma\frac{\partial}{\partial r}H_2 + \gamma\frac{m}{\sin r}H_2 + \cos z(\frac{\partial}{\partial z} - \tan z)H_1 = \sigma\,\cos z\,H_1,$$

$$\gamma(\frac{\partial}{\partial r}H_1 + \frac{\partial}{\partial\sin r}H_3) - \frac{\gamma}{\sin r}[(m - \cos r)H_1 - (m + \cos r)H_3] = \sigma\,\cos z\,H_2,$$

$$\gamma\frac{\partial}{\partial r}H_2 - \gamma\frac{m}{\sin r}H_2 - \cos z(\frac{\partial}{\partial z} - \tan z)H_3 = \sigma\,\cos z\,H_3.$$

$$(15.4.10d)$$

By using the notation in Eq. (15.4.7) it reads simply as

$$0 = \sigma\,\Phi_0, \qquad (15.4.11a)$$

$$+\cos z(\frac{\partial}{\partial z} - \tan z)\Phi_1 = \sigma\,\cos z\Phi_1 - a\Phi_2,$$

$$-b_-\Phi_1 + a_+\Phi_3 = \sigma\,\cos z\,\Phi_2,$$

$$-\cos z(\frac{\partial}{\partial z} - \tan z)\Phi_3 = \sigma \ \cos z\Phi_3 + b \ \Phi_2 \ , \tag{15.4.11b}$$

$$+\cos z(\frac{\partial}{\partial z} - \tan z)E_1 = \sigma \ \cos z \ E_1 - a \ E_2 \ ,$$

$$-b_- E_1 + a_+ E_3 = \sigma \ \cos z \ E_2 \ ,$$

$$-\cos z(\frac{\partial}{\partial z} - \tan z)E_3 = \sigma \ \cos z \ E_3 + b \ E_2 \ , \tag{15.4.11c}$$

$$+\cos z(\frac{\partial}{\partial z} - \tan z)H_1 = \sigma \ \cos z H_1 - a \ H_2 \ ,$$

$$-b_- H_1 + a_+ H_3 = \sigma \ \cos z \ H_2 \ ,$$

$$-\cos z(\frac{\partial}{\partial z} - \tan z)H_3 = \sigma \ \cos z \ H_3 + b \ H_2 \ . \tag{15.4.11d}$$

In taking into account Eqs. (15.4.11a) – (15.4.11d), into Eqs. (15.4.8a) – (15.4.8d), we get

$$-b_- \ E_1 - a_+ \ E_3 - \cos z(\frac{\partial}{\partial z} - 2\tan z)E_2 = M \ \cos z\Phi_0 \ , \tag{15.4.12a}$$

$$i\epsilon \ E_1 + i\sigma \ H_1 = M \ \Phi_1 \ ,$$

$$i\sigma \ H_2 + i\epsilon \ E_2 = M \ \Phi_2 \ ,$$

$$i\epsilon \ E_3 + i\sigma \ H_3 = M \ \Phi_3 \ , \tag{14.4.12b}$$

$$a \ \Phi_0 - i\epsilon \ \cos z\Phi_1 = M \ \cos zE_1 \ ,$$

$$-i\epsilon\Phi_2 - \frac{\partial}{\partial z}\Phi_0 = M \ E_2 \ ,$$

$$b \ \Phi_0 - i\epsilon \ \cos z\Phi_3 = M \ \cos zE_3 \ , \tag{15.4.12c}$$

$$-\sigma \ \Phi_1 = M \ H_1 \ ,$$

$$-i\sigma \ \Phi_2 = M \ H_2 \ ,$$

$$-i\sigma \ \Phi_3 = M \ H_3 \ . \tag{15.4.12d}$$

Next, we will need an explicit form of the Lorentz gauge. Starting from its tensor representation

$$\nabla_\beta(e^{(b)\beta}\Phi_{(b)}^{cart} = 0 \qquad \Longrightarrow$$

$$\frac{\partial\Phi_{(b)cart}}{\partial x^\beta} \ e^{(b)\beta} + \Phi_{(b)}^{cart}\nabla_\beta e^{(b)\beta} = 0 \ , \tag{15.4.13a}$$

or

$$\frac{\partial\Phi_{(b)}^{cart}}{\partial x^\beta} \ e^{(b)\beta} + \Phi_{(b)}^{cart}\frac{1}{\sqrt{-g}} \frac{\partial}{\partial x^\beta}\sqrt{-g}e^{(b)\beta} = 0 \ , \tag{15.4.13b}$$

and further

$$\frac{\partial}{\partial t}\Phi_0^{cart} - \frac{1}{\cos z}(\frac{\partial}{\partial r} + \frac{\cos r}{\sin r})\Phi_1^{cart}$$

$$-\frac{1}{\cos z \sin r}\frac{\partial}{\partial\phi}\Phi_2^{cart} - (\frac{\partial}{\partial z} - 2\tan z)\Phi_3^{cart} = 0.$$

From this, with the substitution of Eq. (15.4.5a), we obtain

$$-i\epsilon\Phi_0^{cart} - \frac{1}{\cos z}\left(\frac{\partial}{\partial r} + \frac{\cos r}{\sin r}\right)\Phi_1^{cart}$$

$$-\frac{im}{\cos z \sin r}\Phi_2^{cart} - \left(\frac{\partial}{\partial z} - 2\tan z\right)\Phi_3^{cart} = 0 . \tag{15.4.13c}$$

To use this relation in these equations, we should transform Eq. (15.4.13c) to the cyclic basis:

$$\Phi_0 = \Phi_0^{cart} , \qquad \Phi_2 = \Phi_3^{cart} ,$$

$$\Phi_3 - \Phi_1 = \sqrt{2}\Phi_1^{cart} , \quad \Phi_3 + \Phi_1 = \sqrt{2}i\,\Phi_2^{cart} ;$$

thus, we have

$$-i\epsilon\Phi_0 - \frac{1}{\cos z}\left(\frac{\partial}{\partial r} + \frac{\cos r}{\sin r}\right)\frac{\Phi_3 - \Phi_1}{\sqrt{2}}$$

$$-\frac{im}{\cos z \sin r}\frac{\Phi_3 + \Phi_1}{\sqrt{2}i} - \left(\frac{\partial}{\partial z} - 2\tan z\right)\Phi_2 = 0 , \tag{15.4.13d}$$

that is

$$-i\epsilon\Phi_0 - \frac{1}{\cos z}\,b_-\,\Phi_1 - \frac{1}{\cos z}\,a_+\,\Phi_3 - \left(\frac{\partial}{\partial z} - 2\tan z\right)\Phi_2 = 0 . \tag{15.4.13e}$$

Now, let us turn to Eqs. (15.4.12a)–(15.4.12d). First, let us consider the case $\sigma \neq 0$, when one must accept from the very beginning such a restriction $\Phi_0 = 0$; correspondingly, this equation become more simple

$$\sigma \neq 0 , \qquad \Phi_0 = 0 ,$$

$$-b_-\,E_1 - a_+\,E_3 - \cos z\left(\frac{\partial}{\partial z} - 2\tan z\right)E_2 = 0 , \tag{15.4.14a}$$

$$i\epsilon E_1 + i\sigma H_1 = M\Phi_1 ,$$

$$i\sigma H_2 + i\epsilon E_2 = M\Phi_2 ,$$

$$i\epsilon E_3 + i\sigma H_3 = M\Phi_3 , \tag{15.4.14b}$$

$$-i\epsilon\Phi_1 = ME_1 , \quad -i\epsilon\Phi_2 = ME_2 , \quad -i\epsilon\Phi_3 = ME_3 , \tag{15.4.14c}$$

$$-\sigma\Phi_1 = MH_1 , \quad -i\sigma\Phi_2 = MH_2 , \quad -i\sigma\Phi_3 = MH_3 . \tag{15.4.14d}$$

Note that by substituting Eq. (15.4.14c) into Eq. (15.4.14a), one gets

$$-b_-\,\Phi_1 - a_+\,\Phi_3 - \cos z\left(\frac{\partial}{\partial z} - 2\tan z\right)\Phi_2 = 0 ,$$

which coincides with the Lorentz condition of Eq. (15.4.13e) when $\Phi_0 = 0$. This condition can be simplified by the following substitutions

$$\Phi_1 = \frac{\varphi_1}{\cos z} , \quad \Phi_3 = \frac{\varphi_3}{\cos z} , \quad \Phi_2 = \frac{1}{\cos^2 z}\varphi_2 ,$$

which results in

$$-b_-\varphi_1 - a_+\varphi_3 - \frac{\partial}{\partial z}\varphi_2 = 0 \; ; \qquad (15.4.15a)$$

in new variables $b_-\varphi_1 = \bar{\varphi}_1$, $a_+\varphi_3 = \bar{\varphi}_3$ it becomes yet simpler

$$\bar{\varphi}_1 + \bar{\varphi}_3 + \frac{\partial}{\partial z}\varphi_2 = 0 \; . \qquad (15.4.15b)$$

The remaining algebraic relations will be fixed-values of σ and relative coefficients of various components

$$\sigma = \pm i\sqrt{\epsilon^2 - M^2} \; , \qquad \Phi_0 = 0 \; ,$$

$$H_j = -i\frac{\sigma}{M}\,\Phi_j \; , \qquad E_j = \frac{i\epsilon}{M}\Phi_j \; . \qquad (15.4.15c)$$

An explicit form of the main functions Φ_j will be found later when exploring the helicity operator equations.

In <u>massless case</u> instead of Eq. (15.4.15c), we have

$$\sigma = \pm i\epsilon \; , \qquad \Phi_0 = 0 \; ,$$

$$H_j = -i\sigma\,\Phi_j \; , \qquad E_j = i\epsilon\,\Phi_j \; . \qquad (15.4.15d)$$

Now, let us consider the case $\underline{\sigma = 0}$, when the system in Eqs. (15.4.12a)–(15.4.12d) becomes

$$-b_-\,E_1 - a_+\,E_3 - \cos z(\frac{\partial}{\partial z} - 2\tan z)E_2 = M\,\cos z\Phi_0 \; , \qquad (15.4.16a)$$

$$i\epsilon E_1 = M\Phi_1, \qquad i\epsilon E_2 = M\Phi_2, \qquad i\epsilon E_3 = M\Phi_3 \; , \qquad (15.4.16b)$$

$$a\,\Phi_0 - i\epsilon\,\cos z\Phi_1 = M\,\cos zE_1 \; ,$$

$$-i\epsilon\Phi_2 - \frac{\partial}{\partial z}\Phi_0 = M\,E_2 \; ,$$

$$b\,\Phi_0 - i\epsilon\,\cos z\Phi_3 = M\,\cos zE_3 \; , \qquad (15.4.16c)$$

$$0 = M\,H_1 \; , \qquad 0 = M\,H_2 \; , \qquad 0 = M\,H_3 \; . \qquad (15.4.16d)$$

Note that when allowing for Eq. (15.4.16b), from Eq. (15.4.16a); it follows

$$-b_-\,\Phi_1 - a_+\,\Phi_3 - \cos z(\frac{\partial}{\partial z} - 2\tan z)\Phi_2 = i\epsilon\,\cos z\Phi_0 \; , \qquad (15.4.16e)$$

which coincides with the Lorentz condition.

Let us introduce these substitutions:

$$\Phi_1 = \frac{\varphi_1}{\cos z} \; , \qquad \Phi_3 = \frac{\varphi_3}{\cos z} \; , \qquad \Phi_2 = \frac{1}{\cos^2 z}\varphi_2 \; ,$$

$$E_1 = \frac{e_1}{\cos z} \; , \qquad E_3 = \frac{e_3}{\cos z} \; , \qquad E_2 = \frac{1}{\cos^2 z}e_2 \; ,$$

$$b_-\varphi_1 = \bar{\varphi}_1 \; , \qquad a_+\varphi_3 = \bar{\varphi}_3 \; ,$$

$$b_-e_1 = \bar{e}_1 \; , \qquad a_+e_3 = \bar{e}_3 \; ,$$

then Eqs. (15.4.16a)–(15.4.16d) will read as:

$$\bar{\varphi}_1 + \bar{\varphi}_3 + \frac{\partial}{\partial z}\varphi_2 = -i\epsilon \, \cos^2 z \Phi_0 \,, \tag{15.4.17a}$$

$$i\epsilon e_1 = M\varphi_1 \,, \quad i\epsilon e_2 = M\varphi_2 \,, \quad i\epsilon e_3 = M\varphi_3 \,, \tag{15.4.17b}$$

$$\Delta \Phi_0 - i\epsilon \bar{\varphi}_1 = M\bar{e}_1 \,, \quad - i\epsilon \varphi_2 - \cos^2 z \frac{\partial}{\partial z}\Phi_0 = Me_2 \,,$$

$$\Delta \Phi_0 - i\epsilon \bar{\varphi}_3 = M\bar{e}_3 \,, \tag{15.4.17c}$$

$$0 = H_1 \,, \qquad 0 = H_2 \,, \qquad 0 = H_3 \,; \tag{15.4.17d}$$

one should take into consideration identity $\Delta = b_-a = a_+b$.

Here, we will show from the helicity operator eigenvalue equation that when $\sigma = 0$ the following relationships must hold:

$$\bar{\varphi}_1 = \bar{\varphi}_3 = \bar{\varphi} \,, \qquad \bar{e}_1 = \bar{e}_3 = \bar{e} \,, \qquad \bar{h}_1 = \bar{h}_3 = \bar{h} \,,$$

$$\Delta \varphi_2 = - \cos^2 z \frac{\partial}{\partial z}\bar{\varphi} \,,$$

$$\Delta e_2 = - \cos^2 z \frac{\partial}{\partial z}\bar{e} \,,$$

$$\Delta h_2 = - \cos^2 z \frac{\partial}{\partial z}\bar{h} \,; \tag{15.4.18}$$

so that from Eqs. (15.4.17a)–(15.4.17d), we get

$$-2\bar{\varphi} - \frac{\partial}{\partial z}\varphi_2 = i\epsilon \, \cos^2 z \, \Phi_0 \,, \tag{15.4.19a}$$

$$\bar{e} = \frac{M}{i\epsilon}\bar{\varphi} \,, \qquad e_2 = \frac{M}{i\epsilon}\varphi_2 \,, \qquad H_j = 0 \,, \tag{15.4.19b}$$

$$(\epsilon^2 - M^2)\varphi_2 - i\epsilon \cos^2 z \frac{\partial}{\partial z}\Phi_0 = 0 \,,$$

$$i\epsilon \Delta \, \Phi_0 + (\epsilon^2 - M^2)\bar{\varphi} = 0 \,. \tag{15.4.19c}$$

Acting on the first equation in Eqs. (15.4.19c) by the operator ∂_z, and excluding the second equation in Eq. (15.4.19c) the variable $\bar{\varphi}$ with the help of Eq. (15.4.19a) – we get

$$(\epsilon^2 - M^2)\frac{\partial}{\partial z}\varphi_2 - i\epsilon \frac{\partial}{\partial z} \cos^2 z \frac{\partial}{\partial z}\Phi_0 = 0 \,,$$

$$2i\epsilon \Delta \, \Phi_0 - (\epsilon^2 - M^2)(\frac{\partial}{\partial z}\varphi_2 + i\epsilon \, \cos^2 z \, \Phi_0) = 0 \,.$$

Summing these two equations, we arrive at the second order equation for Φ_0

$$-i\epsilon \frac{\partial}{\partial z} \cos^2 z \frac{\partial}{\partial z}\Phi_0 + 2i\epsilon \Delta \, \Phi_0 - (\epsilon^2 - M^2)i\epsilon \, \cos^2 z \, \Phi_0 = 0 \,,$$

that is

$$\left(-2\Delta + \frac{\partial}{\partial z} \cos^2 z \frac{\partial}{\partial z} + (\epsilon^2 - M^2) \, \cos^2 z \right) \Phi_0 = 0 \,. \tag{15.4.20a}$$

In this equation, the variables are separated straightforwardly

$$\Phi_0(r, z) = \Phi_0(r)\Phi_0(z) \, , \quad \frac{1}{\Phi_0(r)}(2\Delta)\Phi_0(r) = \Lambda \, ,$$

$$\frac{1}{\Phi_0(z)}\left(\frac{d}{dz}\cos^2 z \frac{d}{dz} + (\epsilon^2 - M^2)\cos^2 z\right)\Phi_0(z) = \Lambda \, . \tag{15.4.20b}$$

In the same manner, through the help of Eq. (15.4.19a) one can exclude the function Φ_0 from second equation in Eq. (15.4.19c)

$$\Delta\left(-2\bar{\varphi} - \frac{\partial}{\partial z}\varphi_2\right) + (\epsilon^2 - M^2)\cos^2 z \, \bar{\varphi} = 0 \, ,$$

and further excluding the variable $\Delta\varphi_2$ with the help of Eq. (15.4.18), we arrive at a second order equation for $\bar{\varphi}$

$$\left(-2\Delta + \frac{\partial}{\partial z}\cos^2 z \frac{\partial}{\partial z} + (\epsilon^2 - M^2)\cos^2 z\right)\bar{\varphi} = 0 \, . \tag{15.4.21a}$$

In this equation, the variables are separated as well:

$$\bar{\varphi}(r, z) = \bar{\varphi}(r)\,\bar{\varphi}(z) \, , \qquad \frac{1}{\bar{\varphi}(r)}(2\Delta)\,\bar{\varphi}(r) = \Lambda \, ,$$

$$\frac{1}{\bar{\varphi}(z)}\left(\frac{d}{dz}\cos^2 z \frac{d}{dz} + (\epsilon^2 - M^2)\cos^2 z\right)\bar{\varphi}(z) = \Lambda \, . \tag{15.4.21b}$$

Note, that from the first equation in Eq. (15.4.19c) it follows an expression for φ_2

$$\varphi_2 = \frac{i\epsilon\,\cos^2 z}{(\epsilon^2 - M^2)}\frac{\partial}{\partial z}\Phi_0 \, . \tag{15.4.22}$$

One can easily verify consistency of the relations obtained. Indeed, let us act on Eq. (15.4.22) by the operator Δ

$$\Delta\varphi_2 = \frac{i\epsilon}{(\epsilon^2 - M^2)}\Delta\cos^2 z\frac{\partial}{\partial z}\Phi_0 = 0 \, .$$

Further, we can allow for Eq. (15.4.18) and get

$$-\cos^2 z\frac{\partial}{\partial z}\bar{\varphi} = \frac{i\epsilon}{(\epsilon^2 - M^2)}\Delta\cos^2 z\frac{\partial}{\partial z}\Phi_0 = 0 \, ,$$

from this, it follows

$$i\epsilon\Delta\,\Phi_0 + (\epsilon^2 - M^2)\bar{\varphi} = 0 \, ,$$

which is an identity

$$-\frac{\partial}{\partial z}\bar{\varphi} \equiv -\frac{1}{(\epsilon^2 - M^2)}(\epsilon^2 - M^2)\frac{\partial}{\partial z}\bar{\varphi} \, .$$

Now, let us turn to equations stemming from diagonalization of helicity operator. In Eqs. (15.4.11a)–(15.4.11d), we notice three similar groups of equations. For instance, equations for H_i are

$$a\, H_2 + \cos z(\frac{\partial}{\partial z} - \tan z)H_1 = \sigma \cos z H_1 \,,$$

$$-b_- H_1 + a_+ H_3 = \sigma \, \cos z \, H_2 \,,$$

$$-b\, H_2 - \cos z(\frac{\partial}{\partial z} - \tan z)H_3 = \sigma \, \cos z \, H_3 \,. \qquad (15.4.23a)$$

Through the help of substitutions

$$H_1 = \frac{1}{\cos z} h_1(r, z) \,,$$

$$H_2 = \frac{1}{\cos^2 z} h_2(r, z) \,,$$

$$H_3 = \frac{1}{\cos z} h_3(r, z) \qquad (15.4.23b)$$

they are simplified

$$a\, h_2 = \cos^2 z(+\sigma - \frac{\partial}{\partial z})\, h_1 \,,$$

$$-b_- h_1 + a_+ h_3 = \sigma \, h_2 \,,$$

$$b\, h_2 = \cos^2 z(-\sigma - \frac{\partial}{\partial z})\, h_3 \,. \qquad (15.4.23c)$$

Let us introduce new variables

$$b_- h_1 = \bar{h}_1 \,, \qquad a_+ h_3 = \bar{h}_3 \,; \qquad (15.4.24a)$$

from Eqs. (15.4.23c), it follows

$$b_- a\, h_2 = \cos^2 z(\sigma - \frac{\partial}{\partial z})\bar{h}_1 \,,$$

$$\bar{h}_3 - \bar{h}_1 = \sigma \, h_2 \,,$$

$$a_+ b\, h_2 = \cos^2 z(-\sigma - \frac{\partial}{\partial z})\, \bar{h}_3 \,. \qquad (15.4.24b)$$

Note that first and third equations contain the same second order operator

$$\Delta = b_- a = a_+ b = \frac{1}{2}\left(-\frac{\partial^2}{\partial r^2} - \frac{\cos r}{\sin r}\frac{\partial}{\partial r} + \frac{m^2}{\sin^2 r}\right). \qquad (15.4.24c)$$

First, let us consider the case $\sigma \neq 0$. Equating the right-hand sides of the first and third equations in Eq. (15.4.24b), we get

$$\sigma(\bar{h}_1 + \bar{h}_3) = -\frac{\partial}{\partial z}(\bar{h}_3 - \bar{h}_1) = -\sigma\frac{\partial}{\partial z}h_2 \,; \qquad (14.4.25a)$$

that is

$$\bar{h}_3 + \bar{h}_1 = -\frac{\partial}{\partial z} h_2 , \qquad \bar{h}_3 - \bar{h}_1 = \sigma \, h_2 .$$

Thus, we arrive at expression for $\bar{h}_1$ and $\bar{h}_3$ through h_2

$$\bar{h}_3 = \frac{1}{2}(+\sigma - \frac{\partial}{\partial z})h_2 , \qquad \bar{h}_1 = \frac{1}{2}(-\sigma - \frac{\partial}{\partial z})h_2 . \tag{15.4.25b}$$

In turn, when substituting Eq. (15.4.25b) into Eq. (15.4.24b); we obtain the same second order equation for h_2

$$b_-ah_2 = \cos^2 z(\sigma - \frac{\partial}{\partial z})\frac{1}{2}(-\sigma - \frac{\partial}{\partial z})h_2 ,$$

$$a_+bh_2 = \cos^2 z(-\sigma - \frac{\partial}{\partial z}) \, \frac{1}{2}(+\sigma - \frac{\partial}{\partial z})h_2 . \tag{15.4.26a}$$

The variables in Eq. (15.4.26a) are separated straightforwardly

$$h_2(r, z) = h_2(r) \, h_2(z) ,$$

$$\frac{1}{h_2(r)} \, (2b_-a) \, h_2(r) = \frac{1}{h_2(z)} \, \cos^2 z(\frac{d^2}{dz^2} - \sigma^2)h_2(z) = \Lambda ,$$

from here, it follows the separated differential equations

$$(2b_-a) \, h_2(r) = \Lambda \, h_2(r) , \tag{15.4.26b}$$

$$(\frac{d^2}{dz^2} - \sigma^2)h_2(z) = \frac{\Lambda}{\cos^2 z} \, h_2(z) . \tag{15.4.26c}$$

Similar results are valid for functions e_i and φ_i:

$$(2b_-a) \, e_2(r) = \Lambda \, e_2(r) ,$$

$$(\frac{d^2}{dz^2} - \sigma^2)e_2(z) = \frac{\Lambda}{\cos^2 z} \, e_2(z) ,$$

$$\bar{e}_1 = \frac{1}{2}(-\sigma - \frac{\partial}{\partial z})e_2, \bar{e}_3 = \frac{1}{2}(+\sigma - \frac{\partial}{\partial z})e_2 ; \tag{15.4.27}$$

$$(2b_-a) \, \varphi_2(r) = \Lambda \, \varphi_2(r) ,$$

$$(\frac{d^2}{dz^2} - \sigma^2)\varphi_2(z) = \frac{\Lambda}{\cos^2 z} \, \varphi_2(z) ,$$

$$\bar{\varphi}_1 = \frac{1}{2}(-\sigma - \frac{\partial}{\partial z})\varphi_2, \bar{\varphi}_3 = \frac{1}{2}(+\sigma - \frac{\partial}{\partial z})\varphi_2 . \tag{15.4.28}$$

Now, let us turn to the system in Eq. (15.4.24b) when $\sigma = 0$; it gives

$$\bar{h}_3 = \bar{h}_1 = \bar{h} ,$$

$$b_-a \, h_2 = - \cos^2 z \frac{\partial}{\partial z} \bar{h} ,$$

$$a_+ b\, h_2 = -\cos^2 z \frac{\partial}{\partial z}\, \bar{h}\,, \qquad (15.4.29a)$$

To clarify, the relations used here start with Eq. (15.4.18).

Let us construct solutions of Eqs. (15.4.28):

$$(2b_- a)\,\varphi_2(r) = \Lambda\,\varphi_2(r)\,,$$

$$\left(\frac{d^2}{dz^2} - \sigma^2\right)\varphi_2(z) = \frac{\Lambda}{\cos^2 z}\varphi_2(z)\,,$$

$$\bar{\varphi}_1 = \frac{1}{2}\left(-\sigma - \frac{\partial}{\partial z}\right)\varphi_2\,, \qquad \bar{\varphi}_3 = \frac{1}{2}\left(+\sigma - \frac{\partial}{\partial z}\right)\varphi_2\,. \qquad (15.4.29b)$$

In the radial equation

$$(2b_- a)\,\varphi_2(r) = \Lambda\,\varphi_2(r)\,,$$

or

$$\left(\frac{d^2}{dr^2} + \frac{\cos r}{\sin r}\frac{d}{dr} - \frac{m^2}{\sin^2 r} + \Lambda\right)\varphi_2(r) = 0, \qquad (15.4.30a)$$

let us introduce a new variable $1 - \cos r = 2\,x$, $x \in [0,\, 1]$:

$$x\,(1-x)\frac{d^2\varphi_2}{dx^2} + (1 - 2\,x)\frac{d\varphi_2}{dx} + \left(\Lambda - \frac{1}{4}\frac{m^2}{x} - \frac{1}{4}\frac{m^2}{1-x}\right)\varphi_2 = 0 \qquad (15.4.30b)$$

and make a substitution $\varphi_2 = x^a\,(1-x)^b\,F_2$; thus we arrive at

$$x\,(1-x)\frac{d^2 F_2}{dx^2} + [2\,a + 1 - (2\,a + 2\,b + 2)\,x]\frac{dF_2}{dx}$$

$$+ \left[-(a+b)\,(a+b+1) + \Lambda + \frac{1}{4}\frac{4\,a^2 - m^2}{x} + \frac{1}{4}\frac{4\,b^2 - m^2}{1-x}\right]F_2 = 0. \qquad (15.4.30c)$$

At a, b, this is taken according to $a = \pm\,|\,m\,|\,/2$, $b = \pm\,|\,m\,|\,/2$, Eq. (15.4.30c) becomes simpler

$$x\,(1-x)\frac{d^2 F_2}{dx^2} + [2\,a + 1 - (2\,a + 2\,b + 2)\,x]\frac{dF_2}{dx} - [(a+b)\,(a+b+1) - \Lambda]\,F_2 = 0$$

it represents the hypergeometric equations with parameters

$$\alpha = a + b + \frac{1}{2} - \frac{1}{2}\sqrt{1 + 4\Lambda}\,, \quad \beta = a + b + \frac{1}{2} + \frac{1}{2}\sqrt{1 + 4\Lambda}\,, \quad \gamma = 2\,a + 1\,.$$

By physical reason for a, b, we take positive values

$$a = +\frac{|\,m\,|}{2}\,, \qquad b = +\frac{|\,m\,|}{2}\,; \qquad (15.4.31a)$$

so the radial function looks as

$$\varphi_2(r) = \left(\sin\frac{r}{2}\right)^{+|m|}\left(\cos\frac{r}{2}\right)^{+|m|} F(\alpha, \beta, \gamma; \sin^2\frac{r}{2})\,; \qquad (15.4.31b)$$

these solutions vanish at the points $r = 0, +\pi$. To have polynomials one should impose the known condition $\alpha = -n_r$; so we get a quantization rule

$$+\frac{\sqrt{1+4\Lambda}}{2} = n_r + \mid m \mid +\frac{1}{2} \ ; \tag{15.4.31c}$$

These corresponding solutions are defined according to

$$\varphi_2 = \left(\sin\frac{r}{2}\right)^{+\mid m \mid} \left(\cos\frac{r}{2}\right)^{+\mid m \mid}$$

$$\times F(-n, 2 \mid m \mid +1+n, \mid m \mid +1; -\sin^2\frac{r}{2}) \ . \tag{15.4.31d}$$

Now, let us solve equation $(15.4.29b)$ for φ_2 in variable z:

$$(\frac{d^2}{dz^2} - \sigma^2)\varphi_2(z) = \frac{\Lambda}{\cos^2 z}\varphi_2(z),$$

$$\sigma^2 = \epsilon^2 - M^2 \ . \tag{15.4.32}$$

A first step is to introduce a new variable (which distinguish between conjugated point $+z$ and $-z$ of spherical space)

$$y = \frac{1+i\tan z}{2} \ , \qquad 1-y = \frac{1-i\tan z}{2} \ ; \tag{15.4.33a}$$

if $z \in [-\pi/2, +\pi/2]$, the variable y belongs to a vertical line in the complex plane

$$y = (\frac{1}{2} - i\infty, \frac{1}{2} + i\infty) \ .$$

To allow for

$$\frac{d}{dz} = \frac{i}{2}\frac{1}{\cos^2 z}\frac{d}{dy} = 2iy(1-y)\frac{d}{dy} \ , \qquad \frac{\Lambda}{\cos^2 z} = 4\Lambda y(1-y) \ ,$$

Eq. $(15.4.32)$ reduces to

$$\left(y(1-y)\frac{d^2}{dy^2} + (1-2y)\frac{d}{dy} + \Lambda - \frac{\epsilon^2 - M^2}{4y(1-y)}\right)\varphi_2 = 0 \ . \tag{15.4.33b}$$

In the region $y \sim 0$, Eq. $(15.4.33b)$ becomes simpler

$$\left(y\frac{d^2}{dy^2} + \frac{d}{dy} - \frac{\epsilon^2 - M^2}{4y}\right)\varphi_2 = 0 \ , \qquad \varphi_2 \sim y^a, \quad a = \pm\frac{\sqrt{\epsilon^2 - M^2}}{2} \ .$$

In the region $y \sim 1$, Eq. $(15.4.33b)$ becomes simpler, as well

$$\left((1-y)\frac{d^2}{dy^2} - \frac{d}{dy} - \frac{\epsilon^2 - M^2}{4(1-y)}\right)\varphi_2 = 0 \ , \qquad \varphi_2 \sim (1-y)^b \ , \quad b = \pm\frac{\sqrt{\epsilon^2 - M^2}}{2} \ .$$

Searching these solutions in the form $\varphi_2(y) = y^a(1-y)^b F(y)$ for $F(y)$ we have

$$y(1-y)F'' + [(2a+1) - y(2a+2b+2)]F'$$

$$+ \left[-(a+b)(a+b+1) + \frac{1}{y}\left(a^2 - \frac{\epsilon^2 - M^2}{4}\right) \right.$$

$$\left. + \Lambda + \frac{1}{1-y}\left(a^2 - \frac{\epsilon^2 - M^2}{4}\right) \right] F = 0 \ . \qquad (15.4.34a)$$

Let it be

$$a = \pm\frac{\sqrt{\epsilon^2 - M^2}}{2} \ , \qquad b = \pm\frac{\sqrt{\epsilon^2 - M^2}}{2} \ , \qquad (15.4.34b)$$

then

$$\varphi_2 = \left(\frac{1 + i\tan z}{2}\right)^a \left(\frac{1 - i\tan z}{2}\right)^b F \ .$$

There are four possibilities depending on a, b

$$a = b = -\frac{\sqrt{\epsilon^2 - M^2}}{2} \ , \qquad \varphi_2 \sim \cos^{-2a} z \ F(z) \ ;$$

$$a = b = +\frac{\sqrt{\epsilon^2 - M^2}}{2} \ , \qquad \varphi_2 \sim \cos^{-2a} z \ F(z) \ ;$$

$$a = -b \ , \qquad \varphi_2 \sim +e^{+2iaz} \ F(z) \ ;$$

$$a = -b \ , \qquad \varphi_2 \sim -e^{-2iaz} \ F(z) \ . \qquad (15.4.35)$$

As relations in Eq. (15.4.34b) hold, Eq. (15.4.34a) takes the form

$$y(1-y)F'' + [(2a+1) - y(2a+2b+2)]F'$$

$$-[(a+b)(a+b+1) - \Lambda] F = 0 \ , \qquad (15.4.36a)$$

which can be recognized as a hypergeometric equation

$$y(1-y)F + [\gamma - (\alpha + \beta + 1)y]F' - \alpha\beta F = 0 \ ,$$

$$\gamma = (2a+1) \ , \quad \alpha = a + b + \frac{1}{2} + \frac{\sqrt{4\Lambda+1}}{2} \ ,$$

$$\beta = a + b + \frac{1}{2} - \frac{\sqrt{4\Lambda+1}}{2} \ . \qquad (15.4.36b)$$

At this point, we should notice that the spectrum for Λ has been clearly found from analyzing the differential equation in the variable r. Therefore, next is to produce a spectrum for energy and consider the cases with $a = b$.

Two possibilities can arise. The first:

$$2a = 2b = \sqrt{\epsilon^2 - M^2} \ , \qquad \beta = -n_z \ ,$$

$$-\sqrt{\epsilon^2 - M^2} = n_z + \frac{1}{2} - \frac{\sqrt{4\Lambda+1}}{2} < 0 \ ,$$

$$\varphi_2 \sim (\cos z)^{-\sqrt{\epsilon^2 - M^2}} \; P_n(\frac{e^{iz}}{2 \cos z}) \; ,$$

$$y = \frac{1 + i \tan z}{2} = \frac{e^{iz}}{2 \cos z} \; ; \qquad (15.4.37)$$

because those solutions tend to infinity at $z = \pm\pi$ they cannot describe physical bound states.

The second:

$$-2a = -2b = +\sqrt{\epsilon^2 - M^2} \; , \qquad \alpha = -n_z \; ,$$

$$+\sqrt{\epsilon^2 - M^2} = n_z + \frac{1}{2} + \frac{\sqrt{4\Lambda + 1}}{2} > 0 \; ,$$

$$\varphi_2 \sim (\cos z)^{+\sqrt{\epsilon^2 - M^2}} \; P_n(\frac{e^{iz}}{2 \cos z}) \; ,$$

$$y = \frac{1 + i \tan z}{2} = \frac{e^{iz}}{2 \cos z} \; . \qquad (15.4.38)$$

These solutions are finite at the points $z = \pm\pi/2$ and they describe bound states.

In the formula for $\sqrt{\epsilon^2 - M^2}$ one must take into account the quantization rule for Λ – thus we arrive at the formula that determine values of energy by two discrete quantum numbers.

$$+\sqrt{\epsilon^2 - M^2} = n_z + n_r + \mid m \mid + 1 \; ; \qquad (15.4.39)$$

remember that these formulas concern the non-zero values for helicity operator $\sigma = \pm i \sqrt{\epsilon^2 - M^2}$.

It remains to specify the energy spectrum for states with $\sigma = 0$ which are determined by the equations

$$\left(-2\Delta + \frac{\partial}{\partial z} \cos^2 z \frac{\partial}{\partial z} + (\epsilon^2 - M^2) \cos^2 z \right) \bar{\varphi} = 0 \; ,$$

$$\bar{\varphi}(r, z) = \bar{\varphi}(r)\bar{\varphi}(z) \; , \quad \frac{1}{\bar{\varphi}(r)} (2\Delta)\bar{\varphi}(r) = \Lambda \; ,$$

$$\frac{1}{\bar{\varphi}(z)} \left(\frac{d}{dz} \cos^2 z \frac{d}{dz} + (\epsilon^2 - M^2) \cos^2 z \right) \bar{\varphi}(z) = \Lambda \; . \qquad (15.4.40)$$

This equation for the variable r has been solved earlier. Next, with the equation for the z variable, we use the substitution $\varphi = \frac{1}{\cos z} f(z)$ reduces to

$$\frac{d^2 f}{dz^2} + \left(\epsilon^2 - M^2 + 1 - \frac{\Lambda}{\cos^2 z} \right) f(z) = 0 \; . \qquad (15.4.41)$$

It coincides with the previous solution:

$$\frac{d^2 \varphi_2}{dz^2} + \left(\epsilon^2 - M^2 - \frac{\Lambda}{\cos^2 z} \right) \varphi_2(z) = 0 \; ,$$

with one formal change

$$\epsilon^2 - M^2 \quad \rightarrow \quad \epsilon^2 - M^2 + 1 \; .$$

Therefore, solutions are written straightforwardly

$$f = \left(\frac{1 + i\tan z}{2}\right)^a \left(\frac{1 - i\tan z}{2}\right)^b F\left(\alpha,\ \beta,\ \gamma;\ \frac{1 + i\tan z}{2}\right), \qquad (15.4.42)$$

where F stands for a hypergeometric function with parameters

$$\alpha = a + b + \frac{1}{2} + \frac{\sqrt{4\Lambda + 1}}{2},$$

$$\beta = a + b + \frac{1}{2} - \frac{\sqrt{4\Lambda + 1}}{2},\ \gamma = (2a + 1). \qquad (15.4.43)$$

The bound states correspond to $a,\ b$, which are defined as

$$a = b = -\frac{\sqrt{\epsilon^2 - M^2 + 1}}{2}.$$

The quantization rule $\alpha = -n_z$ gives

$$+\sqrt{\epsilon^2 - M^2 + 1} = n_z + \frac{1}{2} + \frac{\sqrt{4\Lambda + 1}}{2} > 0. \qquad (15.4.44)$$

Thus, we can allow for quantization for Λ to get the formulas for the energy levels

$$+\sqrt{\epsilon^2 - M^2 + 1} = n_z + n_r + \mid m \mid +1; \qquad (15.4.45)$$

This refers to the case of $\sigma = 0$.

Concluding this section we shall summarize our results. The spin 1 particle is investigated in the 3-dimensional curved space for a constant positive curvature. An extended helicity operator is defined and the variables are separated in a tetrad-based 10-dimensional Duffin–Kemmer equation in a quasi-cylindrical coordinates. The problem is solved exactly in hypergeometric functions, the obtained energy spectrum is determined by three discrete quantum numbers. Transition to a massless case of electromagnetic field is performed. The given problem presents some interest for an exactly solvable model that describes composite systems (particles) of spin 1 or for electromagnetic fields in the non-trivial space-time background. This model presents a finite 3-dimensional box.

Chapter 16

Spin 1 Particle in the Coulomb Field

16.1 Introduction and Separation of the Variables

Many years ago, a very peculiar behavior of a spin 1 particle in presence of the external Coulomb field was noticed by Tamm [432]. As far as we have known about this sort situation with this type of system, much has remained the same. In the present paper, we start examining the problem on the basis of the matrix Duffin–Kemmer–Petiau formalism with the use of the tetrad generally covariant tetrad technique. This turns out to be more convenient than a common Proca tensor approach. In choosing a diagonal spherical tetrad according to

$$dS^2 = dt^2 - dr^2 - r^2(d\theta^2 + \sin^2\theta d\phi^2) \,,$$

$$e^{\alpha}_{(0)} = (1, 0, 0, 0) \,, \qquad e^{\alpha}_{(3)} = (0, 1, 0, 0) \,,$$

$$e^{\alpha}_{(1)} = (0, 0, \frac{1}{r}, 0) \,, \quad e^{\alpha}_{(2)} = (1, 0, 0, \frac{1}{r\sin\theta}) \,, \tag{16.1.1}$$

we reduce the main Duffin–Kemmer–Petiau stationary equation to the form

$$\left[\beta^0(\epsilon + \frac{\alpha}{r}) + i \, [\beta^3\partial_r + \frac{1}{r}(\beta^1 j^{31} + \beta^2 j^{32})] + \frac{1}{r}\Sigma_{\theta,\phi} - M \right] \Phi(x) = 0 \,, \tag{16.1.2}$$

where $\epsilon = E/(c\hbar), \alpha = e^2/(c\hbar), M = mc/\hbar$ and $\Sigma_{\theta,\phi}$ stands for an angular operator (its form means that we have here a generalized Schrödinger–Pauli basis [430, 431])

$$\Sigma_{\theta,\phi} = i \, \beta^1 \partial_\theta + \beta^2 \frac{i\partial + i \, j^{12}\cos\theta}{\sin\theta} \,.$$

Spherical waves with (j, m) quantum numbers should be constructed within the following general substitution

$$\Psi(x) = \{ \, \Phi_0(x), \, \vec{\Phi}(x), \, \mathbf{E}(x), \mathbf{H}(x) \, \} \,,$$

$$\Phi_0(x) = e^{-iEt/\hbar} \, \Phi_0(r) \, D_0 \,, \qquad \vec{\Phi}(x) = e^{-iEt/\hbar} \begin{vmatrix} \Phi_1(r) \, D_{-1} \\ \Phi_2(r) \, D_0 \\ \Phi_3(r) \, D_{+1} \end{vmatrix} \,,$$

$$\mathbf{E}(x) = e^{-iEt/\hbar} \begin{vmatrix} E_1(r)\, D_{-1} \\ E_2(r)\, D_0 \\ E_3(r)\, D_{+1} \end{vmatrix}, \quad \mathbf{H}(x) = e^{-iEt/\hbar} \begin{vmatrix} H_1(r)\, D_{-1} \\ H_2(r)\, D_0 \\ H_3(r)\, D_{+1} \end{vmatrix}; \qquad (16.1.3)$$

short notation for Wigner functions is used: $D_\sigma = D^j_{-m,\sigma}(\phi,\ \theta,\ 0)$, $\sigma = 0, +1, -1$. In accordance with the Pauli approach [431] the quantum number j takes values 0,1,2, ... Using recurrent formulas [429]

$$\nu = \sqrt{j(j+1)}, \qquad a = \sqrt{(j-1)(j+2)},$$

$$\partial_\theta D_{-1} = \frac{1}{2}(a\, D_{-2} - \nu\, D_0), \qquad \frac{m-\cos\theta}{\sin\theta} D_{-1} = \frac{1}{2}(a\, D_{-2} + \nu\, D_0),$$

$$\partial_\theta D_0 = \frac{1}{2}(\nu\, D_{-1} - \nu\, D_{+1}), \qquad \frac{m}{\sin\theta} D_0 = \frac{1}{2}(\nu\, D_{-1} + \nu\, D_{+1}),$$

$$\partial_\theta D_{+1} = \frac{1}{2}(\nu\, D_0 - a\, D_{+2}), \qquad \frac{m+\cos\theta}{\sin\theta} D_{+1} = \frac{1}{2}(\nu\, D_0 + a\, D_{+2}),$$

$$(16.1.4)$$

after a simple algebraic calculation, we arrive at the radial equations (for clarity, this corresponds to the Proca tensor equations and are written down as well)

$$D^b \Phi_{ab} = M\,\Phi_a,$$

$$-(\frac{d}{dr} + \frac{2}{r})\, E_2 - \frac{\nu}{r}(E_1 + E_3) = M\Phi_0,$$

$$+i(\epsilon + \frac{\alpha}{r})\, E_1 + i(\frac{d}{dr} + \frac{1}{r})\, H_1 + i\frac{\nu}{r}\, H_2 = M\Phi_1,$$

$$+i(\epsilon + \frac{\alpha}{r})\, E_2 - i\frac{\nu}{r}(H_1 - H_3) = M\Phi_2,$$

$$+i(\epsilon + \frac{\alpha}{r})\, E_3 - i(\frac{d}{dr} + \frac{1}{r})\, H_3 - i\frac{\nu}{r}\, H_2 = M\,\Phi_3; \qquad (16.1.5)$$

$$D_a\Phi_b - D_b\Phi_a = M\,\Phi_{ab},$$

$$-i(\epsilon + \frac{\alpha}{r})\, \Phi_1 + \frac{\nu}{r}\, \Phi_0 - ME_1 = 0,$$

$$-i(\epsilon + \frac{\alpha}{r})\, \Phi_2 - \frac{d}{dr}\, \Phi_0 - ME_2 = 0,$$

$$-i(\epsilon + \frac{\alpha}{r})\, \Phi_3 + \frac{\nu}{r}\, \Phi_0 - ME_3 = 0,$$

$$-i(\frac{d}{dr} + \frac{1}{r})\, \Phi_1 - i\frac{\nu}{r}\, \Phi_2 - MH_1 = 0,$$

$$+i\frac{\nu}{r}(\Phi_1 - \Phi_3) - MH_2 = 0,$$

$$+i(\frac{d}{dr} + \frac{1}{r})\, \Phi_3 + i\frac{\nu}{r}\Phi_2 - MH_3 = 0, \qquad (16.1.6)$$

where $\nu = \sqrt{j(j+1)/2}$ (note the factor $1/\sqrt{2}$).

Concurrently, with $\mathbf{J}^2$, J_3 let us diagonalize an operator of spacial inversion $\hat{\Pi}$. After transition to spherical tetrad basis, and also to cyclic representation for DKP-matrices β^a, for this discrete operator we get

$$\hat{\Pi} = \begin{vmatrix} 1 & 0 & 0 & 0 \\ 0 & \Pi_3 & 0 & 0 \\ 0 & 0 & \Pi_3 & 0 \\ 0 & 0 & 0 & -\Pi_3 \end{vmatrix} \hat{P} , \qquad \Pi_3 = \begin{vmatrix} 0 & 0 & -1 \\ 0 & -1 & 0 \\ -1 & 0 & 0 \end{vmatrix} . \qquad (16.1.7)$$

Eigenvalue equation $\hat{\Pi}\Psi = P\,\Psi$ results in two different in parity states

$$P = (-1)^{j+1}, \qquad \Phi_0 = 0 , \;\; \Phi_3 = -\Phi_1 , \;\; \Phi_2 = 0 ,$$

$$E_3 = -E_1 , \;\; E_2 = 0, \; H_3 = H_1 ; \qquad (16.1.8)$$

$$P = (-1)^j , \qquad \Phi_3 = \Phi_1 , \; E_3 = +E_1 , \; H_3 = -H_1 , \;\; H_2 = 0 . \qquad (16.1.9)$$

Correspondingly, Eqs. (16.1.5) and (16.1.6) give us 4 and 6 equations:

$P = (-1)^{j+1}$,

$$+i(\epsilon + \frac{\alpha}{r})\, E_1 + i(\frac{d}{dr} + \frac{1}{r})H_1 + i\frac{\nu}{r}H_2 = M\Phi_1 ,$$

$$-i(\epsilon + \frac{\alpha}{r})\,\Phi_1 = ME_1 ,$$

$$-i(\frac{d}{dr} + \frac{1}{r})\Phi_1 = MH_1 , \quad 2i\frac{\nu}{r}\Phi_1 = MH_2 , \qquad (16.1.10)$$

excluding E_1, H_1, H_2, we get a second order differential equation for Φ_1

$$\left[\frac{d^2}{dr^2} + \frac{2}{r}\frac{d}{dr} + (\epsilon + \frac{\alpha}{r})^2 - M^2 - \frac{j(j+1)}{r^2} \right] \Phi_1 = 0 , \qquad (16.1.11a)$$

which coincides with the case of a scalar particle in Coulomb potential (it is a known fact that was noted in Tamm's first paper) [432]. Its solution is well known and provides us with the following energy spectrum (in usual units)

$$E = \frac{mc^2}{\sqrt{1 + \alpha^2/N^2}} , \qquad N = n + \frac{1}{2} + \sqrt{(j + 1/2)^2 - \alpha^2} . \qquad (16.1.11b)$$

For states with parity $P = (-1)^j$, we have the system

$P = (-1)^j$,

$$(\frac{d}{dr} + \frac{2}{r})E_2 + 2\frac{\nu}{r}E_1 + M\Phi_0 = 0 ,$$

$$+i(\epsilon + \frac{\alpha}{r})\, E_1 + i(\frac{d}{dr} + \frac{1}{r})H_1 - M\Phi_1 = 0 ,$$

$$+i(\epsilon + \frac{\alpha}{r})E_2 - 2i\frac{\nu}{r}H_1 - M\Phi_2 = 0\,,$$

$$-i(\epsilon + \frac{\alpha}{r})\,\Phi_1 + \frac{\nu}{r}\Phi_0 - ME_1 = 0\,,$$

$$i(\epsilon + \frac{\alpha}{r})\Phi_2 + \frac{d}{dr}\Phi_0 + ME_2 = 0\,,$$

$$i(\frac{d}{dr} + \frac{1}{r})\Phi_1 + i\frac{\nu}{r}\Phi_2 + MH_1 = 0\,. \tag{16.1.12}$$

16.2   The case of Minimal Value $j = 0$

The states with minimal value $j = 0$ can be treated straightforwardly. In this case, we should start with special substitution of the wave function

$$\Phi_0(x) = e^{-iEt/\hbar}\Phi_0(r)\,, \qquad \vec{\Phi}(x) = e^{-i\epsilon t}\begin{vmatrix} 0 \\ \Phi_2(r) \\ 0 \end{vmatrix},$$

$$\mathbf{E}(x) = e^{-iEt/\hbar}\begin{vmatrix} 0 \\ E_2(r) \\ 0 \end{vmatrix}, \qquad \mathbf{H}(x) = e\begin{vmatrix} 0 \\ H_2(r) \\ 0 \end{vmatrix}. \tag{16.2.1}$$

The $\Sigma_{\theta,\phi}$ acts on this function as a zero operator, and parity $P = (-1)^{0+1} = -1$. Now, the radial system is (for eliminating imaginary unit i, we use slightly different variables): $\Phi_0 = \varphi_0$, $-i\Phi_1 = \varphi_1$, $-i\Phi_2 = \varphi_2$

$$H_2 = 0\,, \qquad -(\frac{d}{dr} + \frac{2}{r})E_2 = M\varphi_0\,,$$

$$(\epsilon + \frac{\alpha}{r})E_2 = M\varphi_2\,, \qquad (\epsilon + \frac{\alpha}{r})\varphi_2 - \frac{d}{dr}\varphi_0 = ME_2\,. \tag{16.2.2}$$

From this, it follows the second order equation for E_2

$$\left[\frac{d^2}{dr^2} + \frac{2}{r}\frac{d}{dr} - \frac{2}{r^2} + (\epsilon + \frac{\alpha}{r})^2 - M^2\right]E_2 = 0\,. \tag{16.2.3}$$

By the substitution $E_2(r) = r^{-1}f(r)$, we get

$$\frac{d^2}{dr^2}f + \left(\epsilon^2 - M^2 + \frac{2\alpha\epsilon}{r} - \frac{2-\alpha^2}{r^2}\right)f = 0\,. \tag{16.2.4}$$

In the dimensionless variables

$$x = r\epsilon = \frac{rE}{c\hbar}\,, \quad \frac{M^2}{\epsilon^2} = \frac{m^2c^4}{E^2} = \lambda^2\,, \tag{16.2.5}$$

it will read as

$$\frac{d^2}{dx^2}f + (1 - \lambda^2 + \frac{2\alpha}{x} - \frac{2-\alpha^2}{x^2})f = 0\,. \tag{16.2.6}$$

Using the substitution $f(x) = x^a e^{-bx} F(x)$, for F, we obtain

$$xF'' + (2a - 2bx)F' + [\ \frac{a(a-1) + \alpha^2 - 2}{x} + (b^2 + 1 - \lambda^2)x + (2\alpha - 2ab)\ ]F = 0\ .$$

Requiring

$$a = \frac{1 \pm \sqrt{9 - 4\alpha^2}}{2}\ ,\quad b = \pm\sqrt{\lambda^2 - 1} = \pm\frac{\sqrt{m^2 c^4 - E^2}}{E}\ , \tag{16.2.7}$$

the choice of upper signs in the formulas provides us with good parameters for bound states, we get

$$x\,F'' + 2(a - bx)\,F' + 2(\alpha - ab)\,F = 0\ . \tag{16.2.8}$$

This equation is solved in confluent hypergeometric functions. Let us specify this solution in detail. Where we shall use an expansion for $F(x)$

$$F(x) = \sum_{k=0}^{\infty} C_k\,x^k\ ,\quad F' = \sum_{k=1}^{\infty} kC_k\,x^{k-1}\ ,\quad F'' = \sum_{k=2}^{\infty} k(k-1)C_k\,x^{k-2}\ ,$$

we get

$$\sum_{k=2}^{\infty} k(k-1)C_k\,x^{k-1} + 2a\sum_{k=1}^{\infty} kC_k\,x^{k-1} - 2b\sum_{k=1}^{\infty} kC_k\,x^k + 2(\alpha - ab)\sum_{k=0}^{\infty} c_k\,x^k = 0\ ,$$

from this, it follows

$$[\ 2aC_1 + 2(\alpha - ab)\ ]x^0\ +\ [\ 2C_2 + 2a\,2C_2 - 2b\,C_1 + 2(\alpha - ab)C_1\ ]\,x$$

$$+ \sum_{n=2}^{\infty} [\ n(n+1)C_{n+1} + 2a(n+1)C_{n+1} - 2bnC_n + 2(\alpha - ab)C_n\ ]\,x^n = 0\ .$$

Therefore, we arrive at recurrent formulas

$$C_1 = -(\alpha - ab)\,C_0 = 0\ ,$$

$$C_2\,2\,(1 + 2a) = 2\,[b - (\alpha - ab)]\,C_1 = 0,\ n = 2, 3, 4, ...,$$

$$C_{n+1}\,(n+1)\,(n+2a) = 2\,[n\,b\ -\ (\alpha - ab)\,]\,C_n = 0\ . \tag{16.2.9}$$

To get the polynomial, we require

$$C_{N+1} = 0 \implies [\ N\,b\ -\ (\alpha - ab)\ ] = 0\ .$$

This quantization rule gives

$$\frac{\alpha - ab}{b} = N\ ,\quad a = \frac{1 + \sqrt{9 - 4\alpha^2}}{2}\ ,\quad b = \frac{\sqrt{m^2 c^4 - E^2}}{E}\ , \tag{16.2.10}$$

so that

$$\frac{2\alpha\epsilon - (1 + \sqrt{9 - 4\alpha^2})\sqrt{m^2 c^4 - E^2}}{2\sqrt{m^2 c^4 - E^2}} = N\ ; \tag{16.2.11}$$

its solution is

$$E = \frac{mc^2}{\sqrt{1 + \alpha^2/(\Gamma + N)^2}}\ ,\quad 2\Gamma = 1 + \sqrt{9 - 4\alpha^2}\ . \tag{16.2.12}$$

16.3   Lorentz Gauge in Presence of Coulomb Potential

As known for a massive spin 1 particle there must exist a generalized Lorentz condition. Let us specify it for this problem under consideration. In the Proca tensor equations for the vector particle

$$D_\alpha \, \Psi_\beta - D_\beta \, \Psi_\alpha = M \, \Psi_{\alpha\beta} \, , \qquad D^\alpha \, \Psi_{\alpha\beta} = M \, \Psi_\beta \, , \qquad (16.3.1)$$

where $D_\alpha = \nabla_\alpha + i(e/c)A_\alpha$. Acting by the operator D_α on the second equation in Eq. (16.3.1), we get

$$(\nabla_\alpha + i\frac{e}{c}A_\alpha) \, \Psi^\alpha = \frac{ie}{2cM} \, F_{\alpha\beta} \, \Psi^{\alpha\beta} \, . \qquad (16.3.2)$$

This Lorentz condition should be translated to tetrad form. To this end, instead of Ψ^α and $\Psi^{\alpha\beta}$ one should introduced their tetrad components

$$\Psi^\alpha = e^{(a)\alpha} \, \Psi_{(a)} \, , \quad \Psi^{\alpha\beta} = e^{(a)\alpha} e^{(b)\beta} \, \Psi_{(a)(b)} \, .$$

Correspondingly, Eq. (16.3.2) takes the form

$$(e^{(a)\alpha}_{;\alpha} \, \Psi_a + e^{(a)\alpha} \, \partial_\alpha) \, \Psi_{(a)}$$

$$+i\frac{e}{c}A^{(a)} \, \Psi_{(a)} = i \, \frac{e}{2cM} \, F^{(a)(b)} \, \Psi_{(a)(b)} \, . \qquad (16.3.3)$$

The Coulomb field $A_0 = e/r$, $F_{r0} = -e/r^2$ in the tetrad description looks

$$A^{(0)} = e^{(0)0} A_0 = \frac{e}{r} \, , \qquad F^{30} = e^{(3)r}e^{(0)0} \, F_{r0} = -\frac{e}{r^2} \, . \qquad (16.3.4)$$

After a simple calculation, we get

$$e^{(0)\alpha}_{;\alpha} = 0 \, , \qquad e^{(1)\alpha}_{;\alpha} = -\frac{\cos\theta}{r\sin\theta} \, , \qquad e^{(2)\alpha}_{;\alpha} = 0 \, , \qquad e^{(3)A}_{;\alpha} = -\frac{2}{r} \, . \qquad (16.3.5)$$

The components functions $\Psi_{(a)}$ and $\Psi_{(a)(b)}$ in Eq. (16.3.3) can be related with components of the DKP-column as follows (transition between cyclic and Cartesian representations; $c \equiv 1/\sqrt{2}$)

$$
\begin{vmatrix}
\Psi_{(0)} \\
\Psi_{(1)} \\
\Psi_{(2)} \\
\Psi_{(3)} \\
\Psi_{(0)(1)} \\
\Psi_{(0)(2)} \\
\Psi_{(0)(3)} \\
\Psi_{(2)(3)} \\
\Psi_{(3)(1)} \\
\Psi_{(1)(2)}
\end{vmatrix}
=
\begin{vmatrix}
1 & 0 & 0 & 0 & 0 & 0 & 0 & 0 & 0 & 0 \\
0 & -c & 0 & +c & 0 & 0 & 0 & 0 & 0 & 0 \\
0 & -ic & 0 & -ic & 0 & 0 & 0 & 0 & 0 & 0 \\
0 & 0 & 1 & 0 & 0 & 0 & 0 & 0 & 0 & 0 \\
0 & 0 & 0 & 0 & -c & 0 & +c & 0 & 0 & 0 \\
0 & 0 & 0 & 0 & -ic & 0 & -ic & 0 & 0 & 0 \\
0 & 0 & 0 & 0 & 0 & 1 & 0 & 0 & 0 & 0 \\
0 & 0 & 0 & 0 & 0 & 0 & 0 & -c & 0 & +c \\
0 & 0 & 0 & 0 & 0 & 0 & 0 & -ic & 0 & -ic \\
0 & 0 & 0 & 0 & 0 & 0 & 0 & 0 & 1 & 0
\end{vmatrix}
=
\begin{vmatrix}
\Phi_0 \, D_0 \\
\Phi_1 \, D_{-1} \\
\Phi_2 \, D_0 \\
\Phi_3 \, D_{+1} \\
E_1 \, D_{-1} \\
E_2 D_0 \\
E_3 \, D_{+1} \\
H_1 \, D_{-1} \\
H_2 D_0 \\
H_3 \, D_{+1}
\end{vmatrix}
\, .
$$

We need only Φ_0, $\Psi_{(1)}$, $\Psi_{(2)}$, $\Psi_{(3)}$, $\Psi_{(0)(3)}$:

$$\Psi_{(1)} = e^{-iEt/\hbar}\frac{1}{\sqrt{2}} (-\Phi_1 \, D_{-1} + \Phi_3 \, D_{+1}) \, ,$$

$$\Psi_{(2)} = e^{-iEt/\hbar} \frac{i}{\sqrt{2}} \left(-\Phi_1 \, D_{-1} - \Phi_3 \, D_{+1}\right),$$

$$\Psi_{(0)} = e^{-iEt/\hbar} \Phi_0 \, D_0, \qquad \Psi_{(3)} = e^{-iEt/\hbar} \Phi_2 \, D_0,$$

$$\Psi_{(0)(3)} = e^{-iEt/\hbar} \, E_2 \, D_0. \tag{16.3.6}$$

Through the help of Eq. (16.3.6), Eq. (16.3.3) gives

$$\frac{1}{\sqrt{2}} \, r \, \Phi_1 \left(\partial_\theta \, D_{-1} - \frac{M - \cos\theta}{\sin\theta} \, D_{-1}\right)$$

$$-\frac{1}{\sqrt{2}} \, r \, \Phi_3 \left(\partial_\theta \, D_{+1} + \frac{M + \cos\theta}{\sin\theta} \, D_{+1}\right)$$

$$+D_0 \left(-\frac{2}{r} \, \Phi_2 \, - \, i\epsilon \, \Phi_0 \, - \, \frac{d}{dr} \, \Phi_2\right) = i \, \frac{\alpha}{2Mr^2} \, E_2 D_0.$$

Now, by taking into account the recurrent formulas [429]

$$\partial_\theta \, D_{-1} - \frac{M - \cos\theta}{\sin\theta} \, D_{-1} = -\sqrt{(j+1)j} \, D_0,$$

$$\partial_\theta \, D_{+1} - \frac{M + \cos\theta}{\sin\theta} \, D_{+1} = -\sqrt{(j+1)j} \, D_0,$$

we arrive at the Lorentz condition in radial form

$$-i \, \epsilon \, \Phi_0 \, - \, \left(\frac{d}{dr} + \frac{2}{r}\right) \Phi_2 \, - \, \frac{\nu}{r} \, (\Phi_1 + \Phi_3) = \frac{i\alpha}{2Mr^2} \, E_2. \tag{16.3.7}$$

For states with parity $P = (-1)^{j+1}$, this condition is satisfied identically. For states with parity $P = (-1)^j$, it gives a more simple relation

$$-i \, \epsilon \, \Phi_0 \, - \, \left(\frac{d}{dr} + \frac{2}{r}\right) \Phi_2 \, - \, \frac{2\nu}{r} \, \Phi_1 = \frac{i\alpha}{2Mr^2} \, E_2. \tag{16.3.8}$$

From Eq. (16.3.8), a very important relationship can be established. To this end, from Eq. (16.3.8) let us exclude Φ_2 with the help of the third equation in Eq. (16.1.12)

$$i \, \epsilon M \, \Phi_0 + i\left(\epsilon + \frac{\alpha}{r}\right)\left(\frac{d}{dr} + \frac{2}{r}\right) E_2$$

$$-\frac{2i\nu}{r}\left(\frac{d}{dr} + \frac{1}{r}\right) H_1 + \frac{2\nu M}{r} \, \Phi_1 = i \, \frac{\alpha}{2r^2} \, E_2.$$

From transforming the second and third terms with the help of 1-st and 2-nd equations in the system of Eq. (16.1.12), we arrive at

$$E_2 = -2Mr \, \Phi_0. \tag{16.3.9}$$

16.4   The Main Function of the First Type, the Confluent Heun Equation

Let us examine the Eqs. (16.1.12), now with the use of the Lorentz condition in Eq. (16.3.8) and its consequence in Eq. (16.3.9). From the 1-st equation in Eq. (16.1.12), we produce

$$E_1 = \frac{Mr}{2\nu}(5 + 2r\frac{d}{dr})\Phi_0 . \qquad (16.4.1)$$

By the help of Eq. (16.3.9), the fourth equation in Eq. (16.1.12) gives

$$\Phi_1 = \frac{-i}{\epsilon + \alpha/r}\left(\frac{\nu}{r} - \frac{5M^2}{2\nu}r - \frac{r^2M^2}{\nu}\frac{d}{dr}\right)\Phi_0 . \qquad (16.4.2)$$

Using the help of Eq. (16.3.9), the fifth equation in Eq. (16.1.12) gives

$$\Phi_2 = \frac{i}{\epsilon + \alpha/r}\left(\frac{d}{dr} - 2M^2r\right)\Phi_0 . \qquad (16.4.3)$$

The sixth equation in Eq. (16.1.12) provides us with the representation for H_1 in term of Φ_0 by means of second order differential operator

$$-MH_1 = \left[(\frac{d}{dr} + \frac{1}{r})\frac{1}{\epsilon + \alpha/r}\left(\frac{\nu}{r} - \frac{5M^2}{2\nu}r - \frac{r^2M^2}{\nu}\frac{d}{dr}\right)\right.$$
$$\left. -\frac{\nu}{r}\frac{1}{\epsilon + \alpha/r}\left(\frac{d}{dr} - 2M^2r\right)\right]\Phi_0. \qquad (16.4.4)$$

Recall that from the third equation in Eq. (16.1.12) one can obtain another representation for H_1 that uses only the first order operator

$$H_1 = -\frac{Mr}{2\nu}\frac{1}{\epsilon + \alpha/r}\left[\frac{d}{dr} + 2r\left[(\epsilon + \frac{\alpha}{r})^2 - M^2\right]\right]\Phi_0 . \qquad (16.4.5)$$

These two representations for H_1 must be consistent with each other.

Now, let us focus on the Lorentz condition

$$-i\,\epsilon\,\Phi_0 - (\frac{d}{dr} + \frac{2}{r})\,\Phi_2 - \frac{2\nu}{r}\,\Phi_1 = i\,\frac{\alpha}{2Mr^2}\,E_2 , \qquad (16.4.6)$$

one can readily derive a second order differential equation for Φ_0:

$$\frac{d^2\Phi_0}{dr^2} + \frac{1}{r}\left(3 - \frac{\epsilon}{\epsilon + \alpha/r}\right)\frac{d\Phi_0}{dr}$$
$$+ \left(\epsilon^2 - \frac{\alpha^2}{r^2} - 3M^2 + 2M^2\frac{\epsilon}{\epsilon + \alpha/r} - \frac{2\nu^2}{r^2}\right)\Phi_0 = 0 . \qquad (16.4.7)$$

The function Φ_0 will be termed as a main function, the remaining ones, $E_2, E_1, \Phi_1, \Phi_2, H_1$, are expressed through it

$$E_2 = -2Mr\,\Phi_0 ,$$

$$E_1 = \frac{Mr}{2\nu}(5 + 2r\frac{d}{dr})\Phi_0 \, , \qquad 2$$

$$\Phi_1 = \frac{-i}{\epsilon + \alpha/r}\left(\frac{\nu}{r} - \frac{5M^2}{2\nu}r - \frac{r^2M^2}{\nu}\frac{d}{dr}\right)\Phi_0 \, , \qquad 3$$

$$\Phi_2 = \frac{i}{\epsilon + \alpha/r}\left(\frac{d}{dr} - 2M^2 r\right)\Phi_0 \, , \qquad 4$$

$$H_1 = -\frac{Mr}{2\nu}\frac{1}{\epsilon + \alpha/r}\left[\frac{d}{dr} + 2r\left[(\epsilon + \frac{\alpha}{r})^2 - M^2\right]\right]\Phi_0 \, . \qquad 5$$

Changing the variable

$$x = -\frac{\epsilon}{\alpha}r \, < 0 \, , \qquad r = -\frac{\alpha}{\epsilon}x \, ; \qquad (16.4.8)$$

Eq. (16.4.7) is reduced to the form

$$\frac{d^2\Phi_0}{dx^2} + \left(\frac{3}{x} - \frac{1}{x-1}\right)\frac{d\Phi_0}{dx} + \left(\alpha^2 - \Lambda^2 - \frac{\alpha^2 + 2\nu^2}{x^2} + \frac{2\Lambda^2}{x-1}\right)\Phi_0 = 0 \, , \quad (16.4.9)$$

where dimensionless parameters are used

$$\Lambda^2 = \alpha^2\,\lambda^2 \, , \qquad \lambda = \frac{mc^2}{E} > 1 \, .$$

Let us consider the behavior of the main function near the point $x = 0$:

$$\frac{d^2\Phi_0}{dx^2} + \frac{3}{x}\frac{d\Phi_0}{dx} - \frac{\alpha^2 + 2\nu^2}{x^2}\Phi_0 = 0 \, ,$$

$$\Phi_0 \sim \text{const } x^A, \qquad A(A-1) + 3A - \alpha^2 - 2\nu^2 = 0 \, ,$$

$$A = -1 - \sqrt{1 + \alpha^2 + 2\nu^2} \, , \quad A = -1 + \sqrt{1 + \alpha^2 + 2\nu^2} \, ; \qquad (16.4.10)$$

to bound states there correspond positive values of A. In the region near $x = +\infty$, the main equation gives

$$\frac{d^2\Phi_0}{dx^2} + \frac{2}{x}\frac{d\Phi_0}{dx} + \left(\alpha^2 - \Lambda^2\right)\Phi_0 = 0 \, ,$$

$$\Phi_0 = e^{+\sqrt{\Lambda^2 - \alpha^2}\,x} = e^{-\sqrt{m^2c^4 - E^2}\,r/\hbar c} \, ; \qquad (16.4.11)$$

to bound states correspond solutions that are vanishing at infinity.

Now, let us introduce substitution $\Phi_0(x) = x^A e^{Bx} f(x)$, Eq. (16.4.9) gives

$$\frac{d^2 f}{dx^2} + \left[2B + \frac{2A+3}{x} + \frac{1}{1-x}\right]\frac{df}{dx} + \left[B^2 + \alpha^2 - \Lambda^2 + \frac{2AB + A + 3B}{x}\right.$$

$$\left. + \frac{A(A-1) + 3A - \alpha^2 - 2\nu^2}{x^2} + \frac{A + B - 2\Lambda^2}{1-x}\right]f(x) = 0 \, . \qquad (16.4.12)$$

In the restrictions on A and B:

$$A(A-1) + 3A - \alpha^2 - 2\nu^2 = 0 \implies A = -1 + \sqrt{1 + 2\nu^2 + \alpha^2} \, ,$$

$$B^2 + \alpha^2 - \Lambda^2 = 0 \quad \Longrightarrow \quad B = +\sqrt{\Lambda^2 - \alpha^2} , \qquad (16.4.13)$$

Eq. (16.4.12) takes the form

$$\frac{d^2 f}{dx^2} + \left[2B + \frac{2A+3}{x} - \frac{1}{x-1} \right] \frac{df}{dx}$$

$$+ \left[\frac{2AB + A + 3B}{x} + \frac{A + B - 2\Lambda^2}{1-x} \right] f = 0 . \qquad (16.4.14)$$

It can be recognized as the confluent Heun equation

$$f = f(a, b, c, d, h; z) , \qquad \frac{d^2 f}{dx^2} + \left(a + \frac{b+1}{x} + \frac{c+1}{x-1} \right) \frac{df}{dx}$$

$$- \frac{[-2d + a(-b-c-2)]x + a(1+b) + b(-1-c) - c - 2h}{2x(x-1)} f = 0 \qquad (16.4.15)$$

with parameters given by

$$a = +2\sqrt{\Lambda^2 - \alpha^2} , \qquad b = +2\sqrt{1 + 2\nu^2 + \alpha^2} ,$$

$$c = -2 , \qquad d = 2\Lambda^2 , \qquad h = +2 . \qquad (16.4.16)$$

One of the known conditions for polynomial solutions is

$$d = -a \left(n + \frac{b+c+2}{2} \right) , \qquad (16.4.17)$$

This will give the following quantization rule

$$\frac{\Lambda^4}{\Lambda^2 - \alpha^2} = (n + \sqrt{1 + 2\nu^2 + \alpha^2})^2 . \qquad (16.4.18)$$

Its solution provides us with energy spectrum

$$E^2 = M^2 c^4 \frac{2\alpha^2}{N^2 - \sqrt{N^4 - 4\alpha^2 N^2}} . \qquad (16.4.19)$$

When N increases to infinity, we get ($N \to \infty$)

$$E^2 = M^2 c^4 \frac{2\alpha^2}{N^2 - N^2\sqrt{1 - 4\alpha^2 N^{-2}}} \approx M^2 c^4 \frac{2\alpha^2}{2\alpha^2} = M^2 c^4 . \qquad (16.4.20)$$

To obtain a nonrelativistic approximation, one must impose a special restriction

$$N = n + \sqrt{1 + 2\nu^2 + \alpha^2} = n + \sqrt{1 + j(j+1) + \alpha^2} \approx n + j , \qquad (16.4.21)$$

which correlates with the known nonrelativistic procedure

$$M + \epsilon + \frac{\alpha}{r} \approx M + \epsilon .$$

In taking into account Eq. (16.4.21), one can derive

$$E^2 = M^2 c^4 \frac{1}{2} \left(1 + \sqrt{1 - \frac{4\alpha^2}{N^2}}\right) \approx M^2 c^4 \left(1 - \frac{\alpha^2}{N^2}\right), \qquad N \approx n + j \, ,$$

that is

$$E = Mc^2 \left(1 - \frac{\alpha^2}{2N^2}\right) = Mc^2 + E' \, . \tag{16.4.22}$$

Thus, the nonrelativistic energy levels are given by

$$E' = -\frac{\alpha^2 M c^2}{2N^2} = -\frac{M e^4}{\hbar^2 N^2} \, , \tag{16.4.23}$$

which coincides with the known exact result (described later in the chapter **16**).

16.5   The Main Radial Function of the Second Type

From general considerations, we may expect two linearly independent solutions for 6-equation system for state with parity $P = (-1)^j$. The Eq. (16.4.6) provides us with only one class. This has made us look for another class (possibly with some different main function).

In this connection, let us turn back to Eq. (16.4.5) multiplied by $-M$ and compare it with Eq. (16.4.4), from that it follows a second order differential equation Φ_0, which is different from Eq. (16.4.7):

$$\frac{d^2 \Phi_0}{dr^2} + \frac{1}{r} \left(6 + \frac{\alpha}{r(\epsilon + \alpha/r)}\right) \frac{d\Phi_0}{dr} + \left[\epsilon^2 - M^2 + \frac{2\epsilon^2 \alpha}{\epsilon r + \alpha} - \frac{\alpha \nu^2}{r^4 M^2 (\epsilon r + \alpha)}\right.$$

$$\left. - \frac{1}{2} \frac{\alpha(-15 + 4\nu^2 - 2\alpha^2)}{r^2(\epsilon r + \alpha)} - \frac{\epsilon(-5 + 2\nu^2 - 3\alpha^2)}{r(\epsilon r + \alpha)}\right] \Phi_0 = 0 \, . \tag{16.5.1}$$

In the variable x, will read as

$$\frac{d^2}{dx^2} \Phi_0 + \frac{1}{x} \left(6 - \frac{x}{x - 1}\right) \frac{d}{dx} \Phi_0 + \left[(1 - \lambda^2)\alpha^2 - \frac{2\alpha^2}{x - 1} + \frac{\nu^2}{\alpha^2 \lambda^2 x^4 (x - 1)}\right.$$

$$\left. + \frac{(-15 + 4\nu^2 - 2\alpha^2)}{2x^2(x - 1)} - \frac{(5 + 2\nu^2 - 3\alpha^2)}{x(x - 1)}\right] \Phi_0 = 0 \, . \tag{16.5.2}$$

By means of the coordinate transformation $y = x^{-1}$, Eq. (16.5.2) becomes

$$\frac{d^2 \Phi_0}{dy^2} + \frac{4y - 3}{y(1 - y)} \frac{d\Phi_0}{dy} + \left[\frac{(1 - \lambda^2)\alpha^2}{y^4} - \frac{2\alpha^2}{y^3(1 - y)} + \frac{\nu^2 y}{\alpha^2 \lambda^2 (1 - y)}\right.$$

$$\left. - \frac{(15 - 4\nu^2 + 2\alpha^2)}{2y(1 - y)} - \frac{(5 + 2\nu^2 - 3\alpha^2)}{y^2(1 - y)}\right] \Phi_0 = 0 \, . \tag{16.5.3}$$

Both differential equations, Eqs. (16.5.2) and (16.5.3) are very complex. We might expect that they can describe some third class of solutions; however, any proof of this does not exist now.

16.6   Nonrelativistic Approximation, Exact Energy Spectrum

First, let us consider the simpler system for states with parity $P = (-1)^{j+1}$:

$$P = (-1)^{j+1},$$

$$i(\epsilon + \frac{\alpha}{r})E_1 + i(\frac{d}{dr} + \frac{1}{r})H_1 + i\frac{\nu}{r}H_2 = M\Phi_1,$$

$$-i(\epsilon + \frac{\alpha}{r})\Phi_1 = ME_1,$$

$$-i(\frac{d}{dr} + \frac{1}{r})\Phi_1 = MH_1, \quad 2i\frac{\nu}{r}\Phi_1 = MH_2. \tag{16.6.1}$$

Here, the H_1, H_2 represent non-dynamical variables, by excluding these; we obtain

$$i(\epsilon + \frac{\alpha}{r})E_1 + \frac{1}{M}(\frac{d}{dr} + \frac{1}{r})^2\Phi_1 - \frac{2\nu^2}{Mr^2}\Phi_1 = M\Phi_1,$$

$$-i(\epsilon + \frac{\alpha}{r})\Phi_1 = ME_1. \tag{16.6.2}$$

Now, we should make special substitution, to introduce big and small constituents ($\Psi_1(r)$ and $\psi_1(r)$, respectively)

$$\Phi_1 = \Psi_1 + \psi_1, \qquad iE_1 = \Psi_1 - \psi_1; \tag{16.6.3}$$

Correspondingly, Eqs. (16.6.1) take the form

$$(\epsilon + \frac{\alpha}{r})(\Psi_1 - \psi_1) + \frac{1}{M}(\frac{d}{dr} + \frac{1}{r})^2(\Psi_1 + \psi_1) - \frac{2\nu^2}{Mr^2}(\Psi_1 + \psi_1) = M(B_1 + \psi_1),$$

$$(\epsilon + \frac{\alpha}{r})(\Psi_1 + \psi_1) = M(B_1 - \psi_1). \tag{16.6.4}$$

Summing and subtracting these two relations, we get:

$$2(\epsilon + \frac{\alpha}{r})\Psi_1 + \frac{1}{M}(\frac{d}{dr} + \frac{1}{r})^2(\Psi_1 + \psi_1) - \frac{2\nu^2}{Mr^2}(\Psi_1 + \psi_1) = 2MB_1, \tag{16.6.5a}$$

$$-2(\epsilon + \frac{\alpha}{r})\psi_1 + \frac{1}{M}(\frac{d}{dr} + \frac{1}{r})^2(\Psi_1 + \psi_1) - \frac{2\nu^2}{Mr^2}(\Psi_1 + \psi_1) = 2M\psi_1. \tag{16.6.5b}$$

Now, we should separate a rest energy by a formal change $\epsilon \Longrightarrow \epsilon + M$; which results in

$$2(\epsilon + \frac{\alpha}{r})\Psi_1 + \frac{1}{M}(\frac{d}{dr} + \frac{1}{r})^2(\Psi_1 + \psi_1) - \frac{2\nu^2}{Mr^2}(\Psi_1 + \psi_1) = 0,$$

$$-2(\epsilon + \frac{\alpha}{r})\psi_1 + \frac{1}{M}(\frac{d}{dr} + \frac{1}{r})^2(\Psi_1 + \psi_1) - \frac{2\nu^2}{Mr^2}(\Psi_1 + \psi_1) = 4M\psi_1.$$

Thus, we produce an equation for a big $\Psi_1(r)$ and small $\psi_1(r)$ components

$$2(\epsilon + \frac{\alpha}{r})\Psi_1 + \frac{1}{M}(\frac{d}{dr} + \frac{1}{r})^2\Psi_1 - \frac{j(j+1)}{Mr^2}\Psi_1 = 0, \tag{16.6.6a}$$

$$(\frac{d}{dr} + \frac{1}{r})^2 \Psi_1 - \frac{j(j+1)}{r^2} \Psi_1 = 4M^2 \psi_1 \ . \tag{16.6.6b}$$

Equation for the big component can be written as Schrödinger equation for a scalar particle

$$\left[\frac{d^2}{dr^2} + \frac{2}{r}\frac{d}{dr} + 2M(\epsilon + \frac{\alpha}{r}) - \frac{j(j+1)}{r^2} \right] \Psi_1 = 0 \ . \tag{16.6.7}$$

Corresponding nonrelativistic 3-dimensional wave function for states with parity $P = (-1)^{j+1}$ is

$$P = (-1)^{j+1} \ , \qquad \Psi = e^{-iEt/\hbar} \begin{vmatrix} +(\Phi_1 + iE_1) \ D_{-1} \\ 0 \\ -(\Phi_1 + iE_1) \ D_{+1} \end{vmatrix} . \tag{16.6.8}$$

Now, let us consider radial equations for states with opposite parity $P = (-1)^j$

$$-(\frac{d}{dr} + \frac{2}{r})E_2 - 2\frac{\nu}{r}E_1 = M\Phi_0 \ , \qquad +i(\epsilon + \frac{\alpha}{r})E_1 + i(\frac{d}{dr} + \frac{1}{r})H_1 = M\Phi_1 \ ,$$

$$+i(\epsilon + \frac{\alpha}{r})E_2 - 2i\frac{\nu}{r}H_1 = M\Phi_2 \ , \qquad -i(\epsilon + \frac{\alpha}{r})\Phi_1 + \frac{\nu}{r}\Phi_0 = ME_1 \ ,$$

$$-i(\epsilon + \frac{\alpha}{r})\Phi_2 - \frac{d}{dr}\Phi_0 = ME_2 \ , \qquad -i(\frac{d}{dr} + \frac{1}{r})\Phi_1 - i\frac{\nu}{r}\Phi_2 = MH_1 \ .$$

$$\tag{16.6.9}$$

Among the four dynamical functions Φ_1, Φ_2, E_1, E_2 separation of big and small constituents is performed as follows:

$$\Phi_1 = \Psi_1 + \psi_1 \ , \qquad \Phi_2 = \Psi_2 + \psi_2 \ ,$$

$$iE_1 = \Psi_1 - \psi_1 \ , \qquad iE_2 = \Psi_2 - \psi_2 \ ; \tag{16.6.10}$$

The nonrelativistic 3-dimensional wave function for states with parity $P = (-1)^j$ is then defined.

$$P = (-1)^j \ , \qquad \Psi = e^{-iEt/\hbar} \begin{vmatrix} (\Phi_1 + iE_1) \ D_{-1} \\ (\Phi_2 + iE_2) \ D_0 \\ (\Phi_1 + iE_1) \ D_{+1} \end{vmatrix} . \tag{16.6.11}$$

Excluding the non-dynamical variables Φ_0, H_1, we obtain the system for the rest energy which can be taken away as well: $\epsilon \Longrightarrow \epsilon + M$)

$$i(\epsilon + M + \frac{\alpha}{r}) \ E_1 + \frac{1}{M} (\frac{d}{dr} + \frac{1}{r}) \ [\ (\frac{d}{dr} + \frac{1}{r})\Phi_1 + \frac{\nu}{r}\Phi_2 \ ] \ = M\Phi_1 \ ,$$

$$i(\epsilon + M + \frac{\alpha}{r})E_2 - \frac{2\nu}{Mr} \ [\ (\frac{d}{dr} + \frac{1}{r})\Phi_1 + \frac{\nu}{r}\Phi_2 \ ] \ = M\Phi_2 \ ,$$

$$-i(\epsilon + M + \frac{\alpha}{r})\Phi_1 + \frac{\nu}{Mr} \ [\ -(\frac{d}{dr} + \frac{2}{r})E_2 - \frac{2\nu}{r}E_1 \ ] \ = ME_1 \ ,$$

$$-i(\epsilon + M + \frac{\alpha}{r})\Phi_2 - \frac{1}{M}\frac{d}{dr} \ [\ -(\frac{d}{dr} + \frac{2}{r})E_2 - \frac{2\nu}{r}E_1 \ ] \ = ME_2 \ .$$

$$(16.6.12)$$

We take into account Eq. (16.6.10), which transforms Eq. (16.6.12) into

$$(\epsilon + M + \frac{\alpha}{r})(\Psi_1 - \psi_1) + \frac{1}{M}(\frac{d}{dr} + \frac{1}{r})^2(\Psi_1 + \psi_1)$$

$$+ \frac{\nu}{M}(\frac{d}{dr} + \frac{1}{r})\frac{1}{r}(\Psi_2 + \psi_2) = M(\Psi_1 + \psi_1) \,,$$

$$(\epsilon + M + \frac{\alpha}{r})(\Psi_2 - \psi_2) - \frac{2\nu}{Mr}(\frac{d}{dr} + \frac{1}{r})(\Psi_1 + \psi_1)$$

$$- \frac{2\nu^2}{Mr^2}(\Psi_2 + \psi_2) = M(\Psi_2 + \psi_2) \,,$$

$$(\epsilon + M + \frac{\alpha}{r})(\Psi_1 + \psi_1) - \frac{\nu}{mr}(\frac{d}{dr} + \frac{2}{r})(\Psi_2 - \psi_2)$$

$$- \frac{2\nu^2}{Mr^2}(\Psi_1 - \psi_1) = M(\Psi_1 - \psi_1) \,,$$

$$(\epsilon + M + \frac{\alpha}{r})(\Psi_2 + \psi_2) + \frac{1}{M}\frac{d}{dr}(\frac{d}{dr} + \frac{2}{r})(\Psi_2 - \psi_2)$$

$$+ \frac{2\nu}{M}\frac{d}{dr}\frac{1}{r}(\Psi_1 - \psi_1) = M(\Psi_2 - \psi_2) \,.$$

From this, we get

$$(\epsilon + \frac{\alpha}{r})(\Psi_1 - \psi_1) + \frac{1}{M}(\frac{d}{dr} + \frac{1}{r})^2(\Psi_1 + \psi_1)$$

$$+ \frac{\nu}{M}(\frac{d}{dr} + \frac{1}{r})\frac{1}{r}(\Psi_2 + \psi_2) = +2M\ \psi_1 \,,$$

$$(\epsilon + \frac{\alpha}{r})(\Psi_1 + \psi_1) - \frac{\nu}{Mr}(\frac{d}{dr} + \frac{2}{r})(\Psi_2 - \psi_2)$$

$$- \frac{2\nu^2}{Mr^2}(\Psi_1 - \psi_1) = -2M\psi_1 \,,$$

$$(\epsilon + \frac{\alpha}{r})(\Psi_2 - \psi_2) - \frac{2\nu}{Mr}(\frac{d}{dr} + \frac{1}{r})(\Psi_1 + \psi_1)$$

$$- \frac{2\nu^2}{Mr^2}(\Psi_2 + \psi_2) = +2M\ \psi_2 \,,$$

$$(\epsilon + \frac{\alpha}{r})(\Psi_2 + \psi_2) + \frac{1}{M}\frac{d}{dr}(\frac{d}{dr} + \frac{2}{r})(\Psi_2 - \psi_2)$$

$$+ \frac{2\nu}{M}\frac{d}{dr}\frac{1}{r}(\Psi_1 - \psi_1) = -2M\ \psi_2 \,.$$

Now, through summing and subtracting the equation within the first couple, and doing the same within second couple, we arrive at

$$(\epsilon + \frac{\alpha}{r})(\Psi_1 - \psi_1) + \frac{1}{M}(\frac{d}{dr} + \frac{1}{r})^2(\Psi_1 + \psi_1) + \frac{\nu}{M}(\frac{d}{dr} + \frac{1}{r})\frac{1}{r}(\Psi_2 + \psi_2)$$

$$+(\epsilon + \frac{\alpha}{r})(\Psi_1 + \psi_1) - \frac{\nu}{Mr}(\frac{d}{dr} + \frac{2}{r})(\Psi_2 - \psi_2) - \frac{2\nu^2}{Mr^2}(\Psi_1 - \psi_1) = 0 \;,$$

$$(\epsilon + \frac{\alpha}{r})(\Psi_1 - \psi_1) + \frac{1}{M}(\frac{d}{dr} + \frac{1}{r})^2(\Psi_1 + \psi_1) + \frac{\nu}{M}(\frac{d}{dr} + \frac{1}{r})\frac{1}{r}(\Psi_2 + \psi_2)$$

$$-(\epsilon + \frac{\alpha}{r})(\Psi_1 + \psi_1) + \frac{\nu}{Mr}(\frac{d}{dr} + \frac{2}{r})(\Psi_2 - \psi_2)$$

$$+ \frac{2\nu^2}{Mr^2}(\Psi_1 - \psi_1) = +4M \; \psi_1 \;,$$

$$(\epsilon + \frac{\alpha}{r})(\Psi_2 - \psi_2) - \frac{2\nu}{Mr}(\frac{d}{dr} + \frac{1}{r})(\Psi_1 + \psi_1) - \frac{2\nu^2}{Mr^2}(\Psi_2 + \psi_2)$$

$$+(\epsilon + \frac{\alpha}{r})(\Psi_2 + \psi_2) + \frac{1}{M}\frac{d}{dr}(\frac{d}{dr} + \frac{2}{r})(\Psi_2 - \psi_2) + \frac{2\nu}{M}\frac{d}{dr}\frac{1}{r}(\Psi_1 - \psi_1) = 0 \;,$$

$$(\epsilon + \frac{\alpha}{r})(\Psi_2 - \psi_2) - \frac{2\nu}{Mr}(\frac{d}{dr} + \frac{1}{r})(\Psi_1 + \psi_1) - \frac{2\nu^2}{Mr^2}(\Psi_2 + \psi_2)$$

$$-(\epsilon + \frac{\alpha}{r})(\Psi_2 + \psi_2) - \frac{1}{M}\frac{d}{dr}(\frac{d}{dr} + \frac{2}{r})(\Psi_2 - \psi_2)$$

$$-\frac{2\nu}{M}\frac{d}{dr}\frac{1}{r}(\Psi_1 - \psi_1) = +4M \; \psi_2 \;.$$

Now, considering Ψ_1, Ψ_2 as big, and ψ_1, ψ_2 as small components, we arrive at two equations for big components, and two equations defining small components through big ones:

$$\frac{1}{M}(\frac{d}{dr} + \frac{1}{r})^2\Psi_1 + \frac{1}{M}(\frac{d}{dr} + \frac{1}{r})\frac{1}{r}\Psi_2 + \frac{\nu}{Mr}(\frac{d}{dr} + \frac{2}{r})\Psi_2 + \frac{2\nu^2}{Mr^2}\Psi_1 = +4M \; \psi_1 \;,$$

$$-\frac{2\nu}{Mr}(\frac{d}{dr} + \frac{1}{r})\Psi_1 - \frac{2\nu^2}{Mr^2}\Psi_2 - \frac{1}{M}\frac{d}{dr}(\frac{d}{dr} + \frac{2}{r})\Psi_2 - \frac{2\nu}{M}\frac{d}{dr}\frac{1}{r}\Psi_1 = +4M \; \psi_2 \;,$$

$$2(\epsilon + \frac{\alpha}{r})\Psi_2 - \frac{2\nu}{Mr}(\frac{d}{dr} + \frac{1}{r})\Psi_1 - \frac{2\nu^2}{Mr^2}\Psi_2 + \frac{1}{M}\frac{d}{dr}(\frac{d}{dr} + \frac{2}{r})\Psi_2 + \frac{2\nu}{M}\frac{d}{dr}\frac{1}{r}\Psi_1 = 0 \;,$$

$$2(\epsilon + \frac{\alpha}{r})\Psi_1 + \frac{1}{M}(\frac{d}{dr} + \frac{1}{r})^2\Psi_1 + \frac{1}{M}(\frac{d}{dr} + \frac{1}{r})\frac{1}{r}\Psi_2 - \frac{\nu}{Mr}(\frac{d}{dr} + \frac{2}{r})\Psi_2 - \frac{2\nu^2}{Mr^2}\Psi_1 = 0 \;.$$

Two last equations provides us with nonrelativistic radial equations – which can be written as

$$r^2\left[\frac{d^2}{dr^2} + \frac{2}{r}\frac{d}{dr} + 2M(\epsilon + \frac{\alpha}{r}) - \frac{2\nu^2}{r^2}\right]\Psi_2 = 2\Psi_2 + 4\nu \; \Psi_1 \;,$$

$$r^2\left[\frac{d^2}{dr^2} + \frac{2}{r}\frac{d}{dr} + 2M(\epsilon + \frac{\alpha}{r}) - \frac{2\nu^2}{r^2}\right]\Psi_1 = 2\nu \; \Psi_2 \;. \tag{6.6.13}$$

It is convenient to present Eqs. (6.6.13) in a matrix form

$$\frac{1}{2}r^2\Delta \left| \begin{array}{c} \Psi_1 \\ \Psi_2 \end{array} \right| = \left| \begin{array}{cc} 0 & \nu \\ 2\nu & 1 \end{array} \right| \left| \begin{array}{c} \Psi_1 \\ \Psi_2 \end{array} \right| . \tag{16.6.14}$$

The right-hand part can be brought to a diagonal form

$$
\begin{vmatrix} f_1 \\ f_2 \end{vmatrix} = \begin{vmatrix} a & c \\ d & b \end{vmatrix} \begin{vmatrix} \Psi_1 \\ \Psi_2 \end{vmatrix} , \qquad r^2 \Delta \begin{vmatrix} f_1 \\ f_2 \end{vmatrix} = \begin{vmatrix} \lambda_1 & 0 \\ 0 & \lambda_2 \end{vmatrix} \begin{vmatrix} f_1 \\ f_2 \end{vmatrix} .
$$

The problem is to solve two systems

$$
\begin{vmatrix} a & c \\ d & b \end{vmatrix} \begin{vmatrix} 0 & \nu \\ 2\nu & 1 \end{vmatrix} = \begin{vmatrix} \lambda_1 & 0 \\ 0 & \lambda_2 \end{vmatrix} \begin{vmatrix} a & c \\ d & b \end{vmatrix} ,
$$

from this, it follows

$$
\begin{cases} \lambda_1 \, a - 2\nu \, c = 0 \\ -\nu \, a + (\lambda_1 - 1) \, c = 0 , \end{cases}
$$

$$
\lambda_1 = \frac{1 + \sqrt{1 + 4j(j+1)}}{2} = j + 1 , \qquad c = \frac{\lambda_1}{2\nu} a ;
$$

$$
\begin{cases} \lambda_2 \, d - 2\nu \, b = 0 \\ -\nu \, d + (\lambda_2 - 1) \, b = 0 , \end{cases}
$$

$$
\lambda_2 = \frac{1 - \sqrt{1 + 4j(j+1)}}{2} = -j , \qquad b = \frac{\lambda_2}{2\nu} d .
$$

The transformation matrix we need is given by

$$
\begin{vmatrix} f_1 \\ f_2 \end{vmatrix} = \begin{vmatrix} a & \lambda_1 a/2\nu \\ d & \lambda_2 d/2\nu \end{vmatrix} \begin{vmatrix} \Psi_1 \\ \Psi_2 \end{vmatrix} . \tag{16.6.15}
$$

Thus, the system in Eqs. (16.6.13) is led to the diagonal form

$$
\left[\frac{d^2}{dr^2} + \frac{2}{r} \frac{d}{dr} + 2M\left(\epsilon + \frac{\alpha}{r}\right) - \frac{2\nu^2}{r^2} - \frac{2\lambda_1}{r^2} \right] f_1 = 0 ,
$$

$$
\left[\frac{d^2}{dr^2} + \frac{2}{r} \frac{d}{dr} + 2M\left(\epsilon + \frac{\alpha}{r}\right) - \frac{2\nu^2}{r^2} - \frac{2\lambda_2}{r^2} \right] f_2 = 0 . \tag{16.6.16}
$$

Note through simple relations

$$
\frac{2\nu^2}{r^2} + \frac{2\lambda_1}{r^2} = \frac{j(j+1) + 2(j+1)}{r^2} = \frac{(j+1)(j+2)}{r^2} ,
$$

$$
\frac{2\nu^2}{r^2} + \frac{2\lambda_2}{r^2} = \frac{j(j+1) - 2j}{r^2} = \frac{(j-1)j}{r^2} .
$$

Thus, we have two problems of the same type (here, f_1 and f_2 correspond to $\nu = j - 1$ and $\nu = j + 1$, respectively)

$$
\left[\frac{d^2}{dr^2} + \frac{2}{r} \frac{d}{dr} + 2M\left(\epsilon + \frac{\alpha}{r}\right) - \frac{\nu(\nu+1)}{r^2} \right] f = 0 . \tag{16.6.17}
$$

In changing the variable $x = 2\sqrt{-2\epsilon M}\, r$

$$
\left[\frac{d^2}{dx^2} + \frac{2}{x} \frac{d}{dx} - \frac{1}{4} - \frac{\alpha M}{x\sqrt{-2\epsilon M}} - \frac{\nu(\nu+1)}{x^2} \right] f(x) = 0 , \tag{16.6.18}
$$

and introducing the substitution $f(x) = x^a e^{-bx} F(x)$, Eq. (16.6.18) is brought to

$$x \frac{d^2 F}{dx^2} + (2a + 2 - 2bx) \frac{dF}{dx}$$

$$+ \left[\frac{a(a+1) - \nu(\nu+1)}{x} - 2b - 2ab + \frac{\alpha M}{\sqrt{-2\epsilon M}} + (b^2 - \frac{1}{4})x \right] F = 0 .$$

$$(16.6.19)$$

When $b = +1/2$, $a = +\nu$, Eq. (16.6.17) is simplified

$$x \frac{d^2 F}{dx^2} + (2\nu + 2 - x) \frac{dF}{dx} - \left[1 + \nu - \frac{\alpha M}{\sqrt{-2\epsilon M}} \right] F = 0 ,$$

what is the confluent hypergeometric equation for $F(A, C; x)$ with parameters given by

$$A = 1 + \nu - \frac{\alpha M}{\sqrt{-2\epsilon M}} , \qquad C = 2\nu + 2 .$$

The quantization condition is $A = -n$, which gives the energy spectrum

$$1 + \nu - \frac{\alpha M}{\sqrt{-2\epsilon M}} = -n \qquad \Longrightarrow$$

$$\epsilon = -\frac{\alpha^2 M}{2(1 + \nu + n)^2} = -\frac{M e^4}{2\hbar^2 (1 + \nu + n)^2} ; \qquad (16.6.20)$$

remember that for linearly independent solutions with parity $P = (-1)^j$ it corresponds to $\nu = j - 1$ and $\nu = j + 1$.

16.7   Conclusions

Quantum-mechanical system – spin 1 particle in external Coulomb field is studied on the basis of the matrix Duffin–Kemmer–Petiau formalism with the use of the tetrad technique. Utlizing the help of a parity operator, the radial 10-equation system is divided into two subsystems of 4 and 6 equations that correspond to parity $P = (-1)^{j+1}$ and $P = (-1)^j$, respectively. The system of 4 equation is reduced to a second order differential equation which coincides with the case of a scalar particle in Coulomb potential. It is shown that the 6-equation system reduces to where there are two different differential equations for a "main" function. One main equation reduces to a confluent Heun equation and provides us with the energy spectrum. Another main equation is a more complex one, and any solutions for this were not constructed. In studying the radial equations, the transition to a nonrelativistic case was performed. In this limit, three types of linearly independent solutions have been constructed in terms of hypergeometric functions.

Chapter 17

Particle with Spin 1 in a Magnetic Field

17.1 Introduction

The problem of a quantum-mechanical particle in an external homogeneous magnetic field is well-known in theoretical physics. In fact, only two cases are considered: a scalar (Schrödinger's) nonrelativistic particle with spin 0, and fermions (nonrelativistic Pauli's and relativistic Dirac's) with spin $1/2$ (the first investigation were done by [433–435]). In this chapter, exact solutions for a spin 1 particle will be constructed in presence of uniform magnetic field.

To treat the problem for a vector particle we take the matrix Duffin–Kemmer–Petiau approach. The main equation in the tetrad form will read as

$$\left[i\,\beta^\alpha(x)\,(\partial_\alpha + B_\alpha - i\frac{e}{\hbar}A_\alpha) \; - \; \frac{Mc}{\hbar} \right] \Psi(x) = 0 \,,$$

$$\beta^\alpha(x) = \beta^a\,e^\alpha_{(a)}(x)\,, \qquad B_\alpha(x) = \frac{1}{2}\,J^{ab}\,e^\beta_{(a)}\nabla_\alpha e_{(b)\beta}\,; \qquad (17.1.1)$$

$e^\alpha_{(a)}(x)$ is a tetrad, J^{ab} stands for generators for a 10-dimensional representation of the Lorentz group referred to as a 4-vector and anti-symmetric tensor (for brevity, we note $Mc/\hbar$ as M). The homogeneous magnetic field $\mathbf{B} = (0,0,B)$ corresponds to 4-potential $A^a = (\,0, \frac{1}{2}\,\mathbf{B} \times \mathbf{r}\,)$; for the cylindric coordinates, the last is given by

$$dS^2 = c^2 dt^2 - dr^2 - r^2\,d\phi^2 - dz^2 \,, \qquad A_\phi = -Br^2/2 \,. \qquad (17.1.2)$$

Choosing a diagonal cylindrical tetrad

$$e^\alpha_{(0)} = (1,0,0,0)\,, \quad e^\alpha_{(1)} = (0,1,0,0)\,, \quad e^\alpha_{(2)} = (0,0,\frac{1}{r},0)\,, \quad e^\alpha_{(3)} = (0,0,0,1)\,, \qquad (17.1.3)$$

after performing the simple calculations, the main equation in Eq. (17.1.1) reduces to the form

$$\left[i\beta^0 \partial_0 + i\beta^1 \partial_r + i\frac{\beta^2}{r}(\partial_\phi + \frac{ieB}{2\hbar}r^2 + J^{12}) + i\beta^3 \partial_z - M \right] \Psi = 0 \,. \qquad (17.1.4)$$

For brevity, let us note $(eB/2\hbar)$ as B. It is better to use the matrices β^a in the so-called cyclic basis, where the generator J^{12} has a diagonal structure.

17.2   Separation of Variables

By utilizing a special substitution (it corresponds to diagonalization of the third projections of momentum P_3 and angular momentum J_3 for a particle with spin 1, as specified for the cylindric tetrad basis)

$$\Psi = e^{-i\epsilon t} e^{im\phi} e^{ikz} \begin{vmatrix} \Phi_0 \\ \vec{\Phi} \\ \mathbf{E} \\ \mathbf{H} \end{vmatrix} , \qquad (17.2.1)$$

the main equation reads

$$\left[\epsilon\beta^0 + i\beta^1 \partial_r - \frac{\beta^2}{r}(m + Br^2 - S_3) - k\beta^3 - M \right] \begin{vmatrix} \Phi_0 \\ \vec{\Phi} \\ \mathbf{E} \\ \mathbf{H} \end{vmatrix} = 0 ; \qquad (17.2.2)$$

after calculation, we arrive at the radial system of 10 equations

$$-\hat{b}_{m-1}\, E_1 - \hat{a}_{m+1}\, E_3 - ik\, E_2 = M\Phi_0 ,$$

$$-i\hat{b}_{m-1}\, H_1 + i\hat{a}_{m+1}\, H_3 + i\epsilon\, E_2 = M\Phi_2 ,$$

$$i\hat{a}_m\, H_2 + i\epsilon\, E_1 - k\, H_1 = M\Phi_1 ,$$

$$-i\hat{b}_m\, H_2 + i\epsilon\, E_3 + k\, H_3 = M\Phi_3 , \qquad (17.2.3a)$$

$$\hat{a}_m\, \Phi_0 - i\epsilon\, \Phi_1 = ME_1 , \qquad -i\hat{a}_m\, \Phi_2 + k\, \Phi_1 = MH_1 ,$$

$$\hat{b}_m\, \Phi_0 - i\epsilon\, \Phi_3 = ME_3 , \qquad i\hat{b}_m\, \Phi_2 - k\, \Phi_3 = MH_3 ,$$

$$-i\epsilon\, \Phi_2 - ik\, \Phi_0 = ME_2 , \qquad i\hat{b}_{m-1}\, \Phi_1 - i\hat{a}_{m+1}\, \Phi_3 = MH_2 ,$$

$$(17.2.3b)$$

where special abbreviations were used for first order differential operators:

$$\frac{1}{\sqrt{2}}\Big(\frac{d}{dr} + \frac{m + Br^2}{r}\Big) = \hat{a}_m , \qquad \frac{1}{\sqrt{2}}\Big(-\frac{d}{dr} + \frac{m + Br^2}{r}\Big) = \hat{b}_m . \qquad (17.2.4)$$

From Eqs. (17.2.3a) and (17.2.3b), it follows 4 equations for the components Φ_a

$$(-\hat{b}_{m-1}\, \hat{a}_m - \hat{a}_{m+1}\hat{b}_m - k^2 - M^2)\, \Phi_0 - \epsilon k\, \Phi_2$$

$$+i\epsilon\,(\, \hat{b}_{m-1}\Phi_1 + \hat{a}_{m+1}\Phi_3\,) = 0 ,$$

$$(\, -\hat{b}_{m-1}\hat{a}_m - \hat{a}_{m+1}\hat{b}_m + \epsilon^2 - M^2\,)\, \Phi_2 + \epsilon k\, \Phi_0$$

$$-ik\,(\,\hat{b}_{m-1}\Phi_1 + \hat{a}_{m+1}\Phi_3\,) = 0\,,$$

$$(\,-\hat{a}_m\hat{b}_{m-1} + \epsilon^2 - k^2 - M^2\,)\,\Phi_1 + \hat{a}_m\hat{a}_{m+1}\,\Phi_3$$

$$+i\epsilon\,\hat{a}_m\,\Phi_0 + ik\,\hat{a}_m\Phi_2 = 0\,,$$

$$(\,-\hat{b}_m\hat{a}_{m+1} + \epsilon^2 - M^2 - k^2\,)\,\Phi_3 + \hat{b}_m\hat{b}_{m-1}\Phi_1$$

$$+i\epsilon\,\hat{b}_m\Phi_0 + ik\,\hat{b}_m\Phi_2 = 0\,. \qquad (17.2.5)$$

17.3   General Analysis of the Radial Equations

Equations (17.2.5) can be transformed to the form

$$[-\hat{b}_{m-1}\hat{a}_m - \hat{a}_{m+1}\hat{b}_m + \epsilon^2 - M^2 - k^2]\,(k\Phi_0 + \epsilon\Phi_2) = 0\,,$$

$$[-\hat{b}_{m-1}\hat{a}_m - \hat{a}_{m+1}\hat{b}_m + \epsilon^2 - k^2 - M^2](\epsilon\Phi_0 + k\Phi_2)$$

$$= (\epsilon^2 - k^2)[(\epsilon\Phi_0 + k\Phi_2) - (i\hat{b}_{m-1}\Phi_1 + i\hat{a}_{m+1}\Phi_3)]\,;$$

$$(17.3.1a)$$

$$(-\hat{a}_m\hat{b}_{m-1} + \epsilon^2 - k^2 - M^2)\Phi_1 + \hat{a}_m\hat{a}_{m+1}\,\Phi_3 + i\epsilon\hat{a}_m\Phi_0 + ik\hat{a}_m\Phi_2 = 0\,,$$

$$(-\hat{b}_m\hat{a}_{m+1} + \epsilon^2 - M^2 - k^2)\Phi_3 + \hat{b}_m\hat{b}_{m-1}\Phi_1 + i\epsilon\hat{b}_m\Phi_0 + ik\hat{b}_m\Phi_2 = 0\,.$$

$$(17.3.1b)$$

Let us introduce new variables

$$F(r) = k\,\Phi_0(r) + \epsilon\,\Phi_2(r)\,, \qquad G(r) = \epsilon\,\Phi_0(r) + k\,\Phi_2(r)\,, \qquad (17.3.2)$$

then Eqs. $(17.3.1a)$ and $(17.3.1b)$ will read as

$$[\,-\hat{b}_{m-1}\,\hat{a}_m - \hat{a}_{m+1}\hat{b}_m + \epsilon^2 - M^2 - k^2\,]\,F = 0\,,$$

$$[\,-\hat{b}_{m-1}\hat{a}_m - \hat{a}_{m+1}\hat{b}_m - M^2\,]\,G$$

$$= -(\epsilon^2 - k^2)\,(i\hat{b}_{m-1}\Phi_1 + i\hat{a}_{m+1}\Phi_3)\,,$$

$$(17.3.3a)$$

$$(\,-\hat{a}_m\hat{b}_{m-1} + \epsilon^2 - k^2 - M^2\,)\,\Phi_1 + \hat{a}_m\hat{a}_{m+1}\,\Phi_3 + i\hat{a}_m\,G = 0\,,$$

$$(\,-\hat{b}_m\hat{a}_{m+1} + \epsilon^2 - M^2 - k^2\,)\,\Phi_3 + \hat{b}_m\hat{b}_{m-1}\Phi_1 + i\,\hat{b}_m\,G = 0\,.$$

$$(17.3.3b)$$

For Eqs. (17.3.3b), let us multiply the first one (from the left) by $\hat{b}_{m-1}$ and the second one by the $\hat{a}_{m+1}$; this results in

$$-\hat{b}_{m-1}\hat{a}_m(\hat{b}_{m-1}\Phi_1) + (\epsilon^2 - k^2 - M^2))(\hat{b}_{m-1}\Phi_1)$$

$$+\hat{b}_{m-1}\hat{a}_m(\hat{a}_{m+1}\Phi_3) + i\hat{b}_{m-1}\hat{a}_m G = 0\,,$$

$$-\hat{a}_{m+1}\hat{b}_m(\hat{a}_{m+1}\Phi_3) + (\epsilon^2 - M^2 - k^2)(\hat{a}_{m+1}\Phi_3)$$

$$+\hat{a}_{m+1}\hat{b}_m(\hat{b}_{m-1}\Phi_1) + i\,\hat{a}_{m+1}\hat{b}_m\,G = 0\,. \qquad (17.3.3c)$$

Again, let us introduce two new field variables

$$\hat{b}_{m-1}\Phi_1 = Z_1 , \qquad \hat{a}_{m+1}\Phi_3 = Z_3 ; \tag{17.3.4}$$

So, Eqs. (17.3.3c) will read as follows

$$-\hat{b}_{m-1}\hat{a}_m Z_1 + (\epsilon^2 - k^2 - M^2)Z_1 + \hat{b}_{m-1}\hat{a}_m Z_3 + i\hat{b}_{m-1}\hat{a}_m G = 0 ,$$

$$-\hat{a}_{m+1}\hat{b}_m Z_3 + (\epsilon^2 - M^2 - k^2)Z_3 + \hat{a}_{m+1}\hat{b}_m Z_1 + i\,\hat{a}_{m+1}\hat{b}_m\,G = 0 .$$

$$\tag{17.3.5}$$

Utilizing the help of the new functions $f(r), g(r)$

$$Z_1 = \frac{f+g}{2} , \quad Z_3 = \frac{f-g}{2} , \qquad Z_1 + Z_3 = f , \quad Z_1 - Z_3 = g , \tag{17.3.6}$$

the system in Eq. (17.3.5) is transformed to the following form

$$-\hat{b}_{m-1}\hat{a}_m\,g + (\epsilon^2 - k^2 - M^2)\frac{f+g}{2} + i\hat{b}_{m-1}\hat{a}_m G = 0 ,$$

$$\hat{a}_{m+1}\hat{b}_m\,g + (\epsilon^2 - M^2 - k^2)\frac{f-g}{2} + i\,\hat{a}_{m+1}\hat{b}_m\,G = 0 . \tag{17.3.7}$$

Combining these equations, we get:

$$[-\hat{b}_{m-1}\hat{a}_m - \hat{a}_{m+1}\hat{b}_m + \epsilon^2 - k^2 - M^2]\,g + i(\hat{b}_{m-1}\hat{a}_m - \hat{a}_{m+1}\hat{b}_m)\,G = 0 ,$$

$$(-\hat{b}_{m-1}\hat{a}_m + \hat{a}_{m+1}\hat{b}_m)\,g + (\epsilon^2 - k^2 - M^2)\,f + i(\hat{b}_{m-1}\hat{a}_m + \hat{a}_{m+1}\hat{b}_m)\,G = 0 . \tag{17.3.8}$$

In turn, Eqs. (17.3.3a) can be presented as

$$(\,-\hat{b}_{m-1}\,\hat{a}_m - \hat{a}_{m+1}\hat{b}_m + \epsilon^2 - M^2 - k^2\,)\,F = 0 ,$$

$$(\,-\hat{b}_{m-1}\hat{a}_m - \hat{a}_{m+1}\hat{b}_m - M^2\,)\,G = -i(\epsilon^2 - k^2)\,f . \tag{17.3.9}$$

Further, through the use of the identities

$$-\hat{b}_{m-1}\,\hat{a}_m - \hat{a}_{m+1}\hat{b}_m = \Delta , \qquad -\hat{b}_{m-1}\,\hat{a}_m + \hat{a}_{m+1}\hat{b}_m = 2B . \tag{17.3.10}$$

Equations (17.3.8) and (17.3.9) can be written as follows

$$(\Delta + \epsilon^2 - M^2 - k^2)\,F = 0 ,$$

$$\Delta\,G = M^2 G - i(\epsilon^2 - k^2)\,f ,$$

$$(\Delta + \epsilon^2 - k^2 - M^2)\,g = 2iB\,G ,$$

$$(\epsilon^2 - k^2 - M^2)\,f - i\Delta\,G + 2B\,g = 0 . \tag{17.3.11}$$

By using the help of the second equation, from the forth equation; it follows

$$f = -i\,G + \frac{2B}{M^2}\,g . \tag{17.3.12}$$

Now, one excludes the function f in the second equation in Eq. (17.3.11) and gets

$$(\Delta + \epsilon^2 - k^2 - M^2)\, G = -i(\epsilon^2 - k^2)\frac{2B}{M^2}\, g\,. \qquad (17.3.13)$$

Thus, the general problem is reduced to the system of four equations

$$(\Delta + \epsilon^2 - M^2 - k^2)\, F = 0\,,$$

$$f = -i\, G + \frac{2B}{M^2}\, g\,,$$

$$(\Delta + \epsilon^2 - k^2 - M^2)\, g = 2i B\, G\,,$$

$$(\Delta + \epsilon^2 - k^2 - M^2)\, G = -2i B\, \frac{\epsilon^2 - k^2}{M^2}\, g\,. \qquad (17.3.14)$$

The structure of this system allows us to separate an evident linearly independent solution as follows

$$f(r) = 0\,, \qquad g(r) = 0\,, \qquad H(r) = 0\,,$$

$$F(r) \neq 0\,, \qquad (\Delta - k^2 - M^2 + \epsilon^2)\, F = 0\,. \qquad (17.3.15)$$

Corresponding functions and energy spectrum are known. We solve the system of the two last equations in Eq. (17.3.14); in matrix form it reads (let $\gamma = (\epsilon^2 - k^2)/M^2$)

$$(\Delta + \epsilon^2 - M^2 - k^2) \begin{vmatrix} g \\ G \end{vmatrix} = \begin{vmatrix} 0 & 2iB \\ -2iB\gamma & 0 \end{vmatrix} \begin{vmatrix} g \\ G \end{vmatrix}\,. \qquad (17.3.16)$$

Let us construct the transformation changing the matrix on the right to a diagonal form

$$(\Delta + \epsilon^2 - M^2 - k^2) \begin{vmatrix} g' \\ G' \end{vmatrix} = \begin{vmatrix} \lambda_1 & 0 \\ 0 & \lambda_2 \end{vmatrix} \begin{vmatrix} g' \\ G' \end{vmatrix}\,,$$

$$\begin{vmatrix} g' \\ G' \end{vmatrix} = S \begin{vmatrix} g \\ G \end{vmatrix}\,, \qquad S = \begin{vmatrix} s_{11} & s_{12} \\ s_{21} & s_{22} \end{vmatrix}\,. \qquad (17.3.17)$$

The problem reduces to linear systems

$$\begin{cases} -\lambda_1\, s_{11} - 2iB\gamma\, s_{12} = 0\,, \\ 2iB\, s_{11} - \lambda_1\, s_{12} = 0\,, \end{cases} \qquad \begin{cases} -\lambda_2\, s_{21} - 2iB\gamma\, s_{22} = 0\,, \\ 2iB\, s_{21} - \lambda_2\, s_{22} = 0\,. \end{cases}$$

The values of λ_1 and λ_2 are given by

$$\lambda_1 = +2B\sqrt{\gamma}\,, \qquad \lambda_2 = -2B\sqrt{\gamma}\,,$$

$$i\, s_{11} - \sqrt{\gamma}\, s_{12} = 0\,, \qquad i\, s_{21} + \sqrt{\gamma}\, s_{22} = 0\,,$$

$$s_{12} = 1,\ s_{22} = 1\,, \qquad S = \begin{vmatrix} -i\,\sqrt{\gamma} & 1 \\ +i\,\sqrt{\gamma} & 1 \end{vmatrix}\,. \qquad (17.3.18)$$

In the new (primed) basis, Eqs. (17.3.16) takes the form of two separated differential equations

$$(\Delta + \epsilon^2 - k^2 - M^2 - 2B\,\sqrt{\gamma}\,)\, g' = 0\,,$$

$$\left(\Delta + \epsilon^2 - k^2 - M^2 + 2B \sqrt{\gamma} \right) G' = 0 . \tag{17.3.19}$$

Let us recall the meaning of Δ, to specify the second order differential equation

$$\left(\frac{d^2}{dr^2} + \frac{1}{r}\frac{d}{dr} - \frac{(m + Br^2)^2}{r^2} + \lambda^2 \right) \varphi(r) = 0 ,$$

$$\lambda^2 = \epsilon^2 - k^2 - M^2 \pm 2B \sqrt{\gamma}, \qquad \sqrt{\gamma} = \frac{\sqrt{\epsilon^2 - k^2}}{M} . \tag{17.3.20}$$

It is convenient to introduce a new variable $x = Br^2$, then Eq. (17.3.20) will read [1]

$$x\frac{d^2\varphi}{dx^2} + \frac{d\varphi}{dx} - \left(\frac{m^2}{4x} + \frac{x}{4} + \frac{m}{2} - \frac{\lambda^2}{4B} \right) \varphi = 0 . \tag{17.3.21}$$

In the substitution $\varphi(x) = x^A e^{-Cx} f(x)$, for $f(x)$, we get

$$x\frac{d^2 f}{dx^2} + (2A + 1 - 2Cx) \frac{df}{dx}$$

$$+ \left[\frac{A^2 - m^2/4}{x} + (C^2 - \frac{1}{4})x - 2AC - C - \frac{m}{2} + \frac{\lambda^2}{4B} \right] f = 0 .$$

When A, C are taken as $A = + \mid m \mid /2$, $C = +1/2$ the previous equation becomes simpler

$$x\frac{d^2 R}{dx^2} + (2A + 1 - x) \frac{dR}{dx} - \left(A + \frac{1}{2} + \frac{m}{2} - \frac{\lambda^2}{4B} \right) R = 0 ,$$

and is of the confluent hypergeometric type

$$x\, Y'' + (\gamma - x)Y' - \alpha Y = 0 ,$$

$$\alpha = \frac{\mid m \mid}{2} + \frac{1}{2} + \frac{m}{2} - \frac{\lambda^2}{4B} , \qquad \gamma = \mid m \mid + 1 .$$

To obtain polynomials we must impose an additional condition $\alpha = -n$; which provides us with the following quantization rule for λ^2

$$\lambda^2 = 4B \left(n + \frac{1}{2} + \frac{\mid m \mid + m}{2} \right) . \tag{17.3.22}$$

Thus, we have arrived at two formulas for the energy spectrum

$$\sqrt{\epsilon^2 - k^2} = \frac{+B + \sqrt{B^2 + M^2(M^2 + \lambda^2)}}{M} ,$$

$$\sqrt{\epsilon^2 - k^2} = \frac{-B + \sqrt{B^2 + M^2(M^2 + \lambda^2)}}{M} . \tag{17.3.23}$$

In turn, the energy spectrum for the case in Eq. (17.3.15) is given by

$$\epsilon^2 = M^2 + k^2 + \lambda^2 . \tag{17.3.24}$$

Thus, in the use of general covariant formalism in the Duffin–Kemmer–Petiau theory for the vector particle, the exact solutions for such a particle are constructed in the presence of an external homogeneous magnetic field. There are three separated types of linearly independent solutions, and the corresponding energy spectra was found.

---

[1] For definiteness let us consider B to be positive, which does not affect the generality of the analysis. So, infinite values of r correspond to infinite and positive values of x.

Chapter 18

Particle with Spin 1 in a Magnetic Field on the Hyperbolic Plane

We constructed exact solutions for the quantum-mechanical equation for a spin $S = 1$ particle in 2-dimensional Riemannian space with a constant negative curvature. This hyperbolic plane is in the presence of an external magnetic field with an analog for a homogeneous magnetic field in Minkowski space. A generalized formula for energy levels describes the quantization for the motion of a vector particle in a magnetic field on the 2-dimensional space H_2 has been found, and a nonrelativistic and relativistic equations have been solved.

18.1   Introduction

The quantization of a quantum-mechanical particle in the homogeneous magnetic field belongs to a classical problem in physics [433–435]. In 1985 – 2010, a more general problem focusing on a curved Riemannian background, hyperbolic and spherical planes, and was extensively studied starting with Comtet and Houston [436,437]; see also Aoki [438], Groshe [439], Klauder and Onofri [440], Avron and Pnueli [441], Dunne [442], Alimohammadi et al. [443,444], Onofri [445], Negro et al. [446], Drukker et al. [447], and Ghanmi and Intissar [448]. This research has provided us with a new system of intriguing dynamics and symmetry, both on classical and quantum levels. The extension to 3-dimensional hyperbolic and spherical spaces was performed recently. In [449,450], exact solutions for a scalar particle for the extended problem of a particle in external magnetic field on the background of Lobachevsky H_3 and Riemann S_3 spacial geometries were found. A corresponding system in the frames of classical mechanics was examined in [451–453].

In the present chapter, we will construct exact solutions for a vector particle described by 10-dimensional Duffin–Kemmer equation in external magnetic field on the background of 2-dimensional spherical space H_2.

A 10-dimensional Duffin–Kemmer equation for a vector particle in a curved space-time has the form

$$\left\{ \beta^c \left[i \left(e_{(c)}^{\beta} \partial_\beta + \frac{1}{2} J^{ab} \gamma_{abc} \right) + \frac{e}{\hbar c} A_{(c)} \right] - \frac{mc}{\hbar} \right\} \Psi = 0 , \qquad (18.1.1)$$

where γ_{abc} stands for Ricci rotation coefficients, $A_a = e^{\beta}_{(a)} A_{\beta}$ represents tetrad components of electromagnetic 4-vector A_{β}; $J^{ab} = \beta^a \beta^b - \beta^b \beta^a$ are generators of 10-dimensional representation of the Lorentz group. For shortness, we use the notation $e/c\hbar \Longrightarrow e$, $mc/\hbar \Longrightarrow M$.

In the space H_3, we will use the system of cylindric coordinates

$$dS^2 = c^2 dt^2 - \cosh^2 z(dr^2 + \sinh^2 r\, d\phi^2) - dz^2 \,,$$

$$u_1 = \cosh z \sinh r \cos\phi \,, \quad u_2 = \cosh z \sinh r \sin\phi \,,$$

$$u_3 = \sinh z \,, \quad u_0 = \cosh z \cosh r \,, \tag{18.1.2}$$

$$G = \{\, r \in [0, +\infty), \ \phi \in [0, 2\pi], \ z \in (-\infty, +\infty) \,\} \,.$$

Generalized expression for electromagnetic potential for an homogeneous magnetic field in the curved model H_3 is given as follows

$$A_{\phi} = -2B \sinh^2 \frac{r}{2} = -B\,(\cosh r - 1) \,. \tag{18.1.3}$$

We will consider this equation in the presence in the field for the model H_3. Corresponding to cylindric coordinates $x^{\alpha} = (t, r, \phi, z)$ a tetrad can be chosen as

$$e^{\beta}_{(a)}(x) = \begin{vmatrix} 1 & 0 & 0 & 0 \\ 0 & \cosh^{-1} z & 0 & 0 \\ 0 & 0 & \cosh^{-1} z \sinh^{-1} r & 0 \\ 0 & 0 & 0 & 1 \end{vmatrix} . \tag{18.1.4}$$

In taking into account the relations

$$\Gamma^r{}_{jk} = \begin{vmatrix} 0 & 0 & \tanh z \\ 0 & -\sinh r \cosh r & 0 \\ \tanh z & 0 & 0 \end{vmatrix} , \quad \Gamma^{\phi}{}_{jk} = \begin{vmatrix} 0 & \coth r & 0 \\ \coth r & 0 & \tanh z \\ 0 & \tanh z & 0 \end{vmatrix} ,$$

$$\Gamma^z{}_{jk} = \begin{vmatrix} -\sinh z \cosh z & 0 & 0 \\ 0 & -\sinh z \cosh z \sinh^2 r & 0 \\ 0 & 0 & 0 \end{vmatrix} ,$$

$$\gamma_{122} = \frac{1}{\cosh z \tanh r} \,, \quad \gamma_{311} = \tanh z \,, \quad \gamma_{322} = \tanh z \,, \tag{18.1.5}$$

Equation (18.1.1) reduces to the form

$$\left\{ i\beta^0 \frac{\partial}{\partial t} + \frac{1}{\cosh z} \left(i\beta^1 \frac{\partial}{\partial r} + \beta^2 \frac{i\partial_{\phi} - eB(\cosh r - 1) + iJ^{12} \cosh r}{\sinh r} \right) \right.$$

$$\left. + i\beta^3 \frac{\partial}{\partial z} - i\frac{\sinh z}{\cosh z} (\beta^1 J^{13} + \beta^2 J^{23}) - M \right\} \Psi = 0 \,. \tag{18.1.6}$$

To separate the variables in Eq. (18.1.6), we employ an explicit form of the Duffin–Kemmer matrices β^a. The most convenient example to use is the cyclic representation, where the generator J^{12} is the diagonal form (we specify matrices by blocks in accordance with $(1-3-3-3)$-splitting)

$$\beta^0 = \begin{vmatrix} 0 & 0 & 0 & 0 \\ 0 & 0 & i & 0 \\ 0 & -i & 0 & 0 \\ 0 & 0 & 0 & 0 \end{vmatrix} , \qquad \beta^i = \begin{vmatrix} 0 & 0 & e_i & 0 \\ 0 & 0 & 0 & \tau_i \\ -e_i^+ & 0 & 0 & 0 \\ 0 & -\tau_i & 0 & 0 \end{vmatrix} , \qquad (18.1.7)$$

where e_i, $e_i^{\,t}$, τ_i denote

$$e_1 = \frac{1}{\sqrt{2}}(-i,\, 0,\, i) , \qquad e_2 = \frac{1}{\sqrt{2}}(1,\, 0,\, 1) , \qquad e_3 = (0, i, 0) ,$$

$$\tau_1 = \frac{1}{\sqrt{2}} \begin{vmatrix} 0 & 1 & 0 \\ 1 & 0 & 1 \\ 0 & 1 & 0 \end{vmatrix} , \qquad \tau_2 = \frac{1}{\sqrt{2}} \begin{vmatrix} 0 & -i & 0 \\ i & 0 & -i \\ 0 & i & 0 \end{vmatrix} , \qquad \tau_3 = \begin{vmatrix} 1 & 0 & 0 \\ 0 & 0 & 0 \\ 0 & 0 & -1 \end{vmatrix} = s_3 .$$

$$(18.1.8)$$

The generator J^{12} explicitly is read as

$$J^{12} = \beta^1 \beta^2 - \beta^2 \beta^1$$

$$= \begin{vmatrix} (-e_1 e_2^+ + e_2 e_1^+) & 0 & 0 & 0 \\ 0 & (-\tau_1 \tau_2 + \tau_2 \tau_1) & 0 & 0 \\ 0 & 0 & (-e_1^+ \bullet e_2 + e_2^+ \bullet e_1) & 0 \\ 0 & 0 & 0 & (-\tau_1 \tau_2 + \tau_2 \tau_1) \end{vmatrix}$$

$$= -i \begin{vmatrix} 0 & 0 & 0 & 0 \\ 0 & \tau_3 & 0 & 0 \\ 0 & 0 & \tau_3 & 0 \\ 0 & 0 & 0 & \tau_3 \end{vmatrix} = -i S_3 . \qquad (18.1.9)$$

18.2 Restriction to 2-Dimensional Model

Let us restrict ourselves to the 2-dimensional case, hyperbolic 2-plane H_2 (formally it is sufficient in Eq. (18.1.6) to remove the dependence on the variable z, fixing its value by $z = 0$)

$$\left[i\beta^0 \frac{\partial}{\partial t} + i\beta^1 \frac{\partial}{\partial r} + \beta^2 \frac{i\partial_\phi - eB (\cosh r - 1) + i J^{12} \cosh r}{\sinh r} - M \right] \Psi = 0 . \quad (18.2.1)$$

Through the use of substitution

$$\Psi = e^{-i\epsilon t} e^{im\phi} \begin{vmatrix} \Phi_0(r) \\ \vec{\Phi}(r) \\ \mathbf{E}(r) \\ \mathbf{H}(r) \end{vmatrix} , \qquad (18.2.2)$$

Equation (18.2.1) assumes the form (introducing notation $m + B(\cosh r - 1) = \nu(r)$)

$$\left[\epsilon\, \beta^0 + i\beta^1\, \frac{\partial}{\partial r} - \beta^2\, \frac{\nu(r) - \cosh r\; S_3}{\sinh r} - M \right] \left| \begin{matrix} \Phi_0(r) \\ \vec{\Phi}(r) \\ \mathbf{E}(r) \\ \mathbf{H}(r) \end{matrix} \right| = 0 \, . \qquad (18.2.3)$$

Equation (18.2.3) will read as

$$\left[\epsilon \left| \begin{matrix} 0 & 0 & 0 & 0 \\ 0 & 0 & i & 0 \\ 0 & -i & 0 & 0 \\ 0 & 0 & 0 & 0 \end{matrix} \right| + i \left| \begin{matrix} 0 & 0 & e_1 & 0 \\ 0 & 0 & 0 & \tau_1 \\ -e_1^+ & 0 & 0 & 0 \\ 0 & -\tau_1 & 0 & 0 \end{matrix} \right| \frac{\partial}{\partial r} \right.$$

$$\left. - \frac{1}{\sinh r} \left| \begin{matrix} 0 & 0 & e_2 & 0 \\ 0 & 0 & 0 & \tau_2 \\ -e_2^+ & 0 & 0 & 0 \\ 0 & -\tau_2 & 0 & 0 \end{matrix} \right| (\nu - \cosh r\; S_3) - M \right] \left| \begin{matrix} \Phi_0 \\ \vec{\Phi} \\ \mathbf{E} \\ \mathbf{H} \end{matrix} \right| = 0 \, , \qquad (18.2.4)$$

or in a block form

$$i e_1 \partial_r \mathbf{E} - \frac{1}{\sinh r}\, e_2 (\nu - \cosh r\; s_3)\mathbf{E} = M\, \Phi_0 \, ,$$

$$i \epsilon \cosh z\, \mathbf{E} + i\tau_1 \partial_r \mathbf{H} - \frac{\tau_2}{\sinh r}(\nu - \cosh r\; s_3)\mathbf{H} = M\, \vec{\Phi} \, ,$$

$$-i\epsilon \cosh z\, \vec{\Phi} - i e_1^+ \partial_r \Phi_0 + \frac{\nu}{\sinh r}\, e_2^+ \Phi_0 = M\mathbf{E} \, ,$$

$$-i\tau_1 \partial_r \vec{\Phi} + \frac{(\nu - \cosh r\; s_3)}{\sinh r}\, \tau_2 \vec{\Phi} = M\, \mathbf{H} \, . \qquad (18.2.5)$$

After the separation of variables, we get:

$$\gamma\!\left(\frac{\partial E_1}{\partial r} - \frac{\partial E_3}{\partial r} \right) - \frac{\gamma}{\sinh r}\left[(\nu - \cosh r)E_1 + (\nu + \cosh r)E_3 \right] = M\Phi_0 \, ,$$

$$+ i\epsilon \cosh z E_1 + i\gamma \frac{\partial H_2}{\partial r} + i\gamma \frac{\nu}{\sinh r} H_2 = M\Phi_1 \, ,$$

$$+ i\epsilon E_2 + i\gamma\!\left(\frac{\partial H_1}{\partial r} + \frac{\partial H_3}{\partial r} \right) - \frac{i\gamma}{\sinh r}\left[(\nu - \cosh r)H_1 - (\nu + \cosh r)H_3 \right] = M\Phi_2 \, ,$$

$$+ i\epsilon E_3 + i\gamma \frac{\partial H_2}{\partial r} - i\gamma \frac{\nu}{\sinh r} H_2 = M\Phi_3 \, , \qquad (18.2.6)$$

$$-i\epsilon \Phi_1 + \gamma \frac{\partial \Phi_0}{dr} + \gamma \frac{\nu}{\sinh r}\Phi_0 = M E_1 \, ,$$

$$-i\epsilon \Phi_2 = M E_2 \, ,$$

$$-i\epsilon \Phi_3 - \gamma \frac{\partial \Phi_0}{\partial r} + \gamma \frac{\nu}{\sinh r}\Phi_0 = M E_3 \, , \qquad (18.2.7)$$

$$-i\gamma\frac{\partial \Phi_2}{\partial r} - i\gamma\frac{\nu}{\sinh r}\Phi_2 = M\cosh z H_1 \,,$$

$$-i\gamma(\frac{\partial \Phi_1}{\partial r} + \frac{\partial \Phi_3}{\partial r}) + \frac{i\gamma}{\sinh r}[(\nu - \cosh r)\Phi_1 - (\nu + \cosh r)\Phi_3] = MH_2 \,,$$

$$-i\gamma\frac{\partial \Phi_2}{\partial r} + i\gamma\frac{\nu}{\sinh r}\Phi_2 = MH_3 \,. \tag{18.2.8}$$

By the notation

$$\frac{1}{\sqrt{2}}(\frac{\partial}{\partial r} + \frac{\nu - \cosh r}{\sinh r}) = \hat{a}_-, \quad \frac{1}{\sqrt{2}}(\frac{\partial}{\partial r} + \frac{\nu + \cosh r}{\sinh r}) = \hat{a}_+ \,, \quad \frac{1}{\sqrt{2}}(\frac{\partial}{\partial r} + \frac{\nu}{\sinh r}) = \hat{a} \,,$$

$$\frac{1}{\sqrt{2}}(-\frac{\partial}{\partial r} + \frac{\nu - \cosh r}{\sinh r}) = \hat{b}_-, \quad \frac{1}{\sqrt{2}}(-\frac{\partial}{\partial r} + \frac{\nu + \cosh r}{\sinh r}) = \hat{b}_+ \,, \quad \frac{1}{\sqrt{2}}(-\frac{\partial}{\partial r} + \frac{\nu}{\sinh r}) = \hat{b} \,,$$

this system then will read as

$$-\hat{b}_- E_1 - \hat{a}_+ E_3 = M\,\Phi_0 \,,$$

$$-i\hat{b}_- H_1 + i\hat{a}_+ H_3 + i\epsilon E_2 = M\,\Phi_2 \,,$$

$$i\hat{a}H_2 + i\epsilon E_1 = M\,\Phi_1 \,,$$

$$-i\hat{b}H_2 + i\epsilon E_3 = M\,\Phi_3 \,, \tag{18.2.9}$$

$$\hat{a}\Phi_0 - i\epsilon\,\Phi_1 = M\,E_1 \,,$$

$$-i\hat{a}\Phi_2 = M\,H_1 \,,$$

$$\hat{b}\Phi_0 - i\epsilon\,\Phi_3 = M\,E_3 \,,$$

$$i\hat{b}\Phi_2 = M\,H_3 \,,$$

$$-i\epsilon\Phi_2 = M\,E_2 \,,$$

$$i\hat{b}_- \Phi_1 - i\hat{a}_+ \Phi_3 = M\,H_2 \,. \tag{18.2.10}$$

18.3   Nonrelativistic Approximation

Excluding the non-dynamical variables Φ_0, H_1, H_2, H_3 with the help of equations

$$-\hat{b}_- E_1 - \hat{a}_+ E_3 = M\,\Phi_0 \,,$$

$$-i\,\hat{a}\Phi_2 = M\,H_1 \,,$$

$$i\hat{b}_- \Phi_1 - i\hat{a}_+ \Phi_3 = M\,H_2 \,,$$

$$i\,\hat{b}\Phi_2 = M\,H_3 \,, \tag{18.3.1}$$

we get 6 equations (grouping them in pairs)

$$i\hat{a}\,(i\hat{b}_- \Phi_1 - i\hat{a}_+ \Phi_3) + i\epsilon\,ME_1 = M^2\Phi_1 \,,$$

$$\hat{a}\,(-\hat{b}_- E_1 - \hat{a}_+ E_3 - i\epsilon\,M\Phi_1 = M^2 E_1 \,, \tag{18.3.2a}$$

$$-i\hat{b}_- \left(-i\,\hat{a}\,\Phi_2\right) + i\hat{a}_+ (i\,\hat{b}\,\Phi_2) + i\epsilon\,ME_2 = M^2\Phi_2\,,$$

$$-i\epsilon\,M\Phi_2 = M^2 E_2\,, \tag{18.3.2b}$$

$$-i\,\hat{b}\,(i\hat{b}_-\Phi_1 - i\hat{a}_+\Phi_3) + i\epsilon\,ME_3 = M^2\Phi_3\,,$$

$$\hat{b}\,(-\hat{b}_- E_1 - \hat{a}_+ E_3) - i\epsilon\,M\Phi_3 = M^2 E_3\,. \tag{18.3.2c}$$

Now, we introduce big and small constituents

$$\Phi_1 = \Psi_1 + \psi_1\,, \qquad iE_1 = \Psi_1 - \psi_1\,,$$

$$\Phi_2 = \Psi_2 + \psi_2\,, \qquad iE_2 = \Psi_2 - \psi_2\,,$$

$$\Phi_3 = \Psi_3 + \psi_3\,, \qquad iE_3 = \Psi_3 - \psi_3\,; \tag{18.3.3}$$

Here, we also should separate the rest energy by formal change $\epsilon \Longrightarrow \epsilon + M$; summing and subtracting equation within each pair in Eq.(18.3.2) and ignore the small constituents ψ_i. We arrive at three equations for big components

$$\left(-2\,\hat{a}\hat{b}_- + 2\epsilon M\right)\Psi_1 = 0\,,$$

$$\left(-(\hat{b}_-\hat{a} + \hat{a}_+\hat{b}) + 2\epsilon M\right)\Psi_2 = 0\,,$$

$$\left(-2\hat{b}\hat{a}_+ + 2\epsilon M\right)\Psi_3 = 0\,. \tag{18.3.4}$$

It is a needed Pauli-like system for the spin 1 particle. Explicitly, Eqs. (18.3.4) will read as

$$\left[\frac{d^2}{dr^2} + \frac{\cosh r}{\sinh r}\frac{d}{dr} - \frac{1}{\sinh r}\frac{d\nu}{dr} - \frac{1 - 2\nu\cosh r}{\sinh^2 r} - \frac{\nu^2}{\sinh^2 r} + 2\,\epsilon\,M\right]\Psi_1 = 0\,,$$

$$\left[\frac{d^2}{dr^2} + \frac{\cosh r}{\sinh r}\frac{d}{dr} - \frac{\nu^2}{\sinh^2 r} + 2\,\epsilon\,M\right]\Psi_2 = 0\,,$$

$$\left[\frac{d^2}{dr^2} + \frac{\cosh r}{\sinh r}\frac{d}{dr} + \frac{1}{\sinh r}\frac{d\nu}{dr} - \frac{1 + 2\nu\cosh r}{\sinh^2 r} - \frac{\nu^2}{\sinh^2 r} + 2\,\epsilon\,M\right]\Psi_3 = 0\,. \tag{18.3.5}$$

By allowing for $\nu(r) = m + B\,(\cosh r - 1)$, we arrive at

$$\left[\frac{d^2}{dr^2} + \frac{\cosh r}{\sinh r}\frac{d}{dr} - B - \frac{1 - 2\,[m + B\,(\cosh r - 1)]\,\cosh r}{\sinh^2 r}\right.$$

$$\left. - \frac{[m + B\,(\cosh r - 1)]^2}{\sinh^2 r} + 2\,\epsilon\,M\right]\Psi_1 = 0\,,$$

$$\left[\frac{d^2}{dr^2} + \frac{\cosh r}{\sinh r}\frac{d}{dr} - \frac{[m + B\,(\cosh r - 1)]^2}{\sinh^2 r} + 2\,\epsilon\,M\right]\Psi_2 = 0\,,$$

$$\left[\frac{d^2}{dr^2} + \frac{\cosh r}{\sinh r} \frac{d}{dr} + B - \frac{1 + 2\left[m + B\left(\cosh r - 1\right)\right]\cosh r}{\sinh^2 r} \right.$$

$$\left. - \frac{\left[m + B\left(\cosh r - 1\right)\right]^2}{\sinh^2 r} + 2\,\epsilon\,M \right] \Psi_3 = 0 \,. \tag{18.3.6}$$

The first and the third equations are symmetric with respect to formal change $m \Longrightarrow -m$, $B \Longrightarrow -B$. In the new variable $1 - \cosh r = 2\,y$, they appear as

$$y\left(1 - y\right) \frac{d^2 \Psi_1}{dy^2} + \left(1 - 2\,y\right) \frac{d\Psi_1}{dy}$$

$$+ \left[B^2 - B - 2\,\epsilon\,M - \frac{1}{4} \frac{\left(2\,B - m - 1\right)^2}{1 - y} - \frac{1}{4} \frac{\left(m - 1\right)^2}{y} \right] \Psi_1 = 0 \,, \tag{18.3.7a}$$

$$y\left(1 - y\right) \frac{d^2 \Psi_2}{dy^2} + \left(1 - 2\,y\right) \frac{d\Psi_2}{dy}$$

$$+ \left[B^2 - 2\,\epsilon\,M - \frac{1}{4} \frac{\left(2\,B - m\right)^2}{1 - y} - \frac{1}{4} \frac{m^2}{y} \right] \Psi_2 = 0 \,, \tag{18.3.7b}$$

$$y\left(1 - y\right) \frac{d^2 \Psi_3}{dy^2} + \left(1 - 2\,y\right) \frac{d\Psi_3}{dy}$$

$$+ \left[B^2 + B - 2\,\epsilon\,M - \frac{1}{4} \frac{\left(2\,B - m + 1\right)^2}{1 - y} - \frac{1}{4} \frac{\left(m + 1\right)^2}{y} \right] \Psi_3 = 0 \,. \tag{18.3.7c}$$

Equation $(18.3.7a)$ with the substitution

$$\Psi_1 = y^{C_1} \left(1 - y\right)^{A_1} f_1$$

results in

$$y\left(1 - y\right) \frac{d^2 f_1}{dy^2} + \left[2\,C_1 + 1 - \left(2\,A_1 + 2\,C_1 + 2\right) y\right] \frac{df_1}{dy}$$

$$+ \left[B^2 - B - 2\,\epsilon\,M - \left(A_1 + C_1\right)\left(A_1 + C_1 + 1\right) \right.$$

$$\left. + \frac{1}{4} \frac{4\,A_1^2 - \left(2\,B - m - 1\right)^2}{1 - y} + \frac{1}{4} \frac{4\,C_1^2 - \left(m - 1\right)^2}{y} \right] f_1 = 0 \,. \tag{18.3.8}$$

At A_1, C_1, when obeying

$$A_1 = \pm \frac{1}{2} \left(2\,B - m - 1\right) \,, \qquad C_1 = \pm \frac{1}{2} \left(m - 1\right) \,,$$

Eq. $(18.3.8)$ becomes simpler

$$y\left(1 - y\right) \frac{d^2 f_1}{dy^2} + \left[2\,C_1 + 1 - \left(2\,A_1 + 2\,C_1 + 2\right) y\right] \frac{df_1}{dy}$$

$$+ \left[B^2 - B - 2\,\epsilon\,M - \left(A_1 + C_1\right)\left(A_1 + C_1 + 1\right) \right] f_1 = 0 \,, \tag{18.3.9a}$$

as a hypergeometric equation with parameters

$$\alpha_1 = A_1 + C_1 + \frac{1}{2} + \sqrt{B^2 - B - 2\,\epsilon\,M + \frac{1}{4}},$$

$$\beta_1 = A_1 + C_1 + \frac{1}{2} - \sqrt{B^2 - B - 2\,\epsilon\,M + \frac{1}{4}},$$

$$\gamma_1 = 2\,C_1 + 1. \tag{18.3.9b}$$

To have finite and single-valued solutions one must impose restrictions $A_1 < 0, C_1 > 0$. Besides, one must get n-order polynomials and satisfy the inequality $A_1 + C_1 + n < 0$. Four different possibilities for A_1, C_1 are (for definiteness, let it be $B > 0$):

$$\text{1.} \qquad A_1 = -\frac{1}{2}\,(2\,B - m - 1), \qquad C_1 = -\frac{1}{2}\,(m - 1),$$

$$\text{2.} \qquad A_1 = +\frac{1}{2}\,(2\,B - m - 1), \qquad C_1 = -\frac{1}{2}\,(m - 1),$$

$$\text{3.} \qquad A_1 = +\frac{1}{2}\,(2\,B - m - 1), \qquad C_1 = +\frac{1}{2}\,(m - 1),$$

$$\text{4.} \qquad A_1 = -\frac{1}{2}\,(2\,B - m - 1), \qquad C_1 = +\frac{1}{2}\,(m - 1).$$

To describe bound state, only variants 1 and 4 are appropriate:

$$1\,, \qquad m < 0\,,$$

$$\alpha_1 = -B + \frac{3}{2} + \sqrt{B^2 - B - 2\,\epsilon\,M + \frac{1}{4}},$$

$$\beta_1 = -B + \frac{3}{2} - \sqrt{B^2 - B - 2\,\epsilon\,M + \frac{1}{4}},$$

$$\gamma_1 = -m + 2\,,$$

$$\text{spectrum} \qquad \alpha_1 = -n\,, \qquad \sqrt{B^2 - B - 2\,\epsilon\,M + \frac{1}{4}} = B - \frac{3}{2} - n\,,$$

$$\epsilon\,M = B - 1 + n\left(B - \frac{3}{2} - \frac{n}{2}\right);$$

$$\tag{18.3.10a}$$

$$4\,, \qquad 0 < m < B\,,$$

$$\alpha_1 = -B + m + \frac{1}{2} + \sqrt{B^2 - B - 2\,\epsilon\,M + \frac{1}{4}},$$

$$\beta_1 = -B + m + \frac{1}{2} - \sqrt{B^2 - B - 2\,\epsilon\,M + \frac{1}{4}},$$

$$\gamma_1 = m \,,$$

spectrum $\qquad \alpha_1 = -n \,, \qquad \sqrt{B^2 - B - 2\,\epsilon\,M + \frac{1}{4}} = B - \frac{1}{2} - (n + m) \,,$

$$\epsilon\,M = (m + n)\left(B - \frac{1}{2} - \frac{1}{2}\,(m + n)\right) .$$

$$(18.3.10b)$$

Formulas in Eqs. $(18.3.10a, b)$ can be joined into a single one

$$\sqrt{B^2 - B - 2\,\epsilon\,M + \frac{1}{4}} = -n - \frac{1}{2} - \frac{|\,2B - m - 1\,| + |\,m - 1\,|}{2} \,. \qquad (18.3.10c)$$

From Eq. $(18.3.7b)$ with the substitution

$$\Psi_2 = y^{C_2}(1 - y)^{A_2} f_2$$

we get

$$y\,(1 - y)\,\frac{d^2 f_2}{dy^2} + [2\,C_2 + 1 - (2\,A_2 + 2\,C_2 + 2)\,y]\,\frac{df_2}{dy}$$

$$+ \Big[B^2 - 2\,\epsilon\,M - (A_2 + C_2)\,(A_2 + C_2 + 1)$$

$$+ \frac{1}{4}\,\frac{4\,A_2^2 - (2\,B - m)^2}{1 - y} + \frac{1}{4}\,\frac{4\,C_2^2 - m^2}{y} \Big]\,f_2 = 0 \,. \qquad (18.3.11)$$

At

$$A_2 = \pm\frac{1}{2}\,(2\,B - m) \,, \qquad C_2 = \pm\frac{m}{2} \,,$$

Eq. $(18.3.11)$ becomes simpler

$$y\,(1 - y)\,\frac{d^2 f_2}{dy^2} + [2\,C_2 + 1 - (2\,A_2 + 2\,C_2 + 2)\,y]\,\frac{df_2}{dy}$$

$$+ \big[B^2 - 2\,\epsilon\,M - (A_2 + C_2)\,(A_2 + C_2 + 1) \big]\,f_2 = 0 \,,$$

$$(18.3.12a)$$

which is recognized as a hypergeometric type

$$\alpha_2 = A_2 + C_2 + \frac{1}{2} + \sqrt{B^2 - 2\,\epsilon\,M + \frac{1}{4}} \,,$$

$$\beta_2 = A_2 + C_2 + \frac{1}{2} - \sqrt{B^2 - 2\,\epsilon\,M + \frac{1}{4}} \,,$$

$$\gamma_2 = 2\,C_2 + 1 \,. \qquad (18.3.12b)$$

From four variants

$$1. \qquad A_2 = -\frac{1}{2}\,(2\,B - m) \,, \qquad C_2 = -\frac{m}{2} \,,$$

2. $A_2 = +\dfrac{1}{2}\left(2\,B - m\right), \qquad C_2 = -\dfrac{m}{2}\,,$

3. $A_2 = +\dfrac{1}{2}\left(2\,B - m\right), \qquad C_2 = +\dfrac{m}{2}\,,$

4. $A_2 = -\dfrac{1}{2}\left(2\,B - m\right), \qquad C_2 = +\dfrac{m}{2}$

only 1 and 4 seem to be appropriate to describe bound states:

$$1\,, \qquad m < 0\,,$$

$$\alpha_2 = -B + \frac{1}{2} + \sqrt{B^2 - 2\,\epsilon\,M + \frac{1}{4}}\,,$$

$$\beta_2 = -B + \frac{1}{2} - \sqrt{B^2 - 2\,\epsilon\,M + \frac{1}{4}}\,,$$

$$\gamma_2 = -m + 1\,,$$

spectrum $\alpha_2 = -n\,, \qquad \sqrt{B^2 - 2\,\epsilon\,M + \frac{1}{4}} = B - \frac{1}{2} - n\,,$

$$\epsilon\,M = \frac{B}{2} + n\left(B - \frac{1}{2} - \frac{n}{2}\right);$$ (18.3.13a)

$$4\,, \qquad 0 < m < B\,,$$

$$\alpha_2 = -B + m + \frac{1}{2} + \sqrt{B^2 - 2\,\epsilon\,M + \frac{1}{4}}\,,$$

$$\beta_2 = -B + m + \frac{1}{2} - \sqrt{B^2 - 2\,\epsilon\,M + \frac{1}{4}}\,,$$

$$\gamma_2 = m + 1\,,$$

spectrum $\alpha_2 = -n\,, \qquad \sqrt{B^2 - 2\,\epsilon\,M + \frac{1}{4}} = B - \frac{1}{2} - (n + m)\,,$

$$\epsilon\,M = \frac{B}{2} + (m + n)\left(B - \frac{1}{2} - \frac{1}{2}(m + n)\right).$$ (18.3.13b)

Formulas in Eq. $(18.3.13a, b)$ can be joined into a single one

$$\sqrt{B^2 - 2\,\epsilon\,M + \frac{1}{4}} = -n - \frac{1}{2} - \frac{|\,2B - m\,| + |\,m\,|}{2}\,.$$ (18.3.13c)

The region for the allowed values of m for bound states is illustrated by Fig. 18.1.

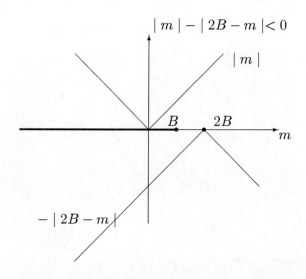

$$|\, m \,| - |\, 2B - m \,| < 0$$

Fig. 18.1 Bound states at $B > 0 : \ m < B$

At $B < 0$, provides a different set of values for m in Fig. 18.2.

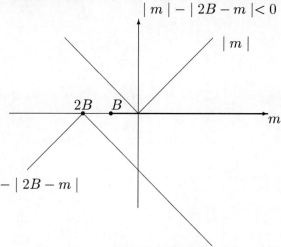

$$|\, m \,| - |\, 2B - m \,| < 0$$

Fig. 18.2 Bound states at $B < 0 : \ B < m$

Similar Figures can be given in connection with the functions Ψ_1 and Ψ_3 as well.

In case of Eq. $(18.3.7c)$, with the substitution

$$\Psi_3 = y^{C_3}(1 - y)^{A_3} f_3 \, ,$$

we will obtain

$$y\,(1-y)\,\frac{d^2 f_3}{dy^2} + \left[2\,C_3 + 1 - (2\,A_3 + 2\,C_3 + 2)\,y\right]\frac{df_3}{dy}$$

$$+ \left[B^2 + B - 2\,\epsilon\,M - (A_3 + C_3)\,(A_3 + C_3 + 1)\right.$$

$$+ \frac{1}{4}\frac{4\,A_3^2 - (2\,B - m + 1)^2}{1 - y} + \frac{1}{4}\frac{4\,C_3^2 - (m+1)^2}{y}\bigg]\,f_3 = 0\,. \qquad (18.3.14)$$

At $A_3,\,C_3$

$$A_3 = \pm\frac{1}{2}\,(2\,B - m + 1)\,, \qquad C_3 = \pm\frac{1}{2}\,(m+1)\,,$$

Equation (18.3.14) will read as

$$y\,(1-y)\,\frac{d^2 f_3}{dy^2} + \left[2\,C_3 + 1 - (2\,A_3 + 2\,C_3 + 2)\,y\right]\frac{df_3}{dy}$$

$$+ \left[B^2 + B - 2\,\epsilon\,M - (A_3 + C_3)\,(A_3 + C_3 + 1)\right] f_3 = 0\,, \qquad (18.3.15a)$$

which is a hypergeometric equation

$$\alpha_3 = A_3 + C_3 + \frac{1}{2} + \sqrt{B^2 + B - 2\,\epsilon\,M + \frac{1}{4}}\,,$$

$$\beta_3 = A_3 + C_3 + \frac{1}{2} - \sqrt{B^2 + B - 2\,\epsilon\,M + \frac{1}{4}}\,,$$

$$\gamma_3 = 2\,C_3 + 1\,. \qquad (18.3.15b)$$

From four possibilities

1. $A_3 = -\frac{1}{2}\,(2\,B - m + 1)\,, \qquad C_3 = -\frac{1}{2}\,(m+1)\,,$

2. $A_3 = +\frac{1}{2}\,(2\,B - m + 1)\,, \qquad C_3 = -\frac{1}{2}\,(m+1)\,,$

3. $A_3 = +\frac{1}{2}\,(2\,B - m + 1)\,, \qquad C_3 = +\frac{1}{2}\,(m+1)\,,$

4. $A_3 = -\frac{1}{2}\,(2\,B - m + 1)\,, \qquad C_3 = +\frac{1}{2}\,(m+1)\,,$

only 1 and 4 are appropriate to describe bound states:

$$1\,, \qquad m < 0\,,$$

$$\alpha_3 = -B - \frac{1}{2} + \sqrt{B^2 + B - 2\,\epsilon\,M + \frac{1}{4}}\,,$$

$$\beta_3 = -B - \frac{1}{2} - \sqrt{B^2 + B - 2\,\epsilon\,M + \frac{1}{4}}\,,$$

$$\gamma_3 = -m\,,$$

spectrum $\quad \alpha_3 = -n\,, \quad \sqrt{B^2 + B - 2\,\epsilon\,M + \dfrac{1}{4}} = B + \dfrac{1}{2} - n\,,$

$$\epsilon\,M = n\left(B + \dfrac{1}{2} - \dfrac{n}{2}\right)\,; \tag{18.3.16a}$$

$$4\,, \quad 0 < m < B\,,$$

$$\alpha_3 = -B + m + \dfrac{1}{2} + \sqrt{B^2 + B - 2\,\epsilon\,M + \dfrac{1}{4}}\,,$$

$$\beta_3 = -B + m + \dfrac{1}{2} - \sqrt{B^2 + B - 2\,\epsilon\,M + \dfrac{1}{4}}\,,$$

$$\gamma_3 = m + 2\,,$$

spectrum $\quad \alpha_3 = -n\,, \quad \sqrt{B^2 + B - 2\,\epsilon\,M + \dfrac{1}{4}} = B - \dfrac{1}{2} - (n + m)\,,$

$$\epsilon\,M = B + (m + n)\left(B - \dfrac{1}{2} - \dfrac{1}{2}\,(m + n)\right)\,. \tag{18.3.16b}$$

Again, formulas in Eq. (18.16a, b) can be joined into a single one

$$\sqrt{B^2 + B - 2\,\epsilon\,M + \dfrac{1}{4}} = -n - \dfrac{1}{2} - \dfrac{|\,2B - m + 1\,| + |\,m + 1\,|}{2}\,. \tag{18.3.16c}$$

18.4   Solution of Radial Equations in Relativistic Case

Let start with Eqs. (18.2.9) and (18.2.10)

$$-\hat{b}_- E_1 - \hat{a}_+ E_3 = M\,\Phi_0\,,$$

$$-i\hat{b}_- H_1 + i\hat{a}_+\,H_3 + i\epsilon\,E_2 = M\,\Phi_2\,,$$

$$i\hat{a}H_2 + i\epsilon\,E_1 = M\,\Phi_1\,,$$

$$-i\hat{b}H_2 + i\epsilon\,E_3 = M\,\Phi_3\,, \tag{18.4.1}$$

$$\hat{a}\Phi_0 - i\epsilon\,\Phi_1 = M\,E_1\,,$$

$$-i\hat{a}\Phi_2 = M\,H_1\,,$$

$$\hat{b}\Phi_0 - i\epsilon\,\Phi_3 = M\,E_3\,,$$

$$i\hat{b}\Phi_2 = M\,H_3\,,$$

$$-i\epsilon\Phi_2 = M\,E_2\,,$$

$$i\hat{b}_- \Phi_1 - i\hat{a}_+\Phi_3 = M\,H_2\,. \tag{18.4.2}$$

Excluding six components E_i, H_i, we derive four second order equations for Φ_a:

$$(-\hat{b}_-\hat{a} - \hat{a}_+\hat{b} + \epsilon^2 - M^2)\Phi_2 = 0\,,$$

$$(-\hat{b}_-\hat{a} - \hat{a}_+\hat{b} - M^2)\Phi_0 + i\epsilon(\hat{b}_-\Phi_1 + \hat{a}_+\Phi_3) = 0 \,,$$

$$(-\hat{a}\hat{b}_- + \epsilon^2 - M^2)\Phi_1 + \hat{a}\hat{a}_+\Phi_3 + i\epsilon\hat{a}\Phi_0 = 0 \,,$$

$$(-\hat{b}\hat{a}_+ + \epsilon^2 - M^2)\Phi_3 + \hat{b}\hat{b}_-\Phi_1 + i\epsilon\hat{b}\Phi_0 = 0 \,. \tag{18.4.3}$$

Once, it should be noted, the existence of a simple solution for the system

$$\Phi_0 = 0 \,, \qquad \Phi_1 = 0 \,, \qquad \Phi_3 = 0 \,,$$

$$(-\hat{b}_-\hat{a} - \hat{a}_+\hat{b} + \epsilon^2 - M^2)\Phi_2 = 0 \,, \tag{18.4.4a}$$

and simple expressions for tensors components

$$E_1 = 0 \,, \qquad H_1 = -iM^{-1}\hat{a}\,\Phi_2 \,,$$

$$E_3 = 0 \,, \qquad H_3 = iM^{-1}\hat{b}\,\Phi_2 \,,$$

$$E_2 = -i\epsilon M^{-1}\Phi_2 \,, \qquad H_2 = 0 \,. \tag{18.4.4b}$$

Let us turn to Eq. (18.4.3) and act on the third equation from the left by operator $\hat{b}_-$, and on the forth equation by operator $\hat{a}_+$. Thus, we introduce the notation

$$\hat{b}_-\Phi_1 = Z_1 \,, \qquad \hat{a}_+\Phi_3 = Z_3 \,,$$

instead of Eq. (18.4.3), we obtain

$$(-\hat{b}_-\hat{a} - \hat{a}_+\hat{b} + \epsilon^2 - M^2)\Phi_2 = 0 \,,$$

$$(-\hat{b}_-\hat{a} - \hat{a}_+\hat{b} - M^2)\Phi_0 + i\epsilon(Z_1 + Z_3) = 0 \,,$$

$$(-\hat{b}_-\hat{a} + \epsilon^2 - M^2)Z_1 + \hat{b}_-\hat{a}Z_3 + i\epsilon\hat{b}_-\hat{a}\Phi_0 = 0 \,,$$

$$(-\hat{a}_+\hat{b} + \epsilon^2 - M^2)Z_3 + \hat{a}_+\hat{b}Z_1 + i\epsilon\hat{a}_+\hat{b}\Phi_0 = 0 \,. \tag{18.4.5}$$

Instead of Z_1, Z_3, let us introduce new functions

$$Z_1 = \frac{f+g}{2} \,, \qquad Z_3 = \frac{f-g}{2} \,,$$

$$Z_1 + Z_3 = f \,, \qquad Z_1 - Z_3 = g \,;$$

This system will then read as

$$(-\hat{b}_-\hat{a} - \hat{a}_+\hat{b} + \epsilon^2 - M^2)\Phi_2 = 0 \,,$$

$$(-\hat{b}_-\hat{a} - \hat{a}_+\hat{b} - M^2)\Phi_0 + i\epsilon\,f = 0 \,,$$

$$-\hat{b}_-\hat{a}\frac{f+g}{2} + (\epsilon^2 - M^2)\frac{f+g}{2} + \hat{b}_-\hat{a}\frac{f-g}{2} + i\epsilon\hat{b}_-\hat{a}\Phi_0 = 0 \,,$$

$$-\hat{a}_+\hat{b}\frac{f-g}{2} + (\epsilon^2 - M^2)\frac{f-g}{2} + \hat{a}_+\hat{b}\frac{f+g}{2} + i\epsilon\hat{a}_+\hat{b}\Phi_0 = 0 \,. \tag{18.4.6}$$

After elementary manipulations with equation 3 and 4, we get

$$(-\hat{b}_-\hat{a} - \hat{a}_+\hat{b} + \epsilon^2 - M^2)\Phi_2 = 0 \,,$$

$$(-\hat{b}_-\hat{a} - \hat{a}_+\hat{b} - M^2)\Phi_0 + i\epsilon\, f = 0\,,$$

$$-\hat{b}_-\hat{a}g + (\epsilon^2 - M^2)\frac{f+g}{2} + i\epsilon\hat{b}_-\hat{a}\Phi_0 = 0\,,$$

$$\hat{a}_+\hat{b}g + (\epsilon^2 - M^2)\frac{f-g}{2} + i\epsilon\hat{a}_+\hat{b}\Phi_0 = 0\,.$$

Now, summing and subtracting equations 3 and 4, we obtain

$$(-\hat{b}_-\hat{a} - \hat{a}_+\hat{b} + \epsilon^2 - M^2)\Phi_2 = 0\,,$$

$$(-\hat{b}_-\hat{a} - \hat{a}_+\hat{b} - M^2)\Phi_0 + i\epsilon\, f = 0\,,$$

$$(-\hat{b}_-\hat{a} + \hat{a}_+\hat{b})g + (\epsilon^2 - M^2)f + i\epsilon(\hat{b}_-\hat{a} + \hat{a}_+\hat{b})\,\Phi_0 = 0\,,$$

$$(-\hat{b}_-\hat{a} - \hat{a}_+\hat{b})g + (\epsilon^2 - M^2)g + i\epsilon(\hat{b}_-\hat{a} - \hat{a}_+\hat{b})\Phi_0 = 0\,. \tag{18.4.7}$$

By taking into account the identities

$$-\hat{b}_-\hat{a} - \hat{a}_+\hat{b} = \Delta_2\,, \qquad -\hat{b}_-\hat{a} + \hat{a}_+\hat{b} = 2B\,, \tag{18.4.8}$$

we arrive at the system

$$(\Delta_2 + \epsilon^2 - M^2)\,\Phi_2 = 0\,, \tag{18.4.9}$$

$$(\Delta_2 - M^2)\,\Phi_0 + i\epsilon\, f = 0\,,$$

$$2B\, g + (\epsilon^2 - M^2)f - i\epsilon\Delta_2\,\Phi_0 = 0\,,$$

$$\Delta_2 g + (\epsilon^2 - M^2)g - 2i\epsilon B\,\Phi_0 = 0\,. \tag{18.4.10}$$

From the second equation, with the use of expression for $\Delta_2\Phi_0$ and according to the first equation, we derive a linear relation between three functions

$$2B\, g - M^2 f - i\epsilon M^2\Phi_0 = 0\,. \tag{18.4.11}$$

Let us exclude f

$$f = \frac{2B}{M^2}\, g - i\epsilon\Phi_0\,,$$

so we get

$$(\Delta_2 + \epsilon^2 - M^2)\, g = 2i\epsilon B\,\Phi_0\,,$$

$$(\Delta_2 + \epsilon^2 - M^2)\,\Phi_0 = -\frac{2i\epsilon B}{M^2}\, g\,. \tag{18.4.12}$$

In using the notation $\gamma = \epsilon^2/M^2$, the system can be presented in a matrix form as follows:

$$(\Delta_2 + \epsilon^2 - M^2)\begin{vmatrix} g \\ \epsilon\,\Phi_0 \end{vmatrix} = \begin{vmatrix} 0 & 2iB \\ -2iB\gamma & 0 \end{vmatrix}\begin{vmatrix} g \\ \epsilon\,\Phi_0 \end{vmatrix}\,, \tag{18.4.13}$$

or symbolically

$$\Delta f = A f \qquad \Delta f' = SAS^{-1} f'\,, \qquad f' = Sf\,.$$

Now, what remains is to find a transformation reducing the matrix A to a diagonal form

$$SAS^{-1} = \begin{vmatrix} \lambda_1 & 0 \\ 0 & \lambda_2 \end{vmatrix}, \qquad S = \begin{vmatrix} a & d \\ c & b \end{vmatrix};$$

the problem is equivalent to the linear systems:

$$-\lambda_1\, a - 2i\gamma B\, d = 0\,, \qquad 2iB\, a - \lambda_1\, d = 0\,;$$
$$-\lambda_2\, c - 2i\gamma B\, b = 0\,, \qquad 2iB\, c - \lambda_2\, b = 0\,.$$

Their solutions can be chosen in the form

$$\lambda_1 = +\frac{2\epsilon B}{M}\,, \qquad \lambda_2 = -\frac{2\epsilon B}{M}\,,$$

$$S = \begin{vmatrix} \epsilon & +iM \\ \epsilon & -iM \end{vmatrix}, \qquad S^{-1} = \frac{1}{-2i\epsilon M}\begin{vmatrix} -iM & -iM \\ -\epsilon & \epsilon \end{vmatrix}. \qquad (18.4.14)$$

New (primed) functions satisfy the following equations

$$1) \qquad \left(\Delta_2 + \epsilon^2 - M^2 - \frac{2\epsilon B}{M}\right) g' = 0\,, \qquad (18.4.15a)$$

$$2) \qquad \left(\Delta_2 + \epsilon^2 - M^2 + \frac{2\epsilon B}{M}\right) \Phi_0' = 0\,; \qquad (18.4.15b)$$

Since, they are independent from each other; there exists two solutions

$$1) \qquad g' \neq 0\,, \qquad \Phi_0' = 0\,; \qquad (18.4.16a)$$

$$2) \qquad g' = 0\,, \qquad \Phi_0' \neq 0\,. \qquad (18.4.16b)$$

The initial functions for these two cases will assume, respectively; the form

$$g = \frac{1}{2\epsilon}g' + \frac{1}{2i\epsilon}\epsilon\Phi_0'\,, \qquad \epsilon\Phi_0 = \frac{1}{2iM}g' - \frac{1}{2iM}\epsilon\Phi_0'\,. \qquad (18.4.17)$$

In cases 1) and 2), they assume, respectively; the form

$$1) \qquad g = \frac{1}{2\epsilon}g'\,, \qquad \epsilon\Phi_0 = \frac{1}{2iM}g'\,; \qquad (18.4.18a)$$

$$2) \qquad g = \frac{1}{2i\epsilon}\epsilon\Phi_0'\,, \qquad \epsilon\Phi_0 = -\frac{1}{2iM}\epsilon\Phi_0'\,. \qquad (18.4.18b)$$

To obtain explicit solutions for these differential equations, we do not need any additional calculations. Instead it suffices to perform some simple formal changes as shown by:

$$\left[\frac{d^2}{dr^2} + \frac{\cosh r}{\sinh r}\frac{d}{dr} - \frac{[m + B(\cosh r - 1)]^2}{\sinh^2 r} + 2\epsilon M\right] f(r) = 0\,,$$

$$\sqrt{B^2 - 2\epsilon M + \frac{1}{4}} = -n - \frac{1}{2} - \frac{|2B - m| + |m|}{2}\,, \qquad (18.4.19)$$

$$2\epsilon M \qquad \Longrightarrow \qquad \begin{cases} (\epsilon^2 - M^2 - \frac{2\epsilon B}{M})\,, \\ (\epsilon^2 - M^2)\,, \\ (\epsilon^2 - M^2 + \frac{2\epsilon B}{M})\,. \end{cases} \qquad (18.4.20)$$

Chapter 19

Particle with Spin 1 in a Magnetic Field on the Spherical Plane

19.1   Introduction

In the present chapter, we will construct exact solutions for a vector particle in external magnetic field on the background of a 2-dimensional spherical space S_2. A 10-dimensional Duffin–Kemmer equation for a vector particle in a curved space-time has the form

$$\left\{ \beta^c [\, i\hbar \,(\, e^\beta_{(c)} \partial_\beta + \frac{1}{2} J^{ab} \gamma_{abc}\,) \,+\, \frac{e}{c} A_c \,] \,-\, mc \right\} \Psi = 0 \,, \qquad (19.1.1)$$

where γ_{abc} stands for Ricci rotation coefficients, $A_a = e^\beta_{(a)} A_\beta$ represent tetrad components of electromagnetic 4-vector A_β; $J^{ab} = \beta^a \beta^b - \beta^b \beta^a$ are generators for a 10-dimensional representation of the Lorentz group. For shortness, we use the notation $e/c\hbar \implies e$, $mc/\hbar \implies M$. In the space S_3, we will use the system of cylindric coordinates

$$dS^2 = c^2 dt^2 - \rho^2 [\cos^2 z (dr^2 + \sin^2 r \, d\phi^2) + dz^2] \,,$$

$$z \in [-\pi/2, +\pi/2], \; r \in [0, +\pi] \,, \; \phi \in [0, 2\pi] \,. \qquad (19.1.2)$$

The generalized expression for a electromagnetic potential on a homogeneous magnetic field in the curved model S_3 is given as follows:

$$A_\phi = -2B \sin^2 \frac{r}{2} = B\,(\cos r - 1)\,. \qquad (19.1.3)$$

We will consider this equation in presence of the field for the model S_3. Corresponding to cylindric coordinates $x^\alpha = (t, r, \phi, z)$, a tetrad can be chosen as

$$e^\beta_{(a)}(x) = \begin{vmatrix} 1 & 0 & 0 & 0 \\ 0 & \cos^{-1} z & 0 & 0 \\ 0 & 0 & \cos^{-1} z \; \sin^{-1} r & 0 \\ 0 & 0 & 0 & 1 \end{vmatrix}. \qquad (19.1.4)$$

Equation (19.1.1) has the form

$$\left\{ i\beta^0 \frac{\partial}{\partial t} + \frac{1}{\cos z}\left(i\beta^1 \frac{\partial}{\partial r} + \beta^2 \frac{i\partial_\phi + eB(\cos r - 1) + iJ^{12}\cos r}{\sin r}\right)\right.$$

$$\left. + i\beta^3 \frac{\partial}{\partial z} + i\frac{\sin z}{\cos z}(\beta^1 J^{13} + \beta^2 J^{23}) - M \right\}\Psi = 0 . \tag{19.1.5}$$

To separate the variables in Eq. (19.1.5), we employ an explicit form of the Duffin–Kemmer matrices β^a. The most convenient method is to use the cyclic representation, where the generator J^{12} is the diagonal form (we specify the matrices by blocks in accordance with $(1 - 3 - 3 - 3)$-splitting)

$$\beta^0 = \begin{vmatrix} 0 & 0 & 0 & 0 \\ 0 & 0 & i & 0 \\ 0 & -i & 0 & 0 \\ 0 & 0 & 0 & 0 \end{vmatrix}, \qquad \beta^i = \begin{vmatrix} 0 & 0 & e_i & 0 \\ 0 & 0 & 0 & \tau_i \\ -e_i^+ & 0 & 0 & 0 \\ 0 & -\tau_i & 0 & 0 \end{vmatrix}, \tag{19.1.6}$$

where e_i, e_i^t, τ_i denote

$$e_1 = \frac{1}{\sqrt{2}}(-i,\, 0,\, i), \qquad e_2 = \frac{1}{\sqrt{2}}(1,\, 0,\, 1), \qquad e_3 = (0, i, 0),$$

$$\tau_1 = \frac{1}{\sqrt{2}}\begin{vmatrix} 0 & 1 & 0 \\ 1 & 0 & 1 \\ 0 & 1 & 0 \end{vmatrix}, \qquad \tau_2 = \frac{1}{\sqrt{2}}\begin{vmatrix} 0 & -i & 0 \\ i & 0 & -i \\ 0 & i & 0 \end{vmatrix}, \qquad \tau_3 = \begin{vmatrix} 1 & 0 & 0 \\ 0 & 0 & 0 \\ 0 & 0 & -1 \end{vmatrix} = s_3 . \tag{19.1.7}$$

The generator J^{12} explicitly reads

$$J^{12} = \beta^1\beta^2 - \beta^2\beta^1 = -i\begin{vmatrix} 0 & 0 & 0 & 0 \\ 0 & \tau_3 & 0 & 0 \\ 0 & 0 & \tau_3 & 0 \\ 0 & 0 & 0 & \tau_3 \end{vmatrix} = -iS_3 . \tag{19.1.8}$$

19.2   Restriction to 2-Dimensional Model

Let us restrict ourselves to 2-dimensional case, spherical space S_2 (formally it is sufficient in Eq. (19.1.5) to remove the dependence on the variable z, fixing its value by $z = 0$)

$$\left[i\beta^0 \frac{\partial}{\partial t} + i\beta^1 \frac{\partial}{\partial r} + \beta^2 \frac{i\partial_\phi + eB(\cos r - 1) + iJ^{12}\cos r}{\sin r} - M \right]\Psi = 0 . \tag{19.2.1}$$

Using the substitution

$$\Psi = e^{-i\epsilon t} e^{im\phi}\begin{vmatrix} \Phi_0(r) \\ \vec{\Phi}(r) \\ \mathbf{E}(r) \\ \mathbf{H}(r) \end{vmatrix}, \tag{19.2.2}$$

Equation (19.2.1) assumes the form (introducing the notation $m + B(1 - \cos r) = \nu(r)$)

$$\left[\epsilon \, \beta^0 + i\beta^1 \frac{\partial}{\partial r} - \beta^2 \frac{\nu(r) - \cos r \, S_3}{\sin r} - M \right] \begin{vmatrix} \Phi_0(r) \\ \vec{\Phi}(r) \\ \mathbf{E}(r) \\ \mathbf{H}(r) \end{vmatrix} = 0 . \qquad (19.2.3)$$

After the separation of variables, then using the notation

$$\frac{1}{\sqrt{2}} (\frac{\partial}{\partial r} + \frac{\nu - \cos r}{\sin r}) = \hat{a}_- , \quad \frac{1}{\sqrt{2}} (\frac{\partial}{\partial r} + \frac{\nu + \cos r}{\sin r}) = \hat{a}_+ , \quad \frac{1}{\sqrt{2}} (\frac{\partial}{\partial r} + \frac{\nu}{\sin r}) = \hat{a} ,$$

$$\frac{1}{\sqrt{2}} (-\frac{\partial}{\partial r} + \frac{\nu - \cos r}{\sin r}) = \hat{b}_- , \quad \frac{1}{\sqrt{2}} (-\frac{\partial}{\partial r} + \frac{\nu + \cos r}{\sin r}) = \hat{b}_+ , \quad \frac{1}{\sqrt{2}} (-\frac{\partial}{\partial r} + \frac{\nu}{\sin r}) = \hat{b} ,$$

we arrive at the radial system

$$-\hat{b}_- E_1 - \hat{a}_+ E_3 = M \, \Phi_0 , \qquad -i\hat{b}_- H_1 + i\hat{a}_+ H_3 + i\epsilon \, E_2 = M \, \Phi_2 ,$$

$$i\hat{a} H_2 + i\epsilon \, E_1 = M \, \Phi_1 , \qquad -i\hat{b} H_2 + i\epsilon \, E_3 = M \, \Phi_3 ,$$

$$(19.2.4)$$

$$\hat{a}\Phi_0 - i\epsilon \, \Phi_1 = M \, E_1 , \qquad -i\hat{a}\Phi_2 = M \, H_1 , \qquad \hat{b}\Phi_0 - i\epsilon \, \Phi_3 = M \, E_3 ,$$

$$i\hat{b}\Phi_2 = M \, H_3 . \qquad -i\epsilon\Phi_2 = M \, E_2 , \qquad i\hat{b}_- \Phi_1 - i\hat{a}_+ \Phi_3 = M \, H_2 .$$

$$(19.2.5)$$

19.3   Nonrelativistic Approximation

Excluding non-dynamical variables Φ_0, H_1, H_2, H_3 with the help of equations

$$-\hat{b}_- E_1 - \hat{a}_+ E_3 = M \, \Phi_0 , \qquad -i \, \hat{a}\Phi_2 = M \, H_1 ,$$

$$i\hat{b}_- \Phi_1 - i\hat{a}_+ \Phi_3 = M \, H_2 , \qquad i \, \hat{b}\Phi_2 = M \, H_3 , \qquad (19.3.1)$$

we get 6 equations (grouping them in pairs)

$$i\hat{a} \, (i\hat{b}_- \Phi_1 - i\hat{a}_+ \Phi_3) + i\epsilon \, ME_1 = M^2 \Phi_1 ,$$

$$\hat{a} \, (-\hat{b}_- E_1 - \hat{a}_+ E_3 - i\epsilon \, M\Phi_1 = M^2 E_1 , \qquad (19.3.2a)$$

$$-i\hat{b}_- \, (-i \, \hat{a} \, \Phi_2) + i\hat{a}_+ (i \, \hat{b} \, \Phi_2) + i\epsilon \, ME_2 = M^2 \Phi_2 ,$$

$$-i\epsilon \, M\Phi_2 = M^2 E_2 , \qquad (19.3.2b)$$

$$-i \, \hat{b} \, (i\hat{b}_- \Phi_1 - i\hat{a}_+ \Phi_3) + i\epsilon \, ME_3 = M^2 \Phi_3 ,$$

$$\hat{b} \, (-\hat{b}_- E_1 - \hat{a}_+ E_3) - i\epsilon \, M\Phi_3 = M^2 E_3 . \qquad (19.3.2c)$$

Now, we will introduce the big and small constituents

$$\Phi_1 = \Psi_1 + \psi_1 \,, \qquad \Phi_2 = \Psi_2 + \psi_2 \,, \qquad \Phi_3 = \Psi_3 + \psi_3 \,,$$

$$iE_1 = \Psi_1 - \psi_1 \,, \qquad iE_2 = \Psi_2 - \psi_2 \,, \qquad iE_3 = \Psi_3 - \psi_3 \,.$$

Here, we should separate the rest energy by a formal change $\epsilon \implies \epsilon + M$; summing and subtracting the equation within each pair in Eq. (19.3.2) and ignore the small constituents ψ_i. We arrive at three equations for the big components

$$\left(-2\,\hat{a}\hat{b}_- + 2\epsilon M \right) \Psi_1 = 0 \,,$$

$$\left(-\hat{b}_-\hat{a} - \hat{a}_+\hat{b} + 2\epsilon M \right) \Psi_2 = 0 \,,$$

$$\left(-2\hat{b}\hat{a}_+ + 2\epsilon M \right) \Psi_3 = 0 \,. \tag{19.3.3}$$

This is a needed for the Pauli-like system with the spin 1 particle.

To allow for $\nu(r) = m + B\,(1 - \cos r)$ from Eq. (19.3.3), we get radial equations in the form:

$$\left[\frac{d^2}{dr^2} + \frac{\cos r}{\sin r}\frac{d}{dr} - B - \frac{1 - 2\,[m + B\,(1 - \cos r)]\,\cos r}{\sin^2 r} \right.$$
$$\left. - \frac{[m + B\,(1 - \cos r)]^2}{\sin^2 r} + 2\,\epsilon\,M \right] \Psi_1 = 0 \,,$$

$$\tag{19.3.4a}$$

$$\left[\frac{d^2}{dr^2} + \frac{\cos r}{\sin r}\frac{d}{dr} - \frac{[m + B\,(1 - \cos r)]^2}{\sin^2 r} + 2\,\epsilon\,M \right] \Psi_2 = 0 \,,$$

$$\tag{19.3.4b}$$

$$\left[\frac{d^2}{dr^2} + \frac{\cos r}{\sin r}\frac{d}{dr} + B - \frac{1 + 2\,[m + B\,(1 - \cos r)]\,\cos r}{\sin^2 r} \right.$$
$$\left. - \frac{[m + B\,(1 - \cos r)]^2}{\sin^2 r} + 2\,\epsilon\,M \right] \Psi_3 = 0 \,.$$

$$\tag{19.3.4c}$$

The first and the third equations are symmetric with respect to the formal change $m \implies -m$, $B \implies -B$.

Let us consider Eq. (19.3.4a). In the new variable $1 - \cos r = 2\,y$, and with the use of a substitution $\Psi_1 = y^{C_1}(1 - y)^{A_1} f_1$ the differential equation assumes the form:

$$y\,(1 - y)\,\frac{d^2 f1}{dy^2} + [2\,C_1 + 1 - (2\,A_1 + 2\,C_1 + 2)\,y]\,\frac{df_1}{dy}$$

$$+ \left[B^2 + B + 2\,\epsilon\,M - (A_1 + C_1)\,(A_1 + C_1 + 1) \right.$$

$$+\frac{1}{4}\frac{4A_1^2-(2B+m+1)^2}{1-y}+\frac{1}{4}\frac{4C_1^2-(m-1)^2}{y}\Bigg]f_1=0\,. \qquad (19.3.5)$$

At A_1, C_1, and according to

$$A_1=+\frac{1}{2}\mid 2B+m+1\mid,\qquad C_1=+\frac{1}{2}\mid m-1\mid, \qquad (19.3.6)$$

Eq. (19.3.5) is recognized as a hypergeometric equation with parameters

$$\alpha_1=A_1+C_1+\frac{1}{2}-\sqrt{B^2+B+2\,\epsilon\,M+\frac{1}{4}}\,,$$

$$\beta_1=A_1+C_1+\frac{1}{2}+\sqrt{B^2+B+2\,\epsilon\,M+\frac{1}{4}}\,,$$

$$\gamma_1=2\,C_1+1\,,\qquad \Psi_1=y^{C_1}\,(1-y)^{A_1}\,F\,(\alpha_1,\,\beta_1,\,\gamma_1;\,y)\,. \qquad (19.3.7)$$

To get polynomials, we need to impose the restriction $\alpha_1=-n$. From this, it follows

$$\Psi_1=y^{C_1}(1-y)^{A_1}\,F(\alpha_1,\beta_1,\gamma_1;\,y)\,,$$

$$\sqrt{B^2+B+2\,\epsilon\,M+\frac{1}{4}}=n+\frac{1}{2}+\frac{\mid 2B+m+1\mid+\mid m-1\mid}{2}\,. \qquad (19.3.8)$$

In a similar manner, we can construct the solutions for Eq. (19.3.4b)

$$\Psi_2=y^{C_2}(1-y)^{A_2}\,F(\alpha_2,\beta_2,\gamma_2;\,y)\,,$$

$$A_2=\pm\frac{1}{2}\,(2\,B+m)\,,\quad C_2=\pm\frac{m}{2}\,,\quad \gamma_2=2\,C_2+1\,,$$

$$\alpha_2=A_2+C_2+\frac{1}{2}-\sqrt{B^2+2\,\epsilon\,M+\frac{1}{4}}\,,$$

$$\beta_2=A_2+C_2+\frac{1}{2}+\sqrt{B^2+2\,\epsilon\,M+\frac{1}{4}}\,; \qquad (19.3.9)$$

from the quantization condition for $\alpha_2=-n$, it follows the energy spectrum

$$\sqrt{B^2+2\,\epsilon\,M+\frac{1}{4}}=n+\frac{1}{2}+\frac{\mid 2B+m\mid+\mid m\mid}{2}\,. \qquad (19.3.10)$$

Finally, we construct solutions for Eq. (19.3.4c):

$$\Psi_3=y^{C_3}(1-y)^{A_3}\,F(\alpha_3,\beta_3,\gamma_3;\,y)\,,$$

$$A_3=\pm\frac{1}{2}\,(2\,B+m-1)\,,\quad C_3=\pm\frac{1}{2}\,(m+1)\,,\quad \gamma_3=2\,C_3+1\,,$$

$$\alpha_3=A_3+C_3+\frac{1}{2}-\sqrt{B^2-B+2\,\epsilon\,M+\frac{1}{4}}\,,$$

$$\beta_3=A_3+C_3+\frac{1}{2}+\sqrt{B^2-B+2\,\epsilon\,M+\frac{1}{4}}\,; \qquad (19.3.11)$$

and requiring $\alpha_3=-n$, we obtain

$$\sqrt{B^2-B+2\,\epsilon\,M+\frac{1}{4}}=n+\frac{1}{2}+\frac{\mid 2B+m-1\mid+\mid m+1\mid}{2}\,. \qquad (19.3.12)$$

19.4   Solution of Radial Equations in Relativistic Case

Let start with Eqs. (19.2.4) and (19.2.5). Excluding six components E_i, H_i with the help of Eq. (19.2.5), we derive four second order equations for Φ_a:

$$(-\hat{b}_-\hat{a} - \hat{a}_+\hat{b} + \epsilon^2 - M^2)\Phi_2 = 0 ,$$

$$(-\hat{b}_-\hat{a} - \hat{a}_+\hat{b} - M^2)\Phi_0 + i\epsilon(\hat{b}_-\Phi_1 + \hat{a}_+\Phi_3) = 0 ,$$

$$(-\hat{a}\hat{b}_- + \epsilon^2 - M^2)\Phi_1 + \hat{a}\hat{a}_+\Phi_3 + i\epsilon\hat{a}\Phi_0 = 0 ,$$

$$(-\hat{b}\hat{a}_+ + \epsilon^2 - M^2)\Phi_3 + \hat{b}\hat{b}_-\Phi_1 + i\epsilon\hat{b}\Phi_0 = 0 . \qquad (19.4.1)$$

Once, it should be noted, the existence of a simple solution for the system

$$\Phi_0 = 0 , \qquad \Phi_1 = 0 , \qquad \Phi_3 = 0 ,$$

$$(-\hat{b}_-\hat{a} - \hat{a}_+\hat{b} + \epsilon^2 - M^2)\Phi_2 = 0 , \qquad (19.4.2a)$$

at this; from Eq. (19.2.5) it follows:

$$E_1 = 0 , \qquad E_2 = -i\epsilon M^{-1}\Phi_2 , \qquad E_3 = 0 ,$$

$$H_1 = -iM^{-1}\hat{a}\,\Phi_2 , \qquad H_2 = 0 , \qquad H_3 = iM^{-1}\hat{b}\,\Phi_2 . \qquad (19.4.2b)$$

Lets us turn to Eq. (19.4.1) and act on the third equation from the left by operator $\hat{b}_-$. And for the fourth equation use the operator $\hat{a}_+$. Thus, introducing the notation

$$\hat{b}_-\Phi_1 = Z_1 , \qquad \hat{a}_+\Phi_3 = Z_3 ,$$

Instead of Eq. (19.4.1), we obtain

$$(-\hat{b}_-\hat{a} - \hat{a}_+\hat{b} + \epsilon^2 - M^2)\Phi_2 = 0 ,$$

$$(-\hat{b}_-\hat{a} - \hat{a}_+\hat{b} - M^2)\Phi_0 + i\epsilon(Z_1 + Z_3) = 0 ,$$

$$(-\hat{b}_-\hat{a} + \epsilon^2 - M^2)Z_1 + \hat{b}_-\hat{a}Z_3 + i\epsilon\hat{b}_-\hat{a}\Phi_0 = 0 ,$$

$$(-\hat{a}_+\hat{b} + \epsilon^2 - M^2)Z_3 + \hat{a}_+\hat{b}Z_1 + i\epsilon\hat{a}_+\hat{b}\Phi_0 = 0 . \qquad (19.4.3)$$

Instead of Z_1, Z_3, let us use new variables

$$Z_1 = \frac{f+g}{2} , \qquad Z_3 = \frac{f-g}{2} ,$$

$$Z_1 + Z_3 = f , \qquad Z_1 - Z_3 = g .$$

Here, the system assumes the form:

$$(-\hat{b}_-\hat{a} - \hat{a}_+\hat{b} + \epsilon^2 - M^2)\Phi_2 = 0 ,$$

$$(-\hat{b}_-\hat{a} - \hat{a}_+\hat{b} - M^2)\Phi_0 + i\epsilon\, f = 0 ,$$

$$-\hat{b}_-\hat{a}\frac{f+g}{2} + (\epsilon^2 - M^2)\frac{f+g}{2} + \hat{b}_-\hat{a}\frac{f-g}{2} + i\epsilon\hat{b}_-\hat{a}\Phi_0 = 0 ,$$

$$-\hat{a}_+\hat{b}\frac{f-g}{2} + (\epsilon^2 - M^2)\frac{f-g}{2} + \hat{a}_+\hat{b}\frac{f+g}{2} + i\epsilon\hat{a}_+\hat{b}\Phi_0 = 0 .$$

$$(19.4.4)$$

Summing and subtracting equations 3 and 4, we get

$$(-\hat{b}_-\hat{a} - \hat{a}_+\hat{b} + \epsilon^2 - M^2)\Phi_2 = 0 ,$$

$$(-\hat{b}_-\hat{a} - \hat{a}_+\hat{b} - M^2)\Phi_0 + i\epsilon\, f = 0 ,$$

$$(-\hat{b}_-\hat{a} + \hat{a}_+\hat{b})g + (\epsilon^2 - M^2)f + i\epsilon(\hat{b}_-\hat{a} + \hat{a}_+\hat{b})\,\Phi_0 = 0 ,$$

$$(-\hat{b}_-\hat{a} - \hat{a}_+\hat{b})g + (\epsilon^2 - M^2)g + i\epsilon(\hat{b}_-\hat{a} - \hat{a}_+\hat{b})\Phi_0 = 0 .$$

$$(19.4.5)$$

By taking into account the identities

$$-\hat{b}_-\hat{a} - \hat{a}_+\hat{b} = \Delta_2, \qquad -\hat{b}_-\hat{a} + \hat{a}_+\hat{b} = 2B ,\qquad (19.4.6)$$

Equations (19.4.5) reduce to the form

$$(\Delta_2 + \epsilon^2 - M^2)\,\Phi_2 = 0 ,\qquad (19.4.7)$$

$$(\Delta_2 - M^2)\,\Phi_0 + i\epsilon\, f = 0 ,$$

$$2B\, g + (\epsilon^2 - M^2)f - i\epsilon\Delta_2\,\Phi_0 = 0 ,$$

$$\Delta_2 g + (\epsilon^2 - M^2)g - 2i\epsilon B\,\Phi_0 = 0 .\qquad (19.4.8)$$

From the second equation, with the use of expression for $\Delta_2\Phi_0$ and according to the first equation; we derive a linear relation between three functions

$$2B\, g - M^2 f - i\epsilon M^2\Phi_0 = 0 .\qquad (19.4.9)$$

Using the help of Eq. (19.4.9), let us exclude f

$$f = \frac{2B}{M^2}\, g - i\epsilon\Phi_0$$

from equations 2 and 3:

$$(\Delta_2 + \epsilon^2 - M^2)\, g = 2i\epsilon B\,\Phi_0 ,$$

$$(\Delta_2 + \epsilon^2 - M^2)\,\Phi_0 = -\frac{2i\epsilon B}{M^2}\, g .\qquad (19.4.10)$$

In the notation $\gamma = \epsilon^2/M^2$, the system can be written in a matrix form

$$(\Delta_2 + \epsilon^2 - M^2)\begin{vmatrix} g \\ \epsilon\,\Phi_0 \end{vmatrix} = \begin{vmatrix} 0 & 2iB \\ -2iB\gamma & 0 \end{vmatrix}\begin{vmatrix} g \\ \epsilon\,\Phi_0 \end{vmatrix} ,\qquad (19.4.11)$$

or symbolically as

$$\Delta f = Af , \qquad \Delta f' = SAS^{-1}\, f' , \qquad f' = Sf .$$

It remains to find a transformation reducing the matrix A to a diagonal form

$$SAS^{-1} = \begin{vmatrix} \lambda_1 & 0 \\ 0 & \lambda_2 \end{vmatrix}, \qquad S = \begin{vmatrix} a & d \\ c & b \end{vmatrix} ;$$

the problem is equivalent to the linear system

$$-\lambda_1\, a - 2i\gamma B\, d = 0 , \qquad 2iB\, a - \lambda_1\, d = 0 ;$$

$$-\lambda_2\, c - 2i\gamma B\, b = 0 , \qquad 2iB\, c - \lambda_2\, b = 0 .$$

Its solutions can be chosen in the form

$$\lambda_1 = +\frac{2\epsilon B}{M} , \qquad \lambda_2 = -\frac{2\epsilon B}{M} , \; S = \begin{vmatrix} \epsilon & +iM \\ \epsilon & -iM \end{vmatrix} ,$$

$$S^{-1} = \frac{1}{-2i\epsilon M} \begin{vmatrix} -iM & -iM \\ -\epsilon & \epsilon \end{vmatrix} . \tag{19.4.12}$$

New (primed) function satisfy the following equations

$$1) \qquad \left(\Delta_2 + \epsilon^2 - M^2 - \frac{2\epsilon B}{M} \right) g' = 0 , \tag{19.4.13a}$$

$$2) \qquad \left(\Delta_2 + \epsilon^2 - M^2 + \frac{2\epsilon B}{M} \right) \Phi_0' = 0 . \tag{19.4.13b}$$

They are independent from each other. There exists two that are linearly independent

$$1) \qquad g' \neq 0, \qquad \Phi_0' = 0 ,$$

$$2) \qquad g' = 0, \qquad \Phi_0' \neq 0 . \tag{19.4.14}$$

The initial functions for these two cases can respectively assume the form

$$1) \qquad g = \frac{1}{2\epsilon} g' , \qquad \epsilon \Phi_0 = \frac{1}{2iM} g' ;$$

$$2) \qquad g = \frac{1}{2i\epsilon} \epsilon \Phi_0' , \qquad \epsilon \Phi_0 = -\frac{1}{2iM} \epsilon \Phi_0' .$$

In each case, Eqs. (19.4.10) has the same form that coincides with Eqs. (19.4.13a) and (19.4.13b), respectively. To obtain explicit solutions for these differential equations, we do not need any additional calculations. Instead, it suffices to perform simple formal changes as shown:

$$2\,\epsilon\, M \qquad \Longrightarrow \qquad \begin{cases} \left(\epsilon^2 - M^2 - \frac{2\epsilon B}{M} \right) , \\ \left(\epsilon^2 - M^2 \right) , \\ \left(\epsilon^2 - M^2 + \frac{2\epsilon B}{M} \right) . \end{cases} \tag{19.4.15}$$

Chapter 20

Electromagnetic Waves in de Sitter Space-Time

20.1 On the 5-Theory of a Massive Spin 1 Particle in de Sitter Space

In the study of examining the fundamental particle fields on the background of an expanding universe; in particular, for de Sitter and anti de Sitter models there has been a long history of research. This intriguing topic can be found in Dirac [454, 455], Schrödinger [48, 456], Lubanski and Rosenfeld [458], Goto [459], Ikeda [460], Nachtmann [461], Chernikov and Tagirov [462], Börner and Dürr [463], Tugov [464], Fushchych and Krivsky [465], Castagnino [466, 467], Vidal [468], Adler [469], Schnirman and Oliveira [470], Tagirov [471], Pestov, Chernikov, and Shavoxina [472], Candelas and Raine [473], Avis, Isham, and Storey [474], Brugarino [475], Fang and Fronsdal [476], Angelopoulos et al. [477], Burges [478], Deser and Nepomechie [479], Dullemond and Beveren [480], Gazeau [481], Allen [482], Flato, Fronsdal, and Gazeau [483], Allen et al. [484, 485], Sanchez [486], Pathinayake et al. [487], Gazeau and Hans [488], Bros et al. [489], Takook [490], Pol'shin [491–493], Gazeau and Takook [494], Deser and Waldron [495, 496], Spradlin et al. [497], Garidi et al. [498], Rouhani and Takook [499], Behroozi et al. [500], Huguet et al. [501], Garidi et al. [502], Huguet et al. [503], Dehghani et al. [504], Faci et al. [505].

The special value of these geometries consists in their simplicity and high symmetry underlying these groups. This makes us believe in the existence of an exact analytical treatment for some fundamental problems of classical and quantum field theory in curved spaces. In particular, there exist special representations for fundamental wave equations, Dirac's and Maxwell's, which are explicitly invariant under respective symmetry groups $SO(4.1)$ and $SO(3.2)$ for these models.

In this chapter, the wave equation in a 5-dimensional form for a massive particle for a spin 1 in the background of a de Sitter space-time model is solved in static coordinates (t, r, θ, ϕ). This covers part of de Sitter space-time in an event horizon. The spherical 5-dimensional waves $A_a(t, r, \theta, \phi), a = 1, ..., 5$ are constructed into three types, $j, j+1, j-1$. The massless case gives electromagnetic wave solutions that obey the Lorentz condition. For the group-theoretical 5-dimensional form, the equations in the massless case are used to produce a recipe to create electromagnetic wave solutions for the Π, E, M types. The first equation is trivial and can be removed by a gauge transformation. The recipe is applicable in arbitrary coordinates, and specified in static coordinates of the de Sitter space.

It is known that a wave equation for spin 1 field on the background of de Sitter space-time can be presented in a form explicitly invariant under the group $SO(4.1)$. To specify some details of that approach, let us start with covariant Proca equations

$$\nabla_\alpha \Psi_\beta - \nabla_\beta \Psi_\alpha = m \, \Psi_{\alpha\beta} \, , \qquad \nabla^\beta \Psi_{\alpha\beta} = m \, \Psi_\alpha \, , \qquad (20.1.1)$$

from this, it follows the equation for the vector Ψ_α

$$(\nabla^\beta \nabla_\beta + m^2)\Psi_\alpha - \nabla_\alpha(\nabla^\beta \Psi_\beta) - R_{\alpha\beta}\Psi^\beta = 0 \, . \qquad (20.1.2)$$

Because Ψ_α obeys the Lorentz condition

$$\nabla^\beta \Psi_\beta = 0 \, , \qquad (20.1.3)$$

Eq. (20.1.2) gives

$$(\nabla^\beta \nabla_\beta + m^2) \, \Psi^\beta - R_{\alpha\beta} \, \Psi^\beta = 0 \, . \qquad (20.1.4)$$

In the massless case, instead of Eq. (20.1.1) we have

$$\nabla_\alpha \, \Psi_\beta - \nabla_\beta \Psi_\alpha = \Psi_{\alpha\beta} \, , \qquad \nabla^\beta \Psi_{\alpha\beta} = 0 \, ; \qquad (20.1.5)$$

and the second order equation is

$$\nabla^\beta \nabla_\beta \Psi^\alpha - \nabla_\alpha(\nabla^\beta \Psi_\beta) - R_{\alpha\beta}\Psi^\beta = 0 \, . \qquad (20.1.6)$$

Equation (20.1.5) has a class of trivial (gauge) solutions

$$\tilde{\Psi}_\alpha = \nabla_\alpha \, f = \partial_\alpha \, f \, , \qquad \tilde{\Psi}_{\alpha\beta} = 0 \, , \qquad (20.1.7)$$

where $f(x)$ is an arbitrary scalar function. This fact often linked to the gauge symmetry

$$\Psi_\alpha(x) \sim \Psi_\alpha(x) + \partial_\alpha \, f(x) \, . \qquad (20.1.8)$$

Trivial solution $\tilde{\Psi}_\alpha(x)$ obeys the Lorentz condition, if $f(x)$ satisfies

$$\nabla^\alpha \nabla_\alpha \, f(x) \equiv \Delta \, f(x) = 0 \, . \qquad (20.1.9)$$

Now, let us specify the previous equations in conformal coordinates in de Sitter space

$$dS^2 = \frac{1}{\Phi^2} \left[(dx^0)^2 - (dx^1)^2 - (dx^2)^2 - (dx^3)^2 \right] ,$$

$$\Phi = (1 - x^2)/2 \, , \quad x^2 = (x^0)^2 - (x^1)^2 - (x^2)^2 - (x^3)^2 \, ; \qquad (20.1.10)$$

$x^0 = ct/\rho$ and so on. Here, it will be convenient to use coordinates

$$x_\alpha = \eta_{\alpha\beta}x^\beta \, , \quad \eta_{\alpha\beta} = \mathrm{diag}(+1, -1, -1, -1) \, ,$$

$$g_{\alpha\beta} = \frac{1}{\Phi^2}\eta_{\alpha\beta} \; , \; \partial_\alpha \Phi = -x_\alpha \; , \; \partial^\alpha \equiv \eta^{\alpha\beta}\partial_\beta \; . \tag{20.1.11}$$

Christoffel symbols are

$$\Gamma^\rho_{\alpha\beta} = \frac{1}{\Phi^2} \left(\delta^\rho_\alpha \, x_\beta \; - \; \delta^\rho_\beta \, x_\alpha \; - \; x^\rho \eta_{\alpha\beta} \right) ;$$

and Proca equations take the form

$$\partial_\alpha \Psi_\beta - \partial_\beta \Psi_\alpha = m \; \Psi_{\alpha\beta} \; , \qquad \Phi^2 \, \partial^\beta \Psi_{\alpha\beta} = m\Psi_\alpha \; . \tag{20.1.12}$$

In the massless case, we have

$$\partial_\alpha \Psi_\beta - \partial_\beta \Psi_\alpha = \Psi_{\alpha\beta} \; , \qquad \partial^\beta \Psi_{\alpha\beta} = 0 \; . \tag{20.1.13}$$

The Lorentz condition in these coordinates appears as

$$\partial^\alpha \, \Psi_\alpha = - \; \frac{2}{\Phi^2} \, x^\alpha \, \Psi_\alpha \; . \tag{20.1.14}$$

Now, starting with x^α, let us introduce five coordinates ξ^a

$$\xi^\alpha = \frac{x^\alpha}{\Phi} \; , \qquad (\alpha = 0, 1, 2, 3) \; , \qquad \xi^5 = \frac{1 + x^2}{1 - x^2} \; ,$$

$$x^\alpha = \frac{\xi^\alpha}{1 + \xi^5} \; , \qquad \Phi = \frac{1}{1 + \xi^5} \; , \qquad a = \alpha, 5 \; ; \tag{20.1.15}$$

they are characterized by

$$\frac{\partial \xi^\alpha}{\partial x^\beta} = \frac{1}{\Phi^2} \left(\Phi \, \delta^\alpha_\beta \; + \; x^\alpha x_\beta \right) , \qquad \frac{\partial \xi^5}{\partial x^\beta} = \frac{x_\beta}{\Phi^2} \; , \qquad \frac{\partial x^\alpha}{\partial \xi^\beta} = \Phi \, \delta^\alpha_\beta \; ,$$

$$\frac{\partial x^\alpha}{\partial \xi^5} = -\Phi \, x^\alpha \; , \qquad (\xi^0)^2 - (\xi^1)^2 - (\xi^2)^2 - (\xi^3)^2 - (\xi^5)^2 = -1 \; ,$$

$$dS^2 = \frac{1}{\Phi^2} \, \eta_{ab} \, dx^a dx^b = \eta_{\alpha\beta} \, d\xi^\alpha d\xi^\beta \; - \; (d\xi^5)^2 \; . \tag{20.1.16}$$

Therefore, de Sitter space can be identified with a sphere in 5-dimensional pseudo-Euclidean space, and thereby it has 10-parametric symmetry group $SO(4.1)$.

$$\xi^{a'} = S^a_{\ b} \, \xi^b \; , \qquad (S^a_{\ b}) \in SO(4.1) \; . \tag{20.1.17}$$

Instead of a 4-vector $\Psi^\alpha(x)$ (let it be designated as $a^\alpha(x)$) let us introduce 5-vector $A^a(\xi)$

$$A^a(\xi) = \frac{\partial \xi^a}{\partial x^\alpha} \, a^\alpha(x) \; ,$$

$$A^\alpha = \frac{1}{\Phi^2}(\Phi\delta^\alpha_\beta + x^\alpha x_\beta)a^\beta \; , \qquad A^5 = \frac{1}{\Phi^2}x_\alpha a^\alpha \; . \tag{20.1.18}$$

The vector $A^a(\xi)$ transforms as a 5-vector ξ^a under the group $SO(4.1)$

$$A^{a'}(\xi') = \frac{\partial \xi^{a'}}{\partial x^\alpha} \, a^\alpha(x) = [\, \frac{\partial}{\partial x^\alpha}(S^a_{\ b} \, \xi^b) \,] \, a^\alpha(x) = S^a_{\ b} \, A^b(\xi) \; . \tag{20.1.19}$$

The inverse relationship to Eq. (20.1.18) has the form

$$a^\alpha(x) = \frac{\partial x^\alpha}{\partial \xi^a} A^a = \Phi \left(A^\alpha - x^\alpha A^5 \right) . \tag{20.1.20}$$

Five variables $A^a(\xi)$ are not independent – the following condition holds

$$\xi^a A_a = 0 . \tag{20.1.21}$$

A wave equation for the 5-vector $A^a(\xi)$ invariant under the group $SO(4.1)$ should be constructed with the help of the following operator:

$$L_{ab} = \xi_a \frac{\partial}{\partial \xi^b} - \xi_b \frac{\partial}{\partial \xi^a} , \tag{20.1.22}$$

its possible form could be

$$\frac{1}{2} L^{ab} L_{ab} A_c + \kappa \, L_{ca} \, A^a + \sigma \, A_c = 0 , \tag{20.1.23}$$

where κ and σ are constants. It is readily verified that the Lorentz gauge has the following 5-form

$$L_{ab} A^b = A_a ; \tag{20.1.24}$$

Therefore, Eq. (20.1.23) appears as

$$\left(\frac{1}{2} L^{ab} L_{ab} + (\kappa + \sigma) \right) A_c = 0 . \tag{20.1.25}$$

Bearing in mind Eq. (20.1.16), we find

$$L_{\alpha\beta} = x_\alpha \frac{\partial}{\partial x^\beta} - x_\beta \frac{\partial}{\partial x^\alpha} , \quad L_{5\alpha} = -\Phi \frac{\partial}{\partial x^\alpha} + x^\beta \, L_{\alpha\beta} ; \tag{20.1.26}$$

and further

$$\frac{1}{2} L^{ab} L_{ab} = -\Phi^2 \left(\partial^\alpha \partial_\alpha + 2 \, \Phi \, x^\alpha \, \partial_\alpha \right) . \tag{20.1.27}$$

The later coincides with covariant d'Alembert operator in conformal coordinates

$$\frac{1}{2} L^{ab} L_{ab} = -\Delta . \tag{20.1.28}$$

Comparing Eq. (20.1.25) with Eq. (20.1.12), we find expression for $(\kappa + \sigma)$

$$\left(\frac{1}{2} L^{ab} L_{ab} + m^2 + 2 \right) A_c = 0 . \tag{20.1.29}$$

Setting here $m^2 = 0$, we get the wave equation for a massless field. Also, we should remember the solutions in Eqs. (20.1.21) and (20.1.24). Let us derive 5-form for this trivial solution

$$\tilde{a}_\alpha = \frac{\partial}{\partial x^\alpha} f , \quad \Delta f = 0 , \tag{20.1.30a}$$

transforming it to 5-form

$$\tilde{A}_\alpha = \left(\Phi \frac{\partial}{\partial x^\alpha} + x_\alpha x^\beta \frac{\partial}{\partial x^\beta} \right) f , \quad \tilde{A}_5 = -x^\alpha \frac{\partial}{\partial x^\alpha} f , \tag{20.1.30b}$$

or shortly

$$\tilde{A}_a = \left(\frac{\partial}{\partial \xi^a} + \xi_a \xi^b \frac{\partial}{\partial \xi^b} \right) f \equiv m_a f . \tag{20.1.30c}$$

In the following, we will use an identity $\Delta = m^a \, m_a , \quad m_a \, A^a = 0 .$

20.2   Spherical Waves in Static Coordinates, Massive Case

Equations for a vector particle will be solved in static coordinates in de Sitter space

$$(\Delta + m^2 + 2) A^b = 0 , \quad \xi_b A^b = 0 , \quad L_{ab}A^b = A_a , \tag{20.2.1}$$

$$dS^2 = (1 - r^2)dt^2 - \frac{dr^2}{1 - r^2} - r^2(d\theta^2 + \sin^2\theta d\phi^2) . \tag{20.2.2}$$

Coordinates $x^\alpha = (t, r, \theta, \phi)$ and ξ^a are referred by

$$\xi^1 = r\sin\theta\cos\phi , \quad \xi^2 = r\sin\theta\sin\phi , \quad \xi^3 = r\cos\theta ,$$

$$\xi^0 = \sinh\sqrt{1 - r^2} , \quad \xi^5 = \cosh\sqrt{1 - r^2} ,$$

$$t = \text{arccot}\,\frac{\xi^0}{\xi^5} , \quad r = \sqrt{(\xi^1)^2 + (\xi^2)^2 + (\xi^3)^2} ,$$

$$\theta = \text{arccot}\,\frac{\sqrt{(\xi^1)^2 + (\xi^2)^2}}{\xi^3} , \quad \phi = \text{arccot}\,\frac{\xi^2}{\xi^1} . \tag{20.2.3}$$

These coordinates (t, r, θ, ϕ) cover the part of the full space.

$$\xi^5 + \xi^0 \geq 0 , \quad \xi^5 - \xi^0 \geq 0 . \tag{20.2.4}$$

For any representation of the group $SO(4.1)$ on the functions $\Psi(\xi)$, we have relationship

$$\xi' = S\,\xi , \quad \Psi'(\xi') = U\,\Psi(\xi) \quad\Longrightarrow\quad \Psi'(\xi) = U\,\Psi(S^{-1}\,\xi) . \tag{20.2.5}$$

In the case $U \equiv S$ and $\Psi \equiv A$, the $(0 - 5)$-rotation

$$\xi^{0'} = \cosh\omega\,\xi^0 + \sinh\omega\,\xi^5 , \qquad \xi^{5'} = \sinh\omega\,\xi^0 + \cosh\omega\,\xi^5 ,$$

with an infinitesimal parameter $\delta\omega$ gives

$$A'(\xi) = (I + \delta\omega\,J_{50})\,A(\xi) , \qquad J_{50} = L_{50} + \sigma_{50} ,$$

$$L_{50} = \xi_5\frac{\partial}{\partial\xi^0} - \xi^0\frac{\partial}{\partial\xi^5} , \quad \sigma_{50} = \begin{vmatrix} 0 & 0 & 0 & 0 & 1 \\ 0 & 0 & 0 & 0 & 0 \\ 0 & 0 & 0 & 0 & 0 \\ 0 & 0 & 0 & 0 & 0 \\ 1 & 0 & 0 & 0 & 0 \end{vmatrix} . \tag{20.2.6}$$

General expression for generators is (where $g_{nb} = \text{diag}(+1, -1, -1, -1, -1)$)

$$(J_{mn})^a{}_b = L_{mn}\,\delta^a_b + (\sigma_{mn})^a{}_b , \quad (\sigma_{mn})^a{}_b = \delta^\alpha_m\,g_{nb} - \delta^a_n\,g_{mb} . \tag{20.2.7}$$

Let us search solutions for Eqs. (20.2.1) by diagonalizing three operators:

$$(-iJ_{50})^a{}_b\,A^b = \epsilon A^a ,$$

$$(\mathbf{J}^2)^a{}_b A^b = j(j+1)A^a \ , \qquad (J_3)^a{}_b A^b = m A^a \ , \tag{20.2.8}$$

where

$$J_k = -\frac{i}{2}\,\epsilon_{ijk}\,(L_{ij}\,+\,\sigma_{ij}) = l_k\,+\,s_k \ , \qquad s_1 = \begin{vmatrix} 0 & 0 & 0 & 0 & 0 \\ 0 & 0 & 0 & 0 & 0 \\ 0 & 0 & 0 & i & 0 \\ 0 & 0 & -i & 0 & 0 \\ 0 & 0 & 0 & 0 & 0 \end{vmatrix} \ ,$$

$$s_2 = \begin{vmatrix} 0 & 0 & 0 & 0 & 0 \\ 0 & 0 & 0 & -i & 0 \\ 0 & 0 & 0 & 0 & 0 \\ 0 & i & 0 & 0 & 0 \\ 0 & 0 & 0 & 0 & 0 \end{vmatrix} \ , \qquad s_3 = \begin{vmatrix} 0 & 0 & 0 & 0 & 0 \\ 0 & 0 & i & 0 & 0 \\ 0 & -i & 0 & 0 & 0 \\ 0 & 0 & 0 & 0 & 0 \\ 0 & 0 & 0 & 0 & 0 \end{vmatrix} \ .$$

First, the eigenfunction equation $(-iJ_{50})\, A = \epsilon A$ is to be solved. In using the identity $J_{50} = -\partial_t + \sigma_{50}$, we get

$$\mathbf{A} \sim e^{-i\epsilon t} \ , \qquad (A^0 + A^5) \sim e^{(-i\epsilon+1)t} \ ,$$

$$(A^0 - A^5) \sim e^{(-i\epsilon-1)t} \ . \tag{20.2.9}$$

Bearing in mind there are two other equations in Eq. (20.2.8). For the 5-vector A^a, we get the following substitution (three types of spherical vectors are used – see in [429])

$$\mathbf{A} = e^{-i\epsilon t}\left[f(r)\,\mathbf{Y}_{jm}^{(j+1)}\,+\,g(r)\,\mathbf{Y}_{jm}^{(j-1)}\,+\,h(r)\,\mathbf{Y}_{jm}^{(j)}\right] \ ,$$

$$A^0 = Y_{jm}\left[e^{(-i\epsilon+1)t}F(r) + e^{(-i\epsilon-1)t}G(r)\right] \ ,$$

$$A^5 = Y_{jm}\left[e^{(-i\epsilon+1)t}F(r) - e^{(-i\epsilon-1)t}G(r)\right] \ . \tag{20.2.10}$$

Radial functions $f(r)$, $g(r)$, $h(r)$, $F(r)$, $G(r)$ are to be constructed on the basis of Eqs. (20.2.1). Using the form of the operator Δ in variables (t, r, θ, ϕ)

$$\Delta = \frac{1}{1-r^2}\frac{\partial^2}{\partial t^2}\,-\,\frac{1}{r^2}\frac{\partial}{\partial r}r^2(1-r^2)\frac{\partial}{\partial r}\,-\,\frac{1}{r^2}\left(\frac{1}{\sin\theta}\frac{\partial}{\partial\theta}\sin\theta\frac{\partial}{\partial\theta}\,+\,\frac{1}{\sin^2\theta}\frac{\partial^2}{\partial\phi^2}\right) \ , \tag{20.2.11}$$

keep in mind the known action of $\mathbf{l}^2$ on spherical functions and vectors

$$\mathbf{l}^2 = -(\frac{1}{\sin\theta}\frac{\partial}{\partial\theta}\sin\theta\frac{\partial}{\partial\theta}\,+\,\frac{1}{\sin^2\theta}\frac{\partial^2}{\partial\phi^2}) \ ,$$

$$\mathbf{l}^2\,\mathbf{Y}_{jm}^{(\nu)} = \nu(\nu+1)\mathbf{Y}_{jm}^{(\nu)} \ , \quad \mathbf{l}^2\,Y_{jm} = j(j+1)Y_{jm} \ ,$$

for radial functions $f(r)$, $g(r)$, $h(r)$, $F(r)$, $G(r)$ we get equations of the same type

$$\left[\frac{d^2}{dr^2} + \frac{2(1-2r^2)}{r(1-r^2)}\frac{d}{dr} - \frac{\Lambda^2}{(1-r^2)^2} - \frac{m^2+2}{1-r^2} - \frac{\nu(\nu+1)}{r^2(1-r^2)}\right]U_{\Lambda,\nu} = 0 \ ;$$

$$(20.2.12a)$$

radial functions are given by

$$f = f_0\, U_{-i\epsilon,j+1}\ , \qquad g = g_0 U_{-i\epsilon,j-1}\ , \qquad h = h_0 U_{-i\epsilon,j}\ ,$$

$$F = F_0 U_{-i\epsilon+1,j}\ , \qquad G = G_0 U_{-i\epsilon+1,j}\ , \qquad (20.2.12b)$$

where f_0, g_0, h_0, F_0, G_0 are constants. Solutions for Eqs. (20.2.12a) can be expressed in terms of hypergeometric functions – let us write down those which are regular in $r = 0$ (let $z = r^2 = \sin^2 \omega$):

$$U_{-i\epsilon,j} = (\sin\omega)^j\, (\cos\omega)^{-i\epsilon}\, F(a,b,c;z)\ ,$$

$$U_{-i\epsilon,j+1} = (\sin\omega)^{j+1}\, (\cos\omega)^{-i\epsilon}\, F(a+1/2,b+1/2,c+1;z)\ ,$$

$$U_{-i\epsilon,j-1} = (\sin\omega)^{j-1}\, (\cos\omega)^{-i\epsilon}\, F(a-1/2,b-1/2,c-1;z)\ ,$$

$$U_{-i\epsilon+1,j} = (\sin\omega)^j\, (\cos\omega)^{-i\epsilon+1}\, F(a+1/2,b+1/2,c;z)\ ,$$

$$U_{-i\epsilon-1,j} = (\sin\omega)^j\, (\cos\omega)^{-i\epsilon-1}\, F(a-1/2,b-1/2,c;z)\ ;$$

$$(20.2.13a)$$

where

$$a = \frac{3/2 + j + i\sqrt{m^2 - 1/4} - i\epsilon}{2}\ ,$$

$$b = \frac{3/2 + j - i\sqrt{m^2 - 1/4} - i\epsilon}{2}\ , \qquad c = j + 3/2\ . \qquad (20.2.13b)$$

From this additional constraint $\xi_a A^a = 0$, for use of (see [429])

$$\xi\, \mathbf{Y}_{jm}^{(j)} = 0\ , \quad \xi\, \mathbf{Y}_{jm}^{(j+1)} = -\sqrt{\frac{j+1}{2j+1}}\, r\, Y_{jm}\ , \quad \xi\, \mathbf{Y}_{jm}^{(j+1)} = -\sqrt{\frac{j}{2j+1}}\, r\, Y_{jm}\ ,$$

one gets

$$-\sqrt{\frac{j+1}{2j+1}}\, r\, f + \sqrt{\frac{j}{2j+1}}\, r\, g + \sqrt{1-r^2}\, (G - F) = 0\ . \qquad (20.2.14)$$

From the Lorentz condition $L_{ab}\, A^b = A_a$, one gets (all details are omitted)

$$-\sqrt{\frac{j+1}{2j+1}}(\frac{d}{dr} + \frac{j+2}{r})\, f + \sqrt{\frac{j}{2j+1}}(\frac{d}{dr} - \frac{j-1}{r})g$$

$$-\frac{i\epsilon}{\sqrt{1-r^2}}(F + G) = 0\ , \qquad (20.2.15a)$$

$$-\sqrt{\frac{j+1}{2j+1}}\, r\, f + \sqrt{\frac{j}{2j+1}}\, r\, g + \sqrt{1-r^2}\, (F - G) = 0\ ; \qquad (20.2.15b)$$

the last equation coincides with Eq. (20.2.14). It is convenient for the expressions in $f(r)$ and $g(r)$ to separate special j-dependent factors

$$\sqrt{\frac{j+1}{2j+1}}\, f(r) = f_0\, U_{-i\epsilon,j+1}\,, \qquad \sqrt{\frac{j}{2j+1}}\, g(r) = g_0\, U_{-i\epsilon,j-1}\,. \qquad (20.2.16)$$

Keep in mind Eq. (20.2.16), where Eqs. (20.2.15a) and (20.2.15b) are changed to

$$f_0(-\frac{d}{d\omega} + \frac{j+2}{\tan\omega})\, U_{-i\epsilon,j+1} + g_0(\frac{d}{d\omega} - \frac{j-1}{\tan\omega})\, U_{-i\epsilon,j-1}$$

$$-i\epsilon(\, F_0 U_{-i\epsilon+1,j}\, +\, G_0 U_{-i\epsilon-1,j}\,) = 0\,, \qquad (20.2.17a)$$

$$-f_0\, \tan\omega\, U_{-i\epsilon,j+1}\, +\, g_0\, \tan\omega\, U_{-i\epsilon,j-1}$$

$$+\, F_0\, U_{-i\epsilon+1,j}\, -\, G_0 U_{-i\epsilon-1,j} = 0\,. \qquad (20.2.17b)$$

These relations may be resolved as follows

$$2\, F_0 U_{-i\epsilon+1,j} = f_0[-\frac{1}{i\epsilon}(-\frac{d}{d\omega} + \frac{j+2}{\tan\omega}) + \tan\omega]U_{-i\epsilon,j+1}$$

$$+g_0[+\frac{1}{i\epsilon}(\frac{d}{d\omega} - \frac{j-1}{\tan\omega}) - \tan\omega]U_{-i\epsilon,j-1}\,, \qquad (20.2.18a)$$

$$2G_0 U_{-i\epsilon-1,j} = f_0[-\frac{1}{i\epsilon}(-\frac{d}{d\omega} + \frac{j+2}{\tan\omega}) - \tan\omega]U_{-i\epsilon,j+1}$$

$$+g_0[+\frac{1}{i\epsilon}(\frac{d}{d\omega} - \frac{j-1}{\tan\omega}) + \tan\omega]U_{-i\epsilon,j-1}\,. \qquad (20.2.18b)$$

From these expressions, in terms of $U_{-i\epsilon\pm1,j}$ and $U_{-i\epsilon,j\pm1}$ for hypergeometric functions, we arrive at

$$2F_0 F(a+1/2, b+1/2, c; z) = f_0[-\frac{2j+3}{i\epsilon} F(a+1/2, b+1/2, c+1; z)$$

$$-\frac{2z}{i\epsilon}\frac{d}{dz}F(a+1/2, b+1/2, c+1; z)] + g_0\frac{2}{i\epsilon}\frac{d}{dz}F(a-1/2, b-1/2, c-1; z)\,, \qquad (20.2.19a)$$

$$2\, G_0 F(a-1/2, b-1/2, c; z) = f_0(1-z)[-\frac{2j+3}{i\epsilon} F(a+1/2, b+1/2, +1; z)$$

$$-\frac{2z}{1-z}F(a+1/2, b+1/2, c+1; z) - \frac{2z}{i\epsilon}\frac{d}{dz}F(a+1/2, b+1/2, c+1; z)]$$

$$+g_0[2F(a-1/2, b-1/2, c-1; z) + \frac{2z}{i\epsilon}(1-z)\frac{d}{dz}F(a-1/2, b-1/2, c-1; z)]\,. \qquad (20.2.19b)$$

By simplicity reason, let us search for solution types:

$$(j+1) - \text{wave}\,, \qquad f_0 \neq 0\,, \ g_0 = 0\,, \ h_0 = 0\,;$$

$$(j-1) - \text{wave} , \qquad f_0 = 0 , \ g_0 \neq 0 , \ h_0 = 0 ;$$

$$j - \text{wave} , \qquad f_0 = 0 , \ g_0 = 0 , \ h_0 \neq 0 . \qquad (20.2.20)$$

The task is to satisfy Eqs. (20.2.19a) and (20.2.19b) and determine the F_0 and G_0 in dependence of three factors f_0, g_0, h_0. For j-wave, from Eqs. (20.2.19a) and (20.2.19b) it follows $F_0 = 0$ and $G_0 = 0$; that is, in this case, the components A^0 and A^5 vanish. In the case of $(j-1)$-wave, the relation in Eq. (20.2.19a) takes the form

$$2\, F_0\, F(a+1/2, b+1/2, c; z) = g_0\, \frac{2z}{i\epsilon}\, \frac{d}{dz}\, F(a-1/2, b-1/2, c-1; z) ;$$

from this, with the use of the formula

$$\frac{d}{dz}\, F(\alpha, \beta, \gamma; z) = \frac{\alpha\beta}{\gamma}\, F(\alpha+1, \beta+1, \gamma+1; z)$$

we immediately produce

$$(j-1) - \text{wave} , \qquad F_0 = \frac{(a-1/2)\,(b-1/2)}{i\epsilon\,(c-1)} . \qquad (20.2.21a)$$

Equation (20.2.19b) for $(j-1)$-wave gives

$$2G_0\, F(a-1/2, b-1/2, c; z)$$

$$= 2\, g_0\, [\, F(a-1/2, b-1/2, c-1; z) + \frac{1-z}{i\epsilon}\, \frac{d}{dz}\, F(a-1/2, b-1/2, c-1; z)\,] ,$$

and further, by the use of the known relation

$$\frac{d}{dz}\, F(\alpha, \beta, \gamma; z) = \frac{\alpha+\beta+\gamma}{1-z}\, F(\alpha, \beta, \gamma; z) + \frac{(\alpha-\gamma)\,(\beta-\gamma)}{\gamma(1-z)}\, F(\alpha, \beta, \gamma+1; z)$$

we get the factor G_0

$$(j-1) - \text{wave} , \qquad G_0 = \frac{(a-c+1/2)\,(b-c+1/2)}{i\epsilon(c-1)}\, g_0 . \qquad (20.2.21b)$$

For $(j+1)$-wave, relation in Eq. (20.2.19a) gives

$$2\, F_0\, F(a+1/2, b+1/2, c; z)$$

$$= f_0\, [-\frac{2j+3}{i\epsilon}\, F(a+1/2, b+1/2, c+1; z) - \frac{2z}{i\epsilon}\, \frac{d}{dz}\, F(a+1/2, b+1/2, c+1; z)\,] ,$$

and further, with the use of the formula

$$z\, \frac{d}{dz}\, F(\alpha, \beta, \gamma; z) = (\gamma-1)\, [\, F(\alpha, \beta, \gamma-1; z) - F(\alpha, \beta, \gamma; z)\,] , \qquad (20.2.22a)$$

we get the factor F_0

$$(j+1) - \text{wave} , \qquad F_0 = \frac{c}{-i\epsilon}\, f_0 . \qquad (20.2.22b)$$

In turn, relation in Eq. (20.2.19b) takes the form

$$2G_0 F(a - 1/2, b - 1/2, c; z) = f_0(1 - z)\left[-\frac{2j + 3}{i\epsilon}F(a + 1/2, b + 1/2, c + 1; z)\right.$$

$$\left. -\frac{2z}{1 - z}F(a + 1/2, b + 1/2, c + 1; z) - \frac{2z}{i\epsilon}\frac{d}{dz}F(a + 1/2, b + 1/2, c + 1; z)\right],$$

$$(20.2.23a)$$

which with the use of (20.2.22a) gives

$$-i\epsilon\,\frac{G_0}{f_0}\,F(a - 1/2, b - 1/2, c; z)$$

$$= (c - a - b)zF(a + 1/2, b + 1/2, c + 1; z) + c(1 - z)F(a + 1/2, b + 1/2, c; z).$$

Let us differentiate the later example:

$$-i\epsilon\,\frac{G_0}{f_0}\,\frac{(a - 1/2)(b - 1/2)}{c}F(a + 1/2, b + 1/2, c + 1; z)$$

$$= (c - a - b)\,(z\frac{d}{dz} + 1)\,F(a + 1/2, b + 1/2, c + 1; z)$$

$$+ c\,(-1\,+\,(\frac{1 - z}{z})\,\frac{d}{dz}\,)\,F(a + 1/2, b + 1/2, c; z),$$

from this, we get

$$-i\epsilon\,\frac{G_0}{f_0}\,\frac{(a - 1/2)(b - 1/2)}{c}\,F(a + 1/2, b + 1/2, c + 1; z)$$

$$= c(-\frac{c - 1}{z} + 2c - a - b - 2)\,F(a + 1/2, b + 1/2, c; z)$$

$$+ c(c - 1)\,\frac{1 - z}{z}\,F(a + 1/2, b + 1/2, c - 1; z)$$

$$-(1 - c)(c - a - b)\,F(a + 1/2, b + 1/2, c; z). \qquad (20.2.23b)$$

Note, let us use the known identity for hypergeometric functions

$$c[(c - 1) - (2c - a - b - 2)z]F(a + 1/2, b + 1/2, c; z)$$

$$+(c - a - 1/2)(c - b - 1/2)zF(a + 1/2, b + 1/2, c + 1; z)$$

$$-c(c - 1)(1 - z)F(a + 1/2, b + 1/2, c - 1; z) = 0,$$

then Eq. (20.2.23b) results in

$$-i\epsilon\frac{G_0}{f_0}\frac{(a - 1/2)(b - 1/2)}{c}F(a + 1/2, b + 1/2, c + 1; z)$$

$$= [(c - a - 1/2)(c - b - 1/2) + (1 - c)(c - a - b)]F(a + 1/2, b + 1/2, c + 1; z);$$

therefore, the factor G_0 is given by

$$(j+1) - \text{wave} , \qquad G_0 = \frac{c}{-i\epsilon} f_0 . \qquad (20.2.23c)$$

The results obtained for the spherical solutions
$\nu = j$,

$$\mathbf{A} = e^{-i\epsilon t} h_0 \, U_{-i\epsilon,j} \, \mathbf{Y}_{jm}^{(j)}(\theta, \phi) , \ A^0 = 0 , \ A^5 = 0 ; \qquad (20.2.24a)$$

$\nu = j+1$,

$$\mathbf{A} = e^{-i\epsilon t} \mathbf{Y}_{jm}^{(j+1)} \sqrt{\frac{2j+1}{j+1}} , \ F_0 = -\frac{c}{i\epsilon} f_0 , \ G_0 = -\frac{c}{i\epsilon} f_0 ; \qquad (20.2.24b)$$

$\nu = j-1$,

$$\mathbf{A} = e^{-i\epsilon t} \mathbf{Y}_{jm}^{(j-1)} \sqrt{\frac{2j+1}{j}} \, g_0 \, U_{-i\epsilon,j-1} ,$$

$$F_0 = \frac{(a-1/2)(b-1/2)}{i\epsilon(c-1)} g_0 , \ \ G_0 = \frac{(a-c+1/2)(b-c+1/2)}{i\epsilon(c-1)} g_0 ;$$

$$(20.2.24c)$$

In the cases of Eqs. (20.2.24b) and (20.2.24c), the components A^0 and A^5 are to be calculated by the same formulas with different values of F_0 and G_0:

$$A^0 = Y_{jm} \left[e^{(-i\epsilon+1)t} F_0 \, U_{-i\epsilon+1,j} + e^{(-i\epsilon-1)t} G_0 \, U_{-i\epsilon-1,j} \right] ,$$

$$A^5 = Y_{jm} \left[e^{(-i\epsilon+1)t} F_0 \, U_{-i\epsilon+1,j} - e^{(-i\epsilon-1)t} G_0 \, U_{-i\epsilon-1,j} \right] .$$

$$(20.2.24d)$$

20.3   Method to Construct Electromagnetic Π, E, M-Waves

Let us recall the situation in the flat Minkowski space. If a scalar function $\Lambda(x)$ is a solution of the massless wave equation

$$\Delta \, \Lambda(x) = 0 , \ \Delta = \partial^0 \partial_0 - \partial^i \partial_i , \qquad (20.3.1)$$

the three linearly independent solutions of the vector wave equation $\Delta \, A^\alpha(x) = 0$ can be constructed as follows [50]:

$$A_\alpha^{(1)} = \frac{\partial}{\partial x^\alpha} \Lambda(x) , \qquad \mathbf{A}^{(2)} = \mathbf{r} \times \mathbf{A}^{(1)} , \qquad \mathbf{A}^{(3)} = \nabla \times \mathbf{A}^{(2)} . \qquad (20.3.2)$$

If the $\Lambda(x)$ is taken as a spherical wave

$$\Lambda(x) = e^{-i\epsilon t} \, Y_{jm}(\theta, \phi) \, f_j(\epsilon r) , \qquad (20.3.3)$$

$f_j(\epsilon r)$ is a Bessel spherical function, the recipe in Eq. (20.3.2) gives three spherical vector solutions [50]:

$$\mathbf{A}^{(1)} \sim e^{-i\epsilon t} \left[\sqrt{\frac{j}{2j+1}} \, f_{j-1} \, \mathbf{Y}_{jm}^{(j-1)} + \sqrt{\frac{j+1}{2j+1}} \, f_{j+1} \, \mathbf{Y}_{jm}^{(j+1)} \right] ,$$

$$A^{(1)0} \sim i\epsilon e^{-i\epsilon t} \, Y_{jm} f_j , \qquad \partial^\alpha A_\alpha^{(\Pi)} = 0 ;$$

$$\mathbf{A}^{(2)} \sim e^{-i\epsilon t} \, f_j(\epsilon r) \, \mathbf{Y}_{jm}^{(j)}(\theta, \phi) , \qquad A^{(2)0} = 0 , \qquad \text{div} \, \mathbf{A}^{(M)} = 0 ;$$

$$\mathbf{A}^{(3)} \sim e^{-i\epsilon t} \left[\sqrt{\frac{j+1}{2j+1}} \, f_{j-1} \, \mathbf{Y}_{jm}^{(j-1)} - \sqrt{\frac{j}{2j+1}} \, f_{j+1} \, \mathbf{Y}_{jm}^{(j+1)} \right] ,$$

$$A^{(3)0} = 0 , \qquad \text{div} \, \mathbf{A}^{(E)} = 0 , \qquad (20.3.4)$$

which are called, respectively, Π-, M-, E-waves.

The task is to extend this method to de Sitter space starting with 5-form:

$$(\Delta + 2)A^b = 0 , \quad \Delta = -\frac{1}{2} L^{ab} L_{ab} = m^a m_a ,$$

$$L_{ab} A^b = 0 , \quad \xi^a A_a = 0 . \qquad (20.3.5)$$

Evidently, the Π-wave should be determined by the rule

$$A_a^{(\Pi)} = m_a \, \Lambda(x) , \quad \Delta \, \Lambda(x) = 0 . \qquad (20.3.6)$$

Let us demonstrate with the help of the commutative relations that the 5-vector $m_a \Lambda(x)$ satisfies $(\Delta + 2)A^b = 0$. Indeed, keep in mind $[m_a, m_b] = L_{ab}$, one may obtain

$$\Delta m_a \Lambda(x) = [m^b m_b, m_a] \Lambda(x) = (L_{ab} m^b + m^b L_{ab}) \Lambda(x) ;$$

which with the help of $[m^b, L_{ab}] = -4m_a$ reduces to

$$\Delta \, m_a \, \Lambda(x) = (\, 2L_{ab} \, m^b + [\, m^b , \, L_{ab} \,] \,) \, \Lambda(x) = (\, 2L_{ab} \, m^b - 4m_a \,) \, \Lambda(x) ,$$

from this, it follows

$$(\Delta + 2)m_b \Lambda(x) = 0 . \qquad (20.3.7)$$

Let us define M-wave. One might expect the structure

$$\mathbf{A}^{(M)} = \vec{\xi} \times \mathbf{A}^{(\Pi)} = (\vec{\xi} \times \mathbf{m}) \, \Lambda(x) , \qquad \Delta\Lambda(x) = 0 . \qquad (20.3.8)$$

Because the operator $(\vec{\xi} \times \mathbf{m})_i = \epsilon_{ijk}\xi_j(\partial_k + \xi_k \xi^a \partial_a)$ commutes with Δ, the equation $\Delta(\vec{\xi} \times \mathbf{m})\Lambda(x) = 0$ holds. Hence, we can conclude that one must make the starting structure of Eq. (20.3.8) more exact

$$\mathbf{A}^{(M)} = (\vec{\xi} \times \mathbf{m})K(x) , \qquad (\Delta + 2)K(x) = 0 ; \qquad (20.3.9)$$

where the scalar function $K(x)$ satisfies the conformally invariant equation. It remains to prove the structure for E-wave:

$$\mathbf{A}^{(E)} = \mathbf{m} \times (\vec{\xi} \times \mathbf{m}) \, \Lambda(x) \,. \qquad (20.3.10)$$

Indeed, the relationship

$$(\Delta + 2) A_n^{(E)} = (\Delta + 2)\epsilon_{nij} m_i \epsilon_{jkl} \xi_k m_l \Lambda(x) = \frac{1}{2}\epsilon_{nij}\epsilon_{jkl}(\Delta + 2) m_i L_{kl} \Lambda(x) \,,$$

with the help of the identity $[m_i, L_{kl}] = (g_{ik}m_l - g_{il}m_k)$, transforms to

$$(\Delta + 2) A_n^{(E)} = \frac{1}{2}\epsilon_{nij}\epsilon_{jkl}(\Delta + 2)(L_{kl}m_i + g_{ik}m_l - g_{il}m_k)\Lambda(x)$$

$$= \frac{1}{2}\epsilon_{nij}\epsilon_{jkl}[L_{kl}(\Delta + 2)m_i + g_{ik}(\Delta + 2)m_l - g_{il}(\Delta + 2)m_k]\Lambda(x) \,;$$

from this, when taking into account $(\Delta + 2) \, m_b \, \Lambda(x) = 0$; one arrives at

$$(\Delta + 2) \, A_n^{(E)} = 0 \,. \qquad (20.3.11)$$

Thus, solutions of the types $\Pi-, M-, E-$ in de Sitter space are determined by

$$A_a^{(\Pi)} = m_a \, \Lambda(x) \,, \qquad \Delta \, \Lambda(x) = 0 \;;$$

$$\mathbf{A}^{(M)} = (\vec{\xi} \times \mathbf{m}) \, K(x) \,, \qquad (\Delta + 2) \, K(x) = 0 \;;$$

$$\mathbf{A}^{(E)} = \mathbf{m} \times (\vec{\xi} \times \mathbf{m}) \, \Lambda(x) \,, \quad \Delta \, \Lambda(x) = 0 \,. \qquad (20.3.12)$$

To obtain A^0 and A^5 for $M-, E-$ waves, one should use the equations

$$L_{ab} \, A^b = 0 \,, \qquad \xi^a \, A_a = 0 \,.$$

20.4   Spherical E, M, Π-Waves in Static Coordinates

Starting with static spherical coordinates in de Sitter space

$$dS^2 = (1 - r^2)dt^2 - \frac{dr^2}{1 - r^2} - r^2 \, (d\theta^2 + \sin^2 \theta d\phi^2) \,, \qquad (20.4.1)$$

a scalar solution for $\Delta \, \Lambda(x) = 0$, let us choose

$$\Delta \, \Lambda(x) = 0 \,, \quad \Lambda(x) = e^{-i\epsilon t} \, Y_{jm}(\theta, \phi) \, f(r) \,,$$

$$f(r) = \sin^j \omega \, (\cos \omega)^{-i\epsilon} \, F(a - 1/2, b + 1/2, c; z) \,,$$

$$a = \frac{j + 1 - i\epsilon}{2} \,, \quad b = \frac{j + 2 - i\epsilon}{2} \,, \quad c = j + 3/2 \,. \qquad (20.4.2)$$

We need an explicit form for m_b in these coordinates (t, r, θ, ϕ)

$$(m^0 + m^5) = e^{+t} \, [\, \frac{1}{\sqrt{1 - r^2}} \, \frac{\partial}{\partial t} + r \, \sqrt{1 - r^2} \, \frac{\partial}{\partial r} \,] \,,$$

$$(m^0 - m^5) = e^{-t} \left[\frac{1}{\sqrt{1-r^2}} \frac{\partial}{\partial t} - r \sqrt{1-r^2} \frac{\partial}{\partial r} \right] ,$$

$$(m^i) = \mathbf{m} = \left(-\nabla + \vec{\xi} \, r \frac{\partial}{\partial r} \right) , \quad \nabla = \frac{\partial}{\partial \vec{\xi}} . \tag{20.4.3}$$

Using the known properties of spherical functions [429], we get a Π-wave representation

$$\Pi - \text{wave} , \qquad (A^0 + A^5)^{\Pi} = e^{(-i\epsilon+1)t} \, Y_{jm} \left(-\frac{i\epsilon}{\cos\omega} + \sin\omega \frac{d}{d\omega} \right) f ,$$

$$(A^0 - A^5)^{\Pi} = e^{(-i\epsilon-1)t} \, Y_{jm} \left(-\frac{i\epsilon}{\cos\omega} - \sin\omega \frac{d}{d\omega} \right) f ,$$

$$\mathbf{A}^{\Pi} = e^{-i\epsilon t} \left[\sqrt{\frac{j+1}{2j+1}} \, \mathbf{Y}_{jm}^{(j+1)} (\cos\omega \frac{d}{d\omega} - \frac{j}{\sin\omega}) f \right.$$

$$\left. - \mathbf{Y}_{jm}^{(j-1)} (\cos\omega \frac{d}{d\omega} + \frac{j+1}{\sin\omega}) f \right] . \tag{20.4.4}$$

For M-wave, using identities [429]

$$\vec{\xi} \times \mathbf{Y}_{jm}^{(j+1)} = i \sqrt{\frac{j}{2j+1}} \, r \, \mathbf{Y}_{jm}^{(j)} , \qquad \vec{\xi} \times \mathbf{Y}_{jm}^{(j-1)} = i \sqrt{\frac{j+1}{2j+1}} \, r \, \mathbf{Y}_{jm}^{(j)} ,$$

we get

$$\mathbf{A}^{(M)} = \frac{i}{\sqrt{j(j+1)}} \, (\vec{\xi} \times \mathbf{m}) \, K(x) = e^{-i\epsilon t} \, \mathbf{Y}_{jm}^{(j)}(\theta, \phi) \, U_{-i\epsilon,j}(z) ; \tag{20.4.5}$$

where $U_{-i\epsilon,j}(z)$ is determined by

$$(\Delta + 2) K(x) = 0 , \quad K(x) = e^{-i\epsilon t} \, Y_{jm}(\theta, \phi) \, U_{-i\epsilon,j}(z) ,$$

$$U_{-i\epsilon,j}(z) = \sin^j \omega \, (\cos\omega)^{-i\epsilon} \, F(a,b,c;z) . \tag{20.4.6}$$

For E-wave, we get

$$\mathbf{A}^{(E)} = \frac{i}{\sqrt{j(j+1)}} \, [\, \mathbf{m} \times (\vec{\xi} \times \mathbf{m}) \,] \, \Lambda(x) = \mathbf{m} \times [\, e^{-i\epsilon t} \, \mathbf{Y}_{jm}^{(j)}(\theta, \phi) \, f(r) \,] , \tag{20.4.7a}$$

and further

$$\mathbf{A}^{(E)} = e^{-i\epsilon t} \left[\sqrt{\frac{j}{2j+1}} \, \mathbf{Y}_{jm}^{(j+1)} (\cos\omega \frac{d}{d\omega} - \frac{j}{\sin\omega}) f \right.$$

$$\left. + \sqrt{\frac{j+1}{2j+1}} \, \mathbf{Y}_{jm}^{(j-1)} (\cos\omega \frac{d}{d\omega} + \frac{j+1}{\sin\omega}) f \right] . \tag{20.4.7b}$$

Suppose that (Π, E)-waves are linear combinations of solutions $(j + 1)$ and $(j - 1)$ constructed in Section **20.2** (at $m = 0$), we reduce the task to study four relations

$$\left(-\frac{i\epsilon}{\cos\omega} + \sin\omega \frac{d}{dr}\right) f = \text{const } U_{-i\epsilon+1,j} , \qquad (20.4.8a)$$

$$\left(-\frac{i\epsilon}{\cos\omega} - \sin\omega \frac{d}{dr}\right) f = \text{const } U_{-i\epsilon-1,j} , \qquad (20.4.8b)$$

$$\left(\cos\omega \frac{d}{d\omega} - \frac{j}{\sin\omega}\right) f = \text{const } U_{-i\epsilon,j+1} , \qquad (20.4.8c)$$

$$\left(\cos\omega \frac{d}{d\omega} + \frac{j+1}{\sin\omega}\right) f = \text{const } U_{-i\epsilon,j-1} ; \qquad (20.4.8d)$$

the four constants are to be found. In the case of Eq. (20.4.8a), we have

$$\left(-\frac{i\epsilon}{\cos\omega} + \sin\omega \frac{d}{dr}\right) f$$

$$= \sin^j \omega \, (\cos\omega)^{-i\epsilon+1} \left[(j - i\epsilon) \, F(a - 1/2, b + 1/2, c; z) \right.$$

$$\left. + 2z \frac{d}{dz} F(a - 1/2, b + 1/2, c; z) \right] ,$$

from this, using the rule

$$z \frac{d}{dz} F(\alpha, \beta, \gamma; z) = \alpha \left[F(\alpha + 1, \beta, \gamma; z) - F(\alpha, \beta, \gamma; z) \right] ,$$

we get

$$\left(-\frac{i\epsilon}{\cos\omega} + \sin\omega \frac{d}{dr}\right) f$$

$$= \sin^j \omega (\cos\omega)^{-i\epsilon+1} (j - i\epsilon) F(a + 1/2, b + 1/2, c; z) = (j - i\epsilon) U_{-i\epsilon+1,j}. \qquad (20.4.9)$$

In the case of Eq. (20.4.8b)

$$\left(-\frac{i\epsilon}{\cos\omega} - \sin\omega \frac{d}{dr}\right) f$$

$$= \sin^j \omega \, (\cos\omega)^{-i\epsilon-1} [-(j - i\epsilon)F + (j + i\epsilon)zF - 2z \frac{d}{dz}F]$$

$$= \sin^j \omega (\cos\omega)^{-i\epsilon-1} [-2i\epsilon \, F(a - 1/2, b + 1/2, c; z)$$

$$-2(a - 1/2)(1 - z)F(a + 1/2, b + 1/2, c; z)] ;$$

and further, with the use of the known identity for hypergeometric functions we get

$$[(\gamma - \alpha - \beta) \, F(\alpha, \beta, \gamma; z)$$

$$+ \alpha \, (1 - z) \, F(\alpha + 1, \beta, \gamma; z) - (\gamma - \beta) \, F(\alpha, \beta - 1, \gamma; z)] = 0$$

(at $\alpha = a - 1/2, \beta = b + 1/2, \gamma = c$); therefore

$$\left(-\frac{i\epsilon}{\cos\omega} + \sin\omega \frac{d}{dr}\right) f$$

$$= -(j + i\epsilon) \sin^j \omega (\cos\omega)^{-i\epsilon+1} F(a - 1/2, b - 1/2, c; z) = -(j + i\epsilon) U_{-i\epsilon-1,j} \; .$$

$$(20.4.10)$$

Let us consider Eq. $(20.4.8c)$:

$$\left(\cos\omega \frac{d}{d\omega} - \frac{j}{\sin\omega} \right) f = \sin^{j+1} \omega \; (\cos\omega)^{-i\epsilon} [(-j + i\epsilon) F + 2(1 - z) \frac{d}{dz} F] \; .$$

Here, we allow for the rule

$$(1 - z) \frac{d}{dz} F(\alpha, \beta, \gamma; z) = [(\alpha + \beta - \gamma) F(\alpha, \beta, \gamma; z)$$

$$+ \frac{(\alpha - \gamma)(\beta - \gamma)}{\gamma} F(\alpha, \beta, \gamma + 1; z) \, ,$$

we arrive at

$$\left(\cos\omega \frac{d}{d\omega} - \frac{j}{\sin\omega} \right) f$$

$$- \frac{j + i\epsilon}{c} \sin^{j+1} \omega \; (\cos\omega)^{-i\epsilon} [\, cF(a - 1/2, b + 1/2, c; z)$$

$$+ (a - 1/2 - c) F(a - 1/2, b + 1/2, c + 1; z)]$$

$$= -\frac{j + i\epsilon}{2} \sin^{j+1} \omega \; (\cos\omega)^{-i\epsilon} \frac{j - i\epsilon}{2} F(a + 1/2, b + 1/2, c + 1; z)] \; ;$$

and further, get

$$\left(\cos\omega \frac{d}{d\omega} - \frac{j}{\sin\omega} \right) f = -\frac{j^2 + \epsilon^2}{2j + 3} U_{-i\epsilon,j+1} \; . \qquad (20.4.11)$$

Finally, for the case of Eq. $(20.4.8d)$; we have

$$\left(\cos\omega \frac{d}{d\omega} + \frac{j + 1}{\sin\omega} \right) f$$

$$= \sin^{j-1} \omega \; (\cos\omega)^{-i\epsilon} [\, (2j + 1) F + z (i\epsilon - j) F + 2z(1 - z) \frac{d}{dz} F] \; ;$$

from this, using the formula

$$z \frac{d}{dz} F(\alpha, \beta, \gamma; z) = (\gamma - 1) [F(\alpha, \beta, \gamma + 1; z) - F(\alpha, \beta, \gamma; z)]$$

we get

$$\left(\cos\omega \frac{d}{d\omega} + \frac{j + 1}{\sin\omega} \right) f = \sin^{j-1} \omega \; (\cos\omega)^{-i\epsilon}$$

$$\times [\, 2(c - 1)(1 - z) F(a - 1/2, b + 1/2, c + 1; z)$$

$$+ 2(c - a - 1/2)z \, F(a - 1/2, b + 1/2, c; z)] \; .$$

The term in square brackets equals to

$$(2j + 1) F(a - 1/2, b - 1/2, c - 1; z) \, ,$$

therefore, we have obtained the following:

$$\left(\cos\omega \frac{d}{d\omega} \;+\; \frac{j+1}{\sin\omega} \right) f = (2j+1)\, U_{-i\epsilon,j-1} \,. \tag{20.4.12}$$

From the results collected, we get spherical waves of three types spherical solutions

$$\Pi - \text{wave}\,, \quad (A^0 + A^5)^\Pi = e^{(-i\epsilon+1)t}\,(+j-i\epsilon)\,U_{-i\epsilon+1,j}\,Y_{jm}\,,$$

$$(A^0 - A^5)^\Pi = e^{(-i\epsilon-1)t}\,(-j-i\epsilon)\,U_{-i\epsilon-1,j}\,Y_{jm}\,,$$

$$\mathbf{A}^\Pi = e^{-i\epsilon t}\,[\sqrt{\frac{j+1}{2j+1}}\,\mathbf{Y}_{jm}^{(j+1)}\,\frac{-(j^2+\epsilon^2)}{2j+3}\,U_{-i\epsilon,j+1}$$

$$- \sqrt{\frac{j}{2j+1}}\,\mathbf{Y}_{jm}^{(j-1)}\,(2j+1)\,]\,U_{-i\epsilon,j-1}\,; \tag{20.4.13a}$$

$$E - \text{wave}\,, \quad \mathbf{A}^{(E)} = e^{-i\epsilon t}\,[\,\sqrt{\frac{j}{2j+1}}\,\mathbf{Y}_{jm}^{(j+1)}\,\frac{-(j^2+\epsilon^2)}{2j+3}\,U_{-i\epsilon,j+1}$$

$$+ \sqrt{\frac{j+1}{2j+1}}\,\mathbf{Y}_{jm}^{(j-1)}\,(2j+1)\,]\,U_{-i\epsilon,j-1}\,; \tag{20.4.13b}$$

$$M - \text{wave}\,, \quad \mathbf{A}^{(M)} = e^{-i\epsilon t}\,\mathbf{Y}_{jm}^{(j)}(\theta,\phi)\,U_{-i\epsilon,j}(z)\,. \tag{20.4.13c}$$

Components A^0 and A^5 for $M-$, E-waves can be found in comparing vector parts $\mathbf{A}$ for these waves; this is given by

$$\nu = j\,, \qquad \mathbf{A} = e^{-i\epsilon t}\,U_{-i\epsilon,j}\,\mathbf{Y}_{jm}^{(j)}(\theta,\phi)\,, \quad h_0 = 1\,, \quad A^0 = 0\,, \quad A^5 = 0\,;$$

$$\nu = j+1\,, \qquad f_0 = \sqrt{\frac{j+1}{2j+1}}\,, \quad \mathbf{A} = e^{-i\epsilon t}\,\mathbf{Y}_{jm}^{(j+1)}\,U_{-i\epsilon,j+1}\,,$$

$$(A^0 \pm A^5) = e^{(-i\epsilon\pm1)t}\,Y_{jm}\,\frac{2j+3}{-i\epsilon}\,\sqrt{\frac{j+1}{2j+1}}\,;$$

$$\nu = j-1\,, \qquad g_0 = \sqrt{\frac{j}{2j+1}}\,, \quad \mathbf{A} = e^{-i\epsilon t}\,\mathbf{Y}_{jm}^{(j-1)}\,U_{-i\epsilon,j-1}\,,$$

$$(A^0 \pm A^5) = e^{(-i\epsilon\pm1)t}\,Y_{jm}\,[\,\frac{(j\mp i\epsilon)\,(j\mp i\epsilon+1)}{i\epsilon(2j+1)}\,\sqrt{\frac{j+1}{2j+1}}\,]\,U_{-i\epsilon\pm1,j}\,.$$

$$\tag{20.4.14}$$

Let us compare these (Π, M, E)- and $(j, j\pm1)$-waves. We see that M- and j-solutions coincide. For Π wave, we see vector parts

$$\mathbf{A}^{(\Pi)} = -\sqrt{\frac{j+1}{2j+1}}\,\frac{j^2+\epsilon^2}{2j+3}\,\mathbf{A}^{(j+1)} \;-\; \sqrt{\frac{j}{2j+1}}\,(2j+1)\,\mathbf{A}^{(j-1)}\,; \tag{20.4.15a}$$

Also these equalities hold

$$(A^0 \pm A^5)^{(\mathrm{II})} = -\sqrt{\frac{j+1}{2j+1}\frac{j^2+\epsilon^2}{2j+3}}\,(A^0 \pm A^5)^{(j+1)}$$

$$-\sqrt{\frac{j}{2j+1}}(2j+1)\,(A^0 \pm A^5)^{(j-1)}\,. \tag{20.4.15b}$$

For E-wave, we have

$$\mathbf{A}^{(E)} = [\,-\sqrt{\frac{j}{2j+1}\frac{j^2+\epsilon^2}{2j+3}}\,\mathbf{A}^{(j+1)} + \sqrt{\frac{j+1}{2j+1}}\,(2j+1)\,\mathbf{A}^{(j-1)}\,]\,; \tag{20.4.16a}$$

from this, it follows A^0 and A^5 for E-wave:

$$(A^0 \pm A^5)^{(E)} = \frac{j-i\epsilon}{i\epsilon}\,\sqrt{j(j+1)}\,e^{(-i\epsilon\pm 1)t}\,Y_{jm}\,U_{-i\epsilon\pm 1,j}\,. \tag{20.4.16b}$$

20.5   Correspondence Principle in de Sitter Model

Let us show that the solutions constructed in de Sitter space are in accordance with the correspondence principle. Namely, they give the known results in the limit of flat Minkowski space. Let us specify the case of Π-wave.

Starting with limiting identities

$$\lim_{\rho\to\infty}[\,r^{j+1}\,U_{-i\epsilon,j+1}\,] = \frac{2^p}{\epsilon}\,\Gamma(1+p)\,\frac{J_p(\epsilon r)}{\sqrt{r}}\,, \qquad p = j+3/2\,,$$

$$\lim_{\rho\to\infty}[\,r^{j-1}\,U_{-i\epsilon,j-1}\,] = \frac{2^{p'}}{\epsilon}\,\Gamma(1+p')\,\frac{J_{p'}(\epsilon r)}{\sqrt{r}}\,, \qquad p' = j-1/2\,, \tag{20.5.1}$$

ρ is the curvature radius, $J_\nu(x)$ stands for the Bessel function we readily produce

$$\lim_{\rho\to\infty}[r^{j-1}\mathbf{A}^{(\mathrm{II})}] = e^{-i\epsilon t}\lim_{\rho\to\infty}[\,-\sqrt{\frac{j}{2j+1}\frac{j^2+\epsilon^2}{\rho^2(2j+3)}}\,\rho^{j+1}\mathbf{Y}_{jm}^{(j+1)}U_{-i\epsilon,j+1}$$

$$-\sqrt{\frac{j}{2j+1}}\rho^{j-1}\mathbf{Y}_{jm}^{(j-1)}U_{-i\epsilon,j-1}] = e^{-i\epsilon t}\,[\,(2j+1)\,\Gamma(j+1/2)\,\frac{2^{j-1/2}}{\epsilon}\,]\,\frac{1}{\sqrt{r}}$$

$$\times\left[\sqrt{\frac{j+1}{2j+1}}\,\mathbf{Y}_{jm}^{(j+1)}\,J_{j+3/2}(\epsilon r) + \sqrt{\frac{j}{2j+1}}\,\mathbf{Y}_{jm}^{(j-1)}\,J_{j-1/2}(\epsilon r)\right]\,; \tag{20.5.2}$$

or with notation $f_\nu(x) = \sqrt{\pi/2x}\,J_{\nu+1/2}(x)$ we get

$$\lim_{\rho\to\infty}[\rho^{j-1}\mathbf{A}^{(\mathrm{II})}] \sim e^{-i\epsilon t}\left[\sqrt{\frac{j+1}{2j+1}}\mathbf{Y}_{jm}^{(j+1)}\,f_{j+1}(\epsilon r) + \sqrt{\frac{j}{2j+1}}\mathbf{Y}_{jm}^{(j-1)}f_{j-1}(\epsilon r)\right]\,,$$

$$\tag{20.5.3}$$

which agrees with the known representation for Π wave in flat space. Besides, keep in mind the identities

$$\lim_{\rho \to \infty} [\, \rho^j \, U_{-i\epsilon \pm 1, j} \,] = \frac{2^p}{\epsilon} \, \Gamma(1+p) \, \frac{J_p(\epsilon r)}{\sqrt{r}} \,, \qquad p = j + 3/2 \,, \qquad (20.5.4)$$

we find the limiting form for $A^{0(\Pi)}$ and $A^{5(\Pi)}$:

$$\lim_{\rho \to \infty} [\, \rho^{j-1}(A^0 \pm A^5)^{(\Pi)} \,]$$

$$= \lim_{\rho \to \infty} \left[\frac{\pm j - i\epsilon\rho}{\rho} \, \exp[(-i\frac{E\rho}{\hbar c} \pm 1)\frac{ct}{\rho}] \, Y_{jm}(\theta, \phi) \, \rho^j \, U_{-i\epsilon \pm 1, j} \right] ; \qquad (20.5.5)$$

from this, it follows

$$\lim_{\rho \to \infty} [\, \rho^{j-1} A^{5(\Pi)} \,] = 0 \,, \quad \lim_{\rho \to \infty} [\, \rho^{j-1} A^{0(\Pi)} \,] \sim e^{-i\epsilon t} \, Y_{jm}(\theta, \phi) \, f_j(\epsilon r) \,, \qquad (20.5.6)$$

this agrees with the known representation for Π wave in flat space [50].

Chapter 21

Spherical Waves of Spin 1 Particle in Anti de Sitter Space-Time

21.1   Introduction

The study to examine the fundamental particle fields on the background of an expanding universe; in particular de Sitter and anti de Sitter models, has a long history. The special value of these geometries consists in their simplicity and underlying high symmetry groups. This makes us believe in the existence of an exact analytical treatment for some fundamental problems of classical and quantum field theory in these curved spaces. In particular, there exists special representations for fundamental wave equations, Dirac's and Maxwell's, which are explicitly invariant under symmetry groups $SO(4.1)$ and $SO(3.2)$ for these models – see references to chapter **20**. In this chapter, we consider the case of spin 1 particle in anti de Sitter model.

21.2   Separation of Variables in DKP-Equation in Anti de Sitter Space

We start our analysis of the spin 1 field with the use of the old and conventional tetrad formalism applied to matrix   Duffin–Kemmer–Petiau formalism. Using a diagonal static spherical tetrad in anti de Sitter space-time $x^\alpha = (t, r, \theta, \phi)$ and corresponding to Ricci coefficients

$$dS^2 = (1 + r^2) \, dt^2 - \frac{dr^2}{1 + r^2} - r^2 (d\theta^2 + \sin^2\theta d\phi^2) \,,$$

$$\Phi = 1 + r^2 \,, \quad g_{\alpha\beta} = \begin{vmatrix} \Phi & 0 & 0 & 0 \\ 0 & -1/\Phi & 0 & 0 \\ 0 & 0 & -r^2 & 0 \\ 0 & 0 & 0 & -r^2 \sin^2\theta \end{vmatrix} \,,$$

$$e^\alpha_{(0)} = (\frac{1}{\sqrt{\Phi}}, 0, 0, 0) \,, \qquad e^\alpha_{(3)} = (0, \sqrt{\Phi}, 0, 0) \,,$$

$$e^\alpha_{(1)} = (0, 0, \frac{1}{r}, 0) \,, \qquad e^\alpha_{(2)} = (1, 0, 0, \frac{1}{r \sin\theta}) \,,$$

$$\gamma_{030} = \frac{r}{\sqrt{\Phi}} \; , \; \gamma_{311} = \frac{\sqrt{\Phi}}{r} \; , \; \gamma_{322} = \frac{\sqrt{\Phi}}{r} \; , \; \gamma_{122} = \frac{\cos\theta}{r \sin\theta} \; , \qquad (21.2.1)$$

we get an explicit form of a matrix Duffin–Kemmer equation for a massive spin 1 particle

$$[\; i\beta^0 \partial_t + i\Phi(\beta^3 \partial_r + \frac{1}{r}(\beta^1 j^{31} + \beta^2 j^{32}) + \frac{\Phi'}{2\Phi}\beta^0 J^{03})$$

$$+ \frac{\sqrt{\Phi}}{r} \, \Sigma_{\theta,\phi} - m\sqrt{\Phi} \;] \; \Phi(x) = 0 \; ,$$

$$\Sigma_{\theta,\phi} = i \, \beta^1 \partial_\theta + \beta^2 \frac{i\partial + i \, j^{12} \cos\theta}{\sin\theta} \; . \qquad (21.2.2)$$

Spherical waves with (j,m) quantum numbers should be constructed within the following general substitution

$$\Phi_{\epsilon jm}(x) = e^{-i\epsilon t} [\; f_1(r) \, D_0, \; f_2(r) \, D_{-1}, \; f_3(r) \, D_0, \; f_4(r) \, D_{+1},$$

$$f_5(r) \, D_{-1}, f_6(r) \, D_0, \; f_7(r) \, D_{+1}, f_8(r) \, D_{-1}, \; f_9(r) \, D_0, \; f_{10}(r) \, D_{+1} \;] \; ;$$

$$(21.2.3)$$

symbol D_σ designates Wigner functions $D^j_{-m,\sigma}(\phi, \theta, 0)$ (we use notation according to the book [429]). Requirement to diagonalize parity operator, $\hat{P} \, \Phi_{\epsilon jm} = P \, \Phi_{\epsilon jm}$, gives

$$P = (-1)^{j+1} \; , \qquad f_1 = f_3 = f_6 = 0 \; , \; f_4 = -f_2 \; ,$$

$$f_7 = -f_5 \; , \; f_{10} = +f_8 \; ;$$

$$P = (-1)^j \; , \qquad f_9 = 0 \; , \; f_4 = +f_2 \; , \; f_7 = +f_5 \; , \; f_{10} = -f_8 \; . \qquad (21.2.4)$$

After the separation of variables, we arrive to the radial systems ($\nu = \sqrt{j(j+1)}/2$):

$$P = (-1)^{j+1} \; ,$$

$$i\epsilon \, f_5 + i\Phi(\frac{d}{dr} + \frac{1}{r} + \frac{\Phi'}{2\Phi}) \, f_8 + i\nu \frac{\sqrt{\Phi}}{r} \, f_9 - m \sqrt{\Phi} \, f_2 = 0 \; ,$$

$$-i\epsilon \, f_2 - m\sqrt{\Phi} \, f_5 = 0 \; , \qquad -i\Phi \, (\frac{d}{dr} + \frac{1}{r}) \, f_2 - m\sqrt{\Phi} \, f_8 = 0 \; ,$$

$$i2\nu \frac{\sqrt{\Phi}}{r} \, f_2 - m \sqrt{\Phi} \, f_9 = 0 \; ; \qquad (21.2.5)$$

$$P = (-1)^j \; ,$$

$$\Phi \, (\frac{d}{dr} + \frac{2}{r}) \, F_6 + \frac{2\nu}{r} \, F_5 + m \, F_1 = 0 \; ,$$

$$i\epsilon \, F_5 + i\Phi \, (\frac{d}{dr} + \frac{1}{r}) \, F_8 - m \, \Phi \, F_2 = 0 \; ,$$

$$i\epsilon F_6 - i2\nu r F_8 - m F_3 = 0 \; , \quad -i\epsilon F_2 + \frac{\nu}{r} F_1 - m F_5 = 0 \; ,$$

$$i\epsilon F_3 + \Phi \frac{d}{dr} F_1 + m \, \Phi \, F_6 = 0 \; ,$$

$$i\Phi \left(\frac{d}{dr} + \frac{1}{r} \right) F_2 + i\frac{\nu}{r} F_3 + m F_8 = 0 ; \tag{21.2.6}$$

in Eq. (21.2.6), we have used substitutions

$$F_1 = \sqrt{\Phi}\, f_1 ,\; F_2 = f_2 ,\; F_3 = \sqrt{\Phi}\, f_3 ,$$

$$F_5 = \sqrt{\Phi}\, f_5 ,\; F_6 = f_6 ,\; F_8 = \sqrt{\Phi}\, f_8 .$$

The case of the minimal value $j = 0$ is to be treated separately, because one must use a special substitution from the very beginning

$$\Phi_{\epsilon jm}(x) = e^{-i\epsilon t} \left(f_1,\, 0,\, f_3,\, 0,\, 0,\, f_6,\, 0,\, 0,\, f_9,\, 0 \right) . \tag{21.2.7}$$

The angular part of the wave operator $\Sigma_{\theta,\phi}$ acts as a zero operator and the main equation takes the form

$$\left[i\beta^0 \partial_t + i\Phi(\beta^3 \partial_r + \frac{1}{r}\,(\beta^1 j^{31} + \beta^2 j^{32}) + \frac{\Phi'}{2\Phi}\beta^0 J^{03}) - m\sqrt{\Phi} \right] \Phi(x) = 0 ; \tag{21.2.8}$$

Correspondingly, we have a very simple radial system

$$-\Phi \left(\frac{d}{dr} + \frac{2}{r} \right) f_6 - m\sqrt{\Phi}\, f_1 = 0 ,\qquad i\epsilon f_6 - m\sqrt{\Phi}\, f_3 = 0 ,$$

$$-i\epsilon f_3 - \Phi(\frac{d}{dr} + \frac{\Phi'}{2\Phi}) f_1 - m\sqrt{\Phi}\, f_6 = 0 ,\qquad f_9 = 0 . \tag{21.2.9}$$

The system in Eq. (21.2.9) describes states with parity $P = (-1)^0 = +1$; states with parity $P = (-1)^{0+1} = -1$ do not exist. The system in Eq. (21.2.9) reduces to second order differential equation for f_6:

$$\frac{d^2}{dr^2} f_6 + \frac{2(1 + 2r^2)}{r(1+r^2)} \frac{d}{dr} f_6 + \left[\frac{\epsilon^2}{(1+r^2)^2} - \frac{m^2 - 2}{1+r^2} - \frac{2}{r^2(1+r^2)} \right] f_6 = 0 , \tag{21.2.10}$$

that is solved in hypergeometric functions

$$f_6(r) = r\,(1+r^2)^{-\epsilon/2}\, F(\alpha, \beta, \gamma, -r^2) ,$$

$$\gamma = 1 + 3/2 ,\qquad \alpha = \frac{3/2 + 1 - \epsilon + \sqrt{m^2 + 1/4}}{2} ,$$

$$\beta = \frac{3/2 + 1 - \epsilon - \sqrt{m^2 + 1/4}}{2} . \tag{21.2.11}$$

The hypergeometric series becomes a polynomial when $\alpha = -n$, $n = 0, 1, 2, ...$; so, we arrive at an energy spectrum

$$\epsilon = N + 3/2 + \sqrt{m^2 + 1/4} ,\qquad N = 2n + 1 \in \{0, 1, 2, ...\} . \tag{21.2.12}$$

21.3   Solutions of Radial Equations at $j > 0$

Let us now turn to Eqs. (21.2.5). Expressing f_5, f_8, f_9 through f_2

$$f_5 = \frac{i}{m\sqrt{\Phi}} \epsilon f_2 , \qquad f_9 = \frac{i}{m} \frac{2\nu}{r} f_2 ,$$

$$f_8 = -\frac{i}{m\sqrt{\Phi}} \Phi \left(\frac{d}{dr} + \frac{1}{r}\right) f_2 , \tag{21.3.1}$$

for f_2, we get

$$\frac{d^2}{dr^2} f_2 + \frac{2(1+2r^2)}{r(1+r^2)} \frac{d}{dr} f_2 + \left[\frac{\epsilon^2}{(1+r^2)^2} - \frac{m^2 - 2}{1+r^2} - \frac{j(j+1)}{r^2(1+r^2)}\right] f_2 = 0 .$$

$$\tag{21.3.2}$$

The following solutions of this type are referred to as j-waves.

Equation (21.3.2) is solved in hypergeometric functions

$$f_2 = U_{\epsilon,j} = (-z)^{j/2}(1-z)^{-\epsilon/2} F(\alpha, \beta, \gamma; z) , \qquad \gamma = j + 3/2 ,$$

$$\alpha = \frac{3/2 + j - \epsilon + \sqrt{m^2 + 1/4}}{2} , \qquad \beta = \frac{3/2 + j - \epsilon - \sqrt{m^2 + 1/4}}{2} . \tag{21.3.3}$$

Restriction $\alpha = -n$, $n = 0, 1, 2, ...$ makes hypergeometric series polynomials, so we get a quantization rule for the energy levels:

$$\epsilon = N + 3/2 + \sqrt{m^2 + 1/4} , \qquad N = 2n + j \in \{0, 1, 2, ...\} . \tag{21.3.4}$$

It is verified easily that at $z \to -\infty$ the radial function $U_{\epsilon,j}(z)$ tends to zero

$$U_{\epsilon,j}(z \to -\infty) \sim z^{j/2} z^{-\epsilon/2} z^n \sim z^{-3/4 - \sqrt{m^2+1/4}/2} .$$

Now, let us turn to Eqs. (21.2.6). Using two different substitutions

$$I. \qquad F_1 = \sqrt{j+1} G_1 , \quad F_2 = i\sqrt{j/2} G_2 , \quad F_3 = i\sqrt{J+1} G_3 ,$$

$$F_5 = \sqrt{j/2} G_5 , \quad F_6 = \sqrt{j+1} G_6 , \quad F_8 = \sqrt{j/2} G_8 ; \tag{21.3.5}$$

$$II. \qquad F_1 = \sqrt{j} G_1 , \quad F_2 = i\sqrt{(j+1)/2} G_2 , \quad F_3 = i\sqrt{j} G_3 ,$$

$$F_5 = \sqrt{(j+1)/2} G_5 , \quad F_6 = \sqrt{j} G_6 , \quad F_8 = \sqrt{(j+1)/2} G_8 , \tag{21.3.6}$$

and expressing G_5, G_6, G_8 through G_1, G_2, G_3; we arrive at three equations, respectively.

$$I.$$

$$\left(\frac{j(j+1)}{r^2} + m^2 - \Phi\left(\frac{d}{dr} + \frac{2}{r}\right)\frac{d}{dr}\right) G_1$$

$$+ \frac{\epsilon j}{r} G_2 + \epsilon \Phi \left(\frac{d}{dr} + \frac{2}{r} \right) \frac{1}{\Phi} G_3 = 0 ,$$

$$\left(\epsilon^2 - m^2 \Phi^2 + \Phi (\frac{d}{dr} + \frac{1}{r}) \Phi (\frac{d}{dr} + \frac{1}{r}) \right) G_2$$

$$+ \frac{\epsilon (j+1)}{r} G_1 + \Phi \frac{j+1}{r} \frac{d}{dr} G_3 = 0 ,$$

$$\left(\frac{\epsilon^2}{\Phi} - \frac{j(j+1)}{r^2} m^2 \right) G_3 - \epsilon \frac{d}{dr} G_1 - \frac{j}{r} \Phi (\frac{d}{dr} + \frac{1}{r}) G_2 = 0 ;$$

$$(21.3.7)$$

II.

$$\left(\frac{j(j+1)}{r^2} + m^2 - \Phi (\frac{d}{dr} + \frac{2}{r}) \frac{d}{dr} \right) G_1$$

$$+ \frac{\epsilon (j+1)}{r} G_2 + \epsilon \Phi (\frac{d}{dr} + \frac{2}{r}) \frac{1}{\Phi} G_3 = 0 ,$$

$$\left(\epsilon^2 - m^2 \Phi^2 + \Phi (\frac{d}{dr} + \frac{1}{r}) \Phi (\frac{d}{dr} + \frac{1}{r}) \right) G_2$$

$$+ \frac{\epsilon j}{r} G_1 + \Phi \frac{j}{r} \frac{d}{dr} G_3 = 0 ,$$

$$(\frac{\epsilon^2}{\Phi} - \frac{j(j+1)}{r^2} - m^2) G_3 - \epsilon \frac{d}{dr} G_1 - \frac{(j+1)}{r} \Phi (\frac{d}{dr} + \frac{1}{r}) G_2 = 0 .$$

$$(21.3.8)$$

To solve Eqs. (21.3.7) and (21.3.8), one can make use of the Lorentz condition. Its explicit form can easily found

$$\frac{-i\epsilon}{\sqrt{\Phi}} f_1 - \sqrt{\Phi} (\frac{d}{dr} + \frac{2}{r} + \frac{\Phi'}{2\Phi}) f_3 - \frac{\nu}{r} (f_2 + f_4) = 0 . \qquad (21.3.9)$$

When $P = (-1)^{j+1}$ in Eq. (21.3.9) holds identically; for substitutions I and II in Eqs. (21.3.7) and (21.3.8) it respectively gives

$$I. \qquad - \epsilon \frac{G_1}{\Phi} = \frac{j}{r} G_2 + (\frac{d}{dr} + \frac{2}{r}) G_3 ,$$

$$II. \qquad - \epsilon \frac{G_1}{\Phi} = \frac{j+1}{r} G_2 + (\frac{d}{dr} + \frac{2}{r}) G_3 . \qquad (21.3.10)$$

In allowing for the relations in Eq. (21.3.10), let us express G_1 through G_2 and G_3, and substitute the results into 2-nd and 3-rd equations in Eqs. (21.3.7) and (21.3.8). Thus, we will respectively get:

I.

$$\left[\frac{d^2}{dr^2} + (\frac{2}{r} + \frac{\Phi'}{\Phi}) \frac{d}{dr} + \frac{\Phi'}{r\Phi} + \frac{\epsilon^2}{\Phi^2} \right.$$

$$\left. - \frac{m^2}{\Phi} - \frac{j(j+1)}{\Phi r^2} \right] G_2 - \frac{2(j+1)}{r^2 \Phi} G_3 = 0 ,$$

$$\left[\frac{d^2}{dr^2} + (\frac{2}{r} + \frac{\Phi'}{\Phi})\frac{d}{dr} + \frac{2\Phi'}{r\Phi} - \frac{2}{r^2} + \frac{\epsilon^2}{\Phi^2}\right.$$

$$\left. - \frac{m^2}{\Phi} - \frac{j(j+1)}{\Phi r^2}\right] G_3 - \frac{2j}{r^2\Phi} G_2 = 0 \ ;$$

$$(21.3.11)$$

II.

$$\left[\frac{d^2}{dr^2} + (\frac{2}{r} + \frac{\Phi'}{\Phi})\frac{d}{dr} + \frac{\Phi'}{r\Phi} + \frac{\epsilon^2}{\Phi^2}\right.$$

$$\left. - \frac{m^2}{\Phi} - \frac{j(j+1)}{\Phi r^2}\right] G_2 - \frac{2j}{r^2\,\Phi} G_3 = 0 \ ,$$

$$\left[\frac{d^2}{dr^2} + (\frac{2}{r} + \frac{\Phi'}{\Phi})\frac{d}{dr} + \frac{2\Phi'}{r\Phi} - \frac{2}{r^2} + \frac{\epsilon^2}{\Phi^2}\right.$$

$$\left. - \frac{m^2}{\Phi} - \frac{j(j+1)}{\Phi r^2}\right] G_3 - \frac{2(j+1)}{r^2\Phi} G_2 = 0 \ .$$

$$(21.3.12)$$

In the case of *I*, and taking $G_3 = +G_2$, from two equations in Eq.(21.3.11); we get the same one

$$I. \qquad G_3 = +G_2 = U_{\epsilon,j+1} \ , \qquad \left[\frac{d^2}{dr^2} + \frac{2(1+2r^2)}{r(1+r^2)}\frac{d}{dr}\right.$$

$$\left. + \frac{\epsilon^2}{(1+r^2)^2} - \frac{M^2 - 2}{1+r^2} - \frac{(j+1)(j+2)}{r^2(1+r^2)}\right] G_2 = 0 \ . \qquad (21.3.13)$$

In the same manner, in the case II, taking $G_3 = -G_2$, we get the same equation (it differs from previous one by the simple formal changing $(j+1)$ into $(j-1)$):

$$II. \qquad G_3 = -G_2 = U_{\epsilon,j-1} \ , \qquad \left[\frac{d^2}{dr^2} + \frac{2(1+2r^2)}{r(1+r^2)}\frac{d}{dr}\right.$$

$$\left. + \frac{\epsilon^2}{(1+r^2)^2} - \frac{M^2 - 2}{1+r^2} - \frac{(j-1)j}{r^2(1+r^2)}\right] G_2 = 0 \ . \qquad (21.3.14)$$

Thus, beside the waves of *j*-type, there also exists two types (all technical details for calculations with hypergeometric function are omitted)

I. $\qquad (j+1) - \text{type} \ ,$

$$G_3 = G_2 = U_{\epsilon,j+1} \ , \qquad -\epsilon\frac{G_1}{\Phi} = (\frac{d}{dr} + \frac{j+2}{r}) G_2 \ ;$$

$$G_1 = \sqrt{-z} \ U_{\epsilon,j+1} - \frac{2j+3}{\epsilon} \sqrt{1-z} \ U_{\epsilon-1,j} \ ,$$

$$U_{\epsilon,j+1} = (-z)^{(j+1)/2} (1-z)^{-\epsilon/2} F(\alpha,\beta,\gamma; z) \ ,$$

$$U_{\epsilon-1,j} = (-z)^{j/2} (1-z)^{-(\epsilon-1)/2} F(\alpha,\beta,\gamma-1; z) \ ,$$

$$\gamma = j + 1 + 3/2 \, , \qquad \alpha = \frac{3/2 + j + 1 - \epsilon + \sqrt{m^2 + 1/4}}{2} \, ,$$

$$\beta = \frac{3/2 + j + 1 - \epsilon - \sqrt{m^2 + 1/4}}{2} \, ; \qquad (21.3.15)$$

II. $(j - 1) -$ type,

$$-G_3 = G_2 = U_{\epsilon, j-1} \, , \qquad -\epsilon \, \frac{G_1}{\Phi} = \left(-\frac{d}{dr} + \frac{j-1}{r} \right) G_2 \, ,$$

$$G_1 = -\sqrt{-z} \; U_{\epsilon, j-1} - \frac{2}{\epsilon} \frac{\alpha\beta}{\gamma} \sqrt{1 - z} \; U_{\epsilon-1, j} \, ,$$

$$U_{\epsilon, j-1} = (-z)^{(j-1)/2} \, (1 - z)^{-\epsilon/2} \, F(\alpha, \beta, \gamma; \, z) \, ,$$

$$U_{\epsilon-1, j} = (-z)^{j/2} \, (1 - z)^{-(\epsilon-1)/2} \, F(\alpha + 1, \beta + 1, \gamma + 1; \, z) \, ,$$

$$\gamma = j - 1 + 3/2 \, , \qquad \alpha = \frac{3/2 + j - 1 - \epsilon + \sqrt{m^2 + 1/4}}{2} \, ,$$

$$\beta = \frac{3/2 + j - 1 - \epsilon - \sqrt{m^2 + 1/4}}{2} \, . \qquad (21.3.16)$$

Let us now collect our results together. There are constructed solutions of three types (here, only $f_1, \dots, f_4$ are specified):

$j - $ wave ,

$$f_1 = f_3 = 0 \, , \quad f_2 = -f_4 = U_{\epsilon, j} \, ;$$

$(j + 1) - $ wave ,

$$f_1 = \sqrt{j + 1} \, \left[\frac{\sqrt{-z}}{\sqrt{1 - z}} \; U_{\epsilon, j+1} - \frac{2j + 3}{\epsilon} \; U_{\epsilon-1, j} \right] \, ,$$

$$f_2 = +f_4 = +i \, \sqrt{j/2} \; U_{\epsilon, j+1} \, , \qquad f_3 = +i \, \sqrt{j + 1} \, \frac{1}{\sqrt{1 - z}} \; U_{\epsilon, j+1} \, ;$$

$(j - 1) - $ wave ,

$$f_1 = \sqrt{j} \, \left[-\frac{\sqrt{-z}}{\sqrt{1 - z}} \; U_{\epsilon, j-1} - \frac{2}{\epsilon} \frac{\alpha\beta}{\gamma} \; U_{\epsilon-1, j} \right] \, ,$$

$$f_2 = +f_4 = i \, \sqrt{\frac{j + 1}{2}} \; U_{\epsilon, j-1} \, , \qquad f_3 = -i \, \sqrt{j} \frac{1}{\sqrt{1 - z}} \; U_{\epsilon, j-1} \, .$$

$$(21.3.17)$$

Three types of solutions correspond to three possible values for the orbital angular moment with a spin 1 particle at fixed j : $l = j, j + 1, j - 1$.

21.4   Massless Limit for Spin 1 Particle

Let us briefly consider a massless limit. The Duffin–Kemmer–Petiau equation stays the same with only a formal change

$$m\sqrt{\Phi} \;\to\; P_6\sqrt{\Phi}\,, \qquad P_6 = \begin{vmatrix} 0 & 0 & 0 & 0 \\ 0 & 0 & 0 & 0 \\ 0 & 0 & I & 0 \\ 0 & 0 & 0 & I \end{vmatrix}, \qquad (21.4.1)$$

which produces evident alterations in the radial system

$$m\sqrt{\phi}f_i \;\to\; 0\,, \quad \text{at } i = 1,2,3,4\,;$$

$$m\sqrt{\phi}f_i \;\to\; \sqrt{\phi}f_i\,, \quad \text{at } i = 5, ..., 10\,. \qquad (21.4.2)$$

In this massless case, the Lorentz condition must be considered as an external gauge restriction for photon field. Other relations remain the same, instead of using the old parameters for the hypergeometric functions new parameters would be necessary.

$$U_{\epsilon,j}\,, \qquad \alpha = \frac{2+j-\epsilon}{2}\,, \qquad \beta = \frac{1+j-\epsilon}{2}\,, \qquad \gamma = j + 3/2\,; \qquad (21.4.3)$$

the energy quantization rule appears as

$$\epsilon = 2n + j + 2 = N + 2\,, \qquad N = 2n + j \in \{0, 1, 2, ...\}\,. \qquad (21.4.4)$$

The case of minimal value $j = 0$ should be considered separately. The system in Eq. (21.2.9) becomes

$$-\Phi\left(\frac{d}{dr} + \frac{2}{r}\right) f_6 - 0\sqrt{\Phi}\, f_1 = 0\,, \qquad i\epsilon f_6 - 0\sqrt{\Phi}\, f_3 = 0\,,$$

$$-i\epsilon f_3 - \Phi\left(\frac{d}{dr} + \frac{\Phi'}{2\Phi}\right) f_1 - \sqrt{\Phi}\, f_6 = 0\,, \qquad f_9 = 0\,,$$

which is equivalent to

$$g_6 = 0\,, \; f_9 = 0\,, \qquad -i\epsilon f_3 - \Phi\left(\frac{d}{dr} + \frac{\Phi'}{2\Phi}\right) f_1 = 0\,. \qquad (21.4.5)$$

Therefore, for all states of electromagnetic field at $j = 0$ the components of electric and magnetic vectors vanish ($F_{\alpha\beta} = 0$); and non-vanishing $f_1(r)$, $f_3(r)$ correspond to solutions of gradient type $A_\alpha = \nabla_\alpha \Phi$. To have fixed two radial functions in Eq. (21.4.5), one must impose a certain gauge condition. In particular, in taking the Lorentz gauge, we get equations (let it be $f_1 = \Phi^{-1/2}F_1$, $f_3 = \Phi^{-1/2}F_3$):

$$-\frac{i\epsilon}{\Phi} F_3 - \frac{d}{dr} F_1 = 0\,, \qquad -\frac{i\epsilon}{\Phi} F_1 - \left(\frac{d}{dr} + \frac{2}{r}\right) F_3 = 0\,; \qquad (21.4.6)$$

from this, it follows

$$\left[\frac{d^2}{dr^2} + \frac{2(1+2r^2)}{r(1+r^2)}\frac{d}{dr} + \frac{\epsilon^2}{(1+r^2)^2}\right] F_1 = 0\,, \qquad (21.4.7)$$

which represents the $j = 0$-spherical solution of the equation $\nabla^\alpha \nabla_\alpha \Phi = 0$, $\Phi = e^{-i\epsilon t} f(r)$.

21.5   5-Dimensional Form of the Wave Equation

It is well known that the wave equation for a particle with spin 1 in the de Sitter and anti de Sitter spaces can be presented in 5-dimensional form; explicitly invariant under groups $SO(4.1)$ and $SO(3.2)$, respectively. Let us specify the problem of spherical wave solutions in anti de Sitter model for the 5-dimensional formalism. It is convenient to start with conformal flat coordinates

$$dS^2 = \frac{(dx^0)^2 - (dx^1)^2 - (dx^2)^2 - (dx^3)^2}{\Phi^2} \ , \quad \Phi = (1 + x^2)/2 \ ; \tag{21.5.1}$$

the Proca tensor equations read as

$$\partial_\alpha \Psi_\beta - \partial_\beta \Psi_\alpha = m \ \Psi_{\alpha\beta} \ , \qquad \Phi^2 \ \partial^\beta \Psi_{\alpha\beta} = m \Psi_\alpha \ . \tag{21.5.2}$$

Let us introduce five coordinates ξ^a $(a = \alpha, \ 5)$:

$$\xi^\alpha = \frac{x^\alpha}{\Phi} \ , \qquad \xi^5 = \frac{1 - x^2}{1 + x^2} \ ,$$

$$x^\alpha = \frac{\xi^\alpha}{1 + \xi^5} \ , \qquad \Phi = \frac{1}{1 + \xi^5} \ ,$$

$$(\xi^0)^2 - (\xi^1)^2 - (\xi^2)^2 - (\xi^3)^2 + (\xi^5)^2 = +1 \ ,$$

$$dS^2 = \eta_{\alpha\beta} \ d\xi^\alpha d\xi^\beta \ + \ (d\xi^5)^2 \ . \tag{21.5.3}$$

In other words, the anti de Sitter space-time can be considered as a sphere in 5-dimensional pseudo-Euclidean space; therefore, the model permits 10-parametric symmetry group $SO(3.2)$. Instead $\Psi^\alpha(x)$ (in the following designated as $a^\alpha(x)$) let us introduce 5-vector $A^a(\xi)$:

$$A^\alpha = (\ \frac{\delta^\alpha_\beta}{\Phi} \ - \ \frac{x^\alpha x_\beta}{\Phi^2}) \ a^\beta \ , \quad A^5 = -\frac{x_\beta \ a^\beta}{\Phi^2} \ ,$$

$$a^\alpha(x) = \Phi \ (A^\alpha \ - \ x^\alpha A^5) \ . \tag{21.5.4}$$

These five components $A^a(\xi)$ obey an additional restriction

$$A^a \xi_a = A^0 \xi^0 - \mathbf{A} \ \vec{\xi} + A^5 \xi^5 = 0 \ .$$

Invariant with respect to $SO(3.2)$ wave equation for vector $A^a(\xi)$ should be constructed with the help of the operator $L_{ab} = \xi_a \ (\partial/\partial \xi^b) - \xi_b \ (\partial/\partial \xi^a)$ and looks as follows

$$(-\frac{1}{2} L^{ab} L_{ab} + m^2 - 2) A_c = 0 \ , \tag{21.5.5a}$$

$$L_{ab} A^b = A_a \ , \qquad A^a \xi_a = 0 \ . \tag{21.5.5b}$$

21.6   Spherical Waves, Separation of Variables

Let us consider the Eqs. $(21.5.5a, b)$ in static coordinates of the anti de Sitter space

$$\xi^1 = r \sin\theta \cos\phi \ , \ \ \xi^2 = r \sin\theta \sin\phi, \ \ \xi^3 = r \cos\theta \ ,$$

$$\xi^0 = \sin t \sqrt{1 + r^2} \ , \ \ \xi^5 = \cos t \sqrt{1 + r^2} \ ;$$

$$t = \operatorname{arccot} \frac{\xi^0}{\xi^5} \ , \ \ r = \sqrt{(\xi^1)^2 + (\xi^2)^2 + (\xi^3)^2} \ ,$$

$$\theta = \operatorname{arccot} \frac{\sqrt{(\xi^1)^2 + (\xi^2)^2}}{\xi^3} \ , \ \ \phi = \operatorname{arccot} \frac{\xi^2}{\xi^1} \ . \tag{21.6.1}$$

For any representation of the group $SO(3.2)$ on the functions $\Psi(\xi)$, we have

$$\xi' = S \, \xi \ , \ \ \Psi'(\xi') = U \, \Psi(\xi) \ \ \ \ \Longrightarrow \ \ \ \ \Psi'(\xi) = U \, \Psi(S^{-1} \, \xi) \ .$$

In particular, for the case of $U \equiv S$ and $\Psi \equiv A$, with the rotation in the plane $(0-5)$

$$\xi^{0'} = \cos\omega \, \xi^0 - \sin\omega \, \xi^5 \ , \ \ \ \ \xi^{5'} = \sin\omega \, \xi^0 + \cos\omega \, \xi^5 \ ,$$

we get

$$A'(\xi) = (I + \delta\omega \, J_{50}) \, A(\xi) \ , \ \ \ \ J_{50} = \mathcal{L}_{50} + \sigma_{50} \ ,$$

$$\mathcal{L}_{50} = \xi_5 \frac{\partial}{\partial \xi^0} - \xi^0 \frac{\partial}{\partial \xi^5} \ , \ \ \ \ \sigma_{50} = \begin{vmatrix} 0 & 0 & 0 & 0 & -1 \\ 0 & 0 & 0 & 0 & 0 \\ 0 & 0 & 0 & 0 & 0 \\ 0 & 0 & 0 & 0 & 0 \\ +1 & 0 & 0 & 0 & 0 \end{vmatrix} \ . \tag{21.6.2}$$

Generators will be used to construct 5-form for energy operator, and total momentum. From requirements

$$(+i J_{50})^a{}_b \, A^b = \epsilon A^a \ ,$$

$$(\mathbf{J}^2)^a{}_b A^b = j(j+1) \, A^a \ , \ \ \ \ (J_3)^a{}_b \, A^b = m \, A^a \tag{21.6.3}$$

it follows

$$\mathbf{A} = e^{-i\epsilon t} \left[f(r) \, \mathbf{Y}^{j+1}_{jm}(\theta, \phi) + g(r) \, \mathbf{Y}^{j-1}_{jm}(\theta, \phi) + h(r) \, \mathbf{Y}^{j}_{jm}(\theta, \phi) \right] ,$$

$$A^0 = \left[e^{-i(\epsilon-1)t} \, F(r) + i \, e^{-i(\epsilon+1)t} \, G(r) \right] Y_{jm}(\theta, \phi) \ ,$$

$$A^5 = \left[i \, e^{-i(\epsilon-1)t} \, F(r) + e^{-i(\epsilon+1)t)} \, G(r) \right] Y_{jm}(\theta, \phi) \ . \tag{21.6.4}$$

At given $j = 1, 2, \ldots$ there exist three linearly independent spherical vectors $\nu = j+1, j, j-1$, when $j = 0$ there exists only one that

$$j = 0 \ , \ \ \ \ \mathbf{A} = e^{-i\epsilon t} \, f(r) \, \mathbf{Y}^1_{00} \ ,$$

$$\frac{1}{2}(A^0 + iA^5) = iG(r) \, e^{-i(\epsilon+1)t} \ , \ \ \ \ \frac{1}{2} \, (A^0 - iA^5) = F(r) \, e^{-i(\epsilon-1)t} \ . \tag{21.6.5}$$

Radial functions $f(r)$, $g(r)$, $h(r)$, $F(r), G(r)$ should be determined from Eq. (21.5.5a, b). From Eq. (21.5.5a), when taking into account the action of $\mathbf{l}^2$ on spherical vectors [429]

$$\mathbf{l}^2\, \mathbf{Y}^{\nu}_{jm} = \nu(\nu+1)\, \mathbf{Y}^{\nu}_{jm} \; ,$$

$$\mathbf{l}^2\, Y_{jm} = j(j+1)\, Y_{jm} \; , \qquad \nu = j, j+1, j-1 \; ,$$

for radial functions $f(r)$, $g(r)$, $h(r)$, $F(r)$ we get equations of the same type

$$\left[\frac{d^2}{dr^2} + \frac{2(1+2r^2)}{r(1+r^2)}\frac{d}{dr} + \frac{\Lambda^2}{(1+r^2)^2} - \frac{\nu(\nu+1)}{r^2(1+r^2)} - \frac{(m^2-2)}{1+r^2}\right] U_{\Lambda,\nu} = 0 \; ; \quad (21.6.6)$$

so we know these five functions have numerical constants f_0, g_0, h_0, F_0, G_0

$$f = f_0\, U_{\epsilon,j+1} \; , \quad g = g_0\, U_{\epsilon,j-1} \; , \quad h = h_0\, U_{\epsilon,j} \; ,$$

$$F = F_0\, U_{\epsilon-1,j} \; , \quad G = G_0\, U_{\epsilon+1,j} \; ;$$

these constants should be obtained from Eqs. (21.5.5b). Solutions of Eq. (21.6.6) are constructed in terms of hypergeometric functions (it suffices to only consider the case $U_{\epsilon,j}$):

$$U_{\epsilon,j} = (-z)^{j/2}\,(1-z)^{-\epsilon/2}\, F(\alpha, \beta, \gamma; z) \; , \qquad z = -r^2, \; \gamma = j+3/2 \; ,$$

$$\alpha = \frac{3/2 + j - \epsilon + \sqrt{m^2+1/4}}{2} \; , \quad \beta = \frac{3/2 + j - \epsilon - \sqrt{m^2+1/4}}{2} \; ; \qquad (21.6.7)$$

we have polynomials when $\alpha = -n$, $n = 0, 1, 2, \ldots$; which results in the quantization rule for energy levels

$$\epsilon = N + 3/2 + \sqrt{m^2+1/4} \; , \qquad N = 2n + j \in \{0, 1, 2, \ldots\} \; . \qquad (21.6.8)$$

From g equations in Eq. (21.5.5b) one can produce relationships between $(G \pm iF)$ and (f, g):

$$G - iF = \frac{1}{\epsilon}\,\sqrt{1+r^2}\,\left[(\frac{d}{dr} + \frac{j+2}{r})\,f - (\frac{d}{dr} - \frac{j-1}{r})\,g\right] \; ,$$

$$G + iF = \frac{-rf + rg}{\sqrt{1+r^2}} \; . \qquad (21.6.9)$$

21.7   Solutions of the Types $(j, j+1, j-1)$

Let us search three linearly independent solutions in the form

$$j - \text{type} \; , \qquad f = 0 \, , \; g = 0 \, , \; h \neq 0 \, ,$$

$$(j+1) - \text{type} \, , \qquad f \neq 0 \, , \; g = 0 \, , \; h = 0 \, ,$$

$$(j-1) - \text{type} \, , \qquad f = 0 \, , \; g \neq 0 \, , \; h = 0 \, .$$

In fact, these requirements are equivalent to diagonalizing the orbital angular operator $\mathbf{l}^2 = \nu(\nu + 1)$, $\nu = j + 1, j, j - 1$. First, let us consider the wave $(j + 1)$

$$f = \sqrt{\frac{2j + 1}{j + 1}} \, f_0 \, U_{\epsilon, j+1} \,, \qquad F = F_0 \, U_{\epsilon-1, j} \,, \qquad G = G_0 \, U_{\epsilon+1, j} \,; \qquad (21.7.1)$$

Eqs. (21.6.9) take the form

$$G + iF = -\frac{r f(r)}{\sqrt{1 + r^2}} \,, \qquad G - iF = \frac{1}{\epsilon} \sqrt{1 + r^2} \left(\frac{d}{dr} + \frac{j + 2}{r} \right) f(r) \,, \qquad (21.7.2)$$

or after transition to the variable $z = -r^2$

$$2 \frac{G_0}{f_0} U_{\epsilon+1, j} = -\frac{\sqrt{-z}}{\sqrt{1 - z}} U_{\epsilon, j+1} + \frac{1}{\epsilon} \sqrt{1 - z} \left(-2\sqrt{-z} \frac{d}{dz} + \frac{j + 2}{\sqrt{-z}} \right) U_{\epsilon, j+1} \,,$$

$$2i \frac{F_0}{f_0} U_{\epsilon-1, j} = -\frac{\sqrt{-z}}{\sqrt{1 - z}} U_{\epsilon, j+1} - \frac{1}{\epsilon} \sqrt{1 - z} \left(-2\sqrt{-z} \frac{d}{dz} + \frac{j + 2}{\sqrt{-z}} \right) U_{\epsilon, j+1} \,.$$

$$(21.7.3)$$

We can allow for explicit forms

$$U_{\epsilon, j+1} = (-z)^{(j+1)/2} (1 - z)^{-\epsilon/2} F(\alpha, \beta, \gamma; z) \,,$$

$$U_{\epsilon+1, j} = (-z)^{j/2} (1 - z)^{-(\epsilon+1)/2} F(\alpha - 1, \beta - 1, \gamma - 1; z) \,,$$

$$U_{\epsilon-1, j} = (-z)^{j/2} (1 - z)^{-(\epsilon-1)/2} F(\alpha, \beta, \gamma - 1; z) \,,$$

$$\gamma = j + 1 + 3/2 \,, \quad \alpha = \frac{3/2 + j + 1 - \epsilon + \sqrt{m^2 + 1/4}}{2} \,,$$

$$\beta = \frac{3/2 + j + 1 - \epsilon - \sqrt{m^2 + 1/4}}{2} \,, \qquad (21.7.4)$$

and using known formulas for hypergeometric functions; we get expressions for G_0, F_0:

$$G_0 = \frac{\gamma - 1}{\epsilon} f_0 = \frac{j + 3/2}{\epsilon} f_0 \,,$$

$$F_0 = i \frac{j + 3/2}{\epsilon} f_0 = i \frac{1 - \gamma}{\alpha + \beta - \gamma} f_0 \,. \qquad (21.7.5)$$

Analogous calculations may be performed for the case of $(j - 1)$-waves:

$$g = \sqrt{\frac{2j + 1}{j}} \, g_0 \, U_{\epsilon, j-1}(z) \,, \quad F = F_0 \, U_{\epsilon-1, j} \,, \quad G = G_0 \, U_{\epsilon+1, j} \,,$$

$$G + iF = \frac{r g(r)}{\sqrt{1 + r^2}} \,, \quad G - iF = -\frac{1}{\epsilon} \sqrt{1 + r^2} \left(\frac{d}{dr} - \frac{j - 1}{r} \right) g(r) \,; $$

$$(21.7.6)$$

or in variable $z = -r^2$

$$2\frac{G_0}{g_0} \, U_{\epsilon+1,j} = \frac{\sqrt{-z}}{\sqrt{1-z}} \, U_{\epsilon,j-1} - \frac{1}{\epsilon} \sqrt{1-z} \left(-2\sqrt{-z}\frac{d}{dz} - \frac{j-1}{\sqrt{-z}} \right) U_{\epsilon,j-1} \, ,$$

$$2i\frac{F_0}{g_0} \, U_{\epsilon-1,j} = \frac{\sqrt{-z}}{\sqrt{1-z}} \, U_{\epsilon,j-1} + \frac{1}{\epsilon} \sqrt{1-z} \left(-2\sqrt{-z}\frac{d}{dz} - \frac{j-1}{\sqrt{-z}} \right) U_{\epsilon,j-1} \, .$$

$$(21.7.7)$$

Using the formulas

$$U_{\epsilon,j-1} = (-z)^{(j-1)/2} \, (1-z)^{-\epsilon/2} \, F(a,b,c; \, z) \, ,$$

$$U_{\epsilon+1,j} = (-z)^{j/2} \, (1-z)^{-(\epsilon+1)/2} \, F(a,b,c+1; \, z) \, ,$$

$$U_{\epsilon-1,j} = (-z)^{j/2} \, (1-z)^{-(\epsilon-1)/2} \, F(a+1,b+1,c+1; \, z) \, ,$$

$$c = j - 1 + 3/2 \, , \quad a = \frac{3/2 + j - 1 - \epsilon + \sqrt{m^2 + 1/4}}{2} \, ,$$

$$b = \frac{3/2 + j - 1 - \epsilon - \sqrt{m^2 + 1/4}}{2}$$

we arrive at

$$G_0 = \frac{(a-c)(b-c)}{\epsilon \, c} \, g_0 \, , \qquad F_0 = i \, \frac{ab}{\epsilon \, c} \, g_0 \, . \qquad (21.7.8)$$

Here, we collect the results together

j-wave $j = 1, 2, 3, ...$

$$\mathbf{A} = e^{-i\epsilon t} h_0 \, U_{-i\epsilon,j}(r) \, \mathbf{Y}_{jm}^j(\theta, \phi) \, , \qquad A^0 = 0 \, , \, A^5 = 0 \, ;$$

$$(21.7.9)$$

quantization rule $\epsilon = 2n + j + 3/2 + \sqrt{m^2 + 1/4}$.

$(j-1)$-wave $j = 1, 2, 3, ...$

$$\mathbf{A} = e^{-i\epsilon t} \sqrt{\frac{2j+1}{j}} \, f(r) \, \mathbf{J}_{jm}^{j-1}(\theta, \phi) \, , \qquad f(r) = f_0 \, U_{\epsilon,j-1} \, ,$$

$$\frac{1}{2}(A^0 + iA^5) = i \, G(r) \, e^{-i(\epsilon+1)t} \, Y_{jm} \, , \qquad \frac{1}{2}(A^0 - iA^5) = F(r) \, e^{-i(\epsilon-1)t} \, Y_{jm} \, ,$$

$$G(r) = \frac{(a-c)(b-c)}{\epsilon \, c} \, g_0 \, U_{\epsilon+1,j} \, , \qquad F(r) = i \, \frac{ab}{\epsilon \, c} \, g_0 \, U_{\epsilon-1,j} \, ,$$

$$(21.7.10)$$

quantization rule $\epsilon = 2n + j - 1 + 3/2 + \sqrt{m^2 + 1/4}$.

$(j+1)$-**wave**, $j = 0, 1, 2, 3, \ldots$

$$\mathbf{A} = e^{-i\epsilon t} \sqrt{\frac{2j+1}{j+1}} \, f(r) \, \mathbf{J}_{jm}^{j+1}(\theta, \phi) \,, \qquad f(r) = f_0 \, U_{\epsilon, j+1} \,,$$

$$\frac{1}{2}(A^0 + iA^5) = i \, G(r) \, e^{-i(\epsilon+1)t} \, Y_{jm} \,, \qquad G(r) = \frac{j+3/2}{\epsilon} \, f_0 \, U_{\epsilon+1, j} \,,$$

$$\frac{1}{2}(A^0 - iA^5) = F(r) \, e^{-i(\epsilon-1)t} \, Y_{jm} \,, \qquad F(r) = i \, \frac{j+3/2}{\epsilon} \, f_0 \, U_{\epsilon-1, j} \,,$$

$$(21.7.11)$$

quantization rule $\epsilon = 2n + j + 1 + 3/2 + \sqrt{m^2 + 1/4}$. Degeneration of the energy levels can be clarified by the table

	j − type	$(j-1)$ − type	$(j+1)$ − type
$N = 1$	$n = 0, \, j = 1$	$n = 0, \, j = 2$	$n = 0, \, j = 0$
$N = 2$	$n = 0, \, j = 2$	$n = 0, \, j = 3$	$n = 0, \, j = 1$
		$n = 0, \, j = 1$	
$N = 3$	$n = 0, \, j = 3$	$n = 0, \, j = 4$	$n = 0, \, j = 2$
	$n = 1, \, j = 1$	$n = 1, \, j = 2$	$n = 1, \, j = 0$
$N = 4$	$n = 0, \, j = 4$	$n = 0, \, j = 5$	$n = 0, \, j = 3$
	$n = 1, \, j = 2$	$n = 1, \, j = 3$	$n = 1, \, j = 1$
$N = 5$	$n = 0, \, j = 5$	$n = 0, \, j = 6$	$n = 0, \, j = 4$
	$n = 1, \, j = 3$	$n = 1, \, j = 4$	$n = 1, \, j = 2$
	$n = 2, \, j = 1$	$n = 2, \, j = 2$	$n = 2, \, j = 0$
$N = 6$	$n = 0, \, j = 6$	$n = 0, \, j = 7$	$n = 0, \, j = 5$
	$n = 1, \, j = 4$	$n = 1, \, j = 5$	$n = 1, \, j = 3$
	$n = 2, \, j = 2$	$n = 2, \, j = 3$	$n = 2, \, j = 1$
		$n = 3, \, j = 1$	
$N = 7$	$n = 0, \, j = 7$	$n = 0, \, j = 8$	$n = 0, \, j = 6$
	$n = 1, \, j = 5$	$n = 1, \, j = 6$	$n = 1, \, j = 4$
	$n = 2, \, j = 3$	$n = 2, \, j = 4$	$n = 2, \, j = 2$
	$n = 3, \, j = 1$	$n = 3, \, j = 2$	$n = 3, \, j = 0$
$N = 8$	$n = 0, \, j = 8$	$n = 0, \, j = 9$	$n = 0, \, j = 7$
	$n = 1, \, j = 6$	$n = 1, \, j = 7$	$n = 1, \, j = 5$
	$n = 2, \, j = 4$	$n = 2, \, j = 5$	$n = 2, \, j = 3$
	$n = 3, \, j = 2$	$n = 3, \, j = 3$	$n = 3, \, j = 1$
		$n = 4, \, j = 1$	

where energy spectrum is given by (at $j = 0$, we have $\nu = j + 1 = 1$)

$$\epsilon = N + \frac{3}{2} + \sqrt{m^2 + 1/4} \,, \qquad N = 2n + \nu \,,$$

$$j = 1, 2, 3, \ldots, \quad \nu = j, j-1, j+1 \,. \qquad (21.7.12)$$

21.8   Riemann–Silberstein–Majorana–Oppenheimer Approach

Maxwell equations in Riemann space can be presented in Riemann–Silberstein–Majorana–Oppenheimer basis as one matrix equation

$$\alpha^c \left(e^\rho_{(c)} \partial_\rho + \frac{1}{2} j^{ab} \gamma_{abc} \right) \Psi = J(x) ,$$

$$\alpha^0 = -iI , \qquad \Psi = \begin{vmatrix} 0 \\ \mathbf{E} + ic\mathbf{B} \end{vmatrix} , \qquad J = \frac{1}{\epsilon_0} \begin{vmatrix} \rho \\ i\mathbf{j} \end{vmatrix} , \qquad (21.8.1)$$

or

$$-i(e^\rho_{(0)} \partial_\rho + \frac{1}{2} j^{ab} \gamma_{ab0}) \Psi + \alpha^k (e^\rho_{(k)} \partial_\rho + \frac{1}{2} j^{ab} \gamma_{abk}) \Psi = J(x) . \qquad (21.8.2)$$

By allowing the notation

$$e^\rho_{(0)} \partial_\rho = \partial_{(0)} , \qquad e^\rho_{(k)} \partial_\rho = \partial_{(k)} , \qquad a = 0, 1, 2, 3 ,$$

$$(\gamma_{01a}, \gamma_{02a}, \gamma_{03a}) = \mathbf{v}_a , \qquad (\gamma_{23a}, \gamma_{31a}, \gamma_{12a}) = \mathbf{p}_a , \qquad (21.8.3)$$

Equation (21.8.2) in the absence of sources, is reduced to

$$-i[\partial_{(0)} + \mathbf{s}(\mathbf{p}_0 + i\mathbf{v}_0)]\Psi + \alpha^k [\partial_{(k)} + \mathbf{s}(\mathbf{p}_k + i\mathbf{v}_k)] \Psi = 0 , \qquad (21.8.4)$$

where

$$s_1 = \begin{vmatrix} 0 & 0 \\ 0 & \tau_1 \end{vmatrix} , \quad s_2 = \begin{vmatrix} 0 & 0 \\ 0 & \tau_1 \end{vmatrix} , \quad s_3 = \begin{vmatrix} 0 & 0 \\ 0 & \tau_1 \end{vmatrix} ,$$

$$\tau_1 = \begin{vmatrix} 0 & 0 & 0 \\ 0 & 0 & -1 \\ 0 & 1 & 0 \end{vmatrix} , \tau_2 = \begin{vmatrix} 0 & 0 & 1 \\ 0 & 0 & 0 \\ -1 & 0 & 0 \end{vmatrix} , \tau_3 = \begin{vmatrix} 0 & -1 & 0 \\ 1 & 0 & 0 \\ 0 & 0 & 0 \end{vmatrix} . \qquad (21.8.5)$$

By the use of a spherical tetrad in the anti de Sitter space the main equation takes the form

$$[-\frac{i\partial_t}{\sqrt{\Phi}} + \sqrt{\Phi}(\alpha^3 \partial_r + \frac{\alpha^1 s_2 - \alpha^2 s_1}{r} + \frac{r}{\Phi} s_3) + \frac{1}{r}\Sigma_{\theta,\phi}] \begin{vmatrix} 0 \\ \psi \end{vmatrix} = 0 ,$$

$$\Sigma_{\theta,\phi} = \frac{\alpha^1}{r}\partial_\theta + \alpha^2 \frac{\partial_\phi + s_3 \cos\theta}{\sin\theta} . \qquad (21.8.6)$$

It is convenient to have the spin matrix s_3 as diagonal, which is reached by a simple linear transformation to the known cyclic basis

$$\Psi' = U_4 \Psi , \qquad U_4 = \begin{vmatrix} 1 & 0 \\ 0 & U \end{vmatrix} ,$$

$$U = \begin{vmatrix} -1/\sqrt{2} & i/\sqrt{2} & 0 \\ 0 & 0 & 1 \\ 1/\sqrt{2} & i/\sqrt{2} & 0 \end{vmatrix} , U^{-1} = \begin{vmatrix} -1/\sqrt{2} & 0 & 1/\sqrt{2} \\ -i/\sqrt{2} & 0 & -i/\sqrt{2} \\ 0 & 1 & 0 \end{vmatrix} , \qquad (21.8.7)$$

so that

$$U\tau_1 U^{-1} = \frac{1}{\sqrt{2}} \begin{vmatrix} 0 & -i & 0 \\ -i & 0 & -i \\ 0 & -i & 0 \end{vmatrix} = \tau_1' , \qquad j'^{23} = s_1' = \begin{vmatrix} 0 & 0 \\ 0 & \tau_1' \end{vmatrix} ,$$

$$U\tau_2 U^{-1} = \frac{1}{\sqrt{2}} \begin{vmatrix} 0 & -1 & 0 \\ 1 & 0 & -1 \\ 0 & 1 & 0 \end{vmatrix} = \tau'_2, \qquad j'^{31} = s'_2 = \begin{vmatrix} 0 & 0 \\ 0 & \tau'_2 \end{vmatrix},$$

$$U\tau_3 U^{-1} = -i \begin{vmatrix} +1 & 0 & 0 \\ 0 & 0 & 0 \\ 0 & 0 & -1 \end{vmatrix} = \tau'_3 \qquad j'^{12} = s'_3 = \begin{vmatrix} 0 & 0 \\ 0 & \tau'_3 \end{vmatrix},$$

$$\alpha'^1 = \frac{1}{\sqrt{2}} \begin{vmatrix} 0 & -1 & 0 & 1 \\ 1 & 0 & -i & 0 \\ 0 & -i & 0 & -i \\ -1 & 0 & -i & 0 \end{vmatrix}, \alpha'^2 = \frac{1}{\sqrt{2}} \begin{vmatrix} 0 & -i & 0 & -i \\ -i & 0 & -1 & 0 \\ 0 & 1 & 0 & -1 \\ -i & 0 & 1 & 0 \end{vmatrix},$$

$$\alpha'^3 = \begin{vmatrix} 0 & 0 & 1 & 0 \\ 0 & -i & 0 & 0 \\ -1 & 0 & 0 & 0 \\ 0 & 0 & 0 & +i \end{vmatrix}.$$

Equation (21.8.7) becomes

$$[-\frac{i\partial_t}{\sqrt{\Phi}} + \sqrt{\Phi}(\alpha'^3 \partial_r + \frac{\alpha'^1 s'_2 - \alpha'^2 s'_1}{r} + \frac{r}{\Phi}s'_3) + \frac{1}{r}\Sigma'_{\theta,\phi}] \begin{vmatrix} 0 \\ \psi' \end{vmatrix} = 0,$$

$$\Sigma'_{\theta,\phi} = \frac{\alpha'^1}{r}\partial_\theta + \alpha'^2 \frac{\partial_\phi + s'_3 \cos\theta}{\sin\theta}. \qquad (21.8.8)$$

21.9 Separating the Variables and Wigner Functions and Solution of the Radial Equations

Let us diagonalize operators $\mathbf{J}^2$, J^3 – corresponding substitution for ψ is

$$\psi = e^{-i\omega t} \begin{vmatrix} 0 \\ f_1(r)D_{-1} \\ f_2(r)D_0 \\ f_3(r)D_{+1} \end{vmatrix}, \qquad (21.9.1)$$

where the shorted notation for Wigner D-functions $D_\sigma = D^j_{-m,\sigma}(\phi,\theta,0)$, $\sigma = -1, 0, +1$; j, m determine the total angular momentum. We use the following recursive relations [429]

$$\partial_\theta D_{-1} = \frac{1}{2}(aD_{-2} - \nu D_0), \qquad \frac{m - \cos\theta}{\sin\theta}D_{-1} = \frac{1}{2}(aD_{-2} + \nu D_0),$$

$$\partial_\theta D_0 = \frac{1}{2}(\nu D_{-1} - \nu D_{+1}), \qquad \frac{m}{\sin\theta}D_0 = \frac{1}{2}(\nu D_{-1} + \nu D_{+1}),$$

$$\partial_\theta D_{+1} = \frac{1}{2}(\nu D_0 - aD_{+2}), \qquad \frac{m + \cos\theta}{\sin\theta}D_{+1} = \frac{1}{2}(\nu D_0 + aD_{+2}),$$

$$\nu = \sqrt{j(j+1)}, \qquad a = \sqrt{(j-1)(j+2)}, \qquad (21.9.2)$$

we get (the factor $e^{-i\omega t}$ is omitted)

$$\Sigma'_{\theta\phi}\Psi' = \frac{\nu}{\sqrt{2}} \begin{vmatrix} (f_1 + f_3)D_0 \\ -i\, f_2 D_{-1} \\ i\,(f_1 - f_3)D_0 \\ +i\, f_2 D_{+1} \end{vmatrix}. \tag{21.9.3}$$

Turning back to Maxwell equation; after a simple calculation, we arrive at the radial system

$$(1) \qquad \sqrt{\Phi}\left(\frac{d}{dr} + \frac{2}{r}\right)f_2 + \frac{1}{r}\frac{\nu}{\sqrt{2}}(f_1 + f_3) = 0\,,$$

$$(2) \qquad \left(-\frac{\omega}{\sqrt{\Phi}} - i\sqrt{\Phi}\frac{d}{dr} - i\frac{\sqrt{\Phi}}{r} - i\frac{r}{\sqrt{\Phi}}\right)f_1 - \frac{i}{r}\frac{\nu}{\sqrt{2}}f_2 = 0\,,$$

$$(3) \qquad -\frac{\omega}{\sqrt{\Phi}}f_2 + \frac{i}{r}\frac{\nu}{\sqrt{2}}(f_1 - f_3) = 0\,,$$

$$(4) \qquad \left(-\frac{\omega}{\sqrt{\Phi}} + i\sqrt{\Phi}\frac{d}{dr} + i\frac{\sqrt{\Phi}}{r} + i\frac{r}{\sqrt{\Phi}}\right)f_3 + \frac{i}{r}\frac{\nu}{\sqrt{2}}f_2 = 0\,. \tag{21.9.4}$$

When combining equations (2) and (4); instead of Eq. (21.9.4), we get

$$(2)+(4), \qquad -\frac{\omega}{\sqrt{\Phi}}(f_1 + f_3) - i\left(\sqrt{\Phi}\frac{d}{dr} + \frac{\sqrt{\Phi}}{r} + \frac{r}{\sqrt{\Phi}}\right)(f_1 - f_3) = 0\,,$$

$$(2)-(4), \quad -\frac{\omega}{\sqrt{\Phi}}(f_1 - f_3) - i\left(\sqrt{\Phi}\frac{d}{dr} + \frac{\sqrt{\Phi}}{r} + \frac{r}{\sqrt{\Phi}}\right)(f_1 + f_3) - \frac{2i}{r}\frac{\nu}{\sqrt{2}}f_2 = 0\,,$$

$$(3) \qquad -\frac{\omega}{\sqrt{\Phi}}f_2 + \frac{i}{r}\frac{\nu}{\sqrt{2}}(f_1 - f_3) = 0\,,$$

$$(1) \qquad \sqrt{\Phi}\left(\frac{d}{dr} + \frac{2}{r}\right)f_2 + \frac{1}{r}\frac{\nu}{\sqrt{2}}(f_1 + f_3) = 0\,.$$

It is easily verified that equation (1) is an identity when we allow for the remaining relations. So independent equations are

$$-\frac{\omega}{\sqrt{\Phi}}f_2 + \frac{i}{r}\frac{\nu}{\sqrt{2}}(f_1 - f_3) = 0\,,$$

$$-\frac{\omega}{\sqrt{\Phi}}(f_1 + f_3) - i\left(\sqrt{\Phi}\frac{d}{dr} + \frac{\sqrt{\Phi}}{r} + \frac{r}{\sqrt{\Phi}}\right)(f_1 - f_3) = 0\,,$$

$$-\frac{\omega}{\sqrt{\Phi}}(f_1 - f_3) - i\left(\sqrt{\Phi}\frac{d}{dr} + \frac{\sqrt{\Phi}}{r} + \frac{r}{\sqrt{\Phi}}\right)(f_1 + f_3) - \frac{2i}{r}\frac{\nu}{\sqrt{2}}f_2 = 0\,. \tag{21.9.5}$$

Let us introduce new functions:

$$f = (f_1 + f_3)/\sqrt{2}\,, \qquad g = (f_1 - f_3)/\sqrt{2}\,,$$

then Eqs. (21.9.5) will appears as

$$f_2 = \frac{i\nu}{\omega} \frac{\sqrt{\Phi}}{r} g = 0 \, , \qquad -\frac{\omega}{\Phi} f - i \left(\frac{d}{dr} + \frac{1}{r} + \frac{r}{\Phi} \right) g = 0 \, ,$$

$$-\frac{\omega^2}{\Phi} g - i\omega \left(\frac{d}{dr} + \frac{1}{r} + \frac{r}{\Phi} \right) f + \frac{\nu^2}{r^2} g = 0 \, . \tag{21.9.6}$$

The system in Eq. (21.9.6) is simplified by substitutions

$$g = \frac{1}{r\sqrt{1+r^2}} G(r) \, , \quad f = \frac{1}{r\sqrt{1+r^2}} F(r) \, ,$$

$$f_2 = \frac{i\nu}{\sqrt{2}\omega} \frac{1}{r^2} G(r) = 0 \, , \qquad i\omega F = \Phi \frac{d}{dr} G \, ,$$

$$i\omega \frac{d}{dr} F + \frac{\omega^2}{\Phi} G - \frac{\nu^2}{r^2} G = 0 \, . \tag{21.9.7}$$

So we have arrived at a single differential equation for $G(r)$:

$$(1+r^2) \frac{d^2 G}{dr^2} + 2r \frac{dG}{dr} + \left(\frac{\omega^2}{1+r^2} - \frac{\nu^2}{r^2} \right) G = 0 \, . \tag{21.9.8}$$

In Eq. (21.9.8), let us introduce a new variable $z = -r^2$, which results in

$$4z(1-z) \frac{d^2 G}{dz^2} + 2(1-3z) \frac{dG}{dz} - \left(\frac{\omega^2}{1-z} + \frac{\nu^2}{z} \right) G = 0 \, , \tag{21.9.9}$$

through the use of substitution $G = z^a(1-z)^b F(z)$. Equation (21.9.9) gives

$$4z(1-z) \frac{d^2 F}{dz^2} + 4 \left[2a + \frac{1}{2} - (2a + 2b + \frac{3}{2})z \right] \frac{dF}{dz}$$

$$+ \left[\frac{4a^2 - 2a - \nu^2}{z} + \frac{4b^2 - \omega^2}{1-z} - 4(a+b)(a+b+\frac{1}{2}) \right] F = 0 \, . \tag{21.9.10}$$

The requirements

$$4a^2 - 2a - \nu^2 = 0 \implies$$

$$a = \frac{1}{4} \pm \frac{1}{4}\sqrt{1 + 4\nu^2} = \frac{1}{4} \pm \frac{1}{2}(j + \frac{1}{2}) = -\frac{j}{2}, +\frac{j+1}{2} \, ,$$

$$4b^2 - \omega^2 = 0 \implies b = \pm \frac{\omega}{2} \, , \, \omega > 0 \, ; \tag{21.9.11}$$

to have solutions vanishing at $r = 0$, one must take positive values $a = (j+1)/2$. Equatiion (21.9.10) takes the form

$$z(1-z) \frac{d^2 F}{dz^2} + [\, 2a + \frac{1}{2} - (2a + 2b + \frac{3}{2})z \,] \frac{dF}{dz}$$

$$-(a+b)(a+b+\frac{1}{2}) F = 0 \, , \tag{21.9.12}$$

this equation is the hypergeometric type

$$\gamma = 2a + \frac{1}{2} , \qquad \alpha + \beta = 2a + 2b + \frac{1}{2} , \qquad \alpha\beta = (a+b)(a+b+\frac{1}{2}) ,$$

that is

$$\alpha = a + b , \qquad \beta = a + b + \frac{1}{2} , \qquad \gamma = 2a + \frac{1}{2} . \qquad (21.9.13)$$

To have polynomials, one must take negative value for $b = -\omega/2$. So, parameters are

$$\alpha = \frac{j+1}{2} - \frac{\omega}{2} , \qquad \beta = \frac{j+1}{2} - \frac{\omega}{2} + \frac{1}{2} , \qquad (21.9.14)$$

and quantization is given by [1]:

$$\alpha = -n , \qquad \omega_{n,j} = 2n + j + 1 + n , \qquad (n = 0, 1, 2, ...) ; \qquad (21.9.15)$$

or in usual units $\omega = (c/\rho)(2n + j + 1)$; ρ is a curvature radius.

---

[1]There exists symmetric variant $\beta = -n \implies \omega_{n,j+1} = 2n + (j+1) + 1$.

Chapter 22

Calculation of the Reflection Coefficient for Particles in de Sitter Space

22.1   Introduction

The problem of the Hawking radiation brings much interest [506–509]. In particular, the radiation in de Sitter space-time that details the penetration of the quantum mechanical particles through the de Sitter horizon have been examined in literature.For further research see Lohiya, Panchapakesan, and Khanal [510–515], Bogush, Otchik, and Red'kov [516, 518], Motolla [517], Mishima and Nakayama [519], Polarski [520], Suzuki, Takasugi, and Umetsu [521–524], Red'kov and Ovsiyuk [525]. Some vagueness remains up to this point, and the present chapter aims to clarify this situation.

In this chapter, exact wave solutions for a particle with spin 0 in the static coordinates of the de Sitter space-time model are examined in detail. The procedure for calculating the reflection coefficient $R_{\epsilon j}$ is analyzed. First, for the scalar particle, two pairs of linearly independent solutions are specified explicitly: running and standing waves. A known algorithm for calculation of the reflection coefficient $R_{\epsilon j}$ on the background of the de Sitter space-time model is analyzed. It is shown that the determination of $R_{\epsilon j}$ requires an additional constrain on the quantum numbers $\epsilon R/\hbar c \gg j$, where R is a curvature radius. When we take this condition into account, the value of $R_{\epsilon j}$ turns out to be precisely zero. It has been claimed that the calculation of the reflection coefficient $R_{\epsilon j}$ is not required at all. The reason for this is because there is no barrier in the effective potential curve on the background of the de Sitter space-time. The same conclusion holds for arbitrary particles with higher spins. This was demonstrated explicitly with the help of the exact solutions for electromagnetic and Dirac fields in [525].

The structure of this chapter is as follows. First, we state the problem in Section **22.2**. More details concerning this approach can be found in [525].

In Section **22.3**, we demonstrate that the basic instructive definition for the calculation of the reflection coefficient in de Sitter model. This model is grounded exclusively on the use of zero order approximation $\Phi^{(0)}(r)$ in the expansion of a particle wave function in a series with the form

$$\Phi(r) = \Phi^{(0)}(r) + \left(\frac{1}{R^2}\right)\Phi^{(1)}(r) + \left(\frac{1}{R^2}\right)^2\Phi^{(2)}(r) + \dots \qquad (22.1.1)$$

What is even more important and we will demonstrate it explicitly; this recipe cannot be extended by accounting for contributions of higher order terms. So, the result $R_{\epsilon j} = 0$ that is obtained here, from examining zero-order term $\Phi^{(0)}(r)$ persists and cannot be improved.

22.2 Reflection Coefficient

Wave equation for a spin 0 particle (M is used instead of $McR/\hbar$, R is the curvature radius) will read as

$$\left(\frac{1}{\sqrt{-g}}\partial_\alpha\sqrt{-g}g^{\alpha\beta}\,\partial_\beta + 2 + M^2\right)\Psi(x) = 0\,, \qquad (22.2.1)$$

and is considered in static coordinates

$$dS^2 = \Phi dt^2 - \frac{dr^2}{\Phi} - r^2(d\theta^2 + \sin^2\theta d\phi^2)\,,$$

$$0 \le r < 1\,, \qquad \Phi = 1 - r^2\,. \qquad (22.2.2)$$

For spherical waves

$$\Psi(x) = e^{-i\epsilon t}f(r)Y_{jm}(\theta,\phi),\ \epsilon = ER/\hbar c\,,$$

the differential equation for $f(r)$ is

$$\frac{d^2f}{dr^2} + \left(\frac{2}{r} + \frac{\Phi'}{\Phi}\right)\frac{df}{dr} + \left(\frac{\epsilon^2}{\Phi^2} - \frac{M^2+2}{\Phi} - \frac{j(j+1)}{\Phi r^2}\right)f = 0\,. \qquad (22.2.3)$$

All solutions are constructed in terms of hypergeometric functions (let $r^2 = z$):
    regular at $r = 0$; standing waves are given as

$$f(z) = z^{j/2}\,(1-z)^{-i\epsilon/2}\,F(a,b,c;z)\,, \qquad (22.2.4)$$

$$\kappa = j/2\,,\quad \sigma = -i\epsilon/2\,,\quad c = j + 3/2\,,$$

$$a = \frac{3/2 + j + i\sqrt{M^2 - 1/4} - i\epsilon}{2}\,,$$

$$b = \frac{3/2 + j - i\sqrt{M^2 - 1/4} - i\epsilon}{2}\,; \qquad (22.2.5)$$

    singular at $r = 0$; standing waves are

$$g(z) = z^{-(j+1)/2}\,(1-z)^{-i\epsilon/2}\,F(\alpha,\,\beta,\,\gamma;\,z)\,,$$

$$\kappa = -(j+1)/2\,,\qquad \sigma = -i\epsilon/2\,,$$

$$c = -j + 1/2 \, , \ \alpha = \frac{1/2 - j + i\sqrt{M^2 - 1/4} - i\epsilon}{2} \, ,$$

$$\beta = \frac{1/2 - j - i\sqrt{M^2 - 1/4} - i\epsilon}{2} \, . \qquad (22.2.6)$$

When using the Kummerrelations, one can expand the standing waves into linear combinations of the running waves

$$f(z) = \frac{\Gamma(c)\Gamma(c - a - b)}{\Gamma(c - a)\Gamma(c - b)} \, U_{run}^{out}(z) \, + \, \frac{\Gamma(c)\Gamma(a + b - c)}{\Gamma(a)\Gamma(b)} \, U_{run}^{in}(z) \, , \qquad (22.2.7)$$

$$U_{run}^{out}(z) = z^{j/2} \, (1 - z)^{-i\epsilon/2} \, F(a, b, a + b - c + 1; 1 - z) \, ,$$

$$U_{run}^{in}(z) = z^{j/2} \, (1 - z)^{+i\epsilon/2} \, F(c - a, c - b, c - a - b + 1; 1 - z) \, ,$$

$$(22.2.8a)$$

$$a^* = (c - a) \, , \ b^* = (c - b) \, , \qquad (a + b - c)^* = -(a + b - c) \, ,$$

$$[\, U_{run}^{out}(z) \,]^* = U_{run}^{in}(z) \, ,$$

$$f(z) = 2 \ \mathrm{Re} \ \frac{\Gamma(c)\Gamma(c - a - b)}{\Gamma(c - a)\Gamma(c - b)} \, U_{out}(z) = 2 \ \mathrm{Re} \ \frac{\Gamma(c)\Gamma(a + b - c)}{\Gamma(a)\Gamma(b)} \, U_{in}(z) \, .$$

$$(22.2.8b)$$

Similarly for $g(z)$

$$g(z) = \frac{\Gamma(\gamma)\Gamma(\gamma - \alpha - \beta)}{\Gamma(\gamma - \alpha)\Gamma(\gamma - \beta)} \, U_{run}^{out}(z) \, + \, \frac{\Gamma(\gamma)\Gamma(\alpha + \beta - \gamma)}{\Gamma(\alpha)\Gamma(\beta)} \, U_{run}^{in}(z) \, ,$$

$$(22.2.9a)$$

$$U_{run}^{out}(z) = z^{j/2} \, (1 - z)^{-i\epsilon/2} F(\alpha + 1 - \gamma, \beta + 1 - \gamma, \alpha + \beta + 1 - \gamma; 1 - z) \, ,$$

$$U_{run}^{in}(z) = z^{j/2} \, (1 - z)^{+i\epsilon/2} F(1 - \alpha, 1 - \beta, \gamma + 1 - \alpha - \beta; 1 - z) \, ,$$

$$(22.2.9b)$$

$$g(z) = 2 \ \mathrm{Re} \ \frac{\Gamma(\gamma)\Gamma(\gamma - \alpha - \beta)}{\Gamma(\gamma - \alpha)\Gamma(\gamma - \beta)} \, U_{run}^{out}(z) = 2 \ \mathrm{Re} \ \frac{\Gamma(\gamma)\Gamma(\alpha + \beta - \gamma)}{\Gamma(\alpha)\Gamma(\beta)} \, U_{run}^{in}(z) \, .$$

$$(22.2.9c)$$

Asymptotic behavior of the running waves is given by the relations

$$U_{run}^{out.}(r \sim 0) \sim \frac{1}{r^{j+1}} \, , \qquad U_{run}^{out}(r \sim 1) \sim (1 - r^2)^{-i\epsilon/2} \, ,$$

$$U_{run}^{in}(r \sim 0) \sim \frac{1}{r^{j+1}} , \qquad U_{run}^{in}(r \sim 1) \sim (1 - r^2)^{+i\epsilon/2} , \qquad (22.2.10a)$$

or in a new radial variable $r^* \in [0, \infty)$:

$$r^* = \frac{R}{2} \ln \frac{1+r}{1-r} , \quad r = \frac{\exp(2r^*/R) - 1}{\exp(2r^*/R) + 1} ,$$

$$U_{run}^{out}(r^* \sim \infty) \sim \left(2^{-iER/\hbar c}\right) \exp(+iEr^*/\hbar c) ,$$

$$U_{run}^{in}(r^* \sim \infty) \sim \left(2^{+iER/\hbar c}\right) \exp(-iEr^*/\hbar c) , \qquad \epsilon = ER/\hbar c .$$

$$(22.2.10b)$$

For the standing waves, we have

$$f(r \sim 0) \sim r^j , \qquad g(r \sim 0) \sim \frac{1}{r^{j+1}} ,$$

$$f(r \sim 1) \sim 2 \, \mathrm{Re} \left[\frac{\Gamma(c)\Gamma(a + b - c)}{\Gamma(a)\Gamma(b)} 2^{+iER/\hbar c} \exp(-iEr^*/\hbar c) \right] ,$$

$$g(r \sim 1) \sim 2 \, \mathrm{Re} \left[\frac{\Gamma(\gamma)\Gamma(\alpha + \beta - \gamma)}{\Gamma(\alpha)\Gamma(\beta)} 2^{+iER/\hbar c} \exp(-iEr^*/\hbar c) \right] .$$

$$(22.2.10c)$$

On can perform the transition to the limit of the flat space-time, in accordance with the following rules:

$$a = \frac{p + 1 - i\epsilon R + i\sqrt{R^2 M^2 - 1/4}}{2} ,$$

$$b = \frac{p + 1 - i\epsilon R - i\sqrt{R^2 M^2 - 1/4}}{2} , \quad p = j + 1/2 ;$$

$$\lim_{R\to\infty} (R^2 z) = R^2, \quad F(a, b, c; z)$$

$$= 1 + \frac{ab}{c} \frac{z}{1!} + \frac{a(a+1)b(b+1)}{c(c+1)} \frac{z^2}{2!} + \cdots , \qquad (22.2.11)$$

and further ($R \to \infty$)

$$\frac{a+n}{R} = \frac{1}{2}\left(\frac{p+1}{R} - i\epsilon + i\sqrt{M^2 - \frac{1}{4R^2}}\right) + \frac{n}{R} \approx \frac{-i\epsilon - iM}{2} ,$$

$$\frac{b+n}{R} = \frac{1}{2}\left(\frac{p+1}{R} - i\epsilon - i\sqrt{M^2 - \frac{1}{4R^2}}\right) + \frac{n}{R} \approx \frac{-i\epsilon + iM}{2} .$$

Let $\epsilon^2 - M^2 \equiv k^2$, then we arrive at

$$\lim_{R\to\infty} F(a, b, c; z) = \Gamma(1 + p) \sum_0^\infty \frac{(-k^2 R^2/4)^n}{n!\Gamma(1 + n + p)} ,$$

$$\lim_{R\to\infty} F(b-c+1, a-c+1, -c+2; z) = \Gamma(1-p) \sum_{0}^{\infty} \frac{(-k^2 R^2/4)^n}{n!\Gamma(1+n-p)} \,.$$

$$(22.2.12)$$

We can allow for the known expansion for Bessel functions

$$J_p(x) = (\frac{x}{2})^p \sum_{0}^{\infty} \frac{(ix/2)^{2n}}{n!\Gamma(1+n+p)} \,,$$

we get (wave amplitude A will be determined later)

$$\lim_{R\to\infty} A\, U_{run}^{out}(z) = \lim_{R\to\infty} A\, \frac{1}{\sqrt{r}}$$

$$\times \left[\frac{\Gamma(-i\epsilon R+1)\Gamma(-p)\,\Gamma(1+p)(2/k)^p R^{-p+1/2}}{\Gamma[\frac{1}{2}(+i\sqrt{R^2 M^2 - 1/4} - i\epsilon R + p + 1)]\Gamma[\frac{1}{2}(-i\sqrt{R^2 M^2 - 1/4} - i\epsilon R + p + 1)]} J_p(kr) \right.$$

$$\left. + \frac{\Gamma(-i\epsilon R+1)\Gamma(+p)\Gamma(1-p)(2/k)^{-p} R^{+p+1/2}}{\Gamma[\frac{1}{2}(+i\sqrt{R^2 M^2 - 1/4} - i\epsilon R - p + 1)]\Gamma[\frac{1}{2}(-i\sqrt{R^2 M^2 - 1/4} - i\epsilon R - p + 1)]} J_{-p}(kr) \right] \,.$$

$$(22.2.13)$$

Performing this limiting procedure, we derive the relation

$$\lim_{R\to\infty} A\, U_{out}(z) \to \frac{1}{i^{j+1}} \sqrt{\frac{2}{kr}}\, H_{j+1/2}^{(1)}(kr) \,, \qquad (22.2.14)$$

where

$$H_{j+1/2}^{(1)}(x) = \frac{ip}{\sin(\pi p)} \left[e^{ip\pi} J_p(x) - J_{-p}(x) \right]$$

stands for Hankel spherical functions.

In connection with the limiting procedure, let us pose a question: When does the relation in Eq. (22.2.14) give us a good approximation? In this case, provided that the curvature radius R is finite. This point is important, since when calculating the reflection coefficient in the de Sitter space; only this approximation in Eq. (22.2.14) was previously used in literature.

To clarify this point, let us compare the radial equation in Minkowski model

$$\left[\frac{d^2}{dr^2} + \frac{2}{r}\frac{d}{dr} + \epsilon^2 - M^2 - \frac{j(j+1)}{r^2} \right] f_{\epsilon j}^0 = 0 \,, \qquad (22.2.15)$$

and the appropriate equation in de Sitter model

$$\left[\frac{d^2}{dr^2} + \frac{2(1-2r^2/R^2)}{r(1-r^2/R^2)}\frac{d}{dr} + \frac{\epsilon^2}{(1-r^2/R^2)^2} - \frac{M^2+2}{1-r^2/R^2} - \frac{j(j+1)}{r^2} \right] f_{\epsilon j} = 0 \,.$$

$$(22.2.16)$$

At the region far from the de Sitter horizon $r \ll R$, the last equation reduces to

$$\left[\frac{d^2}{dR^2} + \frac{2}{R}\frac{d}{dR} + \epsilon^2 - M^2 - \frac{j(j+1)+2}{R^2} - \frac{j(j+1)}{R^2} \right] f_{\epsilon j} = 0 \,. \qquad (22.2.17)$$

So, we immediately conclude that Eq. (22.2.17) coincides with Eq. (22.2.15) for solutions only with quantum numbers obeying the following restriction

$$\epsilon^2 - M^2 \gg \frac{j(j+1)+2}{R^2} . \tag{22.2.18}$$

In usual units, this inequality will read as

$$E = \mu\, mc^2, \qquad \lambda = \frac{\hbar}{mc} , \qquad \frac{\lambda^2}{R^2} \sim 10^{-80} ,$$

$$\mu^2 - 1 \gg \frac{\lambda^2}{R^2} \left[j(j+1)+2 \right] . \tag{22.2.19}$$

Instead, for the massless case; we have

$$\frac{R^2\omega^2}{c^2} \gg \left[j(j+1)+2 \right] \qquad \text{or} \qquad \frac{R^2 4\pi^2}{\lambda^2} \gg \left[j(j+1)+2 \right] . \tag{22.2.20}$$

One additional point should be emphasized here; the radial equation in de Sitter space can be transformed to the form of a Schrödinger like equation with an effective barrierless potential. Indeed, in the variable r^* Eq. (22.2.16) reduces to

$$\left[\frac{d^2}{dr^{*2}} + \epsilon^2 - U(r^*) \right] G(r^*) = 0 ,$$

$$U(r^*) = \frac{1-r^2}{R^2} \left[4(1-r) + \frac{r}{1+r} + m^2 R^2 + \frac{j(j+1)}{r^2} \right] . \tag{22.2.21}$$

It is easily verified that this potential corresponds to repulsive (from the center) force in all space

$$F_{r^*} \equiv -\frac{dU}{dr^*} = \frac{1-r^2}{R^2} + \left[2r \left(\frac{j(j+1)}{r^2} + m^2 R^2 + \frac{r}{1+r} \right. \right.$$

$$+ 4(1-r)) + (1-r^2) \left(\frac{2j(j+1)}{r^3} + 4 - \frac{1}{(1+r)^2} \right) \Bigg] > 0 . \tag{22.2.22}$$

At the de Sitter horizon, $r^* \to \infty$, the potential function $U(r^*)$ tends to zero, so $G(r^*) \sim \exp(\pm i\epsilon r^*)$.

The form of the effective Schrödinger equation modeling a particle in the de Sitter space indicates that the problem of calculation for the reflection coefficient in the system is; this statement should not be even considered. However, in a number of publications such a problem has been treated and solved. So, we should reconsider these calculations and results obtained. Significant steps of our approach will be outlined here.

The existing literature calculations of non-zero reflection coefficients $R_{\epsilon j}$ were based on the usage of the approximate formula for $U_{run}^{out}(R)$ on the region far from de Sitter horizon; Eqs. (22.2.13) and (22.2.14). However, as noted earlier, the formula used is a good approximation only for solutions specified by Eq. (22.2.18), that is when $j \ll \epsilon R$.

It can be shown that the formula existing in the literature gives a trivial result when taking into account this restriction. The scheme of calculation (more detail can found in [11]) will be described later.

The first step consists in the use of the asymptotic formula for the Bessel functions: when $x \gg \nu^2$, we have

$$J(x) \sim \frac{\Gamma(2\nu + 1)\, 2^{-2\nu - 1/2}}{\Gamma(\nu + 1)\, \Gamma(\nu + 1/2)} \frac{1}{\sqrt{x}}$$

$$\times \left[\exp\left(+i(x - \frac{\pi}{2}(\nu + \frac{1}{2}))\right) + \exp\left(-i(x - \frac{\pi}{2}(\nu + \frac{1}{2}))\right) \right],$$

so when $j < j^2 \ll \epsilon R << \epsilon R$ we derive

$$U_{run}^{out}(R) \sim \left[\frac{e^{+i\epsilon R}}{\epsilon R} \left(C_1 \exp(-i\frac{\pi}{2}(p + \frac{1}{2})) + C_2 \exp(-i\frac{\pi}{2}(-p + \frac{1}{2})) \right) \right.$$

$$\left. + \frac{e^{-i\epsilon R}}{\epsilon R} \left(C_1 \exp(+i\frac{\pi}{2}(p + \frac{1}{2})) + C_2 \exp(+i\frac{\pi}{2}(-p + \frac{1}{2})) \right) \right],$$

$$(22.2.23)$$

where C_1 and C_2 are given by

$$C_1 = \frac{\Gamma(a + b + 1 - c)\, \Gamma(1 - c)}{\Gamma(b - c + 1)\, \Gamma(a - c + 1)} \frac{2^{-j-1}\, \Gamma(2p + 1)}{(\epsilon R)^j\, \Gamma(p + 1/2)},$$

$$C_1 = \frac{\Gamma(a + b + 1 - c)\, \Gamma(c - 1)}{\Gamma(a)\, \Gamma(b)} \frac{(\epsilon R)^{j+1}\, \Gamma(-2p + 1)}{2^{-j}\, \Gamma(p + 1/2)}. \qquad (22.2.24)$$

The reflection coefficient $R_{\epsilon j}$ is determined by the coefficients at $e^{-i\epsilon R}/\epsilon R$ and $e^{+i\epsilon R}/\epsilon R$. It is the matter of simple calculation to verify that when $\epsilon R \gg j$, the coefficient $R_{\epsilon j}$ is precisely zero

$$\epsilon R \gg j, \qquad R_{\epsilon j} \equiv 0.$$

This conclusion is consistent with the analysis performed earlier.

22.3 Series Expansion on a Parameter R^{-2} of the Exact Solutions and Calculation of the Reflection Coefficient

In Section **22.3**, we demonstrate the basic instructive definition for the calculation of the reflection coefficient in a de Sitter model. This model is grounded exclusively on the use of zero order approximation $\Phi^{(0)}(r)$ in the expansion of a particle wave function in a series with the form given in Eq. (22.1.1), and cannot be extended by accounting for contributions of higher order terms.

Let us start with the solution, the wave running to the de Sitter horizon

$$U^{out}(z) = \Gamma(a + b - c + 1)\, [\, \alpha\, F(z) + \beta\, G(z)\,],$$

$$\alpha = \frac{\Gamma(1 - c)}{\Gamma(b - c + 1)\Gamma(a - c + 1)}, \qquad \beta = \frac{\Gamma(c - 1)}{\Gamma(a)\Gamma(b)}, \qquad (22.3.1)$$

where

$$F(z) = z^{(p-1/2)/2}(1-z)^{-i\epsilon/2}F(a,b,c;z)\,, \qquad c = j + 3/2 = 1 + p\,,$$

$$a = \frac{1 + p - i\epsilon + i\sqrt{m^2 - 1/4}}{2}\,, \qquad b = \frac{1 + p - i\epsilon - i\sqrt{m^2 - 1/4}}{2}\,,$$

$$(22.3.2)$$

and

$$G(z) = z^{(-p-1/2)/2}(1-z)^{+i\epsilon/2}F(a-c+1,b-c+1,2-c;z)\,, \qquad 2-c = 1-p\,,$$

$$a - c + 1 = \frac{1 - p - i\epsilon + i\sqrt{m^2 - 1/4}}{2}\,, \qquad b - c + 1 = \frac{1 - p - i\epsilon - i\sqrt{m^2 - 1/4}}{2}\,.$$

$$(22.3.3)$$

For the following, we need the expressions for all quantities in usual units. It is convenient to change slightly the designation: now R stands for the curvature radius

$$z = \frac{r^2}{R^2}\,, \qquad \epsilon = \frac{ER}{\hbar c} = \mu\frac{R}{\lambda}\,, \qquad E = \mu\, Mc^2\,,$$

$$m = \frac{McR}{\hbar} = \frac{R}{\lambda}\,, \qquad \lambda = \frac{\hbar}{Mc}\,. \qquad (22.3.4)$$

Now, the relations in Eqs. (22.3.1)–(22.3.3) will read as

$$F(z) = R^{-p+1/2}\,\frac{r^p}{\sqrt{r}}\left(1 - \frac{r^2}{R^2}\right)^{-i\mu R/2\lambda} F(a,b,c;\frac{r^2}{R^2})\,, \qquad c = 1 + p\,,$$

$$a = \frac{1}{2}\left(1 + p - i\mu\frac{R}{\lambda} + i\sqrt{\frac{R^2}{\lambda^2} - \frac{1}{4}}\right)\,, \; b = \frac{1}{2}\left(1 + p - i\mu\frac{R}{\lambda} - i\sqrt{\frac{R^2}{\lambda^2} - \frac{1}{4}}\right)\,;$$

$$(22.3.5)$$

and

$$G(z) = R^{p+1/2}\,\frac{r^{-p}}{\sqrt{r}}\left(1 - \frac{r^2}{R^2}\right)^{+i\mu R/2\lambda} F(a-c+1,b-c+1,2-c;\frac{r^2}{R^2})\,,$$

$$2 - c = 1 - p\,,$$

$$a - c + 1 = \frac{1}{2}\left(1 - p - i\mu\frac{R}{\lambda} + i\sqrt{\frac{R^2}{\lambda^2} - \frac{1}{4}}\right)\,,$$

$$b - c + 1 = \frac{1}{2}\left(1 - p - i\mu\frac{R}{\lambda} - i\sqrt{\frac{R^2}{\lambda^2} - \frac{1}{4}}\right)\,. \qquad (22.3.6)$$

The task consists in obtaining the approximate expressions for $F(r)$ and $G(r)$ in the region far from de Sitter horizon, $r \ll R$. First, let us preserve the leading and next to leading terms (we have a natural small parameter λ/R).

First, let us consider the exponential factors

$$\left(1 - \frac{r^2}{R^2}\right)^{\pm i\mu R/2\lambda} = \exp\left[\pm i\frac{\mu R}{2\lambda}\ln(1 - \frac{r^2}{R^2})\right]$$

$$= \cos\left[\frac{\mu R}{2\lambda}\ln(1 - \frac{r^2}{R^2})\right] \pm i\sin\left[\frac{\mu R}{2\lambda}\ln(1 - \frac{r^2}{R^2})\right]. \qquad (22.3.7)$$

Using the expansion for the logarithmic function

$$\ln(1 - x) = -(x + \frac{x^2}{2} + \frac{x^3}{3} + ...)\,,$$

$$\ln(1 - \frac{r^2}{R^2}) = -\left(\frac{r^2}{R^2} + \frac{1}{2}\frac{r^4}{R^4} + \frac{1}{3}\frac{r^6}{R^6} + ...\right),$$

we get (assuming that $r^2 \ll \lambda R$)

$$\left(1 - \frac{r^2}{R^2}\right)^{\pm i\mu R/2\lambda} = \cos\frac{\mu R}{2\lambda}\left(\frac{r^2}{R^2} + \frac{1}{2}\frac{r^4}{R^4} + \frac{1}{3}\frac{r^6}{R^6} + ...\right)$$

$$\mp i\sin\frac{\mu R}{2\lambda}\left(\frac{r^2}{R^2} + \frac{1}{2}\frac{r^4}{R^4} + \frac{1}{3}\frac{r^6}{R^6} + ...\right)$$

$$\approx \left(1 - \frac{\mu^2 r^4}{8\lambda^2 R^2}\right) \mp i\frac{\mu R}{2\lambda}\left(\frac{r^2}{R^2} + \frac{1}{2}\frac{r^4}{R^4}\right). \qquad (22.3.8a)$$

Terms of Eq. (22.3.8a); this can be written in descending order

$$\left(1 - \frac{r^2}{R^2}\right)^{\pm i\mu R/2\lambda} \approx 1 \mp i\mu\frac{r^2}{2\lambda R} - \frac{\mu^2}{2}\frac{r^4}{4\lambda^2 R^2} \mp i\mu\, X\,\frac{r^4}{4\lambda^2 R^2}\,,$$

$$X = \frac{\lambda}{R} \ll 1\,, \qquad \frac{r^2}{2\lambda R} \ll 1\,. \qquad (22.3.8b)$$

Now, we turn to the hypergeometric function $F(a, b, c; z)$. Because $R \sim 10^{30}, \lambda \sim 10^{-12}$, one can use the approximation of a leading and the next two order terms in the expressions for the following parameters

$$a = \frac{1}{2}\left(1 + p - i\mu\frac{R}{\lambda} + i\frac{R}{\lambda}\sqrt{1 - \frac{\lambda^2}{4R^2}}\right) = \frac{1+p}{2} - i\frac{\mu - 1}{2}\frac{R}{\lambda} - i\frac{\lambda}{16R}\,,$$

$$b = \frac{1}{2}\left(1 + p - i\mu\frac{R}{\lambda} - i\frac{R}{\lambda}\sqrt{1 - \frac{\lambda^2}{4R^2}}\right) = \frac{1+p}{2} - i\frac{\mu + 1}{2}\frac{R}{\lambda} + i\frac{\lambda}{16R}\,. $$

$$(22.3.9)$$

Then, the hypergeometric function is given as

$$F(a, b, c; \frac{r^2}{R^2}) = 1 + \frac{1}{R^2}\frac{ab}{c}r^2 + \frac{1}{2!}\frac{1}{R^4}\frac{a(a+1)b(b+1)}{c(c+1)}(r^2)^2$$

$$+\frac{1}{3!}\frac{1}{R^6}\frac{a(a+1)(a+2)b(b+1)(b+2)}{c(c+1)(c+2)}(r^2)^3+\dots$$

$$+\frac{1}{n!}\frac{1}{R^{2n}}\frac{a(a+1)(a+2)\dots(a+n-1)b(b+1)(b+2)\dots(b+n-1)}{c(c+1)(c+2)\dots(c+n-1)}(r^2)^n+\dots$$

$$(22.3.10)$$

It is convenient to introduce a shortening notation for small quantity $X = \lambda/R$, then a typical term is represented as

$$\frac{1}{R^2}\frac{(a+n)(b+n)}{(c+n)}\approx\frac{1}{p+1+n}$$

$$\times\frac{-i(\mu-1)}{2\lambda}\left(1+i\frac{1+p+2n}{\mu-1}X+\frac{X^2}{8(\mu-1)}\right)$$

$$\times\frac{-i(\mu+1)}{2\lambda}\left(1+i\frac{1+p+2n}{\mu+1}X-\frac{X^2}{8(\mu+1)}\right)$$

$$\approx\frac{1}{p+1+n}\left(-\frac{\mu^2-1}{4\lambda^2}\right)\left[1+\frac{2i\mu}{\mu^2-1}(1+p+2n)X-\frac{(1+p+2n)^2-1/4}{\mu^2-1}X^2\right].$$

$$(22.3.11)$$

Here, we will use the notation

$$k^2=\frac{\mu^2-1}{\lambda^2},$$

then

$$\frac{1}{R^2}\frac{(a+n)(b+n)}{(c+n)}\approx\frac{1}{p+1+n}\left(-\frac{k^2}{4}\right)$$

$$\times\left[1+\frac{2i\mu}{\mu^2-1}(1+p+2n)X-\frac{(1+p+2n)^2-1/4}{\mu^2-1}X^2\right].$$

$$(22.3.12)$$

Thus, we have the following approximate expressions for the first few terms in the series

$$\frac{r^2}{R^2}\frac{ab}{c}\approx\frac{1}{p+1}\left(-\frac{k^2r^2}{4}\right)\left[1+\frac{2i\mu}{\mu^2-1}(1+p)X-\frac{(1+p)^2-1/4}{\mu^2-1}X^2\right],$$

$$\frac{r^2}{R^2}\frac{(a+1)(b+1)}{(c+1)}\approx\frac{1}{p+2}\left(-\frac{k^2r^2}{4}\right)$$

$$\times\left[1+\frac{2i\mu}{\mu^2-1}(1+p+2\times1)X-\frac{(1+p+2\times1)^2-1/4}{\mu^2-1}X^2\right],$$

$$\frac{r^2}{R^2}\frac{(a+2)(b+2)}{(c+2)}\approx\frac{1}{p+3}\left(-\frac{k^2r^2}{4}\right)$$

$$\times\left[1+\frac{2i\mu}{\mu^2-1}(1+p+2\times2)X-\frac{(1+p+2\times2)^2-1/4}{\mu^2-1}X^2\right],$$

$$\frac{r^2}{R^2}\frac{(a+3)(b+3)}{(c+3)}\approx\frac{1}{p+4}\left(-\frac{k^2r^2}{4}\right)$$

$$\times \left[1 + \frac{2i\mu}{\mu^2 - 1}(1 + p + 2 \times 3)\, X - \frac{(1 + p + 2 \times 3)^2 - 1/4}{\mu^2 - 1}\, X^2 \right],$$

$$\frac{r^2}{R^2} \frac{(a+4)(b+4)}{(c+4)} \approx \frac{1}{p+5} \left(-\frac{k^2 r^2}{4} \right)$$

$$\times \left[1 + \frac{2i\mu}{\mu^2 - 1}(1 + p + 2 \times 4)\, X - \frac{(1 + p + 2 \times 4)^2 - 1/4}{\mu^2 - 1}\, X^2 \right],$$

....

$$\frac{r^2}{R^2} \frac{(a+n)(b+n)}{(c+n)} \approx \frac{1}{p+1+n} \left(-\frac{k^2 r^2}{4} \right)$$

$$\times \left[1 + \frac{2i\mu}{\mu^2 - 1}(1 + p + 2n)\, X - \frac{(1 + p + 2n)^2 - 1/4}{\mu^2 - 1}\, X^2 \right].$$

Now, we are ready to write down the expressions for the coefficients of the hypergeometric series preserving only leading and the next two order terms

$$\frac{1}{1!} \frac{r^2}{R^2} \frac{ab}{c} \approx \frac{1}{p+1} \left(-\frac{k^2 r^2}{4} \right) \left[1 + \frac{2i\mu}{\mu^2 - 1}(1 + p)X - \frac{(1 + p)^2 - 1/4}{\mu^2 - 1}X^2 \right],$$

$$\frac{1}{2!} \frac{(r^2)^2}{(R^2)^2} \frac{ab(a+1)(b+1)}{c(c+1)}$$

$$\approx \left(-\frac{k^2 r^2}{4} \right)^2 \frac{1}{2!(p+1)(p+2)} \left\{ 1 + \frac{2i\mu}{\mu^2 - 1}[(1+p) + (1 + p + 2 \times 1)]X \right.$$

$$\left. -X^2 \left[\frac{4\mu^2}{(\mu^2 - 1)^2}(1+p)(1 + p + 2 \times 1) + \frac{(1+p)^2 - 1/4}{\mu^2 - 1} + \frac{(1 + p + 2 \times 1)^2 - 1/4}{\mu^2 - 1} \right] \right\},$$

$$\frac{1}{3!} \frac{(r^2)^3}{(R^2)^3} \frac{ab(a+1)(b+1)(a+2)(b+2)}{c(c+1)(c+2)} \approx \left(-\frac{k^2 r^2}{4} \right)^3 \frac{1}{3!(p+1)(p+2)(p+3)}$$

$$\times \left\{ 1 + X \frac{2i\mu}{\mu^2 - 1}[(1+p) + (1 + p + 2 \times 1) + (1 + p + 2 \times 2)] \right.$$

$$-X^2 \left[\frac{4\mu^2}{(\mu^2 - 1)^2}[(1+p) + (1 + p + 2 \times 1)](1 + p + 2 \times 2) \right.$$

$$\left. \left. +\frac{(1+p)^2 - 1/4}{\mu^2 - 1} + \frac{(1 + p + 2 \times 1)^2 - 1/4}{\mu^2 - 1} + \frac{(1 + p + 2 \times 2)^2 - 1/4}{\mu^2 - 1} \right] \right\},$$

$$\frac{1}{4!} \frac{(r^2)^4}{(R^2)^4} \frac{ab(a+1)(b+1)(a+2)(b+2)(a+3)(b+3)}{c(c+1)(c+2)(c+3)}$$

$$\approx \left(-\frac{k^2 r^2}{4} \right)^4 \frac{1}{4!(p+1)(p+2)(p+3)(p+4)}$$

$$\times \left\{ 1 + X \frac{2i\mu}{\mu^2 - 1}[(1+p) + (1 + p + 2 \times 1) + (1 + p + 2 \times 2) + (1 + p + 2 \times 3)] \right.$$

$$-X^2 \left[\frac{4\mu^2}{(\mu^2-1)^2}[(1+p)+(1+p+2\times1)+(1+p+2\times2)](1+p+2\times3) \right.$$

$$+\frac{(1+p)^2-1/4}{\mu^2-1}+\frac{(1+p+2\times1)^2-1/4}{\mu^2-1}$$

$$\left. +\frac{(1+p+2\times2)^2-1/4}{\mu^2-1}+\frac{(1+p+2\times3)^2-1/4}{\mu^2-1} \right] \Bigg\} ,$$

··

$$\frac{1}{n!}\frac{(r^2)^n}{(R^2)^n}\frac{ab(a+1)(b+1)....(a+n-1)(b+n-1)}{c(c+1)...(c+n)}$$

$$\approx \left(-\frac{k^2r^2}{4} \right)^n \frac{1}{n!(p+1)(p+2)...(p+n)}$$

$$\times \left\{ 1+X\,\frac{2i\mu}{\mu^2-1}[(1+p) \right.$$

$$+(1+p+2\times1)+(1+p+2\times2)+...+(1+p+2\times(n-1))]$$

$$-X^2 \left[\frac{4\mu^2}{(\mu^2-1)^2}[(1+p)+(1+p+2\times1)+... \right.$$

$$+(1+p+2\times(n-2))]\,(1+p+2\times(n-1))$$

$$\left. \left. +\frac{(1+p)^2-1/4}{\mu^2-1}+\frac{(1+p+2\times1)^2-1/4}{\mu^2-1}+...+\frac{(1+p+2\times(n-1))^2-1/4}{\mu^2-1} \right] \right\} .$$

Thus, initial exact hypergeometric function can be approximated by the sum of three series

$$\bar{F} = F(a,b,c;\frac{r^2}{R^2}) = \bar{F}_0(r) + X\,\bar{F}_1(r) + X^2\,\bar{F}_2(r) . \qquad (22.3.13)$$

The leading series $\bar{F}_0(r)$, in fact, reduces to the Bessel function

$$\bar{F}_0(r) = 1 + \frac{(ikr/2)^2}{n!(p+1)} + \frac{(ikr/2)^4}{2!(p+1)(p+2)} + ... + \frac{(ikr/2)^{2n}}{n!(p+1)(p+2)...(p+n)} + ...$$

$$= \Gamma(p+1)\sum_{n=0}^{\infty}\frac{(ikr/2)^{2n}}{n!\Gamma(p+1+n)} = \Gamma(1+p)\left(\frac{kr}{2}\right)^{-p}J_p(kr) ,$$

$$(22.3.14)$$

where

$$J_p(x) = (\frac{x}{2})^p \sum_0^{\infty}\frac{(-x^2/4)^n}{n!\,\Gamma(p+1+n)} , \qquad x = kr .$$

The second series is given by

$$X\,\bar{F}_1(r) = X\,\frac{2i\mu}{\mu^2-1}\left\{(-k^2r^2/4)\frac{p+1}{p+1}+(-k^2r^2/4)^2\frac{[(1+p)+(1+p+2\times1)]}{2!(p+1)(p+2)}\right.$$

$$+(-k^2r^2/4)^3\frac{[(1+p)+(1+p+2\times 1)+(1+p+2\times 2)]}{3!(p+1)(p+2)(p+3)}$$

$$+(-k^2r^2/4)^4\frac{[(1+p)+(1+p+2\times 1)+(1+p+2\times 2)+(1+p+2\times 3)]}{4!(p+1)(p+2)(p+3)(p+4)}+...$$

$$+(-k^2r^2/4)^n\frac{[(1+p)+(1+p+2\times 1)+...+(1+p+2\times(n-1))]}{n!(p+1)(p+2)...(p+n)}+...\Bigg\};$$

$$(22.3.15)$$

that is

$$X\bar{F}_1(r)=X\frac{2i\mu}{\mu^2-1}\left(\frac{-k^2r^2}{4}\right)\left\{1+\left(\frac{-k^2r^2}{4}\right)\frac{[(1+p)+(1+p+2\times 1)]}{2!(p+1)(p+2)}\right.$$

$$+\left(\frac{-k^2r^2}{4}\right)^2\frac{[(1+p)+(1+p+2\times 1)+(1+p+2\times 2)]}{3!(p+1)(p+2)(p+3)}$$

$$+\left(\frac{-k^2r^2}{4}\right)^3\frac{[(1+p)+(1+p+2\times 1)+(1+p+2\times 2)+(1+p+2\times 3)]}{4!(p+1)(p+2)(p+3)(p+4)}+...$$

$$+\left(\frac{-k^2r^2}{4}\right)^n\frac{[(1+p)+(1+p+2\times 1)+...+(1+p+2\times n)]}{(n+1)!(p+1)(p+2)...(p+n+1)}+...\Bigg\};$$

or shorter

$$X\bar{F}_1(r)=X\frac{2i\mu}{\mu^2-1}\Gamma(p+1)\left(\frac{-k^2r^2}{4}\right)$$

$$\times\sum_{n=0}^{\infty}\left(\frac{-k^2r^2}{4}\right)^n\frac{[(1+p)+...+(1+p+2\times n)]}{(n+1)!\,\Gamma(p+2+n)}.$$

$$(22.3.16)$$

Using the known sums

$$(1+p)+...+(1+p+2\times n)=(1+p)(n+1)+2(1+2+3+...n)$$

$$=(p+1)(1+n)+2\frac{(1+n)n}{2}=(n+1)(n+1+p)$$

we derive

$$X\bar{F}_1(r)=X\frac{2i\mu}{\mu^2-1}\,\Gamma(p+1)\left(\frac{-k^2r^2}{4}\right)\sum_{n=0}^{\infty}(-k^2r^2/4)^n\frac{(n+1)(n+1+p)}{(n+1)!\,\Gamma(p+2+n)}$$

$$=X\frac{2i\mu}{\mu^2-1}\,\Gamma(p+1)\left(\frac{-k^2r^2}{4}\right)\sum_{n=0}^{\infty}\frac{(-k^2r^2/4)^n}{n!\,\Gamma(p+1+n)}$$

$$=X\left(\frac{-k^2r^2}{4}\right)\frac{2i\mu}{\mu^2-1}\,\Gamma(p+1)\left(\frac{kr}{2}\right)^{-p}J_p(kr)=X\left(\frac{-k^2r^2}{4}\right)\frac{2i\mu}{\mu^2-1}\,\bar{F}_0(r).$$

Thus, the approximation in Eq. (22.3.13) can be presented as follows

$$\bar{F} = \bar{F}_0(r) + X\bar{F}_1(r) + X^2\bar{F}_2(r)$$

$$= \bar{F}_0(r) + X \frac{2i\mu}{\mu^2 - 1} \left(\frac{-k^2r^2}{4}\right) \bar{F}_0(r) + X^2\bar{F}_2(r) . \qquad (22.3.17)$$

Similar relations can be derived for the hypergeometric series $\bar{G}(r)$:

$$\bar{G} = F(a - c + 1, b - c + 1, 2 - c; \frac{r^2}{R^2})$$

$$= \bar{G}_0(r) + X \bar{G}_1(r) + X^2 \bar{G}_2(r) . \qquad (22.3.18)$$

The leading term again reduces to the Bessel function

$$\bar{G}_0(r) = \Gamma(1 - p) \left(\frac{kr}{2}\right)^{+p} J_{-p}(kr) . \qquad (22.3.19)$$

The next order term is given by

$$X\bar{G}_1(r) = X \left(\frac{-k^2r^2}{4}\right) \frac{2i\mu}{\mu^2 - 1} \bar{G}_0(r) . \qquad (22.3.20)$$

So, the approximation of Eq. (22.3.17) is presented as

$$\bar{G} = \bar{G}_0(r) + X\bar{G}_1(r) + X^2\bar{G}_2(r)$$

$$= \bar{G}_0(r) + X \frac{2i\mu}{\mu^2 - 1} \left(\frac{-k^2r^2}{4}\right) \bar{G}_0(r) + X^2\bar{G}_2(r) . \qquad (22.3.21)$$

Now, let us consider the complete function

$$F(z) = R^{-p+1/2} \frac{r^p}{\sqrt{r}} \left(1 - \frac{r^2}{R^2}\right)^{-i\mu R/2\lambda} F(a, b, c; \frac{r^2}{R^2})$$

$$= R^{-p+1/2} \frac{r^p}{\sqrt{r}} \left[1 + i\mu \frac{r^2}{2\lambda R} - \frac{\mu^2}{2} \frac{r^4}{4\lambda^2 R^2} + i\mu X \frac{r^4}{4\lambda^2 R^2}\right]$$

$$\times \left[\bar{F}_0(r) + \frac{2i\mu}{\mu^2 - 1} \left(\frac{-k^2r^2}{4}\right) X\bar{F}_0(r) + X^2\bar{F}_2(r)\right]$$

$$= R^{-p+1/2} \frac{r^p}{\sqrt{r}} \left[\bar{F}_0(r) + \frac{2i\mu}{\mu^2 - 1} \left(\frac{-k^2r^2}{4}\right) X\bar{F}_0(r) + X^2\bar{F}_2(r)\right.$$

$$+i\mu \frac{r^2}{2\lambda R}\bar{F}_0(r) + i\mu \frac{r^2}{2\lambda R}\frac{2i\mu}{\mu^2 - 1} \left(\frac{-k^2r^2}{4}\right) X\bar{F}_0(r) + i\mu \frac{r^2}{2\lambda R}X^2\bar{F}_2(r)$$

$$-\frac{\mu^2}{2} \frac{r^4}{4\lambda^2 R^2}\bar{F}_0(r) - \frac{\mu^2}{2} \frac{r^4}{4\lambda^2 R^2}\frac{2i\mu}{\mu^2 - 1} \left(\frac{-k^2r^2}{4}\right) X\bar{F}_0(r) - \frac{\mu^2}{2} \frac{r^4}{4\lambda^2 R^2}X^2\bar{F}_2(r)$$

$$+i\mu\, X\, \frac{r^4}{4\lambda^2 R^2}\bar{F}_0(r) + i\mu\, X\, \frac{r^4}{4\lambda^2 R^2}\frac{2i\mu}{\mu^2-1}\,(\frac{-k^2 r^2}{4})\, X\bar{F}_0(r) + i\mu\, X\, \frac{r^4}{4\lambda^2 R^2}X^2\bar{F}_2(r)\Big]$$

$$= R^{-p+1/2}\frac{r^p}{\sqrt{r}}\left[\bar{F}_0(r) - i\mu\,\frac{r^2}{2\lambda R}\,\bar{F}_0(r) + X^2\bar{F}_2(r)\right.$$

$$+i\mu\frac{r^2}{2\lambda R}\bar{F}_0(r) + \mu^2(\frac{r^2}{2\lambda R})^2\bar{F}_0(r) + i\mu\,\frac{r^2}{2\lambda R}\,X^2\bar{F}_2(r)$$

$$-\frac{\mu^2}{2}\,(\frac{r^2}{2\lambda R})^2\,\bar{F}_0(r) + \frac{i\mu^3}{2}(\frac{r^2}{2\lambda R})^3\,\bar{F}_0(r) - \frac{\mu^2}{2}\,(\frac{r^2}{2\lambda R})^2\,X^2\bar{F}_2(r)$$

$$+i\mu\, X\,(\frac{r^2}{2\lambda R})^2\,\bar{F}_0(r) + \mu^2(\frac{r^2}{2\lambda R})^3\bar{F}_0(r) + i\mu\,(\frac{r^2}{2\lambda R})^2 X^3\bar{F}_2(r)\Big].$$

Preserving only first two terms, we have

$$F(z) = R^{-p+1/2}\frac{r^{+p}}{\sqrt{r}}\left[\bar{F}_0(r) + \frac{1}{8}\mu^2\frac{r^2 r^2}{\lambda^2 R^2}\,\bar{F}_0(r) + \frac{\lambda^2}{R^2}\bar{F}_2(r)\right]. \qquad (22.3.22)$$

Similarly, for $G(r)$ we obtain

$$G(z) = R^{+p+1/2}\frac{r^{-p}}{\sqrt{r}}\left[\bar{G}_0(r) + \frac{1}{8}\mu^2\frac{r^2 r^2}{\lambda^2 R^2}\,\bar{G}_0(r) + \frac{\lambda^2}{R^2}\bar{G}_2(r)\right]. \qquad (22.3.23)$$

One should emphasize this feature for the expansions of Eqs. (22.3.22) and (22.3.23): these approximations are real valued, as we must expect.

In the known method of determining and calculating the reflection coefficients in de Sitter model, other authors only used the leading terms in approximations of Eqs. (22.3.22) and (22.3.23), because only these terms allow separate elementary solutions of the form $e^{\pm ikr}$.

Let us perform some additional calculations to clarify the problem. The functions $F(z)$ and $G(z)$ enter the expression for (to horizon) running wave

$$U^{out}(z) = \Gamma(a+b-c+1)[\,\alpha\, F(z) + \beta\, G(z)\,]\,,$$

$$\alpha = \frac{\Gamma(1-c)}{\Gamma(b-c+1)\Gamma(a-c+1)}\,, \qquad \beta = \frac{\Gamma(c-1)}{\Gamma(a)\Gamma(b)}\,, \qquad (22.3.24)$$

where (introducing a very large parameter $Y = R/2\lambda$)

$$\alpha \approx \frac{\Gamma(-p)}{\Gamma[-i(\mu-1)Y + (1-p)/2]\;\Gamma[-i(\mu+1)Y + (1-p)/2]}\,,$$

$$\beta \approx \frac{\Gamma(+p)}{\Gamma[-i(\mu-1)Y + (1+p)/2]\;\Gamma[-i(\mu+1)Y + (1+p)/2]}\,.$$

$$(22.3.25)$$

For all physically reasonable values of quantum numbers j (not very high ones) and values of μ (different from the critical value $\mu = 1$ and not too high ones – they correspond in fact to energies of a particle in units of the rest energy) the argument of Γ-functions in Eq. (22.3.25) are complex-valued with very large imaginary parts.

Let us multiply the given solution in Eq. (22.3.25) by a special factor A which permits us to distinguish small and large parts in this expansion:

$$A = \frac{\Gamma(a - p/2 + 1/4)\Gamma(b - p/2 + 1/4)}{\Gamma(a + b - c + 1)} \, . \tag{22.3.26}$$

So, instead of Eq. (22.3.25), we get

$$AU^{out}(z) = \alpha' \, F(z) + \beta' \, G(z) \, ,$$

$$\alpha' = \Gamma(1 - c)\frac{\Gamma(a - p/2 + 1/4)\Gamma(b - p/2 + 1/4)}{\Gamma(b - c + 1)\Gamma(a - c + 1)} \, ,$$

$$\beta' = \Gamma(c - 1)\frac{\Gamma(a - p/2 + 1/4)\Gamma(b - p/ + 1/4)}{\Gamma(a)\Gamma(b)} \, . \tag{22.3.27}$$

The expressions for α' and β' are

$$\alpha' \approx \Gamma(-p) \, \frac{\Gamma[-i(\mu - 1)Y + 1/4]}{\Gamma[-i(\mu - 1)Y + (1 - p)/2)]} \, \frac{\Gamma[-i(\mu + 1)Y + 1/4]}{\Gamma[-i(\mu + 1)Y + (1 - p)/2)]} \, ,$$

$$\beta' \approx \Gamma(+p) \, \frac{\Gamma[-i(\mu - 1)Y + 1/4]}{\Gamma[-i(\mu - 1)Y + (1 + p)/2)]} \, \frac{\Gamma[-i(\mu + 1)Y + 1/4]}{\Gamma[-i(\mu + 1)Y + (1 + p)/2)]} \, . \tag{22.3.28}$$

We allow for identities

$$\Gamma(-p) = -\frac{\pi}{\sin p\pi}\frac{1}{\Gamma(1 + p)} \, , \qquad \Gamma(p) = +\frac{\pi}{\sin p\pi}\frac{1}{\Gamma(1 - p)} \, , \tag{22.3.29}$$

α' and β' are transformed into

$$\alpha' \approx -\frac{\pi}{\sin p\pi}\frac{1}{\Gamma(1 + p)}\frac{\Gamma[-i(\mu - 1)Y + 1/4]}{\Gamma[-i(\mu - 1)Y + (1 - p)/2)]}\frac{\Gamma[-i(\mu + 1)Y + 1/4]}{\Gamma[-i(\mu + 1)Y + (1 - p)/2)]} \, ,$$

$$\beta' \approx +\frac{\pi}{\sin p\pi}\frac{1}{\Gamma(1 - p)}\frac{\Gamma[-i(\mu - 1)Y + 1/4]}{\Gamma[-i(\mu - 1)Y + (1 + p)/2)]}\frac{\Gamma[-i(\mu + 1)Y + 1/4]}{\Gamma[-i(\mu + 1)Y + (1 + p)/2)]} \, . \tag{22.3.30}$$

Now, we have to take into account the asymptotic formula for Γ-function

$$\frac{\Gamma(z + A)}{\Gamma(z + B)} = z^{A-B}\left(1 + \frac{1}{z}\frac{(A - B)(A + B + 1)}{2} + ...\right) \, ,$$

$$|\arg z| < \pi \, , \qquad |z| \to \infty \, ; \tag{22.3.31}$$

then (remembering that $Y = R/2\lambda$)

$$\frac{\Gamma[-i(\mu - 1)Y + 1/4]}{\Gamma[-i(\mu - 1)Y + (1 - p)/2]}$$

$$\approx \left[-i\frac{(\mu - 1)}{2\lambda}R\right]^{p/2-1/4}\left[1 + \frac{2\lambda}{-i(\mu - 1)R}\frac{(2p - 1)(7 - 2p)}{32}\right],$$

$$\frac{\Gamma[-i(\mu + 1)Y + 1/4]}{\Gamma[-i(\mu + 1)Y + (1 - p)/2]}$$

$$\approx \left[-i\frac{(\mu + 1)}{2\lambda}R\right]^{p/2-1/4}\left[1 + \frac{2\lambda}{-i(\mu + 1)R}\frac{(2p - 1)(7 - 2p)}{32}\right], \quad (22.3.32)$$

and

$$\frac{\Gamma[-i(\mu - 1)Y + 1/4]}{\Gamma[-i(\mu - 1)Y + (1 + p)/2]}$$

$$\approx \left[-i\frac{(\mu - 1)}{2\lambda}R\right]^{-p/2-1/4}\left[1 + \frac{2\lambda}{-i(\mu - 1)R}\frac{(-2p - 1)(7 + 2p)}{32}\right],$$

$$\frac{\Gamma[-i(\mu + 1)Y + 1/4]}{\Gamma[-i(\mu + 1)Y + (1 + p)/2]}$$

$$\approx \left[-i\frac{(\mu + 1)}{2\lambda}R\right]^{-p/2-1/4}\left[1 + \frac{2\lambda}{-i(\mu + 1)R}\frac{(-2p - 1)(7 + 2p)}{32}\right]. \quad (22.3.33)$$

Substituting Eqs. (22.3.32) and (22.3.33) into Eq. (22.3.30); we obtain

$$\alpha' \approx -\left(\frac{\mu^2 - 1}{4\lambda^2}R^2\right)^{p/2-1/4}(-1)^{p/2-1/4}\frac{\pi}{\sin p\pi}\frac{1}{\Gamma(1 + p)}$$

$$\times \left[1 + \frac{2\lambda}{-i(\mu - 1)R}\frac{(2p - 1)(7 - 2p)}{32}\right]\left[1 + \frac{2\lambda}{-i(\mu + 1)R}\frac{(2p - 1)(7 - 2p)}{32}\right],$$

$$\beta' \approx +\left(\frac{\mu^2 - 1}{4\lambda^2}R^2\right)^{-p/2-1/4}(-1)^{-p/2-1/4}\frac{\pi}{\sin p\pi}\frac{1}{\Gamma(1 - p)}$$

$$\times \left[1 + \frac{2\lambda}{-i(\mu - 1)R}\frac{(-2p - 1)(7 + 2p)}{32}\right]\left[1 + \frac{2\lambda}{-i(\mu + 1)R}\frac{(-2p - 1)(7 + 2p)}{32}\right].$$

$$(22.3.34)$$

Preserving the only terms of first two orders; we have

$$\alpha' \approx -\frac{\pi}{\sin p\pi}\frac{1}{\Gamma(1 + p)}\left(\frac{k}{2}\right)^{+p-1/2}R^{+p-1/2}(-1)^{p/2-1/4}$$

$$\times \left(1 + i\frac{(2p - 1)(7 - 2p)}{8}\frac{\mu}{\mu^2 - 1}\frac{\lambda}{R}\right),$$

$$\beta' \approx +\frac{\pi}{\sin p\pi}\frac{1}{\Gamma(1 - p)}$$

$$\times \left(\frac{k}{2}\right)^{-p-1/2} R^{-p-1/2}(-1)^{-p/2-1/4}\left(1 - i\frac{(-2p-1)(7+2p)}{8}\frac{\mu}{\mu^2-1}\frac{\lambda}{R}\right).$$

$$(22.3.35)$$

By substituting expressions of Eq. (22.3.35) into the following expansion

$$AU^{out}(z) = \alpha'\, F(z) + \beta'\, G(z)\,,$$

$$F(r) = R^{-p+1/2}\,\Gamma(1+p)\left(\frac{k}{2}\right)^{-p}\frac{1}{\sqrt{r}}J_p(kr)\,,$$

$$G(r) \approx R^{p+1/2}\,\Gamma(1-p)\left(\frac{k}{2}\right)^{p}\frac{1}{\sqrt{r}}J_{-p}(kr)\,, \qquad (22.3.36)$$

we arrive at the following zero-order approximation

$$AU^{out}(z) = \psi_0^{out}(r) = -\frac{\pi}{\sin p\pi}\frac{1}{\Gamma(1+p)}$$

$$\times \left(\frac{k}{2}\right)^{+p-1/2} R^{+p-1/2}(-1)^{p/2-1/4}R^{-p+1/2}\Gamma(1+p)\left(\frac{k}{2}\right)^{-p}\frac{1}{\sqrt{r}}J_p(kr)$$

$$+\frac{\pi}{\sin p\pi}\frac{1}{\Gamma(1-p)}\left(\frac{k}{2}\right)^{-p-1/2} R^{-p-1/2}(-1)^{-p/2-1/4}R^{p+1/2}\Gamma(1-p)\left(\frac{k}{2}\right)^{p}\frac{1}{\sqrt{r}}J_{-p}(kr),$$

that is

$$AU^{out}(z) = \psi_0^{out}(r)$$

$$= \frac{\pi}{\sin p\pi}\sqrt{\frac{2k}{r}}\left[\ -(-1)^{p/2-1/4}\,J_p(kr) + (-1)^{-p/2-1/4}\,J_{-p}(kr)\ \right]$$

$$= -(-1)^{-p/2-1/4}\sqrt{\frac{2k}{r}}\ \frac{\pi}{\sin p\pi}\left[\ (-1)^{p}\,J_p(kr) - J_{-p}(kr)\ \right]\;;$$

$$(22.3.37)$$

this coincides with a spherical wave propagating in Minkowski space from the origin, and is expressed through the Hankel functions of the first kind

$$H_p^{(1)} = \frac{ip}{\sin p\pi}\left[(-1)^{p}J_p(kr) - J_{-p}(kr)\right]\,. \qquad (22.3.38)$$

22.4   Conclusion

The last, but not the least, mathematical remark should be given. All known quantum mechanical problems with potentials containing one barrier reduced to a second order differential equation with four singular points, is an Heun class equation. In particular, the most popular cosmological problem of that type is a particle in the Schwarzschild space-time background and it reduces to the Heun differential equation. Quantum mechanical problems of the tunneling type are never linked to differential equation of the hypergeometric type, an equation with three singular points. Instead, in the case of de Sitter model with the wave equations for different fields, of spin 0, 1/2, and 1; after the separation of variables are reduced to the second order differential equation with three singular points, and there exists no ground to search in these systems for tunneling class problems.

Chapter 23

On Solutions of Maxwell Equations in the Schwarzschild Space-Time

23.1 Introduction

In this chapter, a general covariant approach to the Maxwell theory is based on the use of the Riemann–Silberstein–Majorana–Oppenheimer complex representation which is used to treat Maxwell field in the background of a Schwarzschild black hole. It is shown that this technique provides us with the possibility when after separating the variables to reduce the problem of a Maxwell field with a differential equation; which is similarly experienced in the case of a scalar field in the Schwarzschild space-time. This differential equation is recognized as a confluent Heun equation.

Another consideration that we have pertains to the electromagnetic field based on a 10-dimensional description. In addition to six components of the strength tensor , one uses a 4 component electromagnetic potential. Such a description of the electromagnetic field is more informative because it includes gauge degrees of freedom. However, this method to describe an electromagnetic field is more complicated. We use it here in a matrix form of Duffin–Kemmer–Petiau. After the separation of the variable, we arrive at a system of 10 radial equations, which can be simplified by the use of additional constraints followed from eigenvalue equation for spacial parity operator $\hat{\Pi}\Psi = P\Psi$; the radial system is divided into two subsystems of 4 and 6 equations, respectively. In our second approach, the problem of electromagnetic field reduces to the confluent Heun differential equation as well. In particular, we explicitly show how our suggested solutions for a complex form are embedded into matrix DKP-formalism, and determine that radial functions are responsible for the gauge degrees of freedom.

23.2   Complex Formalism in Electrodynamics, Curved Space-Time Background

Maxwell equations in Riemann space can be presented in a Riemann–Silberstein–Majorana–Oppenheimer basis as a one matrix equation

$$\alpha^c \left(e^\rho_{(c)} \partial_\rho + \frac{1}{2} j^{ab} \gamma_{abc} \right) \Psi = J(x) \,,$$

$$\alpha^0 = -iI \,, \qquad \Psi = \begin{vmatrix} 0 \\ \mathbf{E} + ic\mathbf{B} \end{vmatrix} \,, \qquad J = \frac{1}{\epsilon_0} \begin{vmatrix} \rho \\ i\mathbf{j} \end{vmatrix} \,, \qquad (23.2.1)$$

or

$$-i(e^\rho_{(0)} \partial_\rho + \frac{1}{2} j^{ab} \gamma_{ab0}) \Psi + \alpha^k (e^\rho_{(k)} \partial_\rho + \frac{1}{2} j^{ab} \gamma_{abk}) \Psi = J(x) \,. \qquad (23.2.2)$$

We can allow for identities

$$\frac{1}{2} j^{ab} \gamma_{ab0} = [s_1(\gamma_{230} + i\gamma_{010}) + s_2(\gamma_{310} + i\gamma_{020}) + s_3(\gamma_{120} + i\gamma_{030})] \,,$$

$$\frac{1}{2} j^{ab} \gamma_{abk} = [s_1(\gamma_{23k} + i\gamma_{01k}) + s_2(\gamma_{31k} + i\gamma_{02k}) + s_3(\gamma_{12k} + i\gamma_{03k})] \,,$$

and by using the notation

$$e^\rho_{(0)} \partial_\rho = \partial_{(0)} \,, \qquad e^\rho_{(k)} \partial_\rho = \partial_{(k)} \,, \qquad a = 0, 1, 2, 3 \,,$$

$$(\gamma_{01a}, \gamma_{02a}, \gamma_{03a}) = \mathbf{v}_a \,, \qquad (\gamma_{23a}, \gamma_{31a}, \gamma_{12a}) = \mathbf{p}_a \,,$$

Equation (23.2.2), in the absence of sources, reduces to

$$-i[\, \partial_{(0)} + \mathbf{s}(\mathbf{p}_0 + i\mathbf{v}_0) \,]\Psi + \alpha^k [\, \partial_{(k)} + \mathbf{s}(\mathbf{p}_k + i\mathbf{v}_k) \,] \Psi = 0 \,. \qquad (23.2.3)$$

Let us consider this matrix equation in the Schwarzschild metric

$$dS^2 = \Phi \, dt^2 - \frac{dr^2}{\Phi} - r^2(d\theta^2 + \sin^2\theta d\phi^2) \,, \quad \Phi = 1 - M/r \,,$$

$$e^\alpha_{(0)} = (\frac{1}{\sqrt{\Phi}}, 0, 0, 0) \,, \qquad e^\alpha_{(3)} = (0, \sqrt{\Phi}, 0, 0) \,,$$

$$e^\alpha_{(1)} = (0, 0, \frac{1}{r}, 0) \,, \qquad e^\alpha_{(2)} = (1, 0, 0, \frac{1}{r\sin\theta}) \,,$$

$$\gamma_{030} = \frac{\Phi'}{2\sqrt{\Phi}} \,, \quad \gamma_{311} = \frac{\sqrt{\Phi}}{r} \,, \quad \gamma_{322} = \frac{\sqrt{\Phi}}{r} \,, \quad \gamma_{122} = \frac{\cos\theta}{r\sin\theta} \,, \qquad (23.2.4)$$

we solve for the explicit form of a matrix equation

$$\left[-\frac{i\partial_t}{\sqrt{\Phi}} + \sqrt{\Phi}(\alpha^3 \partial_r + \frac{\alpha^1 s_2 - \alpha^2 s_1}{r} + \frac{\Phi'}{2\Phi} s_3) + \frac{1}{r} \Sigma_{\theta,\phi} \right] \begin{vmatrix} 0 \\ \psi \end{vmatrix} = 0 \,,$$

$$\Sigma_{\theta,\phi} = \frac{\alpha^1}{r} \partial_\theta + \alpha^2 \frac{\partial_\phi + s_3 \cos\theta}{\sin\theta} \,. \qquad (22.2.5)$$

It is convenient to have the spin matrix s_3 as the diagonal, which is reached by a simple linear transformation in the known cyclic basis; so that

$$j'^{23} = s'_1 = \begin{vmatrix} 0 & 0 \\ 0 & \tau'_1 \end{vmatrix}, \qquad j'^{31} = s'_2 = \begin{vmatrix} 0 & 0 \\ 0 & \tau'_2 \end{vmatrix}, \qquad j'^{12} = s'_3 = \begin{vmatrix} 0 & 0 \\ 0 & \tau'_3 \end{vmatrix},$$

$$\tau'_1 = \frac{1}{\sqrt{2}} \begin{vmatrix} 0 & -i & 0 \\ -i & 0 & -i \\ 0 & -i & 0 \end{vmatrix}, \tau'_2 = \frac{1}{\sqrt{2}} \begin{vmatrix} 0 & -1 & 0 \\ 1 & 0 & -1 \\ 0 & 1 & 0 \end{vmatrix}, \tau'_3 = -i \begin{vmatrix} +1 & 0 & 0 \\ 0 & 0 & 0 \\ 0 & 0 & -1 \end{vmatrix},$$

$$\alpha'^1 = \frac{1}{\sqrt{2}} \begin{vmatrix} 0 & -1 & 0 & 1 \\ 1 & 0 & -i & 0 \\ 0 & -i & 0 & -i \\ -1 & 0 & -i & 0 \end{vmatrix}, \alpha'^2 = \frac{1}{\sqrt{2}} \begin{vmatrix} 0 & -i & 0 & -i \\ -i & 0 & -1 & 0 \\ 0 & 1 & 0 & -1 \\ -i & 0 & 1 & 0 \end{vmatrix},$$

$$\alpha'^3 = \begin{vmatrix} 0 & 0 & 1 & 0 \\ 0 & -i & 0 & 0 \\ -1 & 0 & 0 & 0 \\ 0 & 0 & 0 & +i \end{vmatrix}. \tag{23.2.6}$$

Equation (23.2.5) becomes

$$[-\frac{i\partial_t}{\sqrt{\Phi}} + \sqrt{\Phi}(\alpha'^3\partial_r + \frac{\alpha'^1 s'_2 - \alpha'^2 s'_1}{r} + \frac{\Phi'}{2\Phi}s'_3) + \frac{1}{r}\Sigma'_{\theta,\phi}] \begin{vmatrix} 0 \\ \psi' \end{vmatrix} = 0,$$

$$\Sigma'_{\theta,\phi} = \frac{\alpha'^1}{r}\partial_\theta + \alpha'^2 \frac{\partial_\phi + s'_3\cos\theta}{\sin\theta}. \tag{23.2.7}$$

23.3   Separating the Variables and Wigner Functions

Let us use the diagonalize operators $\mathbf{J}^2$, J^3, with a corresponding substitution for ψ is

$$\psi = e^{-i\omega t} \begin{vmatrix} 0 \\ \varphi_1(r)D_{-1} \\ \varphi_2(r)D_0 \\ \varphi_3(r)D_{+1} \end{vmatrix}, \tag{23.3.1}$$

where the shorted notation for Wigner D-functions $D_\sigma = D^j_{-m,\sigma}(\phi, \theta, 0), \sigma = -1, 0, +1$; j, m determine total angular moment. By using the following recursive relations [429]

$$\partial_\theta D_{-1} = \frac{1}{2}(aD_{-2} - \nu D_0), \quad \frac{m - \cos\theta}{\sin\theta}D_{-1} = \frac{1}{2}(aD_{-2} + \nu D_0),$$

$$\partial_\theta D_0 = \frac{1}{2}(\nu D_{-1} - \nu D_{+1}), \quad \frac{m}{\sin\theta}D_0 = \frac{1}{2}(\nu D_{-1} + \nu D_{+1}),$$

$$\partial_\theta D_{+1} = \frac{1}{2}(\nu D_0 - aD_{+2}), \quad \frac{m + \cos\theta}{\sin\theta}D_{+1} = \frac{1}{2}(\nu D_0 + aD_{+2}),$$

$$\nu = \sqrt{j(j+1)}, \qquad a = \sqrt{(j-1)(j+2)}, \tag{23.3.2}$$

we get (the factor $e^{-i\omega t}$ is omitted)

$$\Sigma'_{\theta\phi}\Psi' = \frac{\nu}{\sqrt{2}} \begin{vmatrix} (\varphi_1 + \varphi_3)D_0 \\ -i\,\varphi_2 D_{-1} \\ i\,(\varphi_1 - \varphi_3)D_0 \\ +i\,\varphi_2 D_{+1} \end{vmatrix} . \tag{23.3.3}$$

Next, in order to simplify the equations, we will change the notation

$$\frac{\nu}{\sqrt{2}} = \sqrt{\frac{j(j+1)}{2}} \implies \nu .$$

Returning back to the Maxwell Eq. (23.2.7), then after a simple calculation, we arrive at the radial system

$$(1) \quad \sqrt{\Phi}\left(\frac{d}{dr} + \frac{2}{r}\right)\varphi_2 + \frac{\nu}{r}\left(\varphi_1 + \varphi_3\right) = 0 ,$$

$$(2) \quad \left(-\frac{\omega}{\sqrt{\Phi}} - i\,\sqrt{\Phi}\frac{d}{dr} - i\frac{\sqrt{\Phi}}{r} - i\frac{\Phi'}{2\sqrt{\Phi}}\right)\varphi_1 - \frac{i\nu}{r}\varphi_2 = 0 ,$$

$$(3) \quad -\frac{\omega}{\sqrt{\Phi}}\varphi_2 + \frac{i\nu}{r}\left(\varphi_1 - \varphi_3\right) = 0 ,$$

$$(4) \quad \left(-\frac{\omega}{\sqrt{\Phi}} + i\,\sqrt{\Phi}\frac{d}{dr} + i\frac{\sqrt{\Phi}}{r} + i\frac{\Phi'}{2\sqrt{\Phi}}\right)\varphi_3 + \frac{i\nu}{r}\varphi_2 = 0 .$$

$$\tag{23.3.4}$$

By combining equations (2) and (4), instead of Eq. (23.3.4) we get

$$(2)+(4)\,, \quad -\frac{\omega}{\sqrt{\Phi}}(\varphi_1 + \varphi_3) - i\left(\sqrt{\Phi}\frac{d}{dr} + \frac{\sqrt{\Phi}}{r} + \frac{\Phi'}{2\sqrt{\Phi}}\right)(\varphi_1 - \varphi_3) = 0 ,$$

$$(2)-(4)\,, \quad -\frac{\omega}{\sqrt{\Phi}}(\varphi_1 - \varphi_3) - i\left(\sqrt{\Phi}\frac{d}{dr} + \frac{\sqrt{\Phi}}{r} + \frac{\Phi'}{2\sqrt{\Phi}}\right)(\varphi_1 + \varphi_3) - \frac{2i\nu}{r}\varphi_2 = 0 ,$$

$$(3) \quad -\frac{\omega}{\sqrt{\Phi}}\varphi_2 + \frac{i\nu}{r}\left(\varphi_1 - \varphi_3\right) = 0 ,$$

$$(1) \quad \sqrt{\Phi}\left(\frac{d}{dr} + \frac{2}{r}\right)\varphi_2 + \frac{\nu}{r}\left(\varphi_1 + \varphi_3\right) = 0 . \tag{23.3.5}$$

It is easily verified that equation (1) is an identity when allowing for those that remain. Therefore, three independent equations are

$$-\frac{\omega}{\sqrt{\Phi}}\varphi_2 + \frac{i\nu}{r}\left(\varphi_1 - \varphi_3\right) = 0 ,$$

$$-\frac{\omega}{\sqrt{\Phi}}(\varphi_1 + \varphi_3) - i\left(\sqrt{\Phi}\frac{d}{dr} + \frac{\sqrt{\Phi}}{r} + \frac{\Phi'}{2\sqrt{\Phi}}\right)(\varphi_1 - \varphi_3) = 0 ,$$

$$-\frac{\omega}{\sqrt{\Phi}}\left(\varphi_1 - \varphi_3\right) - i\left(\sqrt{\Phi}\frac{d}{dr} + \frac{\sqrt{\Phi}}{r} + \frac{\Phi'}{2\sqrt{\Phi}}\right)(\varphi_1 + \varphi_3) - \frac{2i\nu}{r}\,\varphi_2 = 0\,.$$

$$(23.3.6)$$

Let us introduce new functions:

$$f = \varphi_1 + \varphi_3\,, \qquad g = \varphi_1 - \varphi_3\,,$$

then Eqs. (23.3.6) will read as

$$\varphi_2 = \frac{i\nu}{\omega}\frac{\sqrt{\Phi}}{r}g\,, \qquad -\frac{\omega}{\Phi}f - i\left(\frac{d}{dr} + \frac{1}{r} + \frac{\Phi'}{2\Phi}\right)g = 0\,,$$

$$-\frac{\omega^2}{\Phi}g - i\omega\left(\frac{d}{dr} + \frac{1}{r} + \frac{\Phi'}{2\Phi}\right)f + \frac{2\nu^2}{r^2}\,g = 0\,. \qquad (23.3.7)$$

The system in Eq. (23.3.7) is simplified by substitutions

$$g = \frac{1}{r\sqrt{\Phi}}\,G(r)\,, \qquad f = \frac{1}{r\sqrt{\Phi}}\,F(r)\,,$$

so, we have

$$\varphi_2 = \frac{i\nu}{\omega}\frac{1}{r^2}\,G(r)\,, \qquad i\omega\,F = \Phi\frac{d}{dr}G\,,$$

$$+i\omega\,\frac{d}{dr}\,F + \frac{\omega^2}{\Phi}\,G - \frac{2\nu^2}{r^2}\,G = 0\,. \qquad (23.3.8)$$

Finally, we have arrived at a differential equation for $G(r)$:

$$\frac{d^2G}{dr^2} + \frac{\Phi'}{\Phi}\frac{dG}{dr} + \left(\frac{\omega^2}{\Phi^2} - \frac{j(j+1)}{r^2\Phi}\right)G = 0\,, \qquad (23.3.9)$$

its explicit form is

$$\frac{d^2G}{dr^2} + \frac{M}{r(r-M)}\frac{dG}{dr} + \left(\frac{\omega^2 r^2}{(r-M)^2} - \frac{j(j+1)}{r(r-M)}\right)G = 0\,. \qquad (23.3.10)$$

Its mathematically a more convenient form (let us introduce a new variable $x = r/M$) is

$$\frac{d^2G}{dx^2} + \left(\frac{1}{x-1} - \frac{1}{x}\right)\frac{dG}{dx}$$

$$+\left(M^2\omega^2 + \frac{j(j+1)}{x} + \frac{2M^2\omega^2 - j(j+1)}{x-1} + \frac{M^2\omega^2}{(x-1)^2}\right)G = 0\,. \qquad (23.3.11)$$

After using the substitution

$$G = (x-1)^\alpha x^\beta e^{\gamma x} g(x) \qquad (23.3.12)$$

from Eq. (23.3.11), we get

$$\frac{d^2g}{dx^2} + \left[\frac{1 + 2\alpha}{x-1} - \frac{1 - 2\beta}{x} + 2\gamma\right]\frac{dg}{dx} + \left[M^2\omega^2 + \gamma^2 + \frac{M^2\omega^2 + \alpha^2}{(x-1)^2} + \frac{\beta(\beta-2)}{x^2}\right.$$

$$+\frac{j(j+1)+\alpha-\beta-\gamma-2\,\alpha\beta+2\,\beta\gamma}{x}$$

$$+\frac{2M^2\omega^2-j(j+1)-\alpha+\beta+\gamma+2\,\alpha\beta+2\,\alpha\gamma}{x-1}\Bigg]\,g=0\,. \qquad (23.3.13)$$

At the following restrictions on parameters α,β,γ

$$\alpha=\pm iM\omega\,, \qquad \beta=0,\,2\,, \qquad \gamma=\pm iM\omega \qquad (23.3.14)$$

Equation (23.3.11) becomes simpler

$$\frac{d^2g}{dx^2}+\left(\frac{1+2\,\alpha}{x-1}-\frac{1-2\,\beta}{x}+2\gamma\right)\frac{dg}{dx}+\left(\frac{j(j+1)+\alpha-\beta-\gamma-2\,\alpha\beta+2\,\beta\gamma}{x}\right.$$

$$\left.+\frac{2M^2\omega^2-j(j+1)-\alpha+\beta+\gamma+2\,\alpha\beta+2\,\alpha\gamma}{x-1}\right)g=0\,, \qquad (23.3.15)$$

which is recognized as the confluent Heun equation for $G(A,B,C,D,F)$

$$\frac{d^2Z}{dz^2}+\left[A+\frac{1+B}{z}+\frac{1+C}{z-1}\right]\frac{dZ}{dz}+\left[\frac{1}{2}\frac{A-B-C+AB-BC-2F}{z}\right.$$

$$\left.+\frac{1}{2}\frac{A+B+C+AC+BC+2D+2F}{z-1}\right]f=0$$

with parameters given by

$$A=2\gamma\,, \qquad B=2\beta-2\,, \qquad C=2\alpha\,,$$

$$D=2M^2\omega^2\,, \qquad F=1-j(j+1)\,. \qquad (23.3.16)$$

23.4   On Massless Vector Field in Duffin–Kemmer Approach

Let us consider the same problem on the basis of general covariant matrix Duffin–Kemmer formalism. In the Schwarzschild space-time it will read as

$$\left[i\,\beta^0\partial_t\,+\,i\,\Phi\,(\beta^3\partial_r\,+\,\frac{1}{r}\,(\beta^1 j^{31}\,+\,\beta^2 j^{32})\right.$$

$$+\frac{\Phi'}{2\Phi}\,\beta^0 J^{03})\,+\,\frac{\sqrt{\Phi}}{r}\,\Sigma_{\theta,\phi}^\kappa\,-\,I_6\,\sqrt{\Phi}\Bigg]\,\Phi(x)=0\,,$$

$$\Sigma_{\theta,\phi}^\kappa\,=\,i\,\beta^1\partial_\theta\,+\,\beta^2\,\frac{i\partial\,+\,i\,j^{12}\cos\theta}{\sin\theta}\,. \qquad (23.4.1)$$

Through substitution, the solutions of spherical wave type is

$$\Phi_{\omega jm}(x)=e^{-i\omega t}\,[\,f_1(r)\,D_0,\ f_2(r)\,D_{-1},\ f_3(r)\,D_0,\ f_4(r)\,D_{+1},$$

$$f_5(r)\,D_{-1},f_6(r)\,D_0,\ f_7(r)\,D_{+1},\ f_8(r)\,D_{-1},\ f_9(r)\,D_0,\ f_{10}(r)\,D_{+1}\,]\,. \qquad (23.4.2)$$

After the separation of variables, we get 10 radial equations ($\nu = \sqrt{j(j+1)/2}$)

$$-\Phi \left(\frac{d}{dr} + \frac{2}{r}\right) f_6 - \nu \frac{\sqrt{\Phi}}{r} (f_5 + f_7) = 0 \, ,$$

$$i\omega f_5 + i\Phi \left(\frac{d}{dr} + \frac{1}{r} + \frac{\Phi'}{2\Phi}\right) f_8 + i\nu \frac{\sqrt{\Phi}}{r} f_9 = 0 \, ,$$

$$i\omega f_6 + i\nu \frac{\sqrt{\Phi}}{r} (-f_8 + f_{10}) = 0 \, ,$$

$$i\omega f_7 - i\Phi \left(\frac{d}{dr} + \frac{1}{r} + \frac{\Phi'}{2\Phi}\right) f_{10} - i\nu \frac{\sqrt{\Phi}}{r} f_9 = 0 \, ,$$

$$-i\omega f_2 + \nu \frac{\sqrt{\Phi}}{r} f_1 - \sqrt{\Phi} f_5 = 0 \, ,$$

$$-i\omega f_3 - \Phi \left(\frac{d}{dr} + \frac{\Phi'}{2\Phi}\right) f_1 - \sqrt{\Phi} f_6 = 0 \, ,$$

$$-i\omega f_4 + \nu \frac{\sqrt{\Phi}}{r} f_1 - \sqrt{\Phi} f_7 = 0 \, ,$$

$$-i\Phi \left(\frac{d}{dr} + \frac{1}{r}\right) f_2 - i\nu \frac{\sqrt{\Phi}}{r} f_3 - \sqrt{\Phi} f_8 = 0 \, ,$$

$$i\nu \frac{\sqrt{\Phi}}{r} (f_2 - f_4) - \sqrt{\Phi} f_9 = 0 \, ,$$

$$i\Phi \left(\frac{d}{dr} + \frac{1}{r}\right) f_4 + i\nu \frac{\sqrt{\Phi}}{r} f_3 - \sqrt{\Phi} f_{10} = 0 \, . \qquad (23.4.3)$$

Diagonalyzing additionally the spacial inversion operator

$$\hat{P}^{cycl.}_{sph.} = \begin{vmatrix} 1 & 0 & 0 & 0 & 0 & 0 & 0 & 0 & 0 & 0 \\ 0 & 0 & 0 & 1 & 0 & 0 & 0 & 0 & 0 & 0 \\ 0 & 0 & 1 & 0 & 0 & 0 & 0 & 0 & 0 & 0 \\ 0 & 1 & 0 & 0 & 0 & 0 & 0 & 0 & 0 & 0 \\ 0 & 0 & 0 & 0 & 0 & 0 & 1 & 0 & 0 & 0 \\ 0 & 0 & 0 & 0 & 0 & 1 & 0 & 0 & 0 & 0 \\ 0 & 0 & 0 & 0 & 1 & 0 & 0 & 0 & 0 & 0 \\ 0 & 0 & 0 & 0 & 0 & 0 & 0 & 0 & 0 & -1 \\ 0 & 0 & 0 & 0 & 0 & 0 & 0 & 0 & -1 & 0 \\ 0 & 0 & 0 & 0 & 0 & 0 & 0 & -1 & 0 & 0 \end{vmatrix} \hat{P} \, ,$$

from the eigenvalues equation $\hat{P}^{cycl.}_{sph.} \, \Phi_{jm} = P \, \Phi_{jm}$ we obtain two possible eigenvalues and the corresponding restrictions on radial functions:

$$P = (-1)^{j+1} \, ,$$

$$f_1 = f_3 = f_6 = 0 \, , \quad f_4 = -f_2 \, , \quad f_7 = -f_5 \, , \quad f_{10} = +f_8 \, ; \qquad (23.4.4)$$

$$P = (-1)^j \, ,$$

$$f_9 = 0 \, , \quad f_4 = +f_2 \, , \quad f_7 = +f_5 \, , \quad f_{10} = -f_8 \, . \tag{23.4.5}$$

By taking account of Eqs. (23.4.4) and (23.4.5), we get a more simple system

$$P = (-1)^{j+1} \, ,$$

$$i\omega \, f_5 \, + \, i\Phi \left(\frac{d}{dr} + \frac{1}{r} + \frac{\Phi'}{2\Phi} \right) f_8 \, + \, i\nu \frac{\sqrt{\Phi}}{r} \, f_9 = 0 \, , \qquad -i\omega \, f_2 \, - \, \sqrt{\Phi} \, f_5 = 0 \, ,$$

$$-i\Phi \left(\frac{d}{dr} + \frac{1}{r} \right) f_2 \, - \, \sqrt{\Phi} \, f_8 = 0 \, , \qquad i2\nu \frac{\sqrt{\Phi}}{r} \, f_2 \, - \, \sqrt{\Phi} \, f_9 = 0 \, ; \tag{23.4.6a}$$

from this, it follows a second order differential equation for f_2:

$$\frac{d^2 f_2}{dr^2} + \left(\frac{\Phi'}{\Phi} + \frac{2}{r} \right) \frac{df_2}{dr} + \left(\frac{\omega^2}{\Phi^2} - \frac{2\nu^2}{r^2 \Phi} + \frac{\Phi'}{\Phi} \frac{1}{r} \right) f_2 = 0 \, . \tag{23.4.6b}$$

By the substitution of $f_2 = r^{-1} F_2$, it transforms into

$$\frac{d^2 F_2}{dr^2} + \frac{\Phi'}{\Phi} \frac{dF_2}{dr} + \left(\frac{\omega^2}{\Phi^2} - \frac{j(j+1)}{r^2 \Phi} \right) F_2 = 0 \, , \tag{23.4.6c}$$

which coincides with the equation for G in Eq. (23.3.9).

The case $P = (-1)^j$ gives

$$\Phi \left(\frac{d}{dr} + \frac{2}{r} \right) F_6 \, + \, \frac{2\nu}{r} \, F_5 \, = 0 \, , \qquad i\omega \, F_5 \, + \, i\Phi \left(\frac{d}{dr} + \frac{1}{r} \right) F_8 = 0 \, ,$$

$$i\omega \, F_6 \, - \, i \frac{2\nu}{r} \, F_8 = 0 \, , \qquad -i\omega \, F_2 \, + \, \frac{\nu}{r} \, F_1 \, - \, F_5 = 0 \, ,$$

$$i\omega \, F_3 \, + \, \Phi \frac{d}{dr} F_1 \, + \, \Phi \, F_6 = 0 \, ,$$

$$i\Phi \, (\frac{d}{dr} + \frac{1}{r}) \, F_2 \, + \, i \frac{\nu}{r} \, F_3 \, + \, F_8 = 0 \, , \tag{23.4.7}$$

where

$$F_1 = \sqrt{\Phi} \, f_1 \, , \quad F_2 = f_2 \, , \quad F_3 = \sqrt{\Phi} \, f_3 \, ,$$

$$F_5 = \sqrt{\Phi} \, f_5 \, , \quad F_6 = f_6 \, , \quad F_8 = \sqrt{\Phi} \, f_8 \, . \tag{23.4.8}$$

23.5   Relation Between Two Formalisms

To proceed further, let us use relationships between Duffin–Kemmer and complex formalisms. Starting with these identities

$$E_{(1)} = F_{(0)(1)} \, , \qquad E_{(2)} = F_{(0)(2)} \, , \qquad E_{(3)} = F_{(0)(3)} \, ,$$

$$B_{(1)} = -F_{(2)(3)} \, , \qquad B_{(2)} = -F_{(3)(1)} \, , \qquad B_{(3)} = -F_{(1)(2)} \, , \tag{23.5.1}$$

and the known structures of 3-vector and antisymmetric tensor

$$\Psi = e^{-i\omega t} \begin{vmatrix} 0 \\ \varphi_1 D_{-1} \\ \varphi_2 D_0 \\ \varphi_3 D_{+1} \end{vmatrix}, \quad \Phi = e^{-i\omega t} \left[f_1 \, D_0, \, f_2 \, D_{-1}, \, f_3 \, D_0, \, f_4 D_{+1}, \right.$$

$$\left. f_5 \, D_{-1}, \, f_6 \, D_0, \, f_7 \, D_{+1}, \, f_8 \, D_{-1}, \, f_9 \, D_0, \, f_{10} \, D_{+1} \right], \tag{23.5.2}$$

we arrive at three equations

$$e^{-i\omega t} \, D_0 \varphi_2(r) = E_{(2)} + i B_{(2)} = F_{(0)(2)} - i F_{(3)(1)} = e^{-i\omega t} (f_6(r) \, D_0 + i f_9(r) \, D_0) \,,$$

$$e^{-i\omega t} D_{-1} \varphi_1(r) = E_{(1)} + i B_{(1)} = F_{(0)(1)} - i F_{(2)(3)} = e^{-i\omega t} (f_5(r) \, D_{-1} - i \, f_8(r) \, D_{-1}) \,,$$

$$e^{-i\omega t} \, D_{+1} \varphi_3(r) = E_{(3)} + i B_{(3)} = F_{(0)(3)} - i F_{(1)(2)} = e^{-i\omega t} (f_7(r) \, D_{+1} - i f_{10}(r) \, D_{+1}) \,.$$

From this, it follows

$$\varphi_2 = f_6 + i \, f_9 \,, \qquad \varphi_1 = f_5 - i \, f_8 \,, \qquad \varphi_3 = f_7 - i \, f_{10} \,. \tag{23.5.3}$$

When taking into account the constraints related to parity (see Eqs. (23.4.4) and (23.4.5)); we obtain

$$\underline{P = (-1)^{j+1}} \,, \qquad f_6 = 0 \,, \ \ f_7 = -f_5 \,, \ f_{10} = +f_8 \qquad \Longrightarrow$$

$$\varphi_2 = +i \, f_9 \,, \qquad \varphi_1 = f_5 - i \, f_8 \,, \qquad \varphi_3 = -f_5 - i \, f_8 \,; \tag{23.5.4}$$

$$\underline{P = (-1)^{j}} \,, \qquad f_9 = 0 \,, \ \ f_7 = +f_5 \,, \ f_{10} = -f_8 \qquad \Longrightarrow$$

$$\varphi_2 = f_6 \,, \qquad \varphi_1 = f_5 - i \, f_8 \,, \qquad \varphi_3 = f_5 + i \, f_8 \,. \tag{23.5.5}$$

They give the inverse relations

$$\underline{P = (-1)^{j+1}} \,, \qquad f_9 = -i \varphi_2 \,, \qquad f_8 = \frac{i}{2} (\varphi_1 + \varphi_3) \,, \qquad f_5 = \frac{1}{2} (\varphi_1 - \varphi_3) \,;$$

$$\tag{23.5.6}$$

$$\underline{P = (-1)^{j}} \,, \qquad F_6 = \varphi_2 \,, \qquad F_5 = \frac{\sqrt{\Phi}}{2} (\varphi_1 + \varphi_3) \,, \qquad F_8 = i \frac{\sqrt{\Phi}}{2} (\varphi_1 - \varphi_3) \,.$$

$$\tag{23.5.7}$$

23.6   Radial System at $P = (-1)^j$

First, let us examine the case $P = (-1)^j$. To this end, in six equations of Eq. (23.4.7) let us take into account the Eq. (23.5.7). Three first equations (they contain only functions F_5, F_6, F_8)

$$ i\omega F_6 - i\frac{2\nu}{r} F_8 = 0 , \qquad \Phi \left(\frac{d}{dr} + \frac{2}{r} \right) F_6 + \frac{2\nu}{r} F_5 = 0 , $$

$$ i\omega F_5 + i\Phi \left(\frac{d}{dr} + \frac{1}{r} \right) F_8 = 0 , \tag{23.6.1} $$

after translating to the new variables, it will read as

$$ i\omega \, \varphi_2 - i\frac{\nu}{r} \, i\sqrt{\Phi} \, (\varphi_1 - \varphi_3) = 0 , $$

$$ \Phi \left(\frac{d}{dr} + \frac{2}{r} \right) \varphi_2 + \frac{\nu}{r} \, \sqrt{\Phi} \, (\varphi_1 + \varphi_3) = 0 , $$

$$ i\omega \, \sqrt{\Phi} \, (\varphi_1 + \varphi_3) + i\Phi \left(\frac{d}{dr} + \frac{1}{r} \right) i\sqrt{\Phi} \, (\varphi_1 - \varphi_3) = 0 . \tag{23.6.2} $$

Equations (23.6.2) may be compared with Eq. (23.3.5)

$$ (3) \qquad -\frac{\omega}{\sqrt{\Phi}} \varphi_2 + \frac{i\nu}{r} \, (\varphi_1 - \varphi_3) = 0 , $$

$$ (1) \qquad \sqrt{\Phi} \left(\frac{d}{dr} + \frac{2}{r} \right) \varphi_2 + \frac{\nu}{r} \, (\varphi_1 + \varphi_3) = 0 , $$

$$ (2) + (4) \qquad -\frac{\omega}{\sqrt{\Phi}}(\varphi_1 + \varphi_3) - i \left(\sqrt{\Phi}\frac{d}{dr} + \frac{\sqrt{\Phi}}{r} + \frac{\Phi'}{2\sqrt{\Phi}} \right) (\varphi_1 - \varphi_3) = 0 , $$

$$ (2) - (4) \qquad -\frac{\omega}{\sqrt{\Phi}}(\varphi_1 - \varphi_3) - i \left(\sqrt{\Phi}\frac{d}{dr} + \frac{\sqrt{\Phi}}{r} + \frac{\Phi'}{2\sqrt{\Phi}} \right) (\varphi_1 + \varphi_3) - \frac{2i\nu}{r} \, \varphi_2 = 0 . \tag{23.6.3} $$

The first three equations coincide; the last equation in Eq. (23.6.3), (2)–(4) has no counterpart, but it is a consequence of first three in Eq. (23.3.10); and therefore, it can be ignored. Remember that the three functions $\varphi_1, \varphi_2, \varphi_3$ are defined uniquely by G

$$ \varphi_2 = -\frac{1}{i\omega} \frac{\nu}{r^2} \, G(r) , $$

$$ \varphi_1 + \varphi_3 = \frac{1}{i\omega}\frac{\sqrt{\Phi}}{r}\frac{d}{dr}G , \qquad \varphi_1 - \varphi_3 = \frac{1}{r\sqrt{\Phi}} \, G(r) , \tag{23.6.4} $$

where G is determined by the equation

$$ \frac{d^2 G}{dr^2} + \frac{M}{r(r-M)}\frac{dG}{dr} + \left(\frac{\omega^2 r^2}{(r-M)^2} - \frac{j(j+1)}{r(r-M)} \right) G = 0 . \tag{23.6.5} $$

The three remaining equations in Eq. (23.6.3) relate radial functions of the electromagnetic 4-vector, F_1, F_2, F_3, with radial functions of the electromagnetic tensor, F_5, F_6, F_8 as follows

$$i\omega\, F_3 \,+\, \Phi\, \frac{d}{dr}F_1 \,+\, \Phi\, F_6 = 0 \,,$$

$$-i\omega\, F_2 \,+\, \frac{\nu}{r}\, F_1 \,-\, F_5 = 0 \,,$$

$$i\Phi\, \left(\frac{d}{dr} + \frac{1}{r}\right) F_2 \,+\, i\frac{\nu}{r}\, F_3 \,+\, F_8 = 0 \,; \qquad (23.6.6)$$

from this, we obtain

$$i\omega\, F_3 \,+\, \Phi\, \frac{d}{dr}F_1 \,+\, \Phi\, \varphi_2 = 0 \,,$$

$$-i\omega\, F_2 \,+\, \frac{\nu}{r}\, F_1 \,-\, \frac{\sqrt{\Phi}}{2}(\varphi_1 + \varphi_3) = 0 \,,$$

$$i\Phi\, \left(\frac{d}{dr} + \frac{1}{r}\right) F_2 \,+\, i\frac{\nu}{r}\, F_3 + i\frac{\sqrt{\Phi}}{2}(\varphi_1 - \varphi_3) = 0 \,. \qquad (23.6.7)$$

From Eq. (23.6.7), that can allow for Eq. (23.6.4); we get

$$\frac{i\omega}{\Phi}\, F_3 \,+\, \frac{d}{dr}F_1 \,-\, \frac{1}{i\omega}\,\frac{\nu}{r^2}\, G(r) = 0 \,,$$

$$-i\omega\, F_2 \,+\, \frac{\nu}{r}\, F_1 \,-\, \frac{1}{i\omega}\frac{\Phi}{\sqrt{2}\, r}\,\frac{d}{dr}G(r) = 0 \,,$$

$$\Phi\, \left(\frac{d}{dr} + \frac{1}{r}\right) F_2 \,+\, \frac{\nu}{r}\, F_3 + \frac{1}{\sqrt{2}\, r}\, G(r) = 0 \,. \qquad (23.6.8)$$

This system is equivalent to

$$F_3 = \frac{i\Phi}{\omega}\frac{dF_1}{dr} \,-\, \frac{\Phi\,\nu}{\omega^2 r^2}\, G(r) \,,$$

$$F_2 = -\frac{i\nu}{\omega r}\, F_1 + \frac{\Phi}{\sqrt{2}\,\omega^2 r}\,\frac{dG(r)}{dr} \,,$$

$$\frac{d^2 G(r)}{dr^2} + \frac{\Phi'}{\Phi}\frac{dG(r)}{dr} - \frac{\sqrt{2}\,\nu^2}{r^2\,\Phi}\, G(r) + \frac{\omega^2}{\Phi^2}G(r) = 0 \,. \qquad (23.6.9)$$

Note that all the terms with F_1 cancel out each other. It means that the function F_1 can be arbitrary. This fact may be considered as a manifestation of the gauge symmetry in a electromagnetic 4-potential for a fixed electromagnetic tensor.

23.7   Radial System at $P = (-1)^{j+1}$

Now, let us consider the case $P = (-1)^{j+1}$. To this end, in Eq. (23.4.6a)

$$P = (-1)^{j+1} , \qquad -i\omega \, f_2 - \sqrt{\Phi} \, f_5 = 0 ,$$

$$i\omega \, f_5 + i\Phi \left(\frac{d}{dr} + \frac{1}{r} + \frac{\Phi'}{2\Phi} \right) f_8 + i\nu \frac{\sqrt{\Phi}}{r} f_9 = 0 ,$$

$$-i\Phi \left(\frac{d}{dr} + \frac{1}{r} \right) f_2 - \sqrt{\Phi} \, f_8 = 0 ,$$

$$i2\nu \frac{\sqrt{\Phi}}{r} f_2 - \sqrt{\Phi} \, f_9 = 0 ; \qquad\qquad (23.7.1)$$

let us take into account Eqs. (23.5.6); they give

$$\frac{\omega}{\sqrt{\Phi}} \varphi_2 + \frac{i\nu}{r} (\varphi_1 - \varphi_3) = 0 ,$$

$$-\frac{\omega}{\sqrt{\Phi}} (\varphi_1 - \varphi_3) - i\sqrt{\Phi} \left(\frac{d}{dr} + \frac{1}{r} + \frac{\Phi'}{2\Phi} \right) (\varphi_1 + \varphi_3) + \frac{2i\,\nu}{r} \varphi_2 = 0 ,$$

$$-\sqrt{\Phi} \left(\frac{d}{dr} + \frac{2}{r} \right) \varphi_2 + \frac{\nu}{r} (\varphi_1 + \varphi_3) = 0 , \qquad f_2 = -\frac{r}{2\,\nu} \varphi_2 . \qquad (23.7.2)$$

Note that the function f_2 to determine a electromagnetic 4-vector, is uniquely defined by tensor components. This means that for this solution class no gauge freedom exists.

Let us compare Eqs. (23.7.2) with the first three equations in Eq. (23.3.5) (the last equation in Eq. (23.3.5), noted as (2)+(4), may be ignored as their consequence)

$$(3) \quad -\frac{\omega}{\sqrt{\Phi}}\varphi_2 + \frac{i\nu}{r} (\varphi_1 - \varphi_3) = 0 ,$$

$$(1) \quad \sqrt{\Phi} \left(\frac{d}{dr} + \frac{2}{r} \right) \varphi_2 + \frac{\nu}{r} (\varphi_1 + \varphi_3) = 0 ,$$

$$(2) - (4) , \qquad -\frac{\omega}{\sqrt{\Phi}}(\varphi_1 - \varphi_3) - i \left(\sqrt{\Phi}\frac{d}{dr} + \frac{\sqrt{\Phi}}{r} + \frac{\Phi'}{2\sqrt{\Phi}} \right) (\varphi_1 + \varphi_3) - \frac{2i\nu}{r} \varphi_2 = 0 .$$

$$(23.7.3)$$

The difference between Eqs. (23.7.2) and (23.7.3) consists in the sign 'minus' at the function φ_2.

Let us introduce new variables $f = \varphi_1 + \varphi_3$, $g = \varphi_1 - \varphi_3$, then Eqs. (23.7.2) takes the form

$$\varphi_2 = -\frac{i\nu}{\omega} \frac{\sqrt{\Phi}}{r} g , \qquad i\sqrt{\Phi} \left(\frac{d}{dr} + \frac{2}{r} \right) \frac{\sqrt{\Phi}}{r} g + \frac{\omega}{r} f = 0 ,$$

$$-\frac{\omega^2}{\Phi} g - i\omega \left(\frac{d}{dr} + \frac{1}{r} + \frac{\Phi'}{2\Phi} \right) f + \frac{2\nu^2}{r^2} \, g = 0 . \qquad (23.7.4)$$

The system in Eq. (23.7.3) is simplified by the following substitution

$$g = \frac{1}{r\sqrt{\Phi}}\, G(r) \,, \qquad f = \frac{1}{r\sqrt{\Phi}}\, F(r) \,,$$

so, we arrive at

$$\varphi_2 = -\frac{i\nu}{\omega r^2}\, G \,, \qquad \omega F = -i\Phi\, \frac{d}{dr} G \,,$$

$$-\frac{\omega^2}{\Phi} G - i\frac{d}{dr}\omega F + \frac{2\nu^2}{r^2} G = 0 \,. \qquad (23.7.5)$$

Thus, the problem reduces to a single second order differential equation

$$\frac{d^2 G}{dr^2} + \frac{M}{r(r-M)}\frac{dG}{dr} + \left(\frac{\omega^2 r^2}{(r-M)^2} - \frac{j(j+1)}{r(r-M)}\right) G = 0 \,. \qquad (23.7.6)$$

Remember the two relationships

$$P = (-1)^{j+1} \,, \qquad \varphi_2 = -i\frac{\nu}{\omega r^2} G \,,$$

$$\varphi_1 + \varphi_3 = -i\frac{\sqrt{\Phi}}{\omega r}\frac{d}{dr}G \,, \qquad \varphi_1 - \varphi_3 = \frac{1}{r\sqrt{\Phi}}\, G(r) \,,$$

in turn, for the wave with opposite parity; we have

$$P = (-1)^{j} \,, \qquad \varphi_2 = +i\,\frac{\nu}{\omega r^2}\, G \,,$$

$$\varphi_1 + \varphi_3 = -i\frac{\sqrt{\Phi}}{\omega r}\frac{d}{dr}G \,, \qquad \varphi_1 - \varphi_3 = \frac{1}{r\sqrt{\Phi}}\, G(r) \,. \qquad (23.7.7)$$

Relation between two linearly independent solutions found in Majorana–Oppenheimer approach and Duffin–Kemmer–Petiau approach (with different parities) can be summarized in the following way:

$$\begin{array}{llllll}
\text{Majorana–Oppenheimer} & \Longrightarrow & \varphi_2, & +\varphi_1 & \varphi_3 \,, \\[2mm]
P = (-1)^{j} \,, \ \textbf{Duffin–Kemmer} & \Longrightarrow & +\varphi_2, & \varphi_1 \,, & \varphi_3 \,, \\[2mm]
P = (-1)^{j+1} \,, \ \textbf{Duffin–Kemmer} & \Longrightarrow & -\varphi_2, & \varphi_1 \,, & \varphi_3 \,.
\end{array}$$

$$(23.7.8)$$

Due to $(\mathbf{E} + i\mathbf{B}) \sim \vec{\varphi} = (\varphi_i)$, differences in M-O representation by a simple linear transformation provide us with linearly independent solutions of the Maxwell equations.

23.8   Gauge Symmetry

Maxwell equations allow purely gauge (trivial) solutions – for these we must assume the following substitution

$$\Phi_{\omega jm}(x) = e^{-i\omega t} \left[f_1(r)\, D_0,\ f_2(r)\, D_{-1},\ f_3(r)\, D_0,\ f_4(r)\, D_{+1},\ 0,\ 0,\ 0,\ 0,\ 0,\ 0 \right].$$
(23.8.1)

For states with parity $P = (-1)^{j+1}$, we have

$$P = (-1)^{j+1}, \qquad f_1 = f_3 = 0,\ f_4 = -f_2,$$

$$i\omega\, 0 + i\Phi \left(\frac{d}{dr} + \frac{1}{r} + \frac{\Phi'}{2\Phi} \right) 0 + i\nu \frac{\sqrt{\Phi}}{r} 0 = 0, \qquad -i\omega\, f_2 - \sqrt{\Phi}\, 0 = 0,$$

$$-i\Phi \left(\frac{d}{dr} + \frac{1}{r} \right) f_2 - \sqrt{\Phi}\, 0 = 0, \qquad i2\nu \frac{\sqrt{\Phi}}{r} f_2 - \sqrt{\Phi}\, 0 = 0;$$

(23.8.2)

from this, it follows $f_2 = 0$. In other words, no purely gauge solutions exist when parity equals $P = (-1)^{j+1}$.

For states with parity $P = (-1)^j$, we obtain

$$P = (-1)^j, \qquad f_4 = +f_2,\ (F_1 = \sqrt{\Phi}\, f_1,\ F_2 = f_2,\ F_3 = \sqrt{\Phi}\, f_3),$$

$$\Phi \left(\frac{d}{dr} + \frac{2}{r} \right) 0 + \frac{2\nu}{r} 0 = 0, \qquad i\omega\, 0 + i\Phi \left(\frac{d}{dr} + \frac{1}{r} \right) 0 = 0,$$

$$i\omega\, 0 - i\frac{2\nu}{r} 0 = 0, \qquad -i\omega\, F_2 + \frac{\nu}{r} F_1 - 0 = 0,$$

$$i\omega\, F_3 + \Phi \frac{d}{dr} F_1 + \Phi\, 0 = 0, \qquad i\Phi \left(\frac{d}{dr} + \frac{1}{r} \right) F_2 + i\frac{\nu}{r} F_3 + 0 = 0,$$

from this, it follows

$$i\omega\, F_2 = \frac{\nu}{r} F_1, \qquad i\omega\, F_3 = -\Phi \frac{d}{dr} F_1,$$

$$i\Phi \left(\frac{d}{dr} + \frac{1}{r} \right) F_2 + i\frac{\nu}{r} F_3 = 0.$$
(23.8.3)

Excluding the third equation variables in F_2, F_3, we arrive at $0 \equiv 0$. This means that F_1 can be arbitrary, and the two remaining functions are determined by

$$F_1 \text{ is arbitrary}, \qquad i\omega\, F_2 = \frac{\nu}{r} F_1, \qquad i\omega\, F_3 = -\Phi \frac{d}{dr} F_1.$$
(23.8.4)

The last result can be additionally checked with the help of Lorentz gauge condition. Indeed, in radial form it is

$$\frac{-i\omega}{\sqrt{\Phi}} f_1 - \sqrt{\Phi} \left(\frac{d}{dr} + \frac{2}{r} + \frac{\Phi'}{2\Phi} \right) f_3 - \frac{\nu}{r} (f_2 + f_4) = 0.$$
(23.8.5)

It is satisfied identically for states with parity $P = (-1)^{j+1}$; in turn, for those states with parity $P = (-1)^j$, it takes the form

$$\frac{-i\omega}{\sqrt{\Phi}} f_1 - \sqrt{\Phi} \left(\frac{d}{dr} + \frac{2}{r} + \frac{\Phi'}{2\Phi} \right) f_3 - \frac{2\nu}{r} f_2 = 0 \ . \tag{23.8.6}$$

When we take into account Eq. (23.4.8), we get

$$\frac{-i\omega}{\Phi} F_1 - \left(\frac{d}{dr} + \frac{2}{r} \right) F_3 - \frac{2\nu}{r} F_2 = 0 \ . \tag{23.8.7}$$

Now, when we allow for Eq. (23.8.4), we arrive at

$$\left[\frac{d^2}{dr^2} + \left(\frac{2}{r} + \frac{\Phi'}{\Phi} \right) \frac{d}{dr} + \frac{\omega^2}{\Phi^2} - \frac{j(j+1)}{\Phi r^2} \right] F_1 = 0 \ . \tag{23.8.8}$$

Let us compare this with the radial form of the scalar wave $\nabla^\alpha \nabla_\alpha \Psi = 0$. In Schwarzschild metrics, it is given by

$$\left(-\frac{\omega^2}{\Phi} - \frac{1}{r^2} \frac{\partial}{\partial r} r^2 \Phi \frac{\partial}{\partial r} + \frac{j(j+1)}{r^2} \right) f(r) = 0$$

which coincides with Eq. (23.8.8).

Chapter 24

Conclusions and Summaries

In chapter **1**, we treat the (vacuum) Maxwell equations in a curved space-time as the (non-vacuum) Maxwell equations in flat space-time in an effective media with properties determined by metrical structure of the initial curved model $g_{\alpha\beta}(x)$. This metrical structure generates effective constitutive equations for electromagnetic fields:

$$\mathbf{D} = \epsilon_0 \, \epsilon(x) \, \mathbf{E} + \epsilon_0 c \, \alpha(x) \, \mathbf{B} \, , \qquad \mathbf{H} = \epsilon_0 c \, \beta(x) \, \mathbf{E} + \mu_0^{-1} \mu^{-1}(x) \, \mathbf{B} \, ,$$

the form of the four symmetrical tensors $\epsilon^{ik}(x), \alpha^{ik}(x), \beta^{ik}(x), \mu^{ik}(x)$ is found explicitly for the general case of an arbitrary Riemannian space-time geometry $g_{\alpha\beta}(x)$. The main peculiarity in the geometrical generating the effective electromagnetic media characteristics is that four tensors are not independent and obey some additional constraints between them. We provide several examples: the geometrical modeling of the anisotropic media (magnetic crystals) and the geometrical modeling of a uniform media for a moving reference frame in the background of Minkowski electrodynamics. The general problem of transforming arbitrary (linear) material equations has been studied and corresponding formulas have been produced.

In chapter **2**, the Riemann–Silberstein–Majorana–Oppenheimer approach to the Maxwell electrodynamics in presence of electrical sources and arbitrary media is investigated within the matrix formalism. The symmetry of the Maxwell matrix equation under transformations of the complex rotation group $SO(3.C)$ is demonstrated explicitly. For a vacuum case, the matrix form includes four real (4×4) matrices α^b. In presence of media matrix form requires two sets of 4×4 matrices, α^b and β^b – simple and symmetrical realization of them is given. Relations of α^b and β^b to the Dirac matrices in spinor basis is found. Minkowski constitutive relations in case of any linear media are given in a short algebraic form based on the use of complex 3-vector fields and complex orthogonal rotation group $SO(3.C)$. The matrix complex formulation in the Esposito's form, based on the use of two electromagnetic 4-vector is studied and discussed. Extension of the 3-vector complex matrix formalism to arbitrary Riemannian space-time in accordance with the tetrad method by Tetrode–Weyl–Fock–Ivanenko is performed.

In chapter **3**, within the formalism of 16-component Dirac–Kähler field theory, spinor equations for two types of massless vector photon fields with different parities have been derived. The equivalent tensor equations in terms of the field strength tensor F_{ab} and respective 4-vector A_b and 4-pseudo-vector $\tilde{A}_b$ that depend on intrinsic photon parity are studied; they include additional sources, electric 4-vector j_b and magnetic 4-pseudo-vector $\tilde{j}_b$. The equations for two types of photon fields are explicitly uncoupled, their linear combinations through summation or subtraction results in the Maxwell electrodynamics with electric and magnetic charges in 2-potential approach. In this way, the problem of existence for a magnetic charge can be understood as a super selection rule for photon fields with different intrinsic parity. The performed analysis is extended straightforwardly to a curved space-time background. In the frames of this extended Maxwell theory, the known electromagnetic duality is described as a linear transformation mixing field variables referred to photons with different parities. The extended dual transformations concern both field strength tensors and 4-potentials A_b, $\tilde{A}_b$.

In chapter **4**, the metric Duffin–Kemmer–Petiau formalism is elaborated to treat spin 1 particle in curved space-time background. A common view is that the generalization of a wave equation on Riemannian space-time is substantially determined by what a particle is – boson or fermion. As a rule, they say that tensor equations for bosons are extended in a simpler way then spinor equations for fermions. In that context, a very interesting problem is the extension wave equations for boson fields in presence of gravitation. This is the same approach as done by Tetrode–Weyl–Fock–Ivaneneko for spin 1/2 field.

In chapter **5** we relate a generally covariant tensor formalism to a spinor one when these both are applied to a description of the Dirac–Kähler field in a Rimannian space-time. Both methods are taken to be equivalent and the tensor equations are derived from spinor ones. It was shown, that for the characterization of Dirac-Kähler's tensor components, two alternative approaches are suitable: these are whether a tetrad-based tensor classification or a generally coordinate tensor one. By imposing definite restrictions on the the Dirac–Kähler function, we have produced the general covariant form of wave equations for scalar, pseudo-scalar, vector, and pseudo-vector particles. Bosons of spin 0 and 1, with different intrinsic parities, are described by full sets of spinor equations in the frame of the Dirac–Kähler theory. This enables us to obtain the conservation laws for the boson particles with one value of spin by imposing additional linear conditions in the known sixteen conserved currents of the Dirac–Kähler field. In this way for each boson the known conserved quantities, charge vector $j^a(x)$, symmetrical energy-momentum tensor $T^{ab}(x)$, and angular momentum tensor $L^{a[bc]}(x)$, have been found. Additionally, for scalar fields, one conserved current $\nu^a(x)$ has been constructed; it is not zero for only the complex-valued field. For vector particles, two additional currents, $\nu^a(x)$ and $\nu^{ab}(x)$, are found that again do not vanish when fields are complex-valued. Those currents $\nu^a(x)$ and $\nu^{ab}(x)$ have not seemingly show any physical interpretation.

In chapter **6**, we discuss the Pauli–Fierz approach in the description of a massless spin-2 particle. We investigated the framework of a 30-component first order relativistic wave equation theory on a curved space-time background. It is shown that additional gauge symmetry of massless equations established by Pauli–Fierz can be extended only to the curved space-time regions where Ricci tensor vanishes. In all such space-time models the generally covariant S=2 massless wave equation exhibits gauge symmetry property; otherwise, this is not true.

In chapter **7**, a 50-component tensor form of the first order relativistic wave equation for a particle with spin 2 and anomalous magnetic moment is extended in the case of an arbitrary curved space-time geometry. An additional parameter considered in the presence of only a electromagnetic field as related to the anomalous magnetic moment, turns to determine additional interaction terms with external geometrical background through Ricci R_{kl} and Riemann R_{klmn} tensors.

In chapter **8**, tetrad based equation for Dirac–Kähler particle is solved in spherical coordinates in the flat Minkowski space-time. Spherical solutions of the boson type ($J = 0, 1, 2, ...$) are constructed. After performing a special transformation over spherical boson solutions for the Dirac–Kähler equation, (4×4)-matrices $U(x) \implies V(x)$, simple linear expansions of the four rows for a new representative of the Dirac–Kähler field $V(x)$ was established. In this case, in terms of spherical fermion solutions $\Psi_i(x)$ of the four ordinary Dirac equation were derived. However, this fact cannot be interpreted as the possibility not to distinguish between the Dirac–Kähler field and the system four Dirac fermions. The main formal argument is that the special transformation $(I \otimes S(x))$ involved does not belong to the group of tetrad local gauge transformation for Dirac–Kähler field, 2-rank bispinor under the Lorentz group. Therefore, the linear expansions between boson and fermion functions are not gauge invariant under the group of local tetrad rotations.

In chapter **9**, complex formalism of Riemann–Silberstein–Majorana–Oppenheimer in Maxwell electrodynamics is applied to solve the Maxwell equations which are solved exactly on the background of the simplest static cosmological models, spaces of constant curvature of Riemann and Lobachevsky parameterized by spherical coordinates. The separation of variables is realized in the basis of Schrödinger–Pauli type, description of angular dependence in electromagnetic complex 3-vectors is given in terms of Wigner D-functions. In the case of the compact Riemann model, a discrete frequency spectrum for electromagnetic modes that depend on the curvature radius of space and three discrete parameters is found. In the case of hyperbolic Lobachevsky model, no discrete spectrum for frequencies of electromagnetic modes arises.

In chapter **10**, the Duffin–Kemmer form of a massless vector field (Maxwell field) extended to the case of an arbitrary pseudo-Riemannian space-times is applied to solve the Maxwell equations exactly on the background of of the simplest static cosmological model, space of constant curvature of Riemann parameterized by spherical coordinates. The separation of variables is realized in the basis of Schrödinger–Pauli type, description of angular dependence in electromagnetic field functions is given in terms of Wigner D-functions. A discrete frequency spectrum for electromagnetic modes that depend on the curvature radius of space and three discrete parameters is found. The 4-potentials for the spherical electromagnetic waves of the magnetic and electric type have been constructed in explicit form.

In chapters **11** and **12**, complex formalism of Riemann–Silberstein–Majorana–Oppenheimer is applied to solve the Maxwell equations exactly on the background of static cosmological Einstein model, parameterized by special cylindrical coordinates and realized as a Riemann space for a constant positive curvature. A discrete frequency spectrum for electromagnetic modes depend on the curvature radius of space and three parameters is found, and the corresponding basis in electromagnetic solutions have been constructed explicitly. In the case of elliptical model, a part of the constructed solutions should be rejected by continuity considerations. Similar treatment is given for the Maxwell equations in the hyperbolic Lobachevsky model, the complete basis for the electromagnetic solutions which correspond to cylindrical coordinates has been constructed as well. No quantization for frequencies of electromagnetic modes arises here.

In chapter **13** we treat the possibility for Lobachewsky geometry simulate a medium with special constitutive relations, The situation is specified in quasi-cartesian coordinates (x, y, z). Exact solutions of the Maxwell equations in complex 3-vector $\mathbf{E} + i\mathbf{B}$ form, extended to curved space models within the tetrad formalism, have been found in Lobachevsky space. The problem reduces to a second order differential equation which can be associated with an 1-dimensional Schrödinger problem for a particle in external potential field $U(z) = U_0 e^{2z}$. In quantum mechanics, curved geometry acts as an effective potential barrier with reflection coefficient $R = 1$; in electrodynamic context results similar to quantum-mechanical ones arise: the Lobachevsky geometry simulates a medium that effectively acts as an ideal mirror. Penetration of the electromagnetic field into the effective medium, depends on the parameters of an electromagnetic wave, frequency ω, $k_1^2 + k_2^2$, and the curvature radius ρ.

In chapter and **14**, the complex approach to the Maxwell electrodynamics is investigated within the matrix formalism. Within the squaring procedure, we construct four types of formal solutions with the Maxwell equations on the base of scalar d'Alembert solutions. A general problem of separating physical solutions in the linear space $\lambda_0 \Psi^0 + \lambda_1 \Psi^1 + \lambda_2 \Psi^2 + \lambda_3 \Psi^3$ is investigated, the Maxwell equations reduce to a new form which include parameters λ_a. We consider several particular cases for plane waves and cylindrical waves in detail. Possible extension of the technique to a curved space-time models is discussed.

In chapter **15**, spin 1 particle is investigated in 3-dimensional curved spaces for a constant positive and negative curvature. An extended helicity operator is defined and the variables are separated in a tetrad-based 10-dimensional Duffin–Kemmer–Petiau equation. The problem is solved exactly in hypergeometric functions. Transition to a massless case of electromagnetic field is performed. In massive case, the problem is investigated in Pauli non-relativistic approximation as well.

In chapter **16**, a quantum-mechanical system – spin 1 particle in external Coulomb field is studied on the basis of a matrix Duffin–Kemmer–Petiau formalism with the use of the tetrad technique. The separation of the variables is performed with the help of Wigner functions $D^j_{-m\sigma}(\phi, \theta, 0)$; $\sigma = -1, 0, +1$; j and m stand for quantum numbers to determine the square and the third projection of the total angular momentum of the vector particle. Using the help of a parity operator, the radial 10-equation system is divided into two subsystems of 4 and 6 equations that correspond to parity $P = (-1)^{j+1}$ and $P = (-1)^j$, respectively. The 4 equation system is reduced to a second order differential equation which coincides with the case of a scalar particle in Coulomb potential. It is shown that the 6-equation system reduces to two different second order differential equations for a "main" function. One main equation reduces to a confluent Heun equation and provides us with an energy spectrum. Another main equation is more complex, and any solutions for this were not constructed.

In chapters **17**, exact solutions are constructed for the quantum-mechanical equation of a spin $S = 1$ particle in flat Minkowski space-time in presence of external magnertic field.

In chapters **18, 19** we describe spin-1 particle in a 2-dimensional Riemannian space of constant negative and positive curvature in the presence of an external magnetic field, analog of the homogeneous magnetic field in the Minkowski space. A generalized formula for energy levels that describe the quantization of the motion for the vector particle in magnetic field on the 2-dimensional spaces H_2 and $_2$ have been found, non-relativistic and relativistic equations were solved.

In chapter **20**, a 5-dimensional wave equation for a massive spin 1 particle in the background of de Sitter space-time model is solved in static coordinates. The spherical 5-dimensional vectors $A_a, a = 1, ..., 5$ of three types, $j, j + 1, j - 1$ are constructed. In a massless case, they give electromagnetic wave solutions, and obey the Lorentz condition. The 5-form of equations in this massless case is used to produce a recipe to create a electromagnetic wave solution for the types Π, E, M; the first is trivial and can be removed by a gauge transformation. This recipe is specified to produce spherical Π, E, M solutions in static coordinates.

In chapter **21**, three possible techniques to deal with a vector particle in anti de Sitter cosmological model are viewed: Duffin–Kemmer–Petiau matrix formalism is based on the general tetrad recipe, the group theory 5-dimensional approach based on the symmetry group SO(3.2), and a tetrad form of Maxwell equations in a complex Riemann–Silberstein–Majorana–Oppenheimer representation. In the first part, a spin 1 massive field is considered in static coordinates for the anti de Sitter space-time in a tetrad-based approach. The complete set of spherical wave solutions with quantum numbers (ϵ, j, m, l) is constructed; angular dependence in wave functions is described with the help of Wigner functions. The energy quantization rule was found. Transition to the massless case for the electromagnetic field is specified, and electromagnetic solutions in Lorentz gauge were constructed. In the second part, the problem of a particle with a spin 1 is considered on the base of a 5-dimensional wave equation and is specified in the same static coordinates. In the third part, a rarely used approach that is based on a tetrad form of Maxwell equations in complex representation is examined in the anti de Sitter model.

In chapter **22**, the problem of a Hawking radiation in de Sitter space-time is analyzed. In particular, details of penetration for a quantum mechanical particle through the de Sitter horizon has been examined intensively. Yet, there is still some vagueness on this subject. Our aim to clarify this situation. A known algorithm that is used in the calculation of the reflection coefficient $R_{\epsilon j}$ on the background of the de Sitter space-time model is analyzed. It is shown that the determination of $R_{\epsilon j}$ requires an additional constraint on the quantum numbers $\epsilon R/\hbar c >> j$, where R is a curvature radius. When we take this condition into account, the value of $R_{\epsilon j}$ turns out to be precisely zero. As shown, the basic instructive definition for the calculation of the reflection coefficient in de Sitter model is grounded exclusively on the use of a zero order approximation in the expansion of a particle wave function in a series on small parameter $1/R^2$, and it demonstrated that this recipe cannot be extended by an account for using contributions of higher order terms. So, the result $R_{\epsilon j} = 0$ had been obtained through the examination that the zero-order term persists and cannot be improved.

It was claimed that by the calculation of the reflection coefficient $R_{\epsilon j}$, is not required at all, because there is no barrier in the effective potential curve on the background of the de Sitter space-time.

In chapter **23**, we have shown that the generally covariant extended method of Riemann–Silberstein–Majorana–Oppenheimer in electrodynamics, as specified in Schwarzschild metrics, after the separation of variables reduces the problem of electromagnetic solutions to a differential equation that is similar to the case of a scalar field in the Schwarzschild space-time. This differential equation is recognized as a confluent Heun equation. Also, the electromagnetic field is treated on the basis of a 10-dimensional Duffin–Kemmer-Petiau approach. In addition, six components of the strength tensor one uses a 4-component electromagnetic potential. Correspondingly, a system of 10 radial equations is simplified by the use of additional constraints stemming from eigenvalue equation for the spacial parity operator $\hat{\Pi}\Psi = P\Psi$; the radial system is divided into two subsystems of 4 and 6 equations, respectively. In our second approach, the problem of electromagnetic field reduces to the confluent Heun differential equation as well. In particular, we show explicitly how these solutions were found in this complex form and are embedded into the 10-dimensional formalism. We also determine radial functions that are responsible for gauge degrees of freedom.

Bibliography

[1] Lorentz, H.A., De l'influence du mouvement de la terre sur les phénom ènes lu-
mineux. Archives Néerlandaises des Sciences Exactes et Naturelles. 1886. 21. 103–
176; La théorie électromagnétique de Maxwell et son application aux corps mou-
vants. Archives néerlandaises des sciences exactes et naturelles. 1892. 25. 363–552;
Versuch einen Theorie der elektrischen und optischen Erscheinungen in bewegten
Körpern. Leiden: Brill, 1895; Simplified theory of electrical and optical phenomena
in moving systems. Proceedings of the Section of Sciences, Koninklijke Akademie
van Wetenschappen te Amsterdam. 1889. 1. 427–442; Electromagnetic phenomena
in a system moving with any velocity less than that of light. Proceedings of the Section of Sciences, Koninklijke Akademie van Wetenschappen te Amsterdam. 1904.
6. 809–831.

[2] Poincaré, H., Électricité et optique. Paris, Georges Carré, 1890–1891; La théorie
de Lorentz et le principe de réaction. Archives néerlandaises des sciences exactes
et naturelles. 1900. 5. 252–278; La mesure du temps. Revue de métaphysique et
de morale. 1898. 6. 371–384; Électricité et optique: la lumière et les théories
électrodynamiques. Paris: Carré et Naud, 1901; La Science et l'hypothése. Paris:
Flammarion, 1902; Sur la dynamique de l'électron. C. R. Acad. Sci. Paris. 1905.
140. 1504–1508; Sur la dynamique de l'électron. Rendiconti del circolo matematico
di Palermo. 1906. 21. 129–176; La dynamique de l'électron // Revue générale des
sciences pures et appliquées. 1908. 19. 386–402.

[3] Einstein, A., Zur Elektrodynamik der bewegten Körper. Ann. der Phys. 1905. 17.
891–921; Einstein, A.; Laub, J., Über die elektromagnetischen Grundgleichungen
für bewegte Körper. Ann. der Phys. 1908. 26. 532–540; Die im elektromagnetischen
Felde auf ruhende Körper ausgeübten ponderomotorischen Kräfte. Ann. der Phys.
1908. 26. 541–550.

[4] Planck, M., Das Prinzip der Relativität und die rundgleichungen der Mechanik.
Deutsche Physikalische Gesellschaft. Verhandlungen. 1906. 8. 136–141; Zur Dynamik bewegter Systeme. Königlich Preussische Akademie der Wissenschaften
(Berlin). Sitzungsberichte. 1907. 542–570; Reprinted in Ann. Phys. 1908. 26. 1–
34.

[5] Silberstein, L., Elektromagnetische Grundgleichungen in bivectorieller Behandlung.
Ann. der Phys. 1907. 22. no 3. 579–586; Nachtrag zur Abhandlung Über "elektro-

magnetische Grundgleichungen in bivektorieller Behandlung". Ann. der Phys. 1907. 24. no 14. 783–784; The Theory of Relativity. London: Macmillan, 1914.

[6] Minkowski, H., Die Grundlagen für die electromagnetischen Vorgänge in bewegten Kërpern. Nachrichten von der Königlichen Gesellschaft der Wissenschaften zu Göttingen, mathematisch-physikalische Klasse. 1908. 53–111; reprint in Math. Ann. 1910. 68. 472–525; Raum und Zeit. Jahresbericht der deutschen Mathematiker-Vereinigung. 1909. 18. 75–88; Raum und Zeit. Phys. Zeit. 1909. 10. 104–111; Das Relativitätsprinzip. Annalen der Physik. 1915. 47. 927–938.

[7] Von Laue, M., Die Mitführung des Lichtes durch bewegte Körper nach dem Relativitätsprinzip. Ann. Phys. 1907. 23. 989–990.

[8] Von Abraham, M., Zur electromagnetischen Mechanik. Phys. Zeit. 1909. 10. Jahrgang. no 21. 737–741; Die neue Mechanik. Scientia (Rivista di Scienza). 1914. 15. 8–27; Zur Elektrodynamik bewegter Körper. Rend. Circ. Mat. Palermo. 1909. 28. 1; Sull'elettrodinamica di Minkowski. Rendiconti del Circolo Matematico di Palermo. 1910. 30. 33–46.

[9] Lewis, G.N.; Tolman, R.C., The principle of relativity, and non-Newtonian mechanics. Phil. Mag. 1909. 18. 510–523.

[10] Bateman, H., On the conformal transformations of the space of four dimensional and their applications to geometric optics. Proc. London Math. Soc. 1909. 7. 70–92; The transforamation of the electrodynamical equations. Proc. London Math. Soc. 1910. Ser. 8. 223–264; The Mathematical Analysis of Electrical and Optical Wave Motion on the Basis of Maxwell's Equations. 1915, Cambridge; reprinted by Dover, New York, 1955.

[11] Cunningham, E., The principle of relativity in electrodynamics and an extension thereof. Proc. London Math. Soc. 1909. 8. 77–98; The Application of the Mathematical Theory of Relativity to the Electron Theory of Matter. Proc. London Math. Soc. 1911-1912. 10. 116–127; The Principle of Relativity. Cambridge: Cambridge University Press, 1914.

[12] Sommerfeld, A., Zur Relativitätstheorie I: Vierdimensionale Vectoralgebra. Ann. Phys. Lpz. 1910. 32. no 9. 749–776; Zur Relativitätstheorie II: Vierdimensionale Vektoranalysis. Ann. Phys. 1910. 33. no 14. 649–689.

[13] Klein, F., Über die geometrischen Grundlagen der Lorentzgruppe. Jahresbericht der deutschen Mathematiker-Vereinigun. 1910. 19. 281–300; Phys. Zeit. 1911. 12. 17–27.

[14] Lewis, G.N., On Four Dimensional Vector Analysis and its Application in Electrical Theory. Proc. Amer. Acad. Arts and Science. 1910. 46. 165–181; Über vierdimensionale Vektoranalysis und deren Anwendung auf die Elektrizitätstheorie. Jahrbuch der Radioaktivität und Elektronik. 1910. 7. 329–347.

[15] Von Laue., M., Zur Dynamik der Relativitätstheorie. Ann. Phys. 1911. 35. 524–542; Ein Beispiel zur Dynamik der Relativittstheorie. Verhandlungen der Deutschen Physikalischen Gesellschaft. 1911. 13. 513–518; Zur Minkowskischen Elektrodynamik der bewegten Körper. Zeit. Phys. 1950. 128. 387–394; Das Relativitätsprinzip. Braunschweig: Vieweg, 1911.

[16] Alkemade, A.C.V., Transformation equations of the theory of relativity. Annalen der Physik. 1912. 38. no 10. 1035–1042.

[17] Marcolongo, R., Les transformations de Lorentz et les équations de l'électrodynamique. Annales de la Faculté des Sciences de Toulouse. Ser. 3. 1912. 4. 429–468.

[18] Einstein, A.; Grossmann, M., Kovarianzeigenschaften der Feldgleichungen der auf die verallgemeinerte Relativitätstheorie gegründeten Gravitationstheorie. Zeit. für Math. und Phys. 1914. 63. no 1-2. 215–225.

[19] Einstein, A., Prinzipielles zur verallgemeinerten Relativitätstheorie und Gravitationstheorie. 1914. 15. 176–180; Die formale Grundlage der allgemeinen Relativitätstheorie. Kößniglich Preußische Akademie der Wissenschaften (Berlin). Sitzungsberichte. 1914. XLI. 1030–1085; Zur allgemeinen Relativitätstheorie. 1915. XLIV. 778–786; Zur allgemeinen Relativitätstheorie (Nachtrag). 1915. XLVI. 799–801; Erklärung der Perihelbewegung des Merkur aus der allgemeinen Relativitätstheorie. 1915. XLVII. 831–839; Zusammenfassung der Mitteilung "Erklärung der Perihelbewegung des Merkur aus der allgemeinen Relativitätstheorie. 1915. XLVII. 803; Die Feldgleichungen der Gravitation. 1915. XLVIII–XLIX. 844–847; Die Grundlage der allgemeinen Relativitätstheorie. Ann. der Phys. 1916. 49. 769–822.

[20] Hilbert, D., Die Grundlagen der Physik. (Erste Mitteilung). Königliche Gesellschaft der Wissenschaften zu Göttingen. Mathematisch-physikalische Klasse. Nachrichten. 1915. 395–407; Die Grundlagen der Physik (Zweite Mitteilung). Königliche Gesellschaft der Wissenschaften zu Göttingen. Mathematisch-physikalische Klasse. Nachrichten. 1917. 53–76; Die Grundlagen der Physik. Math. Annalen. 1924. 92. 1–32.

[21] Pauli, W., Relativitätstheorie. Leipzig, 1921.

[22] Eddington, A.S., The Mathematical Theory of Relativity. 2d edit. Cambridge. University Press, 1924.

[23] Weyl, H., Raum - Zeit - Materie. Vorlesungen über Allgemeine Relativitätstheorie. Berlin. Springer, 1923.

[24] Tolman, R.C., Relativity, Thermodynamics and Cosmology. Oxford. Clarendon Press, 1934.

[25] Fock, V.A., The theory of Spave, Time and Gravitation. Moscow, 1955 (in Russian).

[26] Schrödinger, E., Space-Time Structure. Cambridge. University Press, 1950.

[27] Schrödinger, E., Expanding Universes. Cambridge. University Press, 1956.

[28] Schmutzer, E., Relativische Physik. Leipzig. Teubner-Verlag, 1968.

[29] Mitzkevich, N.V., Physical Fiels in General Relativity. Moscow, 1969 (in Russian).

[30] Weinberg, S., Gravitation and Cosmology: Principles and Applications of the General Theory of Relativity. Massachusetts, 1971.

[31] Misner, C.W.; Thorne, K.S.; Wheeler, J.A., Gravitation. San Francisco. Freeman, 1973. Vols. I, II, III.

[32] Birrell, N.D.; Davies, P.C.W.; Quantum Fields in Curved Space. Cambridge. University Press, 1982.

[33] Penrose, R.; Rindler, W., Spinors and Space–Time. Vol. I: Two-spinor Calculus and Relativistic Fields. Cambridge: University Press, 1984.

[34] Chandrasekhar, S., The Mathematical Theory of Black Holes. New York. Oxford University Press, 1983.

[35] Gorbatsievich, A.K., Quantum Mechanics in General Relativity. Basic Principles and Elementary Applications. Minsk, 1985 (in Russian).

[36] Gal'tsov, D.V., Particles and Fields in Vicinity of Black Holes. Moscow, 1986 (in Russian).

[37] Landau, L.D.; Lifshitz, E.M., The Classical Theory of Fields. Vol. 2 (4th ed.). Butterworth-Heinemann, 1975.

[38] Landau, L.D.; Lifshitz, E.M., Quantum Mechanics: Non-Relativistic Theory. Vol. 3 (3rd ed.). Pergamon Press, 1977.

[39] Berestetskii, V.B.; Lifshitz, E.M.; Pitaevskii, L.P., Quantum Electrodynamics, 2nd Edition. Pergamon Press, 1982.

[40] Red'kov, V.M., Fields in Riemannian space and the Lorentz Group. Publishing House "Belarusian Science". Minsk, 2009 (in Russian).

[41] Red'kov, V.M., Tetrad Formalism, Spherical Symmetry and Schrödinger Basis. Publishing House "Belarusian Science". Minsk, 2011. (in Russian).

[42] Red'kov, V.M.; Ovsiyuk, E.M., Quantum Mechanics in Spaces of Constant Curvature. Nova Science Publishers. Inc., New York, 2012.

[43] Kottler, F., Maxwell'sche Gleichungen und Metrik. Sitz. Akad. Wien IIa. 1922. 131, 119–146.

[44] Gordon, W., Zur lichtfortp anzung nach der relativität-stheorie. Annalen der Physik. (Leipzig). 1923. 72, 421–456.

[45] Tamm, I.E., Electrodynamics of an anisotropic medium and the special theory of relativity, Zh. Russ. Fiz.-Khim. O-va Chast. Fiz., 1924. 56 no 2-3, 248–262 (in Russian); Crystal optics in the theory of relativity and its relationship to the geometry of a biquadratic form. Zh. Russ. Fiz.-Khim. O-va Chast. Fiz. 1925. 57, no 3-4. 209–240 (in Russian).

[46] Mandelstam, L.I.; Tamm, I.E., Elektrodynamik der anisotropen Medien und der speziallen Relativitatstheorie. Math. Annalen. 1925. 95. 154–160.

[47] Van Dantzig, D., The fundamental equations of electromagnetism, independent of metrical geometry. Proc. Camb. Phil. Soc. 1934. 30. 421–427.

[48] Schrödinger, E., Maxwell's and Dirac's equations in expanding universe. Proc. Roy. Irish. Acad. A. 1940. 46. 25–47.

[49] Tonnelat, M.A., Sur la théorie du photon dans un espace de Riemann. Masson & Cie, 1940.

[50] Stratton, J.A., Electromagnetic Theory. McGraw-Hill book company, inc., 1941

[51] Ueno, Y., On the wave theory of light in general relativity. I. Path of light.; II. Light as the electromagnetic wave. Progr. Theor. Phys. 1953. 10. 442–450 ; 1954. 12, 461–480.

[52] Schrödinger, E. , Space-Time Structure. Cambridge University Press: Cambridge, 1954.

[53] Novacu, V., Introducere in Electrodinamica. Bucharest, Editura Academiei, 1955; Introduction to electrodynamics. Moscow, 1963 (in Russian).

[54] Skrotskii, G.V., The influence of gravitation on the propagation of light. Soviet Phys. Dokl. 1957. 2. 226–229.

[55] Balazs, N.L., The propagation of light rays in moving media. Jour. Optical Soc. Amer. 1955. 45. no 1. 63–64.

[56] Pham Man Quan, Sur les équations de l'électromagné dans la materie. C. R. Acad. Sci. Paris. 1956. 242. 465–467.

[57] Misner, C.W.; Wheeler, J.A., Classical physics as geometry: Gravitation, electro-magnetism, unquantized charge, and mass as properties of curved empty space. Ann. Phys. 1957. 2. no 6. 525–603.

[58] Pham Mau Quan, Inductions électromagnétiques en relativité génélale et principe de Fermat. Archive for Rational Mechanics and Analysis. 1957. 1. 54–80.

[59] Balazs, N.L., Effect of a gravitational field, due to a rotating body, on the plane of polarization of an electromagnetic wave. Phys. Rev. 1958. 110. 236-239.

[60] Tomil'chik, L.M.; Fedorov, A.I., Magnetic anisotropy as metrical property of space. Kristalography. 1959. 4. 498–504 (in Russian).

[61] Dzyaloshinskii, I.E., On the magneto-electrical effect in antiferromagnets. 1959. J. Exp. Theoret. Phys. (USSR). 37. 881–882; Sov. Phys. JETP. 1960. 10. 628–629.

[62] Plebanski, J., Electromagnetic waves in gravitational fields. Phys. Rev. 1960. 118. 1396–1408.

[63] Post, E.J., Formal structure of electrodynamics. General covariance and electromagnetics. North-Holland Pub. Co., 1962.

[64] Winterberg, F., Detection of gravitational waves by stellar scintillation in space. Nuovo Cimento. B. 1968. 53. 264–279.

[65] O'Del, T.H., The electrodynamics of magneto-electric media. North-Holland Pub. Co., 1970.

[66] de Felice, F., On the gravitational field acting as an optical medium. General Relativity and Gravitation. 1971. 2. 347–357.

[67] Bolotowskij, B.M.; Stoliarov, C.N., Contamporain state of electrodynamics of moving medias (unlimited medias). In: Eistein collection. Moscow, 1976. 179–275 (in Russian).

[68] Veselago, V.G., About properties of substances with simultaneously negative values of dielectric and magnetic permeabilities. Soviet Phys. Solid State. 1967. 8. 2853.

[69] Veselago, V.G., The electrodynamics of substances with simultaneously negative values of *epsilon* and μ. Uspekhi Fiz. Nauk. 1967. 92. 517–526 (in Russian); Soviet Phys. Usp.1968, 10. 509–514.

[70] Veselago, V.G., The electromagnetic properties of a mixture of electric and magnetic charges. Soviet Physics JETP. 1967. 25. 680.

[71] Veselago, V.G., Negative Refraction as a Source of Some Pedagogical Problems. Acta Physica Polonica. A. 2007. 112. 777–781.

[72] Fedorov, F.I., Theory of Gyrotropy. Minsk, 1976 (In Rusian).

[73] Barykin, V.N.; Tolkachev, E.A.; Tomilchik, L.M., On symmetry aspects of choice of material equations in micriscopic electrodynamics of moving medias. Vesti AN BSSR, ser. fiz.-mat. 1982. 2. 96–98.

[74] Turov, E.A., Electromagnetic constitutive equations. Moscow, 1983 (in Russian).

[75] Schleich, W.; Scully, M.O., General relativity and modern optics. in Les Houches Session XXXVIII. New trends in atomic physics. Elsevier, Amsterdam, 1984.

[76] Berezin, A.V.; Tolkachev, E.A.; Fedorov, F.I., Dual-invariant constotutive equations for rest hyrotropic medias. Doklady AN BSSR. 1985. 29. 595–597 (in Russian).

[77] Berezin, A.V.; Tolkachev, E.A.; Tregubovic, A.; Fedorov, F.I., Quaternionic constitutive relations for moving hyrotropic medias. Zhurnal Prikladnoj Spektroskopii. 1987. 47. 113–118.

[78] Antoci, S., The electromagnetic properties of material media and Einstein's unified field theory. Progr. Theor. Phys. 1992. 87. 1343–1357.

[79] Antoci, S.; Mihich, L., A forgotten argument by Gordon uniquely selects Abraham's tensor as the energy-momentum tensor for the electromagnetic field in homogeneous, isotropic matter. Nuovo Cimento. B. 1997. 112. 991–1001.

[80] Antoci,S.; Mihich, L., Does light exert Abraham's force in a transparent medium? arXiv:physics/9808002.

[81] Antoci, S.4 Minich, L., One thing that general relativity says about photons in matter. Nuovo Cim. B. 2001. 116. 801–812.

[82] Weiglhofer, W.S., On a medium constraint arising directly from Maxwell's equations. J. Phys. A. 1994. 27. L871–L874.

[83] Weiglhofer, W.S.; Lakhtakia, A., The Post constraint revisited. Arch. Elektron. Ubertrag. 1998. 52. 276–279.

[84] Hillion, P., Constitutive relations and Clifford algebra in electromagnetism. Adv. Appl. Clifford Alg. 1995. 5. 141–158.

[85] Lakhtakia, A.; Weiglhofer, W.S., On a constraint on the electromagnetic constitutive relations of nonhomogeneous linear media. IMA J. of Appl. Math. 1995. 54. 301–306.

[86] Lakhtakia, A.; Weiglhofer, W.S., Lorentz covariance, Occam's razor, and a constraint on linear constitutive relations. Phys. Lett. A. 1996. 213. 107–111.

[87] Mackay, T.G. ; Lakhtakia, A., Negative phase velocity in a uniformly moving, homogeneous, isotropic, dielectric-magnetic medium. J. Phys. A: Math.and Gen. 2004. 37. 5697–5711.

[88] Lakhtakia, A.; Mackay, T.G., Towards gravitationally assisted negative refraction of light by vacuum. J. Phys. A: Math. and Gen. 2004. 37. L505–L510.

[89] Lakhtakia, A., On the genesis of the Post constraint in modern electromagnetism. Optik – International Journal for Light and Electron Optics. 2004. 115. no 4. 151–158.

[90] Lakhtakia, A.; Mackay, T.G.; Setiawan, S., Global and local perspectives of gravitationally assisted negative-phase-velocity propagation of electromagnetic waves in vacuum. Phys. Lett. A. 2005. 336. 89–96.

[91] Mackay, T.G.; Setiawan,S.; Lakhtakia, A., Negative phase velocity of electromagnetic waves and the cosmological constant. Eur Phys J. ser. C. 2005. 41. no S1. 1–4.

[92] Leonhardt, U.; Piwnicki, P., Optics of nonuniformly moving media. Phys. Rev. A. 1999. 60. 4301–4312.

[93] Leonhardt, U.; Piwnicki, P., Relativistic effects of light in moving media with extremely low group velocity. Phys. Rev. Lett. 2000. 84. 822–825.

[94] Leonhardt, U., Space-time geometry of quantum dielectrics. Phys. Rev. A. 2000. 62. 012111 [8 pages].

[95] Leonhardt,U.; Piwnicki, P., Optics of Nonuniformly Moving Media. Phys. Rev. A. 1999. 60. no 6. 4301–4312.

[96] Leonhardt, U.; Piwnicki, P., Relativistic Effects of Light in Moving Media with Extremely Low Group Velocity Phys. Rev. Lett. 2000. 84. 822–825.

[97] Leonhardt, U., Quantum Physics of Simple Optical Instruments. Rept. Prog. Phys. 2003. 66. 1207–1250.

[98] Leonhardt, U.; Philbin, T.G. General Relativity in Electrical Engineering. New J. Phys. 2006. 8. article 247; arXiv:cond-mat/0607418 v2.

[99] Novello, M.; De Lorenci, V.A.; Salim, J.M.; Klippert, R., Geometrical aspects of light propagation in nonlinear electrodynamics. Phys. Rev. D. 2000. 61. no 4. Paper 045001 [10 pages].

[100] Novello, M.; Salim, J.M., Effective electromagnetic geometry. Phys. Rev. D. 2001. 63. Paper 083511 [4 pages].

[101] Novello, M.; Bergliaffa, S.P., Salim, J., Analog black holes in flowing dielectrics. Class. Quant. Grav. 2003. 20. 859–872.

[102] Novello, M.; Bergliaffa, S.P., Effective geometry. AIP Conf. Proc. 2003. 668, 288–300.

[103] de Lange, O.L.; Raab, R.E., Post's constrain for electromagnetic constitutive relations. J. Opt. A. 2001. 3, 23–26.

[104] Raab, R.E.; de Lange, O.L., Symmetry constrains for electromagnetic constitutive relations. J. Opt. A. 2001. 3. 446–451.

[105] Piwnicki, P., Geometrical approach to light in inhomogeneous media. Int. J. Mod. Phys. A. 2002. 17. 1543–1558.

[106] Brevik, I.; Halnes, G., Effective potential for light in moving media. Phys. Rev. D. 2002. 65. Paper 024005 [12 pages].

[107] Vinogradov, A.A., On the form of constiitutive equations in electrodynamics. Uspekhi Fiz. Nauk. 2002. 172. 363–370 (in Russian).

[108] Barkovsky, L.M.; Furs, A.N., Operator methods of description of optical fields in complex medias. Minsk, 2003 (in Russian).

[109] Alexeeva, T.A.; Barkovsky, L.M., To the history of electrodynamics constitutive equations. 7–14 in: Proc. of XII Ann. Sem. Nonlinear Phenomena in Complex Systems. Minsk, 2005.

[110] Obukhov, Y.N.; Hehl, F.W., Spacetime metric from linear electrodynamics. Phys. Lett. B. 1999. 458. 466–470.

[111] de Lorenci, V.A., Klippert, R.; Obukhov, Y.N., On optical black holes in moving dielectrics. Phys. Rev. D. 2003. 68. Paper 061502 [4 pages].

[112] Hehl, F.W.; Obukhov, Y.N., Linear media in classical electrodynamics and the Post constraint. Phys. Lett. A. 2005. 334. 249–259.

[113] Nandi, K.K.; Zhang, Y.Z.; Alsing, P.M.; Evans, P.M.; Bhadra, A., Analog of the Fizeau effect in an effective optical medium. Phys. Rev. D. 2003. 67. Paper 025002 [11 pages].

[114] Vlokh, R., Change of optical properties of space under gravitation field. Ukr. J. Phys. 2004. 5. 27–31.

[115] Vlokh, R., Light absorption under the action of gravitation field. Ukr. J. Phys. 2005. 6. 1–5.

[116] Vlokh, R.; Kostyrko, M., Reflection of light caused by gravitation field of spherically symmentic mass. 2005. Ukr. J. Phys. 6. 125–127.

[117] Vlokh, R.; Kostyrko, M., Estimation of the birefringence change in crystals induced by gravitation field. Ukr. J. Phys. 2005. 6. 120–124.

[118] Vlokh, R.; Kvasnyuk, O., Maxwell equations with accounting of thensor prpperties of time. Ukr. J. Phys. 2007. 8. 125–137.

[119] Matagne, E., Algebraic decomposition of the electromagnetic constitutive tensor. A step toward pre-metric based gravitation. Annalen Phys. 2008. 17. no 1. 17–27.

[120] Boonserm, P.; Cattoen, C.; Faber,T.; Visser, M.; Weinfurtner, S., Effective refractive index tensor for weak field gravity. Class. Quantum Grav. 2005. 22. 1905–1915.

[121] Barcelro, C.; Liberati, S.; Visser, M., Analogue Gravity. Living Reviews in Relativity. 2005. 8. no 12 (2005).

[122] Delphenich, D.H., Symmetries and pre-metric electromagnetism. Annalen der Physik (Leipzig). (2005. 14. 663–704.

[123] Delphenich, D.H., On linear electromagnetic constitutive laws that define almost complex structures. Annalen der Physik (Leipzig). 2007. 16. 207–217.

[124] Delphenich, D.H., Hermitian structures defined by linear electromagnetic constitutive laws. arXiv:07105156

[125] Dereli, T.; Gratus, J.; Tucker,R.W., The covariant description of electromagnetically polarizable media. Phys. Lett. A. 2006. 361. 190–193.

[126] Dereli, T.; Gratus, J.; Tucker, R.W., New perspectives on the relevance of gravitation for the covariant description of electromagnetically polarizable media. J. Phys. A: Math. Theor. 2007. 40. 5695–5715.

[127] Dirac, P.A.M., The Quantum Theory of the Electron. Proc. Roy. Soc. A. 1928. 117. 610–624; The Quantum Theory of the Electron. Part II. Proc. Roy. Soc. A. 1928. 118. 351–361.

[128] Weber, H., Die partiellen Differential-Gleichungen der mathematischen Physik nach Riemann's Vorlesungen. Friedrich Vieweg und Sohn. Braunschweig. 1901. P. 348.

[129] Möglich, F., Zur Quantentheorie des rotierenden Elektrons. Zeit. Phys. 1928. 48. 852–867.

[130] Ivanenko, D.; Landau, L., Zur theorie des magnetischen electrons. Zeit. Phys. 1928. 48. 340–348.

[131] Neumann, J., Einige Bemerkungen zur Diracschen Theorie des relativistischen Dreh-electrons. Zeit. Phys. 1929. 48. 868–881.

[132] van der Waerden, B., Spinoranalyse. Nachr. Akad. Wiss. Gottingen. Math. Physik. Kl. 1929. 100–109 .

[133] Tetrode, H., Allgemein relativistishe Quantentheorie des Elektrons. Zeit. Phys. 1928. 50. no 5-6. 336–346.

[134] Weyl, H., Gravitation and the Electron. Proc. Nat. Acad. Sci. Amer., 1929. 15. 323–334; Gravitation and the Electron. Rice Inst. Pamphlet. 1929. 16. 280–295; Elektron und Gravitation. Zeit. Phys. 1929. 56. 330–352.

[135] Fock, V.; Ivanenko, D., Über eine mögliche geometrische Deutung der relativistis-chen Quantentheorie. Zeit. Phys. 1929. 54. 798–802; Géometrie Quantique linéaire et dÉplacement Parallele. C. R. Acad. Sci. Paris. 1929. 188. 1470–1472.

[136] Fock, V., Geometrisierung der Diracschen Theorie des Elektrons. Zeit. Phys. 1929. 57. 261–277; Sur les Équations de Dirac dans la Théorie de Relativité Générale. C. R. Acad. Sci. Paris. 1929. 189. 25–28; L'équation d'Onde de Dirac et la Geometric de Riemann. J. Phys. Radium. 1929. 10. 392–405.

[137] Juvet, G., Opérateurs de Dirac et Équations de Maxwell. Comm. Math. Helv. 1930. 2. 225–235.

[138] Laporte O., Uhlenbeck G., Application of Spinor Analysis to the Maxwell and Dirac Equations. Phys. Rev. 1931. 37. 1380–1397.

[139] Oppenheimer, J., Note on Light Quanta and the Electromagnetic Field. Phys. Rev. 1931. 38. 725–746.

[140] Majorana, E., Scientific Papers, Unpublished, Deposited at the "Domus Galileana". Pisa, quaderno 2, p. 101/1; 3, p. 11, 160; 15, p. 16;17, p. 83, 159.

[141] de Broglie, L., L'équation d'Ondes du Photon. C. R. Acad. Sci. Paris. 1934. 199. 445–448.

[142] de Broglie, L.; Winter M., Sur le Spin du Photon. C. R. Acad. Sci. Paris. 1934. 199. 813–816.

[143] de Broglie, L., Sur un Cas de Réductibilité en Mécanique Ondulatoire des Particules de Spin 1. C. R. Acad. Sci. Paris. 1939. 208. 1697–1700.

[144] de Broglie, L., Champs Réels et Champs Complexes en Théorie Électromagnétique Quantique du Rayonnement. C. R. Acad. Sci. Paris. 1949. 211. 41–44.

[145] Mercier, A., Expression des équations de l'Électromagnétisme au Moyen des Nombres de Clifford. Thèse de l'Université de Genève. No 953; Arch. Sci. Phys. Nat. Genève. 1935. 17. 1–34.

[146] Petiau, G., University of Paris. Thesis (1936); Acad. Roy. de Belg. Classe Sci. Mem. 1936. 16. no 2

[147] Proca, A., Sur les Equations Fondamentales des Particules Élémentaires. C. R. Acad. Sci. Paris. 1936. 202. 1490–1492.

[148] Proca, A., Sur les Équations Relativistes des Particules Élémentaires. C. R. Acad. Sci. Paris. 1946. 223. 270–272.

[149] Rumer, Yu., Spinor Analysis. Moscow. 1936 (in Russian).

[150] Duffin, R., On the Characteristic Matrices of Covariant Systems. Phys. Rev. 1938. 54. 1114–1114.

[151] Kemmer, N., The Particle Aspect of Meson Theory. Proc. Roy. Soc. London. A. 1939. 173. 91–116.

[152] Kemmer, N., The Algebra of Meson Matrices. Proc. Camb. Phil. Soc. 1943. 39. 189–196.

[153] Kemmer, N., On the Theory of Particles of Spin 1. Helv. Phys. Acta. 1960. 33. 829–838.

[154] Bhabha, H., Classical Theory of Meson. Proc. Roy. Soc. London. A. 1939. 172. 384–409.

[155] Belinfante, F., The Undor Equation of the Meson Field. Physica. 1939. 6. 870–886.

[156] Belinfante, F., Spin of Mesons. Physica. 1939. 6. 887–898.

[157] Taub, A., Spinor Equations for the Meson and their Solution when no Field is Present. Phys. Rev. 1939. 56. 799–810.

[158] Schrödinger, E., Pentads, Tetrads, and Triads of Meson matrices. Proc. Roy. Irish. Acad. A. 1943. 48. 135–146.

[159] Schrödinger, E., Systematics of Meson Matrices. Proc. Roy. Irish. Acad. 1943. 49. 29–42.

[160] Sakata, S.; Taketani M., On the Wave Equation of the Meson. Proc. Phys. Math. Soc. Japan. 1940. 22. 757–770; Reprinted in: Suppl. Progr. Theor. Phys. 1955. 22. 84–97.

[161] Erikson, H.A.S., Vektor-Pseudovektor Meson Theory. Arkiv för matematik, astronomi och fysik. Ark. F. Mat. Ast. Fys. 1942. 29. no 10. [11 pages].

[162] Heitler, W.; Peng, H.W., On the Particle Equation of the Meson. Proc. Roy. Irish. Acad. 1943. 49. 1–28.

[163] Einstein, A.; Bargmann V., Bivector Fields. I, II. Annals of Math. 1944. 45. 1–14; 15–23.

[164] Harish-Chandra. On the Algebra of the Meson Matrices. Proc. Camb. Phil. Soc. 1946. 43. 414–421.

[165] Harish-Chandra, The Correspondence Between the Particle and Wave Aspects of the Meson and the Photon. Proc. Roy. Soc. London. A. 1946. 186. 502–525.

[166] Hoffmann, B., The Vector Meson Field and Projective Relativity. Phys. Rev. 1947. 72. 458–465.

[167] Utiyama, R., On the Interaction of Mesons with the Gravitational Field. Progr. in Theor. Phys. 1947. 2. 38–62.

[168] Schouten, J.A., On meson fields and conformal transformations. Rev. Mod. Phys. 1949. 21. 421–424.

[169] Mercier A., Sur les Fondements de l'Électrodynamique Classique (Méthode Axiomatique). Arch. Sci. Phys. Nat. Genàve. 1949. 2. 584–588.

[170] de Broglie, L.; Tonnelat, M.A., Sur la possibilité d'une structure complexe pour les particules de spin 1. C. R. Acad. Sci. Paris. 1950. 230. 1329–1332

[171] Gupta, S.N., Theory of longitudinal photons in quantum electrodynamics. Proc. Roy. Soc. London. A. 1950. 63, 681–691.

[172] Bleuler, K., Eine neue Methode zur Behandlung der longitudinalen und skalaren Photonen. Helv. Phys. Acta. 1950. 23. 567–586.

[173] O. Brulin,, S. Hjalmars. Wave equations for integer spin particles in gravitational fields. Arkiv for Fysik. 1952. 5. 163–174.

[174] H. Rosen. Special theories of relativity. Amer. J. Phys. 1952. 20. 161–164.

[175] Fujiwara, I., On the Duffin-Kemmer Algebra. Progr. Theor. Phys. 1953. 10. 589–616.

[176] Gürsey, F., Dual Invariance of Maxwell's Tensor. Rev. Fac. Sci. Istanbul. A. 1954. 19. 154–160.

[177] Gupta, S., Gravitation and Electromagnetism. Phys. Rev. 1954. 96. 1683–1685.

[178] Lichnerowicz, A., Théories Relativistes de la Gravitation et de l'Électromagnetisme. Paris, 1955.

[179] Ohmura, T., A New Formulation on the Electromagnetic Field. Prog. Theor. Phys. 1956. 16. 684–685.

[180] Borgardt, A., Matrix Aspects of the Boson Theory. Sov. Phys. JETP. 1956. 30. 334–341.

[181] Borgardt A., Wave Equatons for a Photon. JETF. 1958. 34. 1323–1325.

[182] Fedorov, F., On the Reduction of Wave Equations for Spin-0 and Spin-1 to the Hamiltonian Form. JETP. 1957. 4. 139–141 (in Russian).

[183] Kuohsien, T., Sur les Theories Matricielles du Photon. C. R. Acad. Sci. Paris. 1957. 245. 141–144.

[184] Bludman, S., Some Theoretical Consequences of a Particle Having Mass Zero. Phys. Rev. 1957. 107. 1163–1168.

[185] Good Jr., R., Particle Aspect of the Electromagnetic Field equations. Phys. Rev. 1957. 105. 1914–1919.

[186] Moses, H., A Spinor Representation of Maxwell Equations. Nuovo Cimento Suppl. 1958. 1. 1–18.

[187] Moses, E., Solutions of Maxwell's Equations in Terms of a Spinor Notation: the Direct and Inverse Problems. Phys. Rev. 1959. 113. 1670–1679.

[188] Moses, H., Photon Wave Functions and the Exact Electromagnetic Matrix eEements for Hydrogenic Atoms. Phys. Rev. A. 1973. 8. 1710–1721.

[189] da Silveira, A., Kemmer Wave Equation in Riemann Space. J. Math. Phys. 1960. 1. 489–491.

[190] da Silveira, A., Invariance algebras of the Dirac and Maxwell Equations. Nouvo Cim. A. 1980. 56. 385–395.

[191] Lomont, J., Dirac-Like Wave Equations for Particles of Zero Rest mass and their Quantization. Phys. Rev. 1958. 11. 1710–1716.

[192] Kibble, T., Lorentz Invariance and the Gravitational Field. J. Math. Phys. 1961. 2. 212–221.

[193] Bogush, A.; Fedorov F., On Properties of the Duffin-Kemmer Matrices. Doklady AN BSSR. 1962. 6. 81–85 (in Russian).

[194] Sachs, M.; Schwebel, S., On Covariant Formulations of the Maxwell-Lorentz Theory of Electromagnetism. J. Math. Phys. 1962. 3. 843–848.

[195] Ellis, J., Maxwell's Equations and Theories of Maxwell Form. Ph.D. thesis. University of London, 1964 [417 pages].

[196] Beckers, J.; Pirotte Ch., Vectorial Meson Equations in Relation to Photon Description. Physica. 1968. 39. 205–212.

[197] Casanova, G., Particules Neutre de Spin 1. C. R. Acad. Sci. Paris. A. 1969. 268. 673–676.

[198] Carmeli, M., Group Analysis of Maxwell Equations. J. Math. Phys. 1969. 10. 1699–1703.

[199] Weingarten, D., Complex Symmetries of Electrodynamics. Ann. Phys. 1973. 76. 510–548.

[200] Mignani, R.; Recami, E.; Baldo, M., About a Dirac-Like Equation for the Photon, According to E. Majorana. Lett. Nuovo Cimento. 1974. 11. 568–572.

[201] Newman, E., Maxwell Equations and Complex Minkowski Space. J. Math. Phys. 1973. 14. 102–107.

[202] Frankel, T., Maxwell's Equations. Amer. Math. Mon. 1974. 81. 343–349.

[203] Jackson, J., Classical Electrodynamics. Wiley. New Yor, 1975.

[204] Fushchich, V.I., On the additional invariance of the Dirac and Maxwell equations. Lett. Nuovo Cim. 1974. 11. 508–512.

[205] Edmonds, J., Comment on the Dirac-Like Equation for the Photon. Nuovo Cim. Lett. 1975. 13. 185–186.

[206] Jena, P.; Naik P.; Pradhan T., Photon as the Zero-Mass Limit of DKP Field. J. Phys. A. 1980. 13. 2975–2978.

[207] Venuri, G., A Geometrical Formulation of Electrodynamics. Nuovo Cim. A. 1981. 65. 64–76.

[208] Chow, T., A Dirac-Like Equation for the Photon. J. Phys. A. 1981. 14. 2173–2174.

[209] Fushchich, V.I.; Nikitin, A.G., Symmetries of Maxwell's Equations. Kiev, 1983; Kluwer. Dordrecht. 1987.

[210] Cook, R., Photon Dynamics. Phys. Rev. A. 1982. 25. 2164–2167.

[211] Cook, R., Lorentz Covariance of Photon Dynamics. Phys. Rev. A. 1982. 26. 2754–2760.

[212] Berezin A., Tolkachev E., Fedorov F., Dual-invariant Constitutive Equations for Rest Hyrotropic Medias. Doklady AN BSSR. 1985. 29. 595–597 (in Russian).

[213] Giannetto, E., A Majorana-Oppenheimer Formulation of Quantum Electrodynamics. Lett. Nuovo Cim. 1985. 44. 140–144.

[214] Nunez Yépez H.; Salas Brito A.; Vargas C., Electric and Magnetic Four-Vectors in Classical Electrodynamics. Revista Mexicana de Fisica. 1988. 34. 636.

[215] Kidd, R.; Ardini, J.; Anton, A., Evolution of the Modern Photon. Am. J. Phys. 1989. 57. 27–35.

[216] Recami, E., Possible Physical Meaning of the Photon Wave-Function, According to Ettore Majorana.231–238 in Hadronic Mechanics and Non-Potential Interactions. Nova Sc. Pub., New York, 1990.

[217] Krivsky, I.; Simulik V., Foundations of Quantum Electrodynamics in Field Strengths Terms. Naukova Dumka. Kiev, 1992.

[218] Hillion, P., Spinor Electromagnetism in Isotropic Chiral Media. Adv. Appl. Clifford Alg. 1993. 3. 107–120.

[219] Inagaki, T., Quantum-Mechanical Approach to a Free Photon. Phys. Rev. A. 1994. 49. 2839–2843.

[220] Bialynicki-Birula, I., On the Wave Function of the Photon. Acta Phys. Polon. 1994. 86. 97–116.

[221] Bialynicki-Birula, I., Photon Wave Function. Progress in Optics. 1996. 36. 248–294.

[222] Bialynicki-Birula, I.; Bialynicka-Birula, Z., Heisenberg uncertainty relations for photons Phys. Rev. A. 2012. 86. Paper 022118 [9 pages].

[223] Sipe, J., Photon Wave Functions. Phys. Rev. A. 1995. 52. 1875–1883.

[224] Ghose, P., Relativistic Quantum Mechanics of Spin-0 and Spin-1 Bosons. Found. Phys. 1996. 26. 1441–1455.

[225] Gersten, A., Maxwell Equations as the One-Photon Quantum Equation. Found. Phys. Lett. 1998. 12. 291–298.

[226] Esposito, S., Covariant Majorana Formulation of Electrodynamics. Found. Phys. 1998. 28. 231–244

[227] Torres del Castillo, G.F.; Mercado-Pe'rez, J., Three-dimensional formulation of the Maxwell equations for stationary space-times. J. Math. Phys. 1999. 40. 2882-1891.

[228] Dvoeglazov, V., Speculations on the Neutrino Theory of Light. Annales de la Fondation Louis de Broglie. 1999. 24. 111–127.

[229] Dvoeglazov, V., Historical Note on Relativistic Theories of Electromagnetism. Apeiron. 1998. 5 (1998) 69–88.

[230] Dvoeglazov, V., Generalized Maxwell and Weyl Equations for Massless Particles. Rev. Mex. Fis. 2003. 49. 99–103.

[231] Gsponer, A., On the "equivalence" of the Maxwell and Dirac Equations. Int. J. Theor. Phys. 2002. 41. 689–694.

[232] Ivezić, T., True Transformations Relativity and Electrodynamics. Found. Phys. 2001. 31. 1139–1183.

[233] Ivezić, T., The Invariant Formulation of Special Relativity, or the "True Transformations Relativity", and Electrodynamics. Annales de la Fondation Louis de Broglie. 2002. 27. 287–302.

[234] Ivezić, T., An Invariant Formulation of Special Relativity, or the True Transformations Relativity and Comparison with Experiments. Found. Phys. Lett. 2002. 15. 27–69.

[235] Ivezić, T., The Difference Between the Standard and the Lorentz Transformations of the Electric and Magnetic Fields. Application to Motional EMF. Found. Phys. Lett. 2005. 18. 301–324.

[236] Ivezić, T., Lorentz Invariant Majorana Formulation of the Field Equations and Dirac-like Equation for the Free Photon. EJTP. 2006. 3. 131–142.

[237] Donev, S., Complex Structures in Electrodynamics. arXiv:math-ph/0106008.

[238] Donev, S., From Electromagnetic Duality to Extended Electrodynamics. Annales Fond. L. de Broglie. 2004. 29. 375–392.

[239] Kravchenko, V., On the Relation Between the Maxwell System and the Dirac eEuation. arXiv:mathph/0202009.

[240] Varlamov, V., About Algebraic Foundations of Majorana-Oppenheimer Quantum Electrodynamics and de Broglie-Jordan Neutrino Theory of Light. Ann. Fond. L. de Broglie. 2003. 27. 273–286.

[241] Khan, S., Maxwell Optics: I. An exact matrix Representation of the Maxwell Equations in a Medium. Phys. Scripta. 2005. 71. 440–442; Maxwell Optics: II. An Exact Formalism. arXiv:physics/0205084; arXiv:physics/0205085.

[242] Rollin, S.A. Jr., Spin-1/2 Maxwell fields. Found. Phys. 2004. 34. 815–842.

[243] Kurşunoğlu, B., Complex Orthogonal and Antiorthogonal Representation of Lorentz Group. J. Math. Phys. 1961. 2. 22–32.

[244] Macfarlane, A., On the Restricted Lorentz Group and Groups Homomorphically Related to it. J. Math. Phys. 1962. 3. 1116–1129.

[245] Macfarlane, A., Dirac Matrices and the Dirac Matrix Description of Lorentz Transformations. Commun. Math. Phys. 1966. 2. 133–146.

[246] Fedorov, F.I., The Lorentz group. Moscow, 1978 (in Russian).

[247] Strazhev, V.I.; Tomil'chik, L.M., Electrodynamics with Magnetic Charge. Minsk, 1975 (in Russian).

[248] Kähler, E., Der innere Differentialkalkül. Rendiconti di Matematica. 1962. 21. 425–523.

[249] Cabibbo, N.; Ferrari, E., Quantum Electrodynamics with Dirac Monopoles. Nuovo Cim. 1962. 23. 1147–1154.

[250] Hagen, C.R., Noncovariance of Dirac Monopole. Phys. Rev. B. 1965. 140. 804–810.

[251] Zwanzinger, D., Local-Lagrangian Field Theory ofElectric and Magnetic Charges. Phys. Rev. D. 1971. 3. 880–891.

[252] Vinciarelli, P., Monopole Theory with Potentials Phys. Rev. D. 1972. 6. 3419–3421.

[253] Mignani, R.; Recami, E.: Complex electromagnetic four-potential and the Cabibbo-Ferrari relation for magnetic monopoles. Nuovo Cimento. A. 1975. 30. 533–540.

[254] Strazhev, V.I.; Kruglov, S.I., About Internal Symmetries of the Electromagnetic Field. Acta Physica Polonica. B. 1977. 8. 807–814.

[255] Kresin, Yu.V.; Strazhev, V.I., Two-potential description of the electromagnetic field Teor. Mat. Fis. 1978. 36. 426–429.

[256] Barker, W.; Graziani, F., Quantum Mechanical Formulation of Electron-Monopole Interaction without Dirac Strings. Phys. Rev. D. 1978. 18. 3849–3857.

[257] Barker, W.; Graziani, F.: A Heuristic Potential Theory of Electric and Magnetic Monopoles without Strings. Am. J. Phys. 1978. 46. 1111–1115.

[258] Strazhev, V.I.; Pletyukhov, V.A., Wave equations with multiple representations and a two-potential formulation of electrodynamic. Izvestiya Vysshikh Uchebnykh Zavedenii, Fizika. 1979. 12. 14–18.

[259] Olive, D.I., Exact Electromagnetic Duality. Nucl. Phys. Proc. Suppl. A. 1996. 45. 88–102; Nucl. Phys. Proc. Suppl. A. 1996. 46. 1–15.

[260] Singleton, D., Magnetic Charge as a Hidden Gauge Symmetry. Int. J. Theor. Phys. 1995. 34. 37–46.

[261] Singleton, D., Topological Electric Charge. Int. J. Theor. Phys. 1995. 34 2453–2466.

[262] Singleton, D., Does Magnetic Charge Imply a Massive Photon? Int. J. Theor. Phys. 1996. 35. 2419–2426.

[263] Singleton, D., Electromagnetism with Magnetic Charge and Two Photons. Am. J. Phys. 1996. 64. 452–458.

[264] Berkovits, N., Local Actions with Electric and Magnetic Sources. Phys. Lett. B. 1997. 395. 28–35.

[265] Galvão, C.A.P.; Mignaco, J.A., A Consistent Electromagnetic Duality. [hep-th/0002182]

[266] Ferreira, P.C., Effective Electric and Magnetic Local Actions for Electromagnetism with two Gauge Fields. [hep-th/0510078]

[267] Ferreira, P.C., Explicit Actions for Electromagnetism with Two Gauge Fields with Only one Electric and one Magnetic Physical Fields. J. Math. Phys. 2006. 47. Paper 072902 [17 pages].

[268] Ferreira, P.C., A Pseudo-Photon in Non-Trivial Background Fields Phys. Lett. B. 2007. 651. 74–78.

[269] Kruglov, S.I., Dirac–Kähler equation. Int. J. Theor. Phys. 2002. 41. 653–687.

[270] Strachev, V.I.; Satikov, I.A.; Tsionenko, V.A., Dirac-Kähler equation, classical field. Minsk. Belarus State University, 2007 (in Russian).

[271] Lunardi, J.T.; Pimentel, B.M. ; Teixeira, R.G.; Valverde, J.S., Remarques on the Duffin-Kemmer-Petiau theory and gauge invariance. Phys. Lett. A. 2000. 268. 165–173.

[272] Lunardi, J.T.; Pimentel, B.M.; Valverde, J.S.; Manzoni, L.A., Duffin-Kemmer-Petiau Theory in the Causal Approach. Int. J. Mod. Phys. A. 2002. 17. P. 205–227.

[273] Fainberg, V.Ya.; Pimentel, B.M., Duffin–Kemmer–Petiau and Klein–Gordon–Fock Equations for Electromagnetic, Yang-Mills and external Gravitational Field Interactions: proof of equivalence. Phys. Lett. A. 2000. 271. 16–25.

[274] de Montigny, M.; Khanna, F.C.; Santana, A.E.; Santos, E.S.; Vianna, J.D.M., Galilean covariance and and the Duffin-Kemmer-Petiau equation. J. Phys. A. 2000. 33. L. 273–278.

[275] Kanatchikov, I.V., On the Duffin-Kemmer-Petiau Formulation of the Covariant Hamiltonian Dynamics in Field Theory. Rep. Math. Phys. 2000. 46. 107–112.

[276] Casana, R.; Lunardi, J.T.; Pimentel, B.M.; Teixeira, R.G., Spin 1 fields in Riemann-Cartan space-times via Duffin-Kemmer-Petiau theory. Gen. Rel. Grav. 2002. 34. 1941–1951.

[277] Casana, R.; Fainberg, V.Ya.; Pimentel, B.M.; Lunardi, J.T.; Teixeira, R.G., Massless DKP fields in Riemann-Cartan space-times. Class. Quantum Grav. 2003. 20. 2457–2465.

[278] Yetkin, T.; Havare, A., The Massless DKP Equation and Maxwell Equations in Bianchi Type III Spacetimes. Chin. J. Phys. 2003. 41. 475–485.

[279] Gonen, S.; Havare, A.; Unal, N., Exact solutions of Kemmer equations for Coulomb potential. hep-th/0207087.

[280] Okninski, A., Splitting the Kemmer-Duffin-Petiau equations. Proc. of Institute of Mathematics of NAS of Ukraine. 2004. 50. 902–908.

[281] Bogush, A.A.; Otchik, V.S.; Red'kov, V.M., Generally covariant Duffin-Kemmer formalism and spherical waves for a vector field in de Sitter space. Preprint 426. Institute of Physics. AN BSSR. Minsk, 1986. 45 pages (in Russian).

[282] Bogush, A.A.; Otchik, V.S.; Red'kov, V.M., Vector field in de Sitter space. Vesti AN BSSR. Ser. fiz.-mat. 1986. 1. 58–62 (in Russian).

[283] Red'kov, V.M., Generally relativistical Daffin-Kemmer formalism and behaviour of quantum-mechanical particle of spin 1 in the Abelian monopole field. quant-ph/9812007.

[284] Red'kov, V.M., On discrete symmetry for spin 1/2 and spin 1 particles in external monopole field and quantum-mechanical property of self-conjugacy. quant-ph/ 9812066.

[285] Tokarevskaya, N.G.; Red'kov, V.M., On energy-momentum tensor pf a vector particle in a gravitational field. in: Covariant methods in theoretical physics. Elementary particle physics and relativity theory. Minsk. Iinstiture of Physics. NAN of Belarus, 2001. 143–148 (in Russian).

[286] Tokarevskaya, N.G.; Red'kov, V.M., Duffin–Kemmer formalism and conformal invariance. in: Covariant methods in theoretical physics. Elementary particle physics and relativity theory. Minsk. Iinstitute of Physics, NAN of Belarus, 2001. 149–154 (in Russian).

[287] Bogush, A.A.; Kisel, V.V.; Tokarevskaya, N.G. ; Red'kov, V.M., Non-relativistic approximation in generally covariant of a vector particle. Vesti NANB. Ser. aiz.-mat. 2002. 2. 61–66 (in Russian).

[288] Kisel, V.V.; Tokarevskaya, N.G.; Red'kov, V.M., Scalar particle in Riemannian space, non-minimal interaction, and nonrelativistic approximation. Pages 122–127 in: Proc. of 5th International Conference Bolyai-Gauss-Lobachevsky: Non-Euclidean Geometry In Modern Physics (BGL-5). 10–13 Oct 2006, Minsk, Belarus

[289] Bogush, A.A.; Kisel, V.V.; Tokarevskaya, N.G.; Red'kov, V.M., Duffin–Kemmer–Petiau formalism reexamined: nonrelativistic approximation for spin 0 and spin 1 particles in a Riemannian space-time. Annales de la Fondation Louis de Broglie. 2007. 32. no 2-3. 355–381

[290] Gürsey, F. Reformulation of general relativity in accordance with Mach's principle. Ann. Phys. 1963. 24. no 1-4. 211–244.

[291] Fedorov, F.I.; Pletyukhov, V.A. Wave equations with repeated representations of the Lorentz group. Integer spin. Vesti AN BSSR. Ser. fiz.-mat. nauk. 1969. no 6. 81–88 (in Russian).

[292] Pletyukhov, V.A.; Fedorov, F.I., Wave equations with repeated representations of the Lorentz group for a spin 0 particle. Vesti AN BSSR. Ser. fiz.-mat. nauk. 1970. no 2. 79–85 (in Russian).

[293] Fedorov, F.I.; Pletyukhov, V.A., Wave equations with repeated representations of the Lorentz group. Half-integer spin. Vesti AN BSSR. Ser. fiz.-mat. nauk. 1970. no 3. 78–83 (in Russian).

[294] Pletyukhov, V.A.; Fedorov, F.I., Wave equations with repeated representations of the Lorentz group for a spin 1 particle. Vesti AN BSSR. Ser. fiz.-mat. nauk. 1970, no 3. 84–92 (in Russian).

[295] Capri, A.Z., Non-uniqueness of the spin 1/2 equation. Phys. Rev. 1969. 178. 1811–1815.

[296] Capri, A.Z., First-order wave equations for half-odd-integral spin. Phys. Rev. 1969. 178. no 5. 2427–2433.

[297] Shamaly A.; Capri A.Z., First-order wave equations for integral spin. Nuovo Cim. B. 1971. 2. no 2. 235–253.

[298] Capri, A.Z., Electromagnetic properties of a new spin-1/2 field. Progr. Theor. Phys. 1972. 48. no 4. 1364–1374.

[299] Shamaly, A.; Capri, A.Z., Unified theories for massive spin 1 fields. Can. J. Phys. 1973. 51. no 14. P. 1467 – 1470.

[300] Petras, M., A contribution of the theory of the Pauli-Fierz's equations a particle with spin 3/2. Czech. J. Phys. 1955. 5. no 2. 169–170.

[301] Petras, M., A note to Bhabha's equation for a particle with maximum spin 3/2. Czehc. J. Phys. 1955. 5. no 3. 418–419.

[302] Ulehla, I., Anomalous equations for particles with spin 1/2. JETP. 1957. 33. 473–477 (in Russian).

[303] Formanek, J. On the Ulehla-Petras wave equation. Czehc. J. Phys. B. 1955. 25. no 8. 545–553.

[304] Khalil, M.A.K., Barnacle equivalence structure in relativistic wave equation. Progr. Theor. Phys. 1978. 60. no 5. 1559–1579.

[305] Khalil, M.A.K., An equivalence of relativistic field equations. Nuovo Cimento. A. 1978. 45. no 3. 389–404.

[306] Khalil M.A.K. Reducible relativistic wave equations. J. Phys. A. 1979. Vol. 12, no 5. 649–663.

[307] Bogush, A.A.; Kisel, V.V., Equations with repeated representations of the Lorentz group and Pauli interaction. Vesti AN BSSR. Ser. fiz.-mat. nauk. 1979. no 3. 61–65 (in Russian).

[308] Kisel, V.V., Electric polarizability of a spin 1 particle in the theory of relativistic wave equations. Vesti AN BSSR. Ser. fiz.-mat. nauk. 1982. no 3. 73–78 (in Russian).

[309] Bogush, A.A.; Kisel,V.V., Fedorov, F.I., Description of a free particle by different wave equations. DAN SSSR. 1984. 28. no 8. 702–705 (in Russian).

[310] Bogush, A.A.; Kisel, V.V., Equation for a spin 3/2 particle with anomalous magnetic momentum. Izvestiya Vuzov. Fizika. 1984. no 1. 23–27 (in Russian).

[311] Bogush, A.A.; Kisel, V.V.; Fedorov, F.I., On interpretation of supplementary components of wave functions in presence of electromagnetic interaction. Doklady AN BSSR. 1984. 277. no 2. 343–346 (in Russian).

[312] Wigner, E.P., On unitary representations of the inhomogeneous Lorentz group. Ann. of Math. 1939. 40. 149 – 204.

[313] Pauli, W., Relativistic field theories of elementary particles. Rev. Mod. Phys. 1941. 13. 203–232.

[314] Bhabha, H.J., On the postulational basis of the theory of elementary particles. Rev. Mod. Phys. 1949. 21. no 3. 451–462.

[315] Harish-Chandra, On relativistic wave equation. Phys. Rev. 1947. 71. no 11. 793–805.

[316] Gel'fand, I.M.; Yaglom, A.M., General relativistically invariant equation and infinite infinite-dimensional, representations of the Lorent group JETP. 1948. 18. no 8. 703–733.

[317] Corson, E.M., Introduction to tensors, spinors and relativistic wave equations. London. Blackie and Son, 1953.

[318] Umezawa, H., Quantum Field Theory. Amsterdam. North-Holland, 1956.

[319] Shirokov, Yu.M., Group theory consideration of the bases of relativistic quantum mechanics I. JETP. 1957. 33. no 4. 861–872; II. JETP. 1957. 33. no 5. 1196–1207; III. JETF. 1957. 33. no 5. 1208–1214; IV. JETP. 1958. 34. no 3. 717–724.

[320] Bogush, A.A.; Moroz, L.G., Introduction to klassical field theory. Minsk, Nauka i tekhnika, 1968 (ib Russian).

[321] Lanczos, C. The tensor analytical relationships of Dirac's equation. Zeit. Phys. 1929. 57. 447–473.

[322] Lanczos, C., The covariant formulation of Dirac's equation. Zeit. Phys. 1929. Bd. 57. S. 474 – 483; The conservation laws in the field theoretical representation of Dirac's theory. Zeit. Phys. 1929. 57. 484–493.

[323] Juvet, G., Opérateurs de Dirac et équations de Maxwell. Comm. Math. Helv. 1930. 2. 225–235.

[324] Juvet, G.; Schidlof, A., Sur les nombres hypercomplexes de Clifford et leurs applications à l'analyse vectorielle ordinaire, à l'électromagnétisme de Minkowski et à la théorie de Dirac. Bull. Soc. Sci. Nat. Neuchâtel. 1932. 57. 127–141.

[325] Einstein A.; Mayer V., Semivektoren und Spinoren. Sitz. Ber. Preuss. Akad. Wiss. Berlin. Phys.-Math. Kl. 1932. 522–550.

[326] Einstein, A.; Mayer, W., Die Diracgleichungen für Semivektoren. Proc. Akad. Wet. Amsterdam. 1933. 36. 497–516.

[327] Einstein, A.; Mayer, W., Spaltung der Natürlichsten Feldgleichungen für Semi-Vektoren in Spinor-Gleichungen von Diracschen Tipus. Proc. Akad. Wet. Amsterdam. 1933. 36. 615–619.

[328] Einstein, A.; Mayer, W., Darstellung der Semi-Vektoren als gewöhnliche Vektoren von Besonderem Differentiations Charakter. Ann. of Math. 1934. 35. no 1. 104–110.

[329] Frenkel, Ya.I., Lehrbuch der Elektrodynamik. I. Allgemeine Mechanik der Elektrizität. Verlag J. Springer. Berlin, 1926; II. Makroskopische Elektrodynamik der materiellen Körper. Verlag J. Springer. Berlin, 1928.

[330] Whittaker, E.T., On the relations of the tensor-calculus to the spinor-calculus. Proc. Roy. Soc. London. A. 1937. 158. 38–46.

[331] Proca, A., Sur un article de M.E. Whittaker, intitulé "Les relations entre le calcul tensoriel et le calcul des spineurs". J. Phys. et Radium. 1937. 8. 363–365.

[332] Ruse, H.S., On the geometry of Dirac's equations and their expression in tensor form. Proc. Roy. Soc. Edin. 1936. 57. 97–127.

[333] Taub, A.H., Tensors equations equivalent to the Dirac equations. Ann. Math. 1939. 40. 937.

[334] Taub, A.H., Spinor equations for the meson and their solution when no field is present. Phys. Rev. 1939. 56. no 8. 799–810.

[335] Ivanenko D., Sokolov A. Quantim field theory. Moscow, 1951.

[336] Feshbach, H.; Nickols, W., A wave equation for a particle of maximum spin one. Ann. Phys. N.Y. 1958. 4. no 4. 448–458.

[337] Kähler E. Innerer and äusserer Differentialkalkül. Abh. Dt. Akad. Wiss. Berlin. Kl. Math.-Phys. u. Techn. 1960. N 4.

[338] Kähler, E., Die Dirac-Gleichunung. Abh. Dt. Akad. Wiss. Berlin, Kl. Math.-Phys. u. Techn. 1961. no 1.

[339] Leutwyler, H., Generally covariant Dirac equation and associated boson fields. Nuovo Cimento. 1962. 26. no 5. 1066.

[340] Klauder, J.R., Linear representation of spinor fields by antysymmetric tensors. J. Math. Phys. 1964. 5. no 9. 1204–1214.

[341] Penney, R., Tensorial description of neutrinos. J. Math. Phys. 1965. 6. no 7. 1026–1028.

[342] Cereignani, C., Linear representations of spinors by tensors. J. Math. Phys. 1967. 8. no 3. 417–422.

[343] Streater, R.F.; Wilde, I.F., Fermion states of a boson field. Nucl. Phys. B. 1970. 24. no 3. 561–575.

[344] Pestov, A.B., Connection betwee Dirac and Maxwell equations. Dubna, 1971. 18 pages. Prepeint P2-5798.

[345] Pestov, A.B., Reltivistic equations defined by exterior derivative operators and extended divergence. TMP. 1978. 34. no 1. 48–57.

[346] Pestov, A.B., On the group of internal symmetry of the wave equation defined by exterior derivative operators. Dubna, 1983. Preprin P2-83-506.

[347] Osterwalder, K,. Duality for free Bose fields. Commun. Math. Phys. 1973. 29. no 1. 1–14.

[348] Crumeyrolle, A., Une théorie de Einstein – Dirac en spin maximum 1. Ann. Inst. H. Poincaré. A. 1975. 22. no 1. 43–61.

[349] Durand, E., 16-component theory of the spin-1 particle and its generalization to arbitrary spin. Phys. Rev. D. 1975. 11. no 12. 3405–3416.

[350] Strazhev V.I. On the symmetry group of extended equations for a vector field. Izvestiya Vuzov. Fizika. 1977. no 8. 45–48.

[351] Kruglov, S.I.; Strazhev, V.I., Internal symmetries and conservation lows in ckassical theory of a vector field of general type. Izvestiya Vuzov. Fizika. 1978, no 4. 77–81.

[352] Strazhev, V.I., On dyad symmetry of a vector filed of general type. Acta Phys. Pol. B. 1978. 9. 449–458.

[353] Bogush A.A.; Kruglov S.I.; Strazhev V.I., On the group of internal symmetry of 16-component neory of a cector particles. Doklady AN BSSR. 1978. 22. no 10. 893–895.

[354] Bogush A.A.; Kruglov S.I., On equation of vector field of general type. Procceding of Academy of Sciences of BSSR. ser. phys.-mat. 1978. no 4. 58–65.

[355] Satikov I.A.; Strazhev V.I., On quantum description of the Dirac–Kähler field. TMP. 1987. 73. no 1. 16–25.

[356] Strazhev V.I.; Pletyukhov V.A.; Fedorov F.I., On connection of spin and statistics in the theory of relativistic wave equations with intrinsic degrees of freedom. Minsk, 1988. 36 pages. Preprint no 517. IP AN BSSR.

[357] Strazhev, V.I.; Berezin, A.V.; Satikov, I.A., Dirac–Kähler equations and quantum theory of the Dirac field witn internal symmetry group $SU(2,2)$. Minsk, 1988. 20 pages. Preprint no 522. IP AN BSSR.

[358] Strazhev, V.I.; Tsionenko, D.A., On Dirac–Kähler gauge field teory in a curved space-time. Vestnik BGU. ser. I. fiz.-mat.-inform. 2002. no 2. 15–21.

[359] Tsionenko, D.A., Dirac–Kähler equation in non-Euclidean space-time. Procceding of National Academy of Sciences of Belarus. phys.-mat. 2003. no 1. 81–85.

[360] Strazhev, V.I.; Satikov I.A.; Tsionenko, D.A., Dirac–Kähler equation, classical theory. Minsk, BGU, 2007.

[361] Graf, W., Differential forms as spinors. Ann. Inst. H. Poincaré. A. 1978. 29. no 1. 85–109.

[362] Benn, I.M.; Tucker, R.W., A generation model based on Kähler fermions. Phys. Lett. B. 1982. bf 119. no 4-6. 348–350.

[363] Benn, I.M.; Tucker, R.W., Fermions without spinors. Commun. Math. Phys. 1983. 89. no 3. 341–362.

[364] Benn, I.M.; Tucker, R.W., Kähler fields and five-dimensional Kaluza – Klein theory. J. Phys. A. 1983. 16. no 4. 123–125.

[365] Benn, I.M.; Tucker, R.W., Clifford analysis of exterior forms and Fermi-Bose symmetry. J. Phys. A. 1983. 16. no 17. 4147–4153.

[366] Benn, I.M.; Tucker, R.W., A local right-spin covariant Kähler equation. Phys. Lett. B. 1983. 130. no 3-4. 177–178.

[367] Tucker, R.W.; Benn, I.M., The differential approach to spinors and their symmetries. Nuovo Cim. A. 1985. 88. Ser. 2. no 3. 273–285.

[368] Banks, T.; Dothan, Y.; Horn, D., Geometric fermions. Phys. Lett. B. 1982. 117. no 6. 413–417.

[369] Garbaczewski, P., Quantization of spinor fields. Meaning of "bosonization" in (1+1) and (1+3) dimensions. J. Math. Phys. 1982. 23. no 3. 442–450.

[370] Pletyukhov, V.A.; Strazhev, V.I., On Dirac like wave equation for particles with maximal spin 1. Doklady AN BSSR. 1982. 26. no 8. 691–693.

[371] Pletyukhov, V.A.; Satikov, I.A.; Strazhev, V.I., Relativistic wave equations and massless Dirac–Kähler field. 31 – 35 in: Covariant methods in theoretical physics. Elementary particle physica and relativity theoory. Minsk, Institute of Physics, 1986.

[372] Pletyukhov, V.A.; Strazhev, V.I., On possible extensions of the Dirac–Kähler field. Vesti AN BSSR. ser. fiz.-mat. 1987. no 5. 87–92.

[373] Pletyukhov, V.A.; Strazhev, V.I., Tensorial equations and Dirac particles with internal degrees of freedom. Yadernaya Fizika. 1989. 49. 1505–1514.

[374] Holland, P.R., Tensor conditions for algebraic spinors. J. Phys. A. 1983. 16. no 11. 2363–2374.

[375] Ivanenko, D.D.; Obukhov, Yu.N.; Solodukhin, S.N., On antisymmetric tensor representation. of the Dirac equation. Trieste, 1985. Preprint IC/85/2. ICTP.

[376] Obukhov, Yu.N.; Solodukhin, S.N., Reduction of the Dirac equation and its coonection to Ivanenko–Landau– Kähler equation. TMP. 1993. 94. 276–295.

[377] Bullinaria, J.A., Kähler fermions in arbitrary space-times, their dimensional reduction and relation to spinorial fermions. Ann. Phys. (N.Y.). 1986. 168. no 2. 301–343.

[378] Blau, M., Clifford algebras and Kähler – Dirac spinors. Ph.D. dissertation, Report UWTH Ph 198616. Universitat Wien, 1986. 200 p.

[379] Jourjine, A.N., Space-time Dirac – Kähler spinors. Phys. Rev. D. 1987. 35. no 2. 757–758.

[380] Krolikowski, W., Dirac equation with hidden extra spin: a generalization of kähler equation. I. Acta Phys. Polon. B. 1989. 20. no 10. 849–858.

[381] Krolikowski, W., Dirac equation with hidden extra spin: a generalization of Kähler equation. II. Acta Phys. Polon. B. 1990. 21. no 3. 201–207.

[382] Howe, P.; Penati, S .; Pernici, M; Townsend, P.K., A particle mechanics description of antisymmetric tensor fields. Class. Quant. Grav. 1989. 6. 1125–1140.

[383] Beckers, J.; Debergh, N.; Nikitin, A.G., On parasupersymmetries and relativistic descriptions for spin one particles. I. The free context. Fortschr. Phys. 1995. 43. no 1. 67–80; II. The interacting context (with electromagnetic fields) Fortschr. Phys. 1995. 43. no 1. 81–96.

[384] Kruglov, S.I., Symmetry and electromagnetic interactions of Fields with multispin. N.Y.: Nova Science Pub. Inc., Hauppauge, 2000.

[385] Kruglov, S.I., Dirac – Kähler equations. Intern. J. Theor. Phys. 2002. 41. 653–687.

[386] Kruglov, S.I., On the generalized Dirac equation for fermions with two mass states. Ann. Fond. L. de Broglie. 2004. 29. Hors série 2. 1005–1016.

[387] Marchuk, N.G., Dirac gamma-equation, classical gauge fields and Clifford algebra. Adv. Appl. Clifford Alg. 1998. 8. 181–2242.

[388] Marchuk, N.G., Gauge fields of the matrix Dirac equation. Nuovo Cim. B. 1998. 113. 1287–1295.

[389] Marchuk, N.G., A gauge model with spinor group for a description of local interaction of a fermion with electromagnetic and gravitational fields. Nuovo Cim. B. 2000. 115. 11–25.

[390] Marchuk, N.G., A tensor form of the Dirac equation. Nuovo Cim. B. 2001. 116. no 10. 1225–1248.

[391] Marchuk, N.G., Dirac-type tensor equations with non-Abelian gauge symmetries on pseudo-Riemannian space. Nuovo Cim. B. 2002. 117. 95–120.

[392] Marchuk, N.G., The Dirac equation vs. the Dirac type tensor equation. Nuovo Cim. B. 2002. 117. 511–520.

[393] Krivskij, I.Yu.; Lompej, R.R.; Simulik, B.M., On symmetries of complex Dirac – Kähler equation. TMP. 2005. 143. 64–82.

[394] Pauli, W., Über relativistische Feldleichungen von Teilchen mit beliebigem Spin im elektromagnetishen Feld. Helv. Phys. Acta. 1939. 12. 297–300.

[395] Fierz, M., On relativistic wave equations for particles of arbitrary spin in an electromagnetic field. Proc. Roy. Soc. London. A. 1939. 173. 211–232.

[396] de Broglie, L., Sur l'interprétation de certaines équations dans la théorie des particules de spin 2. C. R. Acad. Sci. Paris. 1941. 212. 657–659.

[397] Fradkin, E.E., To the theory of particle of high spin LETP. 1950. 20. 27–38 (in Russian).

[398] Fedorov, F.I., To the theory of a particle with spin 2. Uchenye zapiski BGU. fiz.-mat. 1951. no 12. 156–173.

[399] Krylov, B.V.; Fedorov, F.I., First order equations for a graviton. DAN BSSR. 1967. 11. no 8. 681–684 (in Russian).

[400] Bogush, A.A.; Krylov, B.V.; Fedorov, F.I., On matrices for equations for particles with spin 2. Vesti AN BSSR, fiz.-mat. 1968. no 1. 74–81 (in Russian).

[401] Fedorov, F.I., First order equations for gravitational field. DAN SSSR. 1968. 179. no 4. 802–805 (in Russian).

[402] Krylov, B.V., On systems of first order equations for a graviton Vesti AN BSSR. fiz.-mat. 1972. no 6. 82–89 (in Russian).

[403] Kisel, V.V., On relativistic wave equations for massive particle with spin 2. Vesti AN BSSR. fiz.-mat. 1986. no 5. 94–99 (in Russian).

[404] Fainberg, V.Ya., To the theory of interaction of particle with higher spns with electromagnetic and meson fields. Trudy FIAN SSSR. 1955. 6. 269–332 (in Russian).

[405] Regge, T., On properties of the particle with spin 2. Nuovo Cimento. 1957. 5. no 2. 325–326.

[406] Buchdahl, H.A., On the compatibility of relativistic wave equations for particles of higher spin in the presence of a gravitational field. Nuovo Cim. 1958. 10. 96–103.

[407] Buchdahl, H.A., On the compatibility of relativistic wave equations in Riemann spaces. Nuovo Cim. 1962. 25. 486–496.

[408] Velo, G.; Zwanziger, D., Noncausality and other defects of interaction Lagrangians for particles with spin one and higher. Phys. Rev. 1969. 188. no 5. 2218–2222.

[409] Velo, G., Anomalous behavior of a massive spin two charged particle in an external electromagnetic field. Nucl. Phys. B. 1972. 43. 389–401.

[410] Hagen, C.R., Minimal electromagnetic coupling of spin-two fields. Phys. Rev. D. 1972. 6. no 4. 984–987.

[411] Hagen, C.R., Scale and conformal transformations in Galilean-covariant field theory. Phys. Rev. D. 1972. 5. no 2. 377–388.

[412] Fedorov, F.I.; Kirilov, A.A. First order equations for gravitational field in vacuum. Acta Physica Polonica. B. 1976. 7. no 3. 161–167 (in Russian).

[413] Cox, W., First-order formulation of massive spin-2 field theories. J. Phys. A. 1982. 15. 253–268.

[414] Barut, A.O.; Xu, B.W., On conformally covariant spin-2 and spin-3/2 equations J. Phys. A. 1982. 15. no 4. 207–210.

[415] Kisel, V.V., Relativistic wave equations with extended sets of representations. Disseration. Institute of Physics. Minsk, 1984 (in Russian).

[416] Loide, R.K., On conformally covariant spin-3/2 and spin-2 equations. J. Phys. A. 1986. 19. no 5. 827–829.

[417] Bogush, A.A.; Kisel, V.V., On description of anomalous magnetic moment of a massive particle with spin 2 in the theory of relativistic wave equations. Izvestia Vuzov. SSSR. Fizika. 1988. 31. no 3. 11–16 (in Russian).

[418] Vasiliev, M.A., More on equations of motion for interacting massless fields of all spins in (3+1)-dimensions. Phys. Lett. B. 1992. 285. 225–234.

[419] Buchbinder, I.L.; Krykhtin, V.A.; Pershin, V.D., On consistent equations for massive spine-2 field coupled to gravity in string theory. Phys. Lett. B. 1999. 466. 216–226.

[420] Buchbinder, I.L.; et al, Equations of motion for massive spin 2 field coupled to gravity Nucl. Phys. B. 2000. 584. 615–640.

[421] Bogush, A.A.; Kisel, V.V.; Tokarevskaya, N.G.; Red'kov, V.M., On equation for a particle wit hspin 2 in external electromagnetic and gravitational fields. Procceedings of National Acadamy of Sciences of Belarus. ser. phys.-mat. 2003. no 1. 62–67 (in Russian).

[422] Red'kov, V.M.; Tokarevskaya, N.G.; Kisel, V.V., Graviton in a curved space-time background and gauge symmetry Nonlinear phenomena in complex systems. 2003. 6. no 3. 772–778.

[423] Kisel, V.V.; Red'kov, V.M., System of tensor equations for a particle with spin 2 and description of an anomalous magnetic mooment. I. Vesti BDPU. named after M. Tank. ser. 3. 2010. no 1(63). 3–6 (in Russian).

[424] Kisel, V.V.; Red'kov, V.M., System of tensor equations for a particle with spin 2 and description of an anomalous magnetic mooment. I. Vesti BDPU. named after M. Tank. ser. 3. 2010. no 2(64). 8–10 (in Russian).

[425] Bogush, A.A.; Kisel, V.V.; Tokarevskaya,N.G.; Red'kov, V.M., Petras theory for a particle with spin 1/2 in a curved space-time. Proccedings of National Acadamy of Sciences of Belarus. ser. phys.-mat. 2002. no 1. 63–68 (in Russian).

[426] Kisel, V.V. ; Tokarevskaya, N.G.; Red'kov, V.M., Petras theory for a particle with spin 1/2 in a curved space-time. Minsk, 2002. 25 pages. Preprint no 737.

[427] Hagen, C.R., Consistency of the anomalous-magnetic-moment coupling of a vector-meson field. Phys. Rev. D. 1974. 9. no 2. 498–499.

[428] Kisel, V.V.; Red'kov, V.M., Shamaly–Capri equation and additional interaction of a vector particle with graviattional field. 107–112 in: Covarian methods in theoretical physics. Physics of elementary particles and relativity theory. Institute of Physics, National Academy of Sciences of Belarus. Minsk, 2001 (in Russian).

[429] Varshalovich, D.A.; Moskalev, A.N.; Hersonskiy, V.K., Quantum theory of angular moment. Nauka, Leningrad, 1975 (in Russian).

[430] Schrödinger, E., The ambiguity of the wave function. Annalen der Physik. 1938. 32. 49–55.

[431] Pauli, W., Über die Kriterium für Ein-oder Zweiwertigkeit der Eigenfunktionen in der Wellenmechanik. Helv. Phys. Acta. 1939. 12. 147–168.

[432] Tamm, I.E.; Motion of mesons in electromagnetic fields. Dokl. Akad. Nauk SSSR. 1940. 29 . no 8-9. 551–554 (in Russian).

[433] Rabi, I.I., Das freie Electron in Homogenen Magnetfeld nach der Diraschen Theorie. Ztshr. Phys. 1928. 49. 507–511.

[434] Landau, L., Diamagnetismus der Metalle, Ztshr. Phys. 1930. 64. 629–637.

[435] Plesset, M.S., Relativistic wave mechanics of the electron deflected by magnetic field. Phys.Rev. 1931. 12. 1728–1731.

[436] Comtet, A.; Houston, P.J., Effective action on the hyperbolic plane in a constant external field. J. Math. Phys. 1985. 26. no 1. 185–191

[437] Comtet, A., On the Landau levels on the hyperbolic plane. Annals of Physics. 1987. 173. 185–209.

[438] Aoki, H., Quantized Hall Effect. Rep. Progr. Phys. 1987. 50. 655–730.

[439] Groshe, C., Path integral on the Poincaré uper half plane with a magnetic field and for the Morse potential. Ann. Phys. (N.Y.), 1988. 187. 110–134.

[440] Klauder, J.R.; Onofri, E., Landau Levels and Geometric Quantization. Int. J. Mod. Phys. A. 1989. 4. 3939–3949.

[441] Avron, J.E.; Pnueli, A., Landau Hamiltonians on Symmetric Spaces. 96 – 117 in: Ideas and methods in mathematical analysis, stochastics, and applications. Vol. II. S. Alverio et al., eds. Cambridge Univ. Press, Cambridge, 1990.

[442] Dunne, G.V., Hilbert Space for Charged Particles in Perpendicular Magnetic Fields. Ann. Phys. (N.Y.) 1992. 215. P. 233 – 263.

[443] Alimohammadi, M.; Shafei Deh Abad, A. Quantum group symmetry of the quantum Hall effect on the non-flat surfaces. J. Phys. A: Math. and Gen. 1996. 29. 559–563.

[444] Alimohammadi, M.; Sadjadi, H.M., Laughlin states on the Poincare half-plane and their quantum group symmetry, J. Phys. A: Math. and Gen. 1996. 29. 5551–5557.

[445] Onofri, E., Landau Levels on a torus. Int. J. Theoret. Phys. 2001. 40. 537–549.

[446] Negro,; del Olmo, M.A.; Rodríguez-Marco, A., Landau quantum systems: an approach based on symmetry. J. Phys. A: Math. and Gen. 2002. 35. 2283–2308.

[447] Drukker, N.; Fiol, B.; Simón, J., Gödel-type Universes and the Landau problem. Journal of Cosmology and Astroparticle Physics. 2004. no 10. Paper 012

[448] Ghanmi, A.; Intissar, A., Magnetic Laplacians of fifferentila forms of the hyperbolic disk and Landau levels. African Journal Of Mathematical Physics. 2004. 1. 21–28.

[449] Bogush, A.A.; Red'kov, V.M.; Krylov, G.G., Schrödinger particle in magnetic and electric fields in Lobachevsky and Riemann spaces. Nonlinear Phenomena in Complex Systems. 2008. 11. no 4. 403–416.

[450] Bogush, A.A.; Red'kov, V.M.; Krylov, G.G., Quantum-mechanical particle in a uniform magnetic field in spherical space S_3. Proceedings of the National Academy of Sciences of Belarus. Ser. fiz.-mat. 2009. 2. 57–63 (in Russian).

[451] Kudryashov, V.V.; Kurochkin, Yu.A.; Ovsiyuk, E.M.; Red'kov, V.M., Motion caused by magnetic field in Lobachevsky space. AIP Conference Proceedings. 2010. 1205. 120–126; Eds. Remo Ruffini and Gregory Vereshchagin. The sun, the stars, the Universe and General relativity. International Conference in Honor of Ya.B. Zeldovich. April 20–23, 2009, Minsk.

[452] Kudryashov, V.V.; Kurochkin, Yu.A.; Ovsiyuk, E.M.; Red'kov, V.M., Motion of a particle in magnetic field in the Lobachevsky space. Doklady Natsionalnoi Akademii Nauk Belarusi. 2009. 53. 50–53 (in Russian).

[453] Kudryashov, V.V.; Kurochkin, Yu.A.; Ovsiyuk, E.M.; Red'kov, V.M., Classical Particle in Presence of Magnetic Field, Hyperbolic Lobachevsky and Spherical Riemann Models. SIGMA. 2010. 6. Paper 004 [34 pages].

[454] Dirac, P.A.M., The electron wave equation in the de Sitter space. Ann. Math. 1935. 36. 657–669.

[455] Dirac, P.A.M., Wave equations in conformal space. Ann. of Math. 1936. 37. 429–442.

[456] Schrödinger, E., The proper vibrations of the expanding universe. Physica. 1939. 6. 899–912.

[457] Schrödinger, E., General theory of relativity and wave mechanics. Wiss. en Natuurkund. 1940. 10. 2–9.

[458] Lubanski, J.K.; Rosenfeld, L., Sur la representation des champs mesoniques dans l'éspace à sinq dimension. Physica. 1942. 9. 117–134.

[459] Goto, K., Wave equations in de Sitter space. Progr. Theor. Phys. 1951. 6. 1013–1014.

[460] Ikeda, M., On a five-dimensional representation of the electromagnetic and electron field equations in a curved space-time. Progr. Theor. Phys. 1953. 10. 483–498.

[461] Nachtmann, O., Quantum theory in de-Sitter space. Commun. Math. Phys. 1967. 6. 1–16.

[462] Chernikov, N.A.; Tagirov, E.A., Quantum theory of scalar field in de Sitter space-time. Ann. Inst. Henri Poincaré. 1968. IX. 109–141.

[463] Börner, G.; Dürr, H.P., Classical and quantum theory in de Sitter space. Nuovo Cim. A. 1969. 64. 669–713.

[464] Tugov, I.I., Conformal covariance and invariant formulation of scalar wave equations. Ann. Inst. Henri Poincaré. A. 1969. 11. 207–220.

[465] Fushchych, W.L.; Krivsky, I.Yu., On representations of the inhomogeneous de Sitter group and equations in five-dimensional Minkowski space. Nucl. Phys. B. 1969. 14. 573–585.

[466] Castagnino, M., Champs de spin entier dans l'espace-temps De Sitter. Ann. Inst. Henri Poincaré. A. 1970. 13. 263–270.

[467] Castagnino, M., Champs spinoriels en Relativité générale; le cas particulier de léspace-temps de De Sitter et les équations d'ond pour les spins éléves Ann. Inst. Yenri Poincaré. A. 1972. 16. 293–341.

[468] Vidal, A., On the consistency of wave equations in de Sitter space. Notas de fi'sica. 1970. 16. no 1 [8 pages].

[469] Adler, S.L., Massless, euclidean quantum electrodynamics on the 5-dimensional unit hypersphere. Phys. Rev. D. 1972. 6. 3445–3461.

[470] Schnirman, E.; Oliveira, C.G., Conformal invariance of the equations of motion in curved spaces. Ann. Inst. Henri Poincaré. A. 1972. 17. 379–397.

[471] Tagirov, E.A., Consequences of field quantization in de Sitter type cosmological models. Ann. Phys. 1973. 76. 561–579.

[472] Pestov, F.B.; Chernikov, N.A., Shavoxina, N.S., Electrodynamical equations in spherical world. Teor. Mat. Fiz. 1975. 25. 327–334.

[473] Candelas, P.; Raine, D.J., General-relativistic quantum field theory: an exactly soluble model. Phys. Rev. D. 1975. 12. 965–974.

[474] Avis, S.J.; Isham, C.J.; Storey, D., Quantum Field Theory In Anti-de Sitter Space-Time. Phys. Rev. D. 1978. 18. 3565–3576.

[475] Brugarino, T., De Sitter-invariant field equations. Ann. Inst. Henri Poincaré. A. 1980. 32. 277–282.

[476] Fang, J.; Fronsdal, C., Massless, half-integer-spin fields in de Sitter space Phys. Rev. D. 1980. 22. 1361–1367.

[477] Angelopoulos, E.; Flato, M.; Fronsdal, C.; Sternheimer D., Massless Particles, Conformal Group, and De Sitter Universe. Phys. Rev. D. 1981. 23. 1278–1289.

[478] Burges, C.J.C., The de Sitter vacuum. Nucl. Phys. B. 1984. 247. 533–543.

[479] Deser, S.; Nepomechie, R.I., Gauge Invariance Versus Masslessness in de Sitter Space. Ann. Phys. 1984. 154. 396–420.

[480] Dullemond, C.; van Beveren, E., Scalar field propagators in anti-de Sitter spacetime. J. Math. Phys. 1985. 26. 2050–2058.

[481] Gazeau, J.P., Gauge fixing and Gupta-Bleuler triplet in de Sitter QED. J. Math. Phys. 1985. 26. 1847–1854.

[482] Allen, B., Vacuum states in de Sitter space. Phys. Rev. D. 1985. 32. 3136–3149.

[483] Flato, M.; Fronsdal, C.; Gazeau, J.P., Masslessness and light-cone propagation in 3+2 de Sitter and 2+1 Minkowski spaces Phys. Rev. D. 1986. 33. 415–420.

[484] Allen, B.; Jacobson, T., Vector two-point functions in maximally symmetric space. Commun. Math. Phys. 1986. 103. 669–692.

[485] Allen, B.; Folacci, A.: The Massless Minimally Coupled Scalar Field In De Sitter Space. Phys. Rev. D. 1987. 35. 3771–3778.

[486] Sánchez, N., Quantum field theory and elliptic interpretation of de Sitter space-time. Nucl. Phys. B. 1987. 294. 1111–1137/

[487] Pathinayake, C.; Vilenkin, A.; Allen, B., Massless scalar and antisymmetric tensor fields in de Sitter space. Phys. Rev. D. 1988. 37. 2872–2877.

[488] Gazeau, J-P.; Hans, M., Integral-spin fields on (3+2)-de Sitter space. J. Math. Phys. 1988. 29. 2533–2552.

[489] Bros J.; Gazeau, J.P.; Moschella, U., Quantum Field Theory in the de Sitter Universe. Phys. Rev. Lett. 1994. 73. 1746–1749.

[490] Takook, M.V., Thèse de l'université Paris VI, 1997. Théorie quantique des champs pour des systèmes élémentaires "massifs" et de "masse nulle" sur l'espace-temps de de Sitter.

[491] Pol'shin, S.A., Group Theoretical Examination of the Relativistic Wave Equations on Curved Spaces. I. Basic Principles. arXiv:gr-qc/9803091.

[492] Pol'shin, S.A., Group Theoretical Examination of the Relativistic Wave Equations on Curved Spaces. II. De Sitter and Anti-de Sitter Spaces. arXiv:gr-qc/9803092

[493] Pol'shin, S.A., Group Theoretical Examination of the Relativistic Wave Equations on Curved Spaces. III. Real reducible spaces. arXiv:gr-qc/9809011

[494] Gazeau, J-P.; Takook, M.V., "Massive" vector field in de Sitter space J. Math.Phys. 2000. 41. 5920–5933.

[495] Deser, S.; Waldron, A., Partial masslessness of higher spins in (A)dS. Nucl. Phys. B. 2001. 607. 577–604.

[496] Deser, S.; Waldron, A., Null propagation of partially massless higher spins in (A)dS and cosmological constant speculations. Phys. Lett. B. 2001. 513. 137–141.

[497] Spradlin, M.; Strominger, A.; Volovich, A., Les Houches lectures on de Sitter space. hep-th/0110007; Prepared for Les Houches Summer School: Session 76: Euro Sum Conference: C01-07-30, p. 423–453.

[498] Garidi, T.; Huguet, E.; Renaud, J., De Sitter Waves and the Zero Curvature Limit Comments. Phys. Rev. D. 2003. 67. Paper 124028 [5 pages].

[499] Rouhani, S.; Takook, M.V., Abelian Gauge Theory in de Sitter Space Mod. Phys. Lett. A. 2005. 20. 2387–2396.

[500] Behroozi, S.; Rouhani, S.; Takook, M.V.; Tanhayi, M.R., Conformally invariant wave-equations and massless fields in de Sitter spacetime. Phys.Rev. D. 2006. 74. Paper 124014 [12 pages].

[501] Huguet, E.; Queva, J.; Renaud, J., Conformally related massless fields in dS, AdS and Minkowski spaces. Phys. Rev. D. 2006. 73. Paper 084025 [7 pages].

[502] Garidi, T.; Gazeau, J.P.; Rouhani, S.; Takook, M.V., Massless vector field in de Sitter Universe. J. Math. Phys. 2008. 49. Paper 032501 [25 ages].

[503] Huguet, E.; Queva, J.; Renaud, J., Revisiting the conformal invariance of the scalar field: From Minkowski space to de Sitter space. Phys. Rev. D. 2008. 77, Paper 044025 [4 pages] (2008)

[504] Dehghani, M.; Rouhani, S.; Takook, M.V.; Tanhayi, M.R., Conformally Invariant "Massless" Spin-2 Field in de Sitter Universe. Phys. Rev. D. 2008. 77. Paper 064028 [12 pages].

[505] Faci, S.; Huguet, E.; Queva, J.; and Renaud, J., Conformally covariant quantization of the Maxwell field in de Sitter space. Phys. Rev. D. 2009. 80. Paper 124005 [13 pages].

[506] Hawking, S.W., Gravitational radiation from colliding black holes. Phys. Rev. Lett. 1971. 26. 1344–1346.

[507] Hawking, S.W., Black hole explositions? Nature. 1974. 248. no 5443. 30–31.

[508] Hawking, S.W., Particle creation by black holes. Commun. Math. Phys. 1975. 43. 199–220/

[509] Hawking, S.W.; Gibbons, G.W., Cosmological event horizons, thermodynamics, and particle creation. Phys. Rev. D. 1977. 15. 2738– 2751.

[510] Lohiya, D.; Panchapakesan, N., Massless scalar field in a de Sitter universe and its thermal flux. J. Phys. A. Math. and Gen. 1978. 11. 1963–1968.

[511] Lohiya, D., Panchapakesan, N.: Particle emission in the de Sitter universe for massless fields with spin. J. Phys. A. Math. and Gen. 1979. 12. 533–539.

[512] Khanal, U.; Panchapakesan, N., Perturbation of the de Sitter-Schwarzchild universe with massless fields. Phys. Rev. D. 1981. 24. 829–834.

[513] Khanal, U.; Panchapakesan, N., Production of massless particles in the de Sitter-Schwarzschild universe. Phys. Rev. D. 1981. 24, 835–838.

[514] Khanal, U., Rotating black hole in asymptotic de Sitter space: Perturbation of the space-time with spin fields Phys. Rev. D. 1983. 28. 1291–1297/

[515] Khanal, U., Further investigations of the Kerr-de Sitter space. Phys. Rev. D. 1985. 32. 879–883.

[516] Otchik, V.S., On the Hawking radiation of spin 1/2 particles in the de Sitter space-time. Class. Quantum Crav. 1985. 2. 539–543.

[517] Motolla, F. Particle creation in de Sitter space. Phys. Rev. D. 1985. 31. 754–766.

[518] Bogush, A.A.; Otchik, V.S.; Red'kov, V.M., Vector field in de Sitter space. Vesti AN NSSR. 1986. no 1. 58–62.

[519] Takashi Mishima, Akihiro Nakayama: Particle production in de Sitter spacetime. Progr. Theor. Phys. 77, 218–222 (1987)

[520] Polarski, D., The scalar wave equation on static de Sitter and anti-de Sitter spaces. Class. Quantum Grav. 1989. 6. 893–900.

[521] Suzuki H.; Takasugi, E., Absorption Probability of De Sitter Horizon for Massless Fields with Spin. Mod. Phys. Lett. A. 1996. 11. 431–436.

[522] Suzuki, H.; Takasugi, E.; Umetsu, H., Perturbations of Kerr-de Sitter Black Hole and Heun's Equations. Prog. Theor. Phys. 1998. 100. 491–505.

[523] Suzuki,H.; Takasugi, E.; Umetsu, H., Analytic Solutions of Teukolsky Equation in Kerr-de Sitter and Kerr-Newman-de Sitter Geometries Prog. Theor. Phys. 1999. 102. 253–272.

[524] Suzuki, H.; Takasugi, E.; Umetsu, H., Absorption rate of the Kerr-de Sitter black hole and the Kerr-Newman-de Sitter black hole Prog. Theor. Phys. 2000. 103. 723–731.

[525] Red'kov, V.M.; Ovsiyuk, E.M., On exact solutions for quantum particles with spin $S = 0, 1/2, 1$ and de Sitter event horizon. Ricerche di matematica. 2011. 60. no 1. 57–88.

[526] Red'kov, V. M.; Ovsiyuk, E.M.; Krylov, G.G., Hawking Radiation in de Sitter Space: Calculation of the Refection Coeficient fo Quantum Particles. Vestnik RUDN. Ser. mat.-inform.-fiz. 2012. no 4. 153–169.

Index

T

U

V

W